MATHEMATICS FOR TECHNICAL OCCUPATIONS

MATHEMATICS FOR TECHNICAL OCCUPATIONS

Dennis Bila
Washtenaw Community College

Ralph Bottorff
Washtenaw Community College

Paul Merritt
Highland Park Community College

Donald Ross
Washtenaw Community College

Winthrop Publishers, Inc., Cambridge, Massachusetts

10 9 8 7 6 5 4 3 2

Library of Congress Cataloging in Publication Data

Main entry under title:

Mathematics for technical occupations.

Includes index.
1. Mathematics—1961– 2. Mathematics—Programmed instruction. I. Bila, Dennis, 1941–
QA39.2.M384 513'.142'077 77-26257
ISBN 0-87626-572-7

TO OUR STUDENTS

CONTENTS

PREFACE

Mathematics for Technical Occupations is designed for a wide range of students with diverse mathematical backgrounds and abilities. It presumes little prior knowledge of mathematics other than the basic operations with whole numbers and is presented in a semiprogrammed worktext format that allows students to proceed at their own pace. Students who have retained some knowledge from previous study in mathematics can go rapidly through the material that is familiar to them, thereby gaining the benefit of a useful review of their skills. Students without this advantage will want to proceed at a slower pace. *Mathematics for Technical Occupations* can be used in a course for the technician who will not continue on to more advanced mathematics. For students who will take further courses in mathematics, but who do not have the necessary background to enter introductory algebra or more advanced courses in technical fields requiring a fundamental knowledge of general mathematics, this text provides all the skills and knowledge necessary to prepare them for such courses.

Mathematics for Technical Occupations is intended for use as the introductory mathematics course in one- and two-year vocational programs in technical areas which do not require intermediate algebra. Such programs might include:

Auto Body Specialist
Auto Body Service Technician
Auto Body Repairman
Automobile Spray Painter
Automotive Service Technician
Automotive Mechanic Program
Construction Specialist
Industrial Drafting Technician
Draftsman-Detailer
Electrical Equipment Repairman
Electronic Service Technician
Hydraulic Assembler
Mechanical-Engineering Technician
Toolroom Machine Operator
Commercial Artist
Technical Illustrator
Photographic Technician
Photographic Assistant
Welding and Fabrication Technician
Combination Welder-Mechanic
Quality Control Inspector
Toolmaker
Tool and Die Maker
Machine Repair
Millwright
Plumber/Pipefitter
Tinsmith/Sheetmetal Worker
Boiler and Powerplant Engineer
Refrigeration Mechanic
Elevator Repairman
Heating and Ventilating Repairman

This list is not complete but simply represents the broad range of programs for which the text is intended.

Although *Mathematics for Technical Occupations* is meant primarily for use in a laboratory with an instructor, its format is also well suited for use in a more traditional classroom.

For many years we have struggled with the problem of providing well-written programmed materials with appropriate applications for those students who wish to learn mathematics relating to their field of interest. Many of these students have had poor experiences in their earlier mathematics courses and are reluctant to take further courses. Our attempts to help them have been hindered by the lack of adequate text material. As a result, we began writing semiprogrammed booklets on single-concept ideas to supplement the material we were using. The positive reactions of students to our materials were encouraging. As a result, we decided to develop a series of semiprogrammed texts written specifically for those who need this level of instruction. *Mathematics for Technical Occupations* is one in this series. Others include *Core Mathematics* (for the general student who has not identified a specific occupational interest), *Mathematics for Health Occupations* (for students interested in allied health fields), *Mathematics for Business Occupations* (for those interested in vocational programs in the business and community service areas), and *Introductory Algebra* and *Intermediate Algebra* (for those pursuing a more traditional course of study leading to precalculus and calculus). We sincerely hope that your experiences with these texts will be as rewarding as ours have been.

Core Mathematics, Introductory Algebra, and *Intermediate Algebra* are all available from Worth Publishers, Inc., N.Y. We would like to thank Worth Publishers for their permission to use excerpts from *Core Mathematics* and *Introductory Algebra* in this text.

We feel that the use of hand-held calculators should be restricted during the study of the first two chapters. Until the basic arithmetic operations are mastered, the use of the hand-held calculator could be a hindrance to the learning of these concepts. However, beyond that point, their use should be permitted, if not encouraged. The use of the calculator during the balance of the study of the text would permit the student to focus attention on the mastery of the content and minimize much of the drudgery associated with the study of the topics included in the text.

The symbols and notation used in the chapters on the metric system are those of the International System of Units (SI). The National Bureau of Standards has adopted the SI in all its research work and publications, except where use of these units would impair communications or reduce the usefulness of the material. Chapter 6 has been reviewed by the National Bureau of Standards. We have used the "metre" and "litre" spelling, though this is still a controversial issue. If conventions are adopted in the United States which conflict with those used in this text, appropriate changes will be made in subsequent editions.

FEATURES OF *Mathematics for Technical Occupations*

FORMAT

Each unit of material (a single concept or a few closely related concepts) is presented in a section. Within a section, short, boxed, numbered frames contain all instructional material, including sample problems and sample solutions. Frames are followed by practice problems, with workspace provided. Answers immediately follow, with numerous solutions and supplementary comments.

EXERCISES

Exercises at the end of each section are quite traditional in nature. However, they are generally shorter than those found in most texts because the student has already done numerous problems and answered many questions during the study of the section. Hence, the exercises serve as a review of the content of the section and as a guide that will help students to recognize whether they have mastered the material. Answers to the problems are provided immediately, permitting the student to advance through the text without having to turn to the back of the text to check answers. Word problems are used throughout.

SAMPLE CHAPTER TESTS

At the end of each chapter we have provided a Sample Chapter Test. It is keyed to each section with the answers provided immediately for student convenience. It may also serve as a pretest for each chapter. However, if an instructor wishes to pretest without the availability of the answers (to the student), one form of the post-test for the chapter (provided in the *Test Manual*) could be used for this purpose.

Upon finishing the chapter, the student should complete the Chapter Sample Test, and the results should be shown to the instructor and discussed with the student. If the instructor and student are confident about mastery of the content of the chapter, a post-test can then be administered. Used in this manner, outstanding results can be anticipated on the first attempt for each post-test.

OBJECTIVES

The Sample Chapter Tests serve as objectives for both student and instructor. The instructor can readily ascertain the objectives of any chapter by examining the sample test at its end, and the student is in a good position to see what is expected of him by examining problems and questions that are going to be asked at the completion of the chapter. We believe that objectives stated verbally are of less benefit to the student than the statement of a problem that must be solved.

GLOSSARY

The glossary at the back of the text provides the student with the pronunciation and the definition of all mathematical words and phrases used in the text. This is particularly important wherever the text is used in a laboratory situation where the student will not hear the words used in class discussions.

SUPPLEMENTS

A *Test Manual* and an *Instructional Kit* are available separately for instructors using *Mathematics for Technical Occupations.* The *Test Manual* contains a reproducible master copy of each of three forms of a post-test for each chapter of the text. Each post-test contains ample workspace and answer blanks. Answers to the tests are given at the end of the *Test Manual.* The *Instructional Kit* discusses the text format in detail and describes at length both the lecture-recitation and laboratory methods of instruction. Sample schedule guidelines and progress records are given, along with a Mathematics Inventory Test which the instructor may give at the start of the course. The kit also includes the author's chapter-by-chapter rationale for the procedures used in the text.

ACKNOWLEDGMENTS

We would like to thank those whose reviews were particularly helpful in the writing of this text: Claude L. Reeves, Tri-County Technical College; Tom Perring, Bunker Hill Community College; Stuart R. Porter, Monroe Community College; and Jeff Slater, North Shore Community College. Since *Mathematics for Technical Occupations* contains material from *Core Mathematics*, we would like to extend another expression of gratitude to the reviewers of that text who originally were so helpful in its preparation.

We would like to express our thanks to our colleagues at Washtenaw Community College and Highland Park Community College for their contributions, with special thanks to Ken Barron,

Robert Bellers, John Mann, Robert Mealing, and Al Robinson for their detailed analysis of course requirements relating to vocational programs in the technical fields. Also to student reviewers Dan Hartmann, Paul Landen, Rich Pickelsimer, Steve Ross, and Terri Ross who worked each and every problem in the text. Their suggestions (and corrections) were invaluable. Certainly we owe a debt of gratitude to all the students who participated in the class testing. Our thanks also to our typists, Shari Brown, Nancy Hayes, Marcia Joiner (now Mrs. Paul Merritt), Marilyn Myers, Carolyn Williams, and Carol Wilson. How they were able to read our handwriting is still a mystery.

Special thanks are due to Ken Tennity, Dorothy Werner, Walt Kirby, and Greg Wood of the Prentice-Hall fieldstaff for their invaluable marketing research assistance.

John Covell, the production editor at Winthrop Publishers, must receive special accolades. His patience, attention to detail, unflappable nature, and support contributed more than words can describe to the successful outcome of this text.

To all those unnamed who participated, we are eternally thankful.

Dennis Bila
Ralph Bottorff
Paul Merritt
Donald Ross

TO THE STUDENT

You are about to brush up on some skills you may already have learned in a previous course; but they may be a bit rusty. You will learn some new techniques, too. This book is designed to help pinpoint those areas where you need to sharpen your skills and to permit you to move quickly through those topics you know well.

If you follow the suggestions below, you will make best use of your time, proceeding through the course as quickly as possible, and you will have mastered all the material in this book.

CHAPTER SAMPLE TESTS

The Chapter Sample Test at the end of each chapter will help you to determine whether you can skip certain sections of the chapter. It is a cumulative review of the entire chapter, and questions are keyed to the individual sections of the chapter. To use this test to best advantage, we suggest this procedure:

1. Work the solutions to all problems neatly on your own paper to the best of your ability. You want to determine what you know about the material without help from someone else. If you receive help in completing this self-test, the result will show another person's knowledge rather than your own.
2. Once you have done as much as you can, check the test with the answers provided. On the basis of errors made, determine the most appropriate course of study. You may need to study all sections in the chapter or you may be able to skip one or two sections.
3. When you have completed the chapter, rework all problems originally missed on the Chapter Sample Test and review the entire test in preparation for the post-test that will cover the entire chapter.

INSTRUCTIONAL SECTIONS

To use each section most effectively:

1. Study the boxed frames carefully.
2. Following each frame are questions based on the information presented in the frame. When completing these questions, do the work in the spaces provided.
3. Use a blank piece of paper as a mask to cover the answers below the divider rule. For example:

Q1 How many eggs can 21 chickens lay in 1 hour, if 1 chicken can lay 7 eggs in $\frac{1}{2}$ hour?

(Work space)

• ### • ### • ### • ### • ### • ### • ### •

Paper Mask

4. After you have written your calculations and response in the workspace, slide the paper mask down, uncovering the correct solution, and check your response with the correct answer given. In this way you can check your progress as you go without accidentally seeing the response before you have completed the necessary thinking or work.

Q1 How many eggs can 21 chickens lay in 1 hour, if 1 chicken can lay 7 eggs in $\frac{1}{2}$ hour?

$7 \times 2 = 14$

$$\begin{array}{r} 14 \\ \times 21 \\ \hline 14 \\ 28 \\ \hline 294 \end{array}$$

• ### • ### • ### • ### • ### • ### • ### •

A1 294: 1 chicken can lay 14 in 1 hour; hence, 21 chickens can lay 21(14) in 1 hour.

Paper Mask

Notice that the answer is often followed by a colon (:). The information following the colon is one of the following:

a. The complete solution,

b. A partial solution, possibly a key step frequently missed by many students,

c. A remark to remind you of an important point, or

d. A comment about the solution.

If your answer is correct, advance immediately to the next step of instruction.

5. Space is provided for you to work in the text. However, you may prefer to use the paper mask to work out your solutions. If you do, be sure to show complete solutions to all problems, clearly numbered, on the paper mask. When both sides of the paper are full, file it in a notebook for future reference.
6. Make all necessary corrections before continuing to the next frame or problem.
7. If something is not clear, talk to your instructor immediately. Do not accept any answers just because the text says so. Make sure you are convinced of the logic behind the work.

8. If difficulty arises when you are studying outside of class, be sure to note the difficulty so that you can ask about it at your earliest convenience.
9. When preparing for a chapter post-test, the frames of each section serve as an excellent review of the chapter.

SECTION EXERCISES

The exercises at the end of each section are provided for additional practice on the content of the section. For your convenience, answers are given immediately following the exercises. However, detailed solutions to these problems are not shown.

You should:

1. Work all problems neatly on your own paper.
2. Check your responses against the answers given.
3. Rework any problem with which you disagree. If you cannot verify the response given, discuss the result with your instructor.
4. In the process of completing the exercise, use the instructional material in the section for review when necessary.

Problems marked with an asterisk (*) are considered more difficult than examples in the instructional section. They are intended to challenge the interested student. Problems of this difficulty will not appear on the Chapter Sample Test or the post-test.

You should realize that the zero is placed to the left of a decimal number in the text to eliminate confusion over the placement of the decimal point. However, it is not incorrect to omit the zero when performing a calculation or expressing an answer, that is, $0.16 = .16$.

This three-part learning plan has proven successful in situations where there have been students with many different backgrounds. With conscientious effort, you will have success with it too.

CHAPTER 1

OPERATIONS WITH FRACTIONS

1.1 EQUIVALENT FRACTIONS

1 A *fraction* is a number that indicates that some whole has been divided into a number of equal parts, and that a portion of the equal parts are counted or represented. For example, the fraction $\frac{2}{3}$ indicates that a whole has been divided into 3 equal parts and that 2 are represented. The bottom number is called the *denominator* and the top number is called the *numerator*. For the fraction $\frac{2}{3}$, the denominator, 3, indicates that a whole is divided into 3 equal parts.

The shaded area represents $\frac{2}{3}$.

The numerator, 2, represents 2 of the 3 equal parts.

Q1 Into how many equal parts has the rectangle been divided?

______

\# \# \# • \# \# \# • \# \# \# • \# \# \# • \# \# \# • \# \# \# • \# \# \# • \# \# \# • \# \# \#

A1 8

Q2 The number of equal parts into which a whole is divided is indicated by the numerator/denominator (circle the correct answer).

\# \# \# • \# \# \# • \# \# \# • \# \# \# • \# \# \# • \# \# \# • \# \# \# • \# \# \# • \# \# \#

A2 denominator

Q3 The number of equal parts represented (shaded) in Q1 is ______.

\# \# \# • \# \# \# • \# \# \# • \# \# \# • \# \# \# • \# \# \# • \# \# \# • \# \# \# • \# \# \#

A3 6

Q4 The number of equal parts represented is called the numerator/denominator (circle the correct answer).

• # # # • # # # • # # # • # # # • # # # • # # # • # # # • # #

A4 numerator

Q5 What portion of the rectangle (Q1) is represented by the shaded area? ______

• # # # • # # # • # # # • # # # • # # # • # # # • # # # • # #

A5 $\frac{6}{8}$

Q6 What fraction is represented by the shaded area?

a. 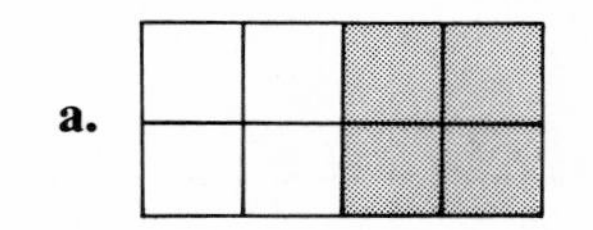 ______ **b.** 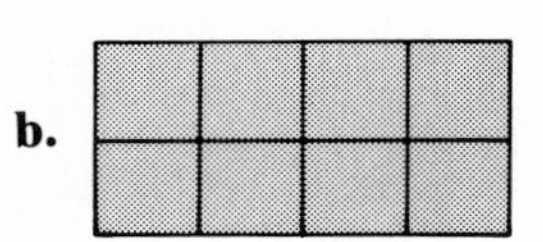 ______

• # # # • # # # • # # # • # # # • # # # • # # # • # # # • # #

A6 **a.** $\frac{4}{8}$ **b.** $\frac{8}{8}$

Q7 What fraction is represented by the shaded area;

a. 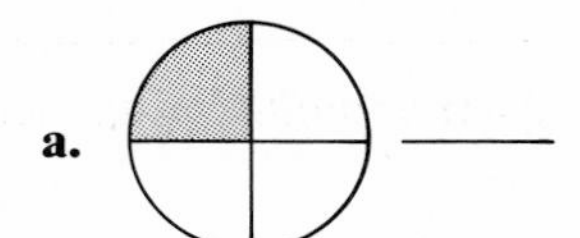 ______ **b.** 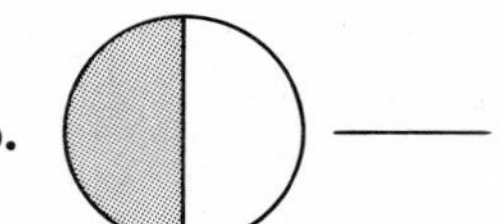 ______ **c.** 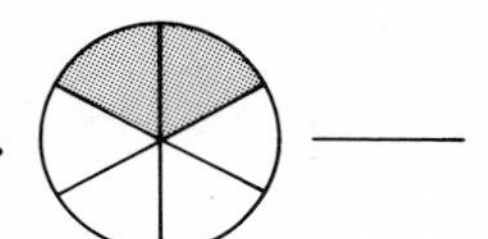 ______ **d.** 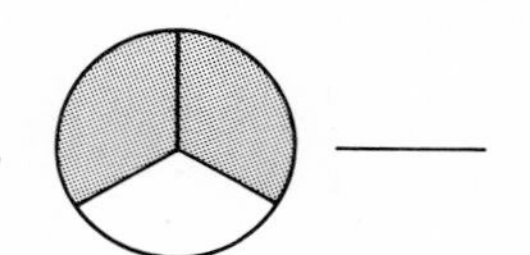 ______

• # # # • # # # • # # # • # # # • # # # • # # # • # # # • # #

A7 **a.** $\frac{1}{4}$ **b.** $\frac{1}{2}$ **c.** $\frac{2}{6}$ **d.** $\frac{2}{3}$

Q8 What fraction does the figure represent? ______

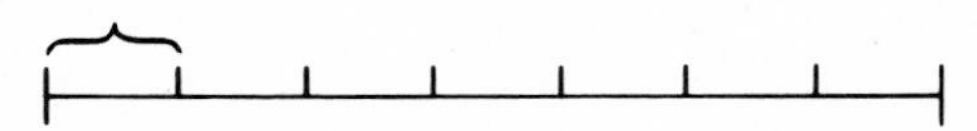

• # # # • # # # • # # # • # # # • # # # • # # # • # # # • # #

A8 $\frac{1}{7}$: The line segment is divided into 7 equal parts (denominator) and 1 equal part is indicated or counted (numerator).

Q9 Shade the circle so that the fraction $\frac{5}{8}$ is indicated.

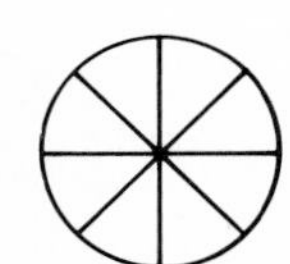

• # # # • # # # • # # # • # # # • # # # • # # # • # # # • # #

A9 or any combination of 5 of the 8 parts

2 The same area may be represented by different fractions. The area on the left is represented by the

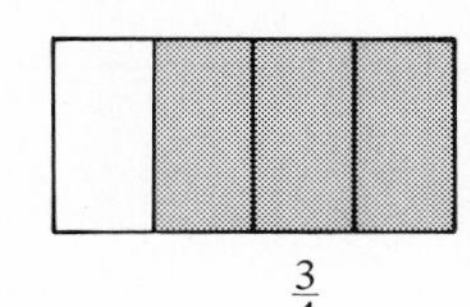

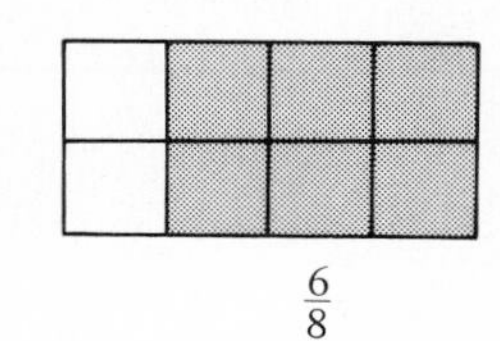

$\frac{3}{4}$ $\frac{6}{8}$

fraction $\frac{3}{4}$; the area on the right is represented by $\frac{6}{8}$. However, both fractions represent the same part of the whole. Therefore, $\frac{3}{4}=\frac{6}{8}$ and they are called *equivalent fractions.*

Q10 What two equivalent fractions can be used to name the shaded area? _______

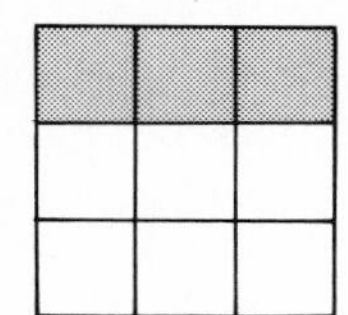

\# # # • # # # • # # # • # # # • # # # • # # # • # # # • # # # • # # #

A10 $\frac{3}{9}, \frac{1}{3}$

3 Equivalent fractions can be obtained by multiplying the numerator and denominator of a fraction by the same number (not zero). For example,

$\frac{3}{4}=\frac{3\times 2}{4\times 2}=\frac{6}{8}$ $\frac{3}{4}=\frac{3\times 3}{4\times 3}=\frac{9}{12}$ $\frac{3}{4}=\frac{3\times 5}{4\times 5}=\frac{15}{20}$

The multiplication of the numerator and the denominator by the same number is usually done mentally. Thus,

$\frac{2}{3}=\frac{8}{12}$

multiplying the numerator and denominator of $\frac{2}{3}$ by 4.

Q11 Change $\frac{5}{7}$ to an equivalent fraction by multiplying numerator and denominator by 6.

\# # # • # # # • # # # • # # # • # # # • # # # • # # # • # # # • # # #

A11 $\frac{30}{42}$: $\frac{5\times 6}{7\times 6}=\frac{30}{42}$

Q12 Change $\frac{2}{3}$ to equivalent fractions by multiplying numerator and denominator by 2, 3, 4, and 5, respectively.

\# # # • # # # • # # # • # # # • # # # • # # # • # # # • # # # • # # #

A12 $\frac{4}{6}, \frac{6}{9}, \frac{8}{12}, \frac{10}{15}$

4 To change $\frac{3}{4}$ to eighths, think

$\frac{3}{4} = \frac{?}{8}$

If the fraction with denominator 8 is to be equivalent to $\frac{3}{4}$, the numerator and denominator of $\frac{3}{4}$ must be multiplied by the same number. Since $8 \div 4 = 2$, the denominator, 4, was multiplied by 2. Likewise, multiply the numerator, 3, by 2, obtaining $\frac{3 \times 2}{4 \times 2} = \frac{6}{8}$.

Q13 What was the denominator multiplied by? $\frac{3}{4} = \frac{\quad}{20}$

\# \# \# • \# \# \# • \# \# \# • \# \# \# • \# \# \# • \# \# \# • \# \# \# • \# \# \# • \# \# \#

A13 5: $20 \div 4 = 5$

Q14 Find the missing value: $\frac{3}{4} = \frac{\quad}{20}$

\# \# \# • \# \# \# • \# \# \# • \# \# \# • \# \# \# • \# \# \# • \# \# \# • \# \# \# • \# \# \#

A14 15: $\frac{3 \times 5}{4 \times 5} = \frac{15}{20}$

Q15 Find the missing value: $\frac{5}{7} = \frac{\quad}{21}$

\# \# \# • \# \# \# • \# \# \# • \# \# \# • \# \# \# • \# \# \# • \# \# \# • \# \# \# • \# \# \#

A15 15: $\frac{5 \times 3}{7 \times 3} = \frac{15}{21}$

Q16 Find the missing value: $\frac{8}{15} = \frac{\quad}{60}$

\# \# \# • \# \# \# • \# \# \# • \# \# \# • \# \# \# • \# \# \# • \# \# \# • \# \# \# • \# \# \#

A16 32: $\frac{8 \times 4}{15 \times 4} = \frac{32}{60}$

Q17 Find the missing value:

a. $\frac{2}{3} = \frac{\quad}{24}$ **b.** $\frac{6}{11} = \frac{\quad}{44}$ **c.** $\frac{3}{4} = \frac{\quad}{28}$ **d.** $\frac{5}{12} = \frac{\quad}{60}$

\# \# \# • \# \# \# • \# \# \# • \# \# \# • \# \# \# • \# \# \# • \# \# \# • \# \# \# • \# \# \#

A17 **a.** 16 **b.** 24 **c.** 21 **d.** 25

5 It is sometimes necessary to represent a whole number* as a fraction. First consider that $8 \div 2$ can be written as $2\overline{)8}$ or $\frac{8}{2}$. All these mean 8 divided by 2. To find the number of 20th's in 5,

*The whole numbers are 0, 1, 2, 3, 4, and so on. The next whole number is formed by adding 1 to the previous whole number.

$5 = \frac{?}{20}$ write 5 as $\frac{5^*}{1}$; hence, $\frac{5}{1} = \frac{?}{20}$

Now, $1 \times 20 = 20$, so the numerator must also be multiplied by 20.

$\frac{5 \times 20}{1 \times 20} = \frac{100}{20}$ or $5 = \frac{100}{20}$

* Recall that (any number) ÷ 1 = that number.

Q18 Find the missing value: $\frac{3}{1} = \frac{\quad}{15}$

\# # # • # # # • # # # • # # # • # # # • # # # • # # # • # # # • # # #

A18 45: $\frac{3 \times 15}{1 \times 15} = \frac{45}{15}$

Q19 Find the missing value: $4 = \frac{\quad}{8}$

\# # # • # # # • # # # • # # # • # # # • # # # • # # # • # # # • # # #

A19 32: $\frac{4 \times 8}{1 \times 8} = \frac{32}{8}$

Q20 Find the missing value:

a. $5 = \frac{\quad}{3}$ **b.** $8 = \frac{\quad}{4}$

\# # # • # # # • # # # • # # # • # # # • # # # • # # # • # # # • # # #

A20 **a.** 15 **b.** 32

6 The equivalent fractions obtained have all been of higher terms (numerator and denominator) than the original fraction. It is often necessary to find an equivalent fraction, but of lower terms. This will be called *reducing* the fraction to a lower term. An equivalent fraction is obtained if the numerator and the denominator of a fraction are divided by the same number (not zero).

Example 1:

$\frac{4 \div 2}{8 \div 2} = \frac{2}{4}$; hence, $\frac{4}{8} = \frac{2}{4}$

Example 2: Reduce $\frac{5}{10}$ by dividing numerator and denominator by 5.

Solution: $\frac{5}{10} = \frac{5 \div 5}{10 \div 5} = \frac{1}{2}$

Q21 Determine the equivalent fraction obtained by dividing the numerator and denominator of $\frac{6}{8}$ by 2.

\# # # • # # # • # # # • # # # • # # # • # # # • # # # • # # # • # # #

A21 $\frac{3}{4}$: $\frac{6 \div 2}{8 \div 2} = \frac{3}{4}$

Q22 Determine equivalent fractions, reducing by the common factor 3:

a. $\frac{3}{9}$ **b.** $\frac{15}{30}$

\# \# \# • \# \# \# • \# \# \# • \# \# \# • \# \# \# • \# \# \# • \# \# \# • \# \# \# • \# \# \#

A22 **a.** $\frac{1}{3}$: $\frac{3 \div 3}{9 \div 3} = \frac{1}{3}$ **b.** $\frac{5}{10}$: $\frac{15 \div 3}{30 \div 3} = \frac{5}{10}$

7 A *common factor* (divisor) of two numbers is a number that will divide both numbers evenly (remainder zero). A common factor of 9 and 12 is 3, because 3 divides both 9 and 12 evenly. A fraction is considered reduced to *lowest terms* if the numerator and denominator have no common factors (divisors). As an example, $\frac{5}{8}$ is reduced to lowest terms, because there are no common factors of 5 and 8.

Q23 Which fractions cannot be reduced (the numerator and denominator have no common factors)?

$\frac{21}{29}, \frac{15}{28}, \frac{12}{27}$ ________

\# \# \# • \# \# \# • \# \# \# • \# \# \# • \# \# \# • \# \# \# • \# \# \# • \# \# \# • \# \# \#

A23 $\frac{21}{29}$ and $\frac{15}{28}$: $\frac{12}{27}$ could be reduced, because 12 and 27 have a common factor of 3.

Q24 Reduce $\frac{12}{27}$.

\# \# \# • \# \# \# • \# \# \# • \# \# \# • \# \# \# • \# \# \# • \# \# \# • \# \# \# • \# \# \#

A24 $\frac{4}{9}$: $\frac{12 \div 3}{27 \div 3} = \frac{4}{9}$

8 Often the terms of a fraction have many common factors. For example, $\frac{12}{18}$ is reduced as follows:

$\frac{12 \div 2}{18 \div 2} = \frac{6}{9}$

However, 6 and 9 have a common factor of 3. Thus, $\frac{6 \div 3}{9 \div 3} = \frac{2}{3}$. $\frac{2}{3}$ is reduced to lowest terms, because 2 and 3 have no common factors. Notice that $\frac{12}{18}$ could have been reduced to lowest terms by first reducing by the common factor 6:

$\frac{12}{18} = \frac{12 \div 6}{18 \div 6} = \frac{2}{3}$

Q25 Reduce $\frac{24}{36}$ to lowest terms.

\# \# \# • \# \# \# • \# \# \# • \# \# \# • \# \# \# • \# \# \# • \# \# \# • \# \# \# • \# \# \#

A25 $\frac{2}{3}$: $\frac{24 \div 12}{36 \div 12} = \frac{2}{3}$ or $\frac{24 \div 2}{36 \div 2} = \frac{12}{18} = \frac{2}{3}$ (reducing by 6)

9 An aid to reducing fractions is to recall some simple *divisibility tests*. That is:

1. A number is divisible by 2 if the last digit is divisible by 2 (that is, if the digit is 0, 2, 4, 6, or 8).

Example: The following numbers are divisible by 2 because the last digit is divisible by 2:

14 38 54 1,026

2. A number is divisible by 3 if the sum of the digits is divisible by 3.

Example: 24 is divisible by 3 because the sum of the digits ($2+4=6$) is divisible by 3. 54 is divisible by 3 because 9 ($5+4$) is divisible by 3. The following numbers are divisible by 3:

15 126 102 390

3. A number is divisible by 5 if the last digit is 5 or 0.

Example: The following numbers are divisible by 5:

85 40 1,080 215

Q26 Show that the following numbers are divisible by 2: 14, 38, 54, 1,026.

• # # # • # # # • # # # • # # # • # # # • # # # • # # # • # #

A26 The last digit is divisible by 2. *Or:*

$14 \div 2 = 7$ $38 \div 2 = 19$
$54 \div 2 = 27$ $1{,}026 \div 2 = 513$

Q27 Show that the following numbers are divisible by 3: 15, 126, 102, 390.

• # # # • # # # • # # # • # # # • # # # • # # # • # # # • # #

A27 The sum of the digits is divisible by 3. *Or:*

$15 \div 3 = 5$ $126 \div 3 = 42$
$102 \div 3 = 34$ $390 \div 3 = 130$

Q28 Show that the following numbers are divisible by 5: 85, 40, 1,080, 215.

• # # # • # # # • # # # • # # # • # # # • # # # • # # # • # #

A28 The last digit is either 5 or 0. *Or:*

$85 \div 5 = 17$ $40 \div 5 = 8$
$1{,}080 \div 5 = 216$ $215 \div 5 = 43$

Q29 Reduce the fractions to lowest terms:

a. $\frac{21}{42}$ **b.** $\frac{15}{39}$

• # # # • # # # • # # # • # # # • # # # • # # # • # # # • # #

A29 **a.** $\frac{1}{2}$ **b.** $\frac{5}{13}$

Q30 Reduce to lowest terms:

a. $\frac{24}{36}$ **b.** $\frac{20}{30}$

• # # # • # # # • # # # • # # # • # # # • # # # • # # # • # #

A30 **a.** $\frac{2}{3}$ **b.** $\frac{2}{3}$

Q31 Reduce to lowest terms:

a. $\frac{21}{28}$ **b.** $\frac{36}{40}$ **c.** $\frac{15}{25}$

• # # # • # # # • # # # • # # # • # # # • # # # • # # # • # #

A31 **a.** $\frac{3}{4}$ **b.** $\frac{9}{10}$ **c.** $\frac{3}{5}$

Q32 Reduce to lowest terms:

a. $\frac{35}{49}$ **b.** $\frac{30}{45}$ **c.** $\frac{27}{36}$ **d.** $\frac{33}{54}$

• # # # • # # # • # # # • # # # • # # # • # # # • # # # • # #

A32 **a.** $\frac{5}{7}$ **b.** $\frac{2}{3}$ **c.** $\frac{3}{4}$ **d.** $\frac{11}{18}$

This completes the instruction for this section.

1.1 EXERCISE

1. What fraction is indicated by the shaded area?

a. 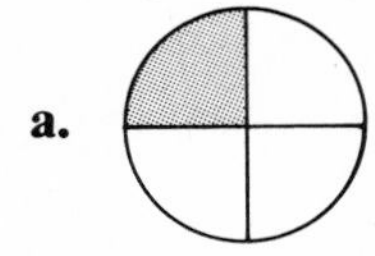

b. 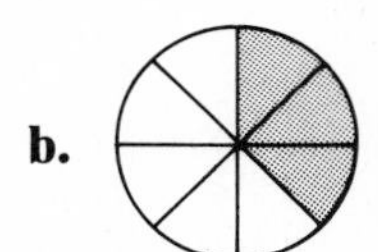

c. 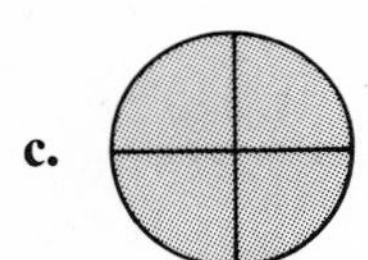

d. 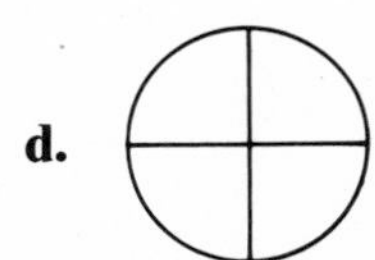

2. Find the missing value:

a. $\frac{3}{4} = \frac{?}{12}$ **b.** $\frac{1}{3} = \frac{?}{9}$ **c.** $\frac{7}{8} = \frac{?}{24}$ **d.** $\frac{6}{7} = \frac{?}{42}$

e. $\frac{3}{8} = \frac{?}{16}$ **f.** $\frac{7}{9} = \frac{56}{?}$ **g.** $\frac{1}{2} = \frac{4}{?}$ **h.** $5 = \frac{?}{15}$

i. $8 = \frac{?}{9}$

3. Reduce to lowest terms:

a. $\frac{6}{8}$ **b.** $\frac{8}{12}$ **c.** $\frac{6}{9}$ **d.** $\frac{12}{16}$ **e.** $\frac{5}{6}$

f. $\frac{56}{63}$ **g.** $\frac{15}{18}$ **h.** $\frac{26}{39}$ **i.** $\frac{43}{86}$ **j.** $\frac{24}{28}$

1.1 EXERCISE ANSWERS

1. **a.** $\frac{1}{4}$ **b.** $\frac{3}{8}$ **c.** $\frac{4}{4}$ or 1 **d.** $\frac{0}{4}$ or 0

2. **a.** 9 **b.** 3 **c.** 21 **d.** 36 **e.** 6
f. 72 **g.** 8 **h.** 75 **i.** 72

3. **a.** $\frac{3}{4}$ **b.** $\frac{2}{3}$ **c.** $\frac{2}{3}$ **d.** $\frac{3}{4}$ **e.** $\frac{5}{6}$
f. $\frac{8}{9}$ **g.** $\frac{5}{6}$ **h.** $\frac{2}{3}$ **i.** $\frac{1}{2}$ **j.** $\frac{6}{7}$

1.2 MIXED NUMBERS AND IMPROPER FRACTIONS

1 A fraction can be defined as the quotient of two whole numbers (the denominator not zero). If the fraction is less than 1, it is called a *proper fraction*. That is, the numerator is less than the denominator. Examples of proper fractions are

$\frac{1}{2}$ $\frac{1}{3}$ $\frac{5}{8}$

Fractions greater than or equal to 1 are called *improper fractions*. That is, the numerator is equal to or greater than the denominator. The following represents the improper fraction $\frac{4}{3}$:

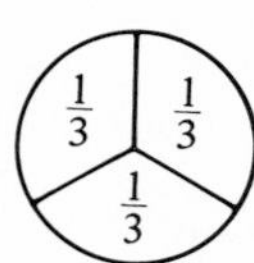

+
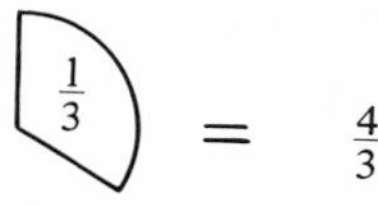

If the numerator and denominator of a fraction are equal, the fraction is equal to 1.

Examples:

$\frac{3}{3} = 1$ $\frac{8}{8} = 1$ $\frac{50}{50} = 1$

Q1 **a.** A fraction greater than 1 is called a (an) ____________ fraction.

b. A fraction less than 1 is called a (an) ____________ fraction.

c. If the numerator is larger than the denominator, the fraction is a (an) ____________ fraction.

d. If the numerator is equal to the demonimator, the fraction is equal to ____________.

\# # # • # # # • # # # • # # # • # # # • # # # • # # # • # # # • # # #

A1 **a.** improper **b.** proper **c.** improper **d.** one

Q2 What improper fraction is represented by the following?

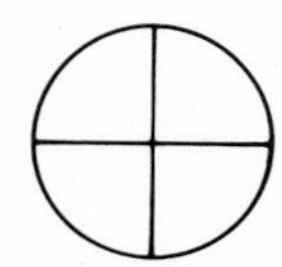
+
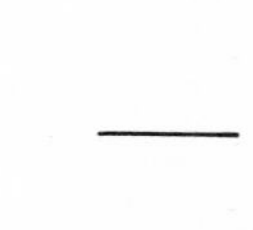
——

\# # # • # # # • # # # • # # # • # # # • # # # • # # # • # # # • # # #

A2 $\frac{6}{4}$

Q3 What improper fraction is represented by the following?

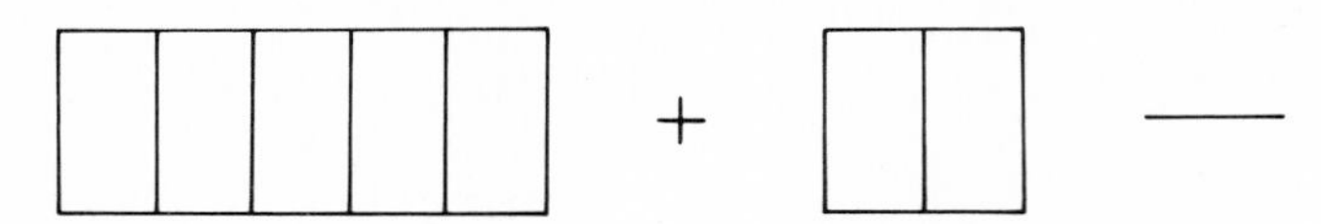

\# # # • # # # • # # # • # # # • # # # • # # # • # # # • # # # • # # #

A3 $\frac{7}{5}$

Q4 Which of the following are improper fractions?

a. $\frac{3}{4}$ **b.** $\frac{7}{2}$ **c.** $\frac{5}{5}$ **d.** $\frac{8}{7}$ ______

\# # # • # # # • # # # • # # # • # # # • # # # • # # # • # # # • # # #

A4 b, c, and d: The numerator is larger than or equal to the denominator.

2 A *mixed number* is a whole number and a proper fraction. The following illustration represents the improper fraction $\frac{4}{3}$ and the mixed number $1\frac{1}{3}$.

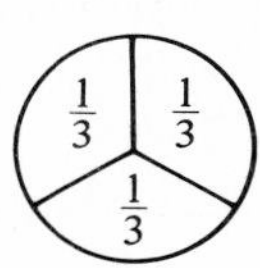

+ $\frac{1}{3}$ = $\frac{4}{3}$ We have $1 + \frac{1}{3} = 1\ \frac{1}{3}$

The plus sign, +, is usually omitted between the whole number and the fraction.

Q5 What mixed number is represented by this illustration?

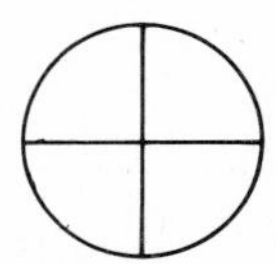 + 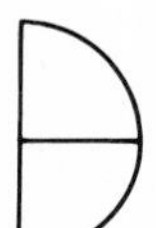 ______

\# # # • # # # • # # # • # # # • # # # • # # # • # # # • # # # • # # #

A5 $1\frac{2}{4}$ or $1\frac{1}{2}$

3 An improper fraction can be written as a mixed number by writing the improper fraction as the sum of a whole number and a common fraction.

Examples:

1. $\frac{7}{4} = \frac{1}{4} + \frac{1}{4} + \frac{1}{4} + \frac{1}{4} + \frac{1}{4} + \frac{1}{4} + \frac{1}{4}$
$= \frac{4}{4} + \frac{3}{4}$
$= 1\frac{3}{4}$

2. $\frac{6}{5} = \frac{5}{5} + \frac{1}{5}$
$= 1 + \frac{1}{5}$
$= 1\frac{1}{5}$

3. $\frac{9}{4} = \frac{4}{4} + \frac{4}{4} + \frac{1}{4}$
$= 1 + 1 + \frac{1}{4}$
$= 2\frac{1}{4}$

Q6 Write the fraction $\frac{5}{4}$ as a mixed number by completing the following steps:

$$\frac{5}{4}=\frac{4}{4}+\frac{1}{4}$$
$$= \_\_\_\_\_\_ + \_\_\_\_\_\_$$
$$= \_\_\_\_\_\_$$

\# # # • # # # • # # # • # # # • # # # • # # # • # # # • # # # • # # #

A6 $= 1+\frac{1}{4}$

$= 1\frac{1}{4}$

Q7 Write $\frac{7}{5}$ as a mixed number.

\# # # • # # # • # # # • # # # • # # # • # # # • # # # • # # # • # # #

A7 $1\frac{2}{5}$: $\frac{7}{5}=\frac{5}{5}+\frac{2}{5}=1\frac{2}{5}$

Q8 Write $\frac{11}{4}$ as a mixed number.

\# # # • # # # • # # # • # # # • # # # • # # # • # # # • # # # • # # #

A8 $2\frac{3}{4}$: $\frac{11}{4}=\frac{4}{4}+\frac{4}{4}+\frac{3}{4}=2\frac{3}{4}$

Q9 Write as mixed numbers:

a. $\frac{8}{3}$ **b.** $\frac{5}{3}$ **c.** $\frac{18}{7}$ **d.** $\frac{9}{2}$

\# # # • # # # • # # # • # # # • # # # • # # # • # # # • # # # • # # #

A9 **a.** $2\frac{2}{3}$ **b.** $1\frac{2}{3}$ **c.** $2\frac{4}{7}$ **d.** $4\frac{1}{2}$

4 An improper fraction can also be changed to a mixed number by dividing the numerator of the fraction by the denominator. The remainder (if any) is written over the denominator and added to the quotient.

Examples:

1. $\frac{5}{3} = 3\overline{)5}$, quotient 1, $\frac{3}{2}$ (remainder 2) $= 1\frac{2}{3}$

2. $\frac{16}{5} = 5\overline{)16}$, quotient 3, $\frac{15}{1}$ (remainder 1) $= 3\frac{1}{5}$

The fractional part of the mixed number should be reduced, if possible. For example,

$\frac{10}{4} = 4\overline{)10}$, quotient 2, $\frac{8}{2}$ (remainder 2) $= 2\frac{2}{4} = 2\frac{1}{2}$

Q10 Change $\frac{15}{4}$ to a mixed number.

\# \# \# • \# \# \# • \# \# \# • \# \# \# • \# \# \# • \# \# \# • \# \# \# • \# \# \# • \# \# \#

A10 $3\frac{3}{4}$: $\frac{15}{4} = 4\overline{)15} = 3\frac{3}{4}$ (quotient 3; $15 - 12 = 3$)

Q11 Change $\frac{42}{12}$ to a mixed number.

\# \# \# • \# \# \# • \# \# \# • \# \# \# • \# \# \# • \# \# \# • \# \# \# • \# \# \# • \# \# \#

A11 $3\frac{1}{2}$: $\frac{42}{12} = 12\overline{)42} = 3\frac{6}{12} = 3\frac{1}{2}$ (quotient 3; $42 - 36 = 6$)

Q12 Change to mixed numbers:

a. $\frac{33}{2}$ **b.** $\frac{73}{7}$ **c.** $\frac{18}{8}$

\# \# \# • \# \# \# • \# \# \# • \# \# \# • \# \# \# • \# \# \# • \# \# \# • \# \# \# • \# \# \#

A12 **a.** $16\frac{1}{2}$ **b.** $10\frac{3}{7}$ **c.** $2\frac{1}{4}$

Q13 Change to mixed or whole numbers:

a. $\frac{44}{4}$ **b.** $\frac{17}{3}$ **c.** $176 \div 24$

\# \# \# • \# \# \# • \# \# \# • \# \# \# • \# \# \# • \# \# \# • \# \# \# • \# \# \# • \# \# \#

A13 **a.** 11 **b.** $5\frac{2}{3}$ **c.** $7\frac{1}{3}$

5 A mixed number can be changed to an improper fraction by writing the whole number as a fraction. For example, $1\frac{2}{3}$ is changed to an improper fraction by writing the 1 as 3 thirds and adding the $\frac{2}{3}$.

That is,

$$1\frac{2}{3} = \frac{3}{3} + \frac{2}{3} = \frac{5}{3}$$

Additional examples are:

1. $3\frac{2}{3} = \frac{9}{3} + \frac{2}{3} = \frac{11}{3}$

2. $4\frac{1}{2} = \frac{8}{2} + \frac{1}{2} = \frac{9}{2}$

Q14 Change $2\frac{2}{3}$ to an improper fraction by completing the steps:

$$2\frac{2}{3} = \frac{6}{3} + \frac{2}{3} = \_\_\_\_$$

• # # # • # # # • # # # • # # # • # # # • # # # • # # # • # #

A14 $\frac{8}{3}$

Q15 Change $1\frac{4}{5}$ to an improper fraction.

• # # # • # # # • # # # • # # # • # # # • # # # • # # # • # #

A15 $\frac{9}{5}$: $1\frac{4}{5} = \frac{5}{5} + \frac{4}{5} = \frac{9}{5}$

Q16 Change to an improper fraction:

a. $1\frac{3}{8}$ **b.** $1\frac{1}{7}$

• # # # • # # # • # # # • # # # • # # # • # # # • # # # • # #

A16 **a.** $\frac{11}{8}$: $1\frac{3}{8} = \frac{8}{8} + \frac{3}{8} = \frac{11}{8}$ **b.** $\frac{8}{7}$: $1\frac{1}{7} = \frac{7}{7} + \frac{1}{7} = \frac{8}{7}$

6 Another method for changing a mixed number to an improper fraction is to multiply the whole number by the denominator and add the product to the existing numerator. This result is placed over the original denominator.

Examples:

1. $2\frac{3}{4} = \frac{2 \times 4 + 3}{4} = \frac{11}{4}$ **2.** $1\frac{3}{7} = \frac{1 \times 7 + 3}{7} = \frac{10}{7}$

Q17 Change $4\frac{2}{5}$ to an improper fraction by completing the following steps:

$$4\frac{2}{5} = \frac{4 \times 5 + 2}{5} = \_\_\_\_$$

• # # # • # # # • # # # • # # # • # # # • # # # • # # # • # #

A17 $\frac{22}{5}$

Q18 Change $5\frac{2}{3}$ to an improper fraction.

• # # # • # # # • # # # • # # # • # # # • # # # • # # # • # #

A18 $\frac{17}{3}$: $5\frac{2}{3} = \frac{5 \times 3 + 2}{3} = \frac{17}{3}$

Q19 Change to an improper fraction:

a. $5\frac{3}{4}$ **b.** $4\frac{7}{8}$

• # # # • # # # • # # # • # # # • # # # • # # # • # # # • # #

A19 **a.** $\frac{23}{4}$: $5\frac{3}{4} = \frac{5 \times 4 + 3}{4} = \frac{23}{4}$ **b.** $\frac{39}{8}$: $4\frac{7}{8} = \frac{4 \times 8 + 7}{8} = \frac{39}{8}$

Q20 Change to an improper fraction:

a. $4\frac{2}{3}$ **b.** $8\frac{7}{8}$ **c.** 8 **d.** $4\frac{3}{10}$

• # # # • # # # • # # # • # # # • # # # • # # # • # # # • # #

A20 **a.** $\frac{14}{3}$ **b.** $\frac{71}{8}$ **c.** $\frac{8}{1}$ **d.** $\frac{43}{10}$

This completes the instruction for this section.

1.2 EXERCISE

1. **a.** A fraction in which the numerator is equal to or larger than the denominator is called a (an) ____________ fraction.
   b. A fraction less than 1 is called a (an) ____________ fraction.
   c. A whole number and a fraction is called a ____________ number.
2. Change to a mixed or whole number:
   a. $\frac{28}{6}$ **b.** $\frac{52}{8}$ **c.** $\frac{48}{9}$ **d.** $\frac{81}{9}$
   e. $222 \div 27$ **f.** $289 \div 51$ **g.** $\frac{102}{13}$ **h.** $\frac{13{,}218}{105}$
3. Change to an improper fraction:
   a. $2\frac{1}{3}$ **b.** $9\frac{2}{3}$ **c.** $6\frac{5}{12}$ **d.** $8\frac{9}{10}$
   e. $5\frac{7}{8}$ **f.** $10\frac{10}{11}$ **g.** $250\frac{7}{8}$ **h.** $42\frac{17}{19}$

1.2 EXERCISE ANSWERS

1. **a.** improper **b.** proper **c.** mixed
2. **a.** $4\frac{2}{3}$ **b.** $6\frac{1}{2}$ **c.** $5\frac{1}{3}$ **d.** 9
   e. $8\frac{2}{9}$ **f.** $5\frac{2}{3}$ **g.** $7\frac{11}{13}$ **h.** $125\frac{31}{35}$
3. **a.** $\frac{7}{3}$ **b.** $\frac{29}{3}$ **c.** $\frac{77}{12}$ **d.** $\frac{89}{10}$
   e. $\frac{47}{8}$ **f.** $\frac{120}{11}$ **g.** $\frac{2{,}007}{8}$ **h.** $\frac{815}{19}$

1.3 ADDITION

Addition of fractions will be easier if you remember that the rules of addition for whole numbers also apply to fractions.

1 When adding whole numbers it was only possible to add numbers that represented the same unit of measure (place value). That is, tens were added to tens, hundreds added to hundreds, and so on. Similarly, when adding fractions it will be possible to add only those fractions expressed as the same unit of measure. Thirds must be added to thirds, fourths to fourths, and so on. Consider the following illustration:

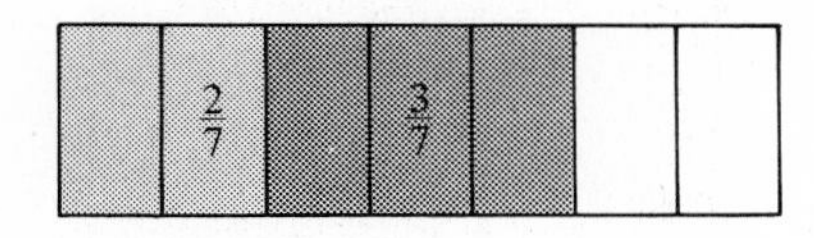

Here 2 equal parts of seven are added to 3 equal parts of seven for a sum of 5 equal parts of seven:

$$\frac{2}{7}+\frac{3}{7}=\frac{5}{7}$$

Two fractions can be combined into a single fraction by addition only if they have the same denominator.

Q1 Find the sum of the fractions represented:

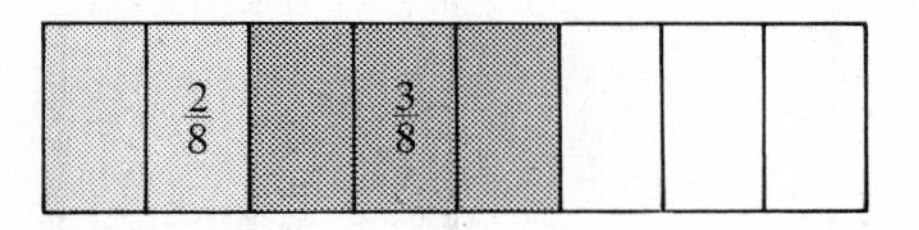

$\frac{2}{8}+\frac{3}{8}$ ______

\# # # • # # # • # # # • # # # • # # # • # # # • # # # • # # # • # # #

A1 $\frac{5}{8}$

2 Notice that the sum of two fractions that have the same denominator can be expressed as a single fraction by writing the sum of the numerators over the common denominator:

$$\frac{2}{7}+\frac{3}{7}=\frac{2+3}{7}=\frac{5}{7}$$

$$\frac{2}{8}+\frac{3}{8}=\frac{2+3}{8}=\frac{5}{8}$$

When possible the sum should be reduced to lowest terms:

$$\frac{3}{12}+\frac{7}{12}=\frac{10}{12}=\frac{5}{6}$$

$$\frac{1}{8}+\frac{3}{8}=\frac{4}{8}=\frac{1}{2}$$

If the sum is an improper fraction, it should be expressed as a mixed number:

$$\frac{2}{5}+\frac{4}{5}=\frac{6}{5}=1\frac{1}{5}$$

$$\frac{5}{6}+\frac{1}{6}=\frac{6}{6}=1$$

Q2 Determine the sum: $\frac{1}{5}+\frac{3}{5}$

\# \# \# • \# \# \# • \# \# \# • \# \# \# • \# \# \# • \# \# \# • \# \# \# • \# \# \# • \# \# \#

A2 $\frac{4}{5}$: $\frac{1}{5}+\frac{3}{5}=\frac{1+3}{5}=\frac{4}{5}$

Q3 Determine the sum: $\frac{1}{6}+\frac{3}{6}$

\# \# \# • \# \# \# • \# \# \# • \# \# \# • \# \# \# • \# \# \# • \# \# \# • \# \# \# • \# \# \#

A3 $\frac{2}{3}$: $\frac{1}{6}+\frac{3}{6}=\frac{4}{6}=\frac{2}{3}$

Q4 Determine the sum: $\frac{7}{3}+\frac{4}{3}$

\# \# \# • \# \# \# • \# \# \# • \# \# \# • \# \# \# • \# \# \# • \# \# \# • \# \# \# • \# \# \#

A4 $3\frac{2}{3}$: $\frac{7}{3}+\frac{4}{3}=\frac{11}{3}=3\frac{2}{3}$

Q5 Determine the sum:

a. $\frac{5}{7}+\frac{6}{7}$ **b.** $\frac{1}{15}+\frac{9}{15}$ **c.** $\frac{11}{15}+\frac{9}{15}$

\# \# \# • \# \# \# • \# \# \# • \# \# \# • \# \# \# • \# \# \# • \# \# \# • \# \# \# • \# \# \#

A5 **a.** $1\frac{4}{7}$ **b.** $\frac{2}{3}$ **c.** $1\frac{1}{3}$

Q6 Determine the sum:

a. $\frac{3}{8}+\frac{5}{8}$ **b.** $\frac{5}{11}+\frac{4}{11}$ **c.** $\frac{5}{48}+\frac{11}{48}$

\# \# \# • \# \# \# • \# \# \# • \# \# \# • \# \# \# • \# \# \# • \# \# \# • \# \# \# • \# \# \#

A6 **a.** 1 **b.** $\frac{9}{11}$ **c.** $\frac{1}{3}$

---

3 In an addition involving mixed numbers, it is usually better to add the whole numbers and fractions separately.

Examples:

1.
$$\begin{array}{r} 3\frac{2}{7} \\ +1\frac{4}{7} \\ \hline 4\frac{6}{7} \end{array}$$

2.
$$\begin{array}{r} 5\frac{7}{12} \\ +3\frac{2}{12} \\ \hline 8\frac{9}{12} \end{array} = 8\frac{3}{4}$$

Q7 Determine the sum:

$$\begin{array}{r} 3\frac{1}{3} \\ +2\frac{1}{3} \\ \hline \end{array}$$

\# \# \# • \# \# \# • \# \# \# • \# \# \# • \# \# \# • \# \# \# • \# \# \# • \# \# \# • \# \# \#

A7 $5\frac{2}{3}$

Q8 Determine the sum:

a. $$\begin{array}{r} 4\frac{1}{4} \\ +3\frac{2}{4} \\ \hline \end{array}$$

b. $$\begin{array}{r} 7\frac{3}{8} \\ +2\frac{2}{8} \\ \hline \end{array}$$

\# \# \# • \# \# \# • \# \# \# • \# \# \# • \# \# \# • \# \# \# • \# \# \# • \# \# \# • \# \# \#

A8 **a.** $7\frac{3}{4}$ **b.** $9\frac{5}{8}$

4 In an addition involving mixed numbers, the sum of the fractions will often be an improper fraction that must be changed to a mixed number and added to the whole-number sum.

Examples:

1. $$\begin{array}{r} 2\frac{5}{8} \\ +3\frac{6}{8} \\ \hline 5\frac{11}{8} \end{array} = 5 + 1\frac{3}{8} = 6\frac{3}{8}$$

2. $$\begin{array}{r} 2\frac{13}{16} \\ +4\frac{15}{16} \\ \hline 6\frac{28}{16} \end{array} = 6 + 1\frac{12}{16} = 7\frac{12}{16} = 7\frac{3}{4}$$

Q9 Determine the sum:

$$\begin{array}{r} 6\frac{1}{2} \\ +7\frac{1}{2} \\ \hline \end{array}$$

\# \# \# • \# \# \# • \# \# \# • \# \# \# • \# \# \# • \# \# \# • \# \# \# • \# \# \# • \# \# \#

A9 14: $6\frac{1}{2} + 7\frac{1}{2} = 13\frac{2}{2} = 13 + 1 = 14$

Q10 Determine the sum:

a. $\begin{array}{r} 1\frac{3}{8} \\ +2\frac{1}{8} \\ \hline \end{array}$

b. $\begin{array}{r} 1\frac{5}{8} \\ +2\frac{7}{8} \\ \hline \end{array}$

\# \# \# • \# \# \# • \# \# \# • \# \# \# • \# \# \# • \# \# \# • \# \# \# • \# \# \# • \# \# \#

A10 **a.** $3\frac{1}{2}$ **b.** $4\frac{1}{2}$: $1\frac{5}{8}+2\frac{7}{8}=3\frac{12}{8}=3+1\frac{4}{8}=4\frac{1}{2}$

Q11 Determine the sum:

a. $15\frac{1}{12}+8\frac{8}{12}$ **b.** $7\frac{1}{8}+2$ **c.** $12+9\frac{1}{3}$

\# \# \# • \# \# \# • \# \# \# • \# \# \# • \# \# \# • \# \# \# • \# \# \# • \# \# \# • \# \# \#

A11 **a.** $23\frac{3}{4}$ **b.** $9\frac{1}{8}$ **c.** $21\frac{1}{3}$

Q12 Determine the sum:

a. $9\frac{3}{4}+4\frac{2}{4}+3\frac{1}{4}$ **b.** $4\frac{5}{8}+\frac{7}{8}+2\frac{6}{8}$

\# \# \# • \# \# \# • \# \# \# • \# \# \# • \# \# \# • \# \# \# • \# \# \# • \# \# \# • \# \# \#

A12 **a.** $17\frac{1}{2}$ **b.** $8\frac{1}{4}$

5 The sum $\frac{1}{2}+\frac{3}{4}$ cannot be written as one fraction as long as the two fractions are in this form. However, since the form of any fraction can be changed by multiplying both the numerator and denominator by the same number (not zero), the forms of the fractions in a sum can be changed so that they have the same denominator. For example, multiply numerator and denominator of $\frac{1}{2}$ by 2 so that both fractions will have the common denominator 4:

$$\frac{1}{2}=\frac{1\times 2}{2\times 2}=\frac{2}{4}$$

Hence,

$$\frac{1}{2}+\frac{3}{4}=\frac{2}{4}+\frac{3}{4}=\frac{5}{4}=1\frac{1}{4}$$

Q13 Multiply numerator and denominator of $\frac{2}{3}$ and $\frac{3}{4}$ by 4 and 3, respectively, so that they have the common denominator 12.

\# \# \# • \# \# \# • \# \# \# • \# \# \# • \# \# \# • \# \# \# • \# \# \# • \# \# \# • \# \# \#

A13 $\frac{2}{3} = \frac{2 \times 4}{3 \times 4} = \frac{8}{12}$ and $\frac{3}{4} = \frac{3 \times 3}{4 \times 3} = \frac{9}{12}$

Q14 The sum $\frac{2}{3} + \frac{3}{4}$ can be expressed as $\frac{}{12} + \frac{}{12}$.

\# \# \# • \# \# \# • \# \# \# • \# \# \# • \# \# \# • \# \# \# • \# \# \# • \# \# \# • \# \# \#

A14 $\frac{8}{12} + \frac{9}{12}$

---

6 To add fractions with unlike denominators, it is necessary first to find a common denominator. Addition of fractions will be simpler if the common denominator is the smallest number possible. This number, the *least common denominator* (*LCD*), is the smallest number that is exactly divisible by each of the original denominators. The LCD of $\frac{2}{3}$ and $\frac{3}{4}$ is 12, because 12 is the smallest number divisible by 3 and 4.

Example: Find the least common denominator for $\frac{1}{9}$ and $\frac{2}{6}$.

Solution: $18 \div 9 = 2$ and $18 \div 6 = 3$

18 is the LCD, because 18 is the smallest number divisible by both 9 and 6.

Q15 Find the LCD for $\frac{1}{4}$ and $\frac{3}{8}$.

\# \# \# • \# \# \# • \# \# \# • \# \# \# • \# \# \# • \# \# \# • \# \# \# • \# \# \# • \# \# \#

A15 8: 8 is the smallest number divisible by both 4 and 8.

Q16 Find the LCD for $\frac{3}{4}$ and $\frac{1}{16}$.

\# \# \# • \# \# \# • \# \# \# • \# \# \# • \# \# \# • \# \# \# • \# \# \# • \# \# \# • \# \# \#

A16 16: 16 is the smallest number divisible by both 4 and 16.

Q17 Find the LCD for $\frac{1}{3}$ and $\frac{5}{6}$.

\# \# \# • \# \# \# • \# \# \# • \# \# \# • \# \# \# • \# \# \# • \# \# \# • \# \# \# • \# \# \#

A17 6: 6 is the smallest number divisible by both 3 and 6.

---

7 It is not always possible to determine the LCD by observation. One method for finding the LCD is to list the multiples of the larger denominator until a multiple is obtained that is divisible by the smaller denominator.

Example: Find the LCD for $\frac{1}{4}$ and $\frac{2}{5}$.

Solution: Multiples of 5 are obtained by multiplying 5 by 1, 2, 3, 4, and so on, respectively.

$5 \times 1 = 5$
$5 \times 2 = 10$
$5 \times 3 = 15$
$5 \times 4 = 20$
$5 \times 5 = 25$
$5 \times 6 = 30$

The smallest multiple of 5 divisible by 4 is 20; hence, 20 is the LCD for $\frac{1}{4}$ and $\frac{2}{5}$.

Q18 List the first five multiples of 8. ____________

\# \# \# • \# \# \# • \# \# \# • \# \# \# • \# \# \# • \# \# \# • \# \# \# • \# \# \# • \# \# \#

A18 8, 16, 24, 32, and 40: $8 \times 1 = 8$, $8 \times 2 = 16$, $8 \times 3 = 24$, and so on.

Q19 Find the LCD for $\frac{1}{6}$ and $\frac{1}{8}$ (find the smallest multiple of 8 divisible by 6).

\# \# \# • \# \# \# • \# \# \# • \# \# \# • \# \# \# • \# \# \# • \# \# \# • \# \# \# • \# \# \#

A19 24: The multiples of 8 are 8, 16, 24, . . ., and 24 is the smallest multiple of 8 divisible by 6. (". . ." means "and so forth.")

Q20 Find the LCD for $\frac{7}{20}$ and $\frac{3}{15}$.

\# \# \# • \# \# \# • \# \# \# • \# \# \# • \# \# \# • \# \# \# • \# \# \# • \# \# \# • \# \# \#

A20 60: 60 is the smallest multiple of 20 divisible by 15 (20, 40, 60, 80, 100, . . .) and $60 \div 15 = 4$.

8 The technique for finding the LCD for three or more fractions is the same as employed with two fractions. That is, find the smallest multiple of the largest denominator that is divisible by the other denominators.

As an example, the LCD for $\frac{1}{3}$, $\frac{1}{4}$, and $\frac{1}{8}$ is 24. Multiples of 8: 8, 16, 24, 32, ...

Note that 16 is divisible by 4 and 8 but is not divisible by 3; hence, 16 is not the LCD. Also, 24 is divisible by 4 and 24 is divisible by 3 ($24 \div 4 = 6$ and $24 \div 3 = 8$); hence, 24 is the LCD for $\frac{1}{3}$, $\frac{1}{4}$, and $\frac{1}{8}$.

Q21 Find the LCD for $\frac{1}{8}$, $\frac{1}{6}$, and $\frac{1}{12}$ (determine the smallest multiple of 12 divisible by 8 and 6).

\# \# \# • \# \# \# • \# \# \# • \# \# \# • \# \# \# • \# \# \# • \# \# \# • \# \# \# • \# \# \#

A21 24: $24 \div 12 = 2$, $24 \div 8 = 3$, and $24 \div 6 = 4$.

Q22 Find the LCD for $\frac{1}{12}$ and $\frac{1}{16}$.

• # # # • # # # • # # # • # # # • # # # • # # # • # # # • # #

A22 48: $48 \div 12 = 4$ and $48 \div 16 = 3$.

Q23 Find the LCD for the following fractions:

a. $\frac{3}{4}, \frac{1}{5}$ **b.** $\frac{7}{15}, \frac{6}{45}$ **c.** $\frac{1}{4}, \frac{5}{6}, \frac{8}{16}$

• # # # • # # # • # # # • # # # • # # # • # # # • # # # • # #

A23 **a.** 20 **b.** 45 **c.** 48

---

9 To add fractions with unlike denominators it will be necessary to rewrite them as equivalent fractions with a common denominator. As an example, add $\frac{2}{3} + \frac{3}{4}$. The LCD is 12, because 12 is the smallest multiple of 4 divisible by 3. It is often easier to add fractions vertically.

$$\frac{2}{3} = \frac{\quad}{12}$$
$$+\frac{3}{4} = \frac{\quad}{12}$$

Write $\frac{2}{3}$ and $\frac{3}{4}$ as equivalent fractions with denominator 12.

$12 \div 3 = 4$; hence, $\frac{2}{3} = \frac{2 \times 4}{3 \times 4}$

$12 \div 4 = 3$; hence, $\frac{3}{4} = \frac{3 \times 3}{4 \times 3}$

$$\frac{2}{3} = \frac{2 \times 4}{3 \times 4} = \frac{8}{12}$$
$$+\frac{3}{4} = \frac{3 \times 3}{4 \times 3} = \frac{9}{12}$$
$$\frac{17}{12} = 1\frac{5}{12}$$

(this step is usually done mentally)

Therefore, $\frac{2}{3} + \frac{3}{4} = 1\frac{5}{12}$.

Q24 Rewrite the fractions as equivalent fractions with the least common denominator.

$$\frac{1}{4} = \_\_\_\_\_$$
$$+\frac{3}{8} = \_\_\_\_\_$$

• # # # • # # # • # # # • # # # • # # # • # # # • # # # • # #

A24 $\frac{1}{4} = \frac{2}{8}$: 8 is the LCD and

$+\frac{3}{8} = \frac{3}{8}$ $8 \div 4 = 2$; multiply $\frac{1 \times 2}{4 \times 2}$

$8 \div 8 = 1$; multiply $\frac{3 \times 1}{8 \times 1}$

Q25 Complete the addition: $\frac{1}{4} = \frac{2}{8}$

$+\frac{3}{8} = \frac{3}{8}$

• # # # • # # # • # # # • # # # • # # # • # # # • # # # • # #

A25 $\frac{5}{8}$

Q26 Rewrite the fractions as equivalent fractions with the LCD:

$\frac{3}{4} =$ ______

$+\frac{1}{5} =$ ______

• # # # • # # # • # # # • # # # • # # # • # # # • # # # • # #

A26 $\frac{3}{4} = \frac{15}{20}$: 20 is the LCD and

$+\frac{1}{5} = \frac{4}{20}$ $20 \div 4 = 5$; multiply $\frac{3 \times 5}{4 \times 5}$

$20 \div 5 = 4$; multiply $\frac{1 \times 4}{5 \times 4}$

Q27 Find the sum of the fractions in A26. ______

• # # # • # # # • # # # • # # # • # # # • # # # • # # # • # #

A27 $\frac{19}{20}$

Q28 Determine the sum: $\frac{1}{2}$

$+\frac{3}{4}$

• # # # • # # # • # # # • # # # • # # # • # # # • # # # • # #

A28 $1\frac{1}{4}$: $\frac{1}{2} = \frac{2}{4}$ $\left(\text{multiply } \frac{1 \times 2}{2 \times 2}\right)$

$+\frac{3}{4} = \frac{3}{4}$

$\frac{5}{4} = 1\frac{1}{4}$

Q29 Determine the sum: $\begin{array}{r} \frac{2}{3} \\ +\frac{1}{9} \\ \hline \end{array}$

\# \# \# • \# \# \# • \# \# \# • \# \# \# • \# \# \# • \# \# \# • \# \# \# • \# \# \# • \# \# \#

A29 $\frac{7}{9}$: $\frac{2}{3} = \frac{6}{9}$ $\left(\text{multiply } \frac{2 \times 3}{3 \times 3}\right)$

$+\frac{1}{9} = \frac{1}{9}$

$\frac{7}{9}$

Q30 Determine the sum: $\begin{array}{r} \frac{5}{6} \\ +\frac{7}{8} \\ \hline \end{array}$

\# \# \# • \# \# \# • \# \# \# • \# \# \# • \# \# \# • \# \# \# • \# \# \# • \# \# \# • \# \# \#

A30 $1\frac{17}{24}$: $\frac{5}{6} = \frac{20}{24}$ $\left(\text{multiply } \frac{5 \times 4}{6 \times 4}\right)$

$+\frac{7}{8} = \frac{21}{24}$ $\left(\text{multiply } \frac{7 \times 3}{8 \times 3}\right)$

$\frac{41}{24} = 1\frac{17}{24}$

Q31 Determine the sum: $\begin{array}{r} \frac{3}{4} \\ +\frac{3}{5} \\ \hline \end{array}$

\# \# \# • \# \# \# • \# \# \# • \# \# \# • \# \# \# • \# \# \# • \# \# \# • \# \# \# • \# \# \#

A31 $1\frac{7}{20}$: $\frac{3}{4} = \frac{15}{20}$ $\left(\text{multiply } \frac{3 \times 5}{4 \times 5}\right)$

$+\frac{3}{5} = \frac{12}{20}$ $\left(\text{multiply } \frac{3 \times 4}{5 \times 4}\right)$

Q32 Determine the sum:

a. $\begin{array}{r} \frac{5}{6} \\ +\frac{5}{12} \\ \hline \end{array}$ **b.** $\begin{array}{r} \frac{2}{3} \\ +\frac{1}{4} \\ \hline \end{array}$

\# \# \# • \# \# \# • \# \# \# • \# \# \# • \# \# \# • \# \# \# • \# \# \# • \# \# \# • \# \# \#

A32 **a.** $1\frac{1}{4}$: $\frac{5}{6} = \frac{10}{12}$, $+\frac{5}{12} = \frac{5}{12}$

b. $\frac{11}{12}$: $\frac{2}{3} = \frac{8}{12}$, $+\frac{1}{4} = \frac{3}{12}$

Q33 Determine the sum:

a. $\frac{3}{4} + \frac{1}{9}$

b. $\frac{3}{5} + \frac{5}{6}$

• ### • ### • ### • ### • ### • ### • ### •

A33 **a.** $\frac{31}{36}$: $\frac{3}{4} = \frac{27}{36}$, $+\frac{1}{9} = \frac{4}{36}$

b. $1\frac{13}{30}$: $\frac{3}{5} = \frac{18}{30}$, $+\frac{5}{6} = \frac{25}{30}$

Q34 Determine the sum:

a. $\frac{1}{6} + \frac{4}{7}$

b. $\frac{2}{3} + \frac{5}{6}$

c. $\frac{1}{12} + \frac{1}{16}$

• ### • ### • ### • ### • ### • ### • ### •

A34 **a.** $\frac{31}{42}$ **b.** $1\frac{1}{2}$ **c.** $\frac{7}{48}$

Q35 Determine the sum:

a. $\frac{3}{4} + \frac{2}{5} + \frac{3}{10}$

b. $\frac{2}{3} + \frac{5}{6} + \frac{4}{9}$

• ### • ### • ### • ### • ### • ### • ### •

A35 **a.** $1\frac{9}{20}$: $\frac{3}{4} = \frac{15}{20}$, $\frac{2}{5} = \frac{8}{20}$, $+\frac{3}{10} = \frac{6}{20}$

b. $1\frac{17}{18}$: $\frac{2}{3} = \frac{12}{18}$, $\frac{5}{6} = \frac{15}{18}$, $+\frac{4}{9} = \frac{8}{18}$

---

10 The sum of mixed numbers with unlike fractions is determined as follows: Add the fractions and whole numbers separately, and then combine the separate sums. For example,

$$2\frac{5}{6} = 2\frac{15}{18}$$
$$+8\frac{4}{9} = 8\frac{8}{18}$$
$$= 10\frac{23}{18} = 10 + 1\frac{5}{18} = 11\frac{5}{18}$$

Q36 Determine the sum:

$$\begin{array}{r} 1\frac{2}{3} \\ +3\frac{1}{4} \\ \hline \end{array}$$

\# \# \# • \# \# \# • \# \# \# • \# \# \# • \# \# \# • \# \# \# • \# \# \# • \# \# \# • \# \# \#

A36 $4\frac{11}{12}$:

$$\begin{array}{rl} 1\frac{2}{3} & = 1\frac{8}{12} \\ +3\frac{1}{4} & = 3\frac{3}{12} \\ \hline & = 4\frac{11}{12} \end{array}$$

Q37 Determine the sum: $1\frac{2}{3}+3\frac{1}{4}+\frac{5}{8}$

\# \# \# • \# \# \# • \# \# \# • \# \# \# • \# \# \# • \# \# \# • \# \# \# • \# \# \# • \# \# \#

A37 $5\frac{13}{24}$

Q38 A table top is $\frac{9}{16}$ inch thick. The legs of this table are $28\frac{3}{8}$ inches high. What is the distance from the floor to the top of the table?

\# \# \# • \# \# \# • \# \# \# • \# \# \# • \# \# \# • \# \# \# • \# \# \# • \# \# \# • \# \# \#

A38 $28\frac{15}{16}$ inches

This completes the instruction for this section.

1.3 EXERCISE

1. Two fractions can be combined into a single fraction by addition only if they have the same ____________.

2. The rule of adding fractions with the same denominator is to ____________.

3. Addition of fractions with unlike denominators is performed by first finding the ____________.

4. Determine the sum:

a. $\begin{array}{r} \frac{3}{16} \\ +\frac{10}{16} \\ \hline \end{array}$ **b.** $\begin{array}{r} \frac{11}{14} \\ +\frac{12}{14} \\ \hline \end{array}$ **c.** $\begin{array}{r} \frac{4}{27} \\ +\frac{5}{27} \\ \hline \end{array}$ **d.** $\frac{5}{6}+\frac{4}{6}$

e. $\begin{array}{r} 3\frac{2}{3} \\ +\frac{1}{3} \\ \hline \end{array}$ **f.** $\begin{array}{r} 2\frac{3}{7} \\ +4\frac{4}{7} \\ \hline \end{array}$

5. Find the sum:

a. $\begin{array}{r}\frac{1}{8}\\ +\frac{3}{4}\\ \hline\end{array}$ **b.** $\begin{array}{r}\frac{3}{4}\\ +\frac{7}{8}\\ \hline\end{array}$ **c.** $\begin{array}{r}\frac{9}{11}\\ +\frac{5}{6}\\ \hline\end{array}$ **d.** $\begin{array}{r}\frac{5}{8}\\ +\frac{7}{9}\\ \hline\end{array}$

e. $\frac{2}{3}+\frac{7}{9}$ **f.** $\frac{1}{8}+\frac{5}{6}$ **g.** $\frac{1}{2}+\frac{5}{9}$ **h.** $\frac{7}{14}+\frac{2}{7}$

i. $\frac{5}{8}+\frac{5}{7}$ **j.** $\frac{1}{6}+\frac{5}{9}$ **k.** $\frac{7}{12}+\frac{1}{18}$ **l.** $4\frac{3}{8}+\frac{7}{16}$

m. $\frac{2}{7}+\frac{1}{3}+\frac{1}{2}$ **n.** $3\frac{2}{3}+2\frac{5}{6}+1\frac{7}{12}$

EXERCISE ANSWERS

1. denominator
2. Add the numerators and place the sum over the denominator.
3. LCD (least common denominator)
4. a. $\frac{13}{16}$ **b.** $1\frac{9}{14}$ **c.** $\frac{1}{3}$ **d.** $1\frac{1}{2}$
e. 4 **f.** 7
5. a. $\frac{7}{8}$ **b.** $1\frac{5}{8}$ **c.** $1\frac{43}{66}$ **d.** $1\frac{29}{72}$
e. $1\frac{4}{9}$ **f.** $\frac{23}{24}$ **g.** $1\frac{1}{18}$ **h.** $\frac{11}{14}$
i. $1\frac{19}{56}$ **j.** $\frac{13}{18}$ **k.** $\frac{23}{36}$ **l.** $4\frac{13}{16}$
m. $1\frac{5}{42}$ **n.** $8\frac{1}{12}$

1.4 PRIME AND COMPOSITE NUMBERS

1 A *prime number* is a whole number greater than 1 divisible by exactly two different factors—itself and 1 only. All other whole numbers are *composite numbers*. For example,

1 is *not* prime, because 1 is not divisible by two different factors.

2 is prime, because 2 is divisible by itself and 1 only.

3 is prime, because 3 is divisible by itself and 1 only.

4 is *not* prime, because 4 is divisible by more than two factors. 4 is divisible by 4, 1, and 2. Therefore, 4 is composite.

Q1 What are the factors (divisors) of 5? ________

• # # # • # # # • # # # • # # # • # # # • # # # • # # # • # #

A1 5 and 1

Q2 Is 5 a prime number? ________ Why? ________________________________

\# # # • # # # • # # # • # # # • # # # • # # # • # # # • # # # • # # #

A2 yes; 5 has exactly two factors, 5 and 1.

Q3 What are the factors (divisors) of 6? __________

\# # # • # # # • # # # • # # # • # # # • # # # • # # # • # # # • # # #

A3 6, 1, 2, and 3

Q4 Is 6 a prime number? ________ Why? ________________________________

\# # # • # # # • # # # • # # # • # # # • # # # • # # # • # # # • # # #

A4 no; 6 has more than two factors (6, 1, 2, and 3).

Q5 Circle the prime numbers:

1 2 3 4 5 6 7 8 9 10 11 12 13 14 15 16 17 18 19 20 21
22 23 24 25

\# # # • # # # • # # # • # # # • # # # • # # # • # # # • # # # • # # #

A5 1 (2) (3) 4 (5) 6 (7) 8 9 10 (11) 12 (13) 14 15 16 (17) 18 (19) 20 21
22 (23) 24 25

Q6 Explain why 9 is not a prime. ________________________________

\# # # • # # # • # # # • # # # • # # # • # # # • # # # • # # # • # # #

A6 9 has more than two factors (9, 1, and 3).

Q7 Explain why 11 is prime. ________________________________

\# # # • # # # • # # # • # # # • # # # • # # # • # # # • # # # • # # #

A7 11 has exactly two factors, itself and 1.

---

2 The number 2 is the first prime number. The multiples of 2 are 2, 4, 6, 8, 10, 12, and so on. Any multiple of 2 is divisible by itself (that is, 2) and by 1; so the multiples of 2, except 2, are composite. For example, 8 is a multiple of 2 and is divisible by 8, 4, 2, and 1; hence, 8 is composite.

Q8 List the first five multiples of 3. ____________

\# # # • # # # • # # # • # # # • # # # • # # # • # # # • # # # • # # #

A8 3, 6, 9, 12, 15

Q9 Which of the multiples of 3 are prime? ______

\# # # • # # # • # # # • # # # • # # # • # # # • # # # • # # # • # # #

A9 3

Q10 List the first five multiples of 5. ____________

\# # # • # # # • # # # • # # # • # # # • # # # • # # # • # # # • # # #

A10 5, 10, 15, 20, 25

Q11 Which of the multiples of 5 are prime? ______

• # # # • # # # • # # # • # # # • # # # • # # # • # # # • # #

A11 5

3 The preceding problems suggest a technique for finding the prime numbers. On the list given in Q12:

1. Circle the first prime, 2, and cross out all other multiples of 2 (every other number counting from 2).
2. Circle the second prime, 3, and cross out all other multiples of 3 (every third number counting from 3).
3. Circle 5 and cross out all other multiples of 5 (every fifth number counting from 5).
4. Circle 7 and cross out all other multiples of 7.

Continue in this manner until only primes remain on your list. This technique for locating primes is called the *sieve of Eratosthenes.*

Q12 Use the technique discussed in Frame 3 to locate all primes in the list below.

	2	3	4	5	6	7	8	9	10
11	12	13	14	15	16	17	18	19	20
21	22	23	24	25	26	27	28	29	30
31	32	33	34	35	36	37	38	39	40
41	42	43	44	45	46	47	48	49	50
51	52	53	54	55	56	57	58	59	60
61	62	63	64	65	66	67	68	69	70
71	72	73	74	75	76	77	78	79	80
81	82	83	84	85	86	87	88	89	90
91	92	93	94	95	96	97	98	99	100

• # # # • # # # • # # # • # # # • # # # • # # # • # # # • # #

A12

	(2)	(3)	~~4~~	(5)	~~6~~	(7)	~~8~~	~~9~~	~~10~~
(11)	~~12~~	(13)	~~14~~	~~15~~	~~16~~	(17)	~~18~~	(19)	~~20~~
~~21~~	~~22~~	(23)	~~24~~	~~25~~	~~26~~	~~27~~	~~28~~	(29)	~~30~~
(31)	~~32~~	~~33~~	~~34~~	~~35~~	~~36~~	(37)	~~38~~	~~39~~	~~40~~
(41)	~~42~~	(43)	~~44~~	~~45~~	~~46~~	(47)	~~48~~	~~49~~	~~50~~
~~51~~	~~52~~	(53)	~~54~~	~~55~~	~~56~~	~~57~~	~~58~~	(59)	~~60~~
(61)	~~62~~	~~63~~	~~64~~	~~65~~	~~66~~	(67)	~~68~~	~~69~~	~~70~~
(71)	~~72~~	(73)	~~74~~	~~75~~	~~76~~	~~77~~	~~78~~	(79)	~~80~~
~~81~~	~~82~~	(83)	~~84~~	~~85~~	~~86~~	~~87~~	~~88~~	(89)	~~90~~
~~91~~	~~92~~	~~93~~	~~94~~	~~95~~	~~96~~	(97)	~~98~~	~~99~~	~~100~~

4 Often a number can be expressed as a product of factors in many different ways. For example,

$20 = 4 \times 5$ $\quad 20 = 10 \times 2$ $\quad 20 = 2 \times 2 \times 5$

However, a number may be expressed as a product of prime factors in only one way:

$20 = 4 \times 5$ $\quad\quad$ $20 = 10 \times 2$

$= 2 \times 2 \times 5$ $\quad\quad$ $= 5 \times 2 \times 2$

By expressing 4 and 10 as a product of primes, 20 has been expressed as a product of primes two times in exactly the same way (remember that the order of the factors is not important).

Q13 Express 8 as a product of primes. ______________

• ### • ### • ### • ### • ### • ### • ### •

A13 $8 = 2 \times 2 \times 2$

Q14 Express 30 as a product of primes. ______________

• ### • ### • ### • ### • ### • ### • ### •

A14 $30 = 2 \times 3 \times 5$

---

5 When expressing a composite number as a product of primes, it will be helpful to recall the following divisibility tests:

1. If the last digit of the number is divisible by 2, the number is divisible by 2.

Example: 304; 4 is divisible by 2, so 304 is divisible by 2. $304 \div 2 = 152$.

2. If the sum of the digits is divisible by 3, the number is divisible by 3.

Example: 411; the sum of the digits, $4 + 1 + 1 = 6$, is divisible by 3, so 411 is divisible by 3. $411 \div 3 = 137$.

3. If the last digit is 0 or 5, the number is divisible by 5.

Example: 85; the last digit is 5, so 85 is divisible by 5. $85 \div 5 = 17$.

The sieve of Eratosthenes can be used to help decide whether or not a number is prime.

Example: Express 45 as a product of primes.

Solution: The last digit is 5, so 45 is divisible by 5.

$45 \div 5 = 9$, hence $45 = 9 \times 5$

$= \underbrace{3 \times 3}\times 5$ (prime factors of 9)

Q15 Express 54 as a product of primes. ______________

• ### • ### • ### • ### • ### • ### • ### •

A15 $54 = 2 \times 3 \times 3 \times 3$: $54 = 2 \times 27$ (divide 54 by 2)

$= 2 \times 3 \times 9$ (divide 27 by 3)

$= 2 \times 3 \times 3 \times 3$

Other combinations of the same prime factors will give the same result.

Q16 Express 110 as a product of primes. ______________

• ### • ### • ### • ### • ### • ### • ### •

A16 $110 = 2 \times 5 \times 11$: $110 = 2 \times 55$

$= 2 \times 5 \times 11$

Q17 Express as a product of primes:

a. 90 ____________ **b.** 53 ____________

\# # # • # # # • # # # • # # # • # # # • # # # • # # # • # # # • # # #

A17 **a.** $90 = 2 \times 3 \times 3 \times 5$ **b.** 53 is prime

6 Prime factors can be used to determine the least common denominator (LCD) for two or more fractions. The denominators are expressed as the product of prime factors. The LCD is the product of these prime factors; *each factor is used as many times as it is found in any one of the denominators.*

Suppose that 12, 15, and 10 are the denominators for three fractions. First, list the prime factors of each:

$12 = 2 \times 2 \times 3$
$15 = 3 \times 5$
$10 = 2 \times 5$

Since 2 is used as a factor twice in the number 12, it is used twice in the LCD. The numbers 3 and 5 are used only once in each of the numbers 12, 10, and 15, so they are used only once in the LCD:

$\text{LCD} = 2 \times 2 \times 3 \times 5$
$= 60$

Q18 To find the sum $\frac{1}{54} + \frac{1}{20} + \frac{1}{12}$, first express the denominators as prime factors.

54 = ____________ 20 = ____________ 12 = ____________

\# # # • # # # • # # # • # # # • # # # • # # # • # # # • # # # • # # #

A18 $54 = 2 \times 3 \times 3 \times 3$, $20 = 2 \times 2 \times 5$, $12 = 2 \times 2 \times 3$

Q19 In the LCD the factor

a. 2 is used how many times? ______ **b.** 3 is used how many times? ______

c. 5 is used how many times? ______

\# # # • # # # • # # # • # # # • # # # • # # # • # # # • # # # • # # #

A19 **a.** twice **b.** three **c.** once

Q20 Find the LCD for $\frac{1}{54}$, $\frac{1}{20}$, and $\frac{1}{12}$.

\# # # • # # # • # # # • # # # • # # # • # # # • # # # • # # # • # # #

A20 540: $\text{LCD} = 2 \times 2 \times 3 \times 3 \times 3 \times 5 = 540$

The factor 2 is used twice, because it is used twice in both 20 and 12. The factor 3 is used 3 times, because it is used 3 times in 54. The factor 5 is used once, because it is used only once in any of the denominators.

Q21 Use prime factors to find the LCD of the following denominators of fractions: 6, 8, and 9.

\# # # • # # # • # # # • # # # • # # # • # # # • # # # • # # # • # # #

A21 72: $6 = 2 \times 3$, $8 = 2 \times 2 \times 2$, $9 = 3 \times 3$; LCD $= 2 \times 2 \times 2 \times 3 \times 3$

Q22 Find the LCD for $\frac{1}{5}, \frac{1}{7}$, and $\frac{1}{10}$.

\# \# \# • \# \# \# • \# \# \# • \# \# \# • \# \# \# • \# \# \# • \# \# \# • \# \# \# • \# \# \#

A22 70: 5 (prime), 7 (prime), $10 = 2 \times 5$; LCD $= 2 \times 5 \times 7$

Q23 Find the LCD for $\frac{1}{12}, \frac{1}{5}$, and $\frac{1}{6}$.

\# \# \# • \# \# \# • \# \# \# • \# \# \# • \# \# \# • \# \# \# • \# \# \# • \# \# \# • \# \# \#

A23 60: $12 = 2 \times 2 \times 3$, 5, $6 = 2 \times 3$; LCD $= 2 \times 2 \times 3 \times 5$

Q24 Find the LCD for $\frac{1}{8}, \frac{1}{12}$, and $\frac{1}{15}$.

\# \# \# • \# \# \# • \# \# \# • \# \# \# • \# \# \# • \# \# \# • \# \# \# • \# \# \# • \# \# \#

A24 120: $8 = 2 \times 2 \times 2$, $12 = 2 \times 2 \times 3$, $15 = 3 \times 5$; LCD $= 2 \times 2 \times 2 \times 3 \times 5$

Q25 Find the sum: $\frac{5}{6} + \frac{11}{15} + \frac{17}{20}$

\# \# \# • \# \# \# • \# \# \# • \# \# \# • \# \# \# • \# \# \# • \# \# \# • \# \# \# • \# \# \#

A25 $2\frac{5}{12}$:

$$\frac{5}{6} = \frac{50}{60}$$
$$\frac{11}{15} = \frac{44}{60}$$
$$+\frac{17}{20} = \frac{51}{60}$$
$$\frac{145}{60} = 2\frac{25}{60}$$

$6 = 2 \times 3$
$15 = 3 \times 5$
$20 = 2 \times 2 \times 5$
LCD $= 2 \times 2 \times 3 \times 5 = 60$

Q26 Find the sum: $\frac{7}{12} + \frac{12}{21} + \frac{18}{35}$

\# \# \# • \# \# \# • \# \# \# • \# \# \# • \# \# \# • \# \# \# • \# \# \# • \# \# \# • \# \# \#

A26 $1\frac{281}{420}$:

$$\begin{aligned} \frac{7}{12} &= \frac{245}{420} \\ \frac{12}{21} &= \frac{240}{420} \\ +\frac{18}{35} &= \frac{216}{420} \\ & \quad \frac{701}{420} \end{aligned}$$

$12 = 2 \times 2 \times 3$
$21 = 3 \times 7$
$35 = 5 \times 7$
$\text{LCD} = 2 \times 2 \times 3 \times 5 \times 7 = 420$

This completes the instruction for this section.

1.4 EXERCISE

1. A prime number is divisible by ______________.
2. List the prime numbers 1–20.
3. If a whole number greater than 1 is not prime, it is ______________.
4. Express the following as a product of primes:
   a. 56 **b.** 78 **c.** 97 **d.** 185
   e. 36 **f.** 100 **g.** 1,000 **h.** 13
5. Find the LCD for the following fractions:
   a. $\frac{1}{18}, \frac{1}{12}, \frac{1}{24}$ **b.** $\frac{1}{14}, \frac{1}{35}, \frac{1}{10}$ **c.** $\frac{1}{20}, \frac{1}{15}, \frac{1}{12}$ **d.** $\frac{1}{9}, \frac{1}{36}, \frac{1}{12}$
   e. $\frac{1}{6}, \frac{1}{8}, \frac{1}{3}$ **f.** $\frac{1}{15}, \frac{1}{14}, \frac{1}{35}$ **g.** $\frac{1}{48}, \frac{1}{75}, \frac{1}{27}$ **h.** $\frac{1}{4}, \frac{1}{36}, \frac{1}{27}$
6. Find the sum:
   a. $\frac{7}{9}+\frac{5}{8}+\frac{9}{14}$ **b.** $\frac{9}{11}+\frac{15}{33}+\frac{17}{30}$ **c.** $\frac{9}{20}+\frac{14}{15}+\frac{7}{8}$

1.4 EXERCISE ANSWERS

1. itself and one only
2. 2, 3, 5, 7, 11, 13, 17, 19
3. composite
4. **a.** $2 \times 2 \times 2 \times 7$ **b.** $2 \times 2 \times 13$ **c.** prime
   d. 5×37 **e.** $2 \times 2 \times 3 \times 3$ **f.** $2 \times 2 \times 5 \times 5$
   g. $2 \times 2 \times 2 \times 5 \times 5 \times 5$ **h.** prime
5. **a.** 72 **b.** 70 **c.** 60
   d. 36 **e.** 24 **f.** 210
   g. 10,800 **h.** 108
6. **a.** $2\frac{23}{504}$ **b.** $1\frac{277}{330}$ **c.** $2\frac{31}{120}$

1.5 SUBTRACTION

1 As in addition, you can subtract one fraction from another if they have the same denominator. Simply subtract numerators and write the difference over the common denominator. For example,

$$\frac{9}{13}-\frac{4}{13}=\frac{9-4}{13}=\frac{5}{13}$$

Q1 Find the difference

a. $\frac{4}{5}-\frac{3}{5}$ **b.** $\frac{5}{8}-\frac{3}{8}$ **c.** $\frac{16}{24}-\frac{9}{24}$

• # # # • # # # • # # # • # # # • # # # • # # # • # # # • # #

A1 **a.** $\frac{1}{5}$: $\frac{4-3}{5}$ **b.** $\frac{1}{4}$: $\frac{5-3}{8}=\frac{2}{8}$ **c.** $\frac{7}{24}$: $\frac{16-9}{24}$

Q2 Find the difference:

a. $\frac{5}{6}-\frac{1}{6}$ **b.** $\frac{11}{12}-\frac{3}{12}$ **c.** $\frac{11}{15}-\frac{4}{15}$

• # # # • # # # • # # # • # # # • # # # • # # # • # # # • # #

A2 **a.** $\frac{2}{3}$ **b.** $\frac{2}{3}$ **c.** $\frac{7}{15}$

2 If two fractions do not have the same denominator, you must change the form of the fractions so that they have a common denominator. As in addition, it will be helpful if the common denominator is the LCD.

Example: Subtract $\frac{3}{4}-\frac{1}{3}$.

Solution:

$$\begin{array}{rcr} \frac{3}{4} & = & \frac{9}{12} \\ -\frac{1}{3} & = & -\frac{4}{12} \\ \hline & & \frac{5}{12} \end{array}$$

LCD = 12

$\frac{3}{4}=\frac{3\times 3}{4\times 3}=\frac{9}{12}$

$\frac{1}{3}=\frac{1\times 4}{3\times 4}=\frac{4}{12}$

Q3 Find the difference:

$$\begin{array}{r} \frac{7}{8} \\ -\frac{3}{4} \\ \hline \end{array}$$

• # # # • # # # • # # # • # # # • # # # • # # # • # # # • # #

A3 $\frac{1}{8}$:

$$\begin{array}{rcr} \frac{7}{8} & = & \frac{7}{8} \\ -\frac{3}{4} & = & -\frac{6}{8} \\ \hline & & \frac{1}{8} \end{array}$$

Q4 Find the difference:

$$\begin{array}{r} \frac{2}{3} \\ -\frac{1}{4} \\ \hline \end{array}$$

\# \# \# • \# \# \# • \# \# \# • \# \# \# • \# \# \# • \# \# \# • \# \# \# • \# \# \# • \# \# \#

A4 $\frac{5}{12}$:
$$\begin{array}{rcr} \frac{2}{3} & = & \frac{8}{12} \\ -\frac{1}{4} & = & -\frac{3}{12} \\ \hline \end{array}$$

Q5 Find the difference: $\frac{11}{16} - \frac{1}{3}$

\# \# \# • \# \# \# • \# \# \# • \# \# \# • \# \# \# • \# \# \# • \# \# \# • \# \# \# • \# \# \#

A5 $\frac{17}{48}$:
$$\begin{array}{rcr} \frac{11}{16} & = & \frac{33}{48} \\ -\frac{1}{3} & = & -\frac{16}{48} \\ \hline \end{array}$$

If you had difficulty finding the LCD, return to Section 2.3 or Section 2.4 and restudy the technique for finding the LCD.

Q6 Find the difference:

$$\begin{array}{r} 8\frac{2}{3} \\ -6\frac{3}{8} \\ \hline \end{array}$$

\# \# \# • \# \# \# • \# \# \# • \# \# \# • \# \# \# • \# \# \# • \# \# \# • \# \# \# • \# \# \#

A6 $2\frac{7}{24}$:
$$\begin{array}{rcr} 8\frac{2}{3} & = & 8\frac{16}{24} \\ -6\frac{3}{8} & = & -6\frac{9}{24} \\ \hline & & 2\frac{7}{24} \end{array}$$

Q7 Find the difference:

a. $\frac{5}{7} - \frac{1}{3}$

b. $\frac{7}{8} - \frac{7}{10}$

\# \# \# • \# \# \# • \# \# \# • \# \# \# • \# \# \# • \# \# \# • \# \# \# • \# \# \# • \# \# \#

A7 **a.** $\frac{8}{21}$

b. $\frac{7}{40}$

Q8 Find the difference:

a. $7\frac{2}{3} - 3\frac{1}{2}$ **b.** $4\frac{5}{16} - 2\frac{1}{4}$

• ### • ### • ### • ### • ### • ### • ### •

A8 **a.** $4\frac{1}{6}$ **b.** $2\frac{1}{16}$

---

3 When subtracting fractions and mixed numbers, it will often be necessary to borrow to find the difference. Consider this example:

$$\begin{array}{rcr} 13\frac{1}{4} & = & 13\frac{2}{8} \\ -6\frac{7}{8} & = & -6\frac{7}{8} \\ \hline \end{array}$$

Since $\frac{2}{8} - \frac{7}{8}$ does not have meaning at this time, it will be necessary to borrow 1 from the 13. The 1 is changed to $\frac{8}{8}$ (any nonzero number divided by itself is 1) and added to $\frac{2}{8}$. The subtraction is then completed.

$$\begin{array}{rcrcr} 13\frac{1}{4} & = & \overset{12}{\cancel{13}}\frac{2}{8} + \frac{8}{8} & = & 12\frac{10}{8} \\ -6\frac{7}{8} & = & -6\frac{7}{8} & = & -6\frac{7}{8} \\ \hline & & & & 6\frac{3}{8} \end{array}$$

Q9 Express 1 with the following denominators:

a. $\frac{\quad}{24}$ **b.** $\frac{\quad}{19}$ **c.** $\frac{\quad}{5}$ **d.** $\frac{\quad}{101}$

• ### • ### • ### • ### • ### • ### • ### •

A9 **a.** $\frac{24}{24}$ **b.** $\frac{19}{19}$ **c.** $\frac{5}{5}$ **d.** $\frac{101}{101}$

Q10 Complete the subtraction problem:

$$\begin{array}{rcr} 21\frac{3}{8} & = & \overset{20}{\cancel{21}}\frac{9}{24} + \frac{24}{24} \\ -17\frac{5}{6} & = & -17\frac{20}{24} \\ \hline \end{array}$$

• ### • ### • ### • ### • ### • ### • ### •

A10 $3\frac{13}{24}$:

$$\begin{array}{r} 20\frac{33}{24} \\ -17\frac{20}{24} \\ \hline 3\frac{13}{24} \end{array}$$

Q11 Find the difference:

$$\begin{array}{r} 3\frac{1}{4} \\ -1\frac{2}{3} \\ \hline \end{array}$$

• # # # • # # # • # # # • # # # • # # # • # # # • # # # • # #

A11 $1\frac{7}{12}$:

$$\begin{array}{rcl} 3\frac{1}{4} = & \overset{2}{\cancel{3}}\frac{3}{12} + \frac{12}{12} = & 2\frac{15}{12} \\ -1\frac{2}{3} = & -1\frac{8}{12} = & -1\frac{8}{12} \\ \hline & & 1\frac{7}{12} \end{array}$$

Q12 Find the difference:

$$\begin{array}{r} 4\frac{1}{5} \\ -1\frac{3}{4} \\ \hline \end{array}$$

• # # # • # # # • # # # • # # # • # # # • # # # • # # # • # #

A12 $2\frac{9}{20}$:

$$\begin{array}{rr} 4\frac{1}{5} = & 3\frac{24}{20} \\ -1\frac{3}{4} = & -1\frac{15}{20} \\ \hline & 2\frac{9}{20} \end{array}$$

4 The difference between a fraction and a whole number can be determined as follows:

$5 - \frac{2}{3}$ is written

$$\begin{array}{r} 5 \\ -\frac{2}{3} \\ \hline \end{array}$$

Borrow 1 from the 5 and change it to $\frac{3}{3}$ and proceed as before:

$$\begin{array}{r} \overset{4}{\cancel{5}}\frac{3}{3} \\ -\frac{2}{3} \\ \hline 4\frac{1}{3} \end{array}$$

Q13 Complete the subtraction problem: $\begin{array}{r} \overset{1}{\not{2}}(\ \ ) \\ -\frac{1}{4} \\ \hline \end{array}$

• # # # • # # # • # # # • # # # • # # # • # # # • # # # • # #

A13 $\begin{array}{r} \overset{1}{\not{2}}\frac{4}{4} \\ -\frac{1}{4} \\ \hline 1\frac{3}{4} \end{array}$

Q14 Find the difference: $\begin{array}{r} 8 \\ -1\frac{2}{3} \\ \hline \end{array}$

• # # # • # # # • # # # • # # # • # # # • # # # • # # # • # #

A14 $6\frac{1}{3}$: $\begin{array}{r} \overset{7}{\not{8}}\frac{3}{3} \\ -1\frac{2}{3} \\ \hline 6\frac{1}{3} \end{array}$

Q15 Find the difference: $\begin{array}{r} 7 \\ -3\frac{2}{5} \\ \hline \end{array}$

• # # # • # # # • # # # • # # # • # # # • # # # • # # # • # #

A15 $3\frac{3}{5}$: $\begin{array}{r} \overset{6}{\not{7}}\frac{5}{5} \\ -3\frac{2}{5} \\ \hline 3\frac{3}{5} \end{array}$

Q16 Find the difference:

a. $3-\frac{5}{8}$ **b.** $6\frac{1}{3}-1\frac{2}{3}$

• # # # • # # # • # # # • # # # • # # # • # # # • # # # • # #

A16 **a.** $2\frac{3}{8}$ **b.** $4\frac{2}{3}$

Q17 Find the difference:

a. $3\frac{4}{5} - 2\frac{1}{2}$

b. $3\frac{5}{8} - 1\frac{11}{12}$

• # # # • # # # • # # # • # # # • # # # • # # # • # # # • # #

A17 a. $1\frac{3}{10}$ b. $1\frac{17}{24}$

Q18 Find the difference:

a. $7\frac{5}{18} - 1\frac{11}{24}$

b. $4\frac{5}{15} - 2\frac{12}{25}$

• # # # • # # # • # # # • # # # • # # # • # # # • # # # • # #

A18 a. $5\frac{59}{72}$ b. $1\frac{64}{75}$

Q19 A board $1\frac{3}{4}$ feet long was sawed from a board $6\frac{1}{2}$ feet. How large was the piece of board left?

• # # # • # # # • # # # • # # # • # # # • # # # • # # # • # #

A19 $4\frac{3}{4}$ feet

This completes the instruction for this section.

1.5 EXERCISE

1. Find the difference:

   a. $\frac{5}{8} - \frac{3}{8}$ b. $\frac{5}{9} - \frac{2}{9}$ c. $\frac{3}{7} - \frac{3}{8}$ d. $\frac{3}{8} - \frac{1}{12}$

   e. $6\frac{3}{4} - 2\frac{1}{8}$ f. $4\frac{5}{9} - 3\frac{5}{12}$

2. Find the difference:

   a. $8 - 7\frac{7}{8}$ b. $3\frac{5}{12} - 1\frac{7}{12}$ c. $7\frac{5}{6} - 2\frac{7}{8}$ d. $3\frac{1}{3} - 2\frac{1}{2}$

   e. $1\frac{3}{10} - \frac{7}{15}$ f. $4\frac{1}{16} - 2\frac{1}{18}$

3. Subtract $\frac{1}{3}$ from $\frac{3}{4}$.

4. Subtract $\frac{2}{5}$ from $\frac{11}{12}$.

5. From $8\frac{1}{2}$ subtract $5\frac{1}{4}$.

6. From $27\frac{9}{16}$ subtract $13\frac{5}{8}$.

1.5 EXERCISE ANSWERS

1. a. $\frac{1}{4}$ b. $\frac{1}{3}$ c. $\frac{3}{56}$

d. $\frac{7}{24}$ **e.** $4\frac{5}{8}$ **f.** $1\frac{5}{36}$

2. a. $\frac{1}{8}$ **b.** $1\frac{5}{6}$ **c.** $4\frac{23}{24}$

d. $\frac{5}{6}$ **e.** $\frac{5}{6}$ **f.** $2\frac{1}{144}$

3. $\frac{5}{12}$ **4.** $\frac{31}{60}$ **5.** $3\frac{1}{4}$ **6.** $13\frac{15}{16}$

1.6 MULTIPLICATION AND DIVISION

1 The product of whole numbers is defined as repeated addition. That is, $4 \times 3 = 3+3+3+3$. The product of a whole number and a fraction can be defined in the same manner.

Examples:

1. $3 \times \frac{2}{7} = \frac{2}{7} + \frac{2}{7} + \frac{2}{7} = \frac{6}{7}$ **2.** $4 \times \frac{1}{2} = \frac{1}{2} + \frac{1}{2} + \frac{1}{2} + \frac{1}{2} = \frac{4}{2} = 2$

Q1 Using repeated addition, find the product: $2 \times \frac{2}{3}$

\# \# \# • \# \# \# • \# \# \# • \# \# \# • \# \# \# • \# \# \# • \# \# \# • \# \# \# • \# \# \#

A1 $1\frac{1}{3}$: $2 \times \frac{2}{3} = \frac{2}{3} + \frac{2}{3} = \frac{4}{3} = 1\frac{1}{3}$

Q2 Using repeated addition, find the product: $\frac{5}{12} \times 3$ $\left(\text{recall that } \frac{5}{12} \times 3 = 3 \times \frac{5}{12}\right)$.

\# \# \# • \# \# \# • \# \# \# • \# \# \# • \# \# \# • \# \# \# • \# \# \# • \# \# \# • \# \# \#

A2 $1\frac{1}{4}$: $3 \times \frac{5}{12} = \frac{5}{12} + \frac{5}{12} + \frac{5}{12} = \frac{15}{12} = 1\frac{3}{12} = 1\frac{1}{4}$

2 The preceding examples suggest a shorter method for finding the product of a whole number and a fraction. In the problem

$$2 \times \frac{3}{7} = \frac{3}{7} + \frac{3}{7}$$
$$= \frac{3+3}{7}$$

the numerator of the sum could be written 2×3. That is, to find the product of a whole number and a fraction, multiply the whole number by the numerator of the fraction and place this product over the denominator.

Examples:

1. $2 \times \frac{3}{7} = \frac{2 \times 3}{7} = \frac{6}{7}$ **2.** $3 \times \frac{5}{12} = \frac{3 \times 5}{12} = \frac{15}{12} = 1\frac{1}{4}$

Q3 Find the product: $3 \times \frac{2}{7}$

\# \# \# • \# \# \# • \# \# \# • \# \# \# • \# \# \# • \# \# \# • \# \# \# • \# \# \# • \# \# \#

A3 $\frac{6}{7}$: $3 \times \frac{2}{7} = \frac{3 \times 2}{7} = \frac{6}{7}$

Q4 Find the product: $\frac{3}{16} \times 5$

\# \# \# • \# \# \# • \# \# \# • \# \# \# • \# \# \# • \# \# \# • \# \# \# • \# \# \# • \# \# \#

A4 $\frac{15}{16}$: $\frac{3}{16} \times 5 = \frac{3 \times 5}{16} = \frac{15}{16}$

Q5 Find the product:

a. $6 \times \frac{2}{15}$ **b.** $4 \times \frac{1}{2}$ **c.** $\frac{1}{3} \times 7$

\# \# \# • \# \# \# • \# \# \# • \# \# \# • \# \# \# • \# \# \# • \# \# \# • \# \# \# • \# \# \#

A5 **a.** $\frac{4}{5}$: $6 \times \frac{2}{15} = \frac{12}{15} = \frac{4}{5}$ **b.** 2: $4 \times \frac{1}{2} = \frac{4}{2} = 2$ **c.** $2\frac{1}{3}$: $\frac{1}{3} \times 7 = \frac{7}{3} = 2\frac{1}{3}$

3 The product of two fractions can be illustrated as follows:

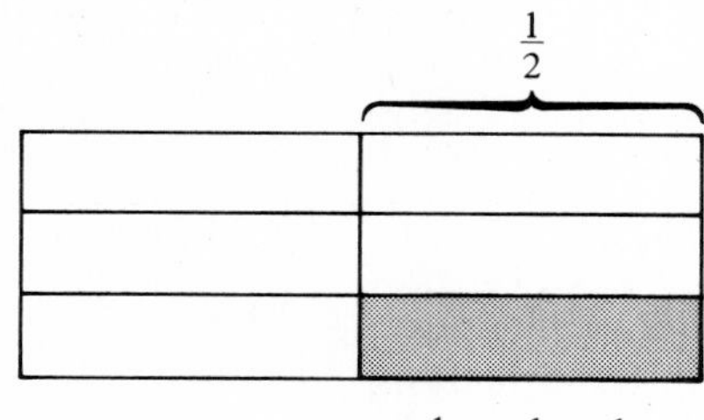

$\frac{1}{3}$ of $\frac{1}{2} = \frac{1}{6}$

The problem $\frac{1}{3}$ of $\frac{1}{2} = \frac{1}{6}$ can be restated as

$\frac{1}{3} \times \frac{1}{2} = \frac{1}{6}$

That is, "of" means to multiply.

Q6 In the following diagram illustrate $\frac{1}{2} \times \frac{2}{3} \cdot \left(\frac{1}{2} \text{ of } \frac{2}{3}\right)$:

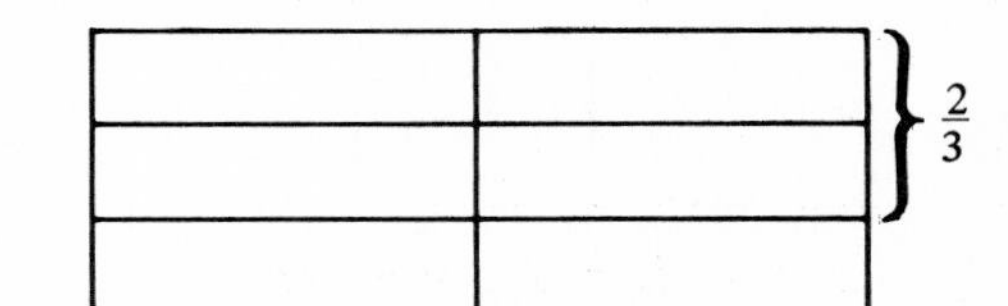

\# \# \# • \# \# \# • \# \# \# • \# \# \# • \# \# \# • \# \# \# • \# \# \# • \# \# \# • \# \# \#

A6

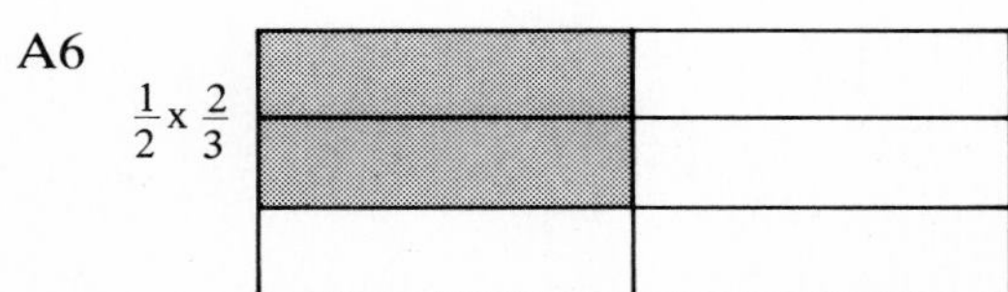

Q7 Use the diagram in A6 to find how many sixths are represented by $\frac{1}{2}\times\frac{2}{3}$. ______

• # # # • # # # • # # # • # # # • # # # • # # # • # # # • # #

A7 $\frac{2}{6}$

4 The preceding examples illustrate the rule for multiplying fractions. The product of two fractions is the product of the numerators over the product of the denominators.

Examples:

1. $\frac{1}{2}\times\frac{2}{3}=\frac{1\times 2}{2\times 3}=\frac{2}{6}=\frac{1}{3}$ **2.** $\frac{5}{7}\times\frac{3}{8}=\frac{5\times 3}{7\times 8}=\frac{15}{56}$

Q8 Complete the multiplication process: $\frac{2}{3}\times\frac{4}{5}=\frac{2\times(\ \ )}{(\ \ )\times(\ \ )}$

• # # # • # # # • # # # • # # # • # # # • # # # • # # # • # #

A8 $\frac{2}{3}\times\frac{4}{5}=\frac{2\times 4}{3\times 5}=\frac{8}{15}$

Q9 Find the product: $\frac{1}{5}\times\frac{1}{3}$

• # # # • # # # • # # # • # # # • # # # • # # # • # # # • # #

A9 $\frac{1}{15}$: $\frac{1}{5}\times\frac{1}{3}=\frac{1\times 1}{5\times 3}$

Q10 Find the product:

a. $\frac{5}{12}\times\frac{7}{9}$ **b.** $\frac{9}{5}\times\frac{3}{4}$

• # # # • # # # • # # # • # # # • # # # • # # # • # # # • # #

A10 **a.** $\frac{35}{108}$: $\frac{5}{12}\times\frac{7}{9}=\frac{5\times 7}{12\times 9}$ **b.** $1\frac{7}{20}$: $\frac{9}{5}\times\frac{3}{4}=\frac{9\times 3}{5\times 4}$

Q11 Find the product:

a. $\frac{4}{5}\times\frac{3}{6}$ **b.** $\frac{8}{7}\times\frac{4}{7}$ **c.** $\frac{8}{5}\times\frac{3}{7}$

• # # # • # # # • # # # • # # # • # # # • # # # • # # # • # #

A11 **a.** $\frac{2}{5}$ **b.** $\frac{32}{49}$ **c.** $\frac{24}{35}$

5 The product of two fractions should be reduced to lowest terms when possible; however, it is often possible to reduce before multiplying. For example, $\frac{3}{4}\times\frac{2}{5}$ can be written as $\frac{3\times 2}{4\times 5}$. Now,

$\frac{3\times 2}{4\times 5}=\frac{2\times 3}{4\times 5}$

Since $\frac{2}{4}=\frac{1}{2}$,

$$\frac{2\times 3}{4\times 5}=\frac{1\times 3}{2\times 5}$$

Hence,

$$\frac{3}{4}\times\frac{2}{5}=\frac{3}{10}$$

This reduction is usually shown as

$$\frac{3}{\underset{2}{\not 4}}\times\frac{\overset{1}{\not 2}}{5}=\frac{3\times 1}{2\times 5}=\frac{3}{10}$$

The example illustrates the following rule: *In multiplying fractions, you can divide any numerator and any denominator by the same nonzero number.*

Q12 Find the product, reducing before multiplying (show all work): $\frac{3}{10}\times\frac{4}{9}$

\# \# \# • \# \# \# • \# \# \# • \# \# \# • \# \# \# • \# \# \# • \# \# \# • \# \# \# • \# \# \#

A12 $\frac{\overset{1}{\not 3}}{\underset{5}{\not 10}}\times\frac{\overset{2}{\not 4}}{\underset{3}{\not 9}}=\frac{2}{15}$

Q13 Find the product: $\frac{11}{3}\times\frac{5}{11}$

\# \# \# • \# \# \# • \# \# \# • \# \# \# • \# \# \# • \# \# \# • \# \# \# • \# \# \# • \# \# \#

A13 $1\frac{2}{3}$: $\frac{\overset{1}{\not 11}}{3}\times\frac{5}{\underset{1}{\not 11}}=\frac{5}{3}$

Q14 Find the product:

a. $\frac{7}{3}\times\frac{3}{8}$

b. $\frac{2}{3}\times\frac{5}{7}$

\# \# \# • \# \# \# • \# \# \# • \# \# \# • \# \# \# • \# \# \# • \# \# \# • \# \# \# • \# \# \#

A14 **a.** $\frac{7}{8}$

b. $\frac{10}{21}$

Q15 Find the product:

a. $\frac{3}{4}\times\frac{1}{9}$

b. $\frac{7}{8}\times\frac{9}{7}$

\# \# \# • \# \# \# • \# \# \# • \# \# \# • \# \# \# • \# \# \# • \# \# \# • \# \# \# • \# \# \#

A15 **a.** $\frac{1}{12}$ **b.** $1\frac{1}{8}$

---

6 The product of mixed numbers can be determined by changing the mixed numbers to improper fractions and multiplying.

Examples:

1. $2\frac{1}{3}\times\frac{2}{5}=\frac{7}{3}\times\frac{2}{5}=\frac{14}{15}$ **2.** $8\frac{1}{2}\times4\frac{2}{3}=\frac{17}{2}\times\frac{14}{3}=\frac{119}{3}=39\frac{2}{3}$

Q16 Find the product: $2\frac{3}{4}\times\frac{5}{6}$

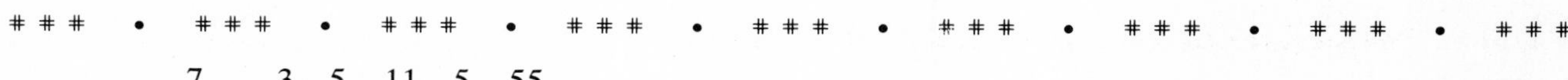

A16 $2\frac{7}{24}$: $2\frac{3}{4}\times\frac{5}{6}=\frac{11}{4}\times\frac{5}{6}=\frac{55}{24}$

Q17 Find the product:

a. $\frac{5}{8}\times2\frac{1}{5}$ **b.** $6\frac{1}{4}\times\frac{4}{5}$

• # # # • # # # • # # # • # # # • # # # • # # # • # # # • # #

A17 **a.** $1\frac{3}{8}$ **b.** 5

Q18 Find the product:

a. $1\frac{2}{3}\times4\frac{1}{5}$ **b.** $1\frac{6}{7}\times2\frac{1}{4}$

• # # # • # # # • # # # • # # # • # # # • # # # • # # # • # #

A18 **a.** 7 **b.** $4\frac{5}{28}$

Q19 John practices hockey every afternoon for $1\frac{3}{4}$ hours. How many hours does he practice in 5 days?

• # # # • # # # • # # # • # # # • # # # • # # # • # # # • # #

A19 $8\frac{3}{4}$ hours $\left(5\times1\frac{3}{4}\right)$

---

7 A division such as $3\div\frac{1}{2}$ can be interpreted as: How many $\frac{1}{2}$ units are contained in 3 units? There are two $\frac{1}{2}$ units in each whole, so there are six $\frac{1}{2}$ units in 3 units.

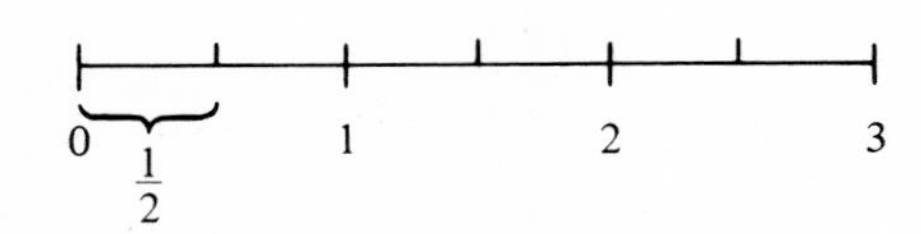

Hence, $3 \div \frac{1}{2} = 6$.

Q20 How many $\frac{1}{3}$ units are contained in a whole? ______

\# \# \# • \# \# \# • \# \# \# • \# \# \# • \# \# \# • \# \# \# • \# \# \# • \# \# \# • \# \# \#

A20 3

Q21 How many $\frac{1}{3}$ units are contained in 6 units? ______

\# \# \# • \# \# \# • \# \# \# • \# \# \# • \# \# \# • \# \# \# • \# \# \# • \# \# \# • \# \# \#

A21 18

Q22 $6 \div \frac{1}{3} =$ ______

\# \# \# • \# \# \# • \# \# \# • \# \# \# • \# \# \# • \# \# \# • \# \# \# • \# \# \# • \# \# \#

A22 18

Q23 How many $\frac{1}{4}$ units are contained in 5 units? ______

\# \# \# • \# \# \# • \# \# \# • \# \# \# • \# \# \# • \# \# \# • \# \# \# • \# \# \# • \# \# \#

A23 20

Q24 $5 \div \frac{1}{4} =$ ______

\# \# \# • \# \# \# • \# \# \# • \# \# \# • \# \# \# • \# \# \# • \# \# \# • \# \# \# • \# \# \#

A24 20

8 The quotient of two fractions can be determined by changing the division process into an equivalent multiplication process. Consider the example $\frac{3}{7} \div \frac{2}{3}$:

$\frac{3}{7} \div \frac{2}{3}$ is equivalent to $\dfrac{\frac{3}{7}}{\frac{2}{3}}$

This fraction can be simplified by multiplying numerator and denominator by $\frac{3}{2}$:

$$\frac{3}{7} \div \frac{2}{3} = \frac{\frac{3}{7}}{\frac{2}{3}} = \frac{\frac{3}{7} \times \frac{3}{2}}{\frac{2}{3} \times \frac{3}{2}} = \frac{\frac{3}{7} \times \frac{3}{2}}{1} = \frac{3}{7} \times \frac{3}{2}$$

The division problem $\frac{3}{7} \div \frac{2}{3}$ has been changed to an equivalent multiplication problem, $\frac{3}{7} \times \frac{3}{2}$.

Q25 Change the division $\frac{2}{3} \div \frac{5}{7}$ into a multiplication problem by completing the missing steps:

$$\frac{2}{3} \div \frac{5}{7} = \frac{\frac{2}{3} \times \frac{(\ )}{(\ )}}{\frac{5}{7} \times \frac{7}{5}} = \frac{\frac{(\ )}{(\ )} \times \frac{(\ )}{(\ )}}{(\ )} = \frac{(\ )}{(\ )} \times \frac{(\ )}{(\ )}$$

\# \# \# • \# \# \# • \# \# \# • \# \# \# • \# \# \# • \# \# \# • \# \# \# • \# \# \# • \# \# \#

A25 $\frac{\frac{2}{3} \times \frac{7}{5}}{\frac{5}{7} \times \frac{7}{5}} = \frac{\frac{2}{3} \times \frac{7}{5}}{1} = \frac{2}{3} \times \frac{7}{5}$

9 The preceding examples verify the rule for dividing fractions: *Invert the divisor and multiply*. For example,

$\frac{2}{3} \div \frac{5}{7}$

(invert the divisor)

$\frac{2}{3} \div \frac{5}{7}$ is equivalent to $\frac{2}{3} \times \frac{7}{5} = \frac{14}{15}$

Hence, $\frac{2}{3} \div \frac{5}{7} = \frac{14}{15}$.

Q26 Which fraction is the divisor?

$\frac{2}{3} \div \frac{7}{8}$ ______

\# \# \# • \# \# \# • \# \# \# • \# \# \# • \# \# \# • \# \# \# • \# \# \# • \# \# \# • \# \# \#

A26 $\frac{7}{8}$

Q27 The rule for dividing fractions is to ______________________________

____

\# \# \# • \# \# \# • \# \# \# • \# \# \# • \# \# \# • \# \# \# • \# \# \# • \# \# \# • \# \# \#

A27 invert the divisor and multiply

Q28 Write the division $\frac{2}{3} \div \frac{7}{8}$ as an equivalent multiplication problem.

\# \# \# • \# \# \# • \# \# \# • \# \# \# • \# \# \# • \# \# \# • \# \# \# • \# \# \# • \# \# \#

A28 $\frac{2}{3} \times \frac{8}{7}$

Q29 Find the quotient: $\frac{2}{3} \div \frac{7}{8}$

• ### • ### • ### • ### • ### • ### • ### •

A29 $\frac{16}{21}$: $\frac{2}{3} \div \frac{7}{8} = \frac{2}{3} \times \frac{8}{7} = \frac{16}{21}$

Q30 Find the quotient: $\frac{4}{7} \div \frac{8}{9}$

• ### • ### • ### • ### • ### • ### • ### •

A30 $\frac{9}{14}$: $\frac{4}{7} \div \frac{8}{9} = \frac{4}{7} \times \frac{9}{8} = \frac{36}{56}$

Q31 Find the quotient: $\frac{4}{5} \div \frac{3}{7}$

• ### • ### • ### • ### • ### • ### • ### •

A31 $1\frac{13}{15}$: $\frac{4}{5} \div \frac{3}{7} = \frac{4}{5} \times \frac{7}{3} = \frac{28}{15}$

Q32 Find the quotient:

a. $\frac{4}{7} \div \frac{3}{5}$ **b.** $\frac{3}{4} \div \frac{5}{8}$

• ### • ### • ### • ### • ### • ### • ### •

A32 **a.** $\frac{20}{21}$ **b.** $1\frac{1}{5}$

Q33 Find the quotient:

a. $\frac{6}{7} \div \frac{9}{10}$ **b.** $\frac{6}{7} \div \frac{12}{13}$

• ### • ### • ### • ### • ### • ### • ### •

A33 **a.** $\frac{20}{21}$ **b.** $\frac{13}{14}$

10 The quotient of $\frac{3}{4} \div 6$ can be determined as follows: $6 = \frac{6}{1}$; hence, $\frac{3}{4} \div 6$ is equivalent to $\frac{3}{4} \div \frac{6}{1}$. Therefore,

$$\frac{3}{4} \div \frac{6}{1} = \frac{3}{4} \times \frac{1}{6} = \frac{1}{8}$$

Q34 $\frac{2}{3} \div 5$ is equivalent to $\frac{2}{3} \div (\quad)$.

\# # # • # # # • # # # • # # # • # # # • # # # • # # # • # # # • # # #

A34 $\frac{5}{1}$

Q35 $\frac{2}{3} \div \frac{5}{1} = \frac{2}{3} \times (\quad)$

\# # # • # # # • # # # • # # # • # # # • # # # • # # # • # # # • # # #

A35 $\frac{1}{5}$

Q36 Find the quotient: $\frac{2}{3} \div 5$

\# # # • # # # • # # # • # # # • # # # • # # # • # # # • # # # • # # #

A36 $\frac{2}{15}$: $\frac{2}{3} \div \frac{5}{1} = \frac{2}{3} \times \frac{1}{5} = \frac{2}{15}$

Q37 Find the quotient: $\frac{2}{5} \div 10$

\# # # • # # # • # # # • # # # • # # # • # # # • # # # • # # # • # # #

A37 $\frac{1}{25}$: $\frac{2}{5} \div \frac{10}{1} = \frac{2}{5} \times \frac{1}{10} = \frac{1}{25}$

Q38 Find the quotient: $3 \div \frac{2}{3}$

\# # # • # # # • # # # • # # # • # # # • # # # • # # # • # # # • # # #

A38 $4\frac{1}{2}$: $3 \div \frac{2}{3} = 3 \times \frac{3}{2} = \frac{9}{2}$

Q39 Find the quotient: $3\frac{3}{4} \div \frac{3}{4}$

\# # # • # # # • # # # • # # # • # # # • # # # • # # # • # # # • # # #

A39 5: $3\frac{3}{4} \div \frac{3}{4} = \frac{15}{4} \div \frac{3}{4} = \frac{15}{4} \times \frac{4}{3} = 5$

Q40 Find the quotient:

a. $\frac{6}{11} \div \frac{5}{11}$ **b.** $9 \div \frac{2}{3}$

\# \# \# • \# \# \# • \# \# \# • \# \# \# • \# \# \# • \# \# \# • \# \# \# • \# \# \# • \# \# \#

A40 **a.** $1\frac{1}{5}$ **b.** $13\frac{1}{2}$

Q41 Find the quotient:

a. $\frac{6}{7} \div 3\frac{1}{4}$ **b.** $4\frac{2}{3} \div 3\frac{1}{4}$

\# \# \# • \# \# \# • \# \# \# • \# \# \# • \# \# \# • \# \# \# • \# \# \# • \# \# \# • \# \# \#

A41 **a.** $\frac{24}{91}$ **b.** $1\frac{17}{39}$

Q42 Find the quotient:

a. $2\frac{1}{2} \div 10$ **b.** $6\frac{1}{2} \div 4\frac{1}{3}$

\# \# \# • \# \# \# • \# \# \# • \# \# \# • \# \# \# • \# \# \# • \# \# \# • \# \# \# • \# \# \#

A42 **a.** $\frac{1}{4}$ **b.** $1\frac{1}{2}$

Q43 Don went on a 500-mile trip this summer. He traveled $\frac{2}{5}$ of the way by train and the rest of the way by bus. How many miles did Don travel by train? ______ By bus? ______

\# \# \# • \# \# \# • \# \# \# • \# \# \# • \# \# \# • \# \# \# • \# \# \# • \# \# \# • \# \# \#

A43 200 miles by train $\left(\frac{2}{5} \times 500\right)$, 300 miles by bus $\left(\frac{3}{5} \times 500\right)$

Q44 Mrs. Pumford cut a large piece of meat weighing $18\frac{2}{3}$ pounds into pieces weighing $1\frac{3}{4}$ pounds each. How many smaller pieces did she get? ______

\# \# \# • \# \# \# • \# \# \# • \# \# \# • \# \# \# • \# \# \# • \# \# \# • \# \# \# • \# \# \#

A44 $10\frac{2}{3}$ pieces: $\left(18\frac{2}{3} \div 1\frac{3}{4}\right)$

This completes the instruction for this section.

1.6 EXERCISE

1. Find the products:

a. $\frac{1}{2}\times\frac{3}{4}$ b. $\frac{1}{7}\times\frac{3}{11}$ c. $16\times\frac{3}{4}$ d. $30\times\frac{2}{15}$

e. $\frac{1}{3}\times\frac{1}{6}$ f. $\frac{1}{8}\times 16$ g. $\frac{4}{9}\times 6$ h. $4\times\frac{9}{10}$

i. $\frac{4}{5}\times\frac{3}{8}$ j. $\frac{3}{4}\times\frac{12}{16}$ k. $\frac{4}{7}\times\frac{3}{8}$ l. $3\frac{1}{3}\times\frac{3}{5}$

m. $2\frac{2}{5}\times\frac{5}{8}$ n. $2\frac{1}{7}\times 1\frac{2}{5}$ o. $1\frac{7}{8}\times 2\frac{1}{5}$

2. Find the quotients:

a. $\frac{8}{9}\div\frac{3}{4}$ b. $\frac{5}{9}\div\frac{3}{4}$ c. $8\div\frac{1}{4}$

d. $6\div\frac{5}{6}$ e. $5\frac{1}{4}\div 3\frac{1}{2}$ f. $6\frac{1}{4}\div 1\frac{3}{5}$

g. $8\div 2\frac{1}{2}$ h. $6\frac{7}{8}\div\frac{11}{14}$ i. $\frac{6}{7}\div 8$

3. A bag contains 25 pounds of fertilizer. How many $1\frac{1}{3}$-pound bags can be filled from the 25-pound bag?

4. A housing development requires $\frac{2}{5}$ of an acre for a house. How many acres will be needed to build 147 houses?

1.6 EXERCISE ANSWERS

1. a. $\frac{3}{8}$ b. $\frac{3}{77}$ c. 12 d. 4

e. $\frac{1}{18}$ f. 2 g. $2\frac{2}{3}$ h. $3\frac{3}{5}$

i. $\frac{3}{10}$ j. $\frac{9}{16}$ k. $\frac{3}{14}$ l. 2

m. $1\frac{1}{2}$ n. 3 o. $4\frac{1}{8}$

2. a. $1\frac{5}{27}$ b. $\frac{20}{27}$ c. 32

d. $7\frac{1}{5}$ e. $1\frac{1}{2}$ f. $3\frac{29}{32}$

g. $3\frac{1}{5}$ h. $8\frac{3}{4}$ i. $\frac{3}{28}$

3. $18\frac{3}{4}$ bags 4. $58\frac{4}{5}$ acres

1.7 APPLICATIONS

1 Numbers that are expressed in terms of a standard unit of measure are called *denominate numbers.* The number 3 by itself is abstract. But 3 weeks, 3 days, 3 dollars, 3 feet, and 3 pounds are denominate numbers; they identify the measures of quantity.

Q1 Which of the following are denominate numbers? ________________

a. 5 ounces **b.** $6\frac{1}{2}$ **c.** 2 seconds **d.** 7 houses **e.** 18,000 **f.** 1 pint

\# \# \# • \# \# \# • \# \# \# • \# \# \# • \# \# \# • \# \# \# • \# \# \# • \# \# \# • \# \# \#

A1 a, c, and f: Houses is not a standard unit of measure.

2 Denominate numbers are generally expressed in the largest unit of measure possible. That is, 68 minutes would be expressed as 1 hour 8 minutes (60 minutes = 1 hour).

Example 1: 90 seconds = 1 minute 30 seconds

Example 2: 26 days = 3 weeks 5 days

Standard units of measure to be used in this section, and their abbreviations, are given below.

1 yr (year) = 52 wk (weeks)
1 yr = 12 mo (months)
1 yr = 365 da (days)
1 yr = 360 da (commonly used for business computations)
1 wk = 7 da
1 da = 24 hr (hours)
1 hr = 60 min (minutes)
1 min = 60 sec (seconds)

Q2 Express in larger units of measure:

a. 83 min **b.** 65 sec

\# \# \# • \# \# \# • \# \# \# • \# \# \# • \# \# \# • \# \# \# • \# \# \# • \# \# \# • \# \# \#

A2 **a.** 1 hr 23 min **b.** 1 min 5 sec

3 When adding or subtracting abstract numbers, it is necessary to add like place values; that is, units to units, tens to tens, hundreds to hundreds, and so on. Likewise, in the addition or subtraction of denominate numbers, the numbers of like denomination (unit of measure) are arranged in vertical columns with the smallest denomination on the right. The like denominations are then added or subtracted a column at a time from right to left.

Example:

7 hr	30 min	17 sec
+3 hr	42 min	50 sec
10 hr	72 min	67 sec

Working from right to left, each denomination should be expressed as a larger denomination, where possible. That is, 67 sec = 1 min 7 sec. The 1 min is added to 72 min leaving a sum of 10 hr 73 min 7 sec. However, 73 min = 1 hr 13 min. The 1 hr is added to 10 hr leaving the final sum 11 hr 13 min 7 sec.

Q3 Add and simplify by writing denominations as larger denominations where possible:

a.

7 hr	23 min
4 hr	58 min

b.

2 wk	3 da	6 hr
1 wk	6 da	23 hr

\# \# \# • \# \# \# • \# \# \# • \# \# \# • \# \# \# • \# \# \# • \# \# \# • \# \# \# • \# \# \#

A3 **a.** 12 hr 21 min:

	7 hr	23 min
	4 hr	58 min
	11 hr	81 min
or	12 hr	21 min

b. 4 wk 3 da 5 hr:

	2 wk	3 da	6 hr
	1 wk	6 da	23 hr
	3 wk	9 da	29 hr
or	3 wk	10 da	5 hr
or	4 wk	3 da	5 hr

Q4 Add:

a.

8 hr	17 min	46 sec
10 hr	39 min	42 sec
26 hr	51 min	35 sec

b.

6 wk	12 da	7 hr
7 wk	35 da	3 hr
8 wk	12 da	16 hr

• # # # • # # # • # # # • # # # • # # # • # # # • # # # • # #

A4 **a.** 1 da 21 hr 49 min 3 sec **b.** 29 wk 4 da 2 hr

---

4 Subtraction of denominate numbers is similar to addition; however, it may be necessary to express a larger denomination as a smaller denomination when borrowing.

Example:

7 hr	23 min
−2 hr	56 min

To subtract 56 min from 23 min it is first necessary to borrow 1 hr (60 min) from 7 hr. The 60 min borrowed is then added to 23 min giving 83 min. The subtraction is now completed:

6 hr	83 min
−2 hr	56 min
4 hr	27 min

Q5 Subtract:

a.

18 min	15 sec
−6 min	45 sec

b.

3 da	14 hr	6 min
−1 da	21 hr	36 min

• # # # • # # # • # # # • # # # • # # # • # # # • # # # • # #

A5 **a.** 11 min 30 sec:

17 min	75 sec
−6 min	45 sec

b. 1 da 16 hr 30 min:

	3 da	13 hr	66 min
	−1 da	21 hr	36 min
or	2 da	37 hr	66 min
	−1 da	21 hr	36 min

Q6 Subtract:

a.

8 hr	15 min
−5 hr	25 min

b.

88 hr	36 min	24 sec
−69 hr	49 min	48 sec

• # # # • # # # • # # # • # # # • # # # • # # # • # # # • # #

A6 **a.** 2 hr 50 min **b.** 18 hr 46 min 36 sec

5 When multiplying a denominate number by an abstract number, first multiply each denomination separately and then change the product to a larger denomination, where possible.

Example:

	3 hr	17 min
	×	6
	18 hr	102 min
or	19 hr	42 min

To divide a denominate number by an abstract number, divide each denomination separately, rewriting partial remainders as smaller denominations. The rewritten smaller denominations are added to like denominations and the division is continued.

Example:

$$\begin{array}{r|l} & 2\text{ hr} \quad 12\text{ min} \\ 6 & 13\text{ hr} \quad 12\text{ min} \\ & \underline{12\text{ hr}} \qquad \updownarrow \text{ add} \\ & 1\text{ hr} = 60\text{ min} \\ & \qquad\quad \overline{72\text{ min}} \\ & \qquad\quad \underline{72\text{ min}} \\ & \qquad\qquad 0 \end{array}$$

Q7 Complete:

a. 12 min 13 sec
× 7

b. $5\overline{)23\text{ hr} \quad 10\text{ min}}$

\# \# \# • \# \# \# • \# \# \# • \# \# \# • \# \# \# • \# \# \# • \# \# \# • \# \# \# • \# \# \#

A7 **a.** 1 hr 25 min 31 sec **b.** 4 hr 38 min

6 Hours worked are sometimes expressed as fractional parts of an hour, for example, $7\frac{3}{4}$ hr. It may be convenient to express these fractional numbers as denominate numbers, in terms of hours and minutes. This is accomplished by recalling that an hour is 60 minutes and $\frac{3}{4}$ of an hour is $\frac{3}{4}$ of 60 minutes.

Example: Change $7\frac{3}{4}$ hr to hours and minutes.

Solution:

$\frac{3}{4}$ of 60 min $= \frac{3}{4} \times \frac{60}{1}$ min $= 45$ min

Hence, $7\frac{3}{4}$ hr $= 7$ hr 45 min.

Q8 Express as hours and minutes:

a. $\frac{1}{4}$ hr **b.** $3\frac{1}{2}$ hr

\# \# \# • \# \# \# • \# \# \# • \# \# \# • \# \# \# • \# \# \# • \# \# \# • \# \# \# • \# \# \#

A8 **a.** 15 min **b.** 3 hr 30 min

Q9 Express as hours and minutes:

a. $7\frac{2}{5}$ hr **b.** $8\frac{3}{10}$ hr

\# # # • # # # • # # # • # # # • # # # • # # # • # # # • # # # • # # #

A9 **a.** 7 hr 24 min **b.** 8 hr 18 min

7 Generally, hours worked are expressed as fractional or decimal numbers. Therefore, it will be necessary to express denominate numbers as fractional numbers.

Example 1: Express 7 hours 25 minutes as a fractional part of an hour.

Solution: It is necessary to express 25 minutes as a fractional part of an hour. Since 1 hour = 60 minutes, 25 minutes is $\frac{25}{60}$ of an hour. Hence, 7 hours 25 minutes = $7\frac{25}{60}$ hr. The fraction is reduced when possible. That is, $7\frac{25}{60}$ hr = $7\frac{5}{12}$ hr.

Example 2: Express 8 hr 40 min as a fractional part of an hour.

Solution: 8 hr 40 min = $8\frac{40}{60}$ hr = $8\frac{2}{3}$ hr

Q10 Express as a fractional part of an hour:

a. 6 hr 15 min **b.** 10 hr 12 min

\# # # • # # # • # # # • # # # • # # # • # # # • # # # • # # # • # # #

A10 **a.** $6\frac{1}{4}$ hr: 6 hr 15 min = $6\frac{15}{60}$ hr = $6\frac{1}{4}$hr **b.** $10\frac{1}{5}$ hr

Q11 Express as a fractional part of an hour:

a. 7 hr 42 min **b.** 5 hr 33 min

\# # # • # # # • # # # • # # # • # # # • # # # • # # # • # # # • # # #

A11 **a.** $7\frac{7}{10}$ hr **b.** $5\frac{11}{20}$ hr

Q12 Sally worked the following hours last week: Mon 7 hr 20 min, Tue 6 hr 35 min, and Wed 8 hr 50 min. Change each denominate number to a fractional part of an hour and find the total hours worked.

\# # # • # # # • # # # • # # # • # # # • # # # • # # # • # # # • # # #

A12 $22\frac{3}{4}$ hr: 7 hr 20 min $= 7\frac{1}{3}$ hr $= 7\frac{4}{12}$ hr

6 hr 35 min $= 6\frac{7}{12}$ hr $= 6\frac{7}{12}$ hr

8 hr 50 min $= 8\frac{5}{6}$ hr $= 8\frac{10}{12}$ hr

Q13 In Q12, find the total hours worked by adding the denominate numbers.

• # # # • # # # • # # # • # # # • # # # • # # # • # # # • # #

A13 22 hr 45 min:

7 hr	20 min
6 hr	35 min
8 hr	50 min
21 hr	105 min

Q14 Nancy worked $6\frac{3}{8}$ hours on Monday, $7\frac{1}{2}$ hours on Tuesday, $8\frac{3}{4}$ hours on Wednesday, $7\frac{3}{8}$ hours on Thursday, and $9\frac{1}{4}$ hours on Friday. How many hours did she work during the week?

• # # # • # # # • # # # • # # # • # # # • # # # • # # # • # #

A14 $39\frac{1}{4}$ hr

Q15 Six employees of the Mayflower Hotel worked the following hours. Find the total hours worked for each employee.

M	*T*	*W*	*Th*	*F*	*S*	*Total Hours*
8	$7\frac{1}{2}$	$7\frac{3}{4}$	$8\frac{1}{4}$	$8\frac{1}{2}$	—	**a.**
$7\frac{1}{4}$	8	$7\frac{1}{4}$	—	$8\frac{1}{4}$	$5\frac{3}{4}$	**b.**
$6\frac{1}{2}$	$6\frac{1}{2}$	$8\frac{1}{4}$	7	8	$2\frac{3}{4}$	**c.**
8	8	$7\frac{1}{4}$	8	$7\frac{1}{4}$	—	**d.**
$7\frac{3}{4}$	8	$7\frac{3}{4}$	$8\frac{1}{2}$	7	—	**e.**
$7\frac{3}{4}$	8	$7\frac{1}{4}$	—	9	$7\frac{1}{2}$	**f.**

• # # # • # # # • # # # • # # # • # # # • # # # • # # # • # #

A15 **a.** 40 **b.** $36\frac{1}{2}$ **c.** 39 **d.** $38\frac{1}{2}$ **e.** 39 **f.** $39\frac{1}{2}$

---

8 Fractions are involved in the solution of many types of technical problems. When solving problems remember to look for the key word or words that indicate the mathematical operation to be used.

Example: Three lamps are connected in a series. Their resistances are $1\frac{2}{3}$ ohms, $\frac{2}{5}$ ohm, and $\frac{1}{2}$ ohm. If the total resistance is the sum of the individual resistances, what is the total resistance in ohms of the three lamps?

Solution: The key word is "sum," which indicates that the individual resistances are to be added. That is,

$$\text{total resistance} = 1\frac{2}{3} + \frac{2}{5} + \frac{1}{2}$$

$$\text{total resistance} = 1\frac{20}{30} + \frac{12}{30} + \frac{15}{30}$$

$$\text{total resistance} = 2\frac{17}{30} \text{ ohms}$$

Q16 Three castings weigh $7\frac{1}{4}$ pounds, $18\frac{2}{3}$ pounds, and $42\frac{3}{8}$ pounds. If $6\frac{11}{12}$ pounds of metal is removed in machining from all three castings, how much weight remains?

• ### • ### • ### • ### • ### • ### • ### •

A16 $61\frac{3}{8}$ pounds: First, find the total weight.

$$7\frac{1}{4} + 18\frac{2}{3} + 42\frac{3}{8} = 68\frac{7}{24} \text{ pounds}$$

Since $6\frac{11}{12}$ pounds were removed, subtract $6\frac{11}{12}$ from the total weight.

$$68\frac{7}{24} - 6\frac{11}{12} = 61\frac{9}{24} \text{ pounds}$$

Q17 An alloy used for bearings is $\frac{11}{19}$ copper, $\frac{7}{19}$ tin, and $\frac{1}{19}$ zinc. How many pounds of each make up 95 pounds of alloy?

• ### • ### • ### • ### • ### • ### • ### •

A17 55 pounds copper: The word "of" indicates multiplication; that is, pounds of copper is $\frac{11}{19}$ of 95 or
35 pounds tin
5 pounds zinc $\frac{11}{19} \times 95$, etc.

Q18 If a motor makes 2,100 revolutions per hour, how many revolutions does it make in 15 minutes $\left(\frac{1}{4}\text{ hour}\right)$?

\# \# \# • \# \# \# • \# \# \# • \# \# \# • \# \# \# • \# \# \# • \# \# \# • \# \# \# • \# \# \#

A18 525 revolutions per $\frac{1}{4}$ hour: $\frac{1}{4}$ of 2,100 or $\frac{1}{4}\times 2{,}100$

Q19 A rod $12\frac{3}{4}$ inches long is to be divided into 8 equal parts. What is the length of each part?

\# \# \# • \# \# \# • \# \# \# • \# \# \# • \# \# \# • \# \# \# • \# \# \# • \# \# \# • \# \# \#

A19 $1\frac{19}{32}$ inches: $12\frac{3}{4}\div 8$

9 *Detailed dimensions* of figures often involve addition and subtraction of fractions. Consider the following example.

Example: Find the length of the bolt.

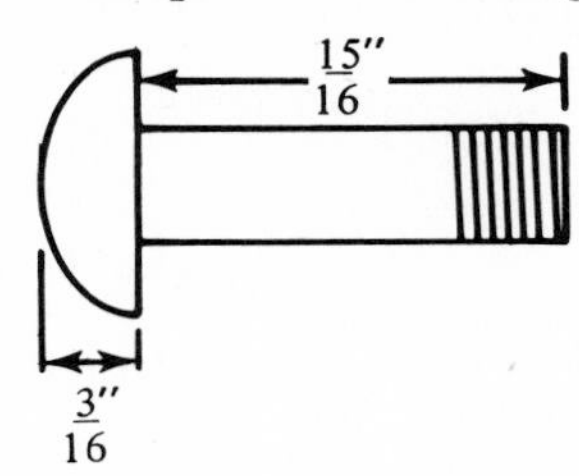

Note: $\frac{15}{16}$ inches can be represented as $\frac{15''}{16}$. Feet would be represented using one raised slash. That is, 8 feet is represented as 8′.

Solution: The length of the bolt is found by adding $\frac{3''}{16}$ and $\frac{15''}{16}$.

$$\frac{3}{16}+\frac{15}{16}=\frac{18}{16}=1\frac{2}{16}=1\frac{1}{8}$$

Hence, the length is $1\frac{1''}{8}$.

Q20 Find the perimeter (distance around).

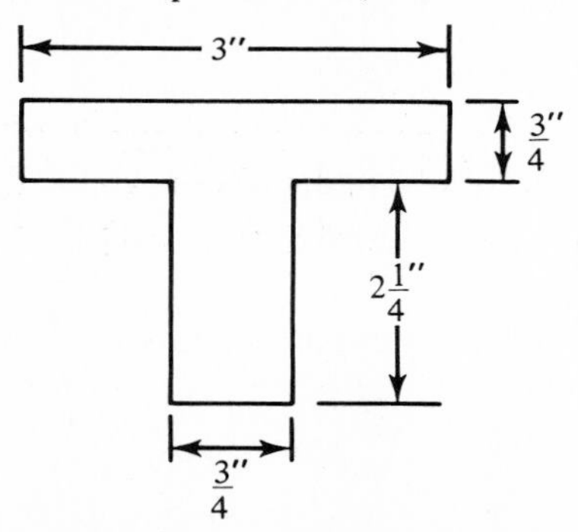

\# \# \# • \# \# \# • \# \# \# • \# \# \# • \# \# \# • \# \# \# • \# \# \# • \# \# \# • \# \# \#

A20 12″:

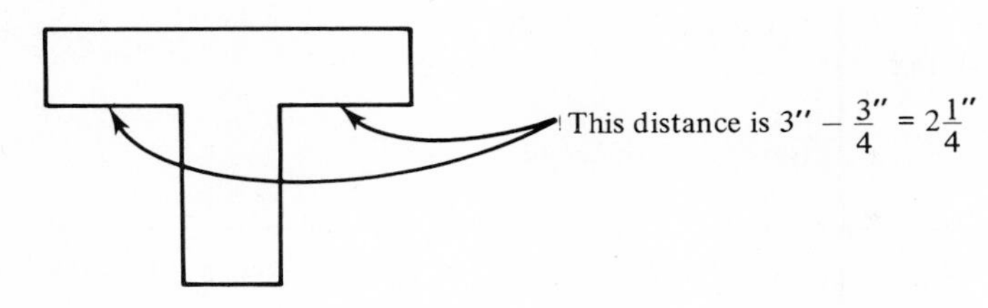

perimeter $= 3+\frac{3}{4}+\frac{3}{4}+\frac{3}{4}+2\frac{1}{4}+2\frac{1}{4}+2\frac{1}{4}$

Q21 Find the total length of the bolt.

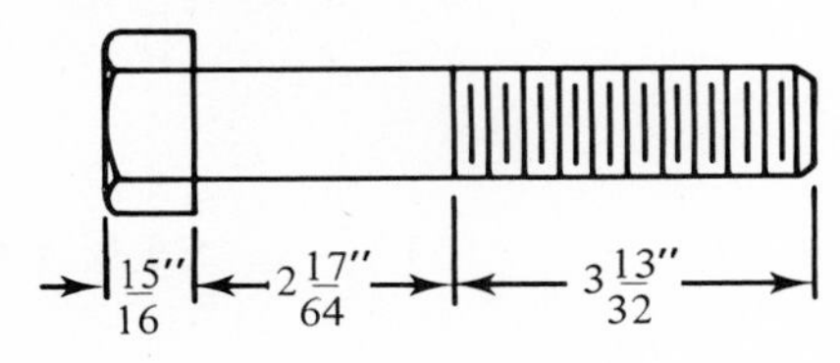

\# \# \# • \# \# \# • \# \# \# • \# \# \# • \# \# \# • \# \# \# • \# \# \# • \# \# \# • \# \# \#

A21 $6\frac{39}{64}''$: $\frac{15}{16}+2\frac{17}{64}+3\frac{13}{32}$

Q22 Find the missing dimension.

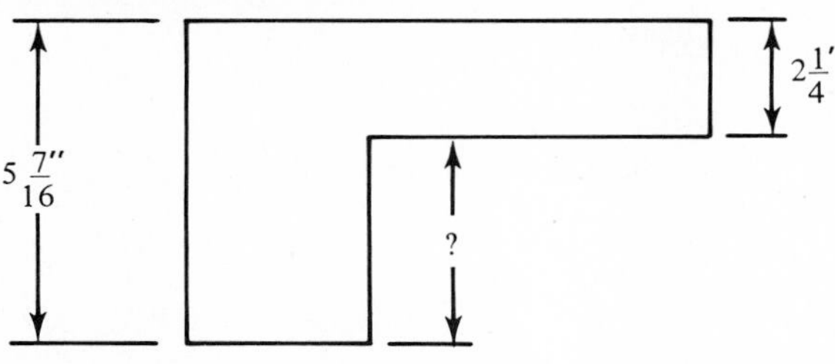

\# \# \# • \# \# \# • \# \# \# • \# \# \# • \# \# \# • \# \# \# • \# \# \# • \# \# \# • \# \# \#

A22 $3\frac{3}{16}''$: $5\frac{7}{16}-2\frac{1}{4}$

Q23 Find the missing dimension if the total length is $1\frac{5}{8}''$.

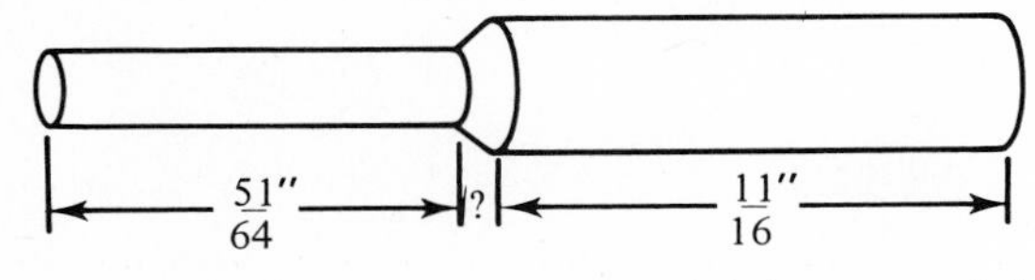

\# \# \# • \# \# \# • \# \# \# • \# \# \# • \# \# \# • \# \# \# • \# \# \# • \# \# \# • \# \# \#

A23 $\frac{9}{64}''$: $1\frac{5}{8}-\left(\frac{51}{64}+\frac{11}{16}\right)=1\frac{5}{8}-1\frac{31}{64}$

10 Another area involving fractions is that of *tapers*. A piece of work that decreases gradually in diameter is said to be tapered. Consider the accompanying diagram. The diameter of the large end is

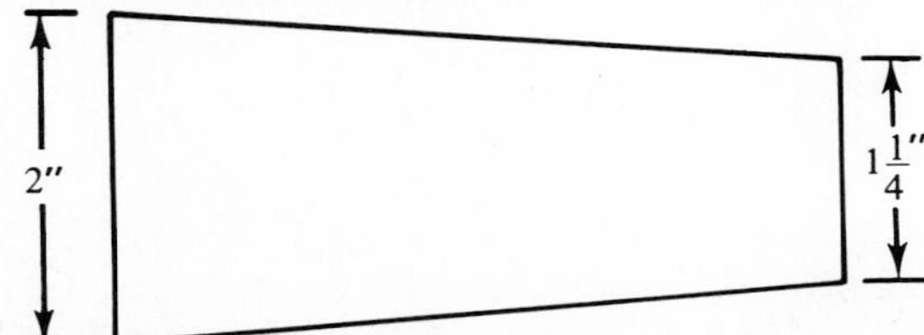

2 inches and the diameter of the small end is $1\frac{1}{4}$ inches. The taper is the difference between the diameters. That is, the taper of the above piece of work is $2'' - 1\frac{1}{4}'' = \frac{3}{4}''$.

Q24 Determine the taper for the following piece of work.

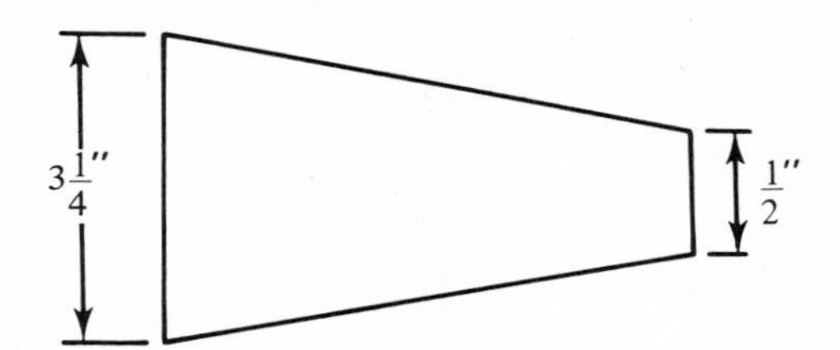

\# \# \# • \# \# \# • \# \# \# • \# \# \# • \# \# \# • \# \# \# • \# \# \# • \# \# \# • \# \# \#

A24 $2\frac{3}{4}''$: $3\frac{1}{4} - \frac{1}{2}$

11 Taper is usually defined as a fraction of an inch for each foot of length or as a fraction of an inch for each inch of length.

Example 1: Find the taper in 6 feet if a piece of work is tapered $\frac{1}{4}''$ per foot.

Solution: For every foot of length the taper is $\frac{1}{4}''$, hence, the taper in 6′ is

$$6 \times \frac{1}{4} = 1\frac{1}{2}$$

The taper in 6 feet is then $1\frac{1}{2}$ inches. That is, the diameter at the end of the 6-foot piece is $1\frac{1}{2}$ inches less than the diameter at the beginning.

Example 2: Find the taper in $8\frac{3}{4}''$ if a piece of work is tapered $\frac{1}{16}''$ per inch of length.

Solution:

$$8\frac{3}{4} \times \frac{1}{16} = \frac{35}{4} \times \frac{1}{16} = \frac{35}{64}$$

The taper is $\frac{35}{64}''$.

Q25 Find the taper in a piece of work 16 feet long if the taper is $\frac{1}{4}''$ per foot.

\# \# \# • \# \# \# • \# \# \# • \# \# \# • \# \# \# • \# \# \# • \# \# \# • \# \# \# • \# \# \#

A25 4″: $16 \times \frac{1}{4}$

Q26 Find the taper in a piece of work $8\frac{1}{2}''$ long if the taper is $\frac{3}{16}''$ per inch.

\# \# \# • \# \# \# • \# \# \# • \# \# \# • \# \# \# • \# \# \# • \# \# \# • \# \# \# • \# \# \#

A26 $1\frac{19}{32}''$: $8\frac{1}{2}\times\frac{3}{16}$

Q27 A piece of work is tapered $\frac{1}{8}''$ per inch of length. What is the diameter at the end of a 12″ piece if the diameter is 6″ in the beginning?

\# \# \# • \# \# \# • \# \# \# • \# \# \# • \# \# \# • \# \# \# • \# \# \# • \# \# \# • \# \# \#

A27 $4\frac{1}{2}''$: First find the taper in 12″: $12\times\frac{1}{8}=1\frac{1}{2}''$. Ending diameter $=6''-1\frac{1}{2}''$

12 If the overall taper of a piece of work is known, the taper per inch can be found by dividing the taper by the length of the piece.

Example: Find the taper per inch in a piece of work 8 inches long if the taper is 2 inches.

Solution: Taper per inch = taper ÷ length $=2''\div 8''=\frac{1}{4}$. Thus, the taper per inch is $\frac{1}{4}''$ per inch.

Q28 Find the taper per inch in a piece of work 15 inches long if the taper is $1\frac{1}{4}$ inches.

\# \# \# • \# \# \# • \# \# \# • \# \# \# • \# \# \# • \# \# \# • \# \# \# • \# \# \# • \# \# \#

A28 $\frac{1}{12}''$ per inch: taper per inch = taper ÷ length $=15''\div 1\frac{3}{4}''$

Q29 Find the taper per inch in the following piece of work. (Hint: First find the taper.)

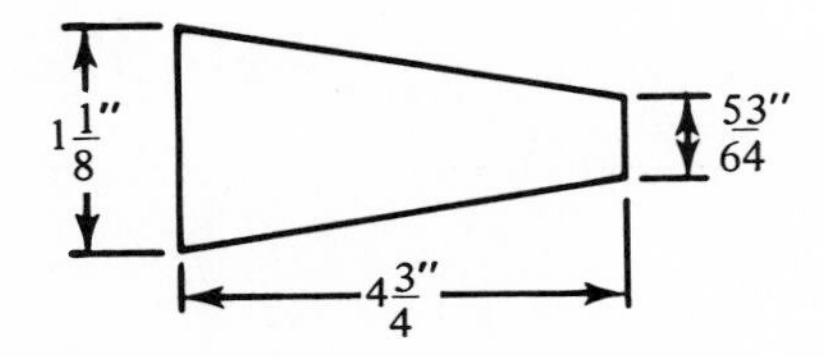

\# \# \# • \# \# \# • \# \# \# • \# \# \# • \# \# \# • \# \# \# • \# \# \# • \# \# \# • \# \# \#

A29 $\frac{1}{16}''$ per inch: taper $=1\frac{1}{8}''-\frac{53''}{64}$

taper per inch $=\frac{19}{64}\div 4\frac{3}{4}$

Q30 Find the taper per foot in Q29 above.

• ### • ### • ### • ### • ### • ### • ### •

A30 $\frac{3''}{4}$ per foot: If the taper per inch is $\frac{1}{16}''$ the taper per foot will be 12 times as great: $= 12 \times \frac{1}{16}$.

This completes the instruction for this section.

1.7 EXERCISE

1. **a.** Change 3 hr 27 min to min.
b. Change 204 min to hr and min.

2. Perform the indicated operation and simplify:

a. 8 hr 10 min + 3 hr 55 min

b. 8 hr 10 min − 3 hr 45 min

c. 4 hr 55 min + 2 hr 37 min + 3 hr 52 min

d. 7 hr 15 min − 2 hr 55 min

e. 3 hr 20 min × 8

f. 15 hr 40 min ÷ 5

3. Change to a fractional part of an hour:
a. 8 hr 10 min **b.** 7 hr 25 min **c.** 6 hr 12 min
d. 15 hr 6 min **e.** 10 hr 33 min **f.** 2 hr 50 min

4. Change to a denominate number in hours and minutes:
a. $8\frac{2}{3}$ hr **b.** $7\frac{1}{10}$ hr **c.** $8\frac{1}{5}$ hr
d. $9\frac{3}{4}$ hr **e.** $8\frac{5}{6}$ hr **f.** $7\frac{3}{10}$ hr

5. Seven employees of Canton Ltd. worked the following hours. Find the total hours worked for each employee.

a. 7	$8\frac{1}{4}$	$7\frac{1}{4}$	$7\frac{1}{2}$	9	—
b. 8	8	7	—	7	7
c. $7\frac{1}{4}$	$7\frac{1}{2}$	$7\frac{1}{4}$	$7\frac{3}{4}$	$6\frac{1}{2}$	—
d. $7\frac{1}{4}$	$7\frac{1}{2}$	$7\frac{1}{2}$	$3\frac{1}{4}$	$8\frac{1}{2}$	$4\frac{1}{2}$
e. 8	8	$7\frac{1}{2}$	—	8	$7\frac{1}{2}$
f. $8\frac{1}{2}$	$7\frac{1}{2}$	$7\frac{3}{4}$	—	7	$7\frac{1}{4}$
g. $7\frac{1}{2}$	$8\frac{1}{2}$	8	7	5	4

6. On Thursday Frank traveled $5\frac{1}{4}$, $6\frac{1}{6}$, and $4\frac{5}{8}$ miles. On Friday he traveled $3\frac{1}{2}$, $\frac{7}{8}$, $3\frac{5}{16}$, and $1\frac{7}{16}$ miles. How many more miles did Frank travel on Thursday?

7. A service station sold $5\frac{1}{8}$ cases of #30 weight oil, $6\frac{3}{4}$ cases of #20 weight, and $13\frac{5}{8}$ cases of #10 weight. How many cases of oil were sold?
8. A man, his wife, and 2 children want to fly to Cleveland. The price of a regular ticket is $72. Before tax, what is the total cost of tickets for the family?

One Way Airlines

Special Family rates for Tickets

Husband	– pays regular price
Wife	— pays $\frac{2}{3}$ of regular price
Child	– pays $\frac{1}{2}$ of regular price

9. A stock which closed on Wednesday at $42\frac{3}{8}$ lost $2\frac{3}{4}$ points on the following day. At what price did it close on Thursday?
10. If a family lives on a budget of $650 per month and two-fifths of its budget is spent for food and rent, how much is spent for food and rent?
11. David Peet purchased $12\frac{1}{2}$ yards of gravel at $6 per yard. What was the cost of his purchase?
12. Richard Wallen ordered $6\frac{1}{2}$ tons of coal. Only $\frac{2}{3}$ of the coal was delivered. How much was delivered?
13. A pipe $61\frac{1}{2}''$ long is divided into 14 equal parts. What is the length of each part?
14. One inch lumber now has an actual thickness of $\frac{25}{32}''$. It is being proposed that this be reduced to $\frac{3}{4}''$. What is the reduction in thickness being proposed?
15. What would the outlet voltage be on a 110 volt line that has a voltage drop of $4\frac{3}{4}$ volts?
16. How long will it take to machine 20 pins if each pin takes $7\frac{1}{2}$ minutes?
17. Find the length of the shaft.

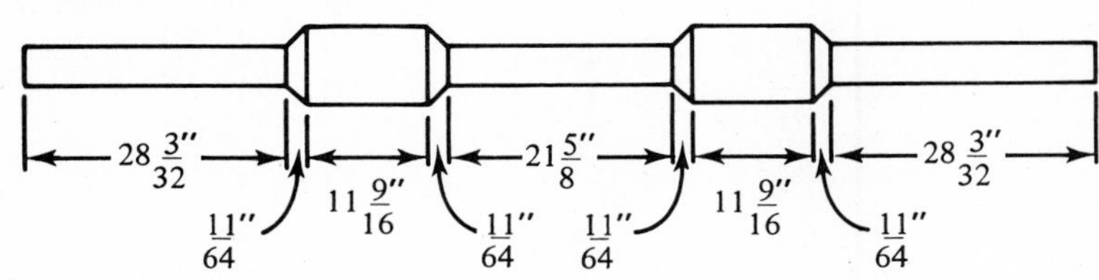

18. Find the missing dimensions:

a. b.

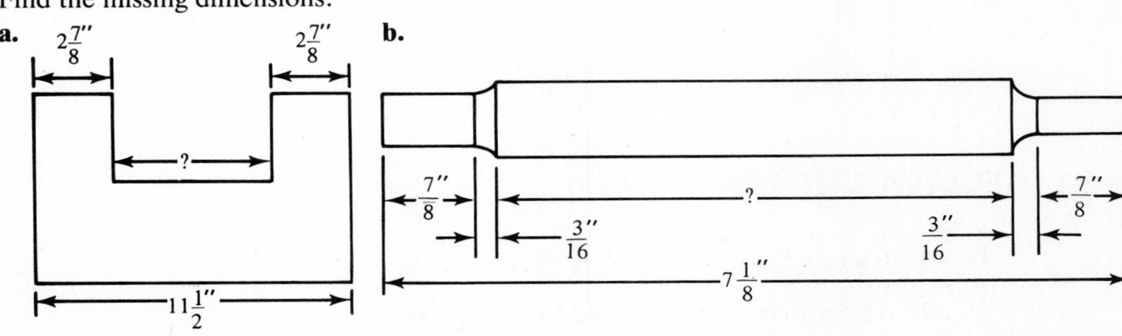

19. Find the taper of the figure below.

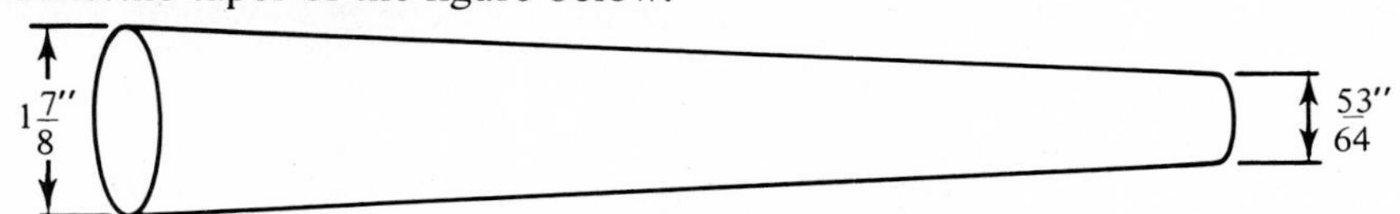

20. If the above figure (problem 19) is 3 feet long, find the taper per foot.

21. Find the taper in a piece of work 7″ long if the taper is $\frac{1}{16}''$ per inch.

22. Find the taper in a piece of work 3 feet long if the taper is $\frac{5}{32}''$ per inch.

1.7 EXERCISE ANSWERS

1. a. 207 **b.** 3 hr 24 min

2. a. 12 hr 5 min **b.** 4 hr 25 min **c.** 11 hr 24 min **d.** 4 hr 20 min **e.** 26 hr 40 min **f.** 3 hr 8 min

3. a. $8\frac{1}{6}$ hr **b.** $7\frac{5}{12}$ hr **c.** $6\frac{1}{5}$ hr **d.** $15\frac{1}{10}$ hr **e.** $10\frac{11}{20}$ hr **f.** $2\frac{5}{6}$ hr

4. a. 8 hr 40 min **b.** 7 hr 6 min **c.** 8 hr 12 min **d.** 9 hr 45 min **e.** 8 hr 50 min **f.** 7 hr 18 min

5. a. 39 hr **b.** 37 hr **c.** $36\frac{1}{4}$ hr **d.** $38\frac{1}{2}$ hr **e.** 39 hr **f.** 38 hr **g.** 40 hr

6. $6\frac{11}{12}$ mi **7.** $25\frac{1}{2}$ cases **8.** $192

9. $39\frac{5}{8}$ **10.** $260 **11.** $75

13. $4\frac{1}{3}$ tons **13.** $4\frac{11}{28}''$ **14.** $\frac{1}{32}''$

15. $105\frac{1}{4}$ volts **16.** $2\frac{1}{2}$ hrs **17.** $101\frac{5}{8}''$

18. a. $5\frac{3}{4}''$ **b.** $4\frac{5}{8}''$

19. $1\frac{3}{64}''$ **20.** $\frac{67}{192}''$ per foot **21.** $\frac{7}{16}''$

22. $5\frac{5}{8}''$

CHAPTER 1 SAMPLE TEST

At the completion of Chapter 1 you should be able to work the following problems.

1.1 EQUIVALENT FRACTIONS

1. Supply the missing value:

a. $\frac{4}{5} = \frac{?}{35}$ **b.** $9 = \frac{?}{8}$

2. Reduce to lowest terms:

a. $\frac{18}{24}$ b. $\frac{34}{51}$

1.2 MIXED NUMBERS AND IMPROPER FRACTIONS

3. a. Change $5\frac{5}{8}$ to an improper fraction.

b. Express $\frac{516}{8}$ as a mixed number.

1.3 ADDITION

4. Add:

a. $\frac{1}{15}+\frac{9}{15}$ b. $3\frac{5}{12}+2\frac{11}{12}$ c. $\frac{1}{2}+\frac{1}{7}$ d. $5\frac{5}{6}+4\frac{7}{9}$

1.4 PRIME AND COMPOSITE NUMBERS

5. List the prime numbers between 20 and 30.

6. Express as a product of primes:

a. 42 b. 31 c. 51 d. 840

1.5 SUBTRACTION

7. Subtract:

a. $\frac{5}{6}-\frac{5}{12}$ b. $4\frac{5}{8}-2\frac{1}{6}$ c. $7-2\frac{3}{5}$ d. $5\frac{7}{12}-3\frac{7}{8}$

1.6 MULTIPLICATION AND DIVISION

8. Multiply:

a. $\frac{3}{4}\times\frac{8}{11}$ b. $1\frac{1}{4}\times 1\frac{2}{3}$ c. $3\frac{1}{2}\times 6$ d. $8\times\frac{1}{16}$

9. Divide:

a. $3\div\frac{1}{3}$ b. $\frac{2}{5}\div\frac{5}{6}$ c. $3\frac{1}{3}\div 2$ d. $2\frac{1}{2}\div\frac{5}{2}$

1.7 APPLICATIONS

10. Three employees worked as follows:

	M	*W*	*F*
Jim	7 hr 30 min	8 hr 10 min	6 hr 15 min
Ladrie	8 hr	7 hr 12 min	8 hr 45 min
Thomas	7 hr 5 min	8 hr 13 min	10 hr 20 min

Find the number of hours that each worked for the days given.

11. Mr. Ackley paid \$182 for $8\frac{2}{3}$ yards of cement. What was the price per yard?

12. Find the missing dimension if the overall length is $12\frac{5}{8}''$.

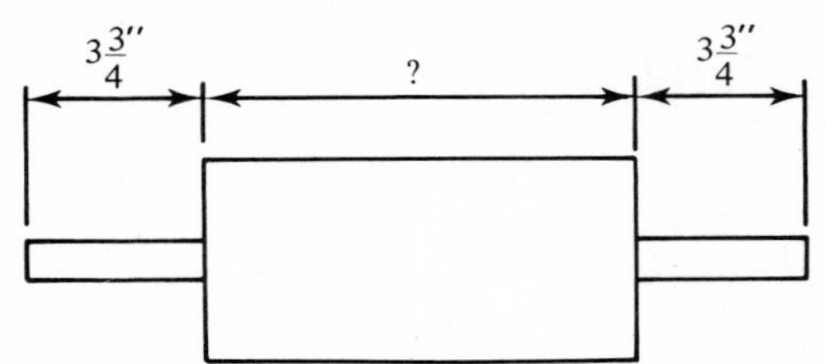

13. A piece of work is tapered $\frac{1}{4}''$ per inch. Find the small diameter of a piece 3″ long if the large diameter is $10\frac{3}{16}''$.

14. Find the taper per inch of a rod 8 inches long if the diameters of the ends are $4\frac{1}{2}''$ and $2\frac{3}{8}''$, respectively.

CHAPTER 1 SAMPLE TEST ANSWERS

1. a. 28 **b.** 72

2. a. $\frac{3}{4}$ **b.** $\frac{2}{3}$

3. a. $\frac{45}{8}$ **b.** $64\frac{1}{2}$

4. a. $\frac{2}{3}$ **b.** $6\frac{1}{3}$ **c.** $\frac{9}{14}$ **d.** $10\frac{11}{18}$

5. 23, 29

6. a. $2\times3\times7$ **b.** prime **c.** 3×17 **d.** $2\times2\times2\times3\times5\times7$

7. a. $\frac{5}{12}$ **b.** $2\frac{11}{24}$ **c.** $4\frac{2}{5}$ **d.** $1\frac{17}{24}$

8. a. $\frac{6}{11}$ **b.** $2\frac{1}{12}$ **c.** 21 **d.** $\frac{1}{2}$

9. a. 9 **b.** $\frac{12}{25}$ **c.** $1\frac{2}{3}$ **d.** 1

10. Jim, 21 hrs 55 min or $21\frac{11}{12}$ hrs; Laurie, 23 hrs 57 min or $23\frac{19}{20}$ hrs; Thomas, 25 hrs 38 min or $25\frac{19}{30}$ hrs

11. \$ 21 **12.** $5\frac{1}{8}''$ **13.** $9\frac{7}{16}''$ **14.** $\frac{17}{64}''$ per inch

CHAPTER 2

OPERATIONS WITH DECIMAL FRACTIONS

A fraction was defined, in Chapter 1, as the quotient of two whole numbers with a nonzero denominator. These types of fractions could be called *common fractions*. The fractions discussed in this chapter are a type of fraction and are called *decimal fractions*.

2.1 PLACE VALUE

1 The Hindu–Arabic number system is called a *decimal system* because there are 10 digits and the place values are based on 10. The place values, from right to left, are ones, tens, hundreds, thousands, and so on.

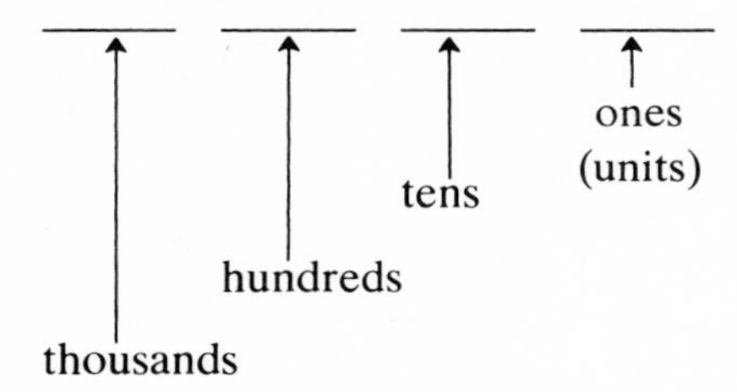

Each place value to the left of another has a value of 10 times the preceding value; that is, the thousands place is 10 times the hundreds place ($1{,}000 = 10 \times 100$), the hundreds place is 10 times the tens place ($100 = 10 \times 10$), and the tens place is 10 times the ones place ($10 = 10 \times 1$). Or, each place value to the right of another is one-tenth $\left(\frac{1}{10}\right)$ of the preceding place. Thus, the value to the right of the thousands place is $\frac{1}{10}$ of 1,000, or 100 (hundreds place, $\frac{1}{10} \times 1{,}000 = 100$).

Q1 The value of the place to the right of the hundreds place is $\frac{1}{10} \times$ _____ = _____.

\# \# \# • \# \# \# • \# \# \# • \# \# \# • \# \# \# • \# \# \# • \# \# \# • \# \# \# • \# \# \#

A1 100, 10

Q2 The value of the place to the right of the tens place is $\frac{1}{10} \times$ _____ = _____.

\# \# \# • \# \# \# • \# \# \# • \# \# \# • \# \# \# • \# \# \# • \# \# \# • \# \# \# • \# \# \#

A2 10, 1

Q3 Continuing with the above pattern, the value of the place to the right of the ones place is $\frac{1}{10} \times$ ______ = ______.

\# \# \# • \# \# \# • \# \# \# • \# \# \# • \# \# \# • \# \# \# • \# \# \# • \# \# \# • \# \# \#

A3 $1, \frac{1}{10}$

Q4 The value of the place to the right of the tenths place is $\frac{1}{10} \times$ ______ = ______.

\# \# \# • \# \# \# • \# \# \# • \# \# \# • \# \# \# • \# \# \# • \# \# \# • \# \# \# • \# \# \#

A4 $\frac{1}{10}, \frac{1}{100}$

2 Question 3 suggests place value to the right of the ones place. This idea can be extended with the introduction of decimal notation. A *decimal fraction* is a fraction whose denominator is 10, 100, 1,000, and so on (a power of 10). Decimal fractions are usually written without denominators, with a *decimal point* representing place value:

$\frac{1}{10}$ is written 0.1 (one tenth)

$\frac{1}{100}$ is written 0.01 (one hundredth)

The place value one place to the right of the ones place is tenths. The digit 1 in 0.1 is one place to the right of the ones place.

The place value one place to the right of the tenths place is hundredths. The digit 1 in 0.01 is one place to the right of the tenths place. Consider:

$\frac{5}{10} = 0.5$ $\quad \frac{8}{100} = 0.08$

$\frac{6}{10} = 0.6$ $\quad \frac{3}{100} = 0.03$

Q5 Write the following decimal fractions using decimal notation:

a. $\frac{3}{10}$ ______ **b.** $\frac{9}{10}$ ______ **c.** $\frac{2}{100}$ ______

\# \# \# • \# \# \# • \# \# \# • \# \# \# • \# \# \# • \# \# \# • \# \# \# • \# \# \# • \# \# \#

A5 **a.** 0.3 **b.** 0.9 **c.** 0.02

Q6 Write each of the following as a fraction:

a. 0.7 ______ **b.** 0.07 ______ **c.** 0.01 ______

\# \# \# • \# \# \# • \# \# \# • \# \# \# • \# \# \# • \# \# \# • \# \# \# • \# \# \# • \# \# \#

A6 **a.** $\frac{7}{10}$ **b.** $\frac{7}{100}$ **c.** $\frac{1}{100}$

Q7 **a.** The place value one place to the right of the ones place is ______________.

b. The place value two places to the right of the ones place is ______________.

\# \# \# • \# \# \# • \# \# \# • \# \# \# • \# \# \# • \# \# \# • \# \# \# • \# \# \# • \# \# \#

A7 **a.** tenths **b.** hundredths

3 Below is a diagram that shows some of the values of the places to the right of the decimal point. It should be used to determine the place value of numbers written in decimal notation. Decimal fractions written using decimal notation are simply called *decimals.*

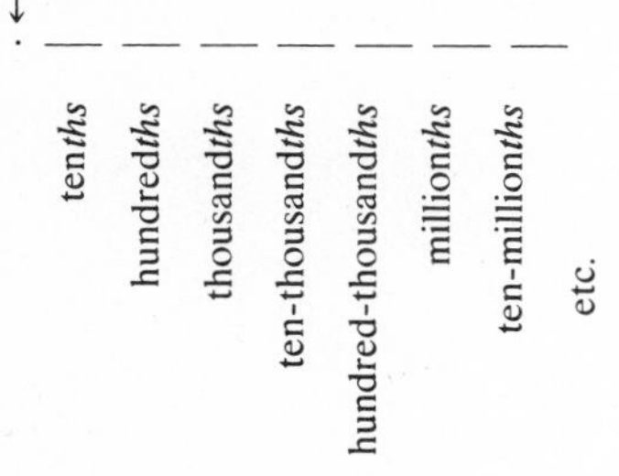

For the numeral 0.18793:

The place value of the 1 is tenths.

The place value of the 8 is hundredths.

The place value of the 7 is thousandths.

The place value of the 9 is ten-thousandths.

The place value of the 3 is hundred-thousandths.

Q8 In the numeral 21.0874, what digit is in the

a. tenths place? ______

b. tens place? ______

c. hundredths place? ______

d. ones place? ______

e. ten-thousandths place? ______

f. thousandths place? ______

• ### • ### • ### • ### • ### • ### • ### •

A8 **a.** 0 **b.** 2 **c.** 8 **d.** 1 **e.** 4 **f.** 7

Q9 Determine the place value of each numeral in 93.1726.

9 ____________ 3 ____________ 1 ____________

7 ____________ 2 ____________ 6 ____________

• ### • ### • ### • ### • ### • ### • ### •

A9 9 (tens), 3 (ones), 1 (tenths), 7 (hundredths), 2 (thousandths), 6 (ten-thousandths)

Q10 **a.** In the decimal 0.28, the 8 means ______.

b. In the decimal 3.017, the 7 means ______.

• ### • ### • ### • ### • ### • ### • ### •

A10 **a.** $\frac{8}{100}$ **b.** $\frac{7}{1{,}000}$

4 The decimal 0.13 can be expressed as

$$0.13=\frac{1}{10}+\frac{3}{100}$$
$$=\frac{1\times 10}{10\times 10}+\frac{3}{100}$$
$$=\frac{10}{100}+\frac{3}{100}$$
$$=\frac{13}{100}$$

The decimal 0.28 can be expressed as

$$0.28=\frac{2}{10}+\frac{8}{100}$$
$$=\frac{2\times 10}{10\times 10}+\frac{8}{100}$$
$$=\frac{20}{100}+\frac{8}{100}$$
$$=\frac{28}{100}$$

The decimal 0.013 can be expressed as

$$0.013=\frac{0}{10}+\frac{1}{100}+\frac{3}{1,000}$$
$$=0+\frac{1\times 10}{100\times 10}+\frac{3}{1,000}$$
$$=\frac{10}{1,000}+\frac{3}{1000}$$
$$=\frac{13}{1,000}$$

Q11 Express the decimal as a fraction by completing the following:

$$0.17=\frac{1}{10}+\frac{7}{100}$$
$$=\frac{1\times(\quad)}{10\times(\quad)}+\frac{7}{100}$$
$$=(\qquad\qquad)+\frac{7}{100}$$
$$=\_\_\_\_\_\_$$

\# \# \# • \# \# \# • \# \# \# • \# \# \# • \# \# \# • \# \# \# • \# \# \# • \# \# \# • \# \# \#

A11
$$0.17=\frac{1}{10}+\frac{7}{100}$$
$$=\frac{1\times 10}{10\times 10}+\frac{7}{100}$$
$$=\frac{10}{100}+\frac{7}{100}$$
$$=\frac{17}{100}$$

Q12 Express the decimals as fractions (do not reduce):

a. 0.06 ______ **b.** 0.27 ______ **c.** 0.81 ______

• ### • ### • ### • ### • ### • ### • ### •

A12 **a.** $\frac{6}{100}$ **b.** $\frac{27}{100}$ **c.** $\frac{81}{100}$

Q13 Express the decimal as a fraction by completing the following:

$$0.273 = \frac{2}{10} + \frac{7}{100} + \frac{3}{(\quad)}$$
$$= \frac{2 \times 100}{10 \times 100} + \frac{7 \times (\quad)}{100 \times (\quad)} + \frac{3}{1{,}000}$$
$$= (\qquad) + (\qquad) + (\qquad)$$
$$= \underline{\qquad}$$

• ### • ### • ### • ### • ### • ### • ### •

A13
$$0.273 = \frac{2}{10} + \frac{7}{100} + \frac{3}{1{,}000}$$
$$= \frac{2 \times 100}{10 \times 100} + \frac{7 \times 10}{100 \times 10} + \frac{3}{1{,}000}$$
$$= \frac{200}{1{,}000} + \frac{70}{1{,}000} + \frac{3}{1{,}000}$$
$$= \frac{273}{1{,}000}$$

Q14 Express the decimals as fractions (do not reduce; do the problem mentally if possible):

a. 0.213 ______ **b.** 0.085 ______ **c.** 0.372 ______

• ### • ### • ### • ### • ### • ### • ### •

A14 **a.** $\frac{213}{1{,}000}$ **b.** $\frac{85}{1{,}000}$ **c.** $\frac{372}{1{,}000}$

Q15 Express the fractions as decimals:

a. $\frac{3}{1{,}000}$ ______ **b.** $\frac{158}{1{,}000}$ ______ **c.** $\frac{29}{1{,}000}$ ______

• ### • ### • ### • ### • ### • ### • ### •

A15 **a.** 0.003* **b.** 0.158 **c.** 0.029

*0.003 = .003. The use of the 0 to the left of the decimal point is a publisher's convention to minimize errors in the placement of the decimal point. In common practice, it is not incorrect to omit it.

5 A mixed number can be represented as a decimal if the fraction is a decimal fraction, and conversely.

Example: Write $2\frac{3}{10}$ as a decimal.

Solution: $2\frac{3}{10} = 2.3$

Two additional examples are

$2\frac{17}{100} = 2.17 \qquad 25\frac{3}{1,000} = 25.003$

Q16 Convert to decimals:

a. $17\frac{7}{10}$ ______ **b.** $12\frac{8}{100}$ ______ **c.** $201\frac{5}{1,000}$ ______

\### • ### • ### • ### • ### • ### • ### • ### • ###

A16 **a.** 17.7 **b.** 12.08 **c.** 201.005

Q17 Convert to fractions (do not reduce):

a. 213.7 ______ **b.** 2.002 ______ **c.** 9.16 ______

\### • ### • ### • ### • ### • ### • ### • ### • ###

A17 **a.** $213\frac{7}{10}$ **b.** $2\frac{2}{1,000}$ **c.** $9\frac{16}{100}$

6 To read a decimal, first read the number involved and then give the place value in which the last digit of the number lies.

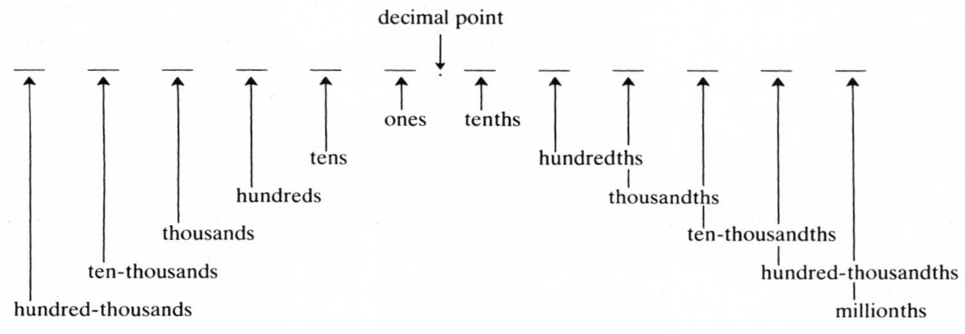

Thus,

0.0134 is read "one hundred thirty-four *ten-thousandths*."

0.85 is read "eighty-five *hundredths*."

0.00203 is read "two hundred three *hundred-thousandths*."

0.045 is read "forty-five *thousandths*."

To read a number consisting of a whole number and a decimal:

Step 1: Read the whole number.

Step 2: Read "and" for the decimal point.

Step 3: Read the decimal as above.

Examples:

1. 12.03 is read "twelve and three hundredths."

2. 5,280.123 is read "five thousand two hundred eighty and one hundred twenty-three thousandths."

Q18 Write the decimal in words:

a. 1.03 ______________________________

b. 42.13 ____________________

c. 12.0439 ____________________

d. 10,800.876 ____________________

____

e. 400.002 ____________________

f. 0.402 ____________________

• ### • ### • ### • ### • ### • ### • ### •

A18 **a.** one and three hundredths
b. forty-two and thirteen hundredths
c. twelve and four hundred thirty-nine ten-thousandths
d. ten thousand eight hundred and eight hundred seventy-six thousandths
e. four hundred and two thousandths
f. four hundred two thousandths

Q19 When reading decimals, the decimal point is read "________."

• ### • ### • ### • ### • ### • ### • ### •

A19 and

---

7 When reading a number, it is important to remember that *and* is used to connect a whole number and a decimal, but it is not used anywhere else in the number:

two hundred thirty-six and fifty-two thousandths

236.052

Q20 Write as decimals:

a. two and six tenths ____________

b. fifteen and seventy-five hundredths ____________

c. one thousand fifty-seven and six thousandths ____________

d. four thousand five hundred seven and forty-three hundred-thousandths ____________

• ### • ### • ### • ### • ### • ### • ### •

A20 **a.** 2.6 **b.** 15.75 **c.** 1,057.006 **d.** 4,507.00043

Q21 Write the decimals in words:

a. 0.3 ____________________

b. 0.32 ____________________

c. 6.35 ____________________

d. 2.003 ____________________

e. 50.052 ____________________

• ### • ### • ### • ### • ### • ### • ### •

A21
a. three tenths
b. thirty-two hundredths
c. six and thirty-five hundredths
d. two and three thousandths
e. fifty and fifty-two thousandths

Q22 Write as decimals:

a. five and six tenths ____________

b. twenty-five ten-thousandths ____________

c. two hundred eight and five hundredths ____________

d. two thousand fifteen ten-thousandths ____________

e. two thousand and fifteen ten-thousandths ____________

• # # # • # # # • # # # • # # # • # # # • # # # • # # # • # #

A22 **a.** 5.6 **b.** 0.0025 **c.** 208.05 **d.** 0.2015 **e.** 2,000.0015

This completes the instruction for this section.

2.1 EXERCISE

1. What is the place value of the 7 in the following numbers?
   a. 13.872 **b.** 702.01 **c.** 0.0007
2. In a number the word "and" represents the ____________ .
3. Write the following decimals as a fraction or mixed number (do not reduce):
   a. 0.704 **b.** 0.0302 **c.** 2.07
   d. 0.00045 **e.** 17.108 **f.** 2.007
4. Write as decimals:
   a. $\frac{19}{1,000}$ **b.** $5\frac{5}{100}$ **c.** $7\frac{9}{10}$
   d. $\frac{72}{100}$ **e.** $\frac{204}{10,000}$ **f.** $\frac{32}{100,000}$
5. Write the decimals in words:
   a. 0.1 **b.** 0.07 **c.** 7.07 **d.** 21.4805
6. Write as a decimal:
   a. three and fifty-seven hundredths
   b. six thousand four ten-thousandths
   c. six thousand and four ten-thousandths
   d. thirty-three and seventy-eight hundredths

2.1 EXERCISE ANSWERS

1. **a.** hundredths **b.** hundreds **c.** ten-thousandths
2. decimal point
3. **a.** $\frac{704}{1,000}$ **b.** $\frac{302}{10,000}$ **c.** $2\frac{7}{100}$
   d. $\frac{45}{100,000}$ **e.** $17\frac{108}{1,000}$ **f.** $2\frac{7}{1,000}$

4. a. 0.019 **b.** 5.05 **c.** 7.9
d. 0.72 **e.** 0.0204 **f.** 0.00032

5. a. one tenth
b. seven hundredths
c. seven and seven hundredths
d. twenty-one and four thousand eight hundred five ten-thousandths

6. a. 3.57 **b.** 0.6004 **c.** 6,000.0004 **d.** 33.78

2.2 ROUNDING OFF

1 Measurements are never exact. In measuring the length of something, you may measure to the nearest inch, the nearest 0.1 inch, the nearest 0.01 inch, or even closer. Each measurement is an approximation of the length. The measurement recorded is usually a number that has been rounded off.

Since a digit several places to the right of the decimal point represents a very small quantity, a number that has several decimal places is frequently "rounded off" to two or three decimal places. *Rounding off* a number is a way of saying that the number is correct to a specified number of decimal places. To round off a decimal number, look at the digit immediately to the right of the digit that you are rounding off to.

1. If this digit is 5 or larger, the digit in the place you are rounding off to is increased by 1 and all digits to the right are replaced by zeros.
2. If this digit is 4 or smaller, the digit in the place you are rounding off to is left as it is and all digits to the right are replaced by zeros. If these zeros are to the right of the decimal point, they are dropped.

Examples: Round off to the nearest hundredth:

1. 27.8463 to the nearest hundredth is 27.8500 or 27.85
   5 or larger
2. 0.7349 to the nearest hundredth is 0.7300 or 0.73
   4 or smaller

Q1 When rounding off 4.763 to the nearest tenth, which digit must be 5 or larger, or 4 or smaller? (Circle the answer.)

• ### • ### • ### • ### • ### • ### • ### •

A1 4.7⑥3

Q2 Round off to the nearest tenth:

a. 4.$\underline{7}$63 ______ **b.** 5.$\underline{0}$31 ______ **c.** 76.$\underline{3}$49 ______

• ### • ### • ### • ### • ### • ### • ### •

A2 **a.** 4.8 **b.** 5.0 **c.** 76.3

Q3 Round off to the nearest tenth:

a. 76.352 ______ **b.** 89.015 ______ **c.** 7.555 ______

• ### • ### • ### • ### • ### • ### • ### •

A3 **a.** 76.4 **b.** 89.0 **c.** 7.6

Q4 When rounding off 5.722 to the nearest hundredth, which digit must be 5 or larger, or 4 or smaller? (Circle the answer.)

• ### • ### • ### • ### • ### • ### • ### •

A4 5.72②

Q5 Round off to the nearest hundredth:

a. 0.385 ______ b. 0.296 ______ c. 5.722 ______

• ### • ### • ### • ### • ### • ### • ### •

A5 a. 0.39 b. 0.30 c. 5.72

Q6 Round off to the nearest hundredth:

a. 0.0782 ______ b. 6.0035 ______ c. 2.8347 ______

• ### • ### • ### • ### • ### • ### • ### •

A6 a. 0.08 b. 6.00 c. 2.83

Q7 Round off the the nearest thousandth:

a. 0.0672 ______ b. 0.05550 ______ c. 0.6338 ______

• ### • ### • ### • ### • ### • ### • ### •

A7 a. 0.067 b. 0.056 c. 0.634

Q8 Round off to the nearest thousandth:

a. 17.36371 ______ b. 0.3999 ______ c. 0.00191 ______

• ### • ### • ### • ### • ### • ### • ### •

A8 a. 17.364 b. 0.400 c. 0.002

Q9 Round off to the nearest thousand:

a. 7,398 ______ b. 62,275 ______ c. 9,872.5 ______

• ### • ### • ### • ### • ### • ### • ### •

A9 a. 7,000 b. 62,000 c. 10,000

Q10 Round off to the nearest whole number (ones place):

a. 479.23 ______ b. 6.872 ______ c. 17.50 ______

• ### • ### • ### • ### • ### • ### • ### •

A10 a. 479 b. 7 c. 18

This completes the instruction for this section.

2.2 EXERCISE

Round off each number as indicated:

1. 4.54716
   a. nearest hundredth b. nearest thousandth c. nearest tenth
2. 8.58637
   a. nearest hundredth b. nearest ten-thousandth c. nearest thousandth

3. 2.07462
  a. nearest hundredth **b.** nearest thousandth **c.** nearest tenth

4. 3.098736
  a. nearest hundredth **b.** three decimals **c.** one decimal

5. 1.70954
  a. two decimals **b.** three decimals **c.** one decimal

2.2 EXERCISE ANSWERS

1. a. 4.55 **b.** 4.547 **c.** 4.5
2. a. 8.59 **b.** 8.5864 **c.** 8.586
3. a. 2.07 **b.** 2.075 **c.** 2.1
4. a. 3.10 **b.** 3.099 **c.** 3.1
5. a. 1.71 **b.** 1.710 **c.** 1.7

2.3 ADDITION AND SUBTRACTION

1 When adding two or more decimals it will be necessary to add digits that represent the same place value. This is illustrated in an example:

$$\begin{array}{r} 0.13 = \frac{1}{10} + \frac{3}{100} \\ +0.24 = \frac{2}{10} + \frac{4}{1000} \\ \hline 0.37 = \frac{3}{10} + \frac{7}{100} \end{array}$$

It is not necessary to write decimals as fractions to find their sum; however, it will be helpful to arrange the numbers in vertical columns so that tenths are added to tenths, hundredths are added to hundredths, and so on.

Q1 Find the sum: $\begin{array}{r} 0.34 \\ +0.23 \\ \hline \end{array}$

• ### • ### • ### • ### • ### • ### • ### •

A1 0.57

Q2 Add 0.316 + 0.472 by arranging vertically.

• ### • ### • ### • ### • ### • ### • ### •

A2 0.788: $\begin{array}{r} 0.316 \\ +0.472 \\ \hline 0.788 \end{array}$

2 If the decimal numbers in an addition problem are arranged correctly the decimal points will lie on a vertical line.

Example: Find the sum 53.7 + 0.036 + 9.12 + 0.0005.

Solution:

```
 53.7
  0.036
  9.12
+0.0005
 62.8565   (decimal points aligned)
```

Q3 Align the decimals correctly: 273.1204 + 7.982

• ### • ### • ### • ### • ### • ### • ### •

A3

```
273.1204
 +7.982
```

Q4 Align the decimals correctly: 0.3 + 0.84 + 0.7 + 1.213

• ### • ### • ### • ### • ### • ### • ### •

A4

```
 0.3
 0.84
 0.7
+1.213
```

3 Zeros are often used to "square off" the columns of decimals as a mean of keeping the column additions correct. The sum 53.7 + 0.036 + 9.12 + 0.0005 could be written as follows by using zeros to "square off" the decimals.

```
 53.7000
  0.0360
  9.1200
+0.0005
 62.8565
```

Q5 Align and square off the decimals: 273.1204 + 7.982

• ### • ### • ### • ### • ### • ### • ### •

A5

```
273.1204
 +7.9820
```

Q6 Align and square off the decimals: 0.3 + 0.84 + 0.7 + 1.213

• ### • ### • ### • ### • ### • ### • ### •

A6
$$\begin{array}{r} 0.300 \\ 0.840 \\ 0.700 \\ +1.213 \\ \hline \end{array}$$

Q7 Find the sum:

a.
$$\begin{array}{r} 0.92 \\ +0.27 \\ \hline \end{array}$$

b.
$$\begin{array}{r} 0.51 \\ 0.30 \\ +0.84 \\ \hline \end{array}$$

c.
$$\begin{array}{r} 1.38 \\ 0.92 \\ +3.54 \\ \hline \end{array}$$

\# \# \# • \# \# \# • \# \# \# • \# \# \# • \# \# \# • \# \# \# • \# \# \# • \# \# \# • \# \# \#

A7 **a.** 1.19 **b.** 1.65 **c.** 5.84

Q8 Find the sum:

a. $4.09+3.987$ **b.** $7.56+2.3+5.879$

\# \# \# • \# \# \# • \# \# \# • \# \# \# • \# \# \# • \# \# \# • \# \# \# • \# \# \# • \# \# \#

A8 **a.** 8.077:
$$\begin{array}{r} 4.090 \\ +3.987 \\ \hline 8.077 \end{array}$$

b. 15.739:
$$\begin{array}{r} 7.560 \\ 2.300 \\ +5.879 \\ \hline 15.739 \end{array}$$

Q9 Find the sum:

a. $8.632+0.234+0.81+0.065$ **b.** $0.38+2.1+3.09+0.075+1.0004$

\# \# \# • \# \# \# • \# \# \# • \# \# \# • \# \# \# • \# \# \# • \# \# \# • \# \# \# • \# \# \#

A9 **a.** 9.741 **b.** 6.6454

Q10 Bob Fogg drove 681 miles on a hunting trip. He bought gas three times and paid \$5.72, \$6.10, and \$5.34. His meals cost \$9.60, and he paid \$12 for a motel room. What was the total of his expenses?

\# \# \# • \# \# \# • \# \# \# • \# \# \# • \# \# \# • \# \# \# • \# \# \# • \# \# \# • \# \# \#

A10 \$38.76

Q11 Linda Fogg bought a dress for \$24.63, a pair of shoes for \$19.99, and a handbag for \$7, and a hat for \$12.33. What was the total cost of these items?

\# \# \# • \# \# \# • \# \# \# • \# \# \# • \# \# \# • \# \# \# • \# \# \# • \# \# \# • \# \# \#

A11 $63.95

4 When subtracting decimals it is especially important to use zeros to "square off" decimals.

Example: Find the difference 5.2 − 3.672.

Solution:
$$\begin{array}{r} 5.200 \\ -3.672 \\ \hline 1.528 \end{array}$$

Q12 Find the difference:

a. $\begin{array}{r} 4.9 \\ -3.6 \\ \hline \end{array}$ **b.** $\begin{array}{r} 15.86 \\ -12.78 \\ \hline \end{array}$

\# \# \# • \# \# \# • \# \# \# • \# \# \# • \# \# \# • \# \# \# • \# \# \# • \# \# \# • \# \# \#

A12 **a.** 1.3 **b.** 3.08

Q13 Find the difference:

a. 2.63 − 1.7 **b.** 9.06 − 0.6

\# \# \# • \# \# \# • \# \# \# • \# \# \# • \# \# \# • \# \# \# • \# \# \# • \# \# \# • \# \# \#

A13 **a.** 0.93 **b.** 8.46

Q14 Find the difference:

a. 0.15 − 0.0367 **b.** 5 − 0.9163

\# \# \# • \# \# \# • \# \# \# • \# \# \# • \# \# \# • \# \# \# • \# \# \# • \# \# \# • \# \# \#

A14 **a.** 0.1133 **b.** 4.0837

Q15 A block of copper 1 foot square and 1 inch thick weighs 45.835 pounds. A block of steel the same size weighs 40.809 pounds. How much heavier is the block of copper?

\# \# \# • \# \# \# • \# \# \# • \# \# \# • \# \# \# • \# \# \# • \# \# \# • \# \# \# • \# \# \#

A15 5.026 pounds: $\begin{array}{r} 45.835 \\ -40.809 \\ \hline \end{array}$

This completes the instruction for this section.

2.3 EXERCISE

1. Find the sum:

a. 6.237 + 1.986
b. 3.819 + 8.23 + 1.7
c. 7.3 + 2.9186 + 1.79
d. 17.086 + 43.509 + 18.762
e. 0.768 + 73.8 + 4.680
f. 453 + 289 + 387.6
g. 0.003 + 600.01 + 10.1
h. 0.408 + 0.2976 + 0.34567

2. Find the difference:

a. 43.62 − 37.96
b. 47 − 8.3
c. 4 − 0.68379
d. 97.8 − 0.4568
e. 52.6 − 9.002
f. 4,006.1 − 969
g. 0.05 − 0.005
h. $45.90 − $3.86

2.3 EXERCISE ANSWERS

1. **a.** 8.223 **b.** 13.749 **c.** 12.0086 **d.** 79.357
   e. 79.248 **f.** 1129.6 **g.** 610.113 **h.** 1.05127
2. **a.** 5.66 **b.** 38.7 **c.** 3.31621 **d.** 97.3432
   e. 43.598 **f.** 3,037.1 **g.** 0.045 **h.** $42.04

2.4 MULTIPLICATION AND DIVISION

1 The multiplication of decimals is carried out in the same way as multiplication of whole numbers, except for the placing of the decimal point in the product. Consider the product 0.2×0.3 in fraction form:

$$0.2 \times 0.3 = \frac{2}{10} \times \frac{3}{10} = \frac{6}{100} = 0.06$$

Notice that the product of a one-place decimal (tenths) by a one-place decimal (tenths) is a two-place decimal (hundredths). That is, tenths $\times$ tenths $=$ hundredths.

In general, *the decimal places of the product will equal the sum of the decimal places of the factors*. For example,

```
   12.13   (2 decimal places)
  ×0.212   (3 decimal places)
  ------
    2426
   1213
  2 426
  ------
 2.57156   (5 decimal places)
```

Q1 How many decimal places will there be in the product 2.68×0.1703? ______

• # # # • # # # • # # # • # # # • # # # • # # # • # # # • # #

A1 6: 2.68 has 2 decimal places, and 0.1703 has 4 decimal places. Hence, the product has $4+2=6$ decimal places.

Q2 How many decimal places will be in the product?

a. 8×7.23 ______ **b.** 0.03×2.9 ______ **c.** 0.005×0.0002 ______

• # # # • # # # • # # # • # # # • # # # • # # # • # # # • # #

A2 **a.** 2 **b.** 3 **c.** 7

Q3 Place the decimal point in the product: $3.25 \times 1.6 = 5200$

• # # # • # # # • # # # • # # # • # # # • # # # • # # # • # #

A3 5.200

Q4 Place the decimal point in the product:
a. $8.7 \times 0.36 = 3132$ **b.** $2.07 \times 0.308 = 63756$

• # # # • # # # • # # # • # # # • # # # • # # # • # # # • # #

A4 **a.** 3.132 **b.** 0.63756

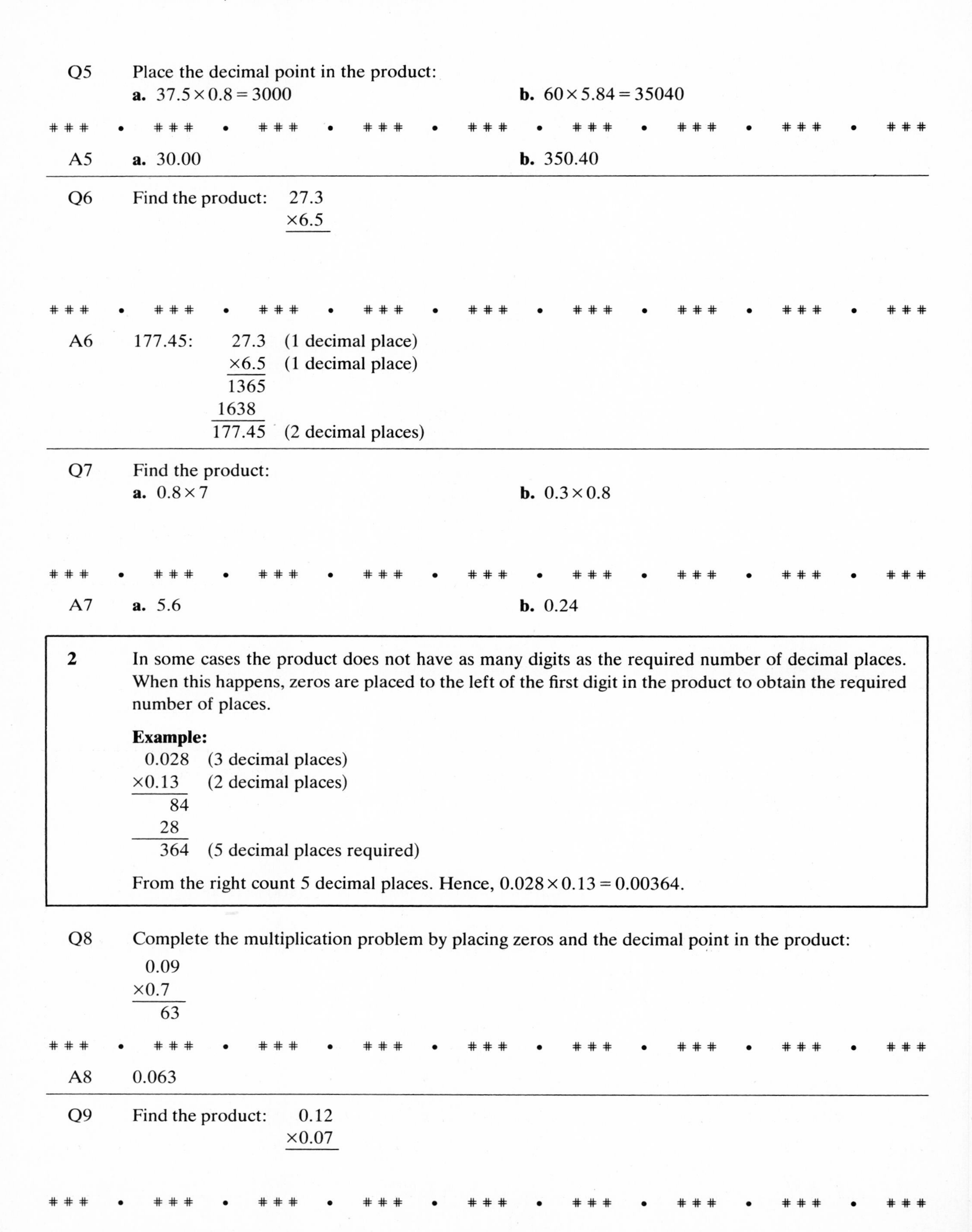

Q5 Place the decimal point in the product:

a. $37.5 \times 0.8 = 3000$ **b.** $60 \times 5.84 = 35040$

\# \# \# • \# \# \# • \# \# \# • \# \# \# • \# \# \# • \# \# \# • \# \# \# • \# \# \# • \# \# \#

A5 **a.** 30.00 **b.** 350.40

Q6 Find the product:

$$\begin{array}{r} 27.3 \\ \times 6.5 \\ \hline \end{array}$$

\# \# \# • \# \# \# • \# \# \# • \# \# \# • \# \# \# • \# \# \# • \# \# \# • \# \# \# • \# \# \#

A6 177.45:

$$\begin{array}{rl} 27.3 & \text{(1 decimal place)} \\ \times 6.5 & \text{(1 decimal place)} \\ \hline 1365 & \\ 1638 & \\ \hline 177.45 & \text{(2 decimal places)} \end{array}$$

Q7 Find the product:

a. 0.8×7 **b.** 0.3×0.8

\# \# \# • \# \# \# • \# \# \# • \# \# \# • \# \# \# • \# \# \# • \# \# \# • \# \# \# • \# \# \#

A7 **a.** 5.6 **b.** 0.24

---

2 In some cases the product does not have as many digits as the required number of decimal places. When this happens, zeros are placed to the left of the first digit in the product to obtain the required number of places.

Example:

$$\begin{array}{rl} 0.028 & \text{(3 decimal places)} \\ \times 0.13 & \text{(2 decimal places)} \\ \hline 84 & \\ 28 & \\ \hline 364 & \text{(5 decimal places required)} \end{array}$$

From the right count 5 decimal places. Hence, $0.028 \times 0.13 = 0.00364$.

Q8 Complete the multiplication problem by placing zeros and the decimal point in the product:

$$\begin{array}{r} 0.09 \\ \times 0.7 \\ \hline 63 \end{array}$$

\# \# \# • \# \# \# • \# \# \# • \# \# \# • \# \# \# • \# \# \# • \# \# \# • \# \# \# • \# \# \#

A8 0.063

Q9 Find the product:

$$\begin{array}{r} 0.12 \\ \times 0.07 \\ \hline \end{array}$$

\# \# \# • \# \# \# • \# \# \# • \# \# \# • \# \# \# • \# \# \# • \# \# \# • \# \# \# • \# \# \#

A9 0.0084:

$$\begin{array}{rl} 0.12 & \text{(2 decimal places)} \\ \underline{\times 0.07} & \text{(2 decimal places)} \\ 0.0084 & \text{(4 decimal places)} \end{array}$$

Q10 Find the product:

a. 0.7×0.07 **b.** 0.008×0.006

\# \# \# • \# \# \# • \# \# \# • \# \# \# • \# \# \# • \# \# \# • \# \# \# • \# \# \# • \# \# \#

A10 **a.** 0.049 **b.** 0.000048

Q11 Find the product:

a. 0.36×0.15 **b.** 3.6×0.15

\# \# \# • \# \# \# • \# \# \# • \# \# \# • \# \# \# • \# \# \# • \# \# \# • \# \# \# • \# \# \#

A11 **a.** 0.054 **b.** 0.54

Q12 Find the product:

a. 2.4×10 **b.** 0.08×10

\# \# \# • \# \# \# • \# \# \# • \# \# \# • \# \# \# • \# \# \# • \# \# \# • \# \# \# • \# \# \#

A12 **a.** 24 **b.** 0.8

Q13 Find the product:

a. 2.4×100 **b.** 0.08×100

\# \# \# • \# \# \# • \# \# \# • \# \# \# • \# \# \# • \# \# \# • \# \# \# • \# \# \# • \# \# \#

A13 **a.** 240 **b.** 8

---

3 A shortcut can be observed for multiplying a decimal by a power of 10 (10, 100, 1,000, etc.). Consider:

$$2.378 \times 10 = 23.78$$

$$2.378 \times 100 = 237.8$$

$$2.378 \times 1{,}000 = 2{,}378$$

$$2.378 \times 10{,}000 = 23{,}780$$

To multiply by a power of 10, the decimal point is moved to the *right* the same number of places as there are zeros in the power of 10. That is,

$2.378 \times \underline{100} = 237.8$
(2 zeros) (decimal point moved 2 places)

Q14 Find the product mentally:

a. 0.072×10 ______ **b.** 0.072×100 ______ **c.** $0.072 \times 1{,}000$ ______

\# \# \# • \# \# \# • \# \# \# • \# \# \# • \# \# \# • \# \# \# • \# \# \# • \# \# \# • \# \# \#

A14 **a.** 0.72 **b.** 7.2 **c.** 72

Q15 Find the product mentally:

a. 23.7×100 ______ **b.** 2.8×10 ______ **c.** $17 \times 1{,}000$ ______

\# \# \# • \# \# \# • \# \# \# • \# \# \# • \# \# \# • \# \# \# • \# \# \# • \# \# \# • \# \# \#

A15 **a.** 2,370 **b.** 28 **c.** 17,000

Q16 Find the product mentally:

a. 0.0017×100 ______ **b.** 5×100 ______ **c.** $0.77 \times 1{,}000$ ______

\# \# \# • \# \# \# • \# \# \# • \# \# \# • \# \# \# • \# \# \# • \# \# \# • \# \# \# • \# \# \#

A16 **a.** 0.17 **b.** 500 **c.** 770

---

4 The division of decimals is carried out in the same way as the division of whole numbers, except for the placing of the decimal point in the quotient.

Example: Find $0.8 \div 2$.

Solution: $0.8 \div 2 = \frac{8}{10} \div 2 = \frac{8}{10} \times \frac{1}{2} = \frac{4}{10} = 0.4$

The problem could have been written:

$$\begin{array}{r} 0.4 \\ 2\overline{)0.8} \\ \underline{8} \end{array}$$

Notice that when the divisor is a whole number, *the decimal point in the quotient is directly above the decimal point in the dividend.*

Q17 Place the decimal point in the quotient (do not divide):

a. $23\overline{)19.218}$ **b.** $7\overline{)16.47}$ **c.** $19\overline{)5.612}$

\# \# \# • \# \# \# • \# \# \# • \# \# \# • \# \# \# • \# \# \# • \# \# \# • \# \# \# • \# \# \#

A17 **a.** $\begin{array}{r}.\\ 23\overline{)19.218}\end{array}$ **b.** $\begin{array}{r}.\\ 7\overline{)16.47}\end{array}$ **c.** $\begin{array}{r}.\\ 19\overline{)5.612}\end{array}$

(In all cases the decimal point is placed directly above the decimal point in the dividend.)

Q18 Find the quotient:

a. $3\overline{)1.8}$ **b.** $8\overline{)3.2}$ **c.** $15\overline{)30.45}$

\# \# \# • \# \# \# • \# \# \# • \# \# \# • \# \# \# • \# \# \# • \# \# \# • \# \# \# • \# \# \#

A18 **a.** 0.6 **b.** 0.4 **c.** 2.03

---

5 When the divisor is a decimal, the division could be completed as follows: $0.168 \div 0.12$ means $\frac{0.168}{0.12}$. The fraction can now be multiplied by $\frac{100}{100}$, making the divisor a whole number.

$$\frac{0.168}{0.12} \times \frac{100}{100} = \frac{16.8}{12} \quad \text{or} \quad \begin{array}{r} 1.4 \\ 12\overline{)16.8} \\ \underline{12} \\ 4\,8 \\ \underline{4\,8} \end{array}$$

The same result could be obtained by using the following steps:

Step 1: Move the decimal point in the divisor to the right to make the divisor a whole number.

Step 2: Move the decimal point in the dividend the same number of places, using zeros as placeholders if necessary.

Step 3: The decimal point in the quotient will be directly above the new decimal point in the dividend.

Example: $\begin{array}{r} 1.4 \\ 0.12_{\wedge}\overline{)0.16_{\wedge}8} \end{array}$ (decimal point moved 2 places to the right)

Q19 Move the decimal point in the divisor and dividend, making the divisor a whole number. (Do no further computation.) $0.02\overline{)14.782}$

• ### • ### • ### • ### • ### • ### • ### •

A19 $0.02_{\wedge}\overline{)14.78_{\wedge}2}$

Q20 Place the decimal point in the quotient correctly:

a. $\begin{array}{r} 5 \\ 5.2\overline{)2.60} \end{array}$ **b.** $\begin{array}{r} 205 \\ 4.36\overline{)0.89380} \end{array}$

• ### • ### • ### • ### • ### • ### • ### •

A20 **a.** $\begin{array}{r} 0.5 \\ 5.2_{\wedge}\overline{)2.6_{\wedge}0} \end{array}$ **b.** $\begin{array}{r} 0.205 \\ 4.36_{\wedge}\overline{)0.89_{\wedge}380} \end{array}$

Q21 Place the decimal point in the quotient correctly:

a. $\begin{array}{r} 7391 \\ 0.02\overline{)14.782} \end{array}$ **b.** $\begin{array}{r} 13 \\ 0.25\overline{)3.25} \end{array}$

• ### • ### • ### • ### • ### • ### • ### •

A21 **a.** 739.1 **b.** 13.

6 Sometimes it may be necessary to place zeros in the quotient, between the decimal point and the first digit of the quotient, to obtain the required number of decimal places. For example,

$$\begin{array}{r} 0.0117 \\ 15\overline{)0.1755} \end{array}$$

Q22 Place the decimal point in the quotient correctly: $\begin{array}{r} 2 \\ 9\overline{)0.18} \end{array}$

• ### • ### • ### • ### • ### • ### • ### •

A22 0.02

Q23 Place the decimal point in the quotient correctly:

a. $\begin{array}{r} 21 \\ 25\overline{)0.525} \end{array}$ **b.** $\begin{array}{r} 3 \\ 90\overline{)0.0270} \end{array}$

• ### • ### • ### • ### • ### • ### • ### •

A23 **a.** 0.021 **b.** 0.0003

Q24 Find the quotient:

a. $7\overline{)39.2}$ **b.** $0.9\overline{)23.4}$

\# \# \# • \# \# \# • \# \# \# • \# \# \# • \# \# \# • \# \# \# • \# \# \# • \# \# \# • \# \# \#

A24 **a.** 5.6 **b.** 26

Q25 Find the quotient:

a. $0.12\overline{)7.92}$ **b.** $0.016\overline{)11.68}$

\# \# \# • \# \# \# • \# \# \# • \# \# \# • \# \# \# • \# \# \# • \# \# \# • \# \# \# • \# \# \#

A25 **a.** 66 **b.** 730: $0.016_{\wedge}\overline{)11.680_{\wedge}}$

---

7 It will often be necessary to round off quotients to a specified decimal place. One method often used when rounding off is to complete the division to one place beyond the place to which you are rounding off.

Example: Find $5.3 \div 8$ (round off to the nearest tenth).

Solution:

$$
\begin{array}{r}
0.66 \\
8\overline{)5.30} \\
\underline{4\;8} \\
50 \\
\underline{48}
\end{array}
$$

Hence, $5.3 \div 8 = 0.7$ rounded off to the nearest tenth.

Q26 Round off the quotients to the nearest tenth:

a. $0.14\overline{)31.9}$ **b.** $0.33\overline{)70.3}$

\# \# \# • \# \# \# • \# \# \# • \# \# \# • \# \# \# • \# \# \# • \# \# \# • \# \# \# • \# \# \#

A26 **a.** 227.9:

$$
\begin{array}{r}
227.85 \\
0.14_{\wedge}\overline{)31.90_{\wedge}00} \\
\underline{28} \\
39 \\
\underline{28} \\
110 \\
\underline{98} \\
120 \\
\underline{112} \\
80 \\
\underline{70}
\end{array}
$$

b. 213.0

Q27 Round off the quotient to the nearest hundredth:

a. $24\overline{)3.2}$ **b.** $0.73\overline{)5.88}$

\# \# \# • \# \# \# • \# \# \# • \# \# \# • \# \# \# • \# \# \# • \# \# \# • \# \# \# • \# \# \#

A27 **a.** 0.13 **b.** 8.05

Q28 Find the quotient:
a. $27.63 \div 10$ **b.** $27.63 \div 100$

\# \# \# • \# \# \# • \# \# \# • \# \# \# • \# \# \# • \# \# \# • \# \# \# • \# \# \# • \# \# \#

A28 **a.** 2.763 **b.** 0.2763

8 As was the case for multiplication, a shortcut can be observed when dividing a decimal by a power of 10:

$2.3 \div 10 = 0.23$
$2.3 \div 100 = 0.023$
$2.3 \div 1{,}000 = 0.0023$

When dividing by a power of 10, the decimal point is moved to the *left* the same number of places as there are zeros in the power of 10.

Q29 Find the quotient:

a. $4.76 \div 10$ ________ **b.** $16.5 \div 1{,}000$ ________

\# \# \# • \# \# \# • \# \# \# • \# \# \# • \# \# \# • \# \# \# • \# \# \# • \# \# \# • \# \# \#

A29 **a.** 0.476 **b.** 0.0165

Q30 Find the quotient:

a. $0.0091 \div 100$ ________ **b.** $0.23 \div 10$ ________

\# \# \# • \# \# \# • \# \# \# • \# \# \# • \# \# \# • \# \# \# • \# \# \# • \# \# \# • \# \# \#

A30 **a.** 0.000091 **b.** 0.023

9 Multiplying and dividing by powers of 10 can often be confusing. The decimal point is often moved the correct number of places but in the wrong direction. It may help to remember that multiplying by a power of 10 will make the number *larger*; hence, the decimal point is moved to the *right.* Dividing by a power of 10 will make the number *smaller*; hence, the decimal point is moved to the *left.*

Q31 Complete:

a. $2.713 \div 100 =$ ________ **b.** $2.713 \times 10 =$ ________

c. $0.018 \times 1{,}000 =$ ________ **d.** $0.018 \div 10 =$ ________

\# \# \# • \# \# \# • \# \# \# • \# \# \# • \# \# \# • \# \# \# • \# \# \# • \# \# \# • \# \# \#

A31 **a.** 0.02713 **b.** 27.13 **c.** 18 **d.** 0.0018

Q32 Bill drove 603 miles and used 35.8 gallons of gasoline. What was his average number of miles per gallon to the nearest tenth? ____________

• ### • ### • ### • ### • ### • ### • ### •

A32 16.8 miles per gallon

This completes the instruction for this section.

2.4 EXERCISE

1. Dividing a decimal by 100 moves the decimal point _____ spaces to the _______.
2. Multiplying a decimal by 10,000 moves the decimal point _____ spaces to the _______.
3. Find the product:
   a. 0.0276×100 **b.** 100×0.0543 **c.** 3.04×10 **d.** $0.0063 \times 1{,}000$
4. Find the quotient:
   a. $0.0276 \div 10$ **b.** $0.0543 \div 100$ **c.** $3.04 \div 10$ **d.** $0.0063 \div 1{,}000$
5. Find the product:
   a. 0.08×0.03 **b.** 6×0.07 **c.** 0.5×8 **d.** 0.025×0.16
   e. 2.4×0.105 **f.** 3.14×2.5
6. Find the quotient:
   a. $0.3 \div 0.4$ **b.** $0.35 \div 0.05$ **c.** $0.084 \div 0.07$ **d.** $0.63 \div 0.7$
   e. $16.308 \div 0.36$ **f.** $1.5072 \div 0.628$
7. Round the quotient to the nearest hundredth:
   a. $1.05 \div 12$ **b.** $7.82 \div 0.047$

2.4 EXERCISE ANSWERS

1. 2, left
2. 4, right
3. **a.** 2.76 **b.** 5.43 **c.** 30.4 **d.** 6.3
4. **a.** 0.00276 **b.** 0.000543 **c.** 0.304 **d.** 0.0000063
5. **a.** 0.0024 **b.** 0.42 **c.** 4 **d.** 0.004
   e. 0.252 **f.** 7.85
6. **a.** 0.75 **b.** 7 **c.** 1.2 **d.** 0.9
   e. 45.3 **f.** 2.4
7. **a.** 0.09 **b.** 166.38

2.5 FRACTIONS AS DECIMALS

1 Fractions whose denominators are a power of 10 are easily converted to a decimal. For example, $\frac{5}{10} = 0.5$, $\frac{7}{100} = 0.07$, and $\frac{23}{100} = 0.23$.

Fractions whose denominators are not powers of 10 can be converted to decimals by dividing the numerator by the denominator. That is, to convert $\frac{3}{4}$ to a decimal, divide 3 by 4. This is accomplished by placing a decimal point and as many zeros as necessary after the 3.

$$\begin{array}{r} 0.75 \\ 4\overline{)3.00} \\ \underline{2\,8} \\ 20 \\ \underline{20} \end{array}$$

Hence, $\frac{3}{4} = 0.75$.

Q1 Convert $\frac{1}{2}$ to a decimal by dividing 1 by 2.

• ### • ### • ### • ### • ### • ### • ### •

A1 0.5: $\begin{array}{r} 0.5 \\ 2\overline{)1.0} \\ \underline{1\,0} \end{array}$

Q2 Convert $\frac{1}{8}$ to a decimal (3 decimal places).

• ### • ### • ### • ### • ### • ### • ### •

A2 0.125: $\begin{array}{r} 0.125 \\ 8\overline{)1.000} \\ \underline{8} \\ 20 \\ \underline{16} \\ 40 \\ \underline{40} \end{array}$

2 If a zero remainder is obtained when converting a fraction to a decimal, the quotient is called a *terminating decimal*. It is given this name because there is a last digit. Often a zero remainder cannot be obtained and the digits of the quotient repeat themselves in a pattern. These decimals are called *repeating decimals*.

Example: Convert $\frac{7}{33}$ to a decimal.

Solution: $\begin{array}{r} 0.2121 = 0.212121 \cdots \\ 33\overline{)7.0000} \\ \underline{6\,6} \\ 40 \\ \underline{33} \\ 70 \\ \underline{66} \\ 40 \\ \underline{33} \\ 7 \end{array}$

At these points the remainder is the same as the original dividend. This means that the same group of digits will repeat continuously. A bar is placed over the group of digits that repeat. Therefore,

$$\frac{7}{33} = 0.\overline{21}.$$

Q3 $0.\overline{31}$ means ______________

\# \# \# • \# \# \# • \# \# \# • \# \# \# • \# \# \# • \# \# \# • \# \# \# • \# \# \# • \# \# \#

A3 0.313131 and so on (the digits 31 repeat continuously).

Q4 Convert $\frac{1}{3}$ to a repeating decimal.

\# \# \# • \# \# \# • \# \# \# • \# \# \# • \# \# \# • \# \# \# • \# \# \# • \# \# \# • \# \# \#

A4 $0.\overline{3}$:

$$\begin{array}{r} 0.3 = 0.333333\cdots \\ 3\overline{)1.0} \\ \underline{9} \\ 1 \end{array}$$

(the remainder is the same as the dividend)

Q5 Convert $\frac{2}{11}$ to a repeating decimal.

\# \# \# • \# \# \# • \# \# \# • \# \# \# • \# \# \# • \# \# \# • \# \# \# • \# \# \# • \# \# \#

A5 $0.\overline{18}$:

$$\begin{array}{r} 0.18 = 0.181818\cdots \\ 11\overline{)2.00} \\ \underline{1\ 1} \\ 90 \\ \underline{88} \\ 2 \end{array}$$

Q6 Convert $\frac{3}{8}$ to a terminating decimal.

\# \# \# • \# \# \# • \# \# \# • \# \# \# • \# \# \# • \# \# \# • \# \# \# • \# \# \# • \# \# \#

A6 0.375: $\begin{array}{r} 0.375 \\ 8\overline{)3.000} \\ \underline{2\,4} \\ 60 \\ \underline{56} \\ 40 \\ \underline{40} \end{array}$

Q7 Convert $\frac{5}{8}$ to a decimal.

\# \# \# • \# \# \# • \# \# \# • \# \# \# • \# \# \# • \# \# \# • \# \# \# • \# \# \# • \# \# \#

A7 0.625: $\begin{array}{r} 0.625 \\ 8\overline{)5.000} \\ \underline{4\,8} \\ 20 \\ \underline{16} \\ 40 \\ \underline{40} \end{array}$

Q8 Convert $\frac{8}{9}$ to a decimal.

\# \# \# • \# \# \# • \# \# \# • \# \# \# • \# \# \# • \# \# \# • \# \# \# • \# \# \# • \# \# \#

A8 $0.\bar{8}$: $\begin{array}{r} 0.8 = 0.888888\cdots \\ 9\overline{)8.0} \\ \underline{7\,2} \\ 8 \end{array}$

Q9 Convert to a decimal:

a. $\frac{3}{20}$ **b.** $\frac{5}{32}$

\# \# \# • \# \# \# • \# \# \# • \# \# \# • \# \# \# • \# \# \# • \# \# \# • \# \# \# • \# \# \#

A9 **a.** 0.15 **b.** 0.15625

Q10 Convert to a decimal:

a. $\frac{4}{9}$ **b.** $\frac{1}{11}$

• ### • ### • ### • ### • ### • ### • ### •

A10 **a.** $0.\overline{4}$ **b.** $0.\overline{09}$

---

3 When the dividend reappears as a remainder, the quotient will be a repeating decimal. In some cases, the dividend does not reappear as a remainder, but a remainder does appear a second time. This means that the digits in the quotient will repeat, starting with the one obtained when the remainder was first used as a new dividend.

Example: Convert $\frac{1}{6}$ to a decimal.

Solution:

$$\begin{array}{r} 0.16 = 0.166666\cdots \\ 6\overline{)1.00} \\ \underline{6} \\ 40 \\ \underline{36} \\ 4 \end{array}$$

(the remainder 4 repeats a second time)

Hence, $\frac{1}{6} = 0.1\overline{6}$.

Q11 Convert $\frac{7}{12}$ to a repeating decimal.

• ### • ### • ### • ### • ### • ### • ### •

A11 $0.58\overline{3}$:

$$\begin{array}{r} 0.583 = 0.5833333\cdots \\ 12\overline{)7.000} \\ \underline{6\,0} \\ 1\,00 \\ \underline{96} \\ 40 \\ \underline{36} \\ 4 \end{array}$$

Q12 Convert $\frac{62}{495}$ to a decimal.

\# \# \# • \# \# \# • \# \# \# • \# \# \# • \# \# \# • \# \# \# • \# \# \# • \# \# \# • \# \# \#

A12 $0.1\overline{25}$

4 Terminating decimals can be converted to fractions by determining the place value indicated and writing the digits over the appropriate power of 10. If possible, the fraction is reduced.

$$\begin{aligned} 0.375 &= \frac{375}{1{,}000} \quad \text{(dividing numerator and denominator by 5 three times)} \\ &= \frac{75}{200} \\ &= \frac{15}{40} \\ &= \frac{3}{8} \end{aligned}$$

Q13 Convert 0.12 to a fraction reduced to lowest terms.

\# \# \# • \# \# \# • \# \# \# • \# \# \# • \# \# \# • \# \# \# • \# \# \# • \# \# \# • \# \# \#

A13 $\frac{3}{25}$: $0.12 = \frac{12}{100} = \frac{3}{25}$

Q14 Convert to a fraction reduced to lowest terms:
a. 0.85 **b.** 0.024 **c.** 0.425

\# \# \# • \# \# \# • \# \# \# • \# \# \# • \# \# \# • \# \# \# • \# \# \# • \# \# \# • \# \# \#

A14 **a.** $\frac{17}{20}$ **b.** $\frac{3}{125}$ **c.** $\frac{17}{40}$

Q15 Convert to a fraction reduced to lowest terms:
a. 0.125 **b.** 0.1875 **c.** 0.203125

\# \# \# • \# \# \# • \# \# \# • \# \# \# • \# \# \# • \# \# \# • \# \# \# • \# \# \# • \# \# \#

A15 **a.** $\frac{1}{8}$ **b.** $\frac{3}{16}$ **c.** $\frac{13}{64}$

5 In some occupations it is often necessary to convert from fractions to decimals. The following table may prove useful in making these conversions.

Fraction	*Decimal*	*Fraction*	*Decimal*	*Fraction*	*Decimal*	*Fraction*	*Decimal*
$\frac{1}{64}$	0.015625	$\frac{17}{64}$	0.26525	$\frac{33}{64}$	0.515625	$\frac{49}{64}$	0.765625
$\frac{1}{32}$	0.03125	$\frac{9}{32}$	0.28125	$\frac{17}{32}$	0.53125	$\frac{25}{32}$	0.78125
$\frac{3}{64}$	0.046875	$\frac{19}{64}$	0.296875	$\frac{35}{64}$	0.546875	$\frac{51}{64}$	0.796875
$\frac{1}{16}$	0.0625	$\frac{5}{16}$	0.3125	$\frac{9}{16}$	0.5625	$\frac{13}{16}$	0.8125
$\frac{5}{64}$	0.078125	$\frac{21}{64}$	0.328125	$\frac{37}{64}$	0.578125	$\frac{53}{64}$	0.828125
$\frac{3}{32}$	0.09375	$\frac{11}{32}$	0.34375	$\frac{19}{32}$	0.59375	$\frac{27}{32}$	0.84375
$\frac{7}{64}$	0.109375	$\frac{23}{64}$	0.359375	$\frac{39}{64}$	0.609375	$\frac{55}{64}$	0.859375
$\frac{1}{8}$	0.125	$\frac{3}{8}$	0.375	$\frac{5}{8}$	0.625	$\frac{7}{8}$	0.875
$\frac{9}{64}$	0.140625	$\frac{25}{64}$	0.390625	$\frac{41}{64}$	0.640625	$\frac{57}{64}$	0.890625
$\frac{5}{32}$	0.15625	$\frac{13}{32}$	0.40625	$\frac{21}{32}$	0.65625	$\frac{29}{32}$	0.90625
$\frac{11}{64}$	0.171875	$\frac{27}{64}$	0.421875	$\frac{43}{64}$	0.671875	$\frac{59}{64}$	0.921875
$\frac{3}{16}$	0.1875	$\frac{7}{16}$	0.4375	$\frac{11}{16}$	0.6875	$\frac{15}{16}$	0.9375
$\frac{13}{64}$	0.203125	$\frac{29}{64}$	0.453125	$\frac{45}{64}$	0.703125	$\frac{61}{64}$	0.953125
$\frac{7}{32}$	0.21875	$\frac{15}{32}$	0.46875	$\frac{23}{32}$	0.71875	$\frac{31}{32}$	0.96875
$\frac{15}{64}$	0.234375	$\frac{31}{64}$	0.484375	$\frac{47}{64}$	0.734375	$\frac{63}{64}$	0.984375
$\frac{1}{4}$	0.250	$\frac{1}{2}$	0.500	$\frac{3}{4}$	0.750	1	1.000

This completes the instruction for this section.

2.5 EXERCISE

1. Convert to a decimal:

a. $\frac{1}{10}$ **b.** $\frac{7}{100}$ **c.** $\frac{1}{2}$ **d.** $\frac{1}{4}$

e. $\frac{1}{5}$ **f.** $\frac{2}{5}$ **g.** $\frac{1}{3}$ **h.** $\frac{2}{3}$

2. Convert to a decimal:

a. $\frac{3}{8}$ **b.** $\frac{5}{6}$ **c.** $\frac{4}{7}$ **d.** $\frac{3}{11}$

e. $\frac{17}{20}$ **f.** $\frac{31}{40}$ **g.** $\frac{27}{20}$ **h.** $\frac{15}{64}$

3. Convert to a fraction reduced to lowest terms:

a. 0.75	**b.** 0.25	**c.** 0.475	**d.** 0.724
e. 0.119	**f.** 0.648	**g.** 0.256	**h.** 0.425

2.5 EXERCISE ANSWERS

1. a. 0.1	**b.** 0.07	**c.** 0.5	**d.** 0.25
e. 0.2	**f.** 0.4	**g.** $0.\overline{3}$	**h.** $0.\overline{6}$
2. a. 0.375	**b.** $0.8\overline{3}$	**c.** $0.\overline{571428}$	**d.** $0.\overline{27}$
e. 0.85	**f.** 0.775	**g.** 1.35	**h.** 0.234375
3. a. $\frac{3}{4}$	**b.** $\frac{1}{4}$	**c.** $\frac{19}{40}$	**d.** $\frac{181}{250}$
e. $\frac{119}{1{,}000}$	**f.** $\frac{81}{125}$	**g.** $\frac{32}{125}$	**h.** $\frac{17}{40}$

2.6 APPLICATIONS

1 In Section 1.7, detailed dimensions were given in fraction form. With the increased use of the metric system, dimensions will often be stated in decimal form. Metric units will be discussed in Chapter 6.

Example: Find the missing dimensions.

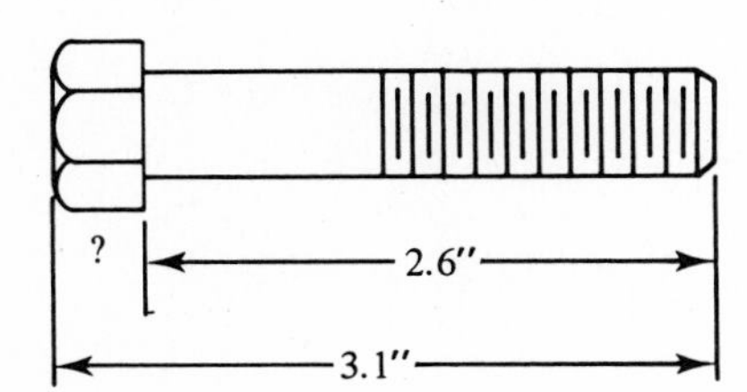

Solution: The missing dimension is found by subtracting 2.6″ from 3.1″.

3.1″ − 2.6″ = 0.5″

Q1 Find the total length of the bolt.

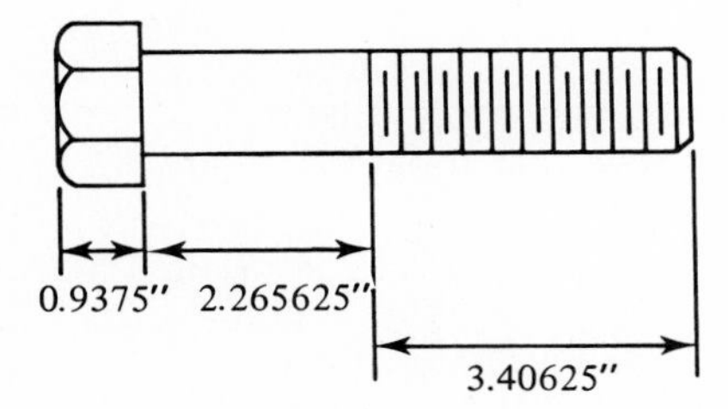

Dimensions would not usually be expressed to the number of decimal places as given here. The dimensions are the decimal equivalents of the fractions in Q21 of Section 1.7.

\# \# \# • \# \# \# • \# \# \# • \# \# \# • \# \# \# • \# \# \# • \# \# \# • \# \# \# • \# \# \#

A1 6.609375″

Q2 Find the missing dimension.

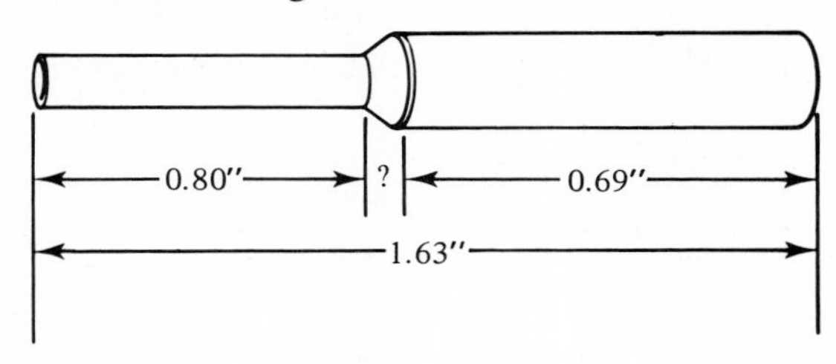

\# \# \# • \# \# \# • \# \# \# • \# \# \# • \# \# \# • \# \# \# • \# \# \# • \# \# \# • \# \# \#

A2 0.14″: 1.63″ − (0.80″ + 0.69″)

Q3 Find the length of the bolt.

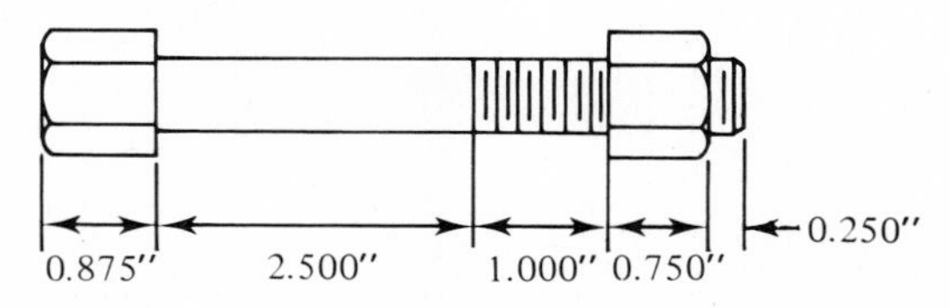

\# \# \# • \# \# \# • \# \# \# • \# \# \# • \# \# \# • \# \# \# • \# \# \# • \# \# \# • \# \# \#

A3 5.375″

2 In Section 1.7 the taper of a piece of work was determined by finding the difference of the diameters of tapered work. While the diameters in 1.7 were represented as fractions, in this section they will be represented as decimals.

Example: Find the taper for the following piece of work.

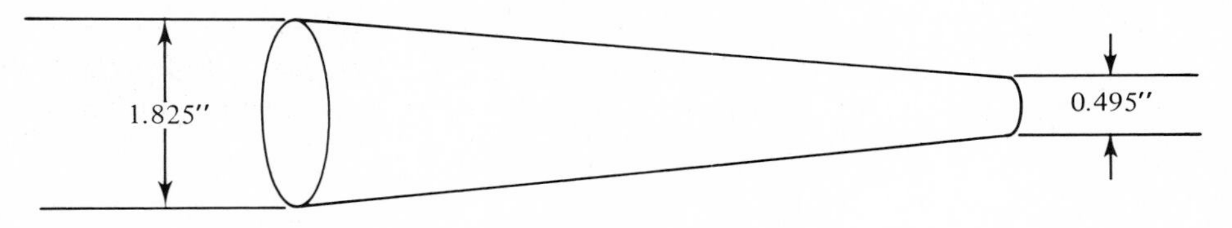

Solution: Taper = 1.825″ − 0.495″ = 1.330″

Q4 If the diameters of a piece of work are 3.025″ and 0.05″, respectively, determine the taper.

\# \# \# • \# \# \# • \# \# \# • \# \# \# • \# \# \# • \# \# \# • \# \# \# • \# \# \# • \# \# \#

A4 2.975″

Q5 Find the taper in 6 feet if a piece of work is tapered 0.25 inches per foot.

\# \# \# • \# \# \# • \# \# \# • \# \# \# • \# \# \# • \# \# \# • \# \# \# • \# \# \# • \# \# \#

A5 1.5 in.: 6×0.25

Q6 Find the taper in 8.75 inches if a piece of work is tapered 0.0625 inches per inch of length (round off to the nearest thousandth).

\# \# \# • \# \# \# • \# \# \# • \# \# \# • \# \# \# • \# \# \# • \# \# \# • \# \# \# • \# \# \#

A6 0.547 in.: 8.75×0.0625

Q7 Find the taper per inch in a piece of work 15 inches long if the taper is 1.25 inches. (Hint: taper per inch = taper ÷ total length.)

\# \# \# • \# \# \# • \# \# \# • \# \# \# • \# \# \# • \# \# \# • \# \# \# • \# \# \# • \# \# \#

A7 $0.08\overline{3}$ in. per inch: $1.25 \div 15$

Q8 The end diameters of a piece of work 4.75 inches in length are 1.125 inches and 0.828 inches, respectively. Find the taper per inch (round off to the nearest thousandth).

\# \# \# • \# \# \# • \# \# \# • \# \# \# • \# \# \# • \# \# \# • \# \# \# • \# \# \# • \# \# \#

A8 0.063 in. per inch: taper $= 1.125'' - 0.828'' = 0.297''$

taper per inch = taper ÷ total length
taper per inch $= 0.297 \div 4.75$

Q9 A local towing service charges \$6.50 per call and \$0.12 per mile towed. The service operator records his beginning odometer reading as 36,008.7 and the ending as 36,026.1. Find the total service charge.

\# \# \# • \# \# \# • \# \# \# • \# \# \# • \# \# \# • \# \# \# • \# \# \# • \# \# \# • \# \# \#

A9 \$8.59: miles driven $= 36{,}026.1 - 36{,}008.7$
service charge $= 17.4 \times \$0.12 + \$6.50$

Q10 An auto repair shop charges its customers \$11.50 per hour labor costs plus cost of parts. Determine the total charges for a tune-up if labor charges were for 1.5 hours (the symbol "@" says "at" and means for each item):

8 spark plugs @ \$1.35 **a.** ____________________

1 air filter @ \$4.75 **b.** ____________________

1 fuel filter @ \$2.16 **c.** ____________________

Total parts **d.** ____________________

Labor **e.** ____________________

Total Charge **f.** ____________________

\# \# \# • \# \# \# • \# \# \# • \# \# \# • \# \# \# • \# \# \# • \# \# \# • \# \# \# • \# \# \#

A10 **a.** \$10.80 **b.** \$4.75 **c.** \$2.16 **d.** \$17.71
e. \$17.25 **f.** \$34.96

Q11 A gasoline company charges an operator \$0.015 per gallon of gasoline pumped. What is the charge for 22,000 gallons of gasoline pumped?

\# \# \# • \# \# \# • \# \# \# • \# \# \# • \# \# \# • \# \# \# • \# \# \# • \# \# \# • \# \# \#

A11 \$330: $22{,}000 \times \$0.015$

Q12 A bolt weighs 0.66 pounds. Approximately how many bolts are in a box weighing 120 pounds?

• # # # • # # # • # # # • # # # • # # # • # # # • # # # • # #

A12 181 bolts: 120 lb ÷ 0.66 lb (Rounded off to the nearest bolt there would be 182; however, because of the box weight, the actual number of bolts would be closer to 181 than 182.)

Q13 A company pays \$3.50 a gross (gross = 12 dozen) for bolts. What is the price of one bolt (round off to the nearest cent)?

• # # # • # # # • # # # • # # # • # # # • # # # • # # # • # #

A13 \$0.02: \$3.50 ÷ 144

Q14 An indoor–outdoor carpet costs \$3.92 a yard. What is the cost of 12.36 yards?

• # # # • # # # • # # # • # # # • # # # • # # # • # # # • # #

A14 \$48.45

Q15 Find the cost of 18 wrenches @ \$7.25.

• # # # • # # # • # # # • # # # • # # # • # # # • # # # • # #

A15 \$130.50

This completes the instruction for this section.

2.6 EXERCISE

1. Find the missing dimension.

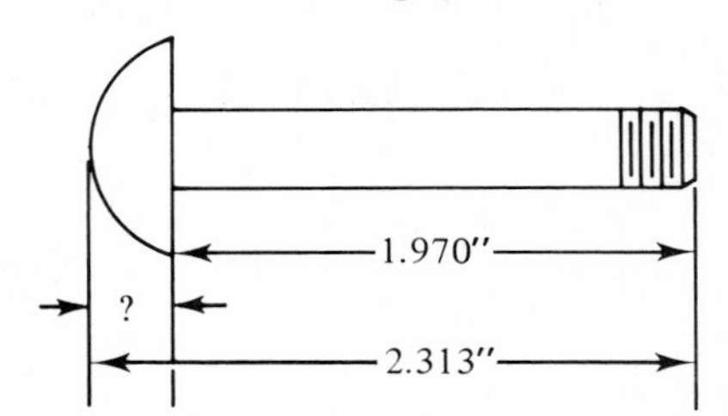

2. Find the missing dimension if the overall length is 12.625 inches.

3. Find the taper of the figure below.

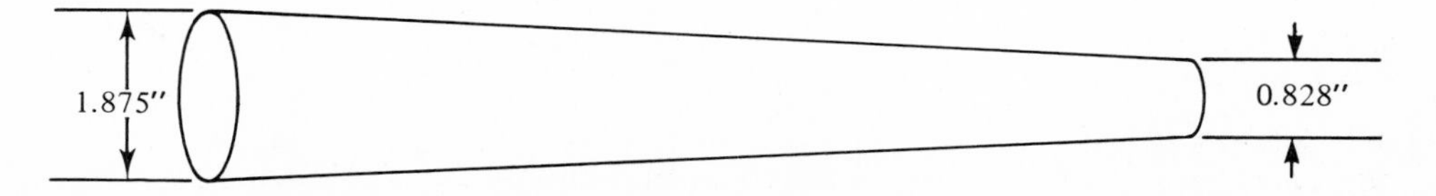

4. Find the taper in 18 feet if the piece of work tapers 0.18 inches per foot.
5. The end diameters of a piece of work are 6.38 inches and 2.72 inches, respectively. Find the taper per inch if the piece is 17.80 inches long (round off to the nearest hundredth).
6. Find the taper in a piece of work 3 feet long if the taper is 0.156″ per inch.
7. How long will it take to machine 40 pins if each pin takes 7.5 minutes?
8. A pipe 16.5 inches long is divided into 14 equal parts. What is the length of each part (round off to the nearest hundredth)?
9. How much do you save if you buy 1 dozen gaskets for $1.80 instead of $0.16 each?
10. A building contractor purchased 231.38 acres of land. If the land is divided into lots of 0.46 acres each, how many whole lots are sold?
11. Complete the following parts list:

Article Number	*Quantity*	*Price*	*Value*
#016	93	@ $0.03	**a.** ______
#108	254	@ 0.02	**b.** ______
#006	78	@ 0.12	**c.** ______
#119	125	@ 0.08	**d.** ______
#253	372	@ 0.23	**e.** ______
#015	18	@ 1.96	**f.** ______
		Total	**g.** ______

12. Complete the cost of painting supplies:

Quantity	*Description*	*Total*
6 gal	pint, outside white @ $5.85	**a.** ______
3 gal	paint, primer @ 4.30	**b.** ______
2	3-in. brushes @ 1.25	**c.** ______
	Total	**d.** ______

13. A piece of angle iron weighs 7.2 pounds per foot. Find the weight of 13.8 feet.
14. Number 2 gauge sheet metal has a thickness of 0.265625 inch and number 28 gauge a thickness of 0.015625 inch. Find the difference in thickness.
15. What will be the height of 115 sheets of number 2 gauge sheet metal (see problem 14)?

2.6 EXERCISE ANSWERS

1. 0.343″ **2.** 5.875″ **3.** 1.047″
4. 3.24 in. **5.** 0.21 in. per inch **6.** 5.616″
7. 300 min or 5 hrs **8.** 1.18 in. **9.** $0.12
10. 503
11. a. $2.79 **b.** $5.08 **c.** $9.36 **d.** $10
e. $85.56 **f.** $35.28 **g.** $148.07
12. a. $35.10 **b.** $12.90 **c.** $2.50 **d.** $50.50
13. 99.36 pounds
14. 0.25 in.
15. 30.546875 in.

CHAPTER 2 SAMPLE TEST

At the completion of Chapter 2, you should be able to work the following problems.

2.1 PLACE VALUE

1. **a.** Write in decimal form: two and seventeen thousandths
b. Write in word form: 17.012
c. Write in decimal form: $\frac{302}{10,000}$
d. What is the place value of the 4 in 2.143?

2.2 ROUNDING OFF

2. Round off the numbers as indicated:
a. 2.416 (nearest hundredth) **b.** 137.054 (nearest *ten*)
c. 60.37 (nearest whole number) **d.** 0.6726 (nearest thousandth)

2.3 ADDITION AND SUBTRACTION

3. Add:
a. $2.076 + 1.76 + 3.7$ **b.** $0.003 + 22 + 3.5$
4. Subtract:
a. $5.06 - 4.982$ **b.** $7 - 3.285$

2.4 MULTIPLICATION AND DIVISION

5. Multiply:
a. 3.27×0.79 **b.** 0.004×0.05 **c.** 0.46×10 **d.** 2.104×100
6. Divide:
a. $0.0448 \div 0.032$ **b.** $1.216 \div 0.04$ **c.** $56.20 \div 100$ **d.** $0.215 \div 10$
7. Round the quotient to the nearest hundredth:
a. $2.24 \div 43$ **b.** $0.4142 \div 0.35$

2.5 FRACTIONS AS DECIMALS

8. Convert to a decimal:
a. $\frac{3}{8}$ **b.** $\frac{1}{11}$ **c.** $\frac{2}{3}$ **d.** $\frac{1}{40}$
9. Convert to a fraction in lowest terms:
a. 0.025 **b.** 0.125 **c.** 0.188 **d.** 0.46

2.6 APPLICATIONS

10. A tire company had total sales of $6,737.36 for their GR78-15. If each tire sold for $31.78, how many tires were sold?

11. Find the missing dimension.

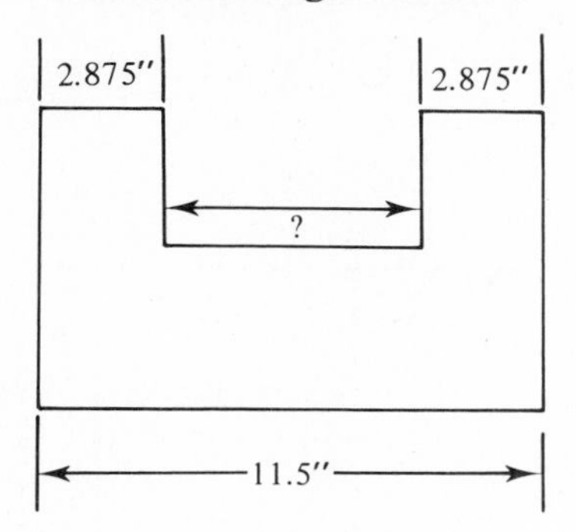

12. Find the taper in a piece of work 3 feet long if the taper is 0.15625" per inch.

13. Calculate the cost of the following building materials:
  a. 803 (2"×4") @ $1.92
  b. 23 yards of concrete @ $33.95
  c. 4,815 feet of flooring @ $0.09
  d. 15,270 bricks @ $0.13

14. A builder purchased 8,722 hinges for $610.54. What was the cost per hinge?

CHAPTER 2 SAMPLE TEST ANSWERS

1. **a.** 2.017 **b.** seventeen and twelve thousandths
c. 0.0302 **d.** hundredths

2. **a.** 2.42 **b.** 140 **c.** 60 **d.** 0.673

3. **a.** 7.536 **b.** 25.503

4. **a.** 0.078 **b.** 3.715

5. **a.** 2.5833 **b.** 0.0002 **c.** 4.6 **d.** 210.4

6. **a.** 1.4 **b.** 30.4 **c.** 0.5620 **d.** 0.0215

7. **a.** 0.05 **b.** 1.18

8. **a.** 0.375 **b.** $0.\overline{09}$ **c.** $0.\overline{6}$ **d.** 0.025

9. **a.** $\frac{1}{40}$ **b.** $\frac{1}{8}$ **c.** $\frac{47}{225}$ **d.** $\frac{23}{50}$

10. 212

11. 5.75 in.

12. 5.625 in.

13. **a.** $1,541.76 **b.** $780.85 **c.** $433.35 **d.** $1,985.10

14. $0.07

CHAPTER 3

OPERATIONS WITH PERCENTS

3.1 PERCENTS

1 In Chapter 2 fractions with a denominator of 100 were called decimal fractions. Fractions with a denominator of 100 are also called *percents*. The word percent means "hundredths." Thus, $\frac{7}{100}$ is equivalent to 7 percent. The symbol % is an abbreviation for the word "percent" and is derived from the denominator of 100. Therefore,

$$\frac{7}{100} = 7\%$$

The symbol 7% is read "7 percent" and means 7 hundredths.

Q1 Write the fractions using the % symbol:

a. $\frac{23}{100} =$ _____ **b.** $\frac{3}{100} =$ _____ **c.** $\frac{25}{100} =$ _____ **d.** $\frac{50}{100} =$ _____

• ### • ### • ### • ### • ### • ### • ### • ### •

A1 **a.** 23% **b.** 3% **c.** 25% **d.** 50%

Q2 Write the percents as fractions:

a. 29% = _____ **b.** 9% = _____ **c.** 53% = _____ **d.** 97% = _____

• ### • ### • ### • ### • ### • ### • ### • ### •

A2 **a.** $\frac{29}{100}$ **b.** $\frac{9}{100}$ **c.** $\frac{53}{100}$ **d.** $\frac{97}{100}$

Q3 Write the percents as fractions and reduce to lowest terms:

a. 25% = _____ **b.** 20% = _____ **c.** 75% = _____ **d.** 44% = _____

• ### • ### • ### • ### • ### • ### • ### • ### •

A3 **a.** $\frac{1}{4}$ **b.** $\frac{1}{5}$ **c.** $\frac{3}{4}$ **d.** $\frac{11}{25}$

2 Some percents can also be expressed as mixed numbers. For example,

$$225\% = \frac{225}{100} = 2\frac{25}{100} = 2\frac{1}{4}$$

Fractional percents can be expressed as a fraction as follows:

$$\frac{1}{2}\% = \frac{\frac{1}{2}}{100} = \frac{1}{2} \div \frac{100}{1} = \frac{1}{2} \times \frac{1}{100} = \frac{1}{200}$$

Q4 Change 250% to a mixed number.

• ### • ### • ### • ### • ### • ### • ### •

A4 $2\frac{1}{2}$: $250\% = \frac{250}{100} = 2\frac{50}{100} = 2\frac{1}{2}$

Q5 Change to a mixed number:

a. 480% **b.** 333%

• ### • ### • ### • ### • ### • ### • ### •

A5 **a.** $4\frac{4}{5}$ **b.** $3\frac{33}{100}$

Q6 Change $\frac{1}{4}\%$ to a fraction.

• ### • ### • ### • ### • ### • ### • ### •

A6 $\frac{1}{400}$: $\frac{1}{4}\% = \frac{\frac{1}{4}}{100} = \frac{1}{400}$

Q7 Change $\frac{2}{3}\%$ to a fraction.

• ### • ### • ### • ### • ### • ### • ### •

A7 $\frac{1}{150}$: $\frac{2}{3}\% = \frac{\frac{2}{3}}{100} = \frac{2}{300}$

3 To change a mixed-number percent to a fraction it is necessary to write the mixed number as an improper fraction over 100. Consider this example:

$$66\frac{2}{3}\% = \frac{66\frac{2}{3}}{100} = \frac{\frac{200}{3}}{100} = \frac{200}{3} \div \frac{100}{1} = \frac{200}{3} \times \frac{1}{100} = \frac{200}{300} = \frac{2}{3}$$

Q8 Change $33\frac{1}{3}\%$ to a fraction.

\# \# \# • \# \# \# • \# \# \# • \# \# \# • \# \# \# • \# \# \# • \# \# \# • \# \# \# • \# \# \#

A8 $\frac{1}{3}$: $33\frac{1}{3}\% = \frac{33\frac{1}{3}}{100} = \frac{\frac{100}{3}}{100} = \frac{100}{300} = \frac{1}{3}$

Q9 Change $12\frac{1}{2}\%$ to a fraction.

\# \# \# • \# \# \# • \# \# \# • \# \# \# • \# \# \# • \# \# \# • \# \# \# • \# \# \# • \# \# \#

A9 $\frac{1}{8}$: $12\frac{1}{2}\% = \frac{12\frac{1}{2}}{100} = \frac{\frac{25}{2}}{100} = \frac{25}{200} = \frac{1}{8}$

Q10 Change to fractions:

a. $16\frac{2}{3}\%$ **b.** $37\frac{1}{2}\%$

\# \# \# • \# \# \# • \# \# \# • \# \# \# • \# \# \# • \# \# \# • \# \# \# • \# \# \# • \# \# \#

A10 **a.** $\frac{1}{6}$ **b.** $\frac{3}{8}$

Q11 Change $3\frac{1}{8}\%$ to a fraction.

\# \# \# • \# \# \# • \# \# \# • \# \# \# • \# \# \# • \# \# \# • \# \# \# • \# \# \# • \# \# \#

A11 $\frac{1}{32}$: $3\frac{1}{8}\% = \frac{3\frac{1}{8}}{100} = \frac{\frac{25}{8}}{100} = \frac{25}{800} = \frac{1}{32}$

4 In Chapter 2 it was shown that fractions with denominator 100 could be written as a decimal. Since percents are written as a fraction with denominator of 100, they can also be expressed as a decimal. For example,

$$18\% = \frac{18}{100} = 0.18$$

Q12 Express 23% as a fraction and a decimal.____________

• ### • ### • ### • ### • ### • ### • ### •

A12 $23\% = \frac{23}{100} = 0.23$

5 A number can be divided by 100 by simply moving the decimal point two places to the left. Hence, to change a percent to a decimal, remove the percent sign and move the decimal point two places to the left.

Examples:

$23\% = 0.23$ because $23\% = 23. \div 100$
$7\% = 0.07$ because $7\% = 7. \div 100$
$17.5\% = 0.175$ because $17.5\% = 17.5 \div 100$

Notice that the decimal point, for a number written without a decimal point, is placed after the last digit. Hence, 8 = 8.

Q13 Express 23.7% as a decimal. ________

• ### • ### • ### • ### • ### • ### • ### •

A13 0.237: 23.7% = 0.237 (drop the percent sign and move the decimal point two places to the left).

Q14 Express as a decimal:

a. 13% = ________ **b.** 75% = ________

• ### • ### • ### • ### • ### • ### • ### •

A14 **a.** 0.13 **b.** 0.75

Q15 Express as a decimal:

a. 19.8% = ________ **b.** 2.75% = ________ **c.** 250% = ________

• ### • ### • ### • ### • ### • ### • ### •

A15 **a.** 0.198 **b.** 0.0275 **c.** 2.50 or 2.5

Q16 Express as a decimal:

a. 700% = ________ **b.** 0.03% = ________
c. 1% = ________ **d.** 5% = ________

• ### • ### • ### • ### • ### • ### • ### •

A16 **a.** 7 **b.** 0.0003 **c.** 0.01 **d.** 0.05

6 To change a decimal to a percent, move the decimal two places to the right and append the percent sign. For example,

$0.28 = 28\%$ because $0.28 = \frac{28}{100} = 28\%$

Examples:

$0.04 = 4\%$
$0.15 = 15\%$
$1.86 = 186\%$
$0.002 = 0.2\%$

Q17 Change 0.617 to a percent. ________

• ### • ### • ### • ### • ### • ### • ### •

A17 61.7%: 0.617 = 61.7% (move the decimal point two places to the right and append the % sign).

Q18 Change to a percent:

a. 0.93 = ________ **b.** 0.0015 = ________

c. 1.72 = ________ **d.** 0.473 = ________

• ### • ### • ### • ### • ### • ### • ### •

A18 **a.** 93% **b.** 0.15% **c.** 172% **d.** 47.3%

7	It may be necessary to use zeros as placeholders when changing a decimal to a percent. For example, change 0.8 to a percent: 0.8 = 0.80 = 80%

Q19 Change 0.7 to a percent. ________

• ### • ### • ### • ### • ### • ### • ### •

A19 70%: 0.7 = 0.70 = 70%

Q20 Change 17 to a percent. ________

• ### • ### • ### • ### • ### • ### • ### •

A20 1,700%: 17 = 17.00 = 1,700%

Q21 Change to a percent:

a. 0.2 = ________ **b.** 5 = ________ **c.** 0.1 = ________ **d.** 213 = ________

• ### • ### • ### • ### • ### • ### • ### •

A21 **a.** 20% **b.** 500% **c.** 10% **d.** 21,300%

8	To change a fraction to a percent, first change the common fraction to a decimal and then change the decimal to a percent. Consider these examples: $\frac{3}{4} = 0.75 = 75\%$ $\frac{1}{8} = 0.125 = 12.5\%$ Recall that to change a fraction to a decimal the numerator is divided by the denominator.

Q22 Change $\frac{3}{8}$ to a percent.

• ### • ### • ### • ### • ### • ### • ### •

A22 37.5%: $8\overline{)3.000}$ with quotient 0.375

Q23 Change to a percent:

a. $\frac{1}{4} = \underset{\text{(decimal)}}{\underline{\hspace{3cm}}} = \underset{\text{(percent)}}{\underline{\hspace{3cm}}}$

b. $\frac{5}{8} = \underset{\text{(decimal)}}{\underline{\hspace{3cm}}} = \underset{\text{(percent)}}{\underline{\hspace{3cm}}}$

• ### • ### • ### • ### • ### • ### • ### •

A23 **a.** 0.25, 25% **b.** 0.625, 62.5%

---

9 A fraction represented by a repeating decimal is usually changed to a percent by completing the division to two decimal places and writing the remainder in fraction form. That is, $\frac{1}{3} = 0.33\frac{1}{3} = 33\frac{1}{3}\%$. The problem could be stated $\frac{1}{3} = 0.33\bar{3} = 33.\bar{3}\%$; however, the first method is preferred.

It should be noted that when moving the decimal point two places to the right, only digits count as places. Fractions do not hold a place. Hence, $0.3\frac{1}{3}$ *does not* equal $3\frac{1}{3}\%$. The division must be carried out to at least two decimal places before the remainder is written as a fraction.

Q24 Express $\frac{1}{6}$ as a percent.

• ### • ### • ### • ### • ### • ### • ### •

A24 $16\frac{2}{3}\%$: $6\overline{)1.00}$ with quotient $0.16\frac{2}{3}$

Q25 Change to a percent:

a. $\frac{2}{3}$ b. $\frac{2}{7}$

• ### • ### • ### • ### • ### • ### • ### •

A25 **a.** $66\frac{2}{3}\%$ **b.** $28\frac{4}{7}\%$

This completes the instruction for this section.

3.1 EXERCISE

Fill the blanks with the proper equivalents, as shown in problem 1.

	Fraction	*Decimal*	*Percent*
1.	$\frac{1}{2}$	0.5	50%
2.	____	0.25	____%
3.	$\frac{1}{3}$	____	____%
4.	____	____	10%

	Fraction	Decimal	Percent
5.	____	0.125	____%
7.	____	____	$66\frac{2}{3}\%$
9.	$\frac{3}{10}$	____	____%
11.	____	0.0625	____%
13.	____	____	80%
15.	____	0.90	____%
17.	$\frac{1}{6}$	____	____%
19.	$\frac{11}{20}$	____	____%
21.	$\frac{3}{5}$	____	____%
23.	____	____	95%
25.	____	0.875	____%
27.	$1\frac{7}{8}$	____	____%
29.	____	____	200%
31.	____	0.08	____%
33.	$4\frac{4}{5}$	____	____%
35.	____	____	180%
37.	____	6.166	____%
39.	$3\frac{3}{4}$	____	____%
41.	$2\frac{1}{2}$	____	____%
43.	____	____	$166\frac{2}{3}\%$
45.	$1\frac{5}{8}$	____	____%

	Fraction	Decimal	Percent
6.	$\frac{2}{5}$	____	____%
8.	____	0.75	____%
10.	____	____	5%
12.	$\frac{3}{8}$	____	____%
14.	____	0.625	____%
16.	____	____	100%
18.	____	____	20%
20.	____	0.375	____%
22.	____	0.166	____%
24.	$\frac{7}{10}$	____	____%
26.	____	____	225%
28.	____	$0.16\frac{2}{3}$	____%
30.	$\frac{1}{100}$	____	____%
32.	____	____	$16\frac{2}{3}\%$
34.	____	5.00	____%
36.	$\frac{1}{50}$	____	____%
38.	____	____	110%
40.	____	0.2	____%
42.	____	4.08	____%
44.	____	2.125	____%
46.	____	3.875	____%

3.1 EXERCISE ANSWERS

1. given

2. $\frac{1}{4}$, 25%

3. $0.33\frac{1}{3}$, $33\frac{1}{3}\%$

4. $\frac{1}{10}$, 0.1 or 0.10

5. $\frac{1}{8}$, 12.5%

6. 0.4 or 0.40, 40%

7. $\frac{2}{3}$, $0.66\frac{2}{3}$

8. $\frac{3}{4}$, 75%

9. 0.3 or 0.30, 30%

10. $\frac{1}{20}$, 0.05
11. $\frac{1}{16}$, 6.25%
12. 0.375, 37.5%
13. $\frac{4}{5}$, 0.8 or 0.80
14. $\frac{5}{8}$, 62.5%
15. $\frac{9}{10}$, 90%
16. 1, 1 or 1.00
17. $0.16\frac{2}{3}$, $16\frac{2}{3}\%$
18. $\frac{1}{5}$, 0.2 or 0.20
19. 0.55, 55%
20. $\frac{3}{8}$, 37.5%
21. 0.6 or 0.60, 60%
22. $\frac{83}{500}$, 16.6%
23. $\frac{19}{20}$, 0.95
24. 0.7 or 0.70, 70%
25. $\frac{7}{8}$, 87.5%
26. $2\frac{1}{4}$, 2.25
27. 1.875, 187.5%
28. $\frac{1}{6}$, $16\frac{2}{3}\%$
29. 2, 2 or 2.00
30. 0.01, 1%
31. $\frac{2}{25}$, 8%
32. $\frac{1}{6}$, $0.16\frac{2}{3}$
33. 4.8 or 4.80, 480%
34. 5, 500%
35. $1\frac{4}{5}$, 1.8 or 1.80
36. 0.02, 2%
37. $6\frac{83}{500}$, 616.6%
38. $1\frac{1}{10}$, 1.1 or 1.10
39. 3.75, 375%
40. $\frac{1}{5}$, 20%
41. 2.5 or 2.50, 250%
42. $4\frac{2}{25}$, 408%
43. $1\frac{2}{3}$, $1.66\frac{2}{3}$
44. $2\frac{1}{8}$, 212.5%
45. 1.625, 162.5%
46. $3\frac{7}{8}$, 387.5%

3.2 WRITING AND SOLVING MATHEMATICAL SENTENCES

1 When such a statement as "2% of 100 is 2" is written in the form $0.02 \times 100 = 2$, it is said to be written mathematically. Written mathematically, 6% of 20 is 1.2 would become: $0.06 \times 20 = 1.2$. Notice that "of" means "multiply" and is replaced by the multiplication symbol, ×. "Is" means "is equal to" and is replaced by the equal sign, =.

Q1 Write "10% of 20 is 2" mathematically. ______________________

• ### • ### • ### • ### • ### • ### • ### •

A1 $0.10 \times 20 = 2$, or $\frac{1}{10} \times 20 = 2$

Q2 Write mathematically:

a. 2% of 14 is 0.28 ______________________

b. 20% of 50 is 10 ______________________

c. 14% of 40 is 5.6 ______________________

d. 200% of 4 is 8 ______

e. $66\frac{2}{3}\%$ of 24 is 16 ______

\# \# \# • \# \# \# • \# \# \# • \# \# \# • \# \# \# • \# \# \# • \# \# \# • \# \# \# • \# \# \#

A2 **a.** $0.02 \times 14 = 0.28$, or $\frac{1}{50} \times 14 = 0.28$ **b.** $0.20 \times 50 = 10$, or $\frac{1}{5} \times 50 = 10$

c. $0.14 \times 40 = 5.6$, or $\frac{7}{50} \times 40 = 5.6$ **d.** $2 \times 4 = 8$

e. $0.66\frac{2}{3} \times 24 = 16$, or $\frac{2}{3} \times 24 = 16$

2 Consider the question: 5% of 12 is what number? Written mathematically: $0.05 \times 12 =$ what number?

When we use letter N to "hold the place" of the missing number, the statement becomes: $0.05 \times 12 = N$? Any letter can be used for the missing (unknown) number. N is a popular choice, because "number" begins with the letter n.

Q3 Write mathematically: 75% of 116 is what number? ______

\# \# \# • \# \# \# • \# \# \# • \# \# \# • \# \# \# • \# \# \# • \# \# \# • \# \# \# • \# \# \#

A3 $0.75 \times 116 = N$? or $\frac{3}{4} \times 116 = N$?

Q4 Write mathematically:

a. 420% of 7 is what number? ______

b. What number is 420% of 7? ______

c. 62% of 19 is what number? ______

d. What number is 62% of 19? ______

\# \# \# • \# \# \# • \# \# \# • \# \# \# • \# \# \# • \# \# \# • \# \# \# • \# \# \# • \# \# \#

A4 **a.** $4.2 \times 7 = N$? **b.** $N = 4.2 \times 7$? **c.** $0.62 \times 19 = N$? **d.** $N = 0.62 \times 19$?

3 In the following question a different part of the problem is missing:

4.1% of what number is 7.3?

$0.041 \times N = 7.3$ (written mathematically)

Notice that "of" is replaced by "×," "what number" by "N," and "is" by "=." ($4.1\% = 0.041$.)

Q5 Write mathematically: 7% of what number is 0.09? ______

\# \# \# • \# \# \# • \# \# \# • \# \# \# • \# \# \# • \# \# \# • \# \# \# • \# \# \# • \# \# \#

A5 $0.07 \times N = 0.09$?

Q6 Write mathematically:

a. 22% of what number is 12? ______

b. 15 is 35% of what number? ______

c. 320% of what number is 400? ____________

d. $37\frac{1}{2}$% of what number is 54? ____________

e. 23.5 is 10% of what number? ____________

• # # # • # # # • # # # • # # # • # # # • # # # • # # # • # #

A6 **a.** $0.22 \times N = 12?$ **b.** $15 = 0.35 \times N?$ **c.** $3.2 \times N = 400?$

d. $0.37\frac{1}{2} \times N = 54?$ or $\frac{3}{8} \times N \times 54?$ **e.** $23.5 = 0.10 \times N?$

---

4 Consider still another type of percent question:

What percent of 90 is 7?
$P \quad \times 90 = 7?$

Notice that the percent is missing; hence, the letter P was used as a reminder that the missing number is a percent.

Q7 Write mathematically: What percent of 70 is 203? ____________

• # # # • # # # • # # # • # # # • # # # • # # # • # # # • # #

A7 $P \times 70 = 203?$

Q8 Write mathematically:

a. What percent of 12 is 0.6? ____________

b. What percent of 9 is 24? ____________

c. 23 is what percent of 19? ____________

d. 14 is what percent of 30? ____________

e. What percent of 30 is 14? ____________

• # # # • # # # • # # # • # # # • # # # • # # # • # # # • # #

A8 **a.** $P \times 12 = 0.6?$ **b.** $P \times 9 = 24?$ **c.** $23 = P \times 19?$ **d.** $14 = P \times 30?$
e. $P \times 30 = 14?$

---

5 In a multiplication problem the numbers that are multiplied together to form the *product* are called the *factors*. In the example $4 \times 3 = 12$, 4 and 3 are the factors and 12 is the product.

Two division problems can be formed from any multiplication problem. For example, $4 \times 3 = 12$ implies that $4 = 12 \div 3$ and $3 = 12 \div 4$.

Q9 Write two division problems that can be formed from $3 \times 5 = 15$.

____________ and ____________

• # # # • # # # • # # # • # # # • # # # • # # # • # # # • # #

A9 $3 = 15 \div 5$, $5 = 15 \div 3$ (either order)

Q10 Write two division problems that can be formed from $5 \times 9 = 45$.

____________ and ____________

• # # # • # # # • # # # • # # # • # # # • # # # • # # # • # #

A10 $5 = 45 \div 9$, $9 = 45 \div 5$ (either order)

6 Generally, a multiplication problem can be represented as: *factor* × *factor* = *product.* From the previous example this statement can be written *one factor* = *product* ÷ *other factor.* For example, if $4 \times 8 = 32$, *what number* = $32 \div 8$ or $N = 32 \div 8$? The answer is that $4 = 32 \div 8$; hence $N = 4$.

Q11 Determine the missing factor represented by the letter N:

a. If $4 \times 8 = 32$, then $N = 32 \div 4$; $N =$ ______

b. If $2 \times 3 = 6$, then $N = 6 \div 3$; $N =$ ______

c. If $45 = 5 \times 9$, then $N = 45 \div 5$; $N =$ ______

d. If $25 \times N = 1{,}200$, then $N = 1{,}200 \div 25$; $N =$ ______

• # # # • # # # • # # # • # # # • # # # • # # # • # # # • # #

A11 **a.** 8 **b.** 2 **c.** 9 **d.** 48

7 Often one of the factors of a product will be unknown. That is, *unknown factor* × *known factor* = *product.* This statement could be written *unknown factor* = *product* ÷ *known factor.* For example, if $N \times 15 = 60$, then $N = 60 \div 15$. N represents the unknown factor, 15 the known factor, and 60 the product.

Q12 Write the unknown factor as the quotient of the product and the known factor:

a. $N \times 7 = 42$ ______________ **b.** $7 \times N = 42$ ______________

• # # # • # # # • # # # • # # # • # # # • # # # • # # # • # #

A12 **a.** $N = 42 \div 7$ **b.** $N = 42 \div 7$

Q13 Solve for N by writing the unknown factor as a quotient of the product and the known factor:
a. $0.03 \times N = 15$ **b.** $2.5 = 0.06 \times N$

• # # # • # # # • # # # • # # # • # # # • # # # • # # # • # #

A13 **a.** 500: $N = 15 \div 0.03 = 500$ **b.** $41\frac{2}{3}$: $N = 2.5 \div 0.06 = 41\frac{2}{3}$

Q14 Solve for N:
a. $0.1 \times N = 2.4$ **b.** $\$1.62 = 0.06 \times N$

• # # # • # # # • # # # • # # # • # # # • # # # • # # # • # #

A14 **a.** 24: $N = 2.4 \div 0.1$ **b.** \$27: $N = \$1.62 \div 0.06$

Q15 Solve for N:

a. $N \times \$21 = \$0.42$

b. $\$0.12 \times N = \$2.40$

• ### • ### • ### • ### • ### • ### • ### •

A15 **a.** 0.02: $N = \$0.42 \div \$21$ **b.** 20: $N = \$2.40 \div \$0.12$

This completes the instruction for this section.

3.2 EXERCISE

1. Write mathematically:

a. 6% of 10 is what number?
b. What number is 17.3% of 92?
c. 520% of what number is 62?
d. What percent of 17 is 9?
e. 17 is 43% of what number?

2. If factor × factor = product, one factor = ________ ÷ ________.

3. If known factor × unknown factor = product, ________ = product ÷ ________.

4. If $75 = N \times 3$, $N =$ ________ ÷ ________.

5. Solve for N or P:

a. $5.2 \times N = 62$
b. $17 = 0.43 \times N$
c. $0.06 \times 10 = N$
d. $\$13 = P \times \$52$
e. $\$0.21 = 0.03 \times N$

3.2 EXERCISE ANSWERS

1. a. $0.06 \times 10 = N$
b. $N = 0.173 \times 92$
c. $5.2 \times N = 62$
d. $P \times 17 = 9$
e. $17 = 0.43 \times N$

2. product, other factor

3. unknown factor, known factor

4. 75, 3

5. a. $11\frac{12}{13}$
b. $39\frac{23}{43}$
c. 0.6
d. 0.25 or 25%
e. \$7

3.3 SOLVING PERCENT PROBLEMS

1 Problems involving percents usually occur in three forms:

1. Finding a percent of a number or quantity. *Example:* What number is 6% of 12?
2. Finding what percent one number is of another. *Example:* 18 is what percent of 36?
3. Finding a number when a certain percent of it is known. *Example:* 12 is 15% of what number?

These questions can be solved by first writing them mathematically.

Example: What number is 6% of 12?

Solution:

$N = 0.06 \times 12$*

$N = 0.72$

$$\begin{array}{r} 12 \\ \times 0.06 \\ \hline 0.72 \end{array}$$

(show work where necessary)

Therefore, 0.72 is 6% of 12.

*It is common to omit the "?" mark.

Q1 What number is 3% of 7?

\# \# \# • \# \# \# • \# \# \# • \# \# \# • \# \# \# • \# \# \# • \# \# \# • \# \# \# • \# \# \#

A1 0.21: $N = 0.03 \times 7$

Q2 0.3% of 63 is what number?

\# • \# \# \# • \# \# \# • \# \# \# • \# \# \# • \# \# \# • \# \# \# • \# \# \# • \# \# \#

A2 0.189: $0.003 \times 63 = N$

Q3 0.4% of 210 is what number?

\# \# \# • \# \# \# • \# \# \# • \# \# \# • \# \# \# • \# \# \# • \# \# \# • \# \# \# • \# \# \#

A3 0.84: $0.004 \times 210 = N$

Q4 What number is 520% of 92?

\# \# \# • \# \# \# • \# \# \# • \# \# \# • \# \# \# • \# \# \# • \# \# \# • \# \# \# • \# \# \#

A4 478.4: $N = 5.2 \times 92$

2 Recall that a mixed-number percent is simplified in the following manner:

$$23\frac{1}{3}\% = \frac{70}{3} \times \frac{1}{100} = \frac{7}{30}$$

Example: $23\frac{1}{3}\%$ of 900 is what number?

Solution:

$$\frac{7}{30} \times 900 = N$$
$$210 = N$$

Therefore, $23\frac{1}{3}\%$ of 900 is 210.

Q5 **a.** What number is 200% of 4? **b.** 24% of 30 is what number?

c. 0.5% of 18 is what number? **d.** What is $15\frac{1}{3}\%$ of 600?

\# \# \# • \# \# \# • \# \# \# • \# \# \# • \# \# \# • \# \# \# • \# \# \# • \# \# \# • \# \# \#

A5 **a.** 8 **b.** 7.2 **c.** 0.09 **d.** 92

3 The second type of percent problem is finding what percent one number is of another.

Example: 18 is what percent of 36?

Solution:

$18 = P \times 36$ (written mathematically)

Recall that *factor* × *factor* = *product* and *unknown factor* = *product* ÷ *known factor*. Hence,

$18 \div 36 = P$ [18 (product) ÷ 36 (known factor)]

$0.5 = P$

$50\% = P$

Therefore, 18 is 50% of 36.

Q6 17 is what percent of 68?

\# \# \# • \# \# \# • \# \# \# • \# \# \# • \# \# \# • \# \# \# • \# \# \# • \# \# \# • \# \# \#

A6 25%: $17 = P \times 68$

$17 \div 68 = P$

$0.25 = P$

Q7 24 is what percent of 30?

\# \# \# • \# \# \# • \# \# \# • \# \# \# • \# \# \# • \# \# \# • \# \# \# • \# \# \# • \# \# \#

A7 80%: $24 = P \times 30$
$24 \div 30 = P$
$0.8 = P$

Q8 21 is what percent of 300?

• # # # • # # # • # # # • # # # • # # # • # # # • # # # • # #

A8 7%: $21 = P \times 300$
$21 \div 300 = P$
$0.07 = P$

Q9 80 is what percent of 64?

• # # # • # # # • # # # • # # # • # # # • # # # • # # # • # #

A9 125%: $80 = P \times 64$
$80 \div 64 = P$
$1.25 = P$

Q10 What percent of 6 is 4?

• # # # • # # # • # # # • # # # • # # # • # # # • # # # • # #

A10 $66\frac{2}{3}$%: $P \times 6 = 4$
$P = 4 \div 6$
$P = 0.66\frac{2}{3}$

Q11 What percent of 25 is 30?

• # # # • # # # • # # # • # # # • # # # • # # # • # # # • # #

A11 120%: $P \times 25 = 30$
$P = 30 \div 25$
$P = 1.2$

Q12 **a.** What percent of 300 is 45?

b. 38 is what percent of 304?

c. What percent of 12 is 4.6?

d. 66.6 is what percent of 74?

\# \# \# • \# \# \# • \# \# \# • \# \# \# • \# \# \# • \# \# \# • \# \# \# • \# \# \# • \# \# \#

A12 **a.** 15% **b.** 12.5% **c.** $38\frac{1}{3}\%$ **d.** 90%

4 The third type of percent problem is finding a number when a certain percent of it is known.

Example: 12 is 15% of what number?

Solution:

$$12 = 0.15 \times N$$
$$12 \div 0.15 = N$$
$$80 = N$$

Therefore, 12 is 15% of 80.

Q13 27 is 5% of what number?

\# \# \# • \# \# \# • \# \# \# • \# \# \# • \# \# \# • \# \# \# • \# \# \# • \# \# \# • \# \# \#

A13 540: $27 = 0.05 \times N$
$27 \div 0.05 = N$

Q14 17 is 4% of what number?

\# \# \# • \# \# \# • \# \# \# • \# \# \# • \# \# \# • \# \# \# • \# \# \# • \# \# \# • \# \# \#

A14 425: $17 = 0.04 \times N$
$17 \div 0.04 = N$

Q15 34 is 125% of what number?

\# \# \# • \# \# \# • \# \# \# • \# \# \# • \# \# \# • \# \# \# • \# \# \# • \# \# \# • \# \# \#

A15 27.2: $34 = 1.25 \times N$
$34 \div 1.25 = N$

Q16 91 is 70% of what number?

\# \# \# • \# \# \# • \# \# \# • \# \# \# • \# \# \# • \# \# \# • \# \# \# • \# \# \# • \# \# \#

A16 130: $91 = 0.7 \times N$
$91 \div 0.7 = N$

Q17 5% of what number is 8?

\# \# \# • \# \# \# • \# \# \# • \# \# \# • \# \# \# • \# \# \# • \# \# \# • \# \# \# • \# \# \#

A17 160: $0.05 \times N = 8$
$N = 8 \div 0.05$

Q18 16% of what number is 10?

\# \# \# • \# \# \# • \# \# \# • \# \# \# • \# \# \# • \# \# \# • \# \# \# • \# \# \# • \# \# \#

A18 62.5: $0.16 \times N = 10$
$N = 10 \div 0.16$

Q19 **a.** 78 is 15.6% of what number? **b.** 8% of what number is 24?

c. 5 is 0.2% of what number? **d.** $33\frac{1}{3}$% of what number is 12.5?

\# \# \# • \# \# \# • \# \# \# • \# \# \# • \# \# \# • \# \# \# • \# \# \# • \# \# \# • \# \# \#

A19 **a.** 500 **b.** 300 **c.** 2,500 **d.** 37.5

This completes the instruction for this section.

3.3 EXERCISE

1. 15 is what percent of 70?
2. 69% of $21 is what number?
3. 6 is 30% of what number?
4. What number is 19% of 20?
5. 152% of 5 is what number?
6. What percent of 320 is 500?
7. 8 is what percent of 15?
8. 32 is 64% of what number?

9. 75% of what number is 4.65?

10. 3.6 is 18% of what number?

11. What number is $33\frac{1}{3}\%$ of 60?

12. 12.5% of 28 is what number?

13. $63\frac{1}{3}\%$ of 3 is what number?

14. 122.5 is what percent of 98?

15. 2 is what percent of 400?

3.3 EXERCISE ANSWERS

1. $21\frac{3}{7}\%$ **2.** $14.49 **3.** 20 **4.** 3.8

5. 7.6 **6.** 156.25% **7.** $53\frac{1}{3}\%$ **8.** 50

9. 6.2 **10.** 20 **11.** 20 **12.** 3.5

13. 1.9 **14.** 125% **15.** 0.5%

3.4 APPLICATIONS

1 The three basic types of percent problems, illustrated in the previous section, will be presented in a variety of situations in this section. A few of the more common applications will be presented first.

One of the more basic types of percent problems is finding the *discount* when the discount is given as a percent. When the price of an article is reduced, the reduction is called a *discount.* Discount is usually expressed as a percent of the *list* (original) price of an article. The percent is called the *rate of discount.* Hence, discount = rate of discount × list price.

Example: Find the discount on a machine priced at $72.60, if it is discounted 15%.

Solution: discount = rate of discount × list price
discount = 15% × $72.60
discount = 0.15 × $72.60
discount = $10.89

Q1 Find the discount on an electric drill if the list price is $17.20 and the rate of discount is 25%.

\# \# \# • \# \# \# • \# \# \# • \# \# \# • \# \# \# • \# \# \# • \# \# \# • \# \# \# • \# \# \#

A1 $4.30: discount = 0.25 × $17.20

Q2 Tires normally selling for $43.95 are being reduced 22%. Find the discount.

\# \# \# • \# \# \# • \# \# \# • \# \# \# • \# \# \# • \# \# \# • \# \# \# • \# \# \# • \# \# \#

A2 $9.67: discount = 0.22 × $43.95

2 When the discount is subtracted from the list price, the result is called the *net or purchase price.* That is, net price = list price − discount.

Example: Find the net price of the tires (Q2) selling for $43.95 discounted at 22%.

Solution: net price = list price − discount
net price = $43.95 − $9.67
net price = $34.28

Q3 Find the net price of an article priced at $173.89 if it is discounted 15%.

• ### • ### • ### • ### • ### • ### • ### •

A3 $147.81: discount = 0.15 × $173.89 = $26.08
net price = $173.89 − $26.08

Q4 Find the net price of machine parts priced at $29.95 if they are reduced 6%.

• ### • ### • ### • ### • ### • ### • ### •

A4 $28.15

3 In the previous problems the list price was decreased by a percentage amount. The list price can also be increased by a certain percentage. The increase is found by multiplying the *rate of increase* by the list price.

Example: A machinist earning $230 per week has his wages increased 15%. What is his weekly salary after the increase?

Solution: increase = rate of increase × original amount
increase = 0.15 × $230
increase = $34.50
increased salary = increase + original amount
increased salary = $34.50 + $230
increased salary = $264.50

Q5 A machine part priced at $8.95 is to be increased 12.5%. Find the new price (round off to the nearest cent).

• ### • ### • ### • ### • ### • ### • ### •

A5 $10.07: increase = 0.125 × $8.95 = $1.12
new price = $1.12 + $8.95

Q6 An article priced at $18.50 is increased by a 4% sales tax. Find the final price.

• ### • ### • ### • ### • ### • ### • ### •

A6 $19.24: increase (sales tax) = 0.04 × $18.50 = $0.74
final price = $0.74 + $18.50

4 Some of the more difficult types of percent problems are those dealing with a *rate of change*. When working rate of change problems it is helpful to remember that the rate of change = the change ÷ original amount.

Example: The price of an article was increased from $6 to $7.50. Find the percent increase.

Solution: Rate of change = the change ÷ original amount. In this case the change represents an increase ($1.50); hence,

rate of increase = increase ÷ original amount
rate of increase = $1.50 ÷ $6
rate of increase = 0.25

Since the rate is usually expressed as a percent, the rate of increase is 25%.

Q7 Rate of change = __________ ÷ __________.

• ### • ### • ### • ### • ### • ### • ### •

A7 the change (increase or decrease) ÷ original amount

Q8 For the following problem complete the blanks:
An article priced at $18.20 is to be remarked $14.56.

a. Original price = ________.

b. The change = ________.

c. Is the change an increase or decrease? ________

• ### • ### • ### • ### • ### • ### • ### •

A8 **a.** $18.20 **b.** $3.64 **c.** decrease

Q9 Determine the rate of decrease in Q8.

• ### • ### • ### • ### • ### • ### • ### •

A9 20%: rate of decrease = decrease ÷ original amount
rate of decrease = $3.64 ÷ $18.20
rate of decrease = 0.2

Q10 Find the rate of decrease if Thompson's income is $13,400 one year and $11,800 the next year (round off the percent to one decimal place).

• ### • ### • ### • ### • ### • ### • ### •

A10 11.9%: rate of decrease = $1,600 ÷ $13,400

Q11 Find the rate of increase when the price of an article changes from $8 to $10.

• ### • ### • ### • ### • ### • ### • ### •

A11 25%: rate of increase = $2 ÷ $8

Q12 What is the rate if increase if production increases from 176 to 192 (round off the percent to one decimal place)?

• ### • ### • ### • ### • ### • ### • ### •

A12 9.1%: rate of increase = $16 \div 176$

5 Common types of percent applications have been presented in the previous pages. These dealt with discounts and rates of change; however, percent applications can be presented in a number of different situations. As stated in Section 3.3, Frame 1, problems involving percents usually occur in three forms:

1. Finding a percent of a number or quantity.
2. Finding what percent one number is of another.
3. Finding a number when a certain percent of it is known.

As discussed in Section 3.3, these types of percent problems can be solved by first writing them mathematically.

Example: Ken saved $576 from his annual income of $6,400. What percent of his income did he save?

Solution: Since he saved $576 and his income is $6,400, the problem could be stated; $576 is what percent of $6,400? Written mathematically:

$$\$576 = P \times \$6{,}400$$

unknown factor = product ÷ known factor

$$P = \$576 \div \$6{,}400$$
$$P = 0.09$$
$$P = 9\%$$

Hence, Ken saved 9% of his income.

Q13 A worker received a $0.45 an hour increase in wages. If his current hourly rate is $8, what is the percent increase in wages (round off the percent to one decimal place)?

• ### • ### • ### • ### • ### • ### • ### •

A13 5.6%: $0.45 is what percent of $8

$$\$0.45 = P \times \$8$$
$$P = \$0.45 \div 8$$

Q14 Out of a production of 1,802 machine parts manufactured, 56 were rejected as imperfect. What percent of the total was rejected (round off the percent to one decimal)?

• ### • ### • ### • ### • ### • ### • ### •

A14 3.1%: 56 is what percent of 1,802

$$56 = P \times 1{,}802$$

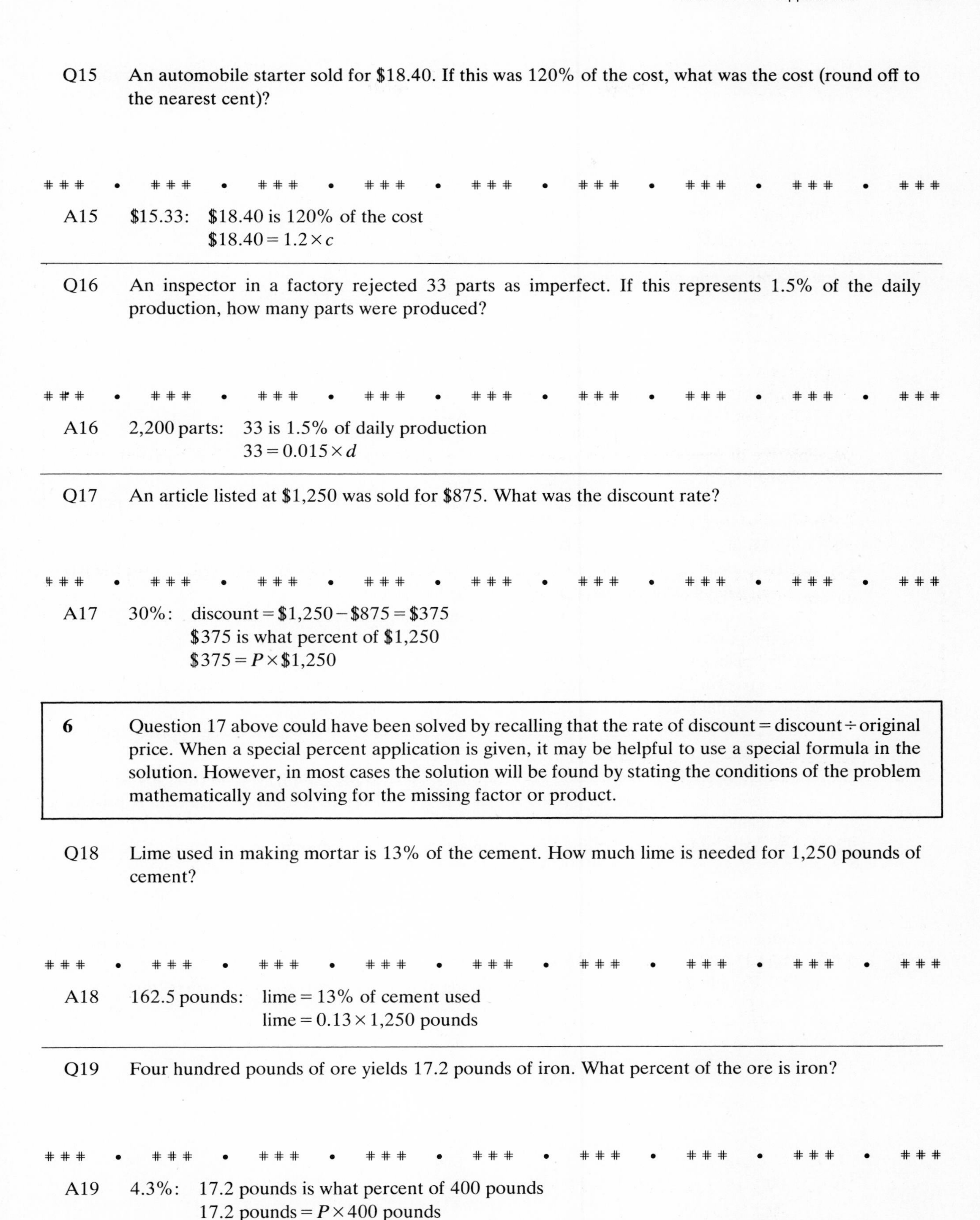

Q15 An automobile starter sold for $18.40. If this was 120% of the cost, what was the cost (round off to the nearest cent)?

• ### • ### • ### • ### • ### • ### • ### •

A15 $15.33: $18.40 is 120% of the cost
$\$18.40 = 1.2 \times c$

Q16 An inspector in a factory rejected 33 parts as imperfect. If this represents 1.5% of the daily production, how many parts were produced?

• ### • ### • ### • ### • ### • ### • ### •

A16 2,200 parts: 33 is 1.5% of daily production
$33 = 0.015 \times d$

Q17 An article listed at $1,250 was sold for $875. What was the discount rate?

• ### • ### • ### • ### • ### • ### • ### •

A17 30%: discount = $1,250 − $875 = $375
$375 is what percent of $1,250
$\$375 = P \times \$1{,}250$

6 Question 17 above could have been solved by recalling that the rate of discount = discount ÷ original price. When a special percent application is given, it may be helpful to use a special formula in the solution. However, in most cases the solution will be found by stating the conditions of the problem mathematically and solving for the missing factor or product.

Q18 Lime used in making mortar is 13% of the cement. How much lime is needed for 1,250 pounds of cement?

• ### • ### • ### • ### • ### • ### • ### •

A18 162.5 pounds: lime = 13% of cement used
lime = $0.13 \times 1{,}250$ pounds

Q19 Four hundred pounds of ore yields 17.2 pounds of iron. What percent of the ore is iron?

• ### • ### • ### • ### • ### • ### • ### •

A19 4.3%: 17.2 pounds is what percent of 400 pounds
17.2 pounds = $P \times 400$ pounds

Q20 A certain ore yields 5% iron. How many pounds of ore are necessary to produce 102 pounds of iron?

• # # # • # # # • # # # • # # # • # # # • # # # • # # # • # #

A20 2,040 pounds: 102 pounds = 5% of ore
102 pounds = $0.05 \times R$

This completes the instruction for this section.

3.4 EXERCISE

1. Find the discount if the original price is $18 and the discount rate is 12%.
2. Find the discount rate if an article priced at $450 is reduced $15 (round off the percent to one decimal place).
3. Find the discount rate if the list price is $15.50 and the discount is $4.50 (round off the percent to one decimal place).
4. If the list price of a machine is $62.50 and the discount rate is 5%, find the net price.
5. In a local election, 411,700 people cast their vote. If this represents a gain of 15% over the last election, what was the number of voters in the last election?
6. An article costs $17.60. If the price is increased 15%, find the new price.
7. Find the percent increase if a price increases from $7.20 to $8 (round off the percent to one decimal place).
8. Frank's pay was decreased from $96 to $91.50. Find the percent decrease (round off the percent to one decimal place).
9. The price of a drill press is $7,800. If this is an increase of 20% over last year's price, find last year's price.
10. An article with 10% tax included sells for $6.49. What is the selling price without the tax?
11. A machine sold at a profit of $17.60 which was 16% of the cost. What was the selling price of the machine?
12. A 4% tax is added to the selling price of an article. Find the amount of tax on an article selling for $18.60.
13. Sixty percent of a certain type of metal is lead. How many pounds of lead are there in 685 pounds of this metal?
14. A factory makes 2,450 machine parts, 12% of which are defective. How many perfect parts are produced?
15. If a ton (2,000 pounds) of ore yields 70 pounds of iron, what percent of the ore is iron?

3.4 EXERCISE ANSWERS

1. $2.16 **2.** 3.3% **3.** 29.0% **4.** $59.37
5. 358,000 **6.** $20.24 **7.** 11.1% **8.** 4.7%
9. $6,500 **10.** $5.90 **11.** $127.60 **12.** $0.74
13. 411 pounds **14.** 2,156 **15.** 3.5%

CHAPTER 3 SAMPLE TEST

At the completion of Chapter 3, you should be able to work the following problems.

3.1 PERCENTS

1. Change to a decimal:
a. 24% **b.** 7.3%

2. Change to a percent:
a. 0.8 **b.** 1.25

3. Change to a fraction:
a. $16\frac{2}{3}\%$ **b.** $12\frac{1}{2}\%$

4. Change to a percent:
a. $\frac{9}{40}$ **b.** $\frac{5}{6}$

3.2 WRITING AND SOLVING MATHEMATICAL SENTENCES

5. Write the following mathematically:
a. 17% of 61 is 10.37 **b.** 5% of 30 is what number?

6. a. If $4 \times 5 = 20$, $5 =$ _____ $\div$ _____.
b. If $0.5 \times A = \$2.70$, $A =$ _____ $\div$ _____.

3.3 SOLVING PERCENT PROBLEMS

7. a. 15 is 30% of what number? **b.** 336 is what percent of 16?
c. What number is 23.8% of 15?

3.4 APPLICATIONS

8. If the list price of an article is $78.20 and the discount is 15%, find the net price.

9. A machine selling for $150 is reduced $35. Find the percent of reduction (round off the percent to one decimal place).

10. It is estimated that 11,250 face bricks are needed for a certain job. How many more bricks are needed if 15% is added for waste (round off to the nearest brick)?

11. Eight hundred fifty pounds of recycled junk metal yields 620.5 pounds of usable metal. The usable metal is what percent of the junk metal?

12. A machine part sells for $8.20 which is 120% of its cost. What is the cost (round off to the nearest cent)?

CHAPTER 3 SAMPLE TEST ANSWERS

1. a. 0.24 **b.** 0.073
2. a. 80% **b.** 125%
3. a. $\frac{1}{6}$ **b.** $\frac{1}{8}$
4. a. 22.5% **b.** $83\frac{1}{3}\%$
5. a. $0.17 \times 61 = 10.37$ **b.** $0.05 \times 30 = N$
6. a. 20, 4 **b.** $2.70, 0.5
7. a. 50 **b.** 2,100% **c.** 3.57
8. $66.47 **9.** 23.3% **10.** 1,688
11. 73% **12.** $6.83

CHAPTER 4

RECOGNITION AND PROPERTIES OF GEOMETRIC FIGURES

Technicians live and work in a geometric world. A knowledge of geometry will make a technician's life and work easier. It will also make the world more pleasing to the eye and intellectually interesting. This chapter will help you obtain that knowledge.

4.1 POINTS, LINES, AND PLANES

1 The most fundamental notion in geometry is a *point.* A point is a location in space and has no thickness. Although a point is represented with a dot on paper, the dot is much too large to actually be a point. The dot helps in thinking about the point.

Q1 Which dot is the best representation of a point? ______

a. • **b.** · **c.** ·

\# # # • # # # • # # # • # # # • # # # • # # # • # # # • # # # • # # #

A1 c: It looks most like a location with no dimensions.

Q2 Is the representation in part **c** of Q1 actually a point? ______

\# # # • # # # • # # # • # # # • # # # • # # # • # # # • # # # • # # #

A2 no: It is still too large and is actually many, many points all very close together.

2 Another geometric object which is so fundamental that other objects depend on its properties is a *line.* A line is a set of points and is represented with a picture such as:

⟵――――――――――――――――⟶

A geometric line is always straight, has no thickness, and extends forever in both directions. The lines you draw on paper help you think about true geometric lines, which really only exist perfectly in your mind. A tightly stretched string is another representation that helps to think of a line in space, but it is still not a perfect model of a geometric line.

Q3 List three important properties of a line. ______________________________

____

\# # # • # # # • # # # • # # # • # # # • # # # • # # # • # # # • # # #

A3 It is straight, has no thickness, and extends forever in both directions.

Q4 The following are sometimes used to represent lines, but are not lines as we speak of them in geometry. Explain why in each case:

a. pencil ________________

b. telephone wire strung on poles ________________

c. a plumb line formed by a weight suspended on a string ________________

d. the "line" separating A1 from Q1 on page 125 ________________

• # # # • # # # • # # # • # # # • # # # • # # # • # # # • # #

A4 **a.** A pencil has thickness, has definite length, and may not be straight.
b. It has thickness. It sags between poles so is not straight and has definite length.
c. The string has thickness and definite length.
d. It has thickness, or you wouldn't see it. It has definite length, and if the page is not a flat surface, the "line" on the page is curved.

3 A *plane* is usually represented by a flat surface such as a table top, wall, or stiff sheet of paper. A geometric plane, however, extends infinitely in every direction. The accompanying figure represents two intersecting planes.

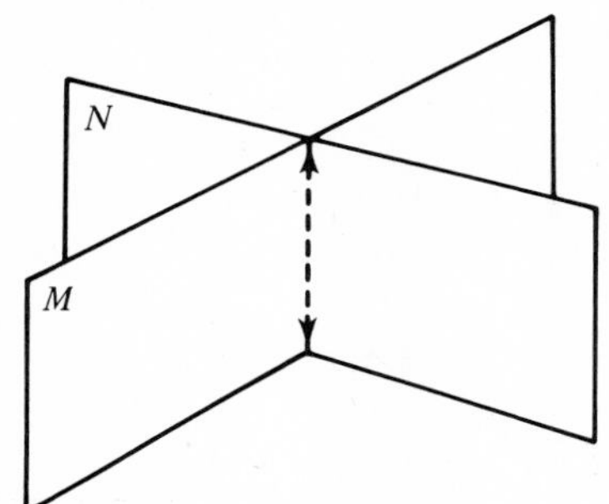

The planes must be drawn in perspective and distortions occur. (For example, it may not look like plane M and plane N meet with square corners.) Keep in mind that all drawings of geometric objects, no matter how carefully they are done, are imperfect. As long as you use the figures to reason correctly, they are serving their purpose.

Q5 Planes M and N are shown intersecting in Frame 3. What geometric figure is formed by the points that are on both planes? ________

• # # # • # # # • # # # • # # # • # # # • # # # • # # # • # #

A5 a line

4 When more than one line is drawn, it is helpful to be able to refer to them. They are sometimes referred to with the notation l_1 and l_2. (This is read l sub one and l sub two.) The 1 and 2 are called

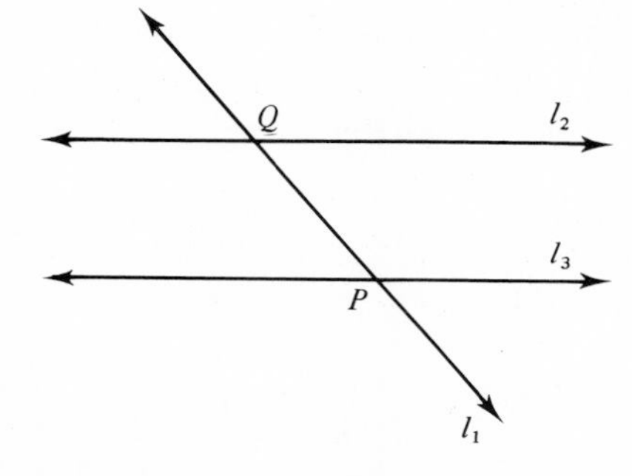

subscripts and are used only to keep track of which l is being discussed. They are *not* exponents. The figure on the previous page contains three lines: l_1, l_2, and l_3. l_3 intersects l_1 at point P, but l_3 does not intersect l_2.

Two lines in one plane may intersect, or they may be parallel. If they are parallel, they have no points in common. If they intersect, they have one and only one point in common.

Intersecting lines

Parallel lines
(write $l_1 \parallel l_2$)

If all four angles formed by intersecting lines are equal, each angle is a *right* angle and the lines are said to be *perpendicular* lines. In the illustration, l_1 is perpendicular to l_2. Write $l_1 \perp l_2$ and place a square where they intersect.

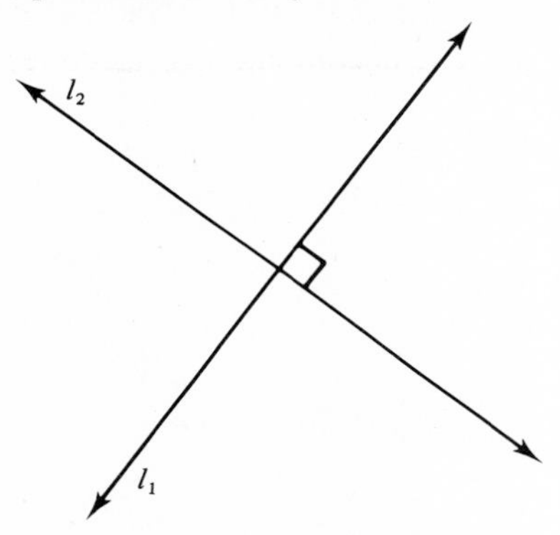

Q6 **a.** From the figure below, write two pairs of lines that appear to be parallel (use the appropriate symbol between each pair of lines). ____________________

b. Write four pairs of lines that appear to be perpendicular (use the appropriate symbol).

____

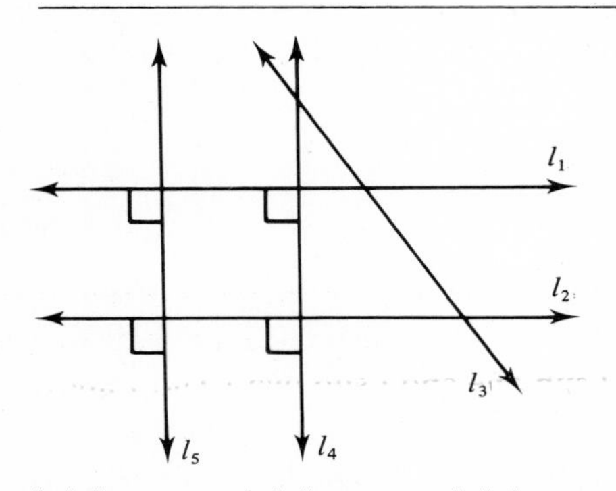

\# \# \# • \# \# \# • \# \# \# • \# \# \# • \# \# \# • \# \# \# • \# \# \# • \# \# \# • \# \# \#

A6 **a.** $l_1 \parallel l_2$ and $l_4 \parallel l_5$

b. $l_1 \perp l_4$, $l_1 \perp l_5$, $l_2 \perp l_4$, and $l_2 \perp l_5$

Q7 How many lines can be drawn through point A? ____________________

$\cdot A$

\# \# \# • \# \# \# • \# \# \# • \# \# \# • \# \# \# • \# \# \# • \# \# \# • \# \# \# • \# \# \#

A7 an infinite number

Q8 How many lines can be drawn that contain both point A and point B? ____________

A · B ·

\# \# \# • \# \# \# • \# \# \# • \# \# \# • \# \# \# • \# \# \# • \# \# \# • \# \# \# • \# \# \#

A8 one

---

5 One and only one line can be drawn through two points. This fact is sometimes stated as: *Two points determine one and only one line.*

Q9 Hold up a finger and thumb on one hand and one finger on the other. How many planes (such as a book cover) touch all three points (two fingers, one thumb)? __________

\# \# \# • \# \# \# • \# \# \# • \# \# \# • \# \# \# • \# \# \# • \# \# \# • \# \# \# • \# \# \#

A9 one

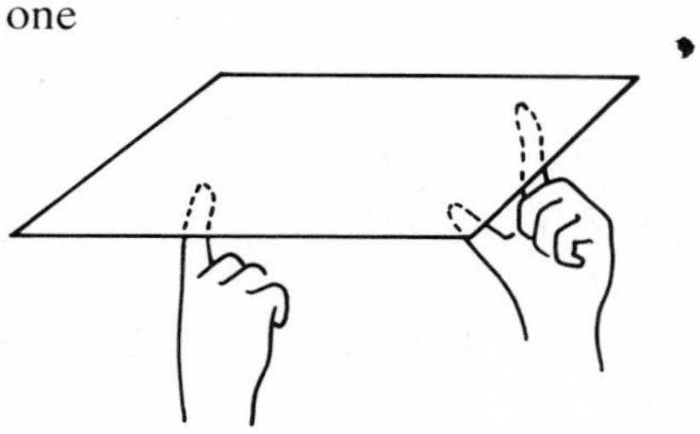

Q10 Suppose there are three points not on one line. How many planes are determined? __________

\# \# \# • \# \# \# • \# \# \# • \# \# \# • \# \# \# • \# \# \# • \# \# \# • \# \# \# • \# \# \#

A10 one

Q11 Hold up one finger on each hand. How many ways can a flat surface touch both points?

____

\# \# \# • \# \# \# • \# \# \# • \# \# \# • \# \# \# • \# \# \# • \# \# \# • \# \# \# • \# \# \#

A11 an infinite number

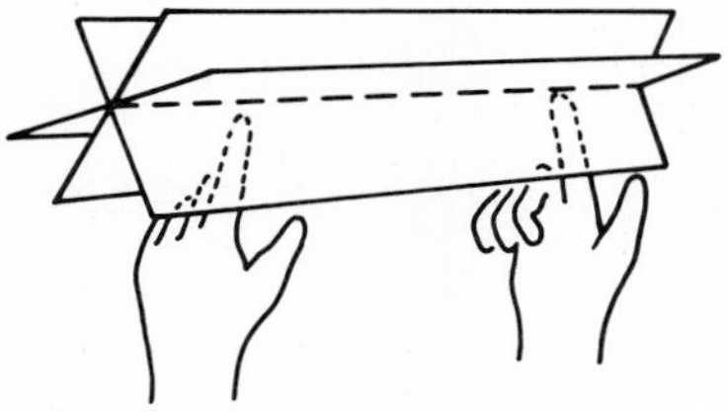

Q12 Suppose that there are two points in space. How many planes contain the two points?

____

\# \# \# • \# \# \# • \# \# \# • \# \# \# • \# \# \# • \# \# \# • \# \# \# • \# \# \# • \# \# \#

A12 an infinite number

---

6 The answers to Q9 through Q12 are examples of two properties of planes: *Two points are contained in an infinite number of planes and three points determine one and only one plane.*

These properties are illustrated with the plane of the surface of a door passing through two points represented by two hinges. The door can be placed in an infinite number of positions. However, when the plane must also contain a third point, such as the tip of your shoe, the door becomes stationary.

Q13 Mark three points on a sheet of paper. Hold your finger above these three points to represent a fourth point.

a. How many ways can a flat surface touch three of the four points? ____________

b. How many lines would be determined by taking two points at a time? ____________

\# \# \# • \# \# \# • \# \# \# • \# \# \# • \# \# \# • \# \# \# • \# \# \# • \# \# \# • \# \# \#

A13 **a.** four planes (including the surface of the paper):

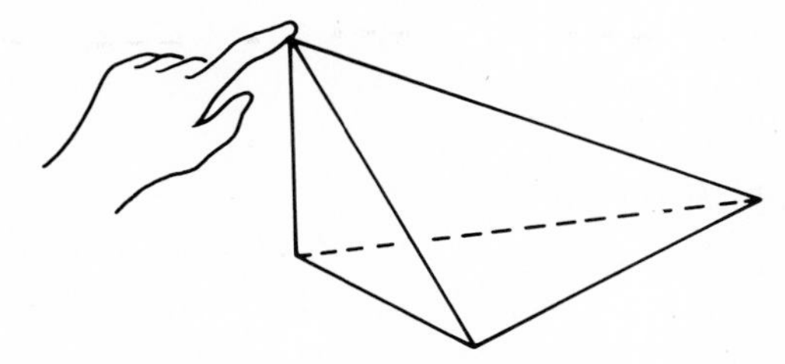

b. six lines: three in the plane of the paper and three connecting the points on the paper with the finger.

7 Points are frequently labeled with capital letters. Lines are referred to by naming two points on the line and drawing a double-headed arrow above them. The following line is represented by $\overleftrightarrow{AB}$:

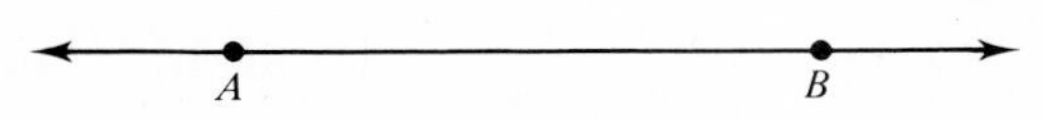

Q14 Label the statements true or false:

__________ **a.** $\overleftrightarrow{AB}$ intersects $\overleftrightarrow{CD}$.

__________ **b.** $\overleftrightarrow{AB} \parallel \overleftrightarrow{CD}$.

__________ **c.** $\overleftrightarrow{CD}$ intersects $\overleftrightarrow{DB}$.

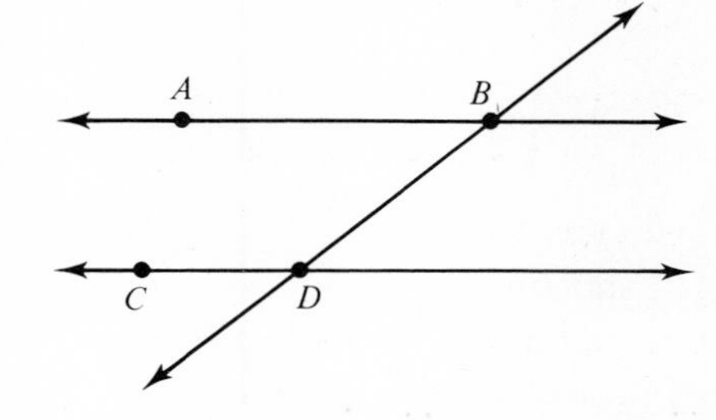

\# \# \# • \# \# \# • \# \# \# • \# \# \# • \# \# \# • \# \# \# • \# \# \# • \# \# \# • \# \# \#

A14 **a.** false **b.** true **c.** true

8 A *line segment* (sometimes referred to as a segment) is the portion of a line between two points including the endpoints. A line segment is referred to with the symbol $\overline{AB}$, where A and B are its endpoints. Segment $\overline{CD}$ is shown as

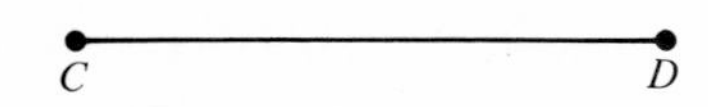

In the next figure $\overleftrightarrow{AB}$ is drawn, but $\overline{AB}$ refers only to the points A and B and those between A and B. The segment is a part of the line.

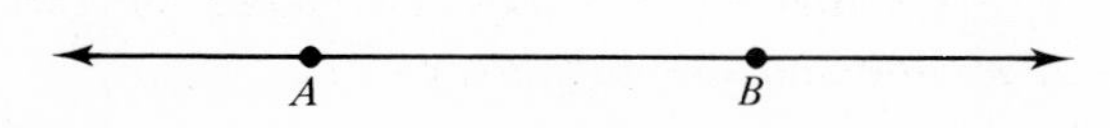

Refer to the figure below for the following examples:

1. P is on $\overleftrightarrow{QR}$.
2. P is not on $\overline{QR}$ (because P is not between Q and R).
3. Q is on $\overleftrightarrow{PR}$.
4. Q is on $\overline{PR}$.
5. R is on $\overleftrightarrow{PQ}$.
6. R is not on $\overline{PQ}$.
7. P is on $\overline{PR}$.

Q15 Label the statements true or false:

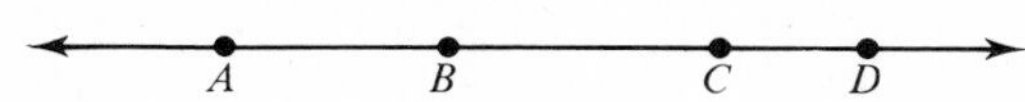

a. A is on $\overleftrightarrow{CD}$. TRUE

b. A is on $\overline{CD}$. FALSE

c. B is on $\overline{AC}$. TRUE

d. B is on $\overleftrightarrow{AC}$. TRUE

e. C is on $\overleftrightarrow{AB}$. TRUE

f. C is on $\overline{AB}$. FALSE

g. C is on $\overline{BC}$. TRUE

\# \# \# • \# \# \# • \# \# \# • \# \# \# • \# \# \# • \# \# \# • \# \# \# • \# \# \# • \# \# \#

A15 **a.** true **b.** false **c.** true **d.** true **e.** true **f.** false **g.** true

Q16 **a.** Name the lines shown in the figure. $\overleftrightarrow{AB}$ $\overleftrightarrow{AC}$ $\overleftrightarrow{CB}$

b. Name the line segments shown in the figure. ______________

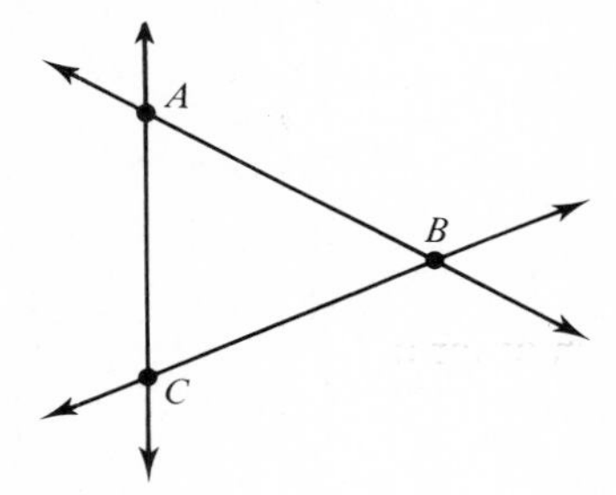

\# \# \# • \# \# \# • \# \# \# • \# \# \# • \# \# \# • \# \# \# • \# \# \# • \# \# \# • \# \# \#

A16 **a.** $\overleftrightarrow{AB}$, $\overleftrightarrow{AC}$, and $\overleftrightarrow{BC}$ **b.** $\overline{AB}$, $\overline{AC}$, and $\overline{BC}$

9 Line segments are equal only if they have the same endpoints. Therefore, there are three different line segments $\overline{AB}$, $\overline{BC}$, and $\overline{AC}$ on the line below.

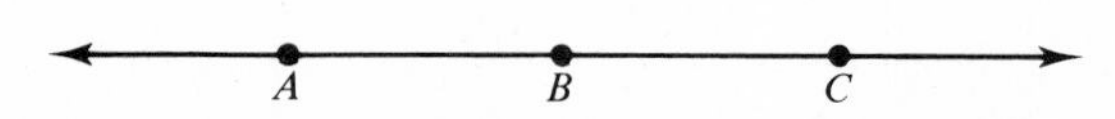

Q17 Name four segments on the figure

____

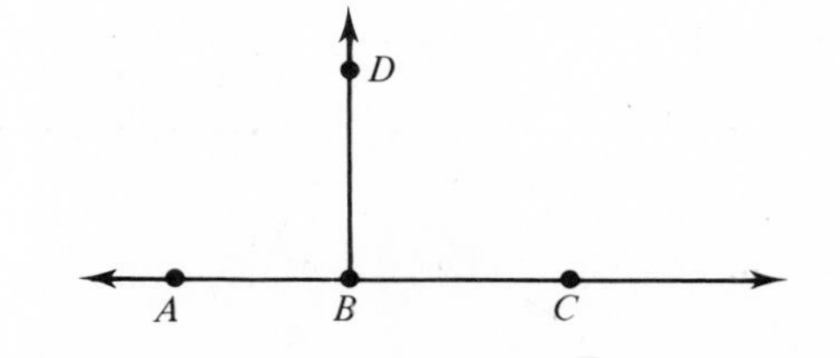

\# \# \# • \# \# \# • \# \# \# • \# \# \# • \# \# \# • \# \# \# • \# \# \# • \# \# \# • \# \# \#

A17 $\overline{AB}$, $\overline{BC}$, $\overline{BD}$, and $\overline{AC}$

10 Two planes are parallel if they do not intersect. An example of parallel planes would be the planes containing the ceiling, table top, and floor in a room. Of course, the physical structures are not planes, but you can think of them being extended forever without ever intersecting.

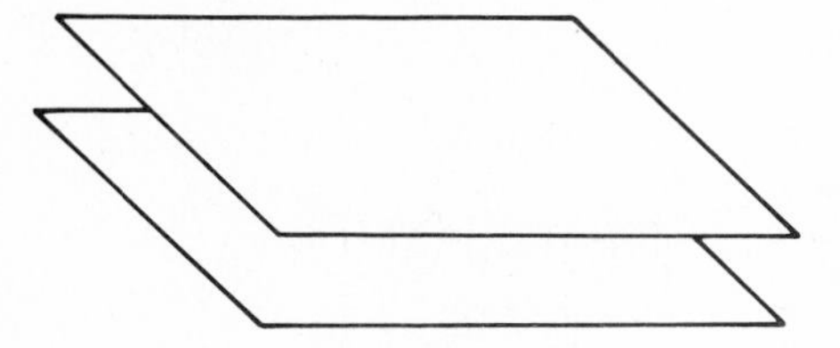

In work using a shaper, *parallels* are used to bolster and level the work in a vise. Each surface is parallel to the surface on the other side.

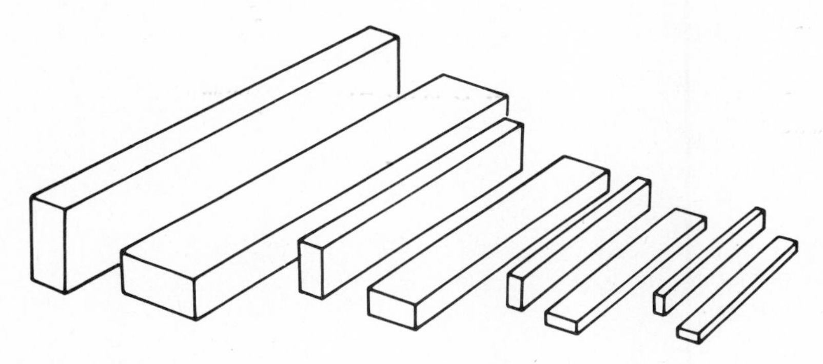

Several sizes of parallels

Another type of parallel is the *angle plate*. They are used in pairs in a vise to hold the work "out of square." As can be seen in the accompanying figure, the inside plane surfaces of the angle plates are parallel with each other in the vise. However, they are not parallel with the outer surfaces of the angle plates that come in contact with the vise.

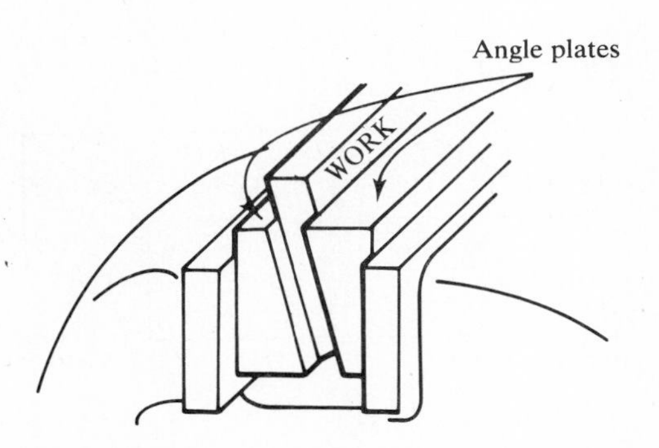

Q18 In the figure at the top of the next page, the work is being held in angle plates in a vise.

a. Name a segment parallel to $\overline{AE}$. ______

b. Name a segment parallel to $\overline{BF}$. ______

c. Name a segment parallel to $\overline{AB}$ and $\overline{CD}$. ______

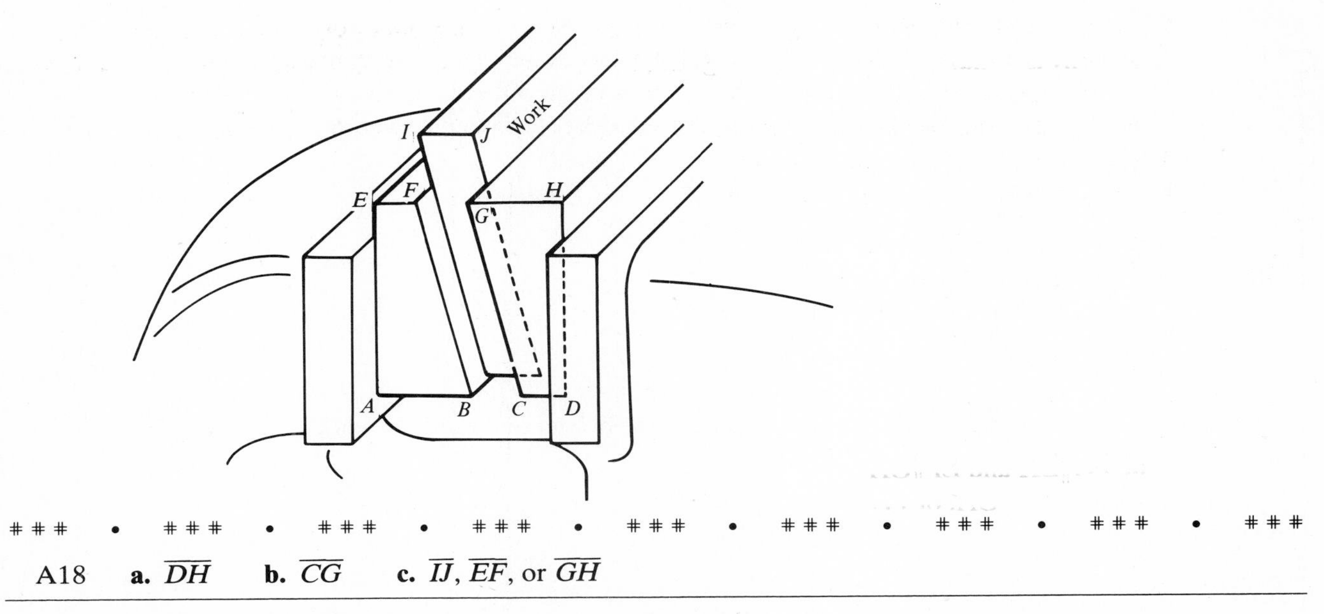

\# \# \# • \# \# \# • \# \# \# • \# \# \# • \# \# \# • \# \# \# • \# \# \# • \# \# \# • \# \# \#

A18 **a.** $\overline{DH}$ **b.** $\overline{CG}$ **c.** $\overline{IJ}$, $\overline{EF}$, or $\overline{GH}$

This completes the instruction for this section.

4.1 EXERCISE

Figure 1

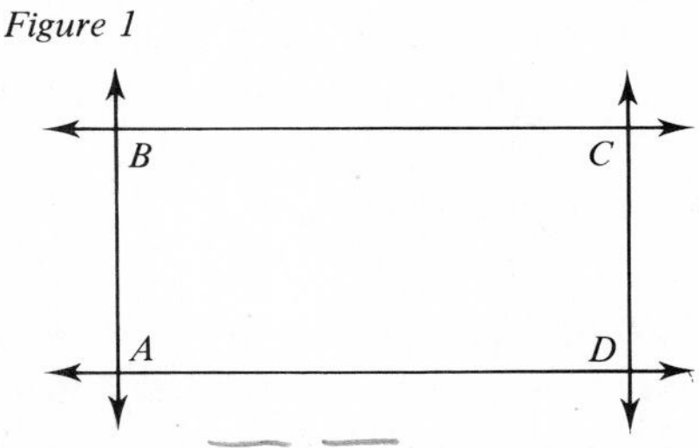

1. Name the four segments that are the four sides of the rectangle $ABCD$.
2. In Figure 1, name two line segments that intersect at B.
3. In Figure 1, name two line segments that appear to be part of parallel lines (there are two possibilities).
4. Figure 2 has the shape of a brick. The dashed lines would not be visible.
   - **a.** Name a pair of edges that have the common point C in the plane closest to you.
   - **b.** Name a pair of edges that appear to be part of parallel lines in the plane farthest from you (there are two possibilities).
   - **c.** Name an edge that is not in the plane of the top or the plane of the bottom of the figure.

Figure 2

5. Use the figure at the right to answer the following:
   - **a.** Name two lines that appear to be parallel.
   - **b.** $\overleftrightarrow{AC}$ and $\overleftrightarrow{BD}$ intersect at what point?
   - **c.** Name eight line segments on the figure.
   - **d.** Are there any perpendicular lines in the figure?

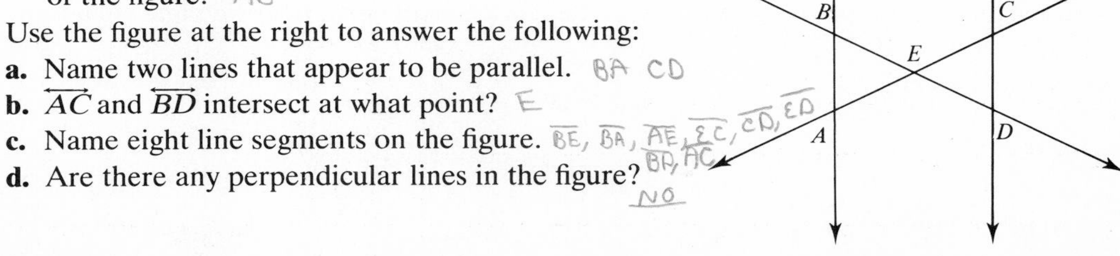

6. What is the best model of a point: the sun, a bowling ball, a grain of sand, or a marble?
7. **a.** Why is it that a table with three legs will always stand firmly on the floor, whereas a table with four legs sometimes rocks?
   b. Is it always possible to adjust a wobbling table with four legs by adjusting exactly one leg?
8. Is there always a plane that contains any three-sided figure?
9. Is there always a plane that contains any four-sided figure?

4.1 EXERCISE ANSWERS

1. $\overline{AB}$, $\overline{BC}$, $\overline{CD}$, and $\overline{DA}$
2. $\overline{AB}$ and $\overline{BC}$
3. $\overline{AB} \| \overline{CD}$ and $\overline{BC} \| \overline{AD}$
4. **a.** $\overline{BC}$ and $\overline{CD}$
   b. $\overline{FG} \| \overline{EH}$ and $\overline{EF} \| \overline{GH}$
   c. $\overline{AB}$, $\overline{CD}$, $\overline{GH}$, or $\overline{FE}$
5. **a.** $\overleftrightarrow{AB} \| \overleftrightarrow{CD}$
   b. E
   c. $\overline{AB}$, $\overline{AE}$, $\overline{BE}$, $\overline{EC}$, $\overline{ED}$, $\overline{CD}$, $\overline{AC}$, and $\overline{BD}$
   d. no
6. A grain of sand. However, something as large as the sun, such as a star, appears to be a point if you are far enough away.
7. **a.** The ends of three legs determine a unique plane that makes the table stable. If the fourth leg is not in the same plane, the table will rock.
   b. yes
8. yes
9. no: For example, think of raising point D in problem 1 above the plane. The figure is still four-sided but not in a plane.

4.2 RAYS AND ANGLES

1 The following geometric figure, which contains A and B and extends forever in each direction, is called a *line*.

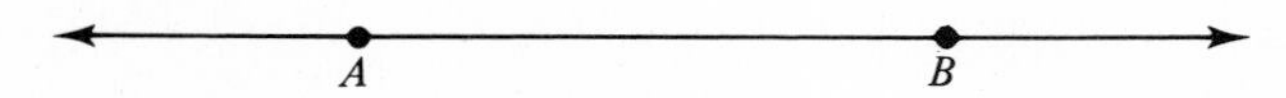

A line is considered to be a set of points and, as the picture demonstrates, is assumed to be straight. A line is labeled by any two points on it. It is thus possible to speak of "line AB" or use the symbol $\overleftrightarrow{AB}$.

The part of the line on one side of a point that includes the endpoint is called a *ray*. A ray is represented by noting its endpoint first along with another point on the ray with a single-headed arrow extending to the right above them. The name of the following rays would be $\overrightarrow{DC}$:

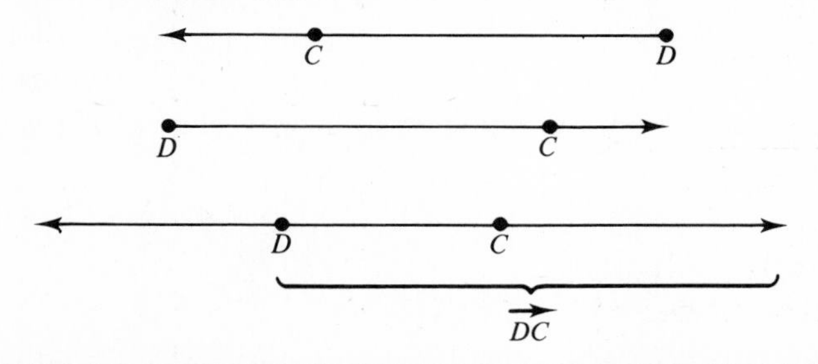

Q1 Name the rays. $\overrightarrow{KL}, \overrightarrow{CD}, \overrightarrow{ED}$ ____________

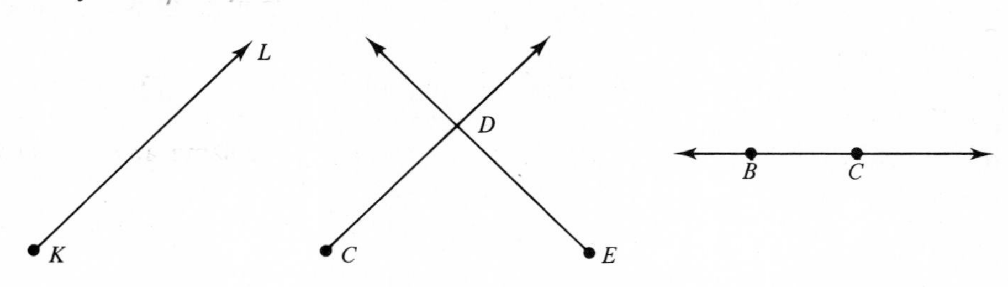

\# \# \# • \# \# \# • \# \# \# • \# \# \# • \# \# \# • \# \# \# • \# \# \# • \# \# \# • \# \# \#

A1 $\overrightarrow{KL}$, $\overrightarrow{CD}$, $\overrightarrow{ED}$, $\overrightarrow{BC}$, and $\overrightarrow{CB}$

2 A good representation of a ray is a flashlight beam or searchlight beam. However, a ray has properties that these representations do not share. A ray has no thickness, extends infinitely in one direction, and has a point (with no dimensions) as its endpoint.

The same ray can be named in several ways. For example, in the figure below $\overrightarrow{BC}$ and $\overrightarrow{BD}$ represent the same ray. We therefore say that $\overrightarrow{BC} = \overrightarrow{BD}$, because they contain the same endpoint and extend along the same line in the same direction.

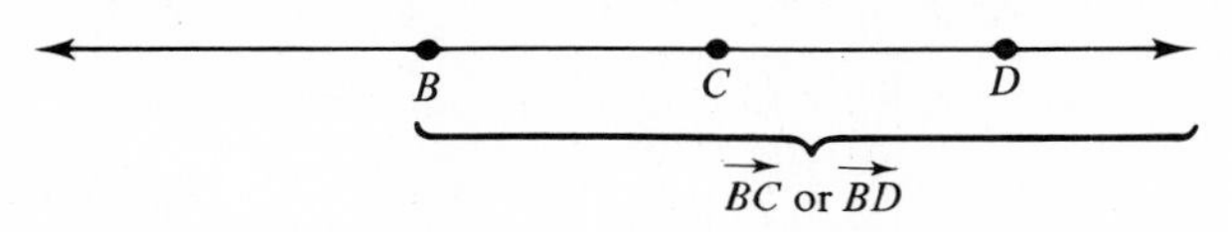

Q2 **a.** List all the rays on the line. ____________

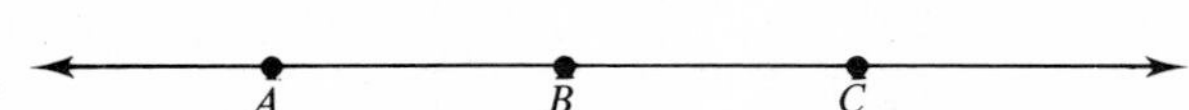

b. Which of the rays in your answer to part **a** are equal? ____________

\# \# \# • \# \# \# • \# \# \# • \# \# \# • \# \# \# • \# \# \# • \# \# \# • \# \# \# • \# \# \#

A2 **a.** $\overrightarrow{AB}$, $\overrightarrow{AC}$, $\overrightarrow{BC}$, $\overrightarrow{CB}$, $\overrightarrow{CA}$, and $\overrightarrow{BA}$ (mathematicians have agreed *not* to use $\overleftarrow{BC}$ to mean the same as $\overrightarrow{CB}$)

b. $\overrightarrow{AB} = \overrightarrow{AC}$ and $\overrightarrow{CB} = \overrightarrow{CA}$

Q3 List six different rays on the figure shown.

____

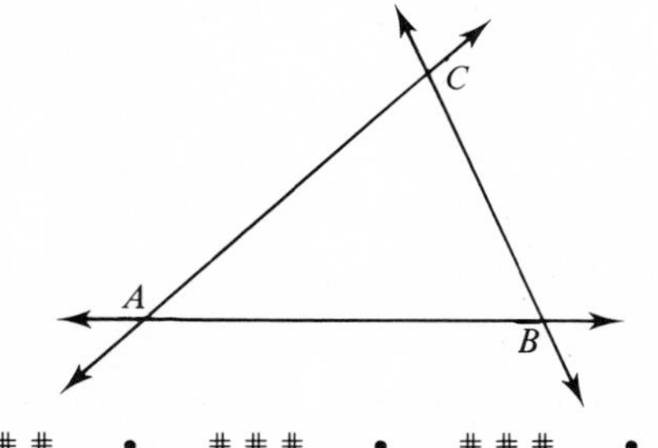

\# \# \# • \# \# \# • \# \# \# • \# \# \# • \# \# \# • \# \# \# • \# \# \# • \# \# \# • \# \# \#

A3 $\overrightarrow{AB}$, $\overrightarrow{AC}$, $\overrightarrow{CB}$, $\overrightarrow{CA}$, $\overrightarrow{BA}$, and $\overrightarrow{BC}$

3 An *angle* consists of two rays with a common endpoint. The common endpoint is called the *vertex* of the angle. The rays are the *sides* of the angle. A representation of an angle is shown. The two rays are $\overrightarrow{BA}$ and $\overrightarrow{BC}$, with the vertex being the point B.

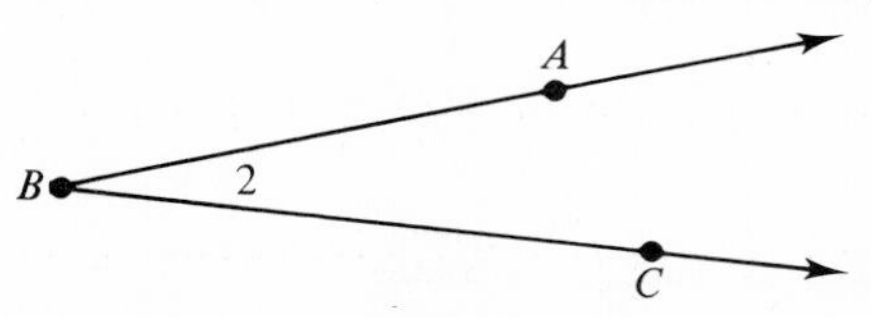

An angle can most easily be named by its vertex or by a number written at the vertex between the sides. The name of the angle is preceded by "$\measuredangle$." Thus, two names for the angle above are $\measuredangle B$ and $\measuredangle 2$.

However, if there is more than one angle at a vertex, a more complicated scheme is used. Another name for the angle shown is $\measuredangle ABC$, where A is on one ray, B (the middle letter) is the vertex, and C is on the other ray. Another way of naming the same angle is $\measuredangle CBA$. However, $\measuredangle ACB$ would not be correct, because C is not the vertex.

Q4 Use three letters to name the three angles in the figure.

$\measuredangle 1 =$ ~~$\measuredangle CAB$~~

$\measuredangle 2 =$ ~~$\measuredangle ABC$~~

$\measuredangle 3 =$ ~~$\measuredangle ACB$~~

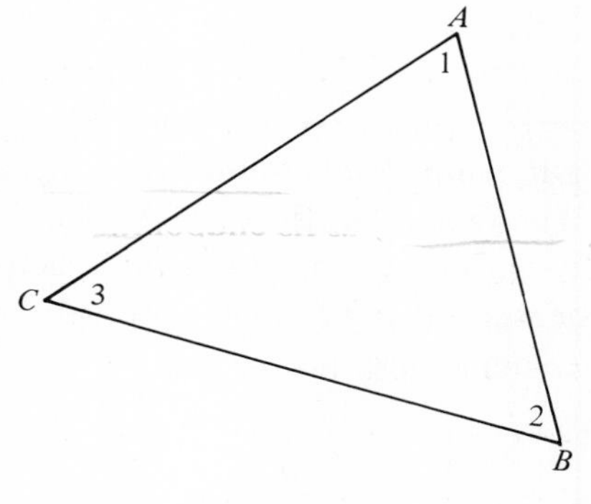

\# \# \# • \# \# \# • \# \# \# • \# \# \# • \# \# \# • \# \# \# • \# \# \# • \# \# \# • \# \# \#

A4 $\measuredangle 1 = \measuredangle CAB$ or $\measuredangle BAC$
$\measuredangle 2 = \measuredangle ABC$ or $\measuredangle CBA$
$\measuredangle 3 = \measuredangle ACB$ or $\measuredangle BCA$

Q5 Name the angles in the following figure by using the easiest method (fewest letters) and the points shown.

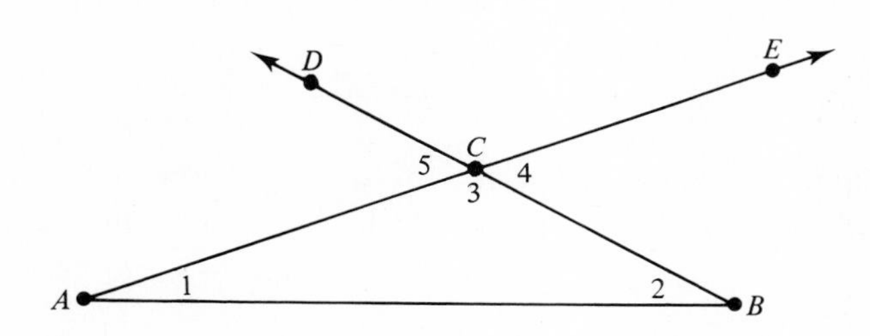

$\measuredangle 1 =$ $\measuredangle A$ $\measuredangle 4 =$ $\measuredangle ECB$

$\measuredangle 2 =$ $\measuredangle B$ $\measuredangle 5 =$ $\measuredangle ACD$

$\measuredangle 3 =$ $\measuredangle ACB$

\# \# \# • \# \# \# • \# \# \# • \# \# \# • \# \# \# • \# \# \# • \# \# \# • \# \# \# • \# \# \#

A5 $\measuredangle 1 = \measuredangle A$ $\measuredangle 4 = \measuredangle ECB$
$\measuredangle 2 = \measuredangle B$ $\measuredangle 5 = \measuredangle DCA$
$\measuredangle 3 = \measuredangle ACB$

Of course, each of the representations could be completely reversed and it represents the same angle; that is, $\measuredangle ACB = \measuredangle BCA$.

Q6 On the figure to the right:

a. Name two angles at E.

______ ______

b. Name two angles at A.

______ ______

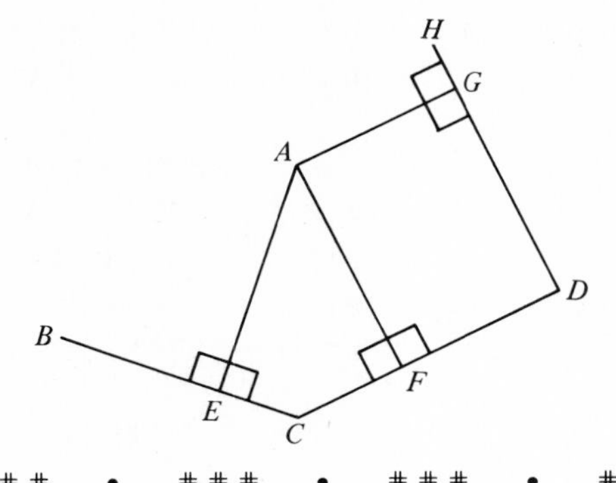

\# \# \# • \# \# \# • \# \# \# • \# \# \# • \# \# \# • \# \# \# • \# \# \# • \# \# \# • \# \# \#

A6 **a.** $\measuredangle BEA$ and $\measuredangle AEC$ **b.** $\measuredangle EAF$ and $\measuredangle FAG$

4 The *measurement of an angle* is a number assigned to the amount of rotation required to swing one ray of the angle to the other ray. The unit of measure will be the *degree*, which is based upon the division of the rotation of 1 complete revolution into 360 parts. Each of these parts is a degree. You should become familiar with the measure of angles so you can tell the approximate size of an angle by inspection. Some frequently used angles are shown.

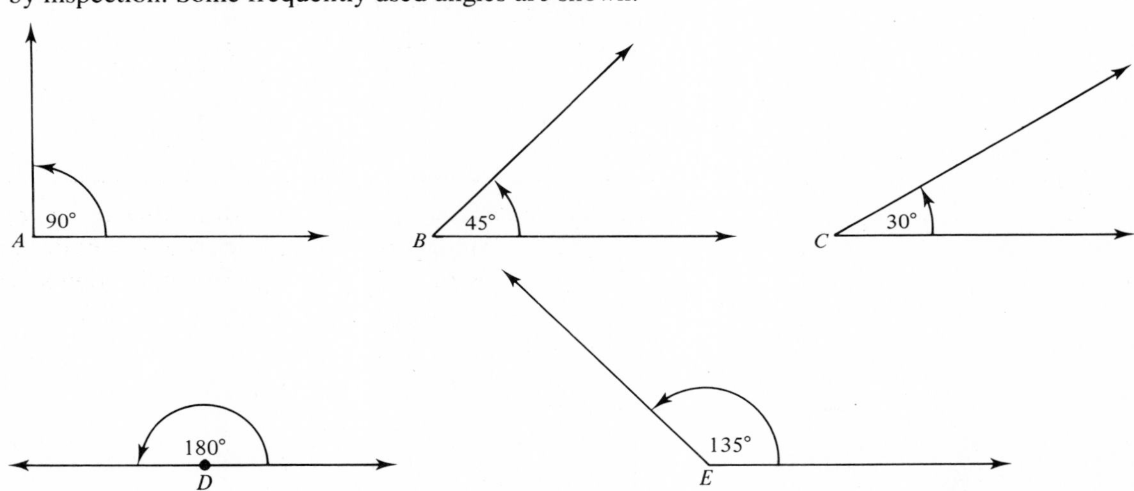

The size of an angle does not depend upon the lengths of the sides of the angle shown since the rays of an angle extend indefinitely. Therefore, angle 1 is considered to be larger than angle 2 in the following figures.

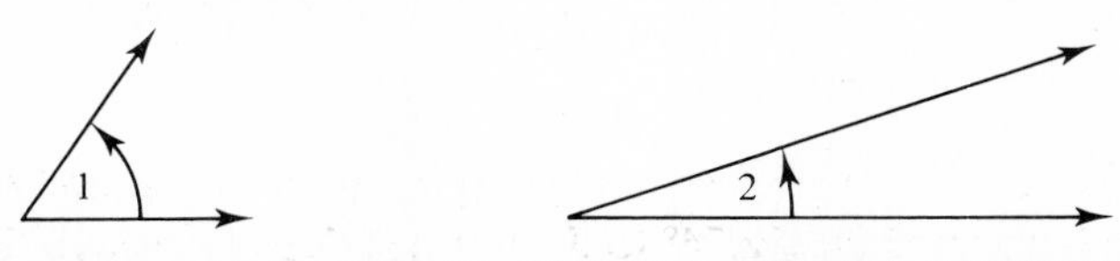

Q7 Compare the angles shown to the examples in Frame 4 to estimate the approximate measure of the angle:

a.

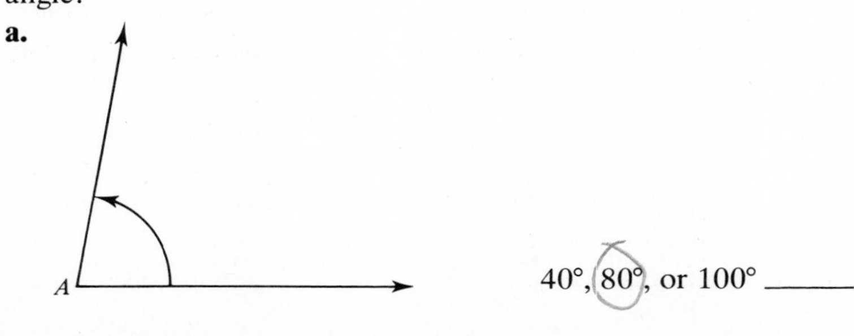

40°, 80°, or 100° ______

b.

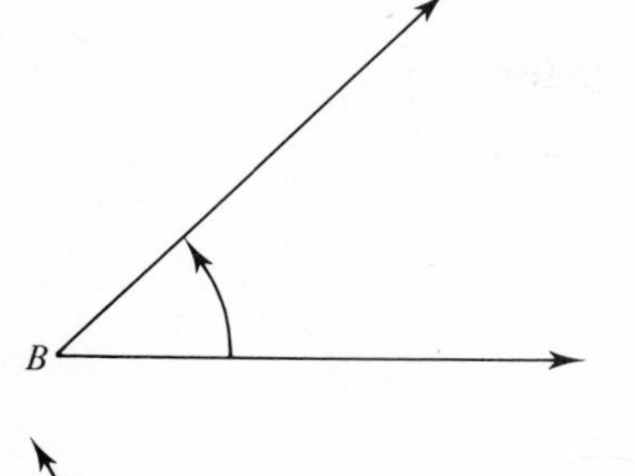

20°, 45°, or 60° ______

c.

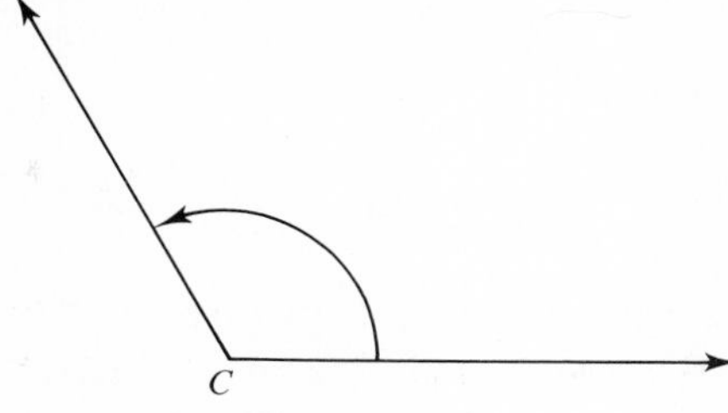

80°, 160°, or 120° ______

d.

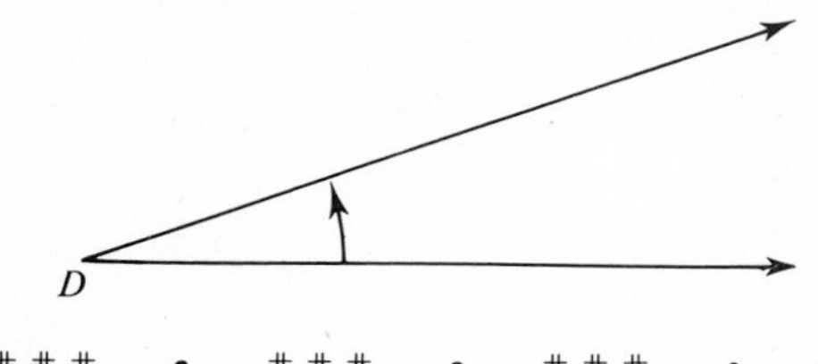

20°, 45°, or 60° ______

\# \# \# • \# \# \# • \# \# \# • \# \# \# • \# \# \# • \# \# \# • \# \# \# • \# \# \# • \# \# \#

A7 **a.** 80° **b.** 45° **c.** 120° **d.** 20°

---

5 The instrument used to measure angles is called a *protractor*. A protractor is needed to complete this section. The protractor is shown as it would be used to measure an angle. The base of the protractor is placed along one ray of the angle with the center mark on the protractor at the vertex. The other

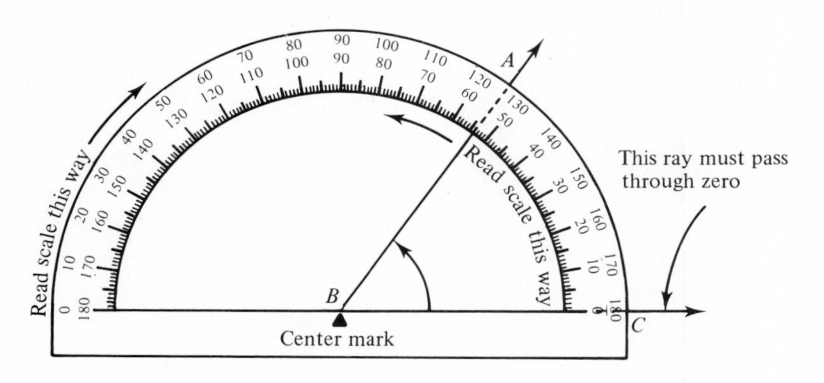

ray of the angle will pass under the scale of the protractor and the degree measure is read from the scale. The measure of the angle shown is approximately 54° to the nearest degree. Notice that each scale mark has two numbers beside it. The numbers larger than 90 are used if an angle over 90° is being measured. Write $m \measuredangle ABC = 54°$ (read "the measure of angle *ABC* is equal to 54 degrees").

Q8 Use the protractor to read the measure of each of the following angles, as shown in the figure at the top of the next page.

a. $m\measuredangle SBG =$ ______ **b.** $m\measuredangle SBA =$ ______

c. $m\measuredangle SBC =$ ______ **d.** $m\measuredangle SBD =$ ______

e. $m\measuredangle SBE =$ ______ **f.** $m\measuredangle SBF =$ ______

g. $m\measuredangle SBR =$ ______

h. $m\measuredangle RBE =$ ______

i. $m\measuredangle RBC =$ ______

j. $m\measuredangle RBA =$ ______

k. $m\measuredangle EBD =$ ______

l. $m\measuredangle GBA =$ ______

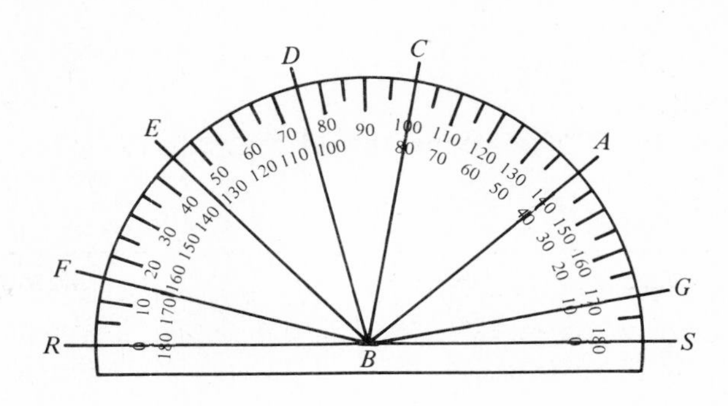

\# \# \# • \# \# \# • \# \# \# • \# \# \# • \# \# \# • \# \# \# • \# \# \# • \# \# \# • \# \# \#

A8 **a.** 10° **b.** 40° **c.** 80° **d.** 105°: You must notice that the angle is larger than 90° and read the appropriate scale.
e. 135° **f.** 165° **g.** 180° **h.** 45°: You must notice that the angle is less than 90° and read the appropriate scale.
i. 100° **j.** 140° **k.** 30°: $75° - 45° = 30°$ **l.** 30°: $40° - 10° = 30°$

Q9 Use a protractor to measure the angles (extend sides of angle if necessary):

a.

C

B A

b.

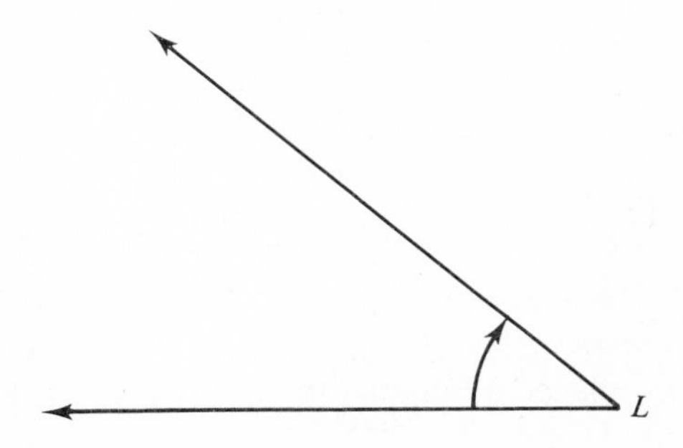

c. K

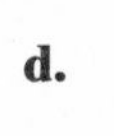

d.

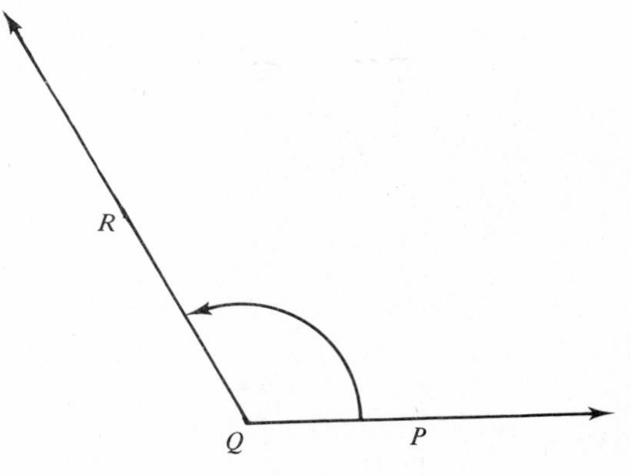

e.

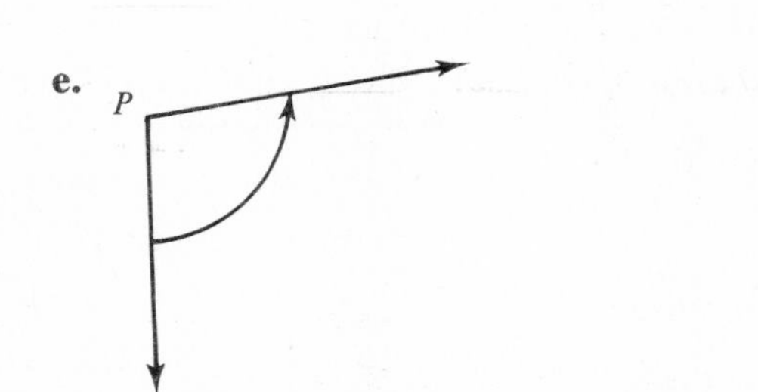

f.

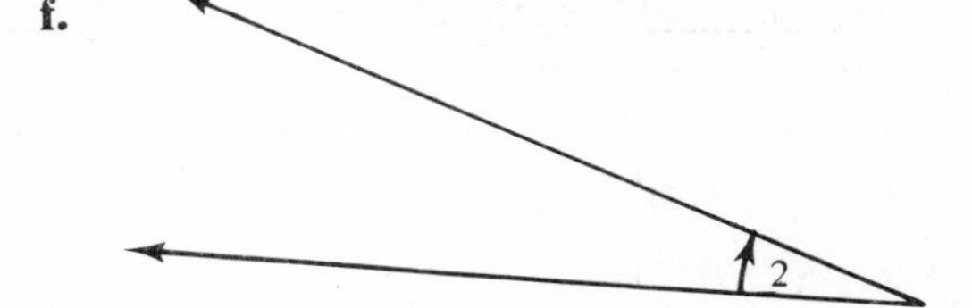

\# \# \# • \# \# \# • \# \# \# • \# \# \# • \# \# \# • \# \# \# • \# \# \# • \# \# \# • \# \# \#

A9 **a.** $m\measuredangle ABC = 67°$ **b.** $m\measuredangle L = 40°$
c. $m\measuredangle K = 40°$ **d.** $m\measuredangle PQR = 118°$
e. $m\measuredangle P = 98°$ (notice that the sides must be extended)
f. $m\measuredangle 2 = 20°$

6 The angle of the cutting surfaces of a chisel is called the *bevel* of a chisel. The correct bevel of a chisel varies from 55° to 90° depending upon the material that is being cut. For general all-round use, 60° to 70° is best. In the illustration two chisels are shown along with the way of measuring the angle. Notice in Figure 2 the lower edge is held flat against the work.

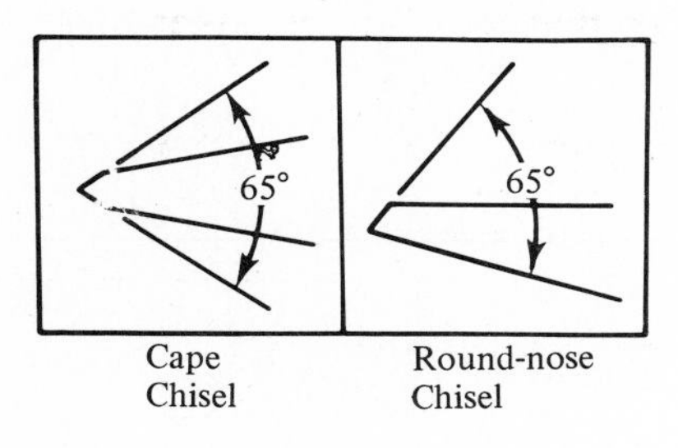

Figure 1

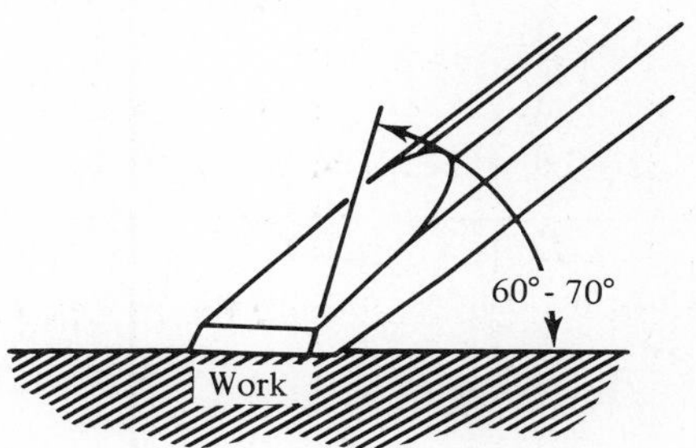

Figure 2

Q10 Measure the bevel of each chisel below:

a.

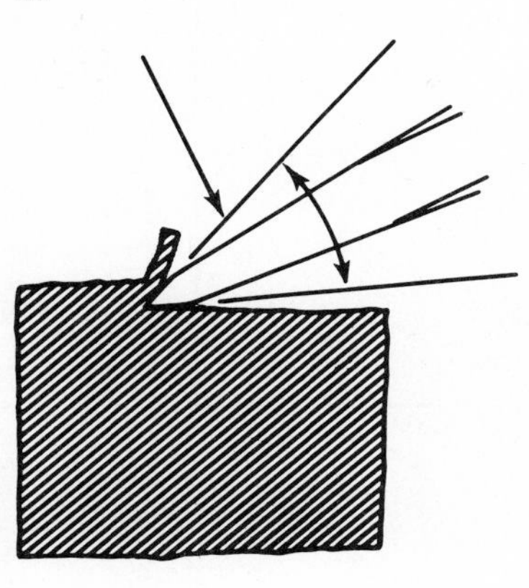

b.

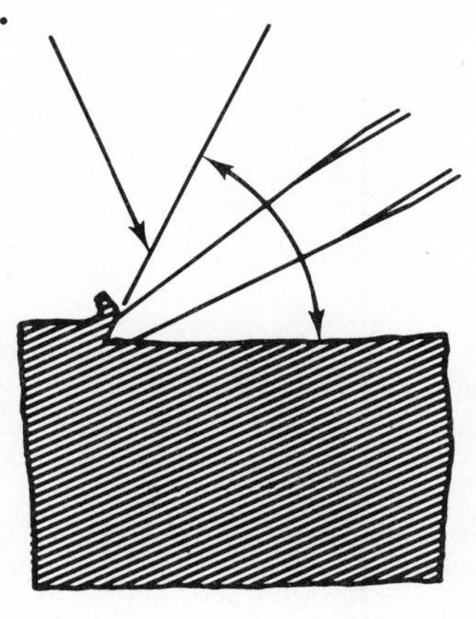

c.

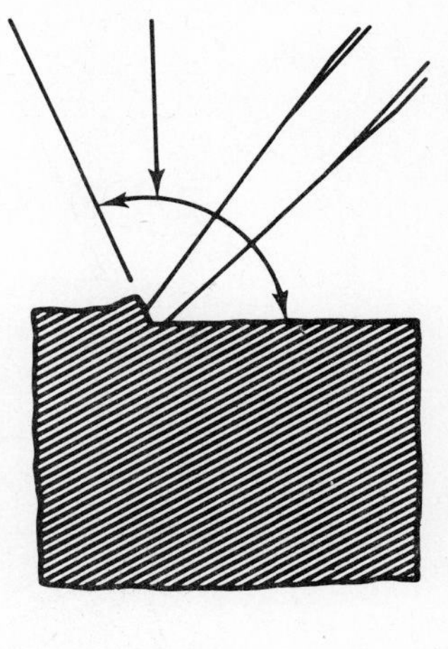

\# # # • # # # • # # # • # # # • # # # • # # # • # # # • # # # • # # #

A10 **a.** 42° **b.** 62° **c.** 114°

Q11 Which chisel in Q10 is beveled correctly for general all-purpose use? __________

\# # # • # # # • # # # • # # # • # # # • # # # • # # # • # # # • # # #

A11 **b**

7 A *truss* is an engineering structure composed of members joined at their ends to form a rigid frame. The rigidity of the triangle makes it the basic element of a truss. You can see trusses in roof structures, bridges, cranes, and towers.

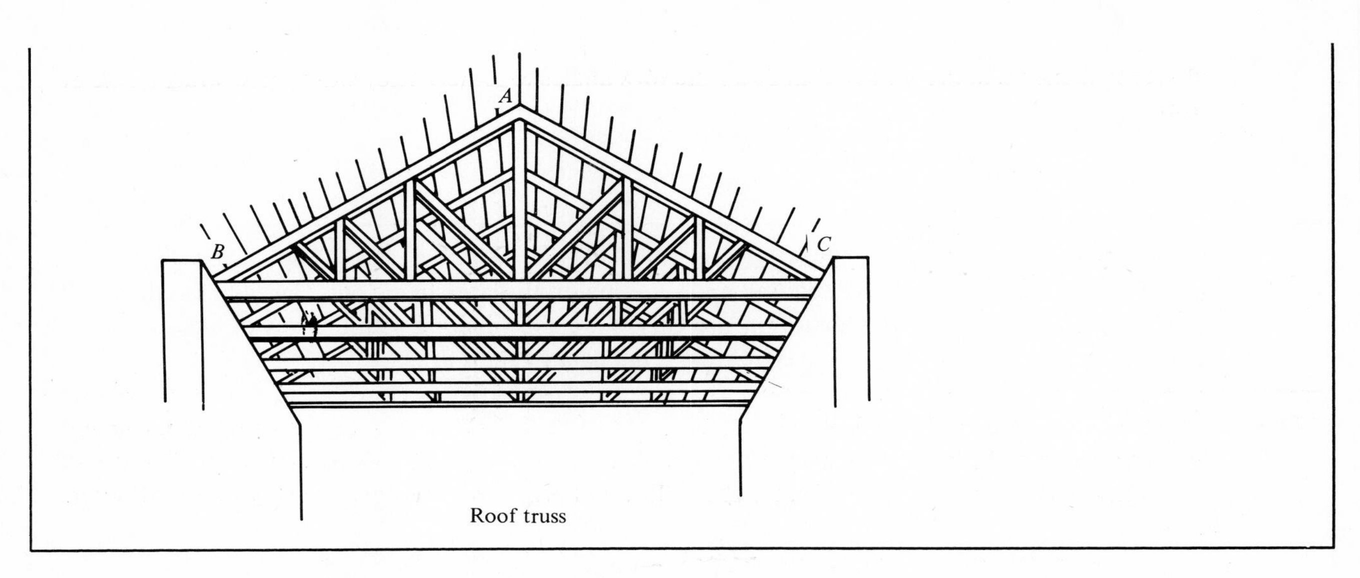

Roof truss

Q12 Measure the angles in the truss shown in Frame 7.

a. $m\measuredangle A =$ ______ **b.** $m\measuredangle B =$ ______ **c.** $m\measuredangle C =$ ______

\# \# \# • \# \# \# • \# \# \# • \# \# \# • \# \# \# • \# \# \# • \# \# \# • \# \# \# • \# \# \#

A12 **a.** 120° **b.** 30° **c.** 30°

Q13 Properties of geometric figures can be discovered by measuring angles (1) Add the measures of the three angles in each triangle shown.

a.

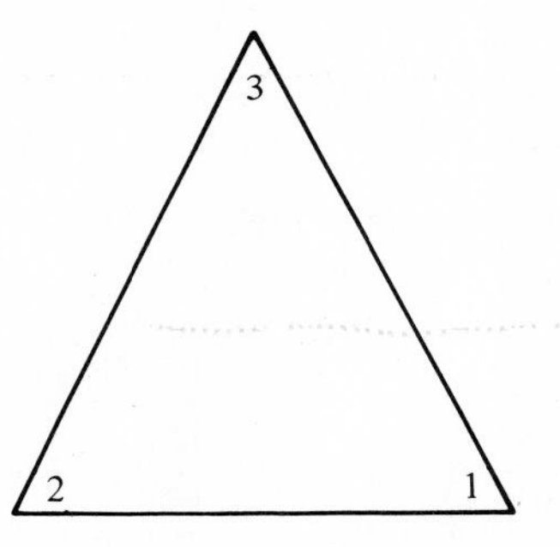

b.

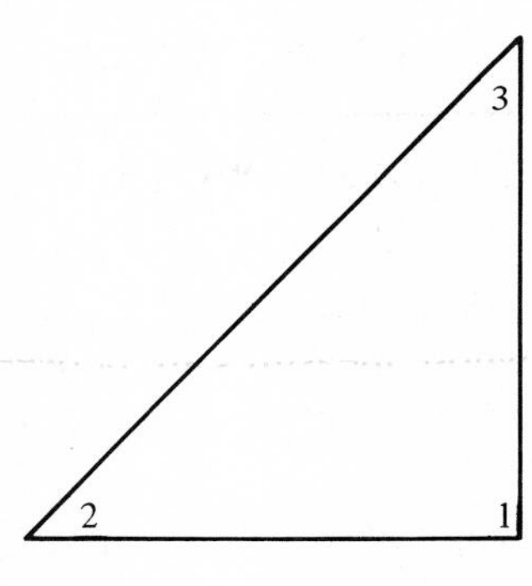

c.

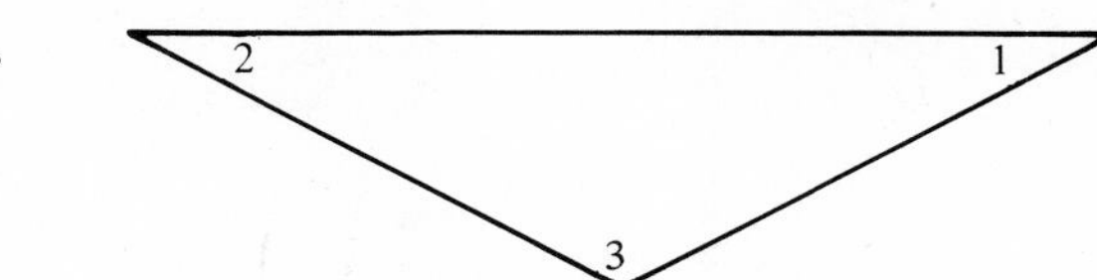

(2) What do you discover? ______________________________

\# \# \# • \# \# \# • \# \# \# • \# \# \# • \# \# \# • \# \# \# • \# \# \# • \# \# \# • \# \# \#

A13 (1) **a.** $63° + 64° + 53° = 180°$
b. $90° + 47° + 43° = 180°$
c. $28° + 124° + 28° = 180°$
(2) It looks as though the sum of the measures of the angles of a triangle is 180°. This is, in fact, true.

8 The results of Q13 may be summarized by saying that the sum of the measures of the angles of a triangle is 180°. This result may be obtained in another way. Cut out a triangle and tear off the three angles as is shown in the illustration. Place the torn angles together. They will form a straight line or 180°.

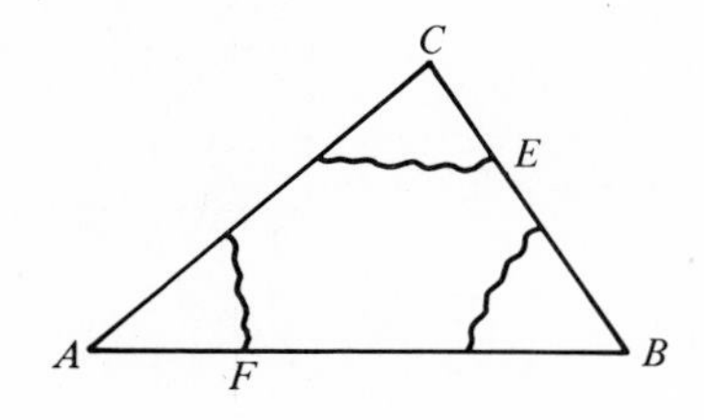

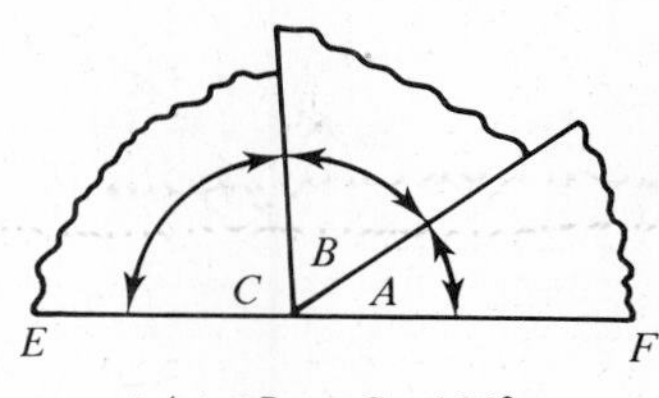

$\angle A + \angle B + \angle C = 180°$

Q14 If two angles of a triangle are known, you can use the fact that the sum of all the angles of a triangle is 180° to obtain the third angle. Suppose angle A and B are on the shore of a river and C is on an island in the river. If $m\measuredangle A = 90°$ and $m\measuredangle B = 35°$, find $m\measuredangle C$ without measuring with a protractor.

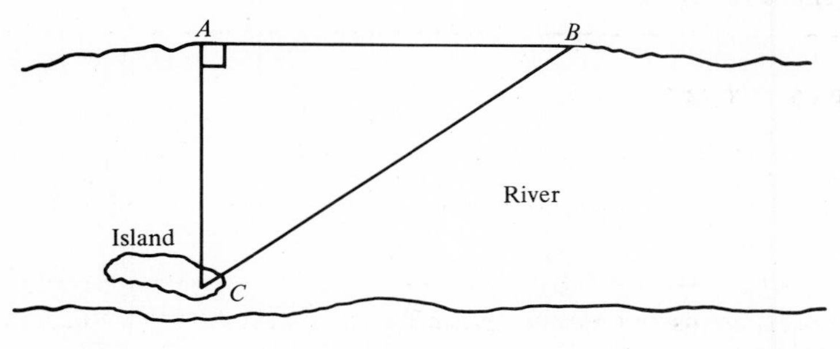

\# \# \# • \# \# \# • \# \# \# • \# \# \# • \# \# \# • \# \# \# • \# \# \# • \# \# \# • \# \# \#

A14 $m\measuredangle C = 55°$: $90° + 35° = 125°$, $180° - 125° = 55°$

9 Angles are classified according to the size of their measures.

1. Angles whose measures are between 0° and 90° are called *acute* angles.
2. Angles whose measures are 90° are called *right* angles.
3. Angles whose measures are between 90° and 180° are called *obtuse* angles.
4. Angles whose measures are 180° are called *straight* angles.

Examples:

1. $\measuredangle BOD$ is acute.
2. $\measuredangle BOE$ is right.
3. $\measuredangle BOC$ is obtuse.
4. $\measuredangle BOA$ is straight.

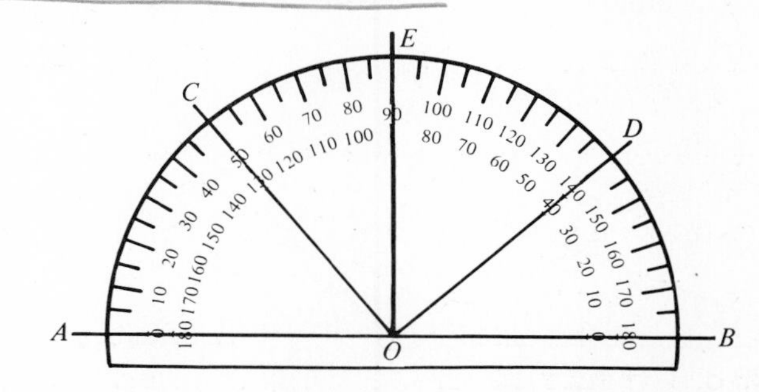

Q15 Write acute, right, or obtuse under each of the following angles:

a.

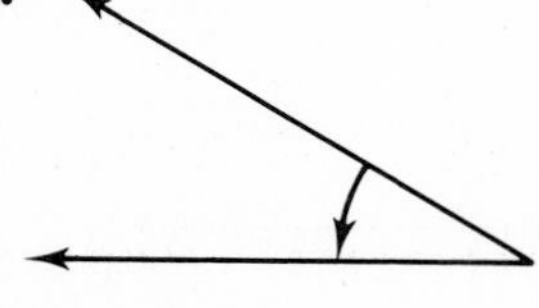

b.

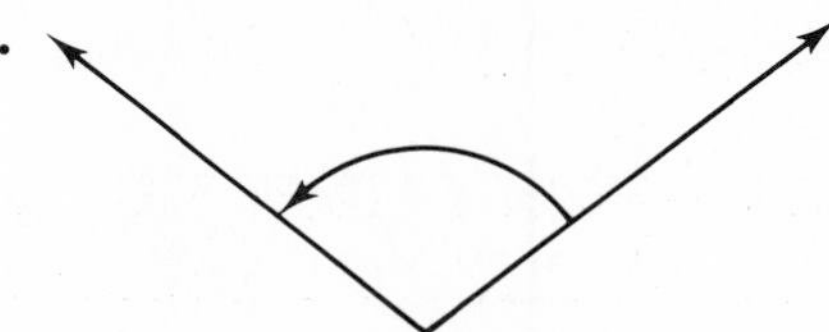

c.

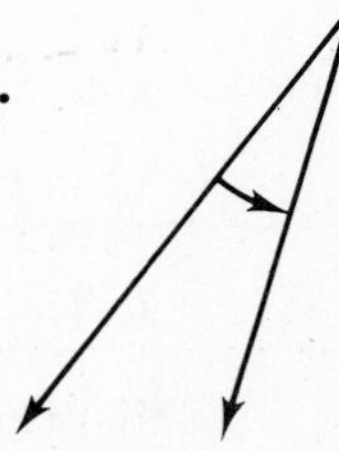

____________ ____________ ____________

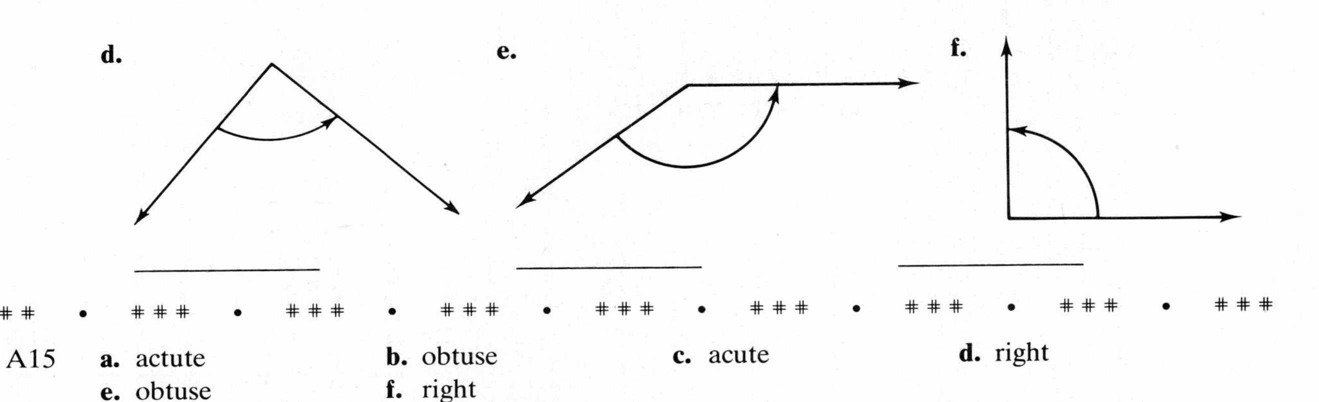

\# \# \# • \# \# \# • \# \# \# • \# \# \# • \# \# \# • \# \# \# • \# \# \# • \# \# \# • \# \# \#

A15 **a.** actute **b.** obtuse **c.** acute **d.** right
e. obtuse **f.** right

Q16 Below is the view from above of a machined part.

a. Name three obtuse angles. ____ ____ ____

b. Name an acute angle. ____

c. Name two right angles. ____ ____

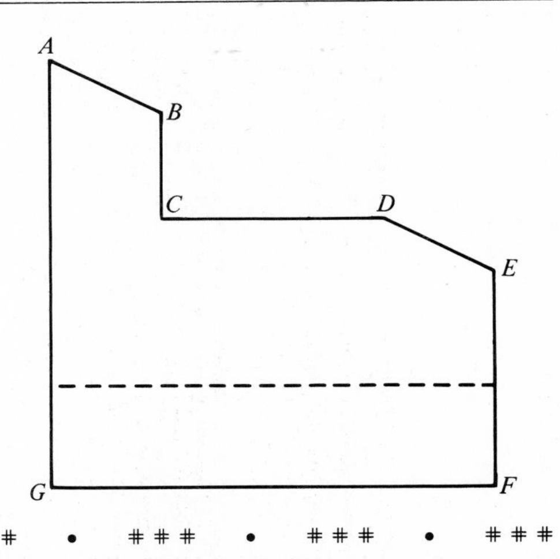

\# \# \# • \# \# \# • \# \# \# • \# \# \# • \# \# \# • \# \# \# • \# \# \# • \# \# \# • \# \# \#

A16 **a.** ∡*B*, ∡*D*, and ∡*E* **b.** ∡*A* **c.** ∡*G* and ∡*F*: ∡*C* forms a right angle outside the figure but not inside the figure.

10 Two angles whose measures added together result in a sum of 90° are said to be *complementary angles*. Each is a complement of the other. For example, if $m∡A = 20°$ and $m∡B = 70°$, angles *A* and *B* are complementary. We say that ∡*A* is the complement of ∡*B* and ∡*B* is the complement of ∡*A*.

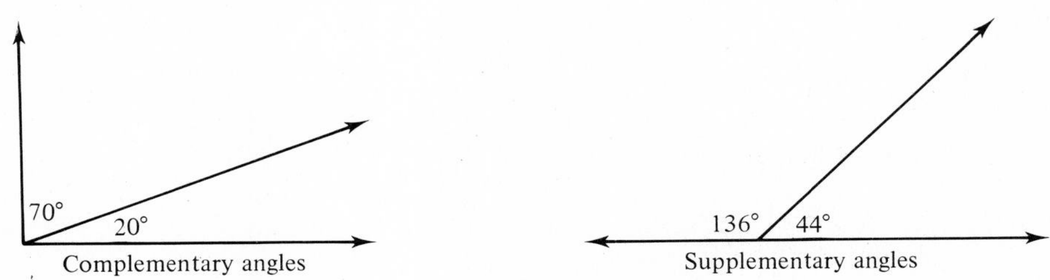

Complementary angles

Supplementary angles

Two angles whose measures added together result in a sum of 180° are said to be *supplementary angles*. Each is a supplement of the other. For example, if $m∡C = 136°$ and $m∡D = 44°$, angles *C* and *D* are supplementary. We say that ∡*C* is the supplement of ∡*D* and ∡*D* is the supplement of ∡*C*.

Q17 **a.** The supplement of an angle of 60° would have what measure? ____

b. The complement of an angle of 60° would have what measure? ____

\# \# \# • \# \# \# • \# \# \# • \# \# \# • \# \# \# • \# \# \# • \# \# \# • \# \# \# • \# \# \#

A17 **a.** 120° **b.** 30°

Q18 **a.** Which of the given angles is the complement of angle X? ________

b. Which of the given angles is the supplement of angle X? ________

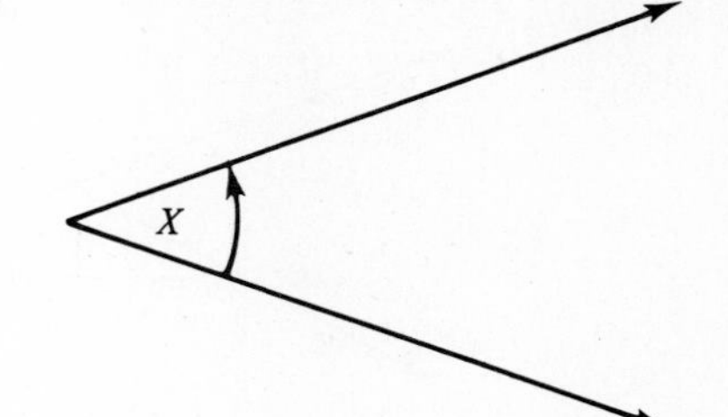

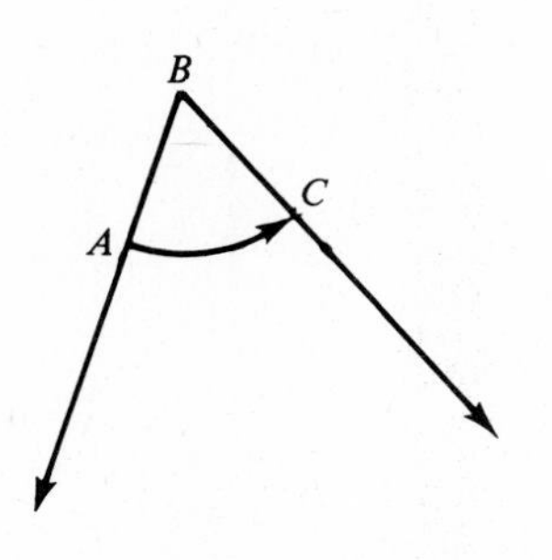

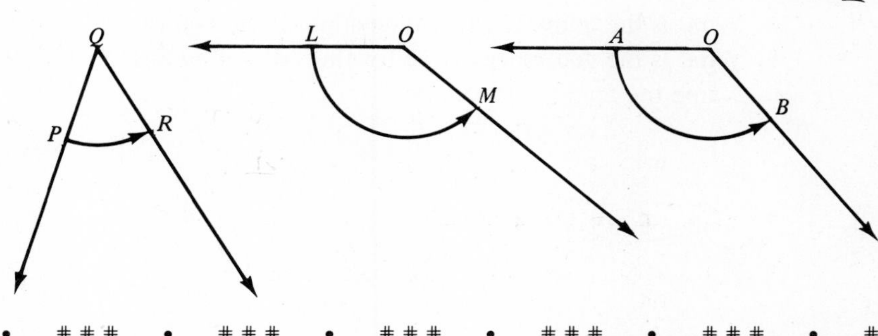

\# \# \# • \# \# \# • \# \# \# • \# \# \# • \# \# \# • \# \# \# • \# \# \# • \# \# \# • \# \# \#

A18 **a.** $\measuredangle PQR$ **b.** $\measuredangle LOM$

Q19 A machinist knows that $\measuredangle ABC$ and $\measuredangle DEF$ on the part he is making are supplementary. If $m\measuredangle ABC$ is 157°, what is the measure of $\measuredangle DEF$?

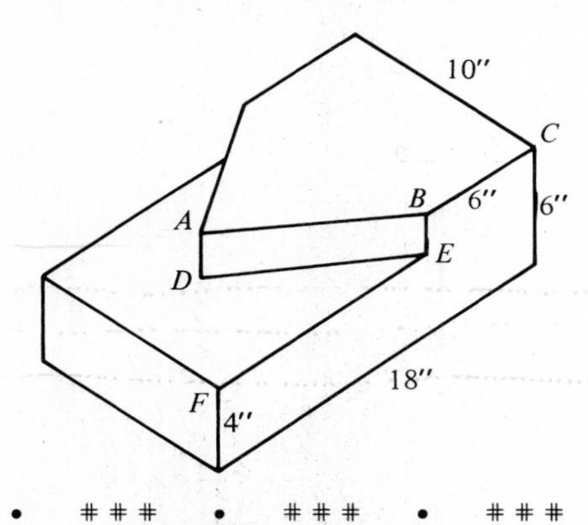

\# \# \# • \# \# \# • \# \# \# • \# \# \# • \# \# \# • \# \# \# • \# \# \# • \# \# \# • \# \# \#

A19 23°: $180° - 157° = 23°$

This completes the instruction for this section.

4.2 EXERCISE

1. Name two rays with endpoints at B.

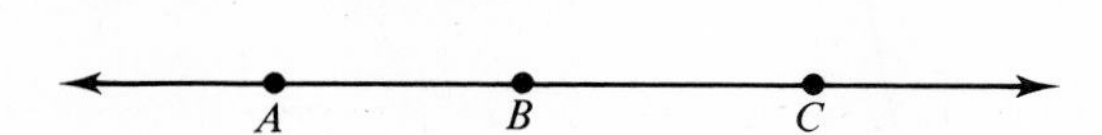

2. Give two other ways of naming the ray AC.

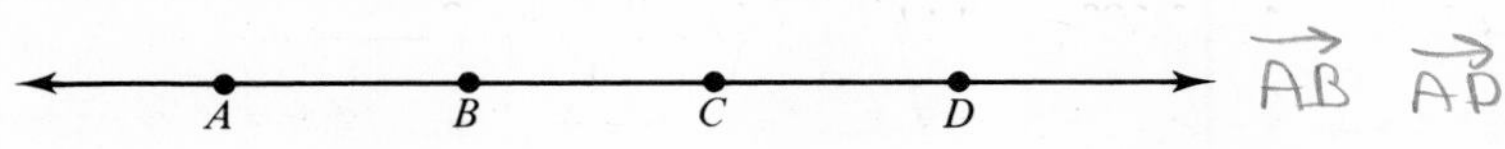

3. Name eight different rays by using the points shown on the intersecting lines.

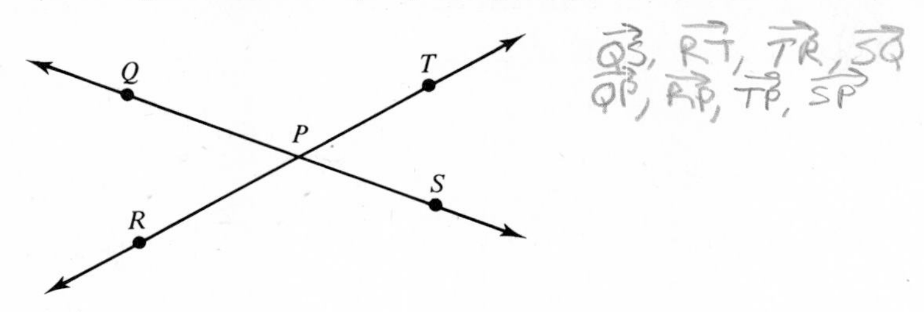

4. What is the name of the intersection of the two rays that make up an angle?

5. What is the geometric name for the sides of an angle?

6. Name the angle in three ways.

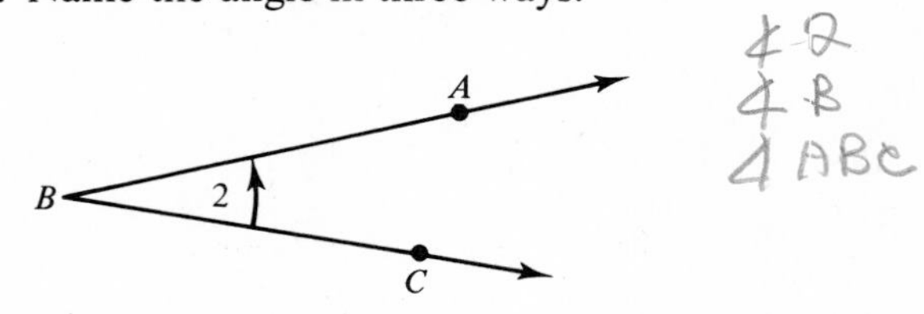

7. Name four angles in the figure.

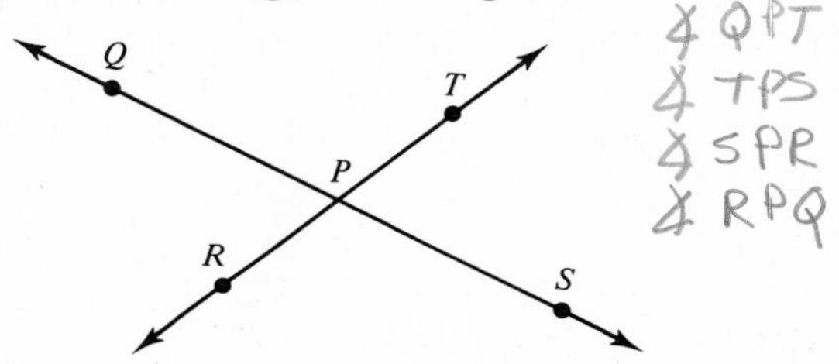

8. How many degrees are in one complete revolution?

9. Without measuring the angles, choose the most likely number of degrees in the angle:

a. **b.**

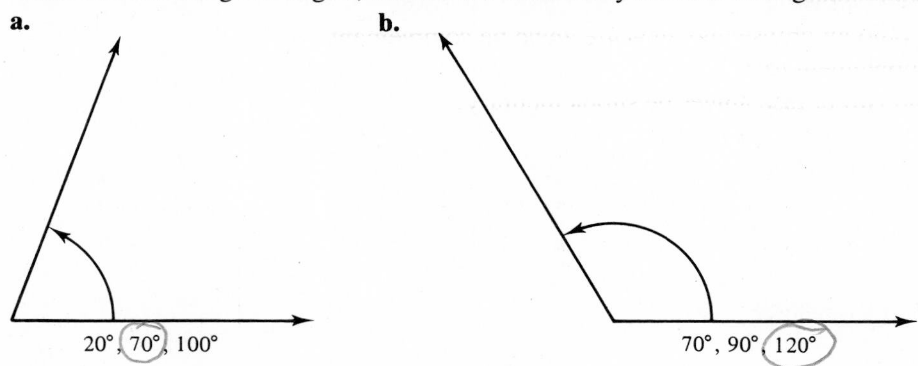

10. What is the name of the instrument used to measure angles?

11. Measure the angles below to the nearest degree:

a. **b.**

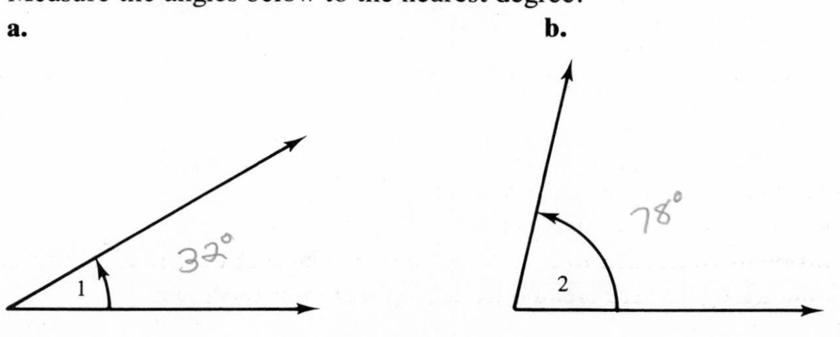

c. **d.**

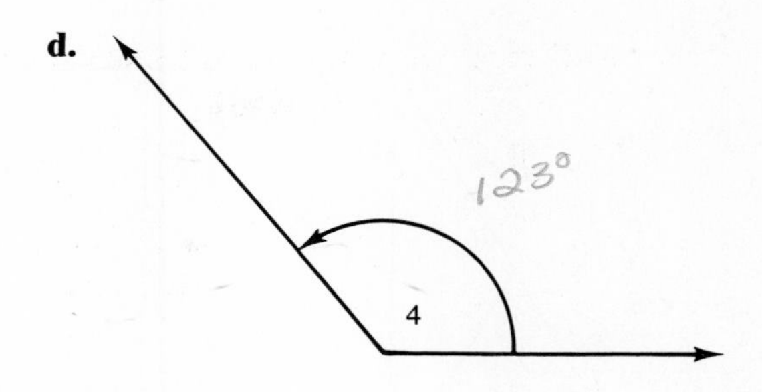

12. Consider the angles and answer the following questions:

a. **b.** **c.** **d.**

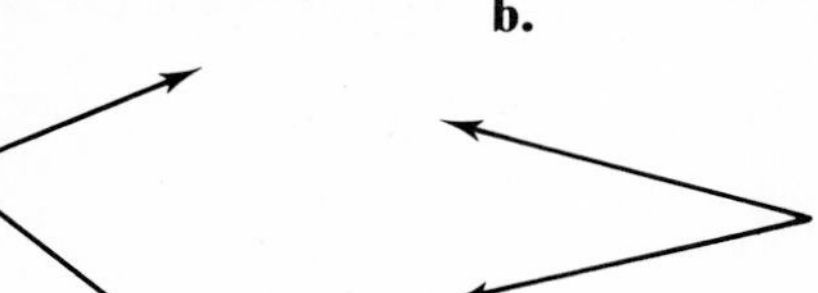

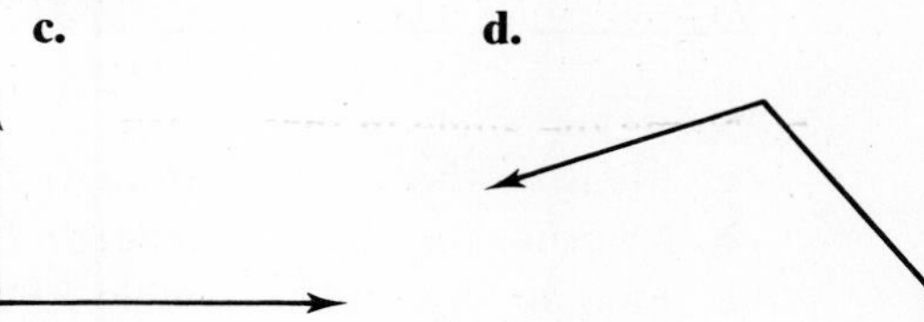

e. **f.** **g.** **h.**

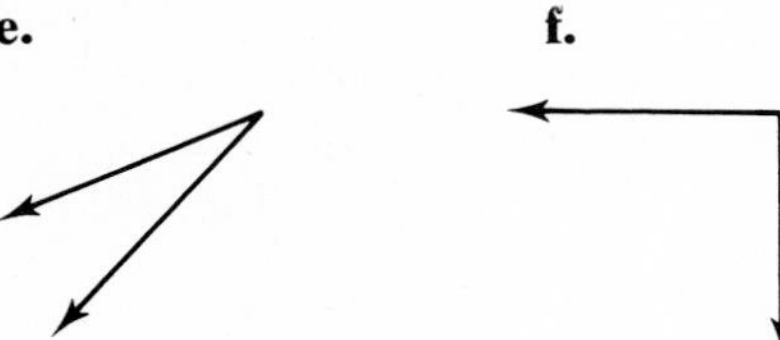

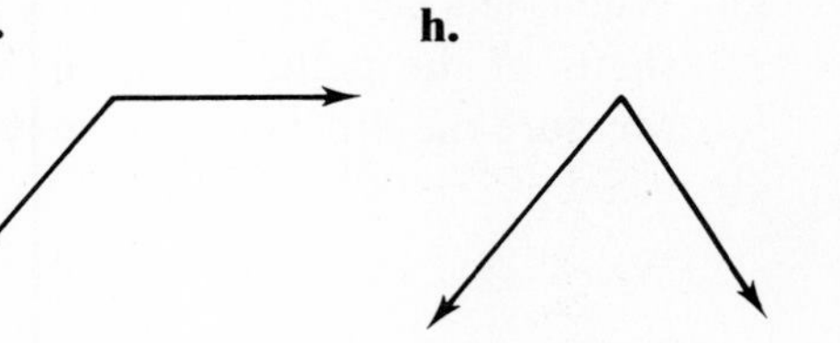

a. Name the letters of the acute angles. B, E, H, A
b. Name the letters of the obtuse angles. D, G
c. Name the letters of the right angles. F, C

13. What does it mean to say that two angles are complementary? = 90°

14. What does it mean to say that two angles are supplementary? = 180°

15. **a.** Could two acute angles be supplementary? NO
b. complementary? YES

16. **a.** Could an obtuse and an acute angle be complementary? NO
b. supplementary? YES

17. Could two obtuse angles be supplementary? NO

18. A bridge truss is shown below:

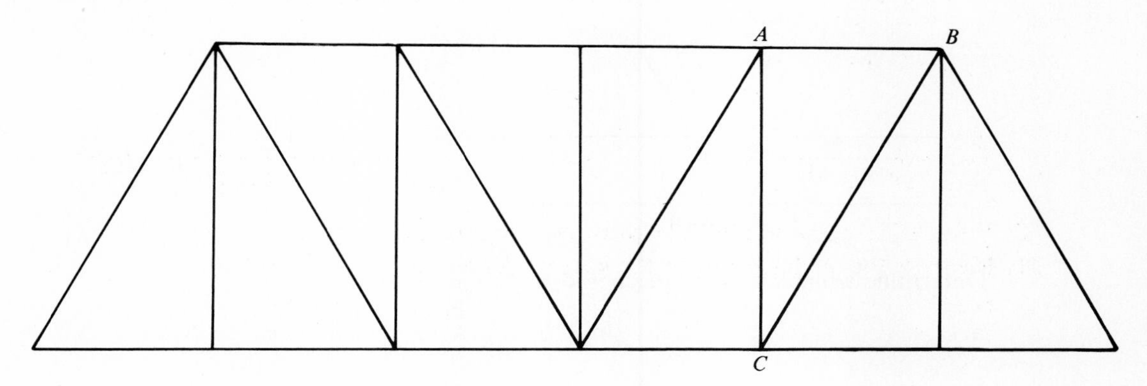

a. Measure the angles in triangle *ABC*. 90°, 60° 30°
b. Are there any complementary angles in triangle *ABC*? ∡ABC, ∡ACB
c. Are there any supplementary angles in triangle *ABC*? NO

19. The bridge in problem 18 needed to be strengthened because of the increase in traffic over the bridge. Additional triangles were created by adding members as shown.

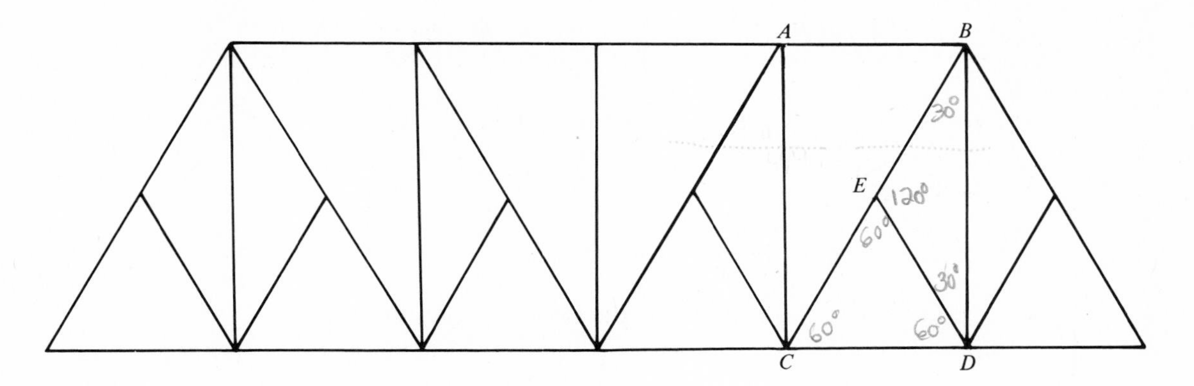

a. Measure the angles in triangle *BED*.

b. Measure the angles in triangle *EDC*.

c. Find an angle from triangle *EBD* and one from triangle *EDC* that are supplementary.

20. When belt layouts are arranged, horizontal belts are best. However, if the line between the shafts of the pulleys is at an angle with the horizontal, that angle should not exceed 45°. Measure the drive arrangements below to see if any exceed that limit.

a.

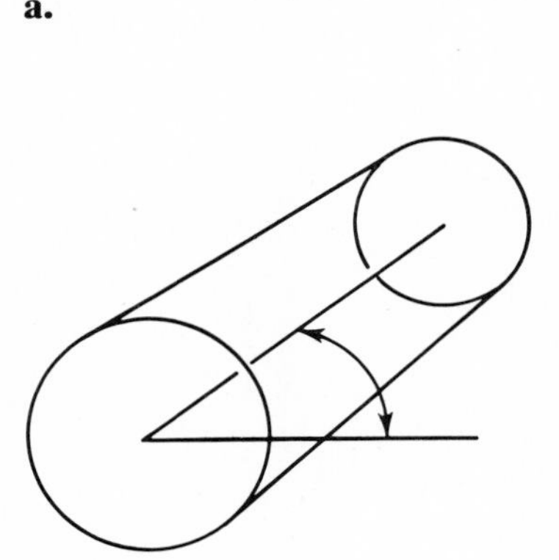

b.

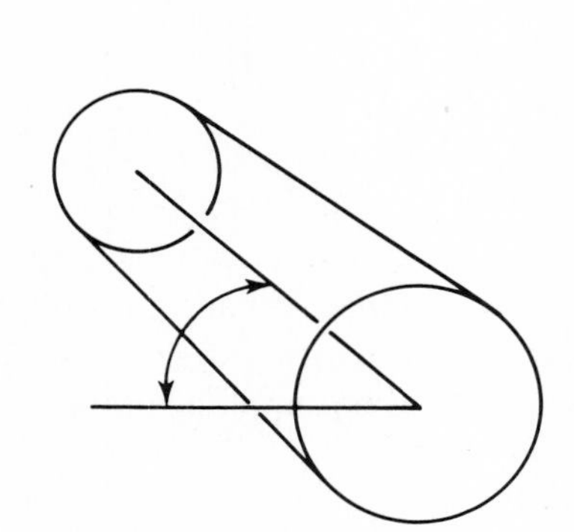

c.

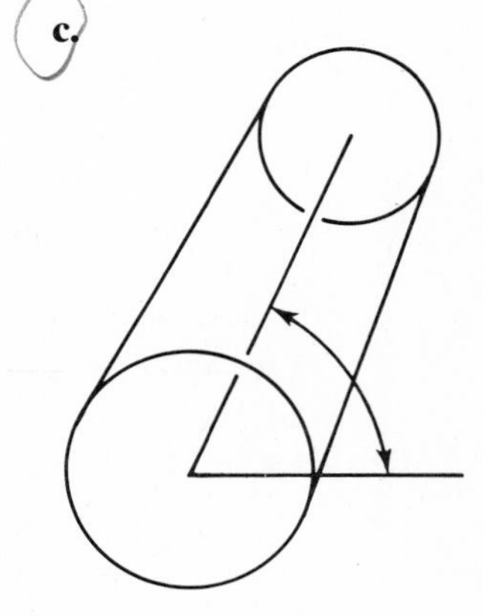

21. A welded single-vee-groove joint is strongest when the work is beveled as below with a 45° minimum measure of angle 1. (This means the angle cannot be smaller than 45°.)

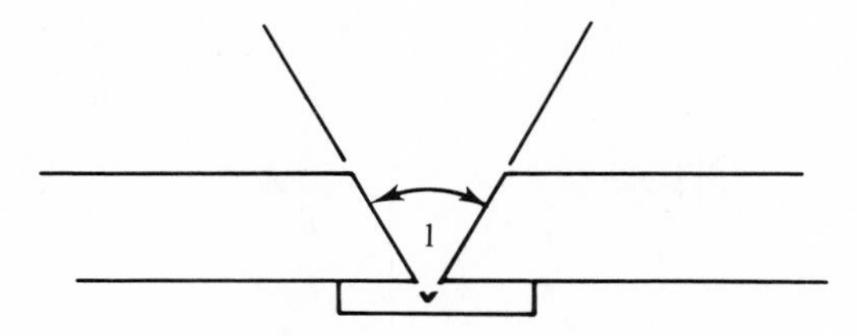

Determine which of the angles below meet the standard.

a.

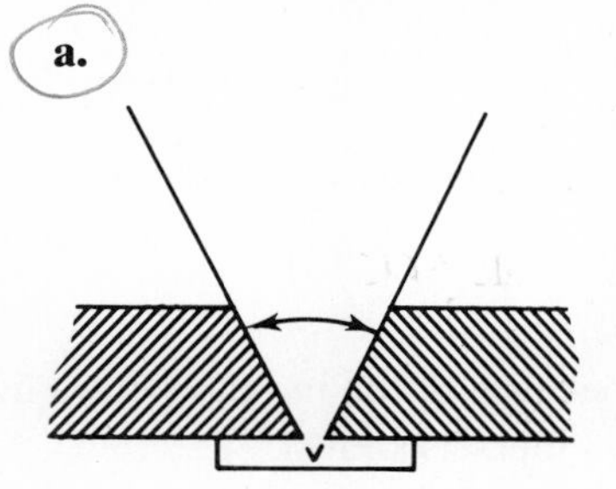

b.

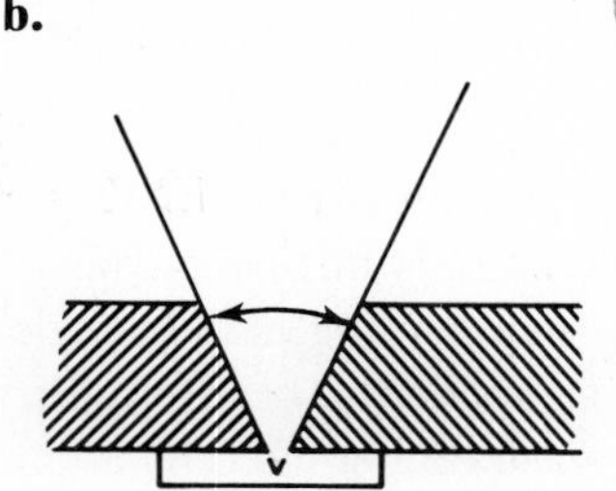

c.

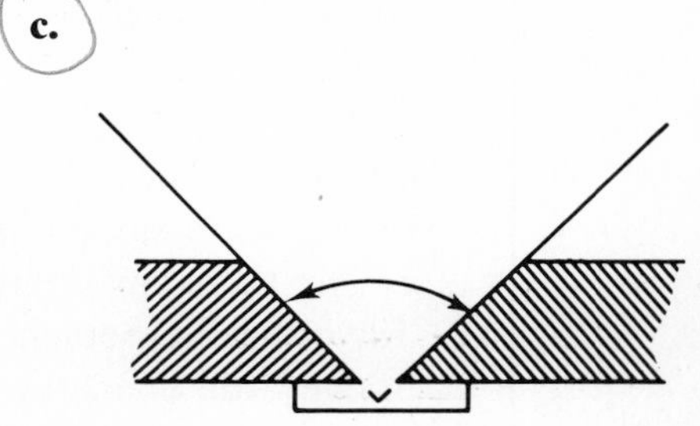

22. In welding a double-vee-groove, the angle 2 should be a minimum of 60°.

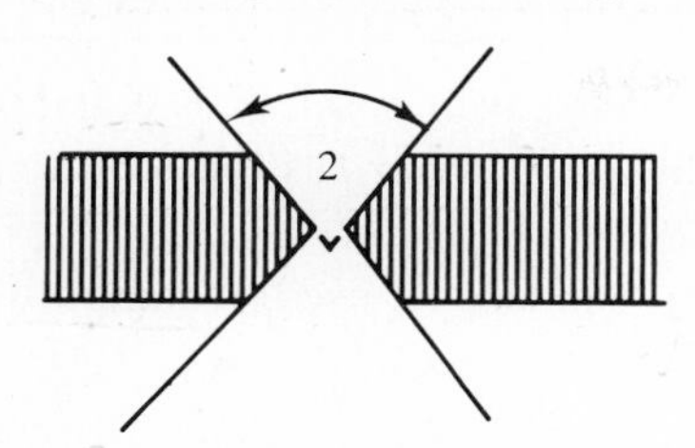

Determine which of the angles below meet the standard.

a.

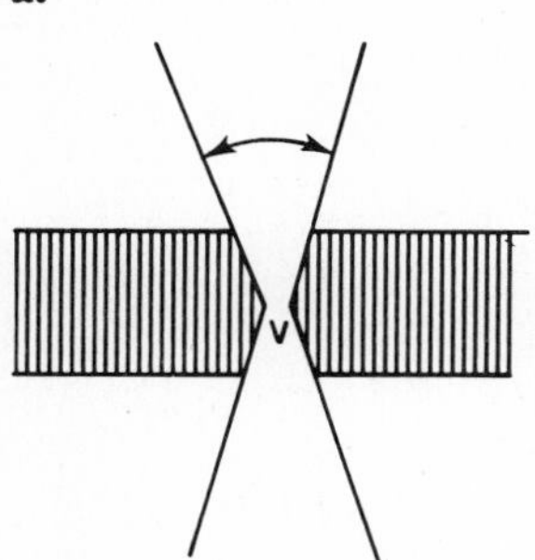

b.

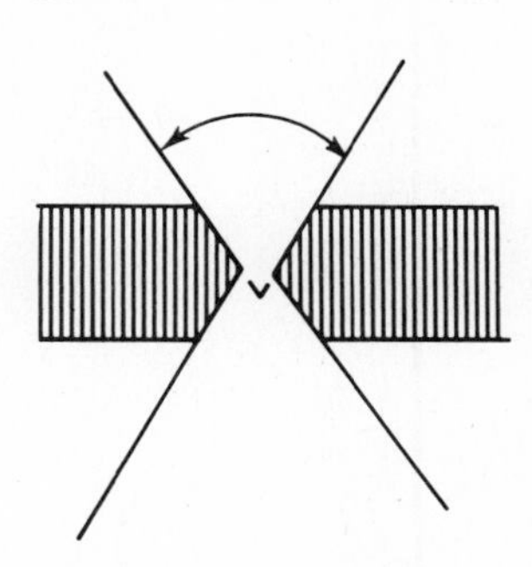

c.

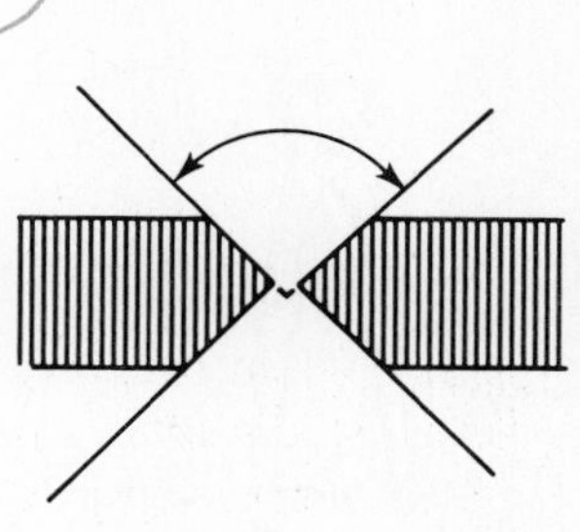

4.2 EXERCISE ANSWERS

1. $\overrightarrow{BC}$ and $\overrightarrow{BA}$
2. $\overrightarrow{AB}$ and $\overrightarrow{AD}$
3. $\overrightarrow{QS}$, $\overrightarrow{PS}$, $\overrightarrow{RT}$, $\overrightarrow{PT}$, $\overrightarrow{SQ}$, $\overrightarrow{PQ}$, $\overrightarrow{TR}$, and $\overrightarrow{PR}$
4. vertex
5. rays
6. $\measuredangle ABC$, $\measuredangle B$, $\measuredangle 2$, or $\measuredangle CBA$
7. $\measuredangle QPR$, $\measuredangle RPS$, $\measuredangle SPT$, and $\measuredangle TPQ$
8. 360°
9. **a.** 70° **b.** 120°
10. protractor
11. **a.** 31° **b.** 78° **c.** 95° **d.** 130°
12. **a.** *a, b, e, h* **b.** *d, g* **c.** *c, f*
13. The sum of their measures is 90°.
14. The sum of their measures is 180°.
15. **a.** no **b.** yes
16. **a.** no **b.** yes
17. no
18. **a.** $m\measuredangle ACB = 30°$. $m\measuredangle A = 90°$, $m\measuredangle ABC = 60°$
b. yes: $\measuredangle ACB$ and $\measuredangle ABC$
c. no
19. **a.** $m\measuredangle EBD = 30°$, $m\measuredangle BDE = 30°$, $m\measuredangle DEB = 120°$
b. $m\measuredangle ECD = 60°$, $m\measuredangle CED = 60°$, $m\measuredangle EDC = 60°$
c. $\measuredangle BED$ is supplementary to each of the angles of triangle *EDC*.
20. *c*
21. *a, b, c*
22. *b, c*

4.3 CIRCLES AND POLYGONS

1 A *circle* is a set of all points in a plane whose distance from a given point, *P*, is equal to a positive number, *r*. The point *P* is the *center* of the circle. The positive number *r* is the *radius* of the circle. Twice the radius is the *diameter* of the circle.

The instrument used to draw circles is the compass. It can also be used to compare distances.

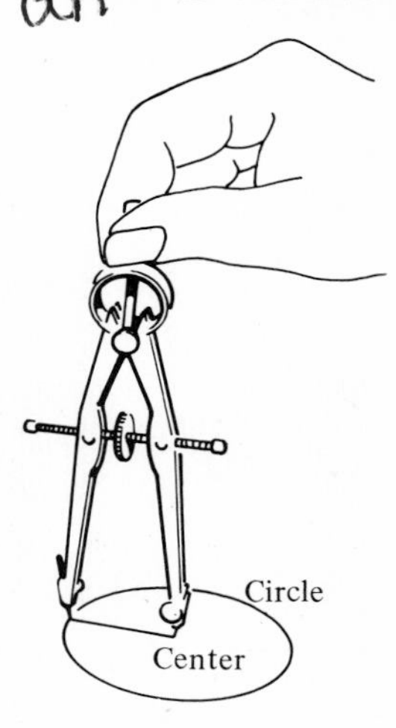

Q1 Use a compass to determine which point is the center of the circle. ______

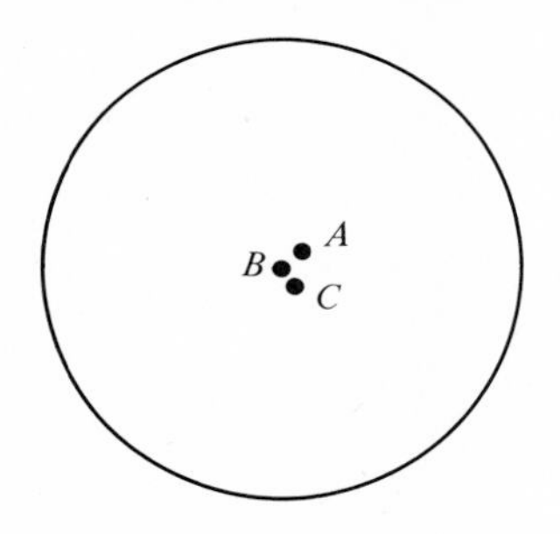

• ### • ### • ### • ### • ### • ### • ### •

A1 *B*

Q2 **a.** Calculate the diameter if the radius of a circle is:
(1) 3 centimetres (2) 8 inches
(3) 2 feet 3 inches (4) 4.25 centimetres

b. Calculate the radius if the diameter of a circle is:
(1) 12 centimetres (2) 4 feet
(3) 7 inches (4) 8 yards 2 feet

• ### • ### • ### • ### • ### • ### • ### •

A2 **a.** (1) 6 centimetres (2) 16 inches (3) 4 feet 6 inches (4) 8.50 centimetres
b. (1) 6 centimetres (2) 2 feet (3) 3.5 inches (4) 4 yards 1 foot

2 The word "radius" (plural form is radii) is also used to refer to a line segment with one endpoint at the center and the other on the circle. Likewise, the word "diameter" has two meanings. It is a line segment with endpoints on the circle containing the center, as well as being used to represent the length of such a line segment. You will be able to tell which meaning is intended from the context of the statement. Because of the definition of a circle, all radii of the same circle are the same length, and all diameters of the same circle are the same length.

Q3 **a.** Identify the line segments that are radii. ______

b. Identify the line segments that are diameters. ______

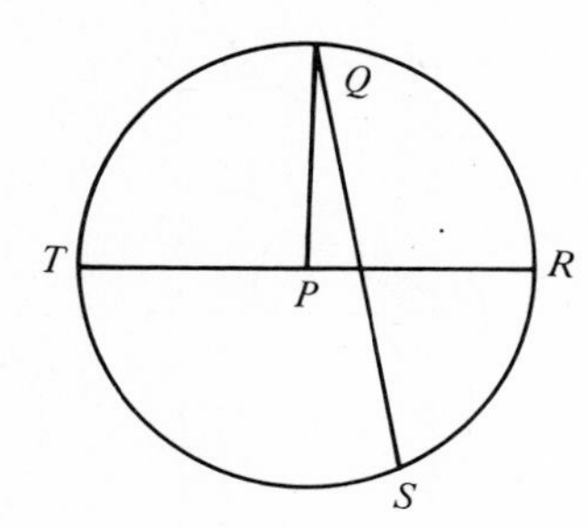

\# # # • # # # • # # # • # # # • # # # • # # # • # # # • # # # • # # #

A3 **a.** $\overline{TP}$, $\overline{PR}$, and $\overline{PQ}$ **b.** $\overline{TR}$

Q4 Suppose $\overline{AB}$ is a radius and is 5 inches long, and $\overline{DE}$ is 1.5 inches (inches is sometimes symbolized with ″):

a. What is the length of AC? ______

b. What is the length of DA? ______

c. What is the length of EA? ______

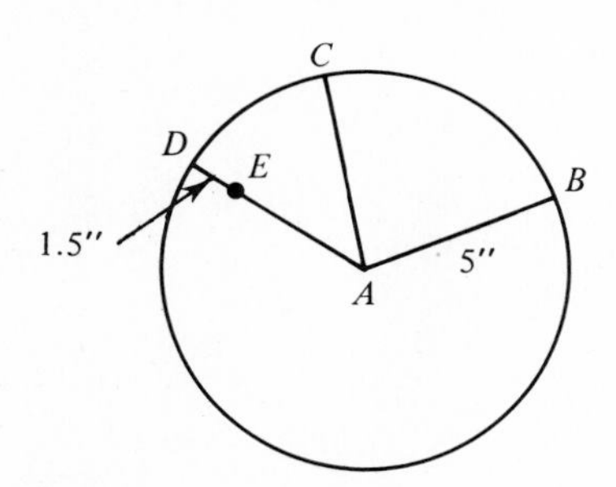

\# # # • # # # • # # # • # # # • # # # • # # # • # # # • # # # • # # #

A4 **a.** 5 inches **b.** 5 inches **c.** 3.5 inches: $5 - 1.5 = 3.5$

Q5 Suppose $\overline{AC}$ is a radius and is 14 centimetres long. How long is diameter $\overline{DB}$?

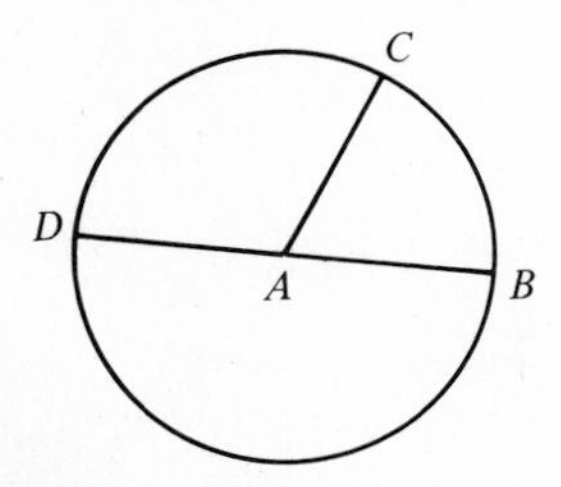

\# # # • # # # • # # # • # # # • # # # • # # # • # # # • # # # • # # #

A5 28 centimetres

Q6 Two parallel shafts are connected by a pair of rolling friction wheels. If the distance between the centers of the shafts is 15 inches and the radius of the larger wheel R is 9.6 inches, what is the radius of the smaller wheel R'? (R' is read "R prime" and represents a second radius.)

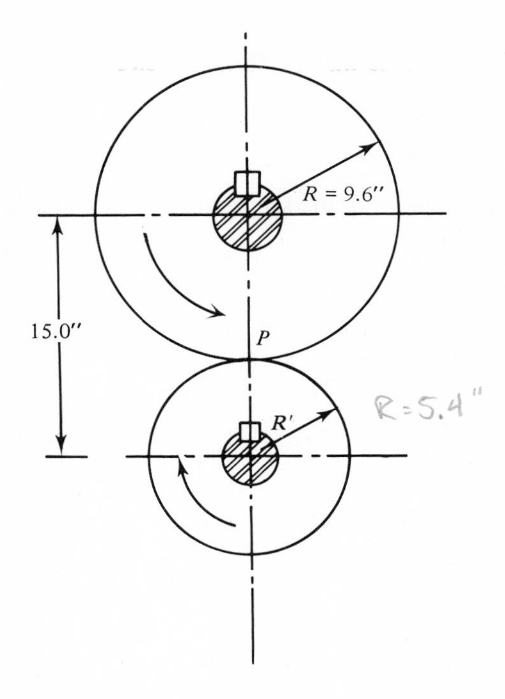

• ### • ### • ### • ### • ### • ### • ### •

A6 5.4 inches: $15.0 - 9.6 = 5.4$

3 The part of a circle connecting two points on a circle is an *arc* of a circle. If the two points are on a diameter of a circle, the arc is called a *semicircle. When naming an arc, move clockwise about the circle.*

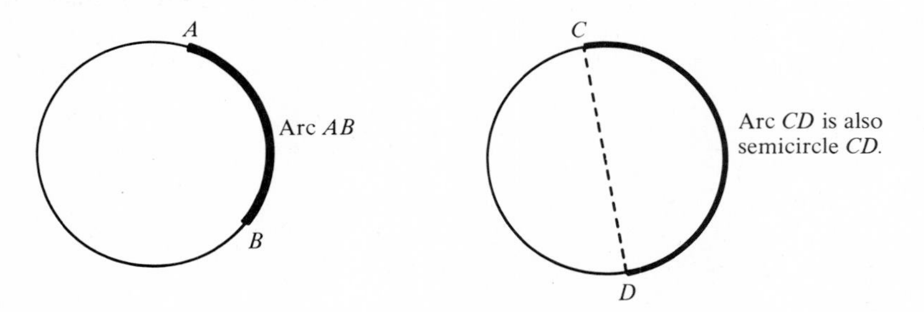

The word arc is used to describe the right and wrong way to remove the shoulder left on a pipe after cutting it with a pipe cutter. The file should be moved in a turning motion, removing excess metal through an *arc* of the circumference. The wrong way results in a series of flat places on the pipe.

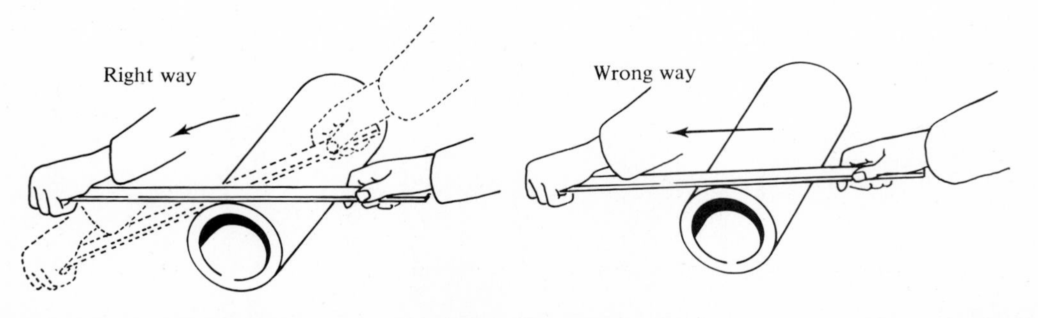

Q7 **a.** Name three arcs on the circle. ________

b. Which of these arcs appears to be a semicircle? ________

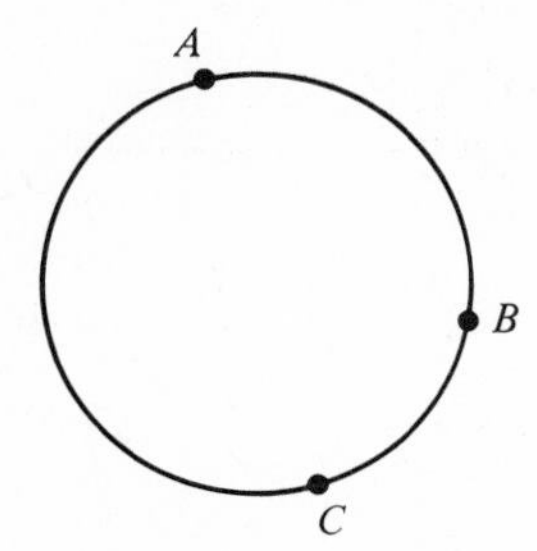

\# \# \# • \# \# \# • \# \# \# • \# \# \# • \# \# \# • \# \# \# • \# \# \# • \# \# \# • \# \# \#

A7 **a.** arc AB, arc BC, arc CA. There are other arcs as well, namely, arc AC, arc BA, and arc CB. Notice that arc $AB \neq$ arc BA since they contain opposite sides of the circle. (Note: The symbol $\neq$ says "is not equal.")

b. arc CA.

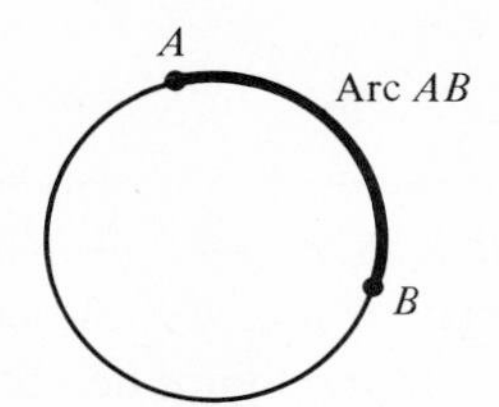

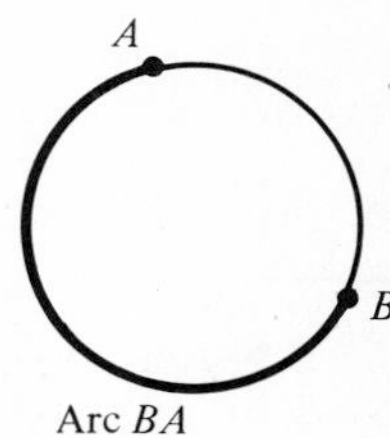

4 A line segment connecting two points on a circle is a *chord.* If the line segment passes through the center of the circle, the chord is on a diameter. If a line intersects only one point on a circle, the line is a *tangent line* and is said to be tangent to the circle. Notice that a chord is a line segment while a tangent is a line.

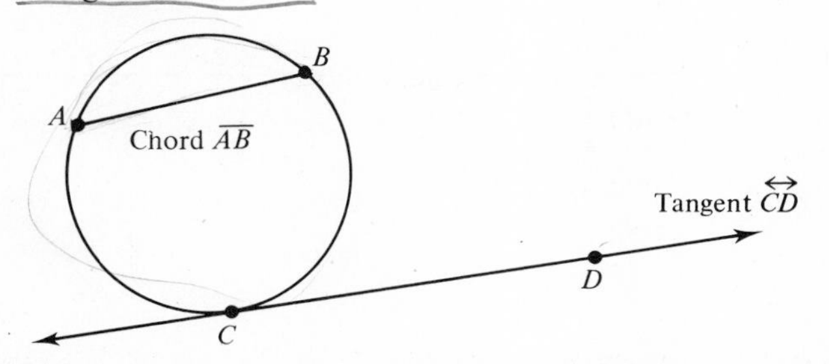

Q8 **a.** Name a tangent to circle 1. ________

b. Name two tangents to circle 2. ________

c. Name two chords of circle 1. ________

d. Name a chord of circle 2. ________

e. Name two diameters. ________

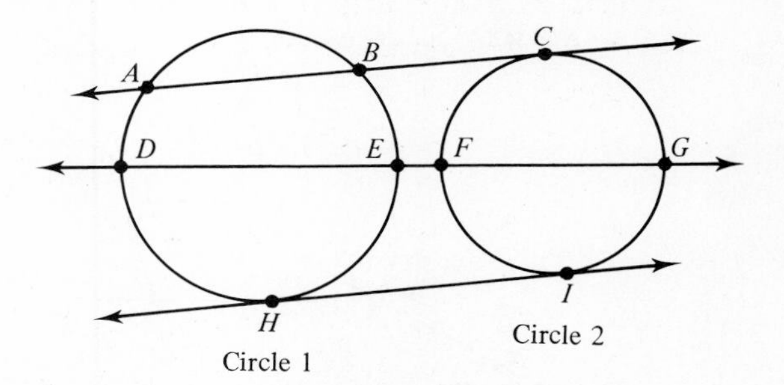

\# \# \# • \# \# \# • \# \# \# • \# \# \# • \# \# \# • \# \# \# • \# \# \# • \# \# \# • \# \# \#

A8 **a.** $\overleftrightarrow{HI}$ **b.** $\overleftrightarrow{HI}$, $\overleftrightarrow{AC}$: $\overleftrightarrow{AC}$ could also be called $\overleftrightarrow{BC}$ or $\overleftrightarrow{AB}$.

c. $\overline{AB}$, $\overline{DE}$ **d.** $\overline{FG}$ **e.** $\overline{DE}$ or $\overline{FG}$

Q9 Two pulleys of different sizes are connected with a belt.

a. Name two tangents.

b. Is arc AD a semicircle?

______

c. Is arc AD larger or smaller than arc DA? ______

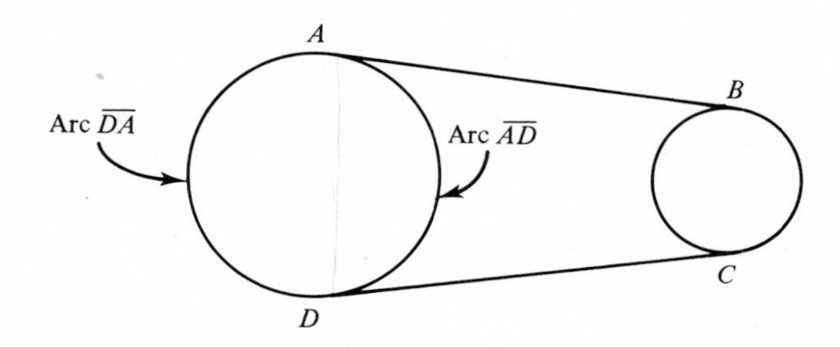

\# \# \# • \# \# \# • \# \# \# • \# \# \# • \# \# \# • \# \# \# • \# \# \# • \# \# \# • \# \# \#

A9 **a.** $\overleftrightarrow{AB}$, $\overleftrightarrow{DC}$ **b.** no **c.** smaller

5 An angle with vertex at the center of a circle is a *central angle*. $\measuredangle ABC$ and $\measuredangle CDE$ are central angles. The figure enclosed by a central angle and its intercepted arc is a *sector* of a circle. A sector is commonly said to be the shape of a piece of pie. A sector CDE is shaded in the figure below.

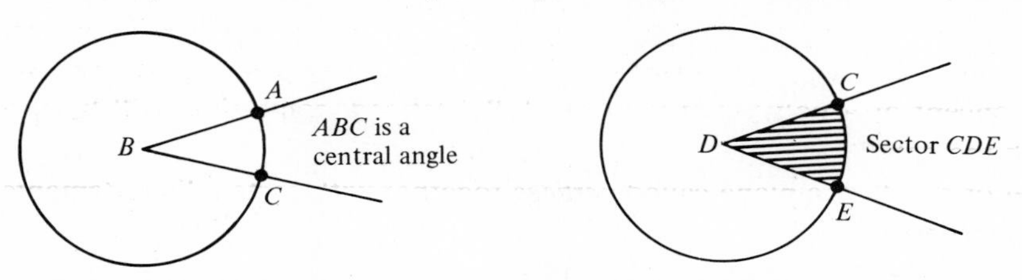

Q10 If an angle is a central angle, indicate yes; if not indicate no:

a. $\measuredangle ABC$ ______

b. $\measuredangle ADE$ ______

c. $\measuredangle CBD$ ______

d. $\measuredangle EDF$ ______

e. $\measuredangle ABD$ ______

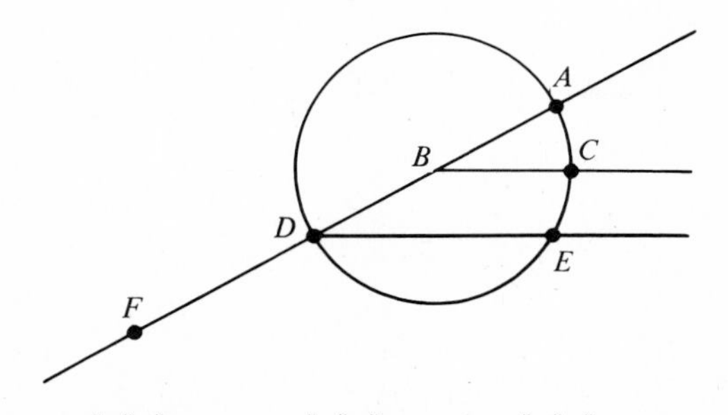

\# \# \# • \# \# \# • \# \# \# • \# \# \# • \# \# \# • \# \# \# • \# \# \# • \# \# \# • \# \# \#

A10 **a.** yes **b.** no **c.** yes **d.** no **e.** yes

Q11 In planning belt-drive arrangements, vertical drive arrangements should be avoided if possible. However, if necessary, the smaller driving pulley is better located at the top. The effective arc of contact is drawn for each arrangement. Measure the central angle for the arc of contact AB in each case.

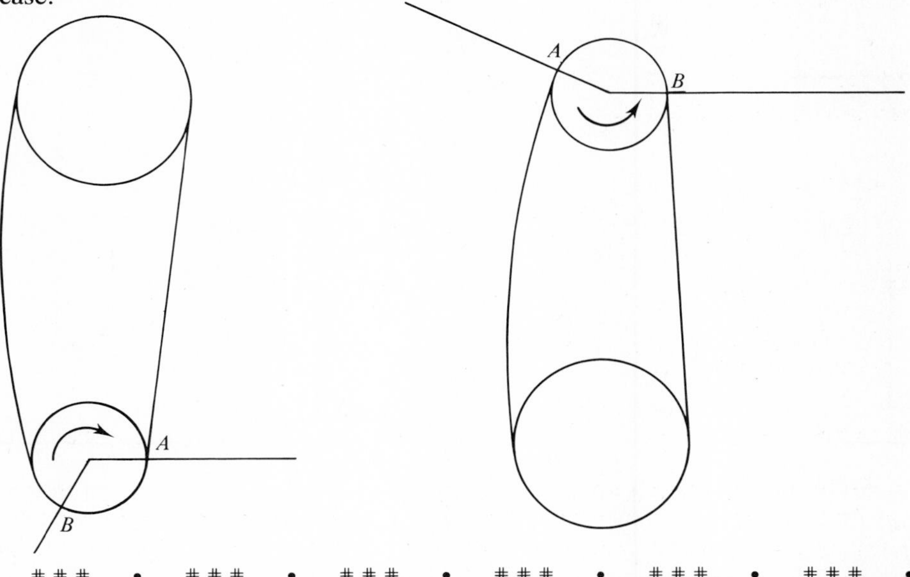

\# \# \# • \# \# \# • \# \# \# • \# \# \# • \# \# \# • \# \# \# • \# \# \# • \# \# \# • \# \# \#

A11 **a.** 120° **b.** 155°

6 You may be familiar with geometric figures called polygons from your past experience. However, since the geometric concept of a polygon is very carefully defined, some time will be spent just becoming familar with the definition.

A *polygon* is a set of points in a plane called *vertices* together with certain line segments called *sides* having these properties:

1. There are a fixed number of vertices (at least three) arranged in a particular order. In particular, there is a first vertex and a last vertex.
2. Each vertex except the last is joined to the next vertex after it by a line segment called a *side*; and the last vertex is joined to the first vertex by a side.
3. Two sides intersect only at their endpoints.
4. No two sides with a common endpoint are in the same straight line.

Some examples of polygons follow:

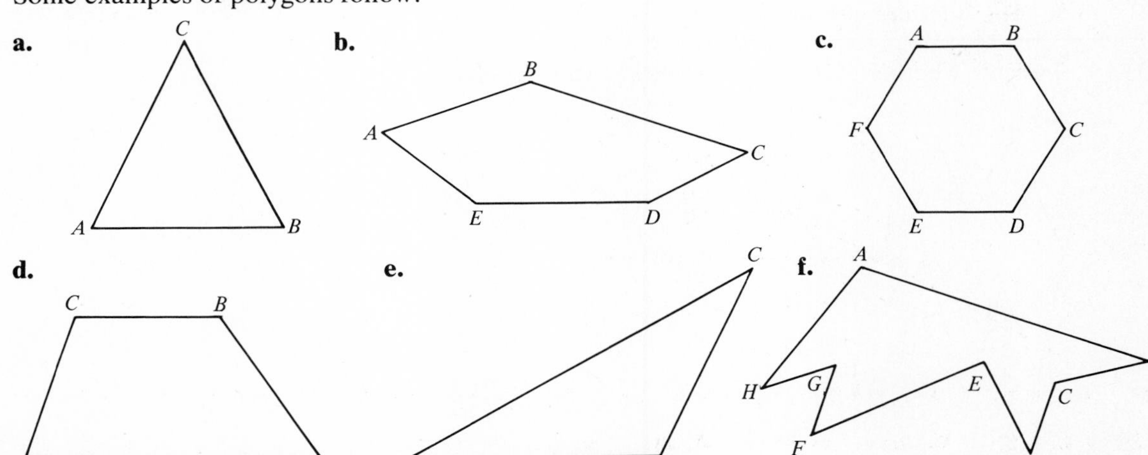

Some examples of geometric figures that are sometimes mistaken for polygons are shown here.

a. b. 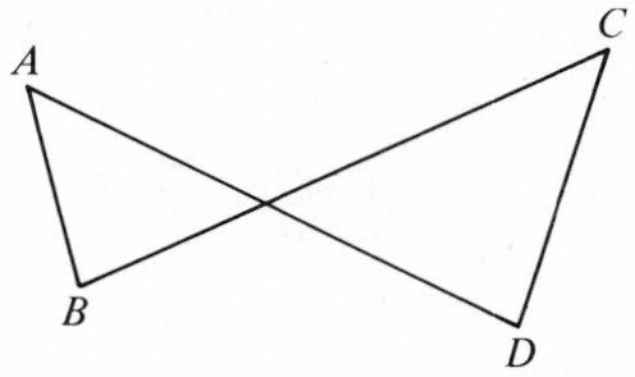 c.

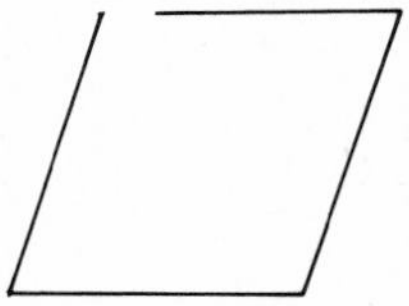

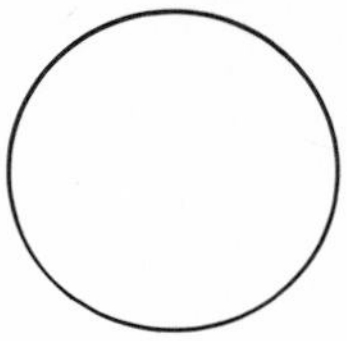

A B C D

d. 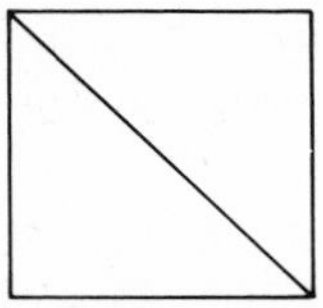 e. 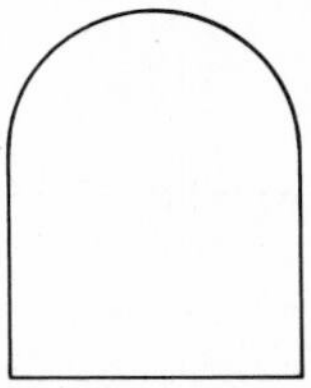 f.

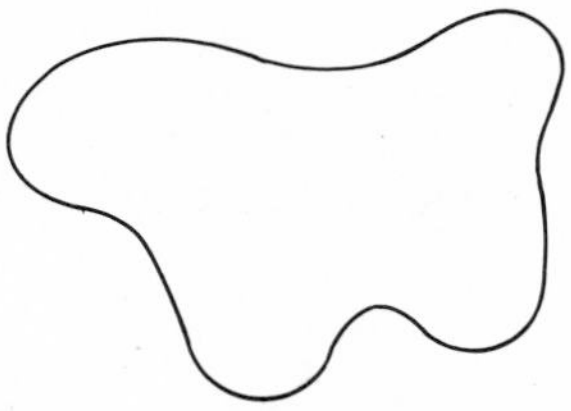

Q12 Explain what part of the definition is not satisfied by each of the geometric figures of Frame 1 that is not a polygon:

a. ______

b. ______

c. ______

d. ______

e. ______

f. ______

• # # # • # # # • # # # • # # # • # # # • # # # • # # # • # #

A12
a. 2; the first vertex is not connected to the last.
b. 3; $\overline{BC}$ intersects $\overline{AD}$ at a point that is not an endpoint.
c. The sides are not line segments.
d. 2; the vertices are not joined consecutively.
e. The sides are not all line segments.
f. The sides are not line segments.

Q13 Indicate whether each of the figures shown is a polygon:

a. b. c.

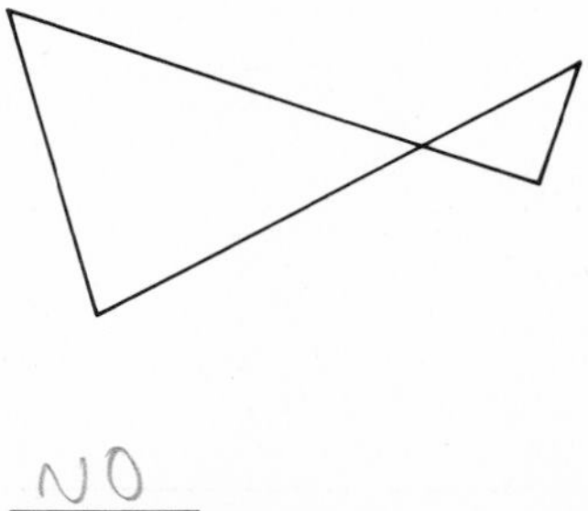

Yes ______ NO ______ NO ______

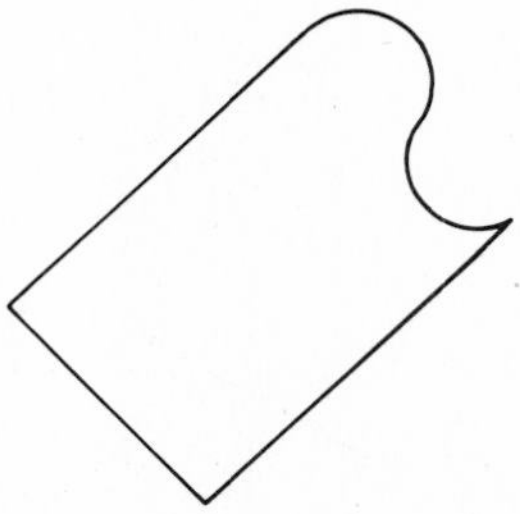

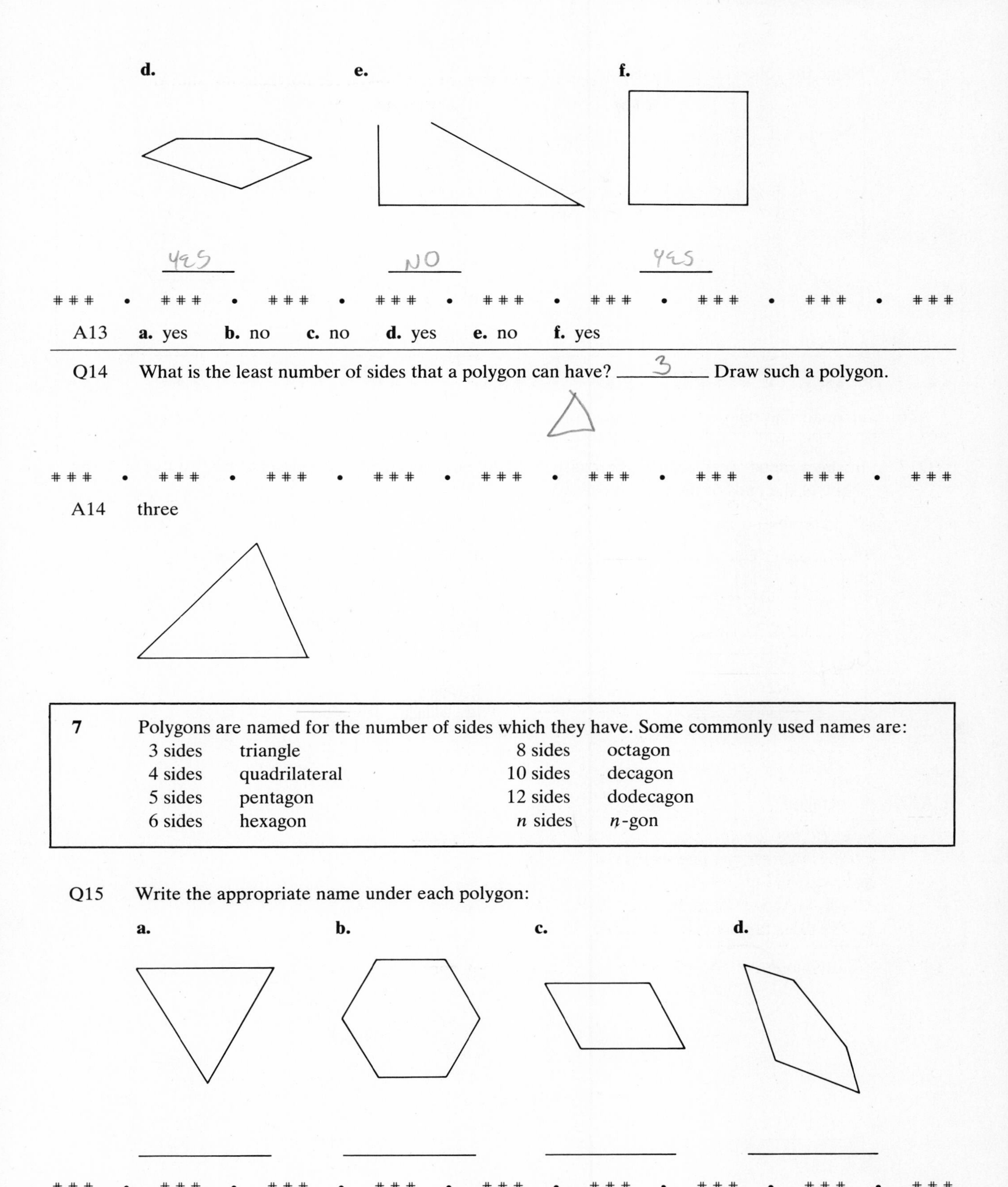

d. **e.** **f.**

\# # # • # # # • # # # • # # # • # # # • # # # • # # # • # # # • # # #

A13 **a.** yes **b.** no **c.** no **d.** yes **e.** no **f.** yes

---

Q14 What is the least number of sides that a polygon can have? ________ Draw such a polygon.

\# # # • # # # • # # # • # # # • # # # • # # # • # # # • # # # • # # #

A14 three

---

7 Polygons are named for the number of sides which they have. Some commonly used names are:

3 sides	triangle	8 sides	octagon
4 sides	quadrilateral	10 sides	decagon
5 sides	pentagon	12 sides	dodecagon
6 sides	hexagon	n sides	n-gon

Q15 Write the appropriate name under each polygon:

a. **b.** **c.** **d.**

\# # # • # # # • # # # • # # # • # # # • # # # • # # # • # # # • # # #

A15 **a.** triangle **b.** hexagon **c.** quadrilateral **d.** pentagon

Q16 Name the following polygons:

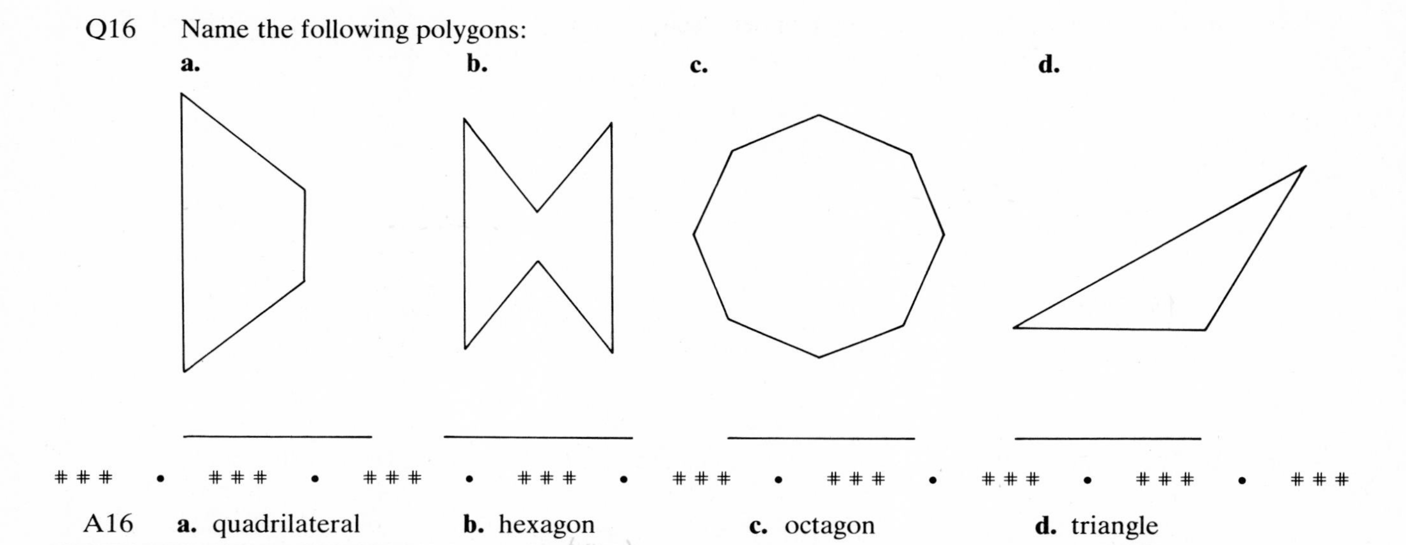

\# \# \# • \# \# \# • \# \# \# • \# \# \# • \# \# \# • \# \# \# • \# \# \# • \# \# \# • \# \# \#

A16 **a.** quadrilateral **b.** hexagon **c.** octagon **d.** triangle

---

Q17 In sheet metal work elbows are constructed to carry water around corners. What polygon is the shape of the ends of the elbows below?

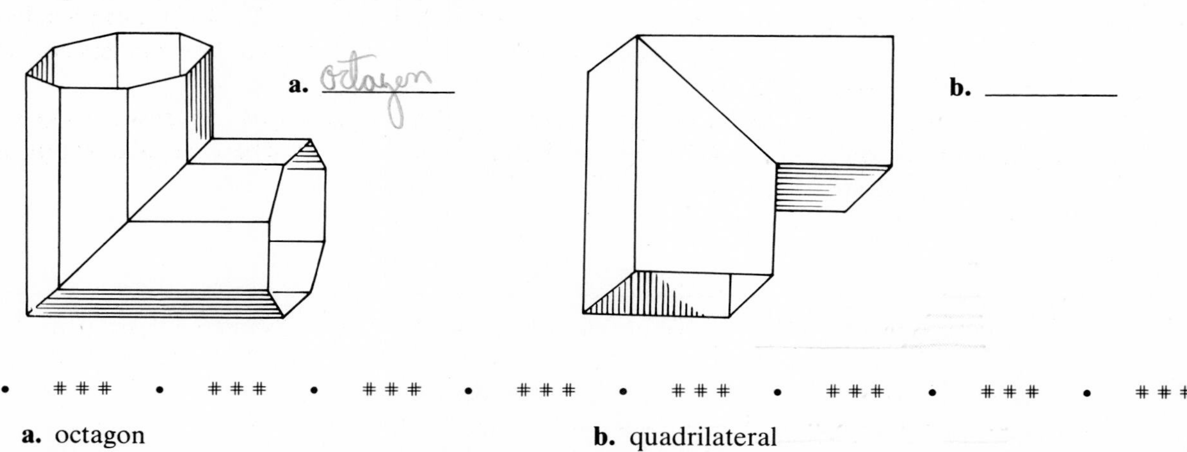

\# \# \# • \# \# \# • \# \# \# • \# \# \# • \# \# \# • \# \# \# • \# \# \# • \# \# \# • \# \# \#

A17 **a.** octagon **b.** quadrilateral

Q18 A bolthead or nut has the shape of a polygon. What shapes are used below?

a.

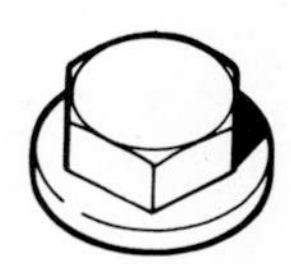

b.

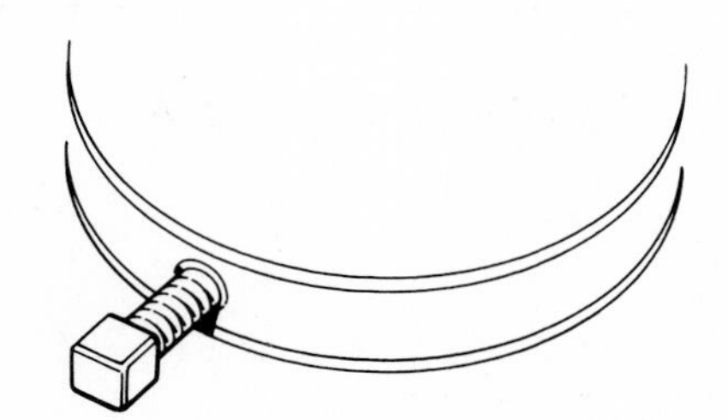

\# \# \# • \# \# \# • \# \# \# • \# \# \# • \# \# \# • \# \# \# • \# \# \# • \# \# \# • \# \# \#

A18 **a.** hexagon **b.** quadrilateral

Q19 A box-end wrench is designed with either a six-point or a twelve-point opening at each end. What n-gon is formed by each wrench below?

a.

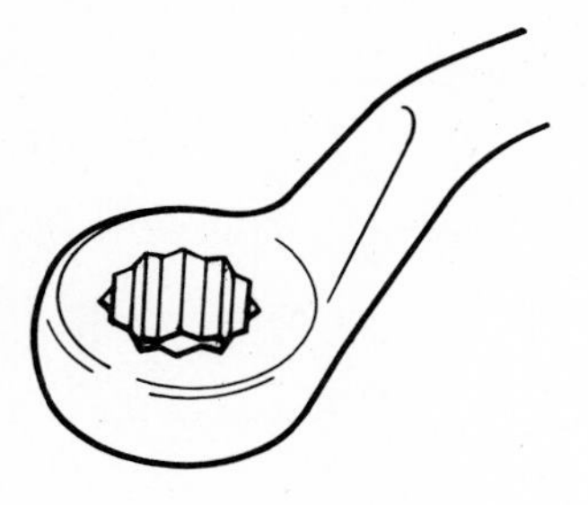

______

b.

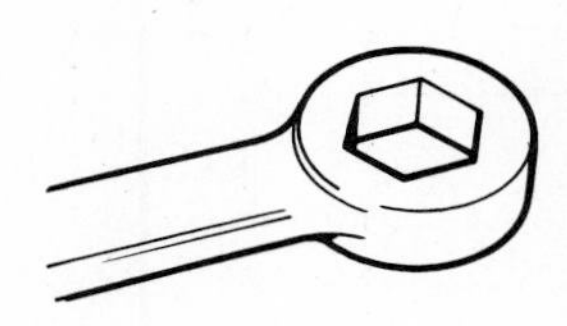

______

\# \# \# • \# \# \# • \# \# \# • \# \# \# • \# \# \# • \# \# \# • \# \# \# • \# \# \# • \# \# \#

A19 **a.** 24-gon **b.** hexagon

8 A polygon divides the points of a plane into three sets—the points of the polygon, the points in the interior of the polygon, and the points in the exterior of the polygon. The plane area bounded by the polygon is the *interior* of the polygon. The *exterior* of a polygon is the set of points that are not in the interior and not on the polygon.

A *triangular region* is a triangle and its interior. A *hexagonal region* is a hexagon together with its interior. Other regions can be described in a similar manner. The triangular region ABC is shaded in the figure.

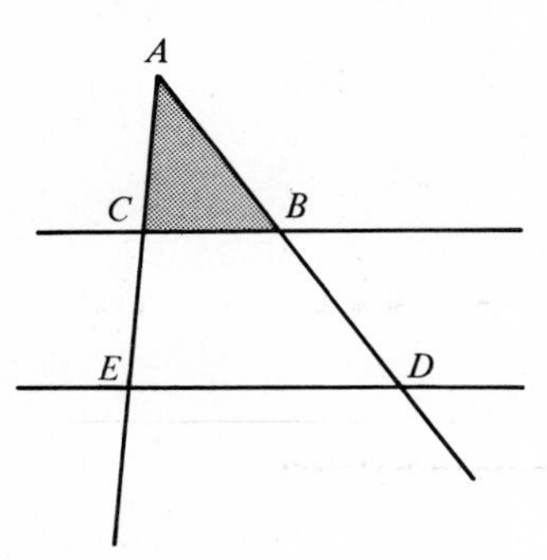

Q20 Shade the triangular region AED in the figure.

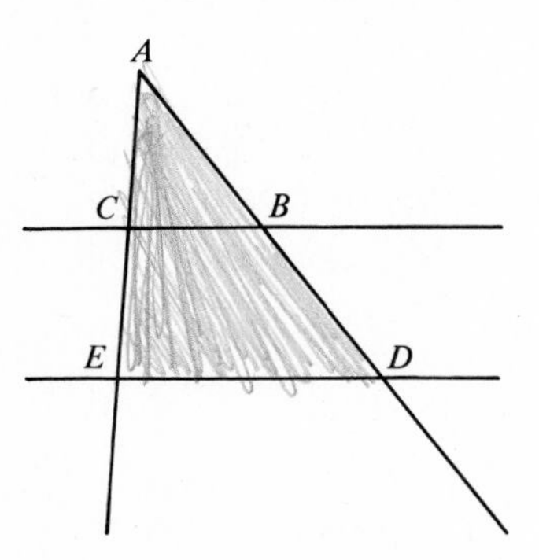

\# \# \# • \# \# \# • \# \# \# • \# \# \# • \# \# \# • \# \# \# • \# \# \# • \# \# \# • \# \# \#

A20

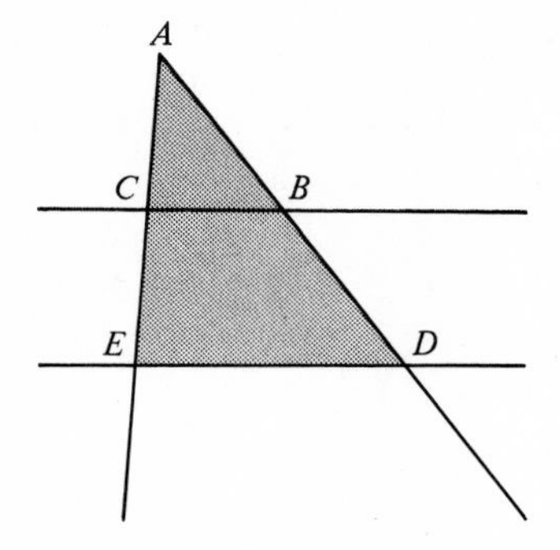

Q21 Shade a quadrilateral region in the figure (there are three possibilities).

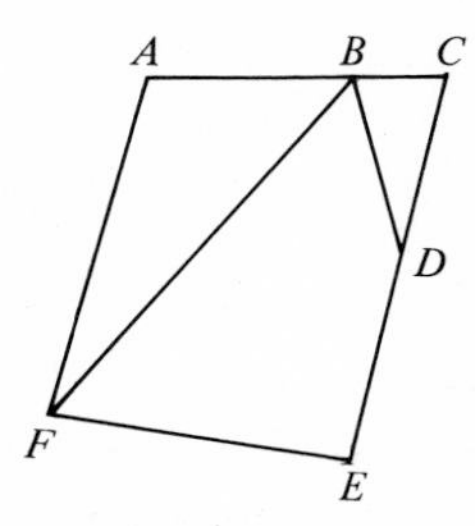

A21 Any one of the following is correct.

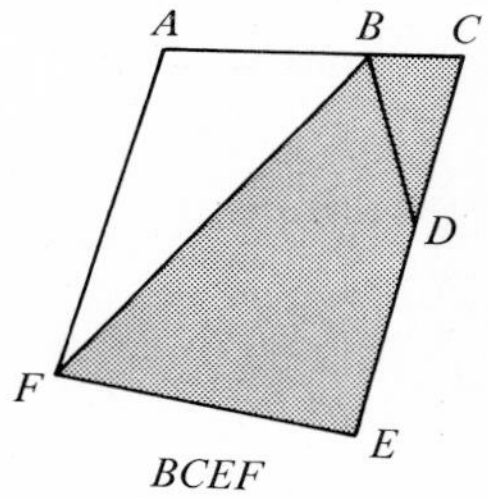

BCEF

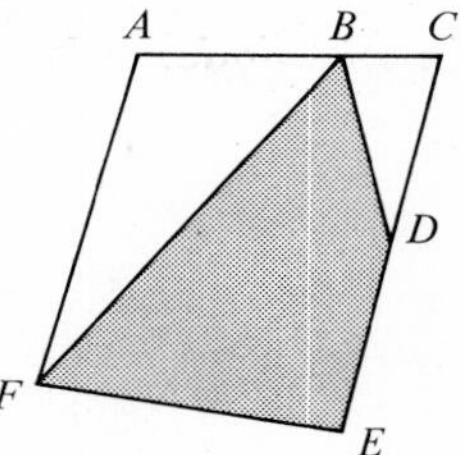

BDEF

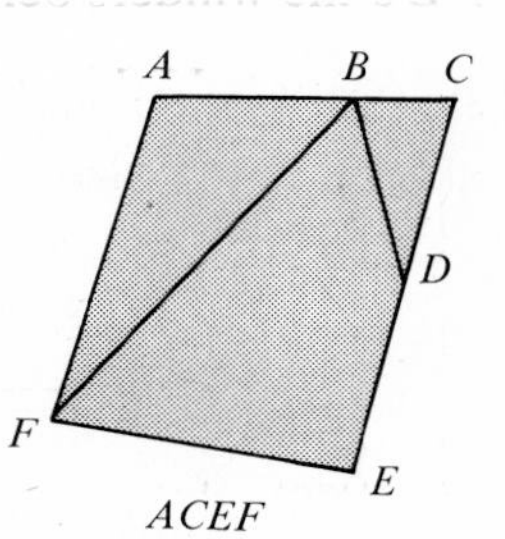

ACEF

Q22 Shade a pentagonal region in the figure.

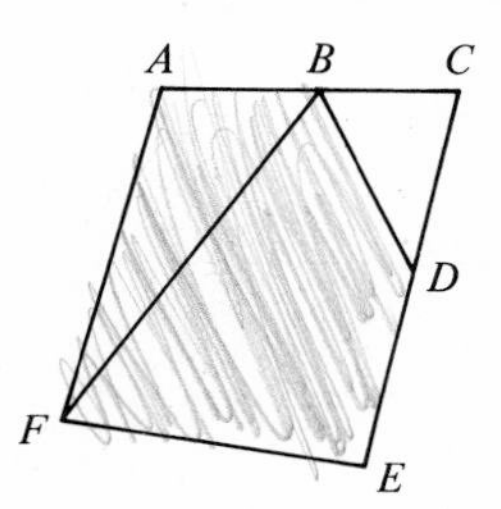

A22 *ABDEF*

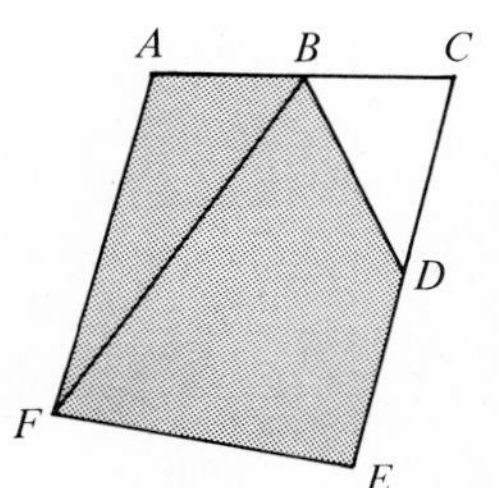

Q23 Is a point on the polygon included in the polygonal region? YES

• ### • ### • ### • ### • ### • ### • ### •

A23 yes

Q24 Place a *P* before an object that resembles a polygon and an *R* before an object that resembles a polygonal region:

P **a.** picture frame

R **b.** picture

______ **c.** door

______ **d.** stop sign

______ **e.** hexagonal sugar cookie

______ **f.** cutting edge of a hexagonal cookie cutter

• ### • ### • ### • ### • ### • ### • ### •

A24 **a.** *P* **b.** *R* **c.** *R* **d.** *R* **e.** *R* **f.** *P*

Q25 The construction of a winding stair sometimes requires steps of different polygonal shapes. These steps are called *winders.*

a. Do the winders below suggest a polygon or a polygonal region? ______________

b. What would the polygonal region of each winder below be called? ______________

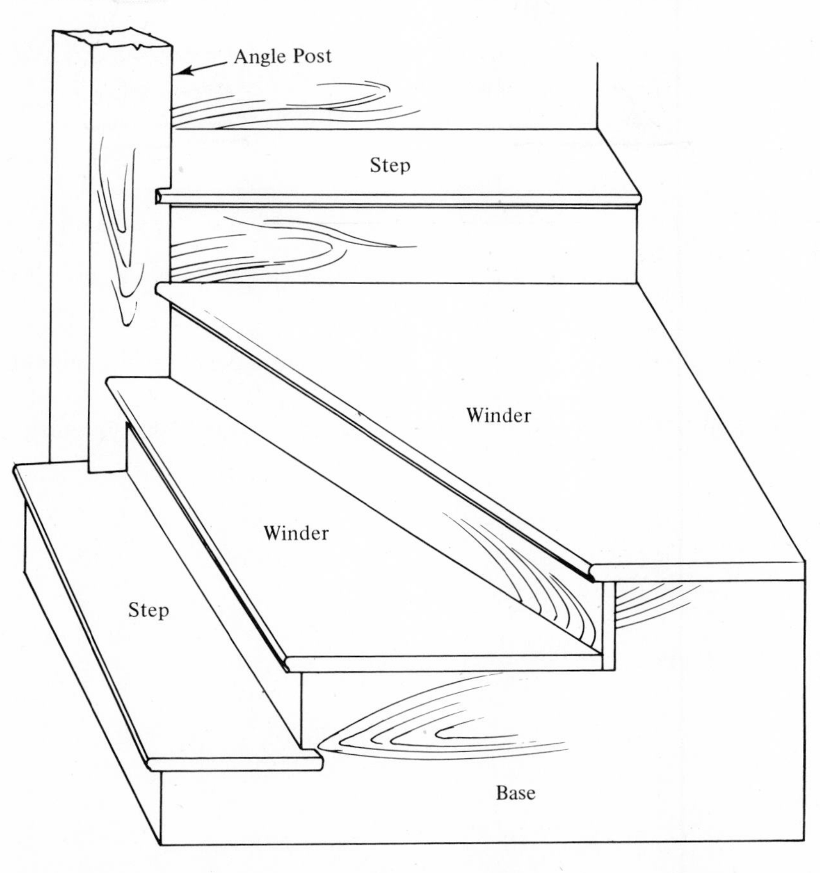

• ### • ### • ### • ### • ### • ### • ### •

A25 **a.** polygonal region **b.** quadrilateral region

9 Triangles are classified by properties of their sides. Some common descriptive names of triangles are given next.

An *equilateral* triangle has three equal sides. An *isosceles* triangle has at least two equal sides. A *scalene* triangle has no pair of sides equal.

Notice that the definition of an isosceles triangle allows an equilateral triangle to also be called isosceles, because it has at least two sides equal. Frequently we indicate the equal sides of a polygon by marking them with a short line through the sides.

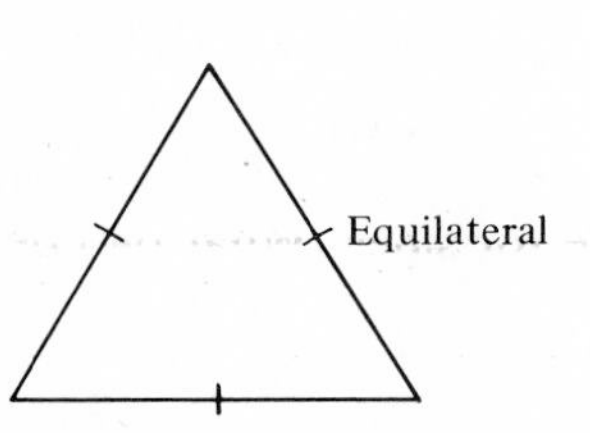

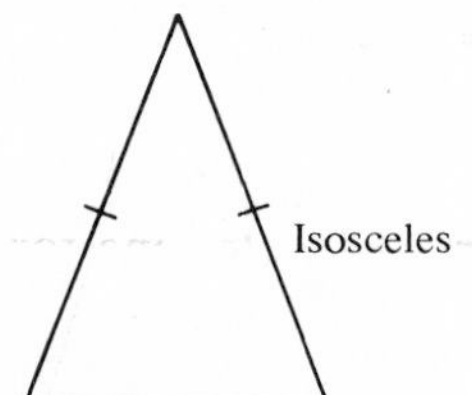

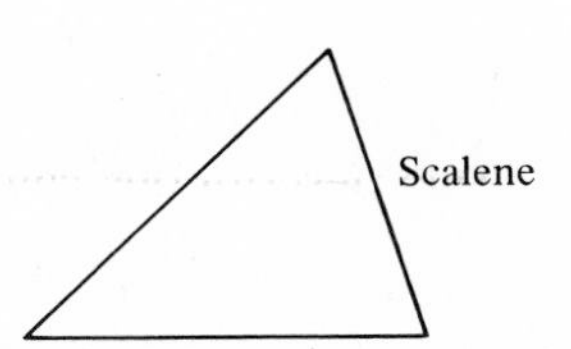

Q26 Write the descriptive name based on the lengths of the sides of each triangle:

a.

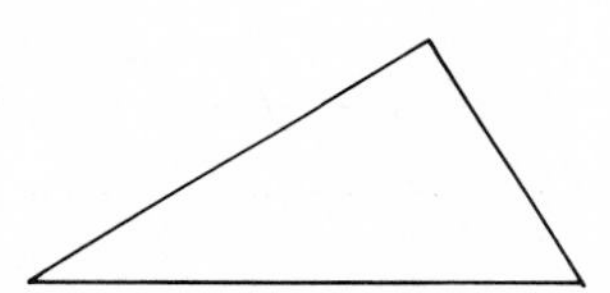

____

b.

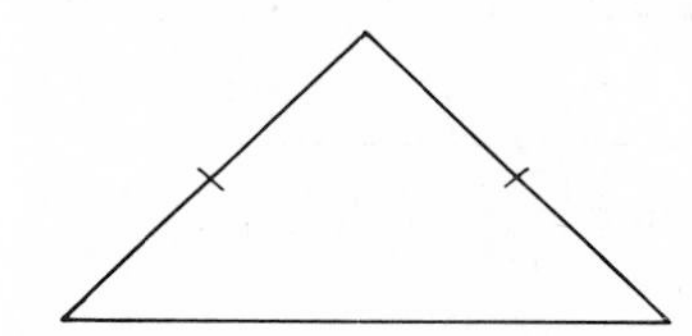

____

c.

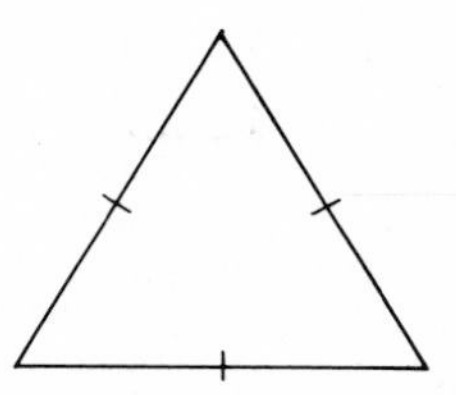

____

\# \# \# • \# \# \# • \# \# \# • \# \# \# • \# \# \# • \# \# \# • \# \# \# • \# \# \# • \# \# \#

A26 **a.** scalene **b.** isosceles **c.** equilateral or isosceles

10 Triangles are also classified by properties of their angles. Some common descriptive names are given here.

A *right triangle* has one angle of 90°

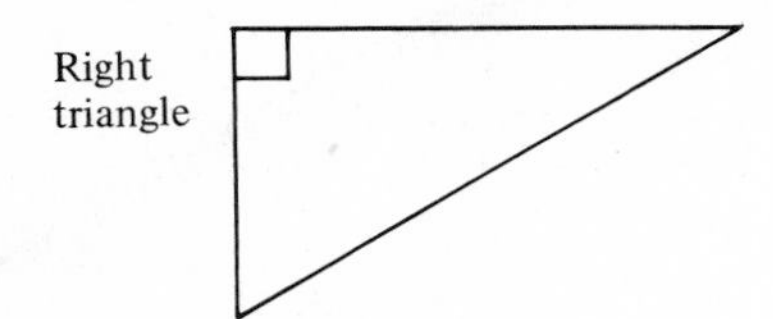

An obtuse triangle has one angle greater than 90°

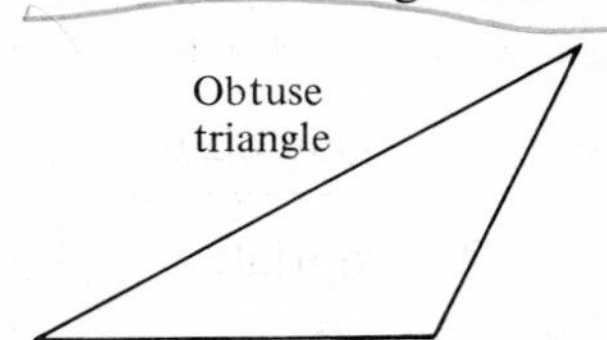

An *acute triangle* has three angles less than 90°

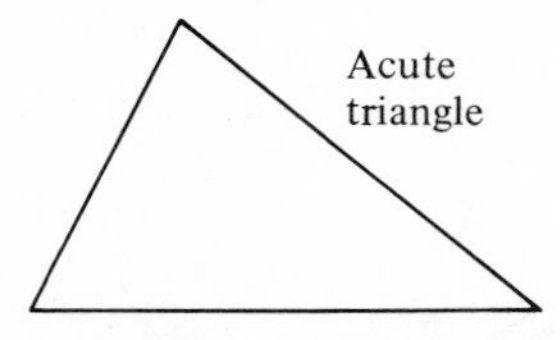

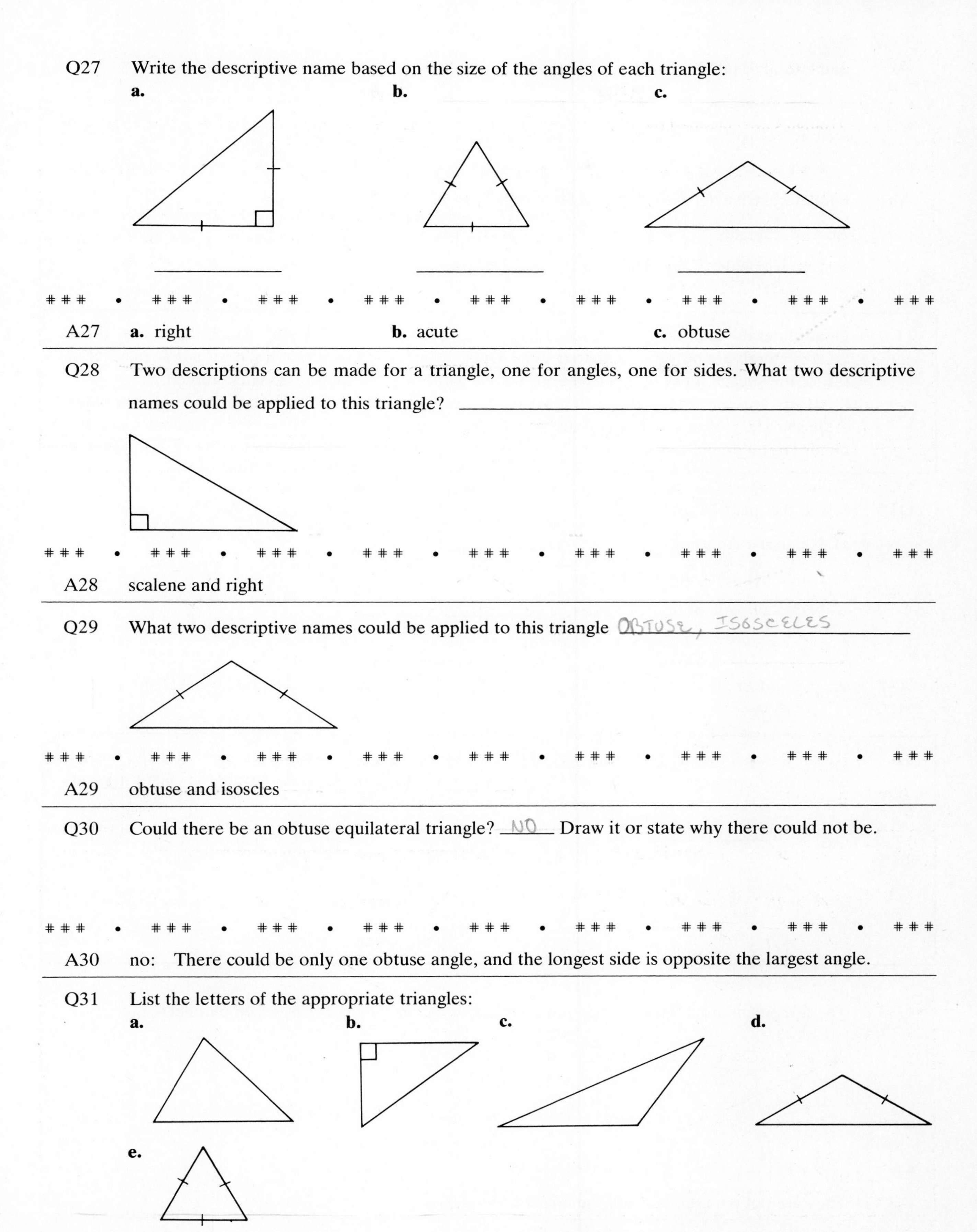

Q27 Write the descriptive name based on the size of the angles of each triangle:

a. **b.** **c.**

__________ __________ __________

\# # # • # # # • # # # • # # # • # # # • # # # • # # # • # # # • # # #

A27 **a.** right **b.** acute **c.** obtuse

Q28 Two descriptions can be made for a triangle, one for angles, one for sides. What two descriptive names could be applied to this triangle? ____________________

\# # # • # # # • # # # • # # # • # # # • # # # • # # # • # # # • # # #

A28 scalene and right

Q29 What two descriptive names could be applied to this triangle OBTUSE, ISOSCELES

\# # # • # # # • # # # • # # # • # # # • # # # • # # # • # # # • # # #

A29 obtuse and isoscles

Q30 Could there be an obtuse equilateral triangle? NO Draw it or state why there could not be.

\# # # • # # # • # # # • # # # • # # # • # # # • # # # • # # # • # # #

A30 no: There could be only one obtuse angle, and the longest side is opposite the largest angle.

Q31 List the letters of the appropriate triangles:

a. **b.** **c.** **d.**

e.

equilateral triangles = ______ right triangles = ______

isosceles triangles = ______ obtuse triangles = ______

scalene triangles = ______ acute triangles = ______

\# \# \# • \# \# \# • \# \# \# • \# \# \# • \# \# \# • \# \# \# • \# \# \# • \# \# \# • \# \# \#

A31 equilateral triangles = **e** right triangles = **b**

isoscles triangles = **d, e** obtuse triangles = **c, d**

scalene triangles = **a, b, c** acute triangles = **a, e**

---

11 Quadrilaterals are used so much that a vocabulary has developed which describes them. Quadrilaterals are polygons with four sides. Opposite sides of a quadrilateral need *not* be parallel. A few of the special types of quadrilaterals will be identified and their properties studied.

First, two terms that deal with quadrilaterals will be defined: adjacent sides and opposite sides. *Adjacent sides* of a quadrilateral are any two sides that intersect at a vertex. *Opposite sides* of a quadrilateral do not intersect.

Q32 Given the quadrilateral *ABCD*:

a. Name the adjacent sides at *B*. ______

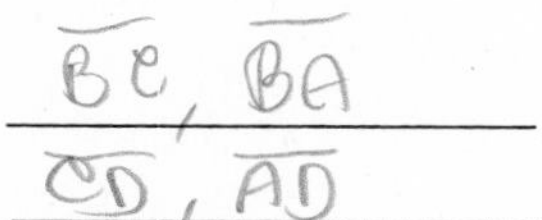

b. Name the adjacent sides at *D*. ______

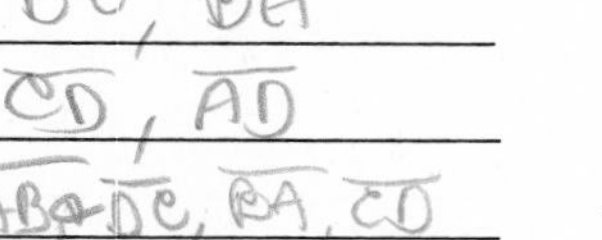

c. Name all pairs of opposite sides. ______

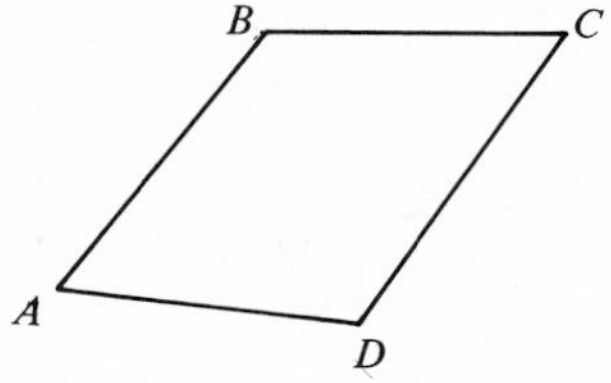

\# \# \# • \# \# \# • \# \# \# • \# \# \# • \# \# \# • \# \# \# • \# \# \# • \# \# \# • \# \# \#

A32 **a.** $\overline{AB}$ and $\overline{BC}$ **b.** $\overline{CD}$ and $\overline{AD}$ **c.** $\overline{AB}$ and $\overline{DC}$, $\overline{AD}$ and $\overline{BC}$

---

12 Recall that parallel lines do not intersect. Two line segments are said to be parallel if the lines of which they are a part are parallel. A quadrilateral with opposite sides parallel is called a *parallelogram*. Below is a parallelogram with $\overline{AB} \| \overline{CD}$ and $\overline{BC} \| \overline{AD}$.

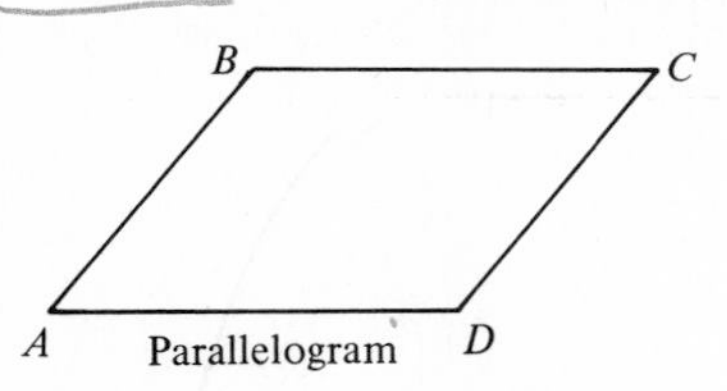

Parallelogram

Q33 The line segments $\overline{AB}$ and $\overline{CD}$ do not intersect. Why are they not considered parallel?

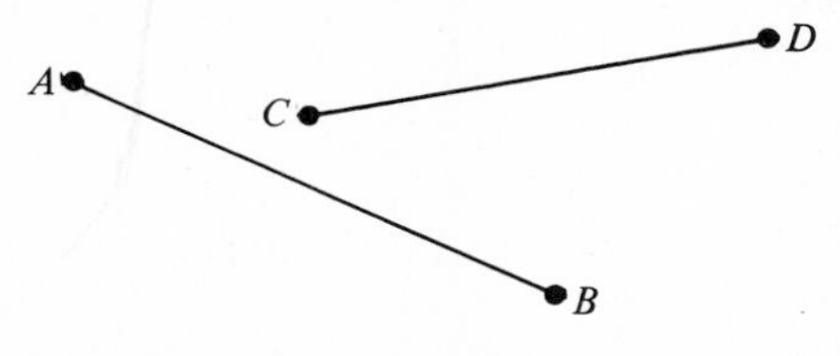

\# \# \# • \# \# \# • \# \# \# • \# \# \# • \# \# \# • \# \# \# • \# \# \# • \# \# \# • \# \# \#

A33 They are not parallel, because the lines of which they are a part do intersect.

Q34 Which of the following figures appear to be parallelograms? ________

a. **b.** **c.** **d.**

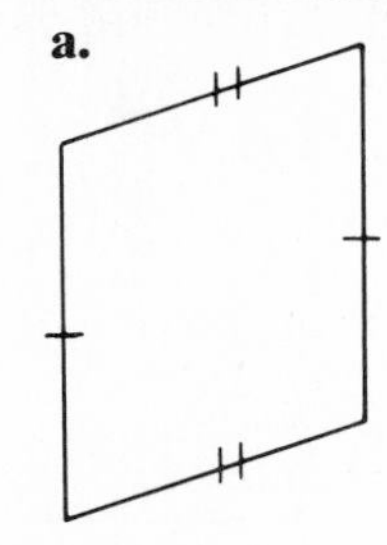

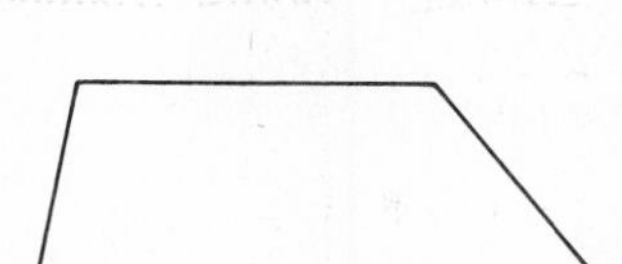
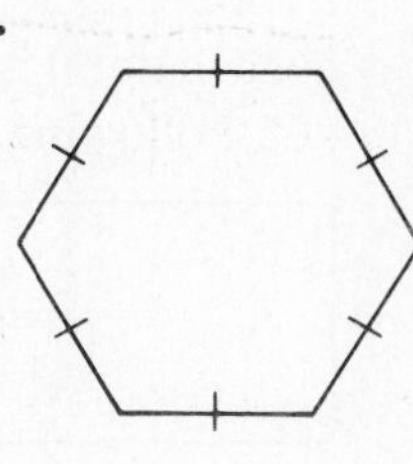

\# # # • # # # • # # # • # # # • # # # • # # # • # # # • # # # • # # #

A34 **a** and **b**: Notice that the opposite sides of figure **d** appear to be parallel. However, a parallelogram must first be a quadrilateral.

13 If the sides of a parallelogram meet at right angles, it is a *rectangle.* If the parallelogram's sides meet at right angles and are of equal length, it is a *square*.

The following quadrilateral is a rectangle as well as being a parallelogram:

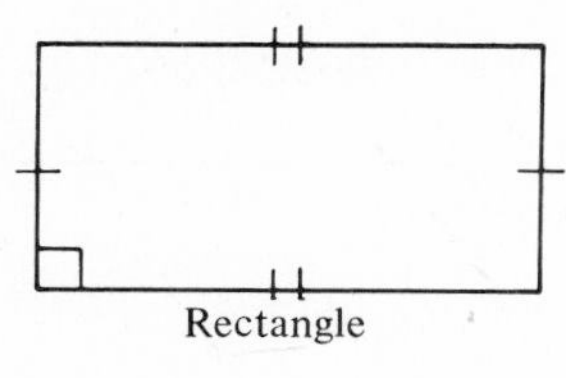
Rectangle

The following quadrilateral is a square as well as being a rectangle and a parallelogram:

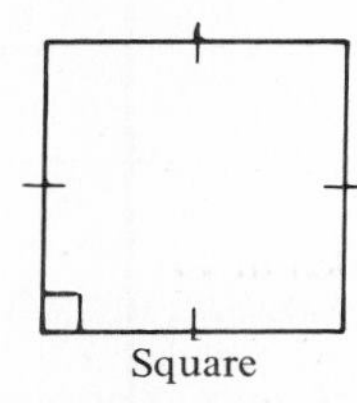
Square

Q35 List the letters of the appropriate geometric figures:

a. **b.** **c.** **d.** **e.**

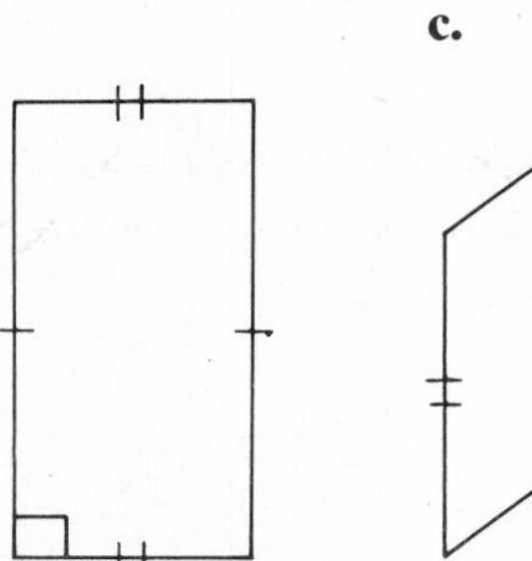
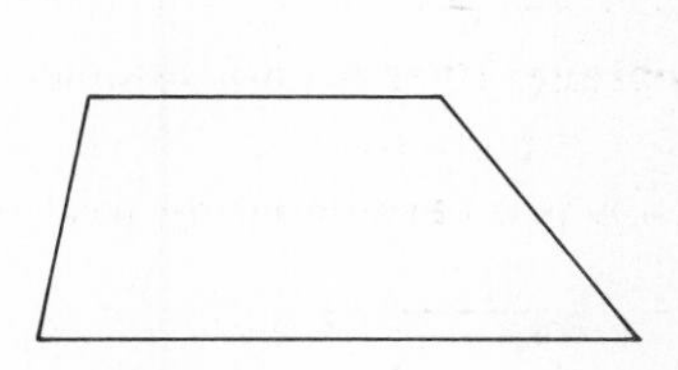
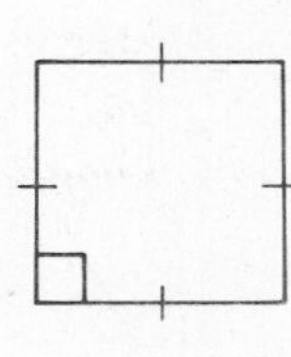

polygons = ________

quadrilaterals = ________ 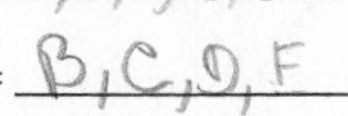

parallelograms = ________ 

rectangles = ________ 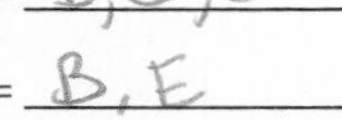

squares = ________ 

\# # # • # # # • # # # • # # # • # # # • # # # • # # # • # # # • # # #

A35 polygons = **a, b, c, d, e**
quadrilaterals = **b, c, d, e**
parallelograms = **b, c, e**
rectangles = **b, e**
squares = **e**

14 If one pair of opposite sides of a quadrilateral are parallel but the other pair are not, the quadrilateral is a *trapezoid*.

Following are two trapezoids:

Q36 Were any of the polygons in Q35 trapezoids? ________ Which? ______

\# # # • # # # • # # # • # # # • # # # • # # # • # # # • # # # • # # #

A36 yes, **d**

15 The *midpoint* is the point on a line segment that divides the line segment into two segments of equal length. The midpoint of line segment $\overline{AB}$ below is C, since the length of $\overline{AC}$ equals the length of $\overline{CB}$.

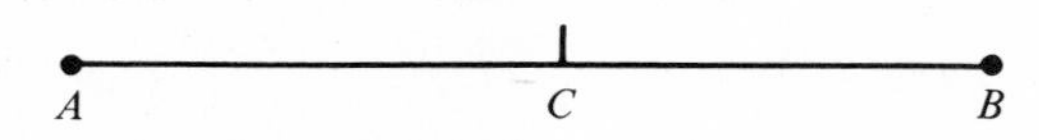

Q37 Connect the midpoints of the quadrilaterals consecutively.

a. What does the figure obtained appear to be? ______________

b. Do you think that always happens? ______________

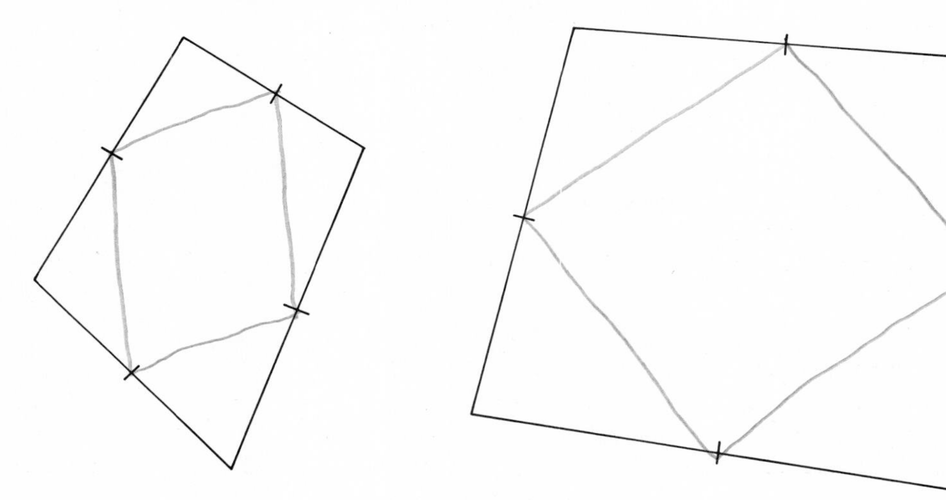

Try drawing some odd-shaped quadrilaterals of your own and test your guess.

\# # # • # # # • # # # • # # # • # # # • # # # • # # # • # # # • # # #

A37 **a.** a parallelogram **b.** yes: They will always be parallelograms.

16 A diagram is sometimes used to classify information. One that is useful for quarilaterals is shown. The diagram can be used to show that all rectangles are parallelograms but not all parallelograms are rectangles.

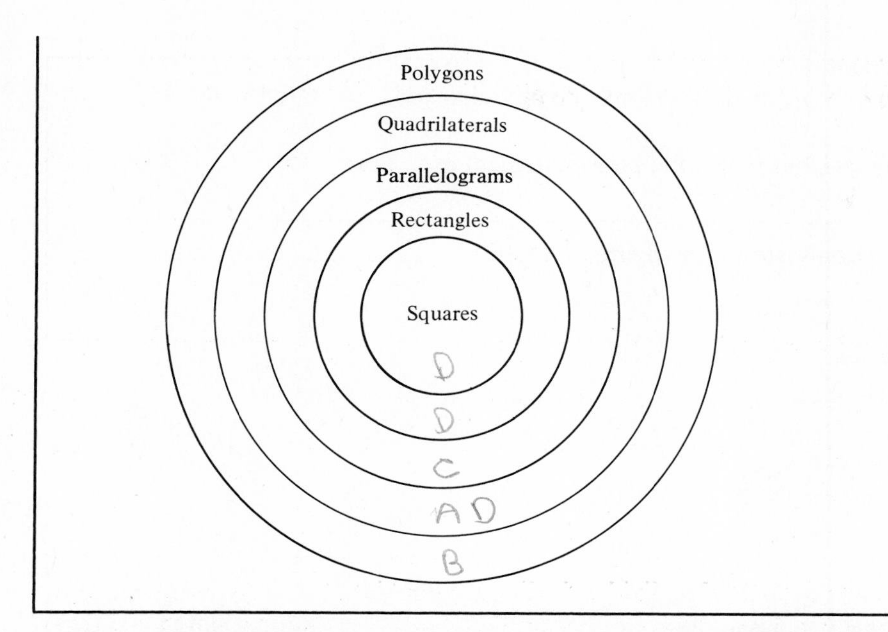

Q38 Place the letter of each of the following in the appropriate space in the diagram of Frame 16:

a. trapezoid **b.** triangle **c.** **d.**

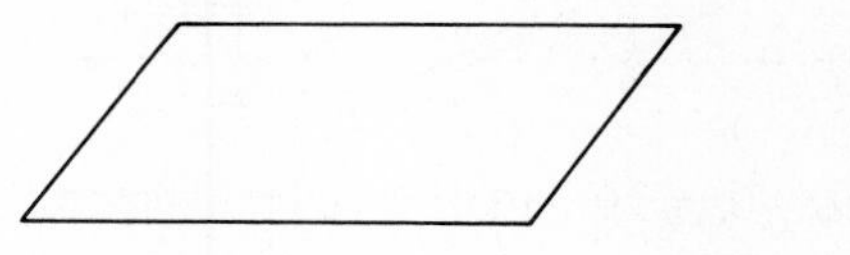

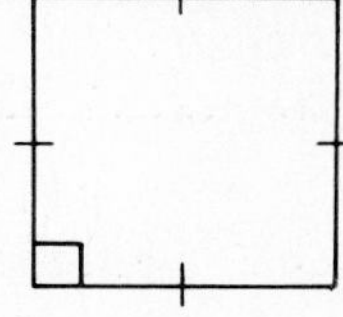

\# # # • # # # • # # # • # # # • # # # • # # # • # # # • # # # • # # #

A38

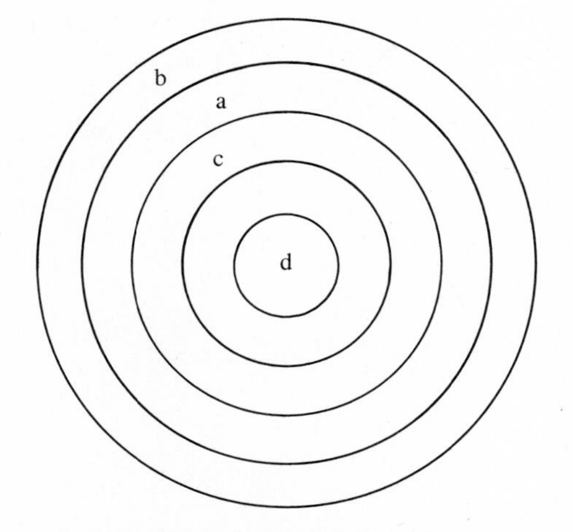

Q39 Use the diagram of Frame 16 to fill in the following five statements:

a. All PARALLELOGRAMS are quadrilaterals but not all quadrilaterals are PARALLELOGRAMS.

b. All SQUARES are quadrilaterals but not all quadrilaterals are SQUARES.

c. All RECTANGLES are quadrilaterals but not all quadrilaterals are RECTANGLES.

d. All RECTANGLES are parallelograms but not all parallelograms are RECTANGLES.

e. All SQUARES are rectangles but not all rectangles are SQUARES.

\# # # • # # # • # # # • # # # • # # # • # # # • # # # • # # # • # # #

A39 **a.** parallelograms, parallelograms
b. rectangles, rectangles — **a**, **b**, and **c** may be interchanged
c. squares, squares
d. squares, squares (or rectangles, but that is the same statement as in Frame 16)
e. squares, squares

Q40 What geometrical polygon predominates in the figure of a stair? ______

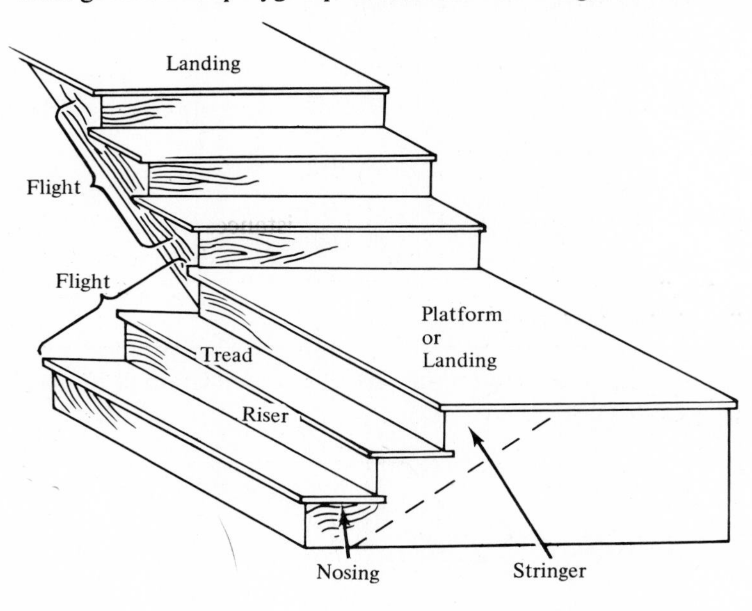

\# \# \# • \# \# \# • \# \# \# • \# \# \# • \# \# \# • \# \# \# • \# \# \# • \# \# \# • \# \# \#

A40 rectangle

This completes the instruction for this section.

4.3 EXERCISE

1. Use the figure to identify the following by name:

   a. P is the ______.

   b. $\overline{PB}$ is a(n) ______.

   c. $\overline{AC}$ is a(n) ______.

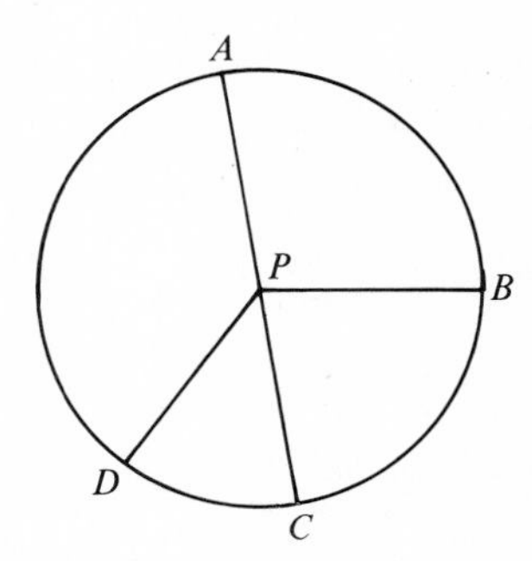

2. How long is the diameter of a circle with a radius of $5\frac{1}{2}$ centimetres?
3. What is the radius of a circle with a diameter of 7 inches?
4. A soil pipe with a 5-inch diameter has what radius?
5. What is the radius of a pulley with a 19.5 centimetre diameter?
6. Consider the pulley system with a crossed belt shown on the next page:
   a. Name a semicircle on the larger pulley.
   b. Name 5 radii on the larger pulley.
   c. Is arc TT' a semicircle?
   d. What is the line $\overleftrightarrow{TS'}$ called?

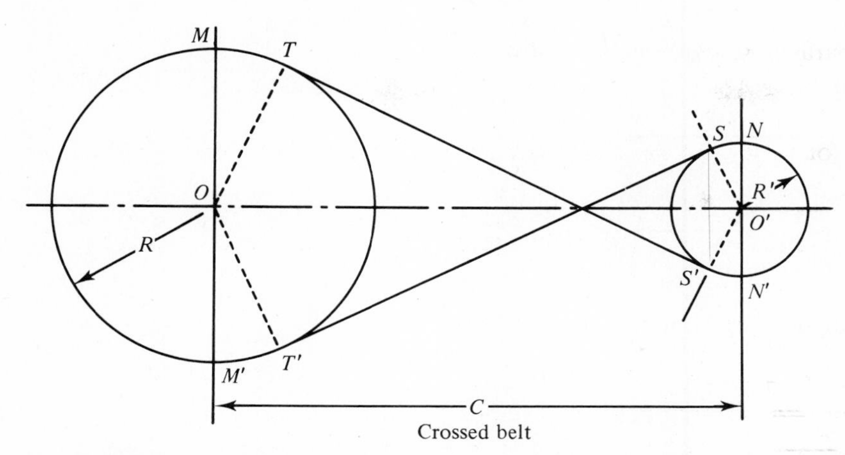

Crossed belt

7. Suppose the figure in problem 6 represented the following distances: arc $MT = 3''$, semicircle $M'M = 18''$, semicircle $NN' = 10''$, arc $SN = 1.5''$, and $\overline{TS'} = 24''$. What length belt would be needed for the pulley system? 83.5"

8. Use the radii of the pulleys below to determine the radius x. 3"

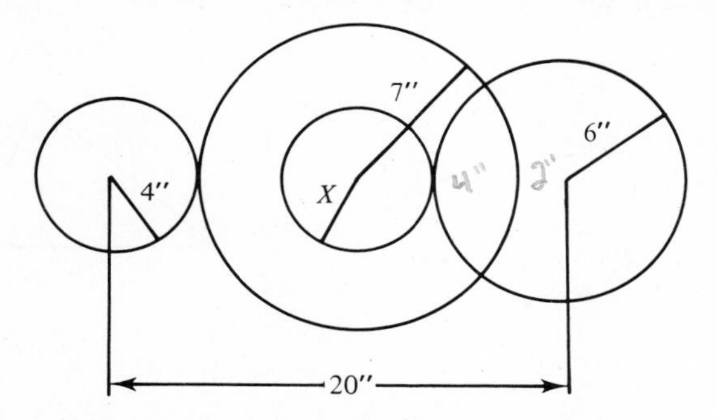

9. Indicate which of the following are polygons:

a. **b.**  **c.** **d.**

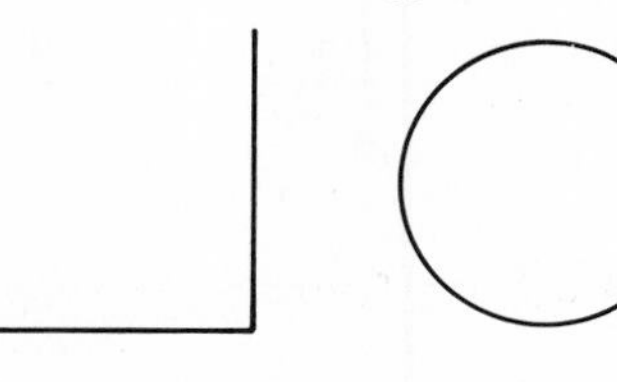

e. **f.** **g.** **h.**

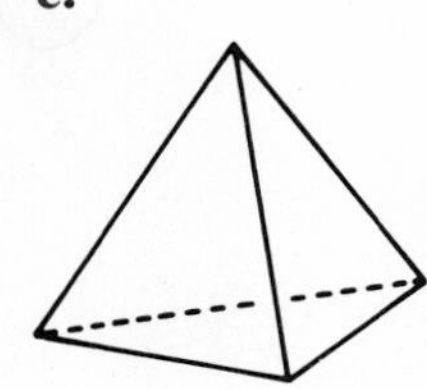

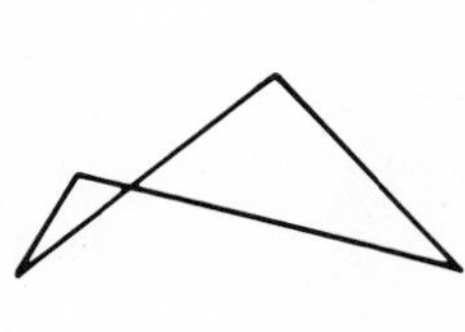

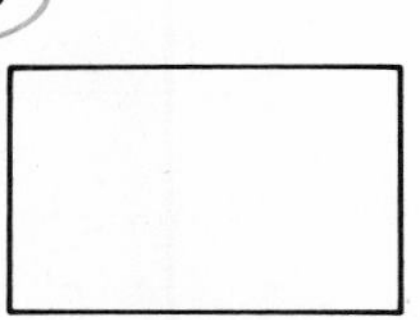

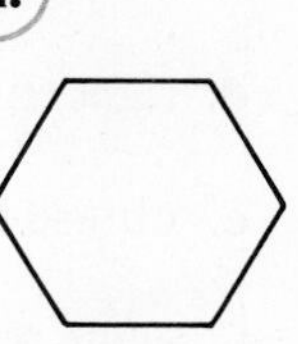

10. Indicate the name of each of the polygons:

a. **b.** **c.** **d.**

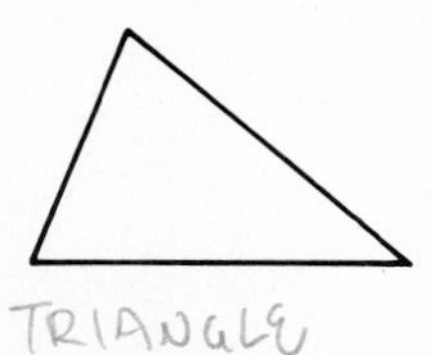
TRIANGLE

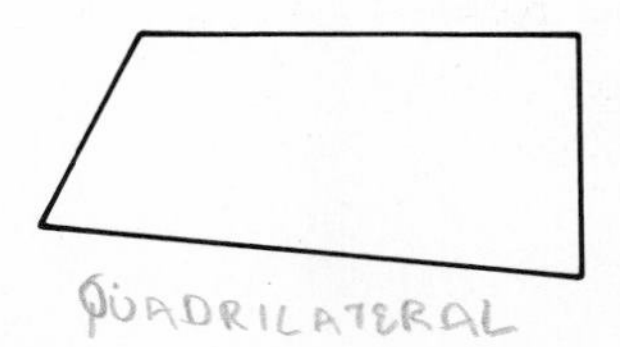
QUADRILATERAL

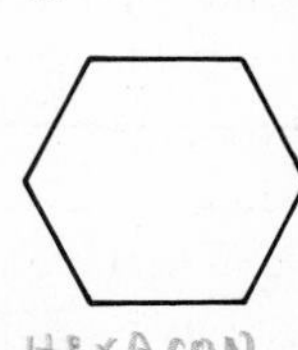
HEXAGON

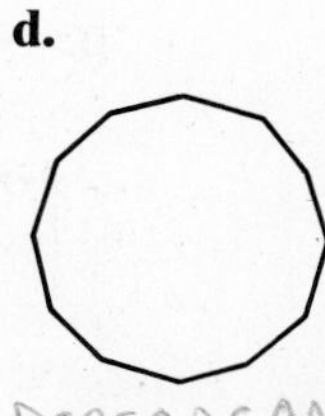
DODECAGON

11. a. Shade a quadrilateral region in the figure.

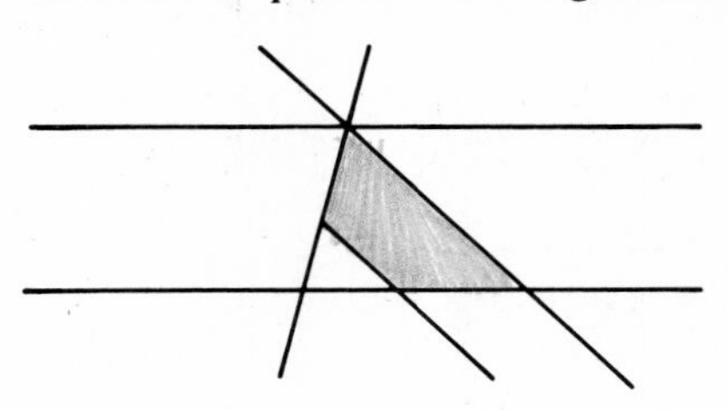

b. Shade a triangular region in the figure (two possible).

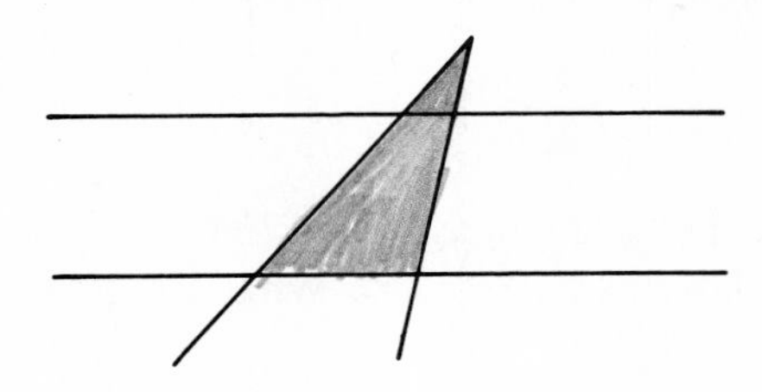

12. List the letters of the appropriate triangles:

a.

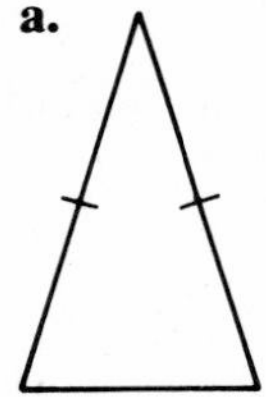

b.

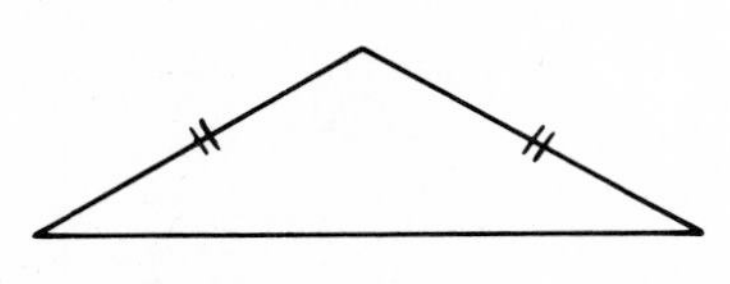

c.

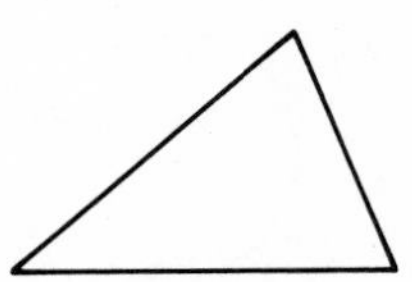

d.

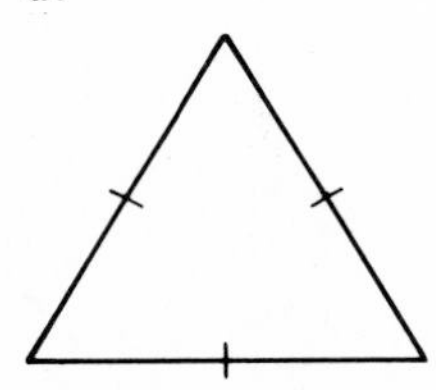

e.

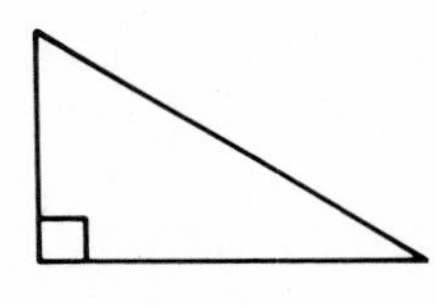

f.

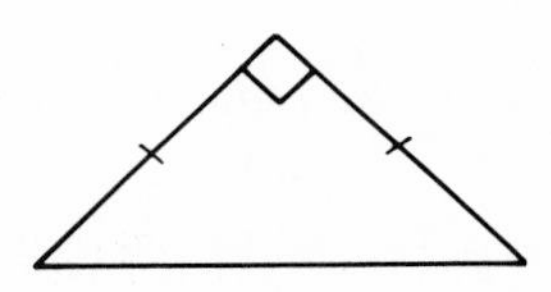

a. equilateral triangles = D

b. isosceles triangles = A, B, D, F

c. scalene triangles = C, E

d. right triangles = E, F

e. obtuse triangles = B

f. acute triangles = A, C, D

13. Consider quadrilateral $ABCD$:

a. Name two sides adjacent to $\overline{DC}$. $\overline{AD}$, $\overline{BC}$

b. Name the side opposite $\overline{AD}$. $\overline{BC}$

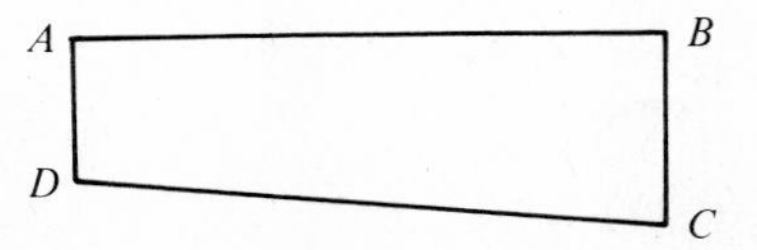

14. Shade a polygonal region that appears to be bounded by a parallelogram.

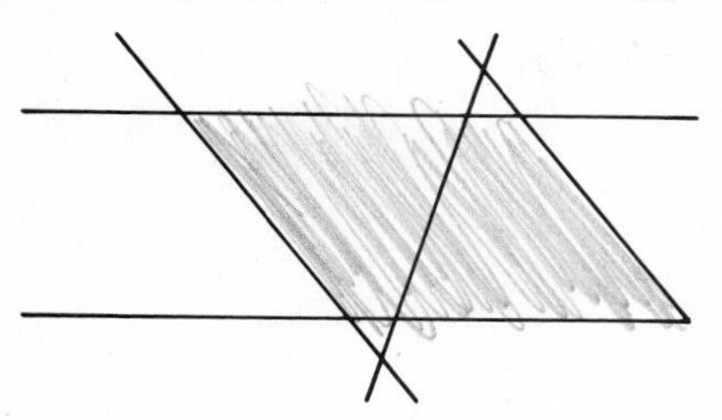

15. How many pairs of parallel sides are contained in a parallelogram? 2
16. Is a square a rectangle? YES
17. Is a rectangle a parallelogram? YES
18. Is a square a parallelogram? YES
19. Is a square a polygon? YES
20. Is a rectangle a square? NO
21. Is a parallelogram a rectangle? NO
22. Is a parallelogram a trapezoid? NO
23. Draw a trapezoid.
24. A square is a rectangle with what further restriction? ALL SIDES MUST BE EQUAL
25. A rectangle is a parallelogram with what further restriction? ALL ANGLES MUST BE RIGHT ANGLES
26. a. Below is a part of the frame of a barn. What polygons predominate the frame? TRIANGLES
b. What kind of triangle appears in each case? RIGHT TRIANGLE

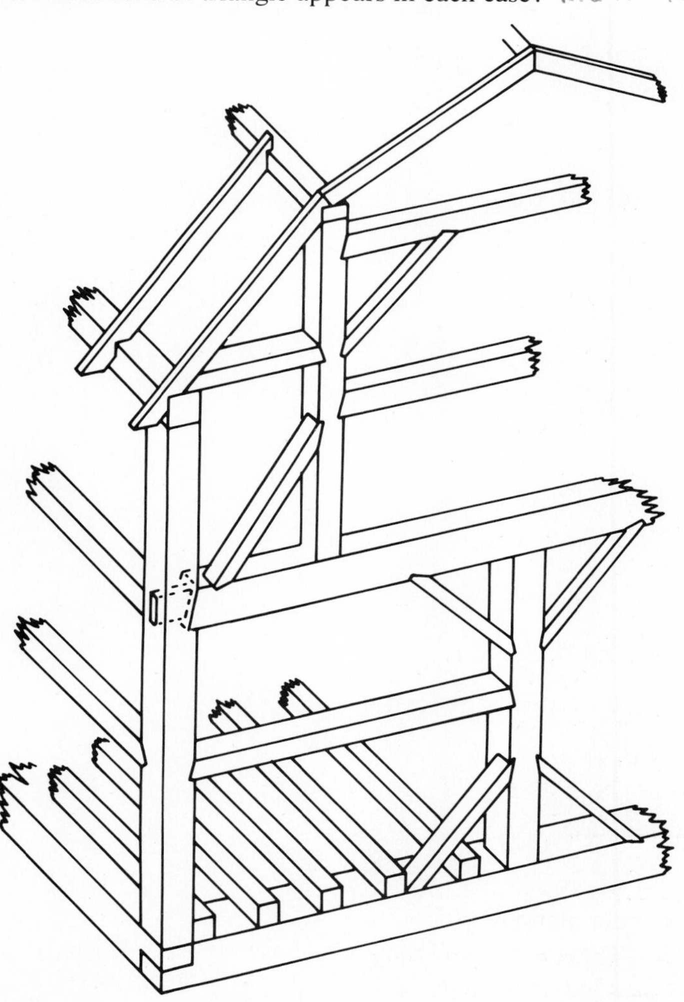

4.3 EXERCISE ANSWERS

1. a. center **b.** radius **c.** diameter
2. 11 centimetres
3. 3.5 inches
4. 2.5 inches
5. 9.75 centimetres
6. a. MM' **b.** $\overline{OM}$, $\overline{OT}$, $\overline{OT'}$, $\overline{OM'}$ and R
c. no **d.** a tangent line
7. 85 inches
8. 3″
9. a, d, g, and h
10. a. triangle **b.** quadrilateral **c.** hexagon **d.** dodecagon
11. a. **b.** or

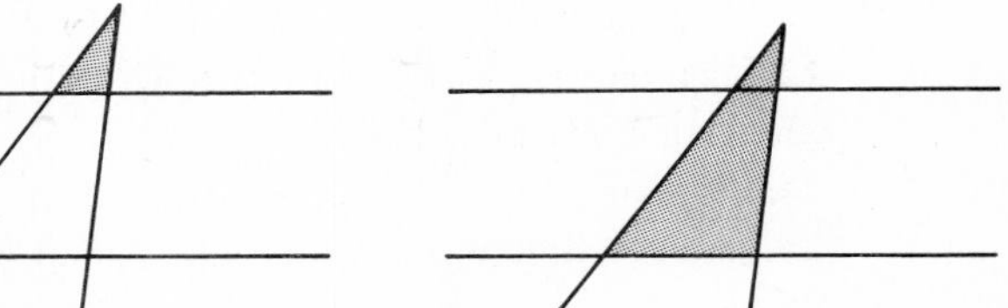

12. a. d **b.** a, b, d, f **c.** c, e **d.** e, f
e. b **f.** a, c, d
13. a. $\overline{AD}$ and $\overline{BC}$ **b.** $\overline{BC}$
14.

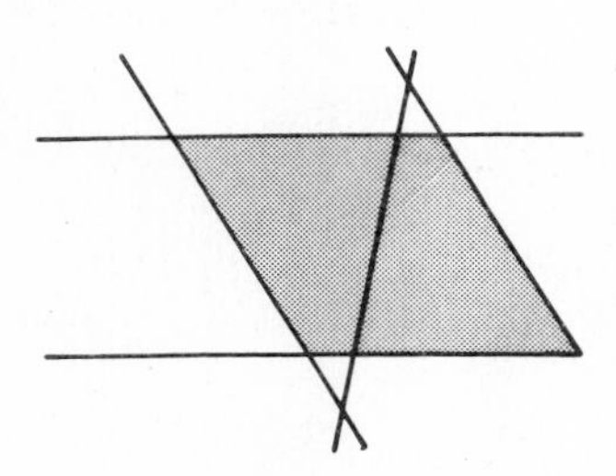

15. 2
16. yes
17. yes
18. yes
19. yes
20. no, not necessarily
21. no, not necessarily
22. no
23.

24. The sides are of equal measure.
25. The angles at the vertices have measure 90°.
26. a. rectangles and triangles: Do you also find the pentagons?
b. right triangles: Many look like they are isosceles.

4.4 POLYHEDRA, PRISMS, AND PYRAMIDS

1 There is a geometric term for solid shapes with flat surfaces. They are called polyhedra. A *polyhedron* is defined as a geometric solid whose surfaces are polygonal regions. Some examples of polyhedra are shown.

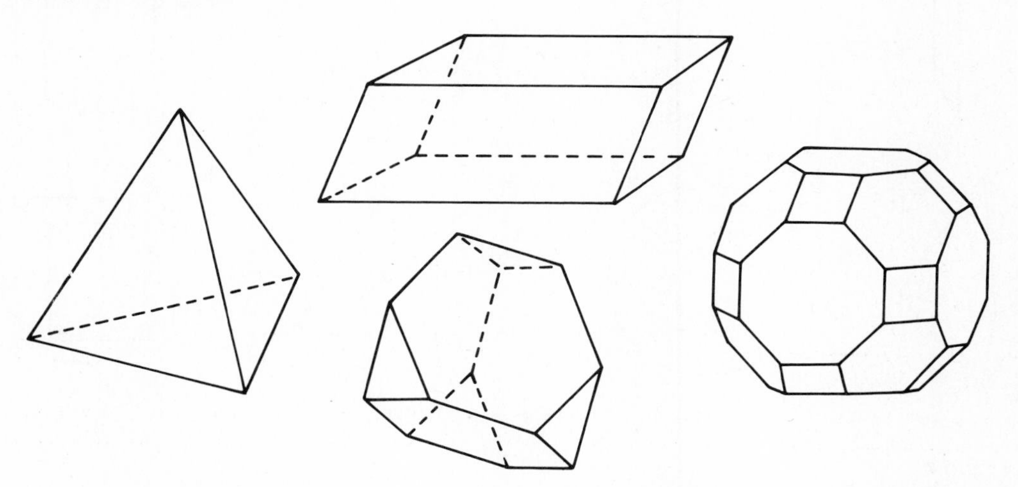

The polygonal regions which form the surface of the polyhedron are called *faces*. The sides of the faces are now called *edges* of the polyhedron. The vertices of the faces are the *vertices* of the polyhedron.

Q1 One common polyhedron is the cube. Examine the cube shown below. You see seven of its eight vertices. Suppose the one you do not see is labeled E. Use your imagination to name all the vertices and edges.

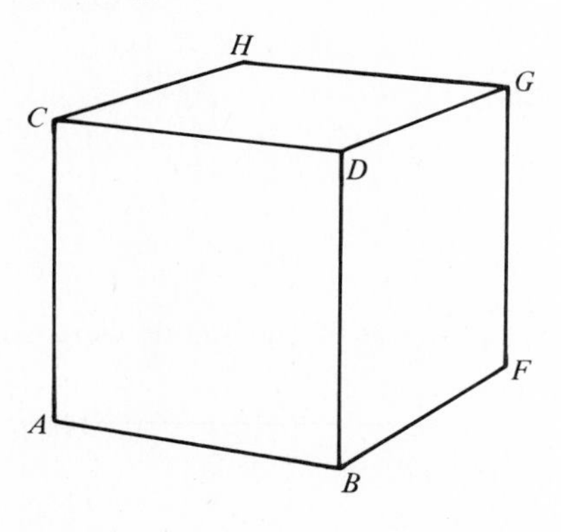

vertices = ____________________

edges = ____________________

\# # # • # # # • # # # • # # # • # # # • # # # • # # # • # # # • # # #

A1 vertices = A, B, C, D, E, F, G, H
edges = $\overline{AB}, \overline{BD}, \overline{DC}, \overline{CA}, \overline{BF}, \overline{DG}, \overline{CH}, \overline{AE}, \overline{EF}, \overline{FG}, \overline{GH}, \overline{HE}$

Q2 How many faces does the cube have in Q1? _____

\# # # • # # # • # # # • # # # • # # # • # # # • # # # • # # # • # # #

A2 6

Q3 Below is an array of solid figures. Some are polyhedra and some are not. List the letters of all figures that are polyhedra. ____________

a.

b.

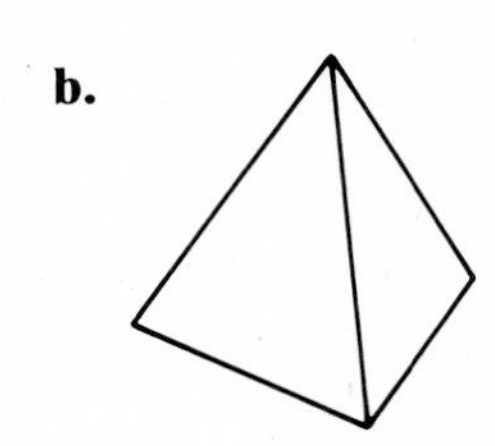

c.

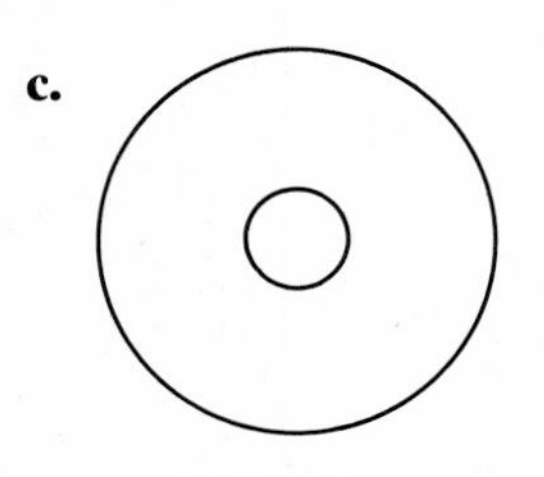

d.

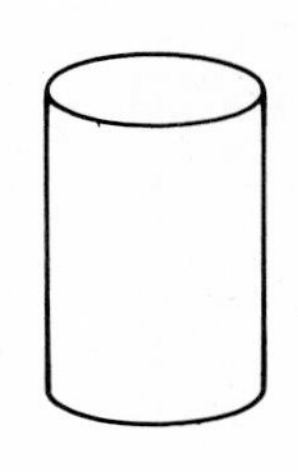

e.

f.

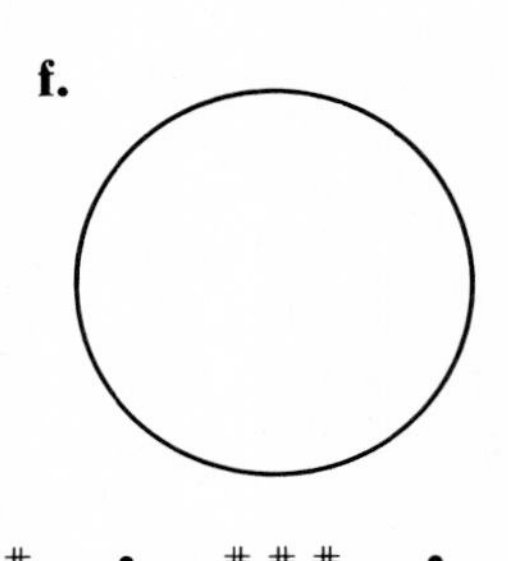

g.

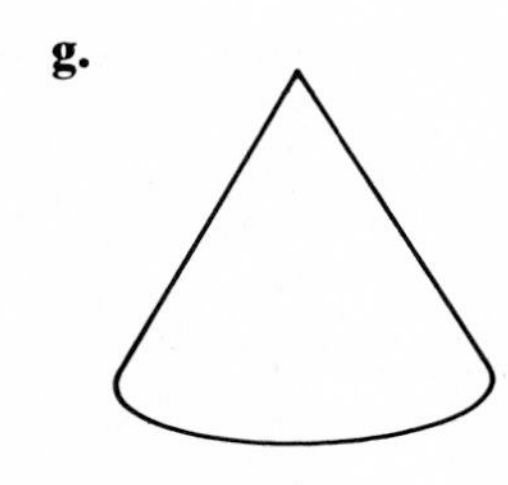

h.

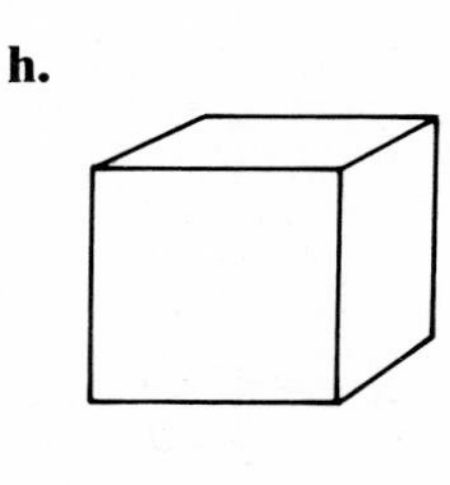

• # # # • # # # • # # # • # # # • # # # • # # # • # # # • # #

A3 **a**, **b**, **e**, and **h**

2 Polyhedra are classified by the number of faces they contain. A *tetrahedron* has four faces, an *octahedron* eight, and a *dodecahedron* twelve. Except for the tetrahedron, polyhedra with a certain number of faces are named with the same prefix as a polygon with that number of sides. For example, a polyhedron with five faces would be called a pentahedron, because a polygon with five sides is called a pentagon. The following polyhedron is an octahedron. Notice that the figure looks

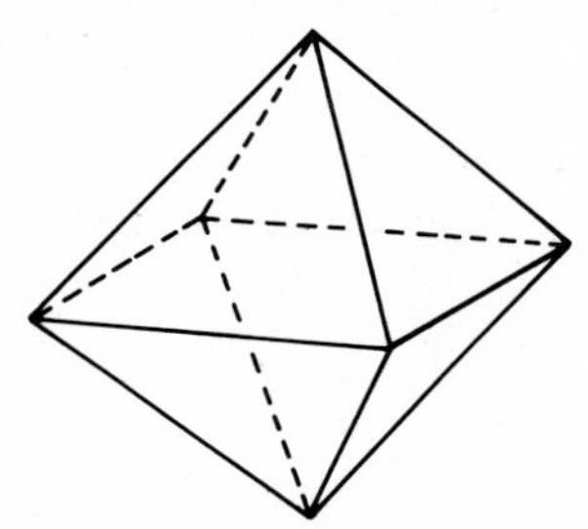

transparent. However, line segments that would actually be hidden if the figure were solid are dashed. This helps a person "view" all the faces.

Q4 Indicate (1) the numbers of faces and (2) the name of each polyhedron:

a.

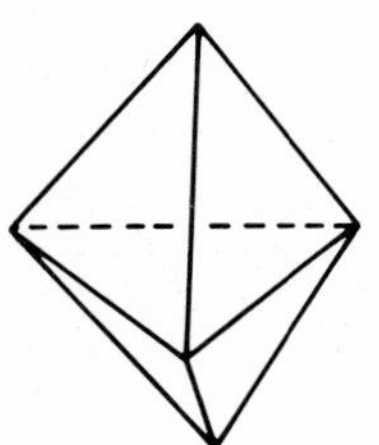

(1) ____________

(2) ____________

b.

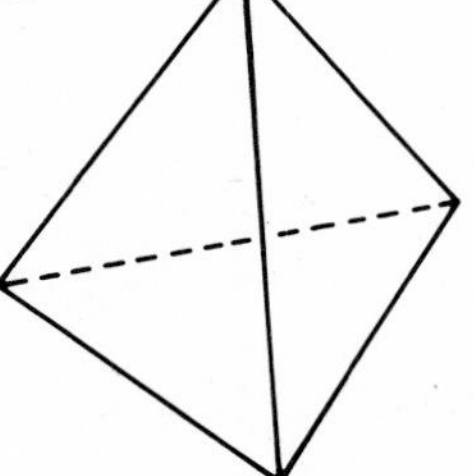

(1) ____________

(2) ____________

c.

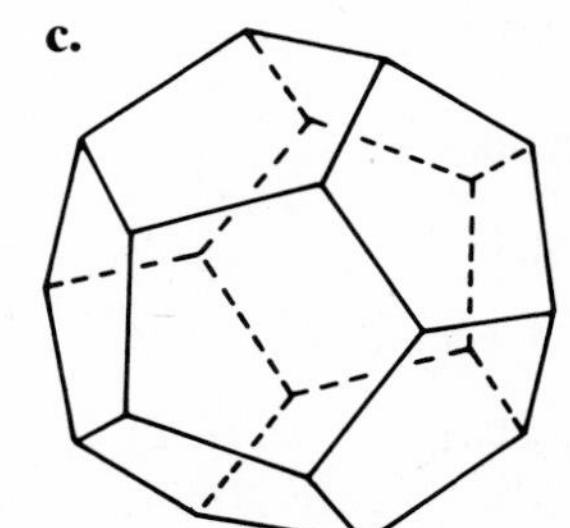

(1) ____________

(2) ____________

• # # # • # # # • # # # • # # # • # # # • # # # • # # # • # #

A4 **a.** (1) 6 faces (2) hexahedron
b. (1) 4 faces (2) tetrahedron
c. (1) 12 faces (2) dodecahedron

---

3 If a polygon has sides and angles of equal measure, it is called a *regular polygon.* A *regular polyhedron* has faces that are regular polygons. Since there are an infinite number of regular polygons, it may be surprising that there are only five regular polyhedra. They are the regular tetrahedron, the cube (actually a regular hexahedron), the regular octahedron, the regular dodecahedron, and the regular icosahedron (20 faces).

Q5 Write the appropriate name under each regular polyhedron:

a.

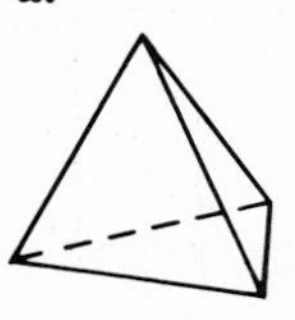

b.

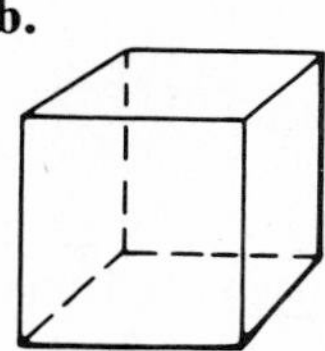

c.

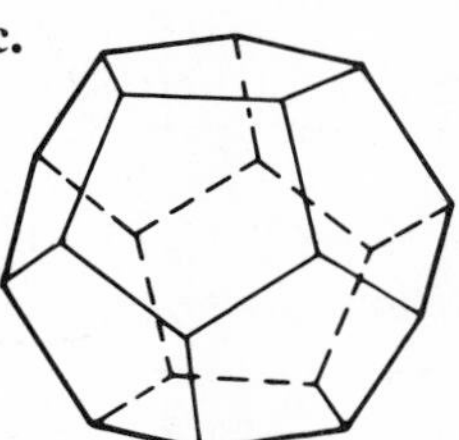

d.

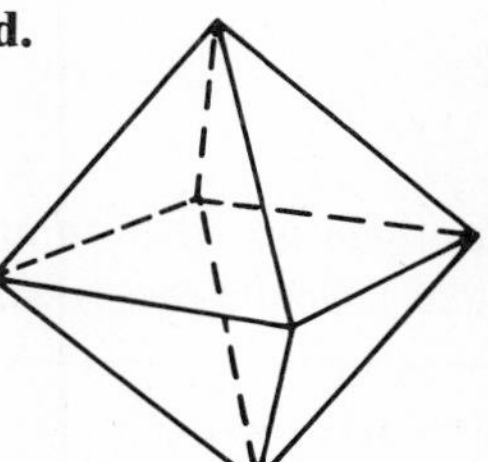

e.

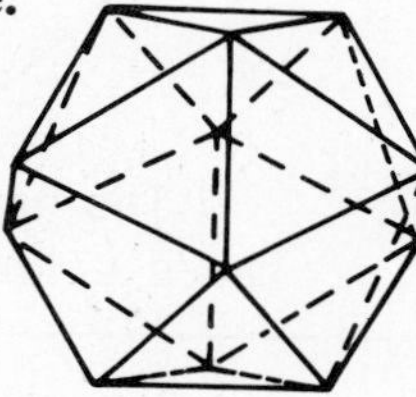

a. ____________ **b.** ____________
c. ____________ **d.** ____________
e. ____________

\# # # • # # # • # # # • # # # • # # # • # # # • # # # • # # # • # # #

A5 **a.** regular tetrahedron **b.** cube
c. regular dodecahedron **d.** regular octahedron
e. regular icosahedron

Q6 It is possible to cut a model of a regular solid on some of its edges and "flatten" it out. Examine the flattened solids here and match them with the regular solids in Q5.

a.

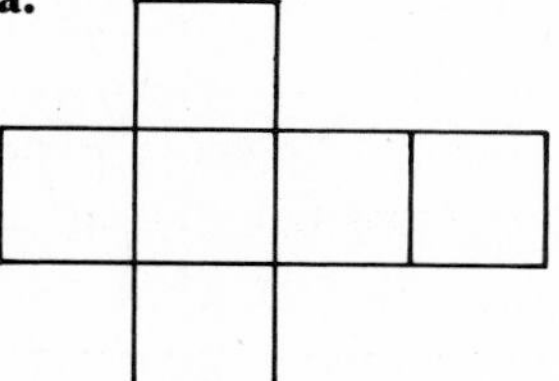

matches ______

b.

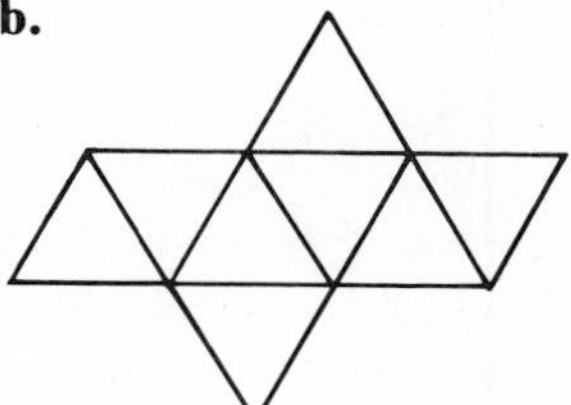

matches ______

c.

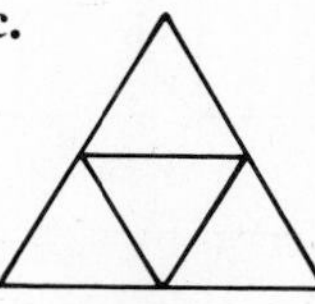

matches ______

d.

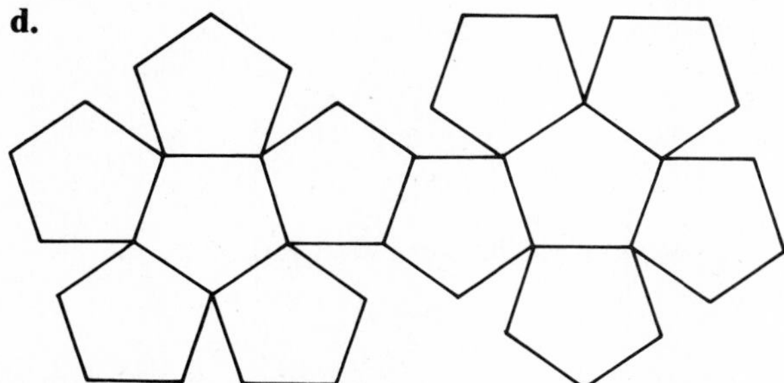

matches ______

e.

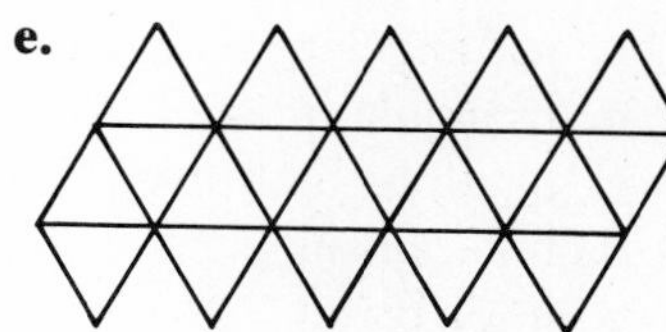

matches ______

\# # # • # # # • # # # • # # # • # # # • # # # • # # # • # # # • # # #

A6 **a.** cube (b) **b.** octahedron (d) **c.** tetrahedron (a)
d. dodecahedron (c) **e.** icosahedron (e)

---

4 A prism is a special kind of polyhedron. A *prism* has two faces which have the same size and shape and which lie in parallel planes; the remaining faces are parallelograms. The two faces that are parallel to each other are called *bases*. The remaining faces are called *lateral* faces. Some examples:

a. **b.** 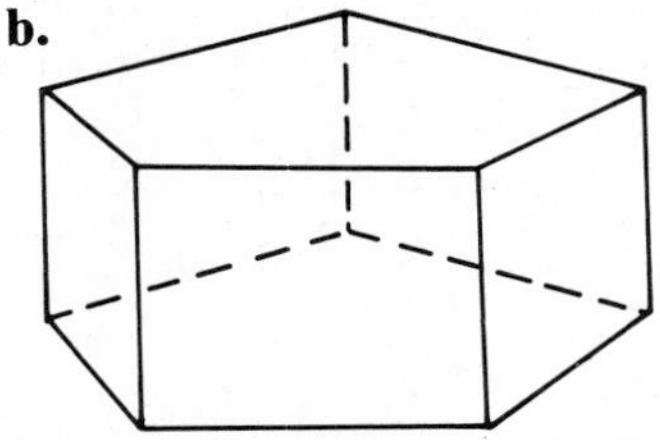 **c.**

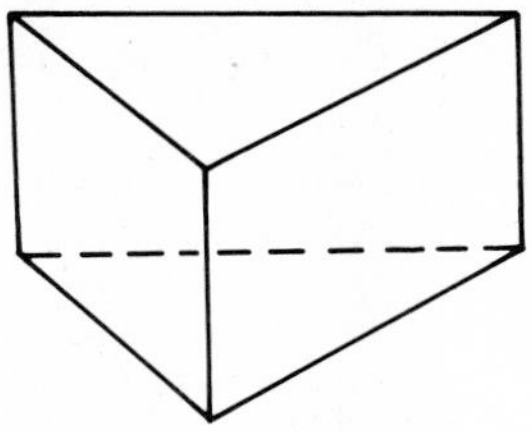

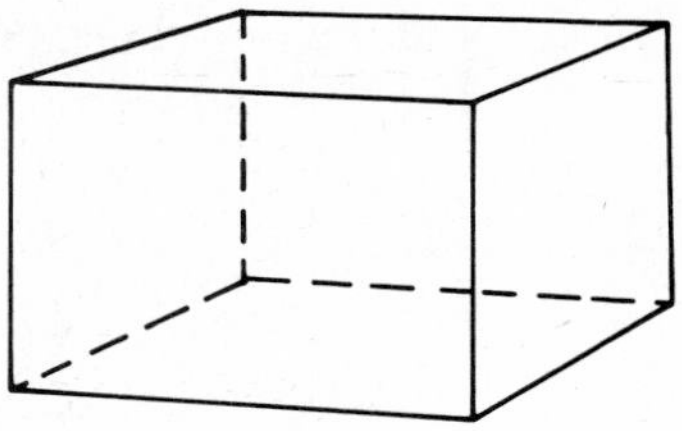

The base determines the name of the prism. A prism whose bases are triangular regions is called a *triangular prism*; a prism whose base is a hexagonal region is a *hexangular prism*; and so on.

Q7 What would the three prisms in Frame 4 be called?

a. ______________________

b. ______________________

c. ______________________

• ### • ### • ### • ### • ### • ### • ### •

A7 **a.** triangular prism **b.** pentangular prism **c.** rectangular prism

---

Q8 Give the descriptive name of each prism; otherwise, indicate "not a prism":

a. **b.** **c.** **d.** **e.** **f.**

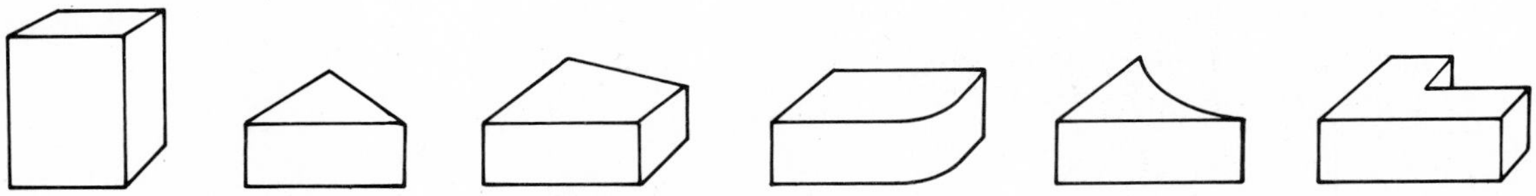

a. ______________________ **b.** ______________________

c. ______________________ **d.** ______________________

e. ______________________ **f.** ______________________

• ### • ### • ### • ### • ### • ### • ### •

A8 **a.** rectangular prism (any of the rectangular faces could be considered the base)
b. triangular prism
c. quadrangular prism
d. not a prism
e. not a prism
f. hexangular prism (the prism bases must be polygons of the same size and shape and lie in parallel planes)

Q9 Machined steel or cast parts are frequently prisms. On the next page a cast iron bedplate is shown.
a. Considering the end of the casting to be the base, what would be the name of the prism? ______________________

b. Think of dividing the prism into three parts at the dotted lines. Three prisms are formed. What would be the name of each? ____________________

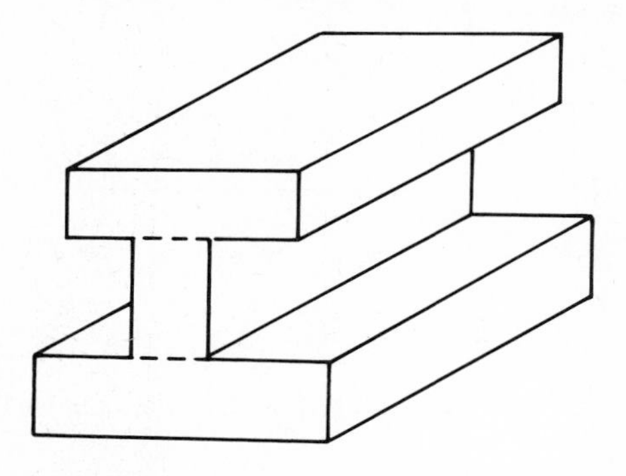

\# # # • # # # • # # # • # # # • # # # • # # # • # # # • # # # • # # #

A9 **a.** dodecagon prism **b.** rectangular prism

Q10 **a.** Consider the end of the V-block below to be the base of the prism. What is the name of the prism? ____________________

b. Consider the V-block to be a prism with another prism removed. What is the name of the original prism? ____________________ What is the name of the prism that has been removed? ____________________

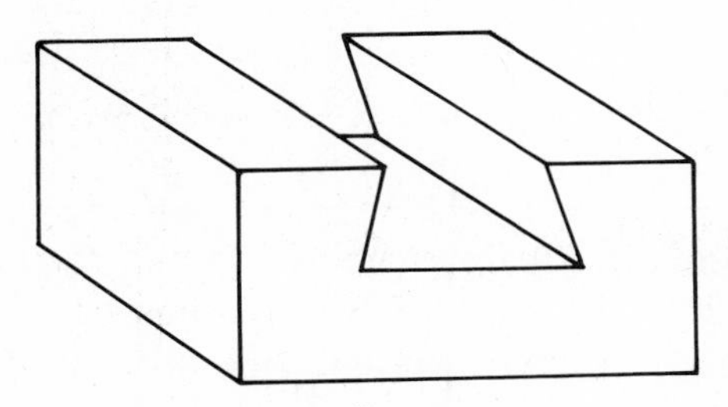

\# # # • # # # • # # # • # # # • # # # • # # # • # # # • # # # • # # #

A10 **a.** octagonal prism **b.** rectangular prism, trapezoidal prism

5 Prisms can be drawn with a small amount of instruction and some practice. To begin with, two views of a common rectangular prism called the cube will be examined. Below is a picture of the cube and a geometric diagram of the cube. The edges that appear on both figures are drawn with solid lines. The edges that are not visible in the picture are called *hidden* edges and are drawn with dashed lines in the diagram.

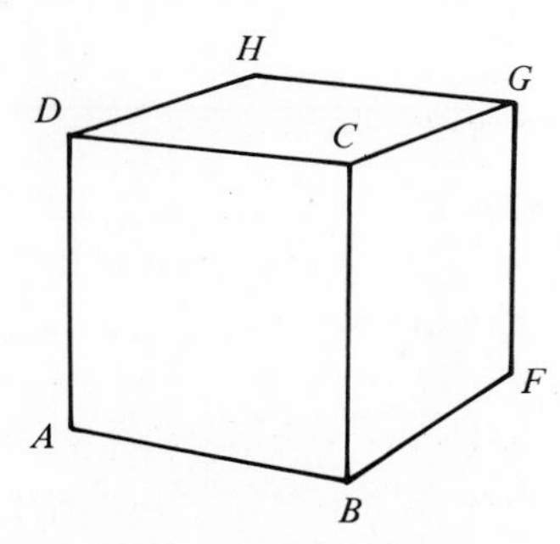

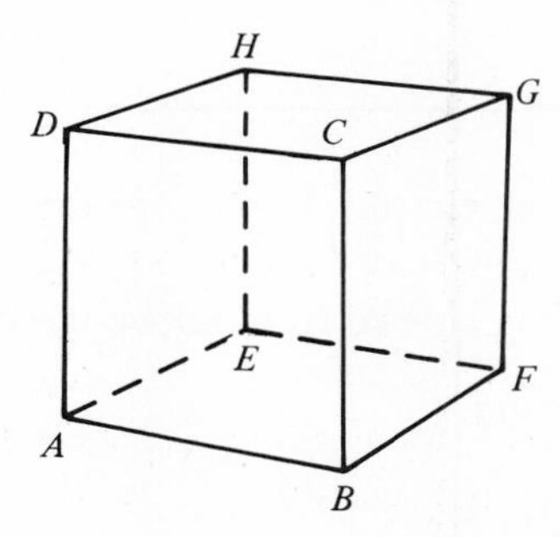

Q11 Consider the cube in Frame 5:

a. What is the one hidden vertex? ____________

b. Name the three hidden edges. ____________

c. One of the hidden faces is *ABFE*. Name two others. ____________

\# # # • # # # • # # # • # # # • # # # • # # # • # # # • # # # • # # #

A11 **a.** *E* **b.** $\overline{AE}$, $\overline{EF}$, and $\overline{EH}$

c. *AEHD* and *EFGH* (The letters should be arranged in order starting with any one of them; that is, $AEHD = EHDA$.)

Q12 Notice that faces on opposite sides of the prism (in Frame 5) have exactly the same shape:

a. Name the front and back lateral faces. ____________

b. Name the left and right lateral faces. ____________

c. Name the top base and then the bottom base. ____________

\# # # • # # # • # # # • # # # • # # # • # # # • # # # • # # # • # # #

A12 **a.** *ABCD* and *EFGH* **b.** *AEHD* and *BFGC* **c.** *DCGH* and *ABFE*

6 Now look at the same cube from a different angle. Suppose that the cube of Frame 5 had face *ABFE* resting on a transparent table. Move your eye down below the table and look up at the cube. Different edges become hidden. Compare this view of the cube with the view in Frame 5 until you "see" in your mind how the angle of sight has changed.

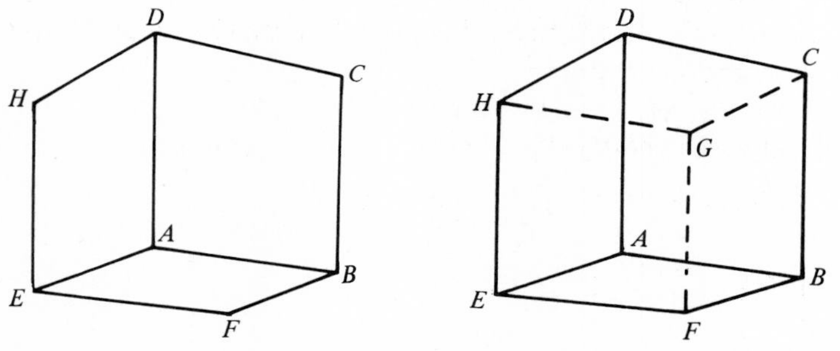

D C H A B E F

D C H G A B E F

Q13 **a.** Is the cube in Frame 5 being viewed from above the cube or from below? ________

b. Is the cube in Frame 6 being viewed from above the cube or from below? ________

\# # # • # # # • # # # • # # # • # # # • # # # • # # # • # # # • # # #

A13 **a.** above **b.** below

Q14 Consider the cube of Frame 6:

a. Name the hidden vertex. G

b. Name the hidden edges. HG, GC, GF

c. Were any of the hidden edges of Frame 6 also hidden in Frame 5? NO

Which ones? ________

d. Name the hidden faces. HDCG, GFBC, HGFE

e. Were any of the hidden faces of Frame 6 also hidden in Frame 5? YES

Which ones? HEFG

\# \# \# • \# \# \# • \# \# \# • \# \# \# • \# \# \# • \# \# \# • \# \# \# • \# \# \# • \# \# \#

A14 **a.** *G* **b.** $\overline{HG}$, $\overline{FG}$, and $\overline{CG}$

c. no, none **d.** *EFGH*, *BCGF*, and *CDHG*

e. yes, *EFGH* (the face on the back of the cube)

Q15 Which sketches are drawn correctly? D

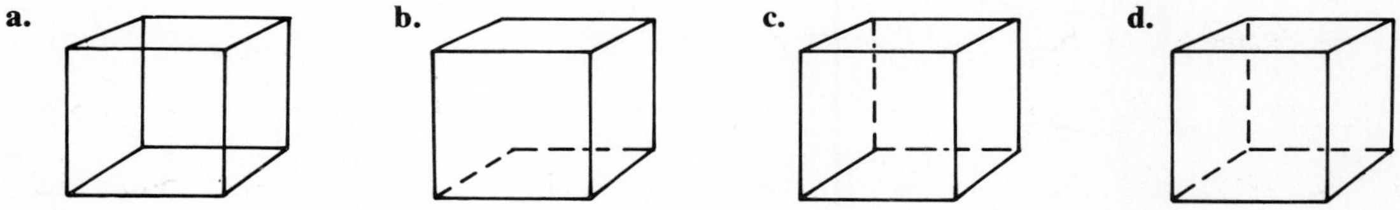

a. **b.** **c.** **d.**

\# \# \# • \# \# \# • \# \# \# • \# \# \# • \# \# \# • \# \# \# • \# \# \# • \# \# \# • \# \# \#

A15 **d**

7 To sketch a cube some of its faces must be distorted in order to "look" right. However, the bottom hidden base is always the same shape as the visible top base. Several views of a cube are shown here. Notice that some of the hidden edges have been sketched. Use these as a guide to complete Q16.

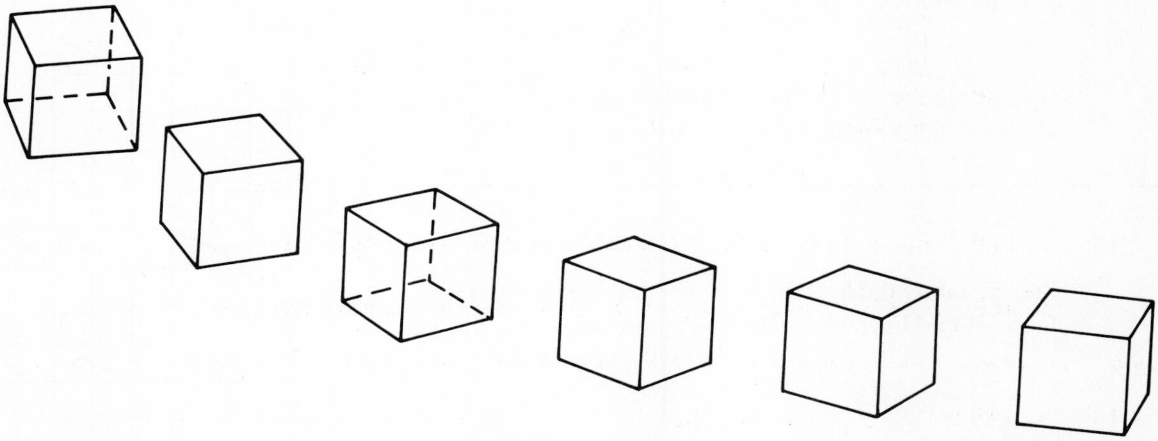

Q16 Sketch the hidden edges in each of these cubes.

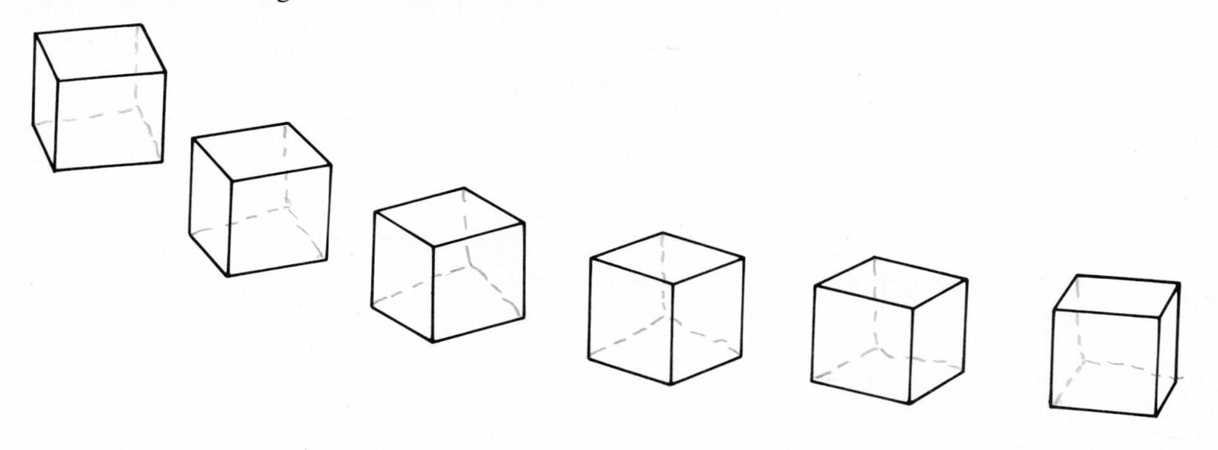

• ### • ### • ### • ### • ### • ### • ### •

A16

8 You have had a chance to interpret sketches of cubes. You now will get an opportunity to make your own sketches. It is helpful to have a background of graph paper on which to draw, although the same procedures may be followed without it. To sketch a cube follow these steps:

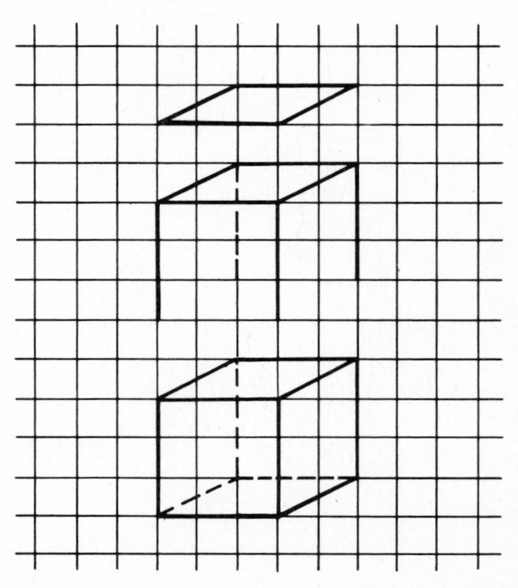

Step 1: Draw the top face. Make opposite edges parallel and the same length.

Step 2: Draw another edge from each vertex as shown. Make them parallel and the same length.

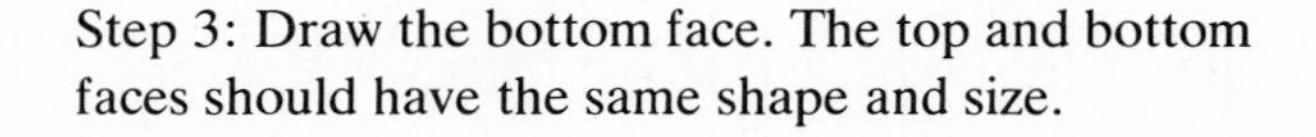

Step 3: Draw the bottom face. The top and bottom faces should have the same shape and size.

Q17 Here is another example. Copy it on the graph provided. Count the squares carefully.

Step 1:

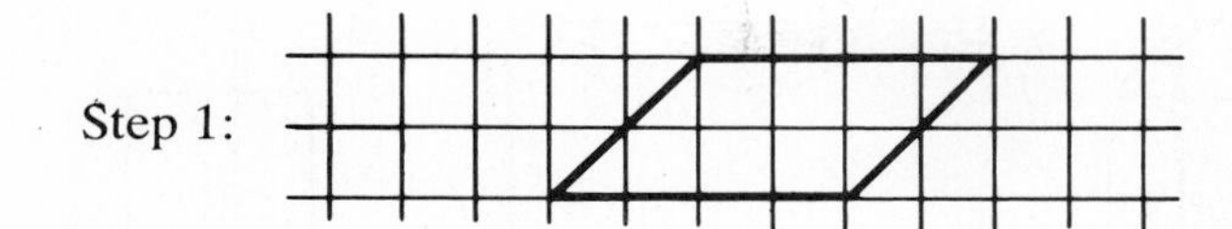

Step 2:

Step 3:

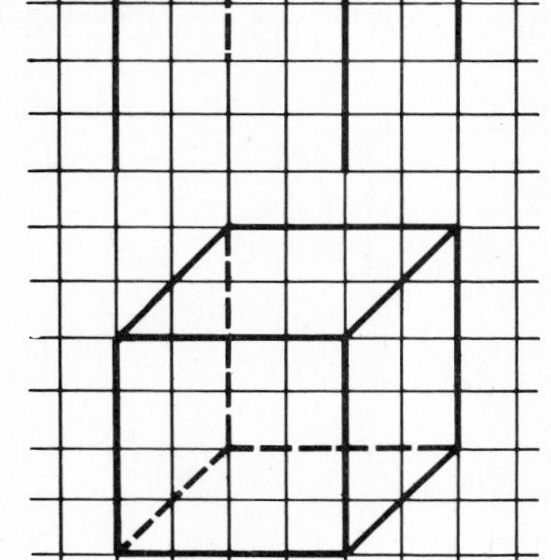

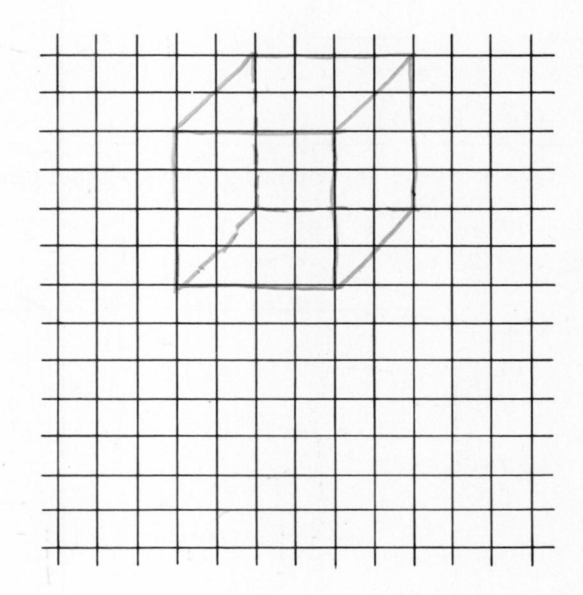

\# # # • # # # • # # # • # # # • # # # • # # # • # # # • # # # • # # #

A17 Compare your sketch with the finished cube.

Q18 Copy each of the following sketches. Count squares to make your drawings the same size and shape.

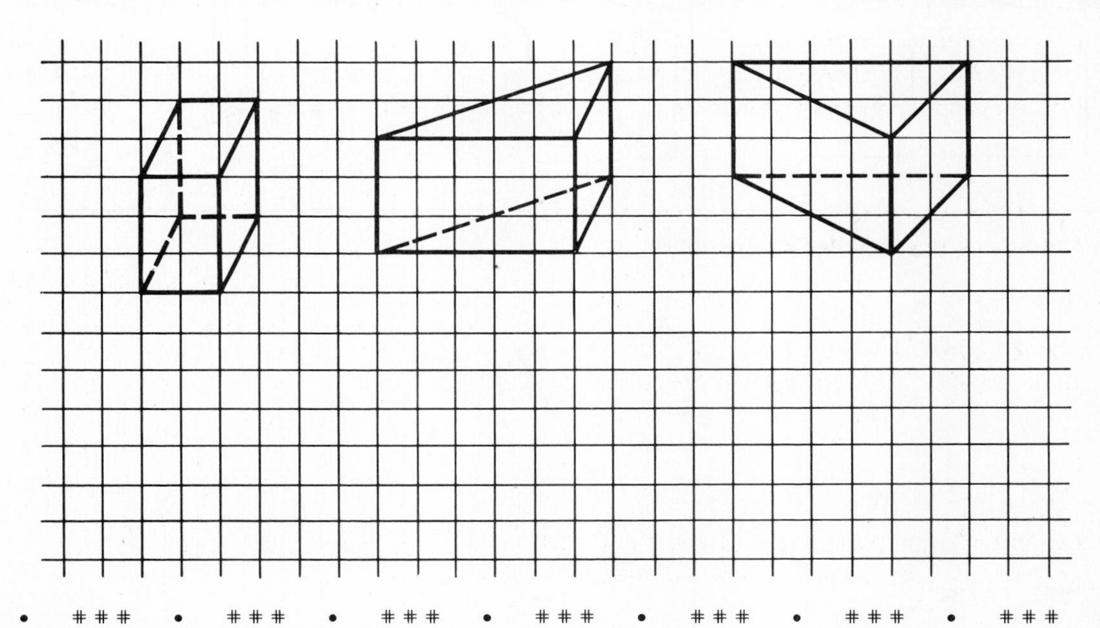

\# # # • # # # • # # # • # # # • # # # • # # # • # # # • # # # • # # #

A18 Compare your sketch with the original prism.

Q19 Which of the following sketches has all the hidden edges drawn correctly?

a.
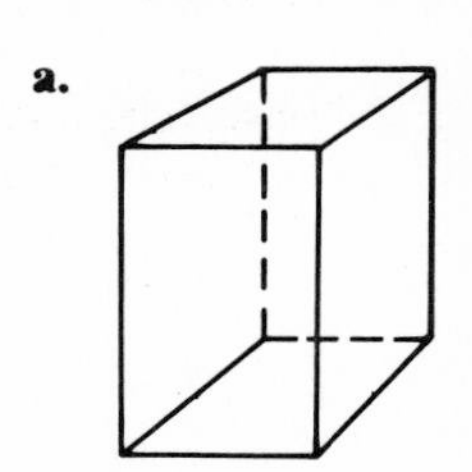
b.
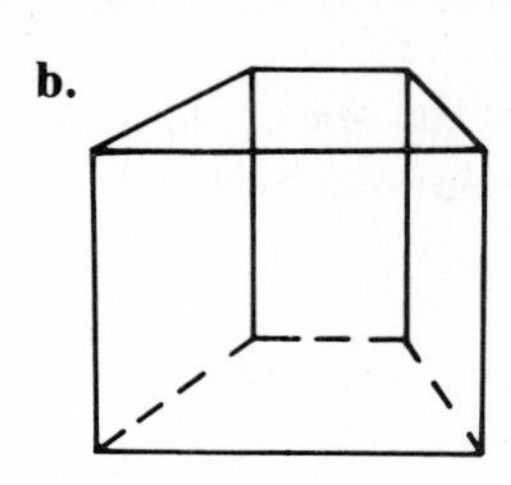
c.
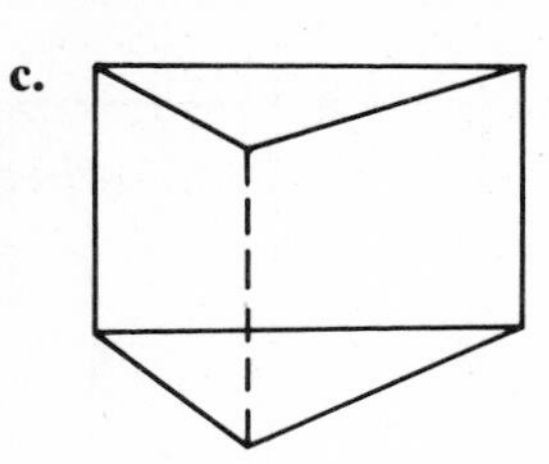
d.
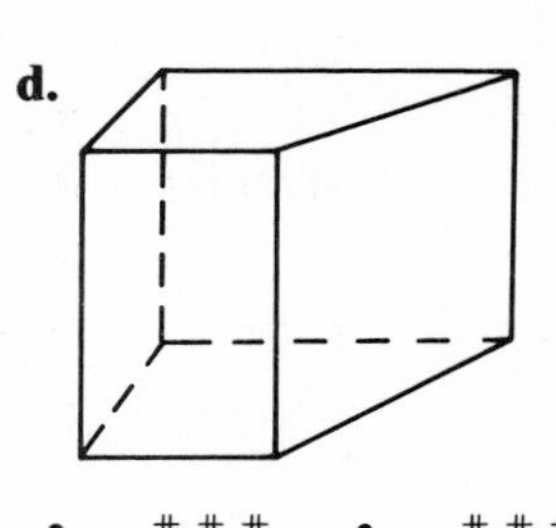

• # # # • # # # • # # # • # # # • # # # • # # # • # # # • # #

A19 **d**

Q20 Sketch the following prisms without using a background of graph paper. Include all hidden edges in your sketch.

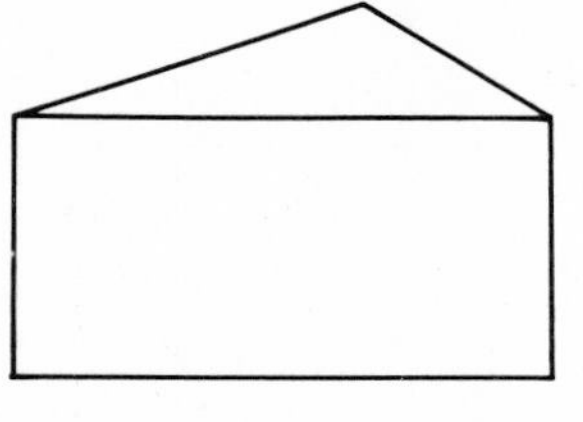
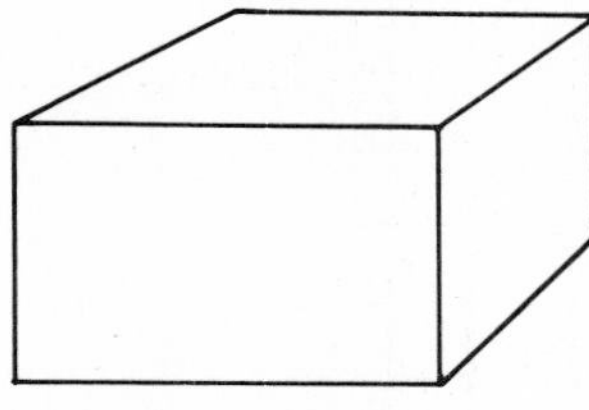
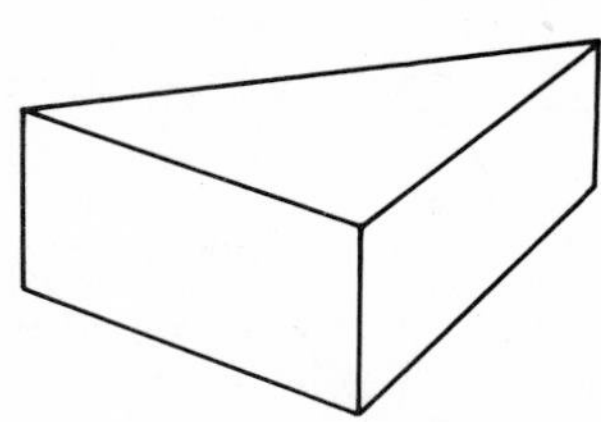

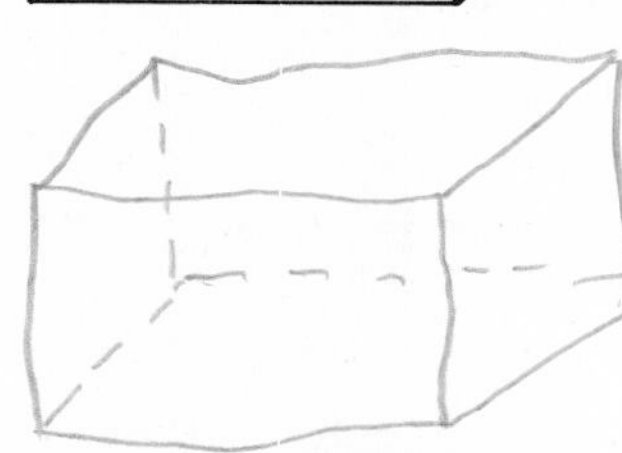
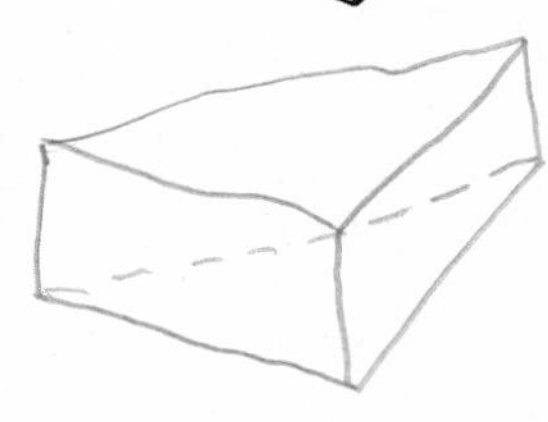

• # # # • # # # • # # # • # # # • # # # • # # # • # # # • # #

A20

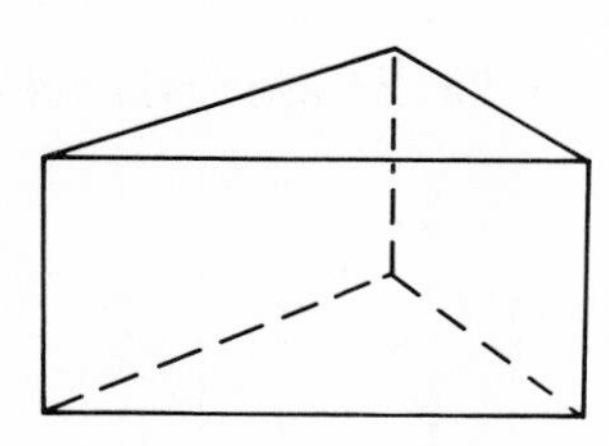
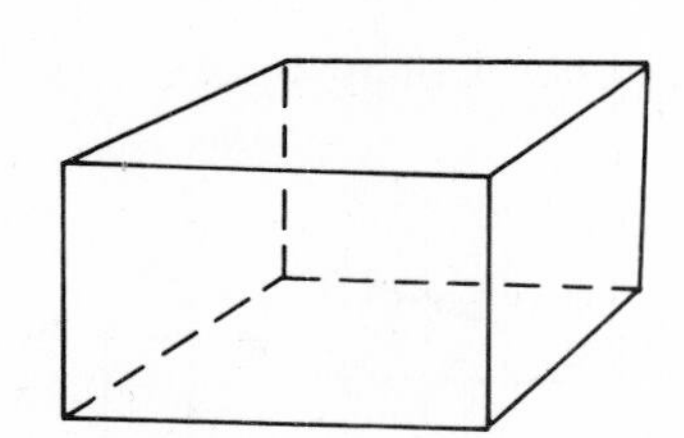
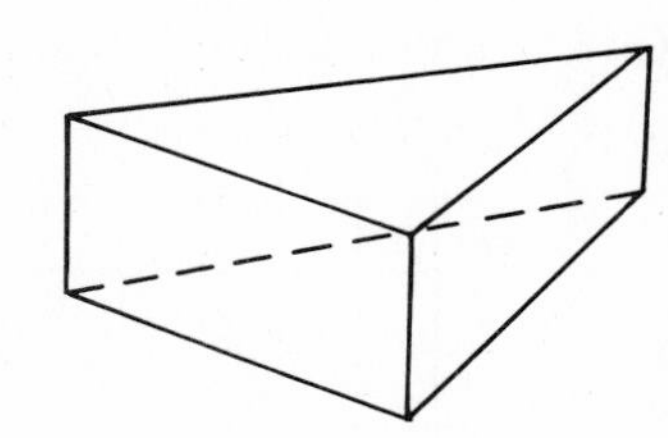

9 The following polyhedra are pyramids:

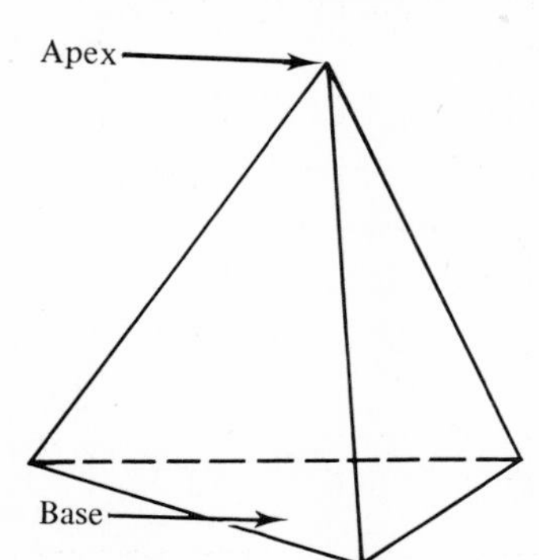

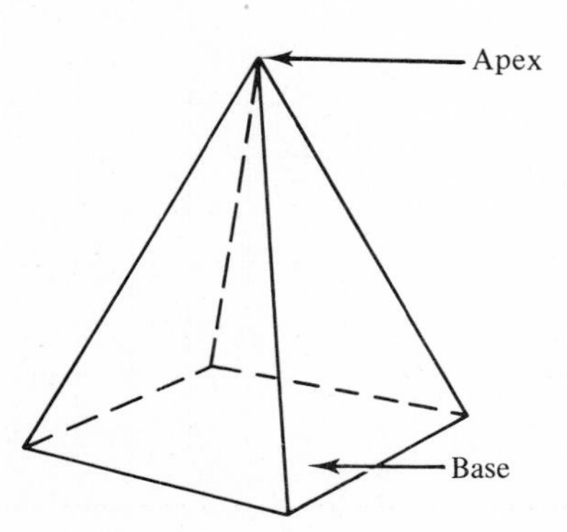

A *pyramid* is formed by a polygonal region called the *base* and a point outside the plane containing the base called the *apex*. A pyramid is the solid formed by all the points in the polygonal region of the base and all triangular regions formed by segments between the vertices of the base and the apex together with all interior points of the figure.

The *altitude* of a pyramid is a segment from the apex perpendicular to the base. The apex along with the vertices of the base are the vertices of the pyramid.

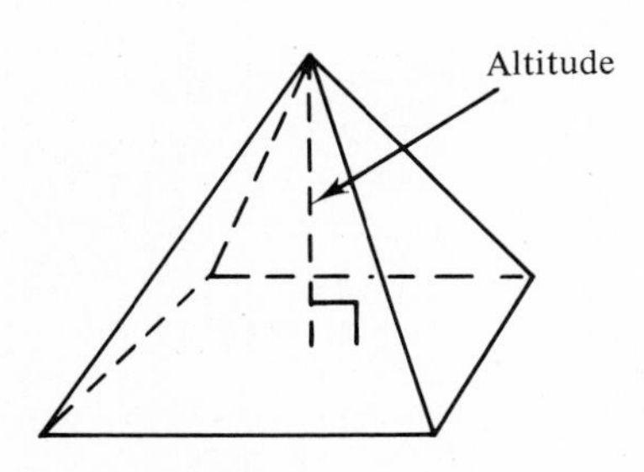

Q21 Choose the sketches with the hidden lines shown correctly: ______

a.

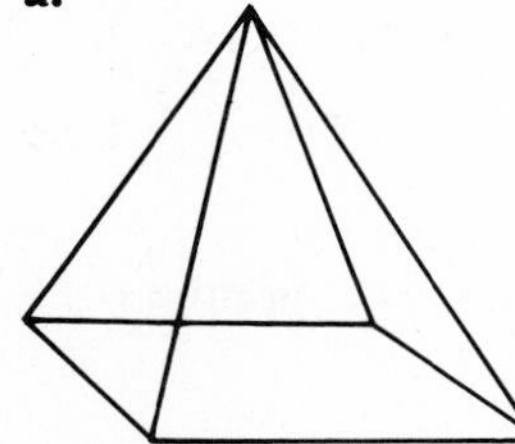

b.

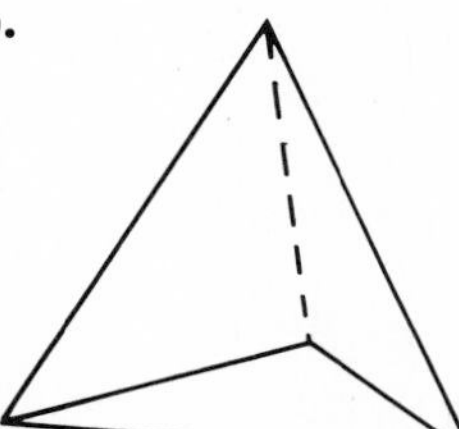

c.

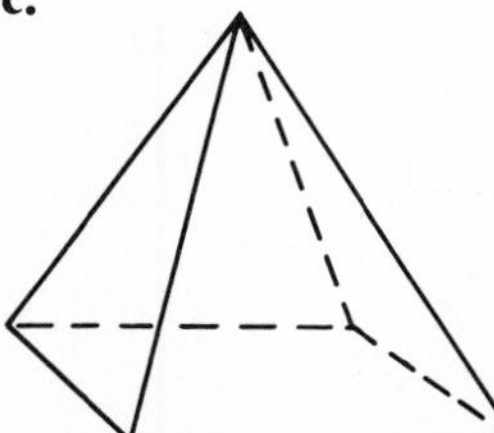

d.

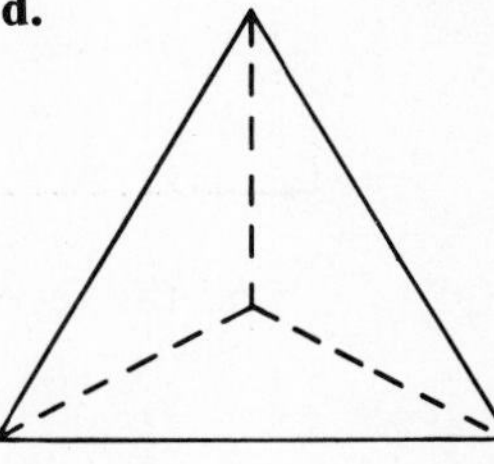

\# # # • # # # • # # # • # # # • # # # • # # # • # # # • # # # • # # #

A21 **c** and **d**

Q22 Use the polygonal regions shown below for the base and the point A for the apex and sketch the pyramids.

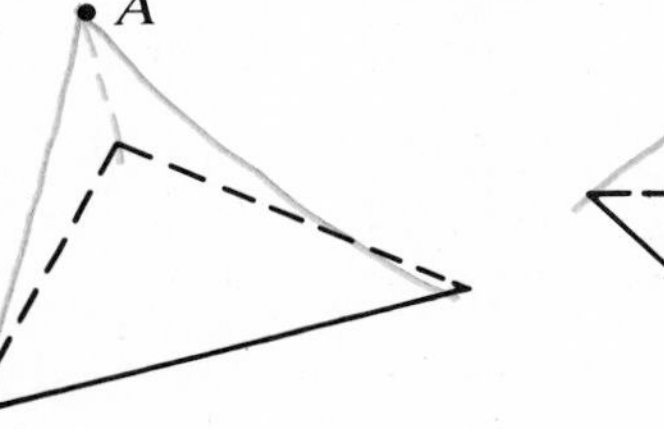

\# # # • # # # • # # # • # # # • # # # • # # # • # # # • # # # • # # #

A22

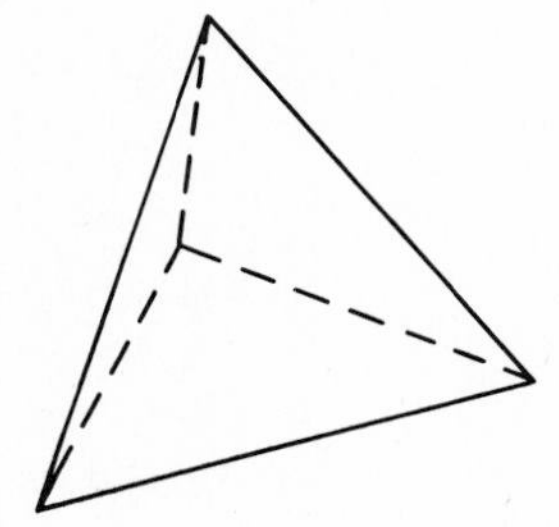

10 Pyramids are classified and named according to the shape of their base. One with a triangular base is a triangular pyramid; one with a rectangular base is a rectangular pyramid; and so on.

Q23 Make a sketch of a triangular pyramid.

a. How many vertices does it have? ______

b. How many edges does it have? ______

c. How many faces does it have? ______

• ### • ### • ### • ### • ### • ### • ### •

A23

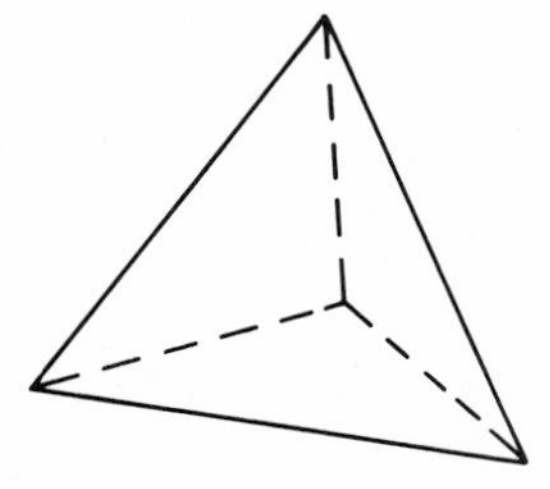

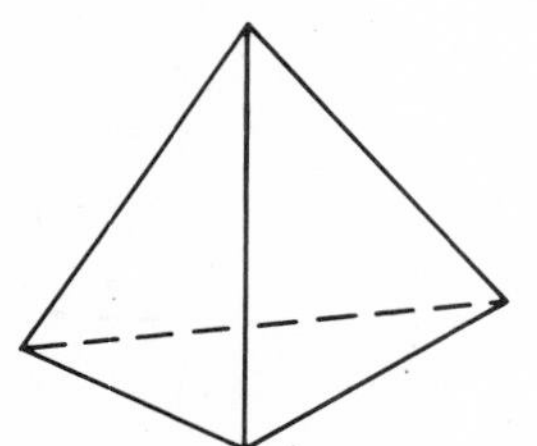

a. 4 **b.** 6 **c.** 4

Q24 If all the faces, including the base, are the same size and shape, a triangular pyramid could also be called a regular ______________.

• ### • ### • ### • ### • ### • ### • ### •

A24 tetrahedron

Q25 Below is a sketch of the pyramids at Gizeh in Egypt. What is the name of these pyramids?

____

• ### • ### • ### • ### • ### • ### • ### •

A25 square pyramids

Q26 The machined steel piece below is made up of two of the solids discussed in this section. What are they? ______________

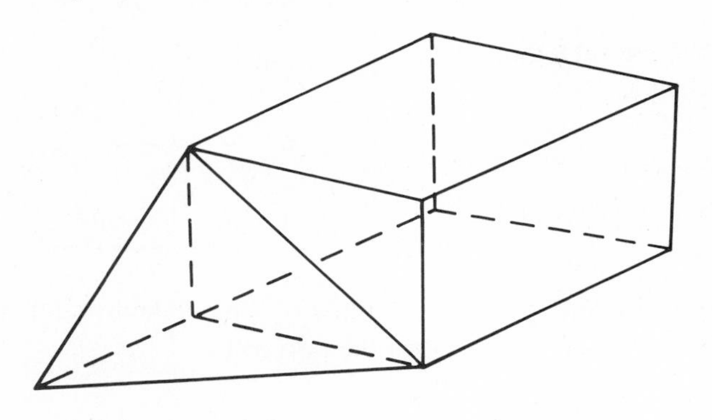

\# # # • # # # • # # # • # # # • # # # • # # # • # # # • # # # • # # #

A26 rectangular prism and triangular pyramid

Q27 If the center of a cube is connected with each of its vertices, how many pyramids are formed? ______

\# # # • # # # • # # # • # # # • # # # • # # # • # # # • # # # • # # #

A27 6

This completes the instruction for this section.

4.4 EXERCISE

1. Consider the polyhedron and answer the questions:

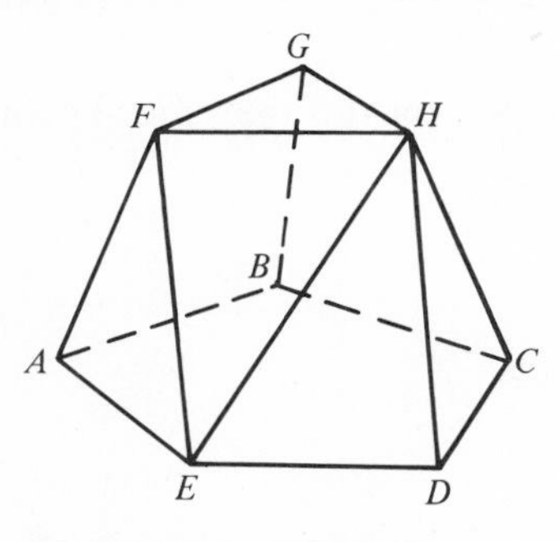

a. Name the "hidden" vertices.
b. Name the "hidden" edges.
c. Name two quadrilateral faces.
d. Name a pentagonal face.
e. Name five triangular faces.
f. What is a more specific name for the polyhedron?

2. How many regular polyhedra are there?
3. Name the regular polyhedra and give the number of faces of each.
4. Use the figure here as an upper base and sketch a cube.

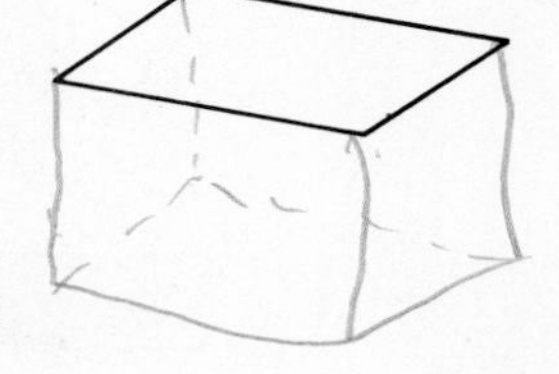

5. a. Use the triangle here as the upper base and sketch a triangular prism.

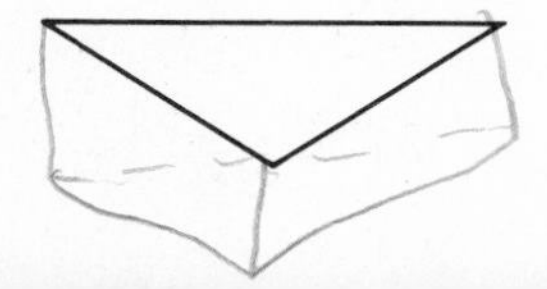

b. How many faces does a triangular prism have?
c. How many edges does it have?
d. How many vertices does it have?

6. Consider the pyramid and answer the following:
a. Name the three hidden edges.
b. Name the triangles that bound the lateral faces.
c. Name the three edges that intersect at B.
d. Name the three angles formed at B.
e. Are any two of the angles formed at B in the same plane?
f. Which vertex is the apex?

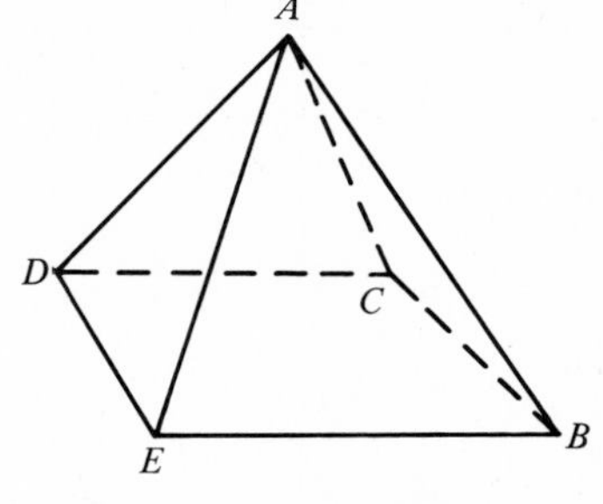

7. Think of a point 1 inch in the air above the polygon shown. Think of the polyhedron formed by connecting each vertex of the polygon with the point above the polygon.

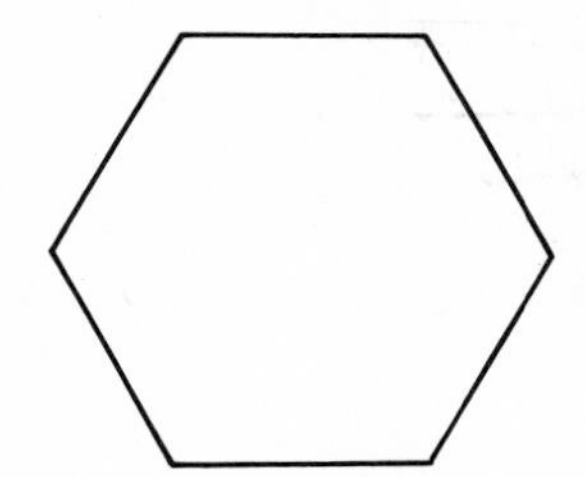

a. What is the name of this polyhedron?
b. How many vertices does it have?
c. How many edges does it have?
d. How many faces does it have?

8. What is the least number of faces that a prism can have?

9. a. Are all prisms polyhedra?
b. Are all polyhedra prisms?

10. Sketch the hidden lines in the following figures:

a.

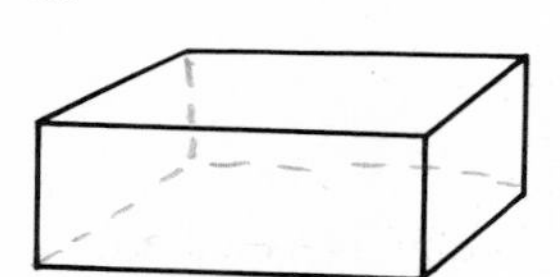

b.

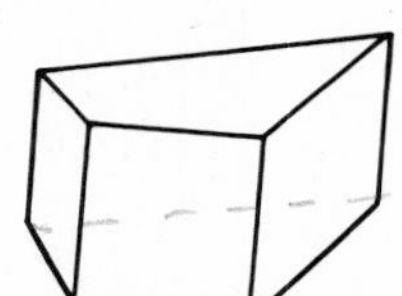

11. Use the polygonal region below for the base and the point A for the apex and sketch a pyramid. Sketch the altitude of the pyramid.

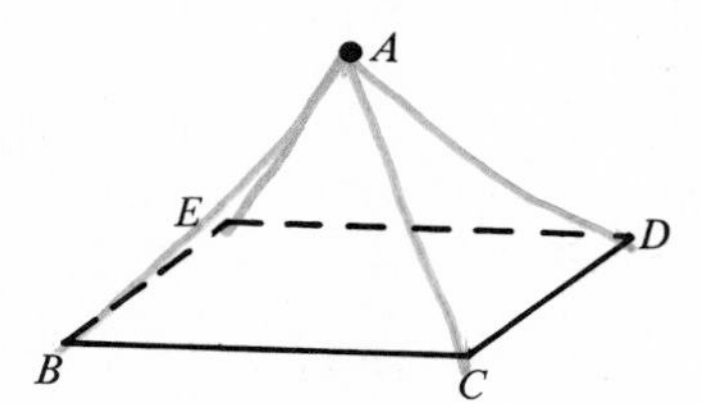

12. a. A miter cut is a straight-blade cut made at an angle of 45° across a board. Is the resulting shaded portion of the board a prism?

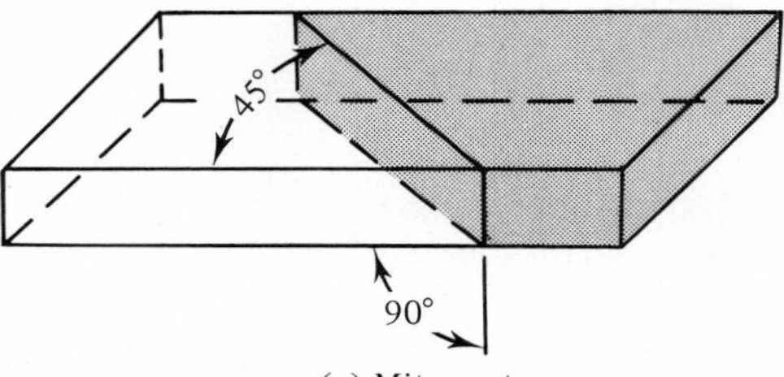

(a) Miter cut

b. A bevel cut is one in which the saw blade is set at an angle of 45° and the cut is made straight across the board. Is the resulting shaded portion a prism?

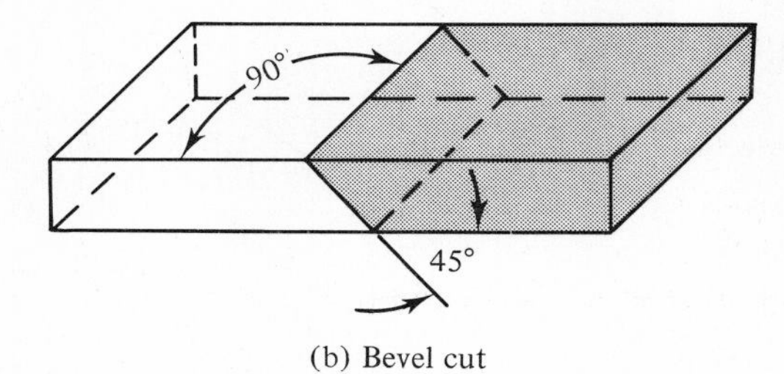

(b) Bevel cut

c. A compound miter or compound bevel cut is one in which the saw blade is set at the required bevel angle and the line of cut across the board is also laid out at an angle. Is the resulting shaded portion a prism?

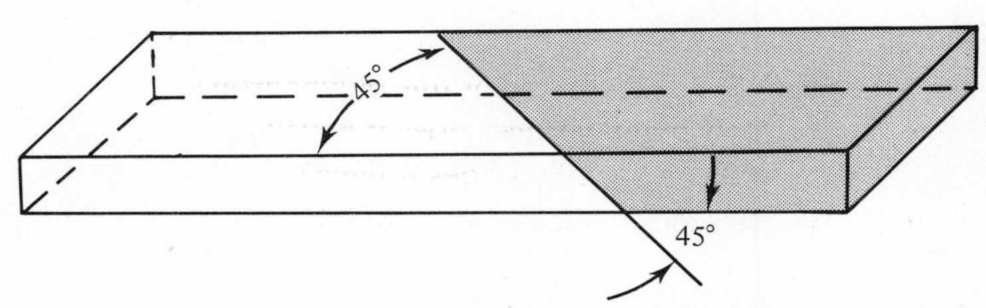

(c) Compound miter or compound bevel cut

4.4 EXERCISE ANSWERS

1. **a.** *B*
b. $\overline{AB}$, $\overline{BC}$, and $\overline{BG}$
c. *ABGF* and *BCHG*
d. *ABCDE*
e. *AEF*, *EHF*, *EDH*, *DCH*, and *FGH*
f. octahedron

2. five

3. regular tetrahedron, 4; cube, 6; regular octahedron, 8; regular dodecahedron, 12; regular icosahedron, 20

4.

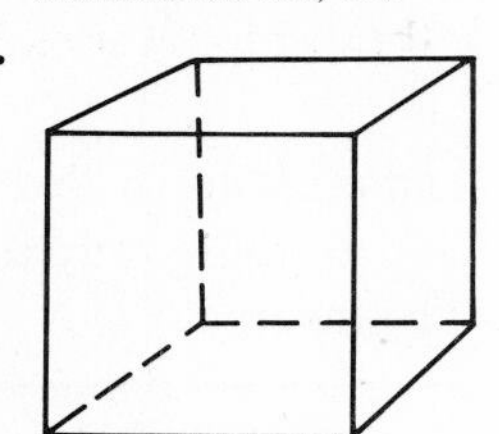

5. **a.**

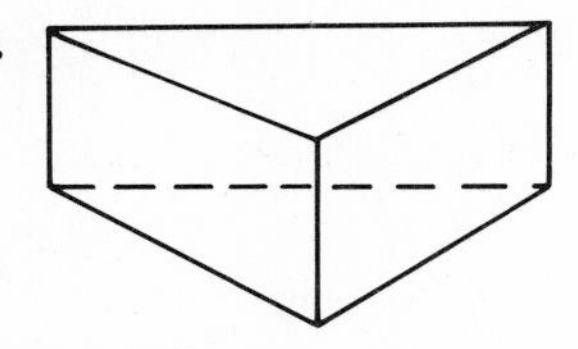

b. 5 faces
c. 9 edges
d. 6 vertices

6. **a.** $\overline{BC}$, $\overline{CD}$, and $\overline{AC}$
b. *ABE*, *AED*, *ADC*, and *ACB*
c. $\overline{BE}$, $\overline{BC}$, and $\overline{BA}$
d. $\measuredangle ABE$, $\measuredangle ABC$, and $\measuredangle CBE$
e. no
f. *A*

7. a. hexagonal pyramid **b.** 7 vertices **c.** 12 edges
d. 7 faces

8. 5 faces

9. a. yes **b.** no

10. a.

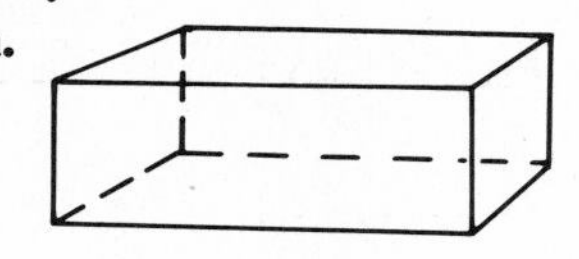

b.

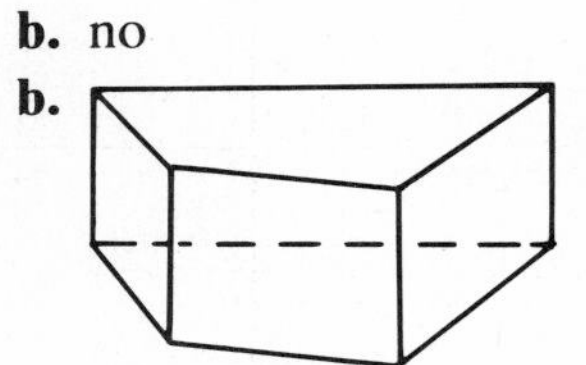

11.

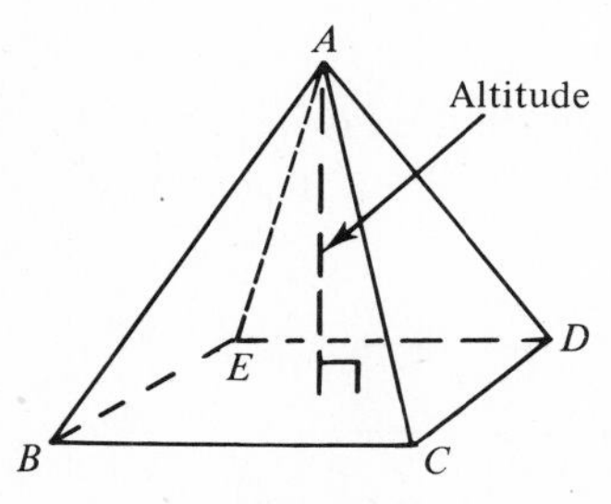

12. a. yes **b.** yes: Consider the front face as the base. **c.** no

4.5 CYLINDERS, CONES, AND SPHERES

1 Circular cylinders are not difficult to recognize. Some examples are shown here.

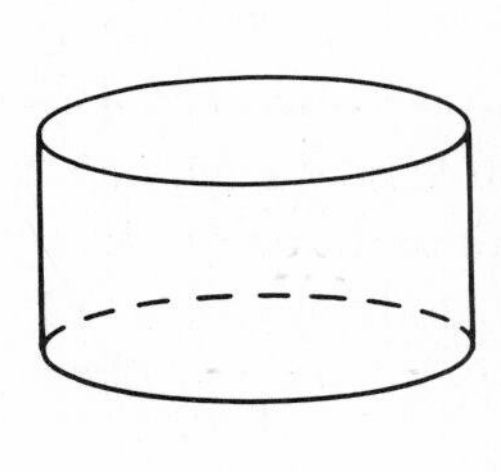

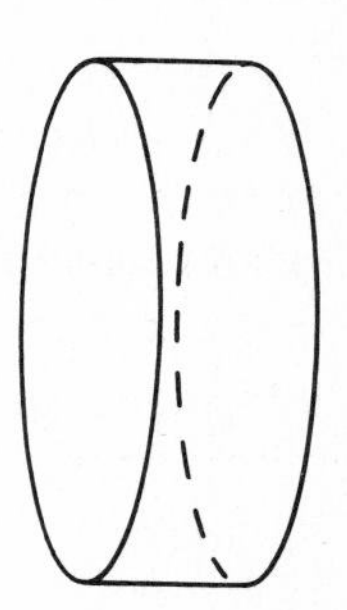

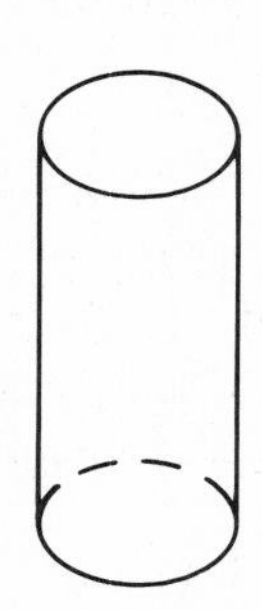

A right-circular *cylinder* consists of two circular regions with the same radius, called *bases*, connected by line segments perpendicular to the planes of the two circles. Because only right-circular cylinders are discussed in this text, we shall refer to them simply as "cylinders."

Q1 Which of the following have a cylindrical shape?

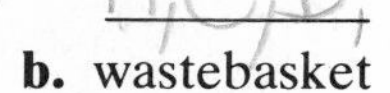

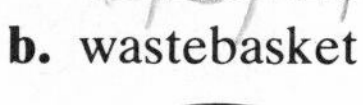

a. round pencil

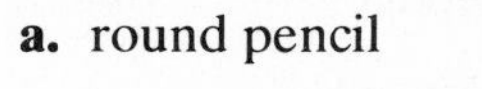

b. wastebasket

c. section of a water pipe

d. beer can

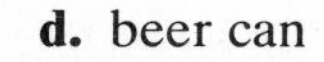

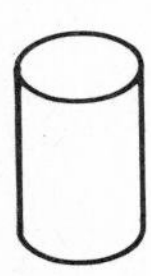

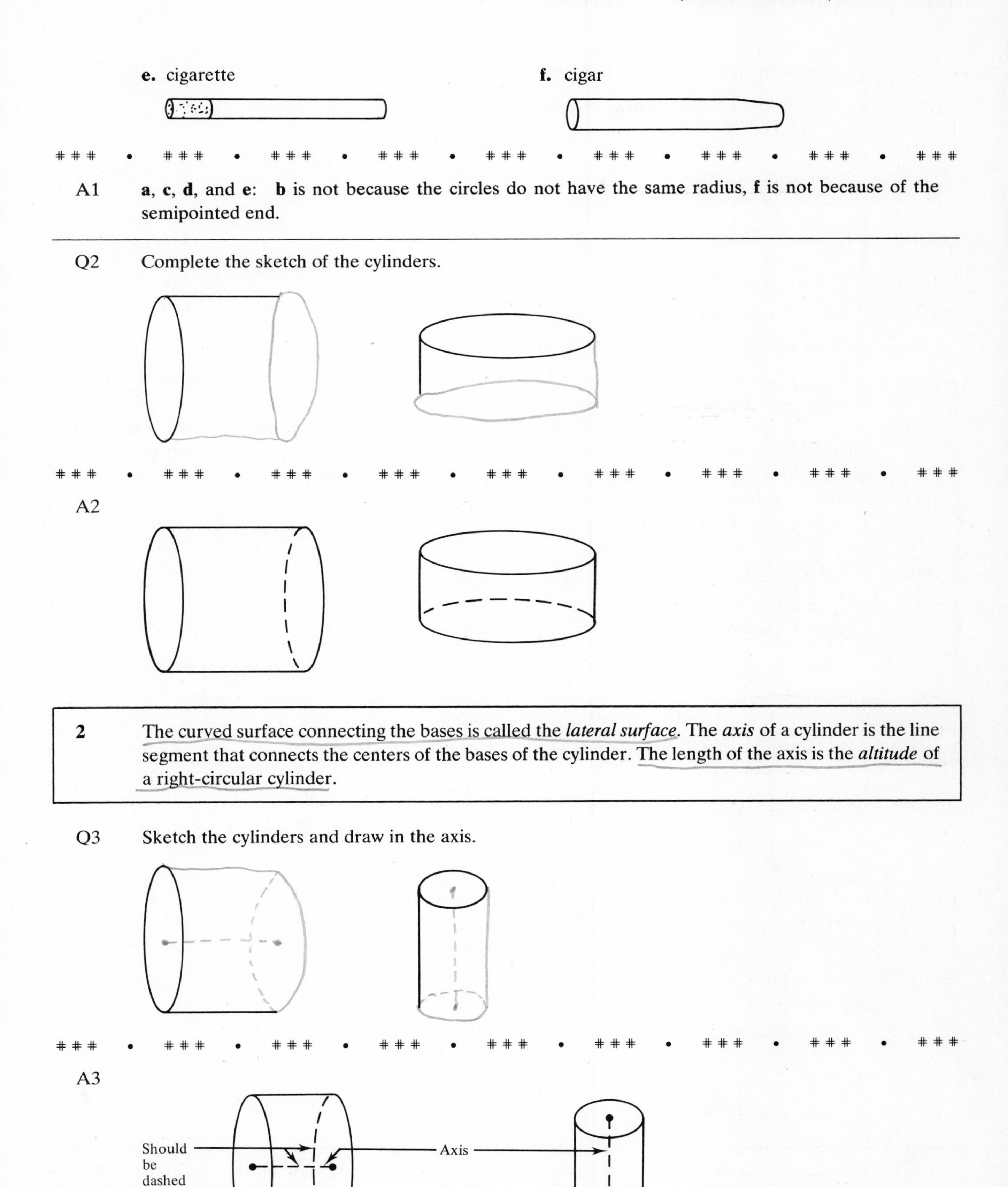

e. cigarette

f. cigar

\# # # • # # # • # # # • # # # • # # # • # # # • # # # • # # # • # # #

A1 **a**, **c**, **d**, and **e**: **b** is not because the circles do not have the same radius, **f** is not because of the semipointed end.

Q2 Complete the sketch of the cylinders.

\# # # • # # # • # # # • # # # • # # # • # # # • # # # • # # # • # # #

A2

2 The curved surface connecting the bases is called the *lateral surface*. The *axis* of a cylinder is the line segment that connects the centers of the bases of the cylinder. The length of the axis is the *altitude* of a right-circular cylinder.

Q3 Sketch the cylinders and draw in the axis.

\# # # • # # # • # # # • # # # • # # # • # # # • # # # • # # # • # # #

A3

Q4 **a.** How many centimetres long is the radius of the cylinder? 46

b. How many centimetres is the altitude of the cylinder? 67

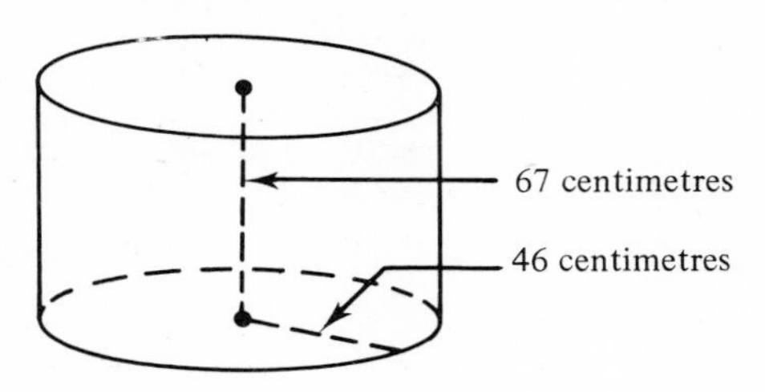

• ### • ### • ### • ### • ### • ### • ### •

A4 **a.** 46 centimetres **b.** 67 centimetres

3 A right-circular *cone* is a circular region and the surface made up of line segments connecting the circle with a point located on a line through the center of the circle and perpendicular to the plane of the circle. The circular region is called the *base*. The point to which the base is connected is the *vertex*. (Be careful that you do not call it the apex.) The line segment connecting the vertex to the center of the base is the *axis*.

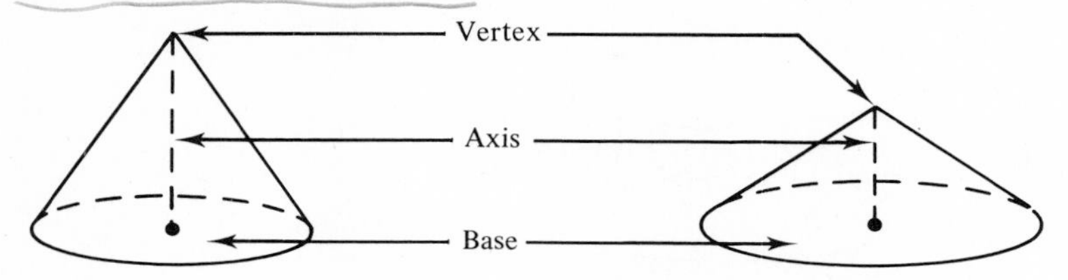

Q5 Make sketches of cones with the base and vertex shown. Include a sketch of the axes.

a.

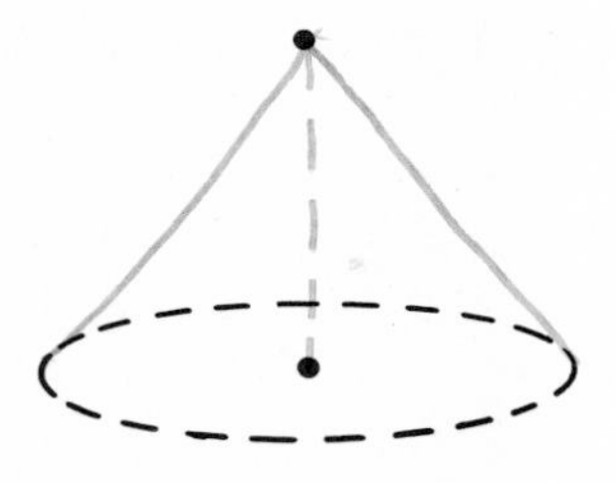

b.

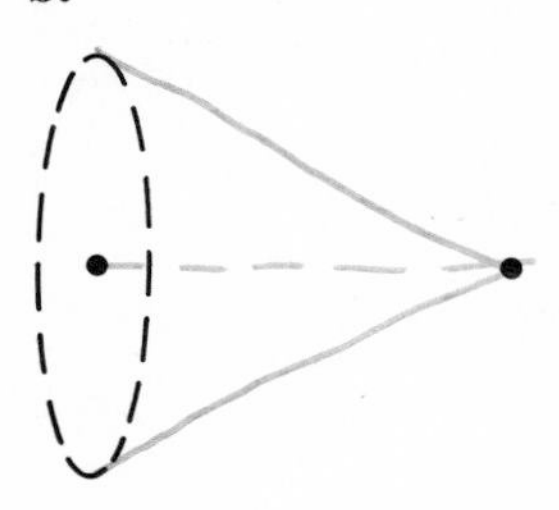

• ### • ### • ### • ### • ### • ### • ### •

A5 **a.**

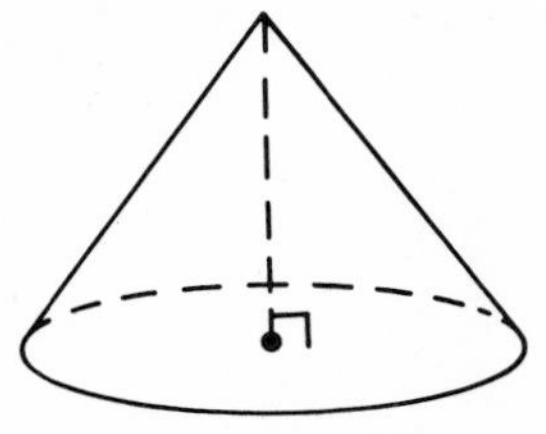

b. Two possible sketches:

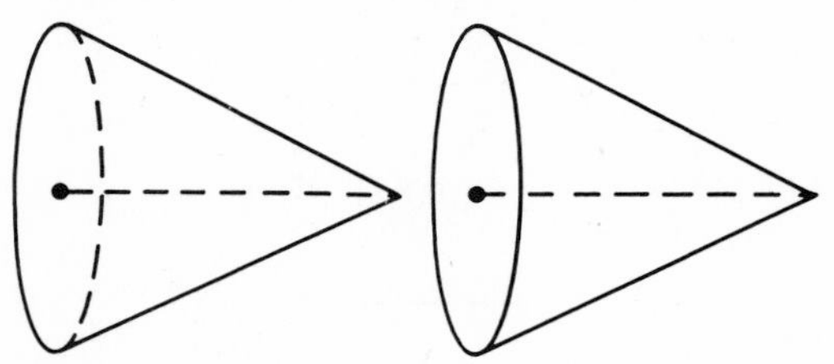

4 The *altitude* of a right-circular cone is the length of its axis. Since the altitude of a cone is a length, it is a number rather than a geometric object. The *slant height* of a right-circular cone is the length of a line segment from the vertex to a point on the circle of the base.

Q6 Use the sketch to complete the sentences below:

a. $\overline{OP}$ is the AXIS of the right-circular cone.

b. The length of $\overline{OP}$ is the ALTITUDE.

c. The length of $\overline{PQ}$ is the SLANT HIGHT of the cone.

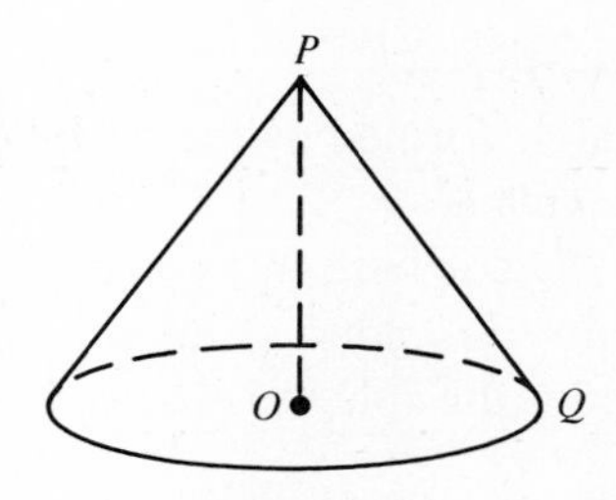

\# \# \# • \# \# \# • \# \# \# • \# \# \# • \# \# \# • \# \# \# • \# \# \# • \# \# \# • \# \# \#

A6 **a.** axis **b.** altitude **c.** slant height

5 A *sphere* is a set of points in space at some fixed distance from a given point. The given point is the *center* of the sphere. A *radius* of a sphere is a line segment from the center to the sphere. The radius also refers to the measure of such a line segment. You will be able to tell which meaning is intended by the context in which the word "radius" is used. A *diameter* of the sphere is a line segment containing the center with endpoints on the sphere. A sketch of a sphere is shown. Notice how the dashed lines are used to give the sketch depth. To draw a sphere use a compass to sketch a circle of the proper radius. Then sketch the elliptical shape in the interior of the circle. Be sure to make the hidden lines dashed.

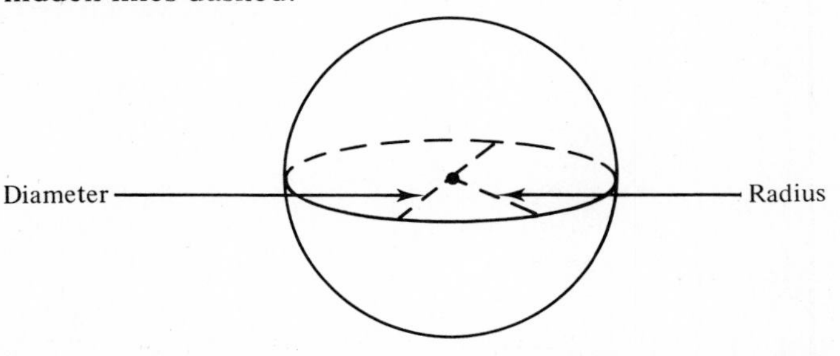

Q7 Sketch a sphere and draw a radius of the sphere.

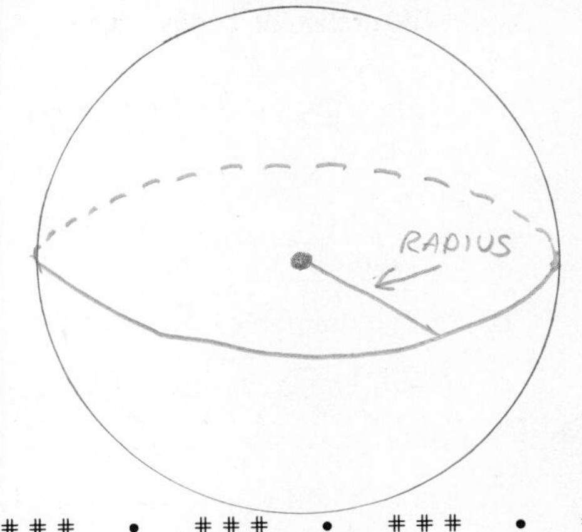

\# \# \# • \# \# \# • \# \# \# • \# \# \# • \# \# \# • \# \# \# • \# \# \# • \# \# \# • \# \# \#

A7

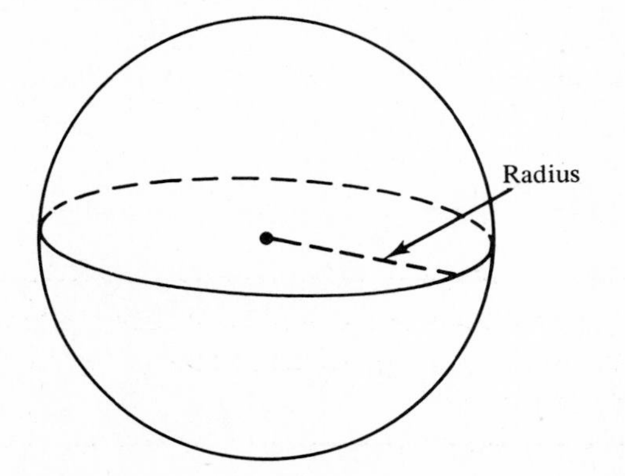

6 A circle formed by the intersection of the sphere and a plane containing the center of the sphere is a *great circle*.

The sphere in Figure 1 has two great circles shown, *AEBF* and *CEDF*. The circle *ABCD* in Figure 2 is not a great circle because it does not contain the center of the sphere. A great circle divides the sphere into two halves each called a *hemisphere*.

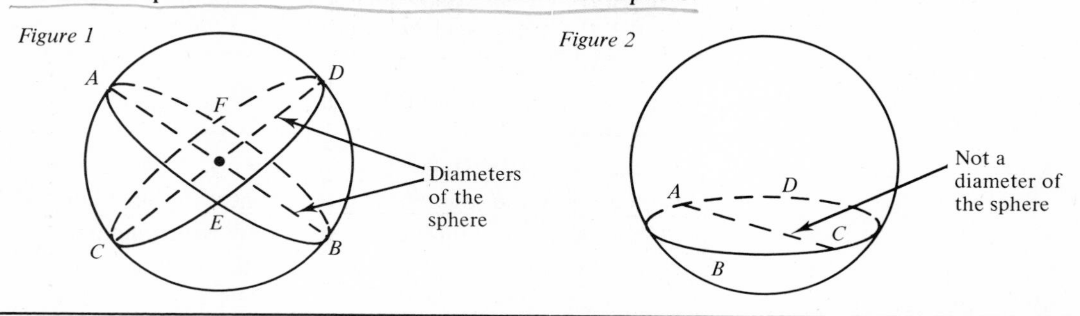

Q8 Use the following figure to answer the questions:

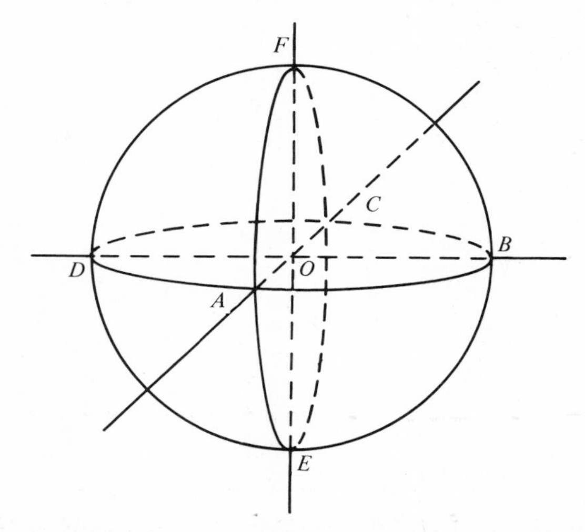

a. $\overline{OC}$ and $\overline{OF}$ are radii of the sphere. Name four more radii. OD, OB, OE, OA

b. $\overline{EF}$ is a diameter of the sphere. Name two other diameters. AC, DB

c. *DEBF* is a great circle. Name two others. DABC, FAEC

• ### • ### • ### • ### • ### • ### • ### •

A8 **a.** $\overline{OB}$, $\overline{OA}$, $\overline{OD}$, and $\overline{OE}$ **b.** $\overline{AC}$ and $\overline{DB}$
c. *AECF* and *ABCD*

Q9 The lines of longitude are shown on a diagram of the earth's surface in Figure 1. Lines of latitude are shown in Figure 2.

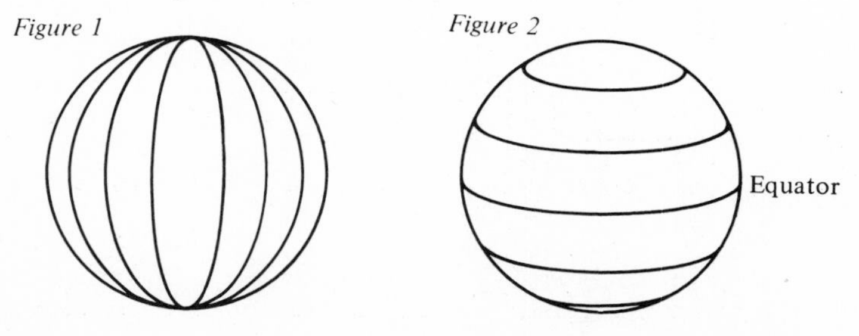

The lines of LONGITUDE are all great circles. Only one of the lines of LATITUDE is a great circle. This great circle is called the EQUATOR.

• ### • ### • ### • ### • ### • ### • ### •

A9 longitude, latitude, equator

Q10 Some farm silos are made up of two solid figures discussed in this section. What are they? CILIDERS, & HEMISPHERE

• ### • ### • ### • ### • ### • ### • ### •

A10 cylinder and hemisphere

Q11 Below is shown a button-head machine bolt with a square nut.

a. What is the geometric figure that makes up the stem? CYLINDER

b. Is the head of the bolt a hemisphere? NO

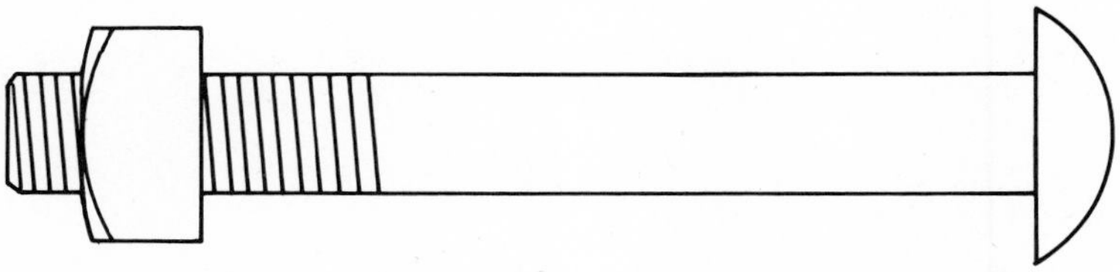

• ### • ### • ### • ### • ### • ### • ### •

A11 **a.** cylinder **b.** no

Q12 A granary finds it does not have the capacity for a crop of wheat. The grain is temporarily stored in a pile formed in a corner where two walls come together. The pile formed has the shape of a fraction of one of the solids of this section. (See the shaded area below.)

a. What is the solid? CONE

b. What fraction is it? 1/4

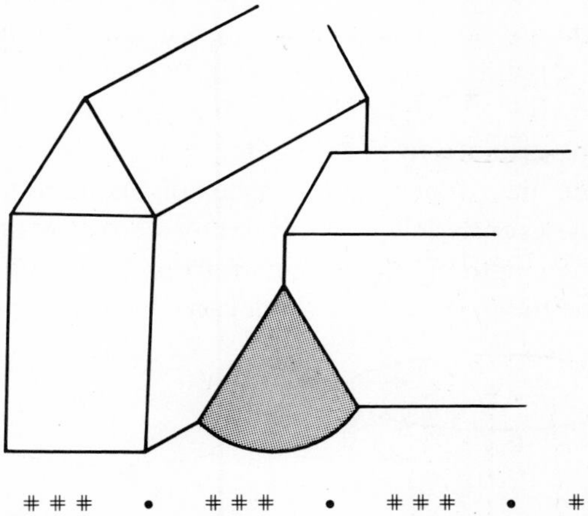

• ### • ### • ### • ### • ### • ### • ### •

A12 **a.** cone **b.** $\frac{1}{4}$

Q13 The geometric solid shown is made up of several solids of this and previous sections.

a. The basic shape is a ______________.

b. The addition on top is half of a ______________.

c. The portion deleted is a ______________.

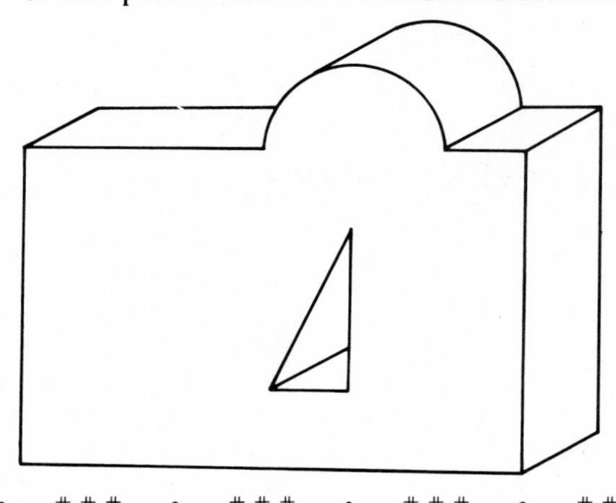

\# # # • # # # • # # # • # # # • # # # • # # # • # # # • # # # • # # #

A13 **a.** rectangular prism **b.** cylinder **c.** triangular prism

---

This completes the instruction for this section.

4.5 EXERCISE

1. Name the geometric figures:

a.

b.

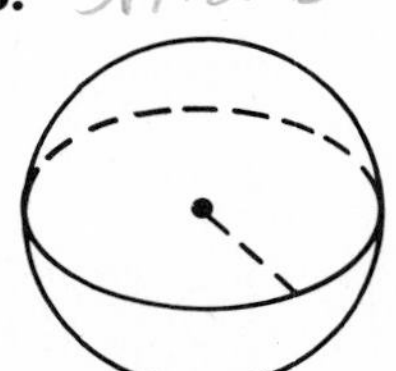

c.

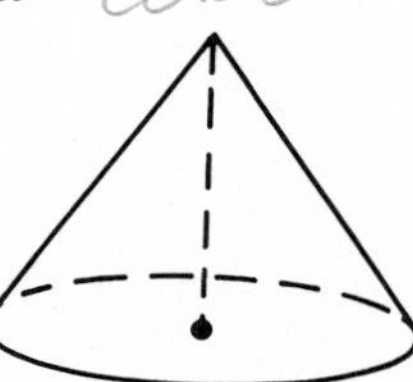

2. Name the term that each of the following describes:

a. the line connecting the centers of the ends of a cylinder

b. the point outside the plane of the base but connected to the base of a cone with line segments

c. the perpendicular distance between bases of a cylinder

d. the distance from the center of a sphere to the sphere

e. the line connecting the vertex of a cone to the center of the base

f. the geometric figure formed by a cross section of a sphere through its center

g. the length of a line segment from the vertex to the base of a right-circular cone.

3. Name the parts of the cylinder to which the arrows point:

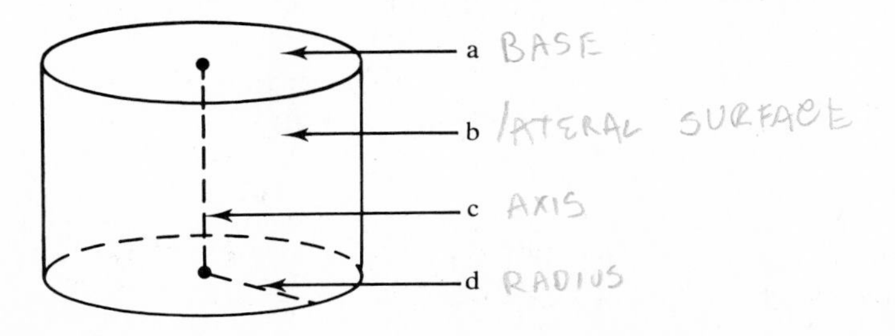

4. Name the parts of the cone to which the arrows point:

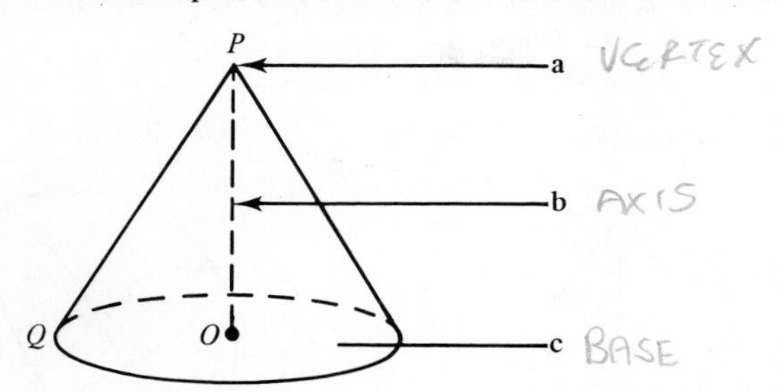

5. Name the line segment whose length is the slant height in problem 4. $\overline{PQ}$

6. Name the part of the sphere to which the arrows point:

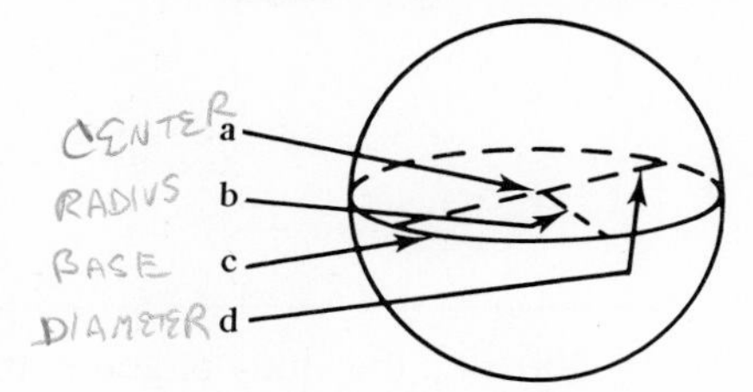

7. Complete the sketch of a cylinder. Sketch the axis.

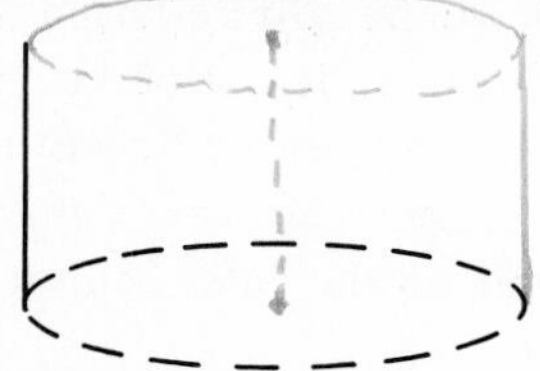

8. Complete the sketch of a cone. Sketch the axis.

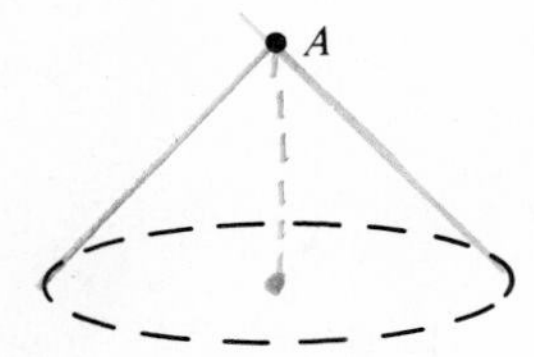

9. Complete the sketch of a sphere. Sketch a radius and two great circles.

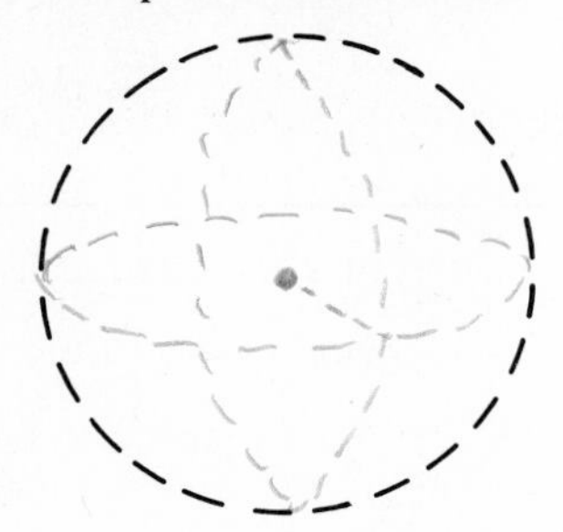

4.5 EXERCISE ANSWERS

1. a. cylinder **b.** sphere **c.** cone

2. a. axis **b.** vertex **c.** altitude **d.** radius

e. axis **f.** great circle **g.** altitude

3. **a.** base **b.** lateral surface **c.** axis **d.** radius
4. **a.** vertex **b.** axis **c.** base
5. $\overline{PQ}$
6. **a.** center **b.** radius **c.** great circle **d.** diameter
7.

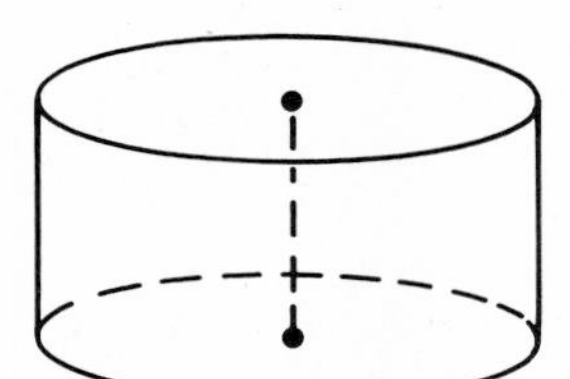

8.

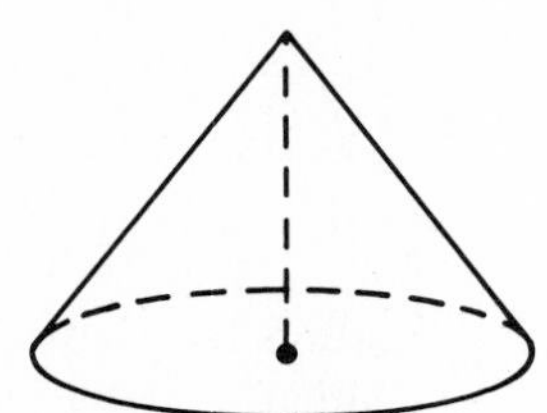

9.

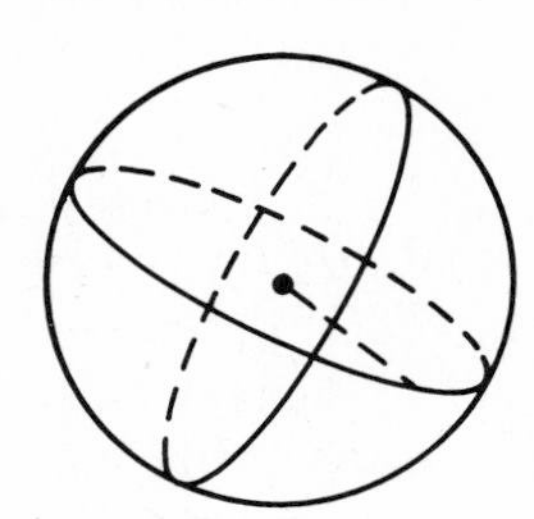

4.6 GEOMETRIC CONSTRUCTIONS

1 To the ancient Greeks, who had developed traditional geometry by 300 B.C., the study of geometric construction was an important subject. They placed very careful restrictions on the tools that could be used in drawing basic geometric forms. These same restrictions will be adhered to in this section.

The tools available were the *compass* and unmarked *straightedge*. A picture of each is shown in the illustrations, and you should have each of these tools in order to complete this section. If you do not have an unmarked straightedge, an inch or centimetre rule will do. However, if the same rules are followed that the Greeks restricted themselves to, the markings on the ruler cannot be used to measure. The straightedge will be used only to draw straight lines.

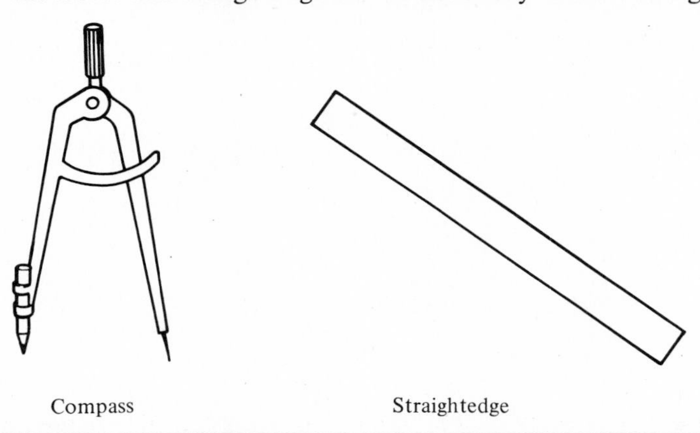

Compass Straightedge

Q1 What civilization developed the original rules for geometric constructions? __________

\# # # • # # # • # # # • # # # • # # # • # # # • # # # • # # # • # # #

A1 Greek

Q2 What are the tools of geometric constructions? ____________ and ____________

\# # # • # # # • # # # • # # # • # # # • # # # • # # # • # # # • # # #

A2 compass, straightedge

2 The Greeks were challenged by the problem of how many geometric objects they could construct with only these tools. However, only a few elementary constructions will be considered here. Many more are possible and you eventually might want to try others.

One of the most elementary constructions would be a circle. When you use the compass to draw a circle, the steel point marks the center and the sharpened pencil marks the circle. For best results use only one hand; hold it by the handle grip and lean the compass in the direction of the turn, as shown below.

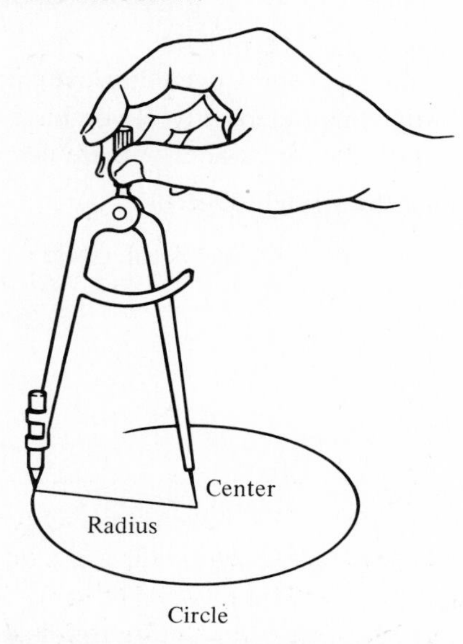

Q3 Use the distance shown below as the length of the radius and draw a circle with a compass.

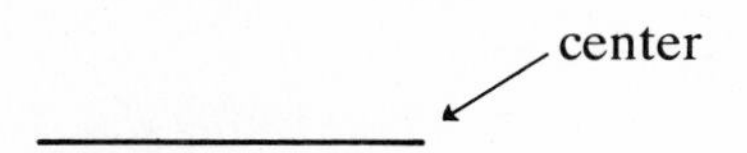

\# \# \# • \# \# \# • \# \# \# • \# \# \# • \# \# \# • \# \# \# • \# \# \# • \# \# \# • \# \# \#

A3

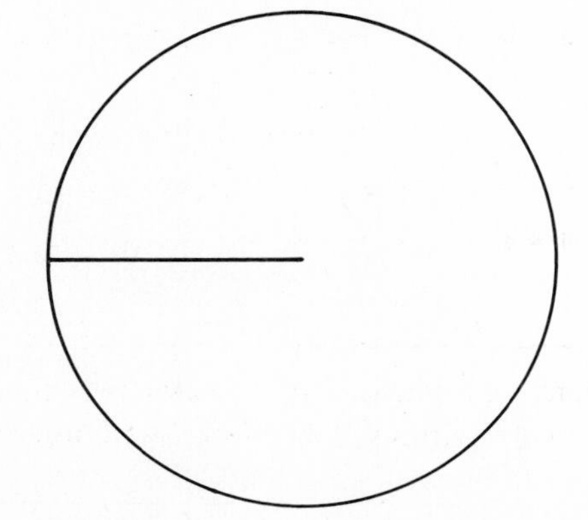

3 The steps for constructing a regular hexagon are given below.

Step 1: Draw a circle with any radius.

Step 2: Using the same compass setting as the radius, and starting at any point on the circle, mark another point on the circle.

Step 3: Using the new point as the starting point, repeat the operation over and over until the original point is marked. If your compass is set correctly it will come out exactly. If you have a slight discrepancy, it is because of the difficulty of being accurate with the instrument you are using.

Step 4: Connect the marked points on the circle in turn using the straightedge.

The resulting figure is a hexagon with sides equal in length to the radius of the circle.

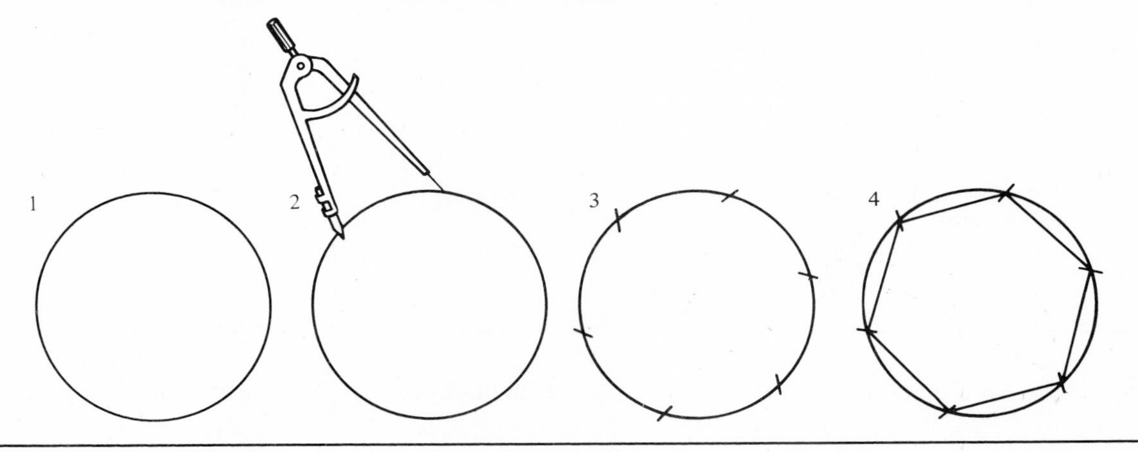

Q4 Construct a hexagon with sides equal in length to the segment shown.

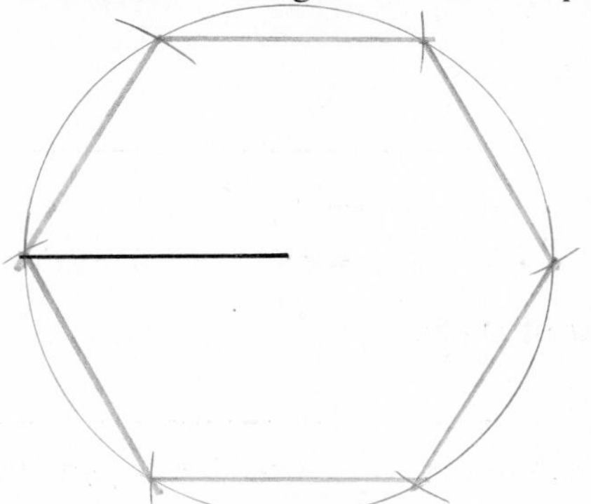

\# # # • # # # • # # # • # # # • # # # • # # # • # # # • # # # • # # #

A4

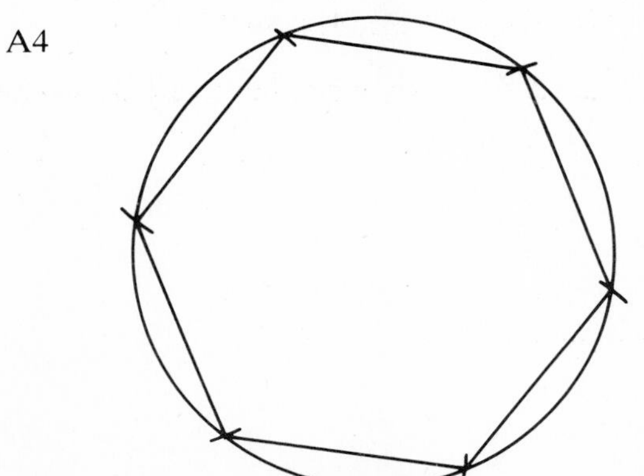

Q5 Construct the figure which results if you follow the same procedure as that of constructing the hexagon but connect each point on the circle with the second point from it. Use the same radius as Q4.

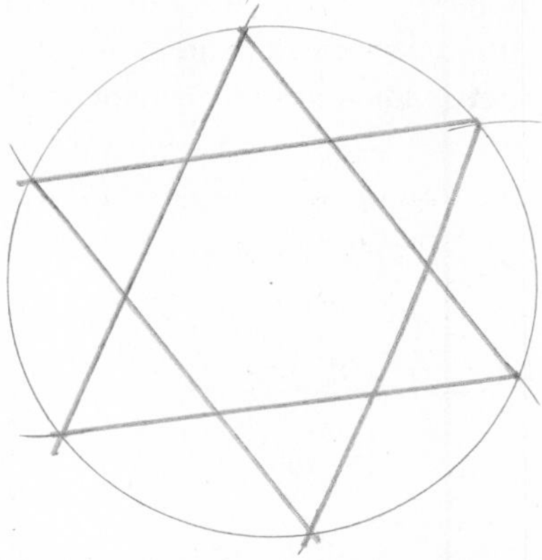

\# \# \# • \# \# \# • \# \# \# • \# \# \# • \# \# \# • \# \# \# • \# \# \# • \# \# \# • \# \# \#

A5

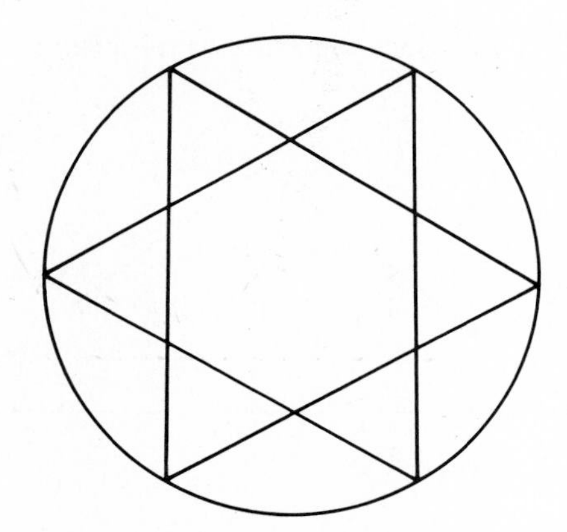

Q6 What would be a good name for the design of Q5? ____________

\# \# \# • \# \# \# • \# \# \# • \# \# \# • \# \# \# • \# \# \# • \# \# \# • \# \# \# • \# \# \#

A6 six-pointed star (Sometimes this design is called the Star of David.)

4 Another elementary construction is the *bisection of a line segment.* To bisect a line segment is to find a point which divides the segment into two segments of equal length.

Step 1: Set the compass at some radius larger than $\frac{1}{2}$ the length of the given segment.

Step 2: Place the point of the compass at an endpoint of the line segment and swing an arc across the segment.

1

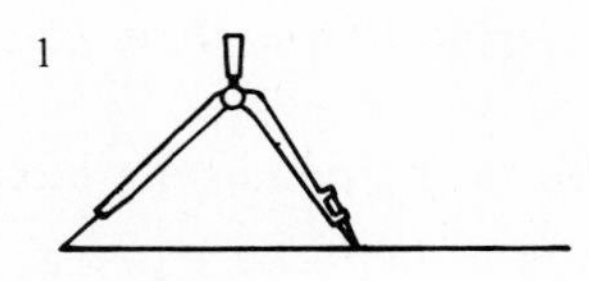

2

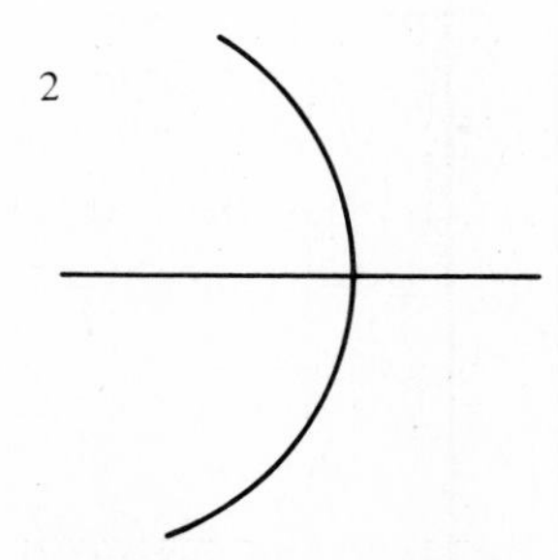

Step 3: Place the point of the compass at the other end of the segment. Swing an arc that intersects the first arc twice, being careful not to change the setting.

Step 4: Connect the intersections of the arcs with the striaightedge. The point where the newly drawn segment intersects the original segment is the point of bisection. The segment connecting the intersections of the arcs is also perpendicular to the segment. It is sometimes called the *perpendicular bisector* of the segment.

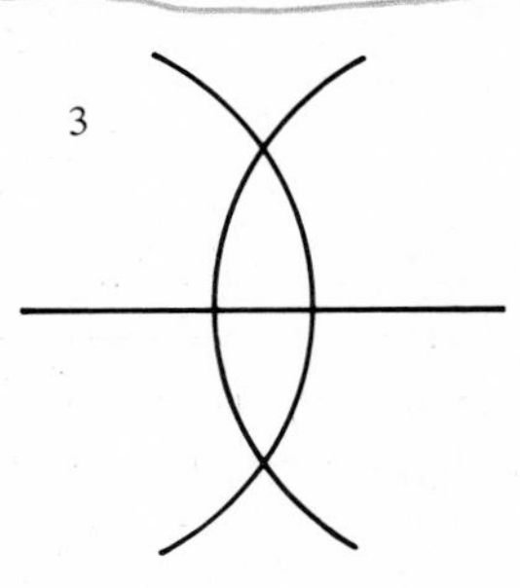

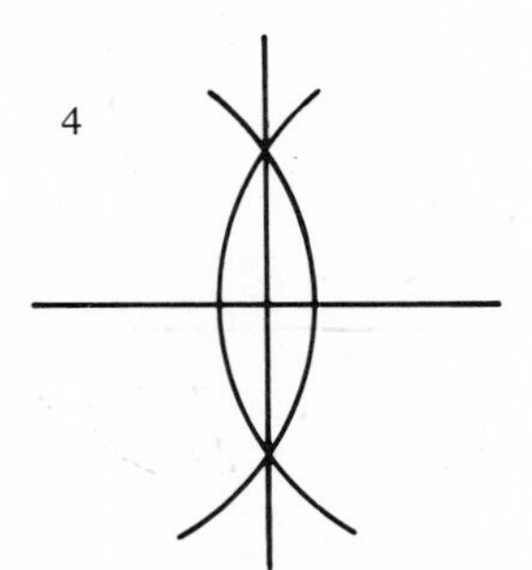

Q7 **a.** Bisect the line segment below.

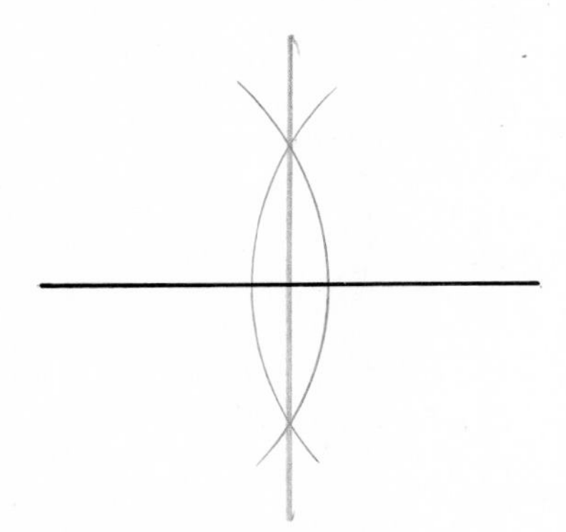

b. Bisect each side of the triangle.

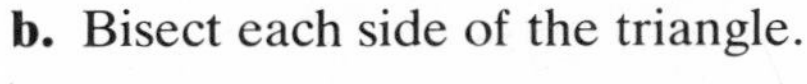

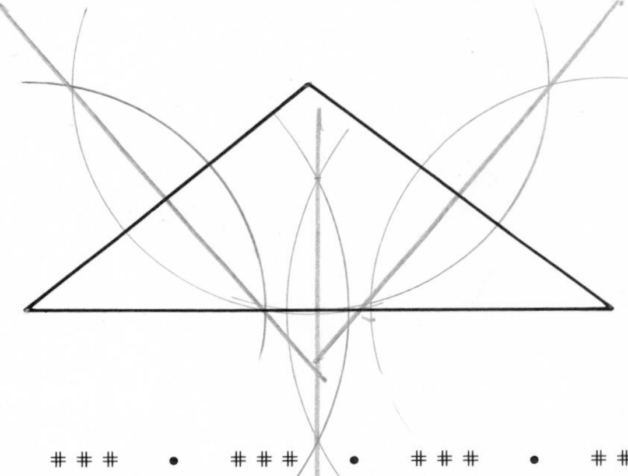

• # # # • # # # • # # # • # # # • # # # • # # # • # # # • # #

A7 **a.**

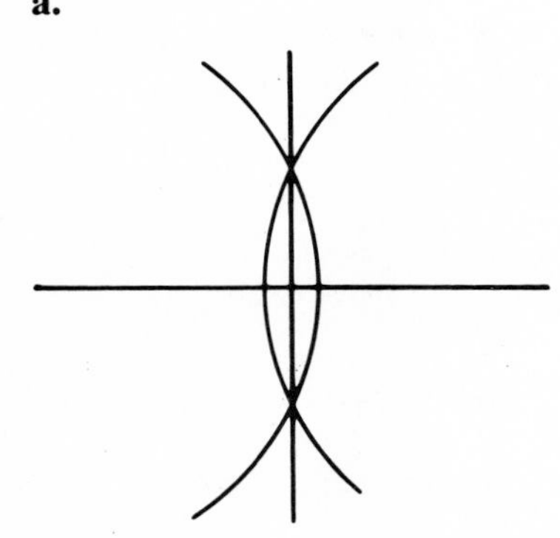

b.

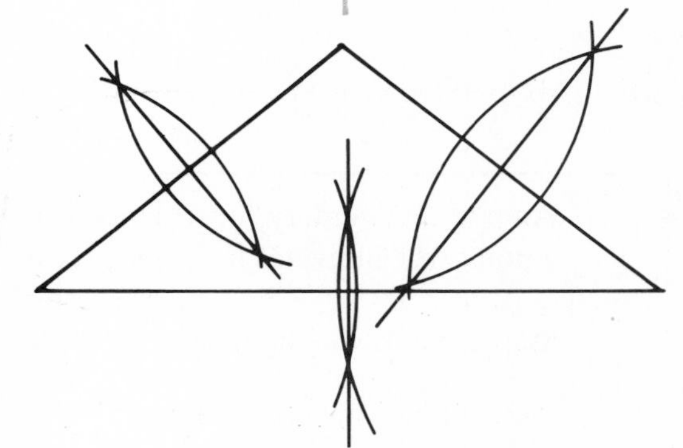

5 *It is a property of triangles that the perpendicular bisectors of the sides meet in a point.* Check the accuracy of your bisectors in Q7 by extending all three until they cross in one point.

It is also a property of geometry that the center of a circle is on the perpendicular bisector of every chord. This property may be used to locate the center of a circle.

Example: Locate the center of the circle.

Solution:

Step 1: Place three points on the circle and draw two chords.

Step 2: Bisect the chords. The center is the intersection of the two perpendicular bisectors.

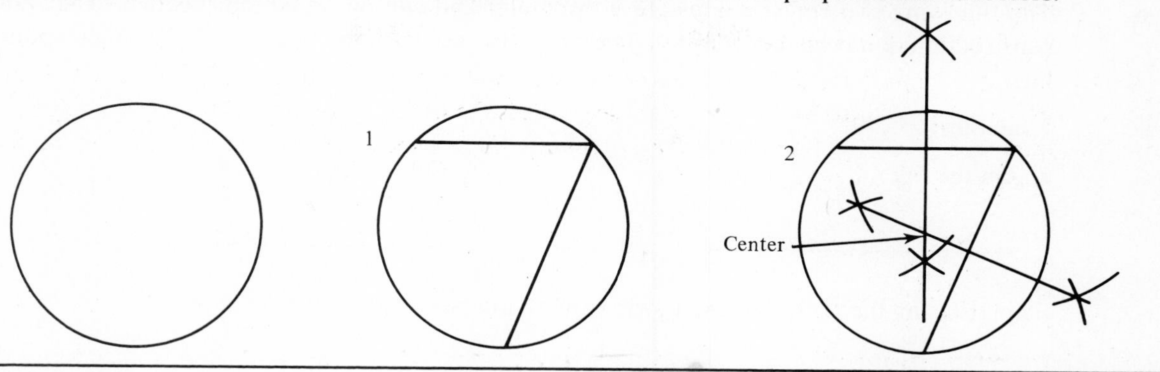

Q8 A technician is given a piece from a broken pulley and is asked to draw a circle of the same size as the pulley. Use the method of Frame 5 to solve the problem. (Hint: You must find the center of the circle.)

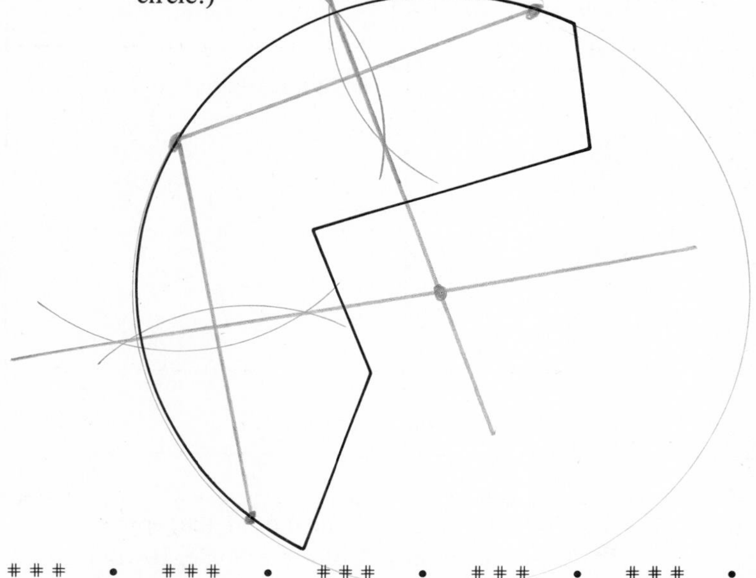

\# # # • # # # • # # # • # # # • # # # • # # # • # # # • # # # • # # #

A8

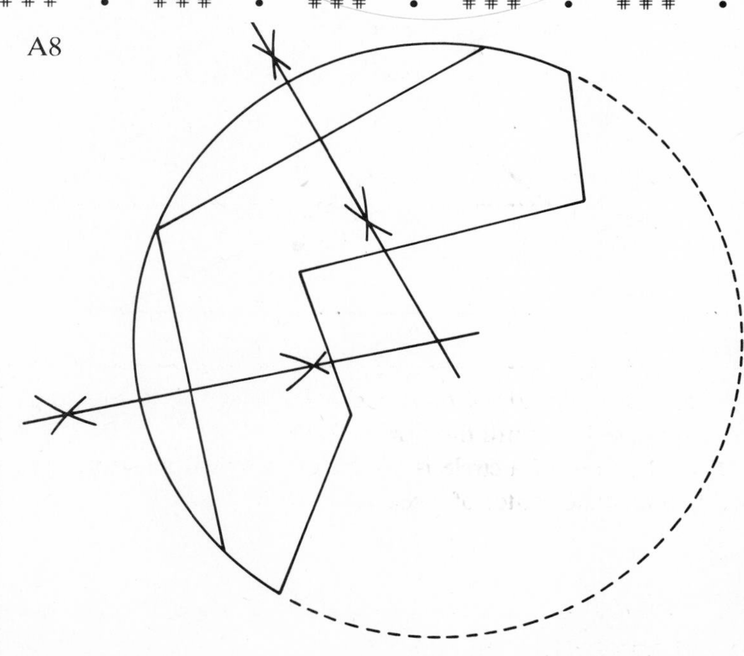

Any three points on the circle could be used to form the chords that are bisected.

6 The following two constructions are very similar and will be presented together. They are (a) the construction of a perpendicular to a line through a point on the line, and (b) through a point off the line.

Examples:

a. On the line

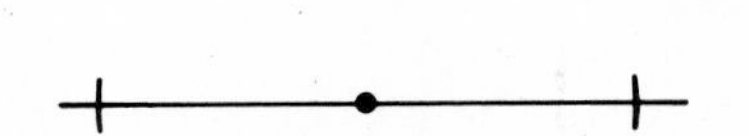

b. Off the line

Step 1: Using the point as center swing an arc intersecting the line twice.

Step 2: Bisect the resulting segment.

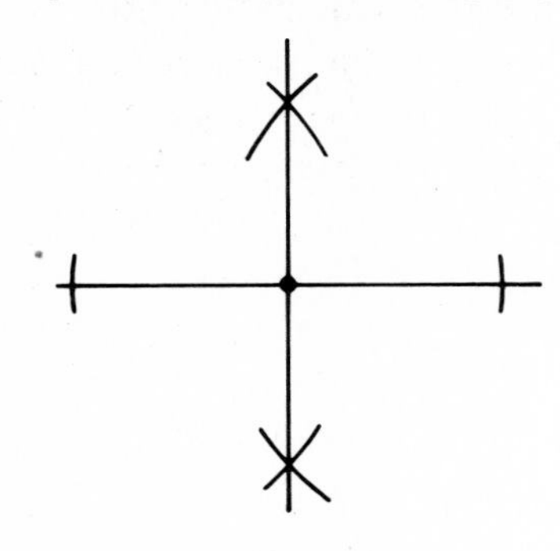

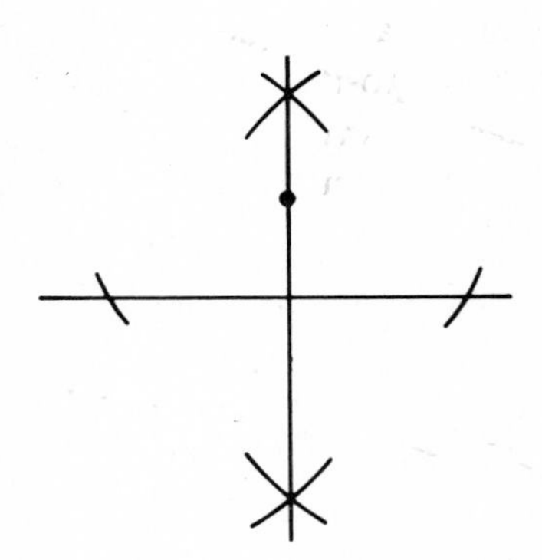

The resulting perpendicular will pass through the original point in both cases.

Q9 **a.** Construct a perpendicular to the line through the point.

b. Construct a perpendicular to the opposite side from vertex A of the triangle. It will be necessary to extend $\overline{BC}$.

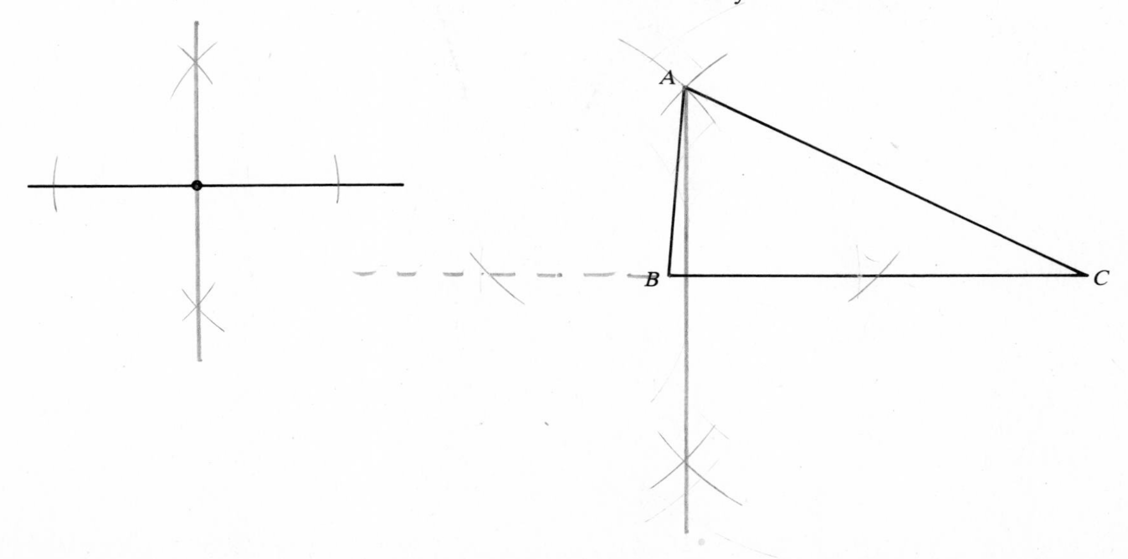

\# \# \# • \# \# \# • \# \# \# • \# \# \# • \# \# \# • \# \# \# • \# \# \# • \# \# \# • \# \# \#

A9 **a.**

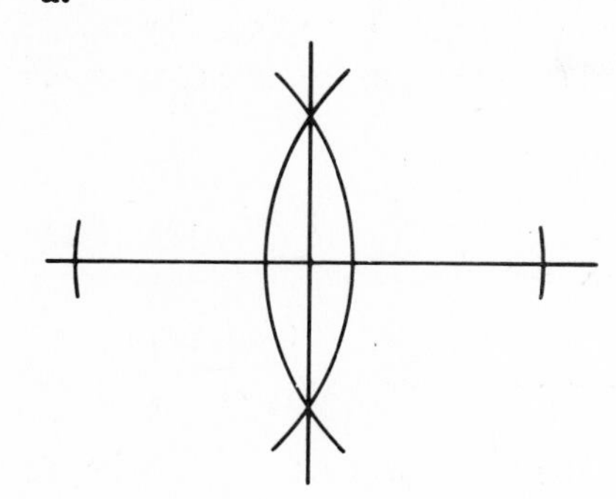

b.

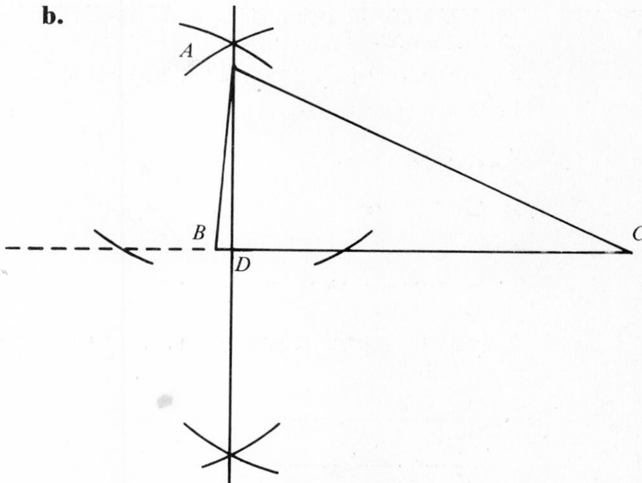

7 The *altitudes* of a triangle are the perpendicular line segments from each vertex to the line containing the opposite side. An altitude, $\overline{AD}$, is shown in A9b. *It is a property of triangles that the three altitudes of a triangle will be contained in lines which interesect in one point.* This property is illustrated below. You will have an opportunity to verify it for another triangle in Q10.

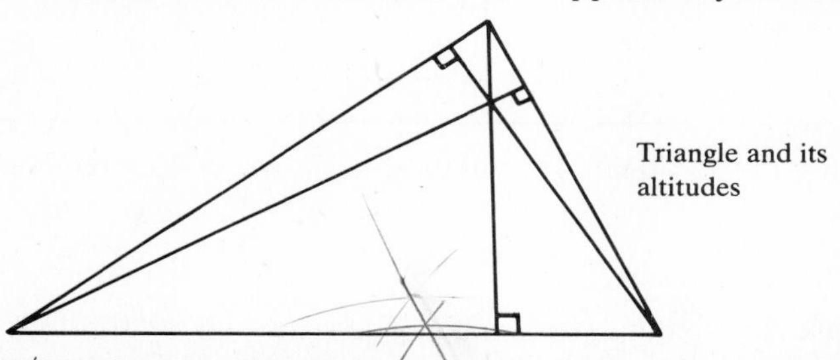

Triangle and its altitudes

Q10 Construct the three altitudes in the triangle and verify that they are contained in lines that intersect in a point. Label the point D.

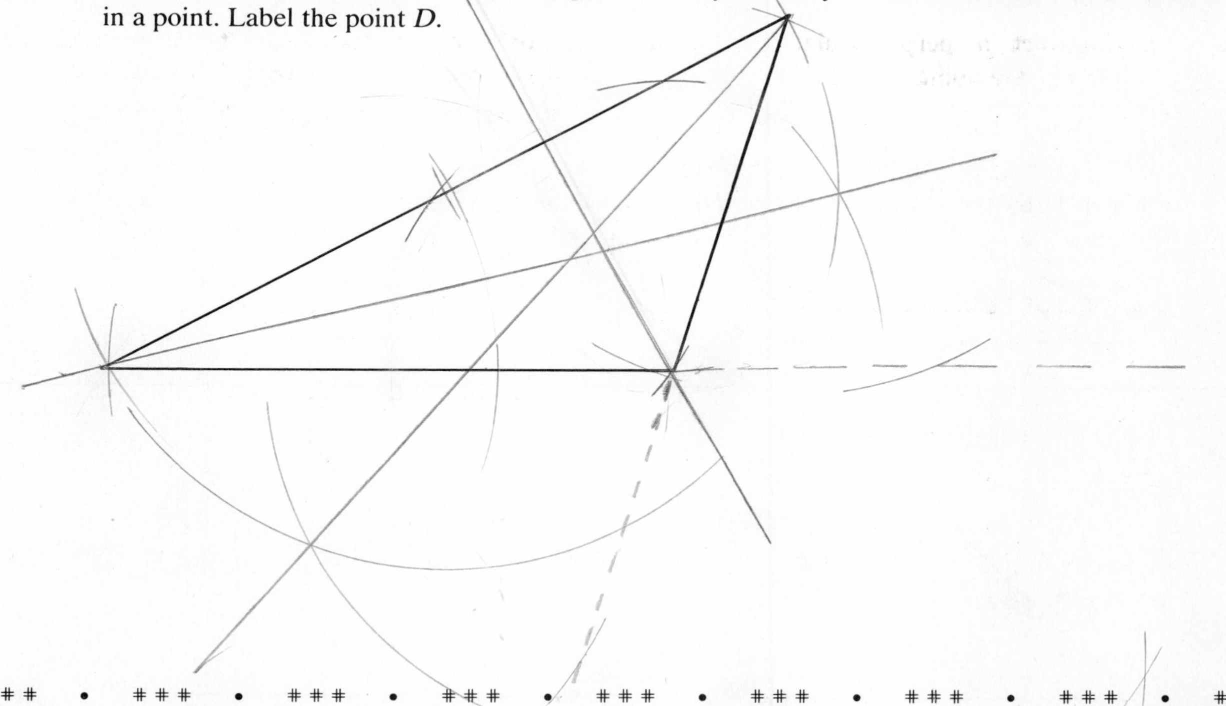

\# \# \# • \# \# \# • \# \# \# • \# \# \# • \# \# \# • \# \# \# • \# \# \# • \# \# \# • \# \# \#

A10 Notice that the point of intersection is sometimes outside of the triangle.

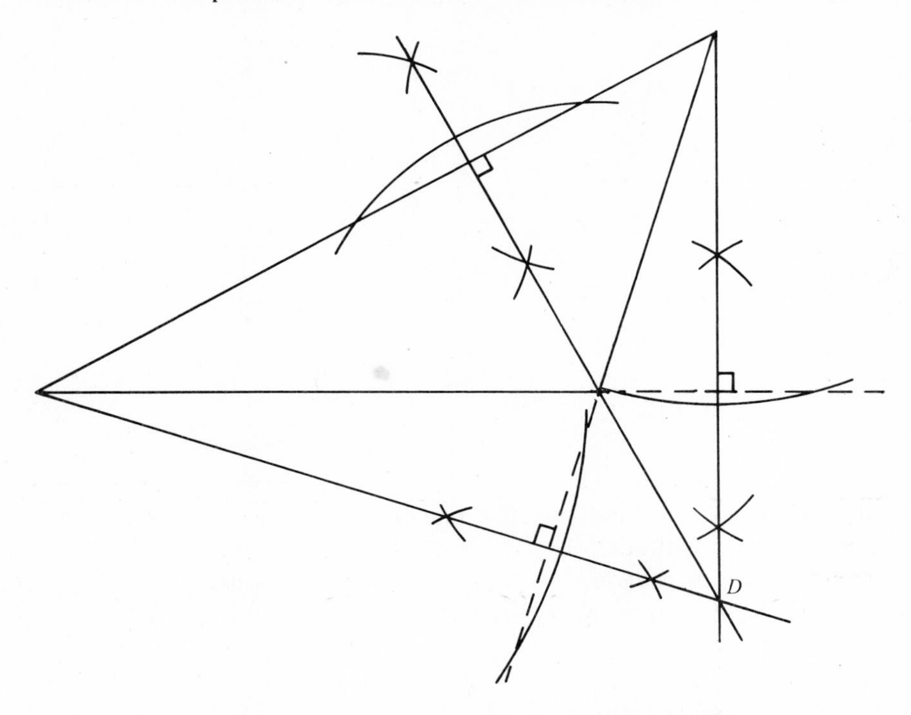

8 Another important construction is to copy an angle. The method is shown in the steps below. To copy the given angle:

Step 1: Draw one side of the angle.

Step 2: Use any radius and draw an arc through the given angle. Then draw an arc through the new side. Use the same radius.

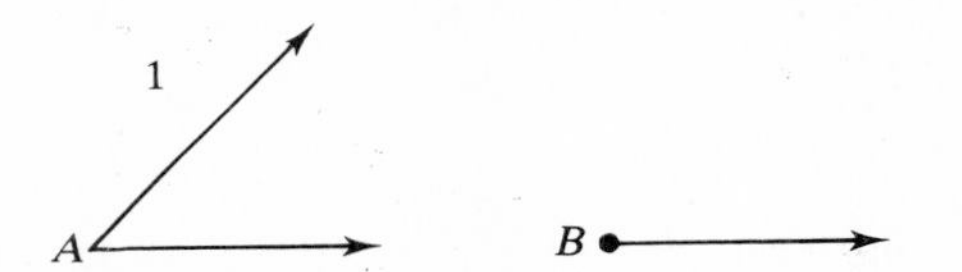

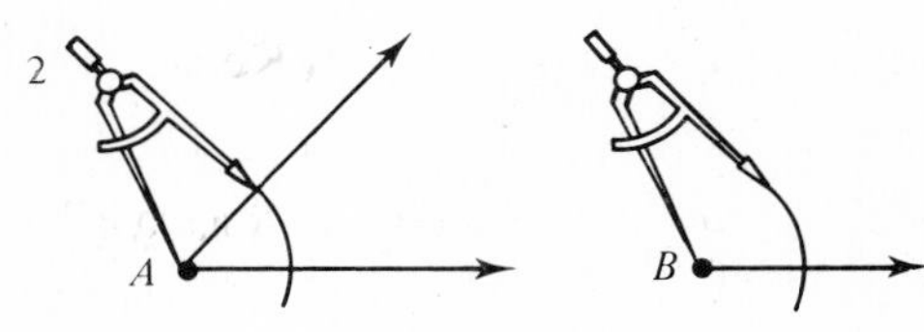

Step 3: Copy the distance between intersections.

Step 4: Draw the second side through the intersection of the arcs.

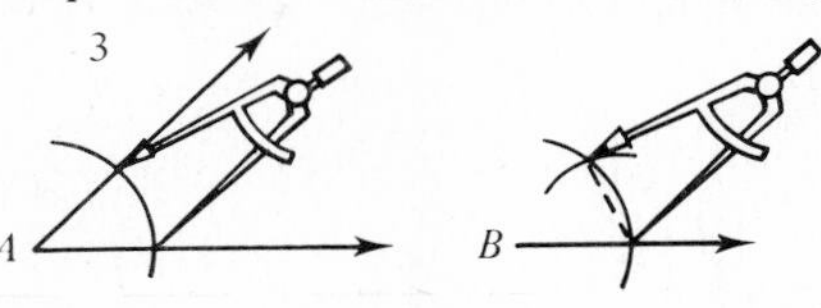

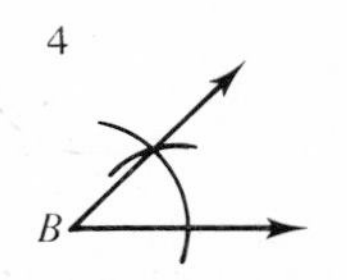

Q11 Copy the angle below at vertex B:

a.

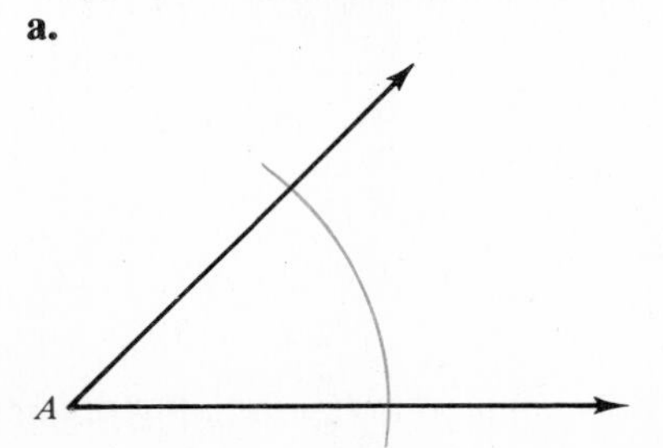

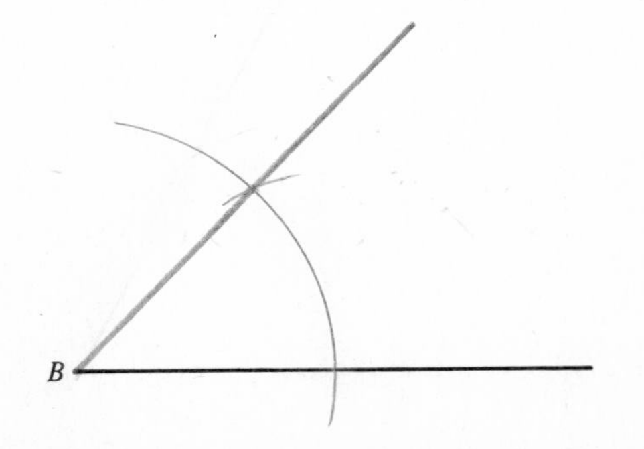

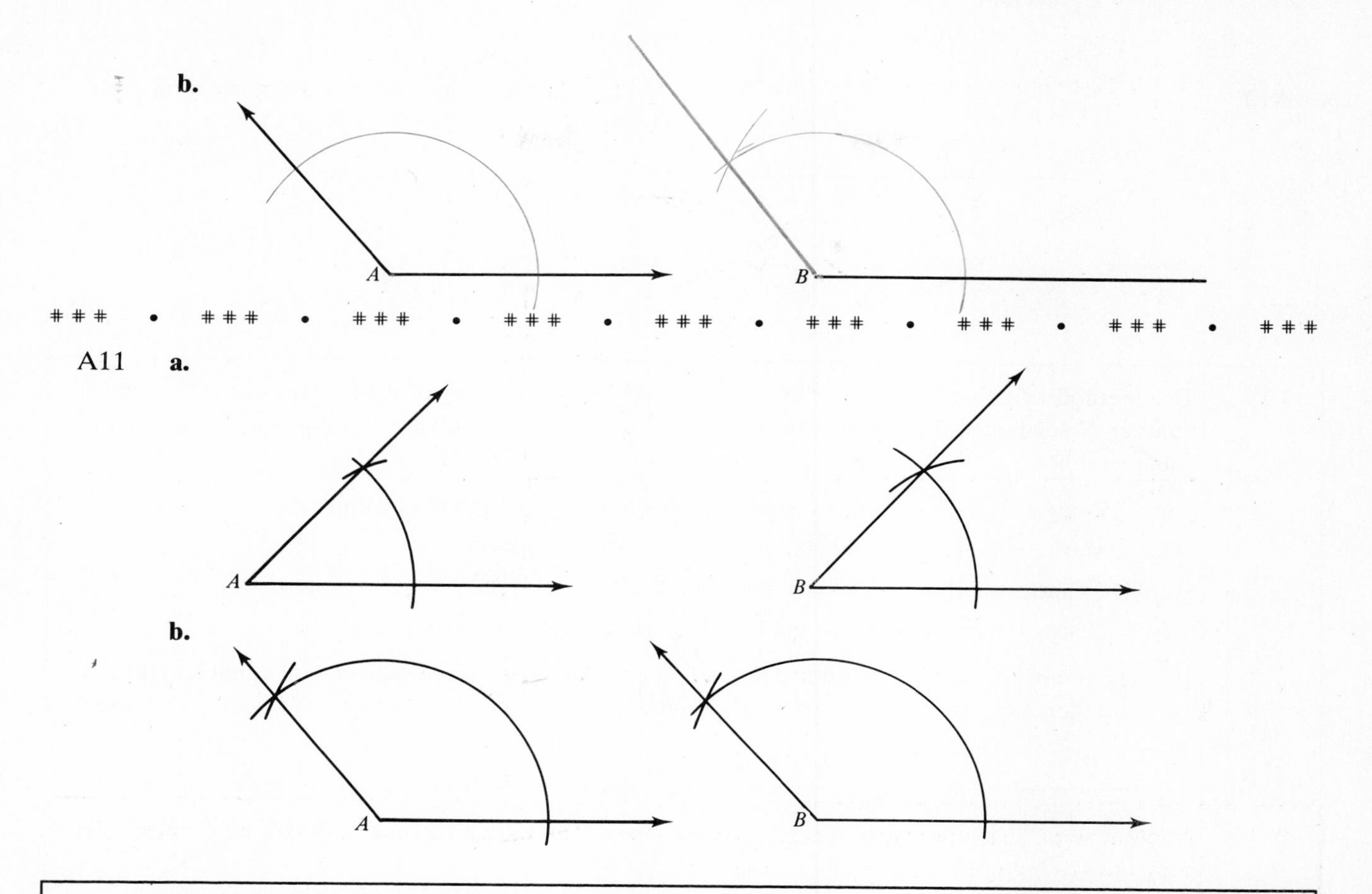

9 After you know how to copy an angle, you can construct a line parallel to a given line through a point not on the given line.

Step 1: Through the point P draw any line intersecting the given line.

Step 2: Draw the arcs shown. Keep the same radius.

Step 3: Copy the angle at P.

Step 4: Draw the parallel line through P.

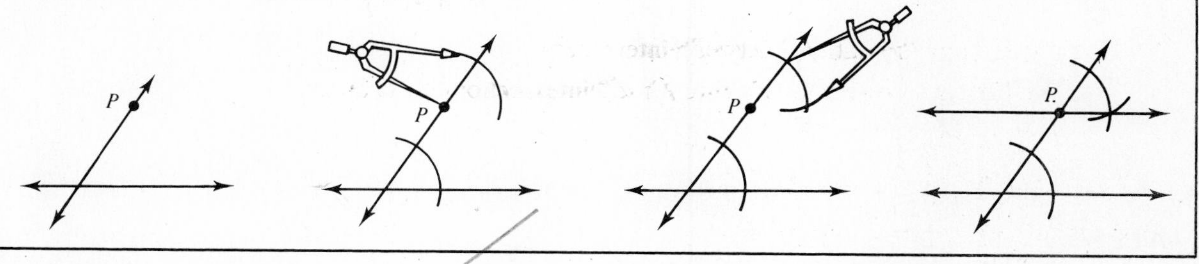

Q12 Use the method of Frame 9 to construct a line parallel to $\overline{AB}$ through point P.

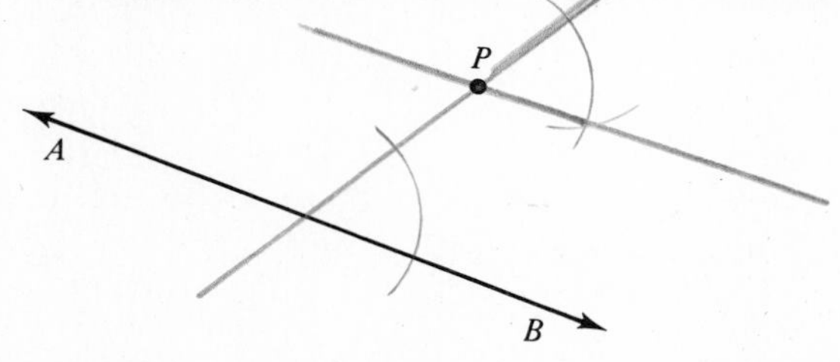

• ### • ### • ### • ### • ### • ### • ### • ### •

A12

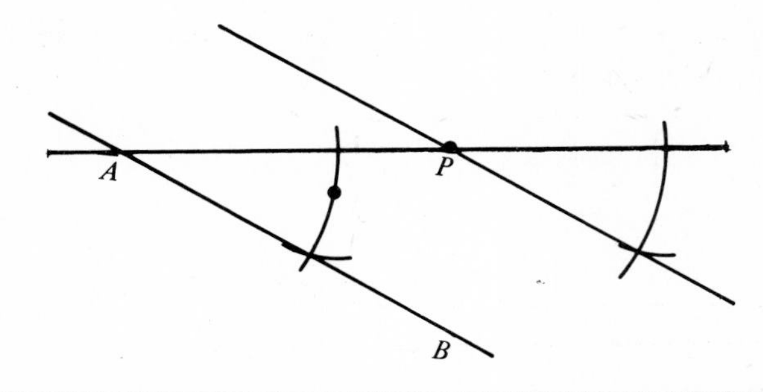

1. Draw any line through P intersecting $\overline{AB}$.
2. Copy $\measuredangle PAB$ at P.

10 The method of constructing parallel lines is used in the dividing of a line segment into a given number of segments of equal length. Follow the steps below to divide $\overline{AB}$ into three segments of equal length.

Step 1: Draw a segment with the endpoint at A of any length and at any angle.

Step 2: Mark 3 equal lengths on $\overline{AC}$ of any size with a compass.

Step 3: Connect the third point with endpoint B.

Step 4: Construct lines parallel to the line through B at the first and second points.

The points of intersection with $\overline{AB}$ divide the segment into three segments of equal length. The same method may be used to divide the segment into a different number of segments of equal length.

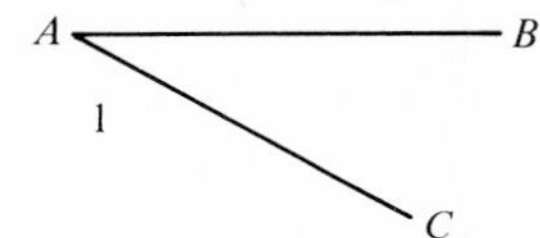

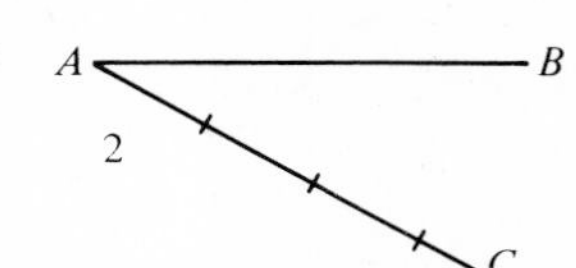

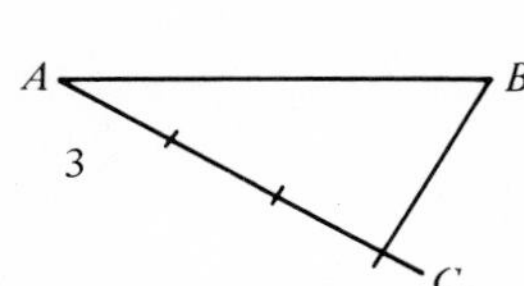

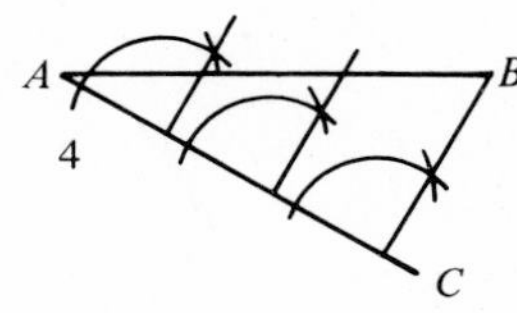

Q13 Divide the segment below into four segments of equal length.

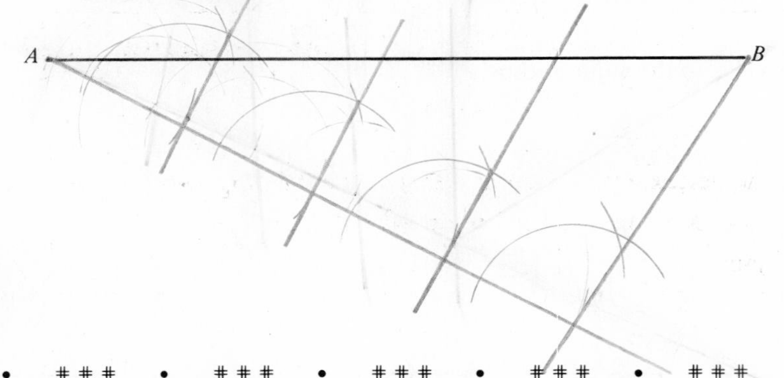

\# \# \# • \# \# \# • \# \# \# • \# \# \# • \# \# \# • \# \# \# • \# \# \# • \# \# \# • \# \# \#

A13

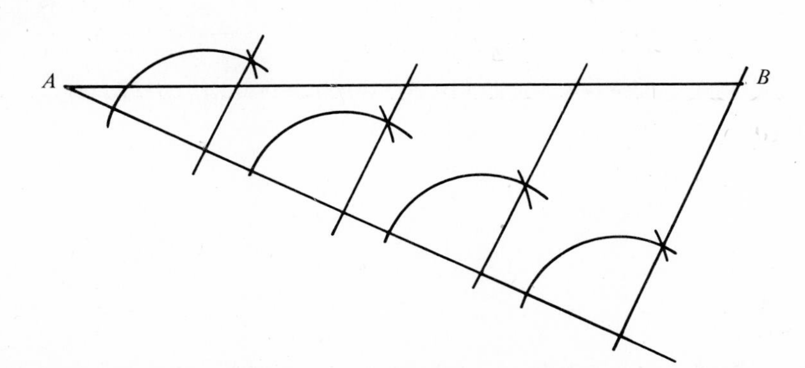

Q14 Four holes are to be equally spaced along the centerline of a steel plate. Locate the centers of the holes by construction. (Divide it into five equal segments.)

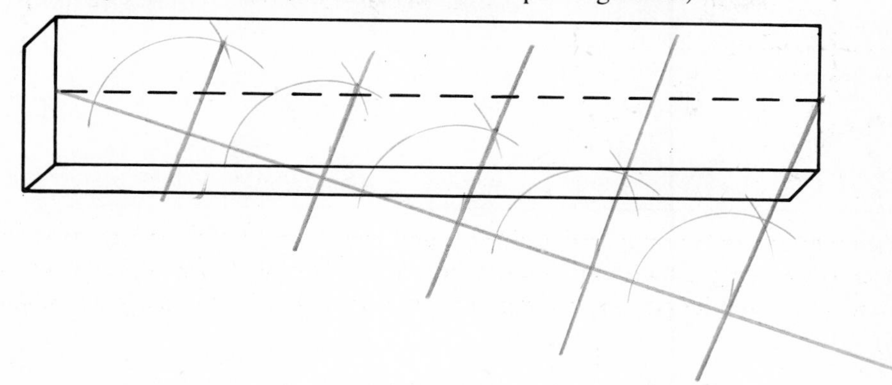

\# # # • # # # • # # # • # # # • # # # • # # # • # # # • # # # • # # #

A14

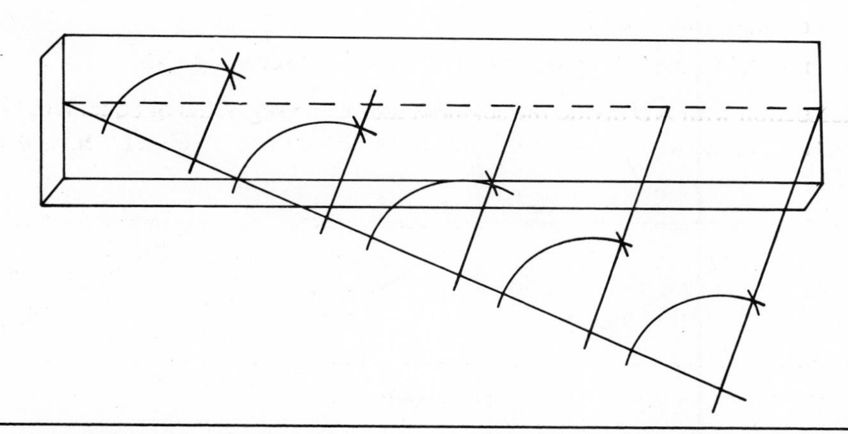

11 An angle is bisected when a ray is located which forms angles of equal measure with the sides of the original angle. The method is shown below.

To bisect ∡ABC follow these steps.

Step 1: Locate the point at the vertex. Use any radius. Draw an arc through the two sides.

Step 2: Draw two intersecting arcs as shown. Keep the same radius.

Step 3: Draw the angle bisector.

1

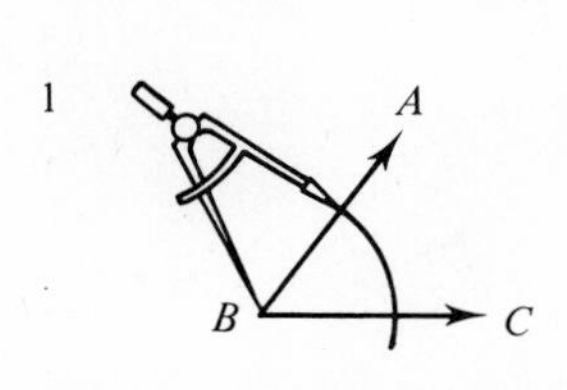

2

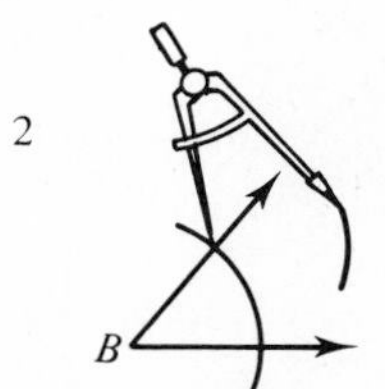

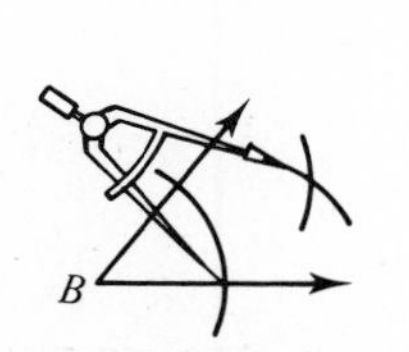

3

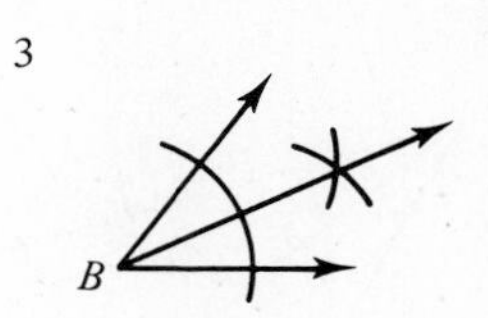

Q15 Use the method of Frame 10 to bisect the angle.

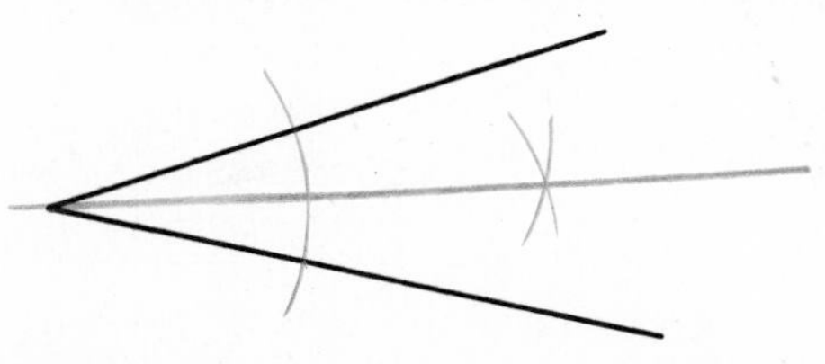

\# # # • # # # • # # # • # # # • # # # • # # # • # # # • # # # • # # #

A15

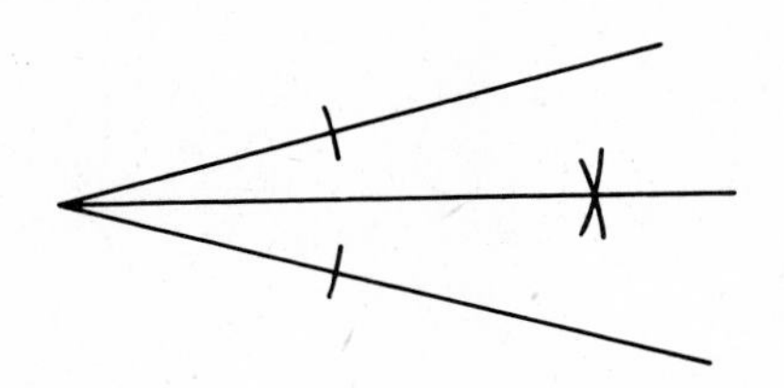

12 The angle bisectors of the angles of any triangle meet in one point. This fact is illustrated below.

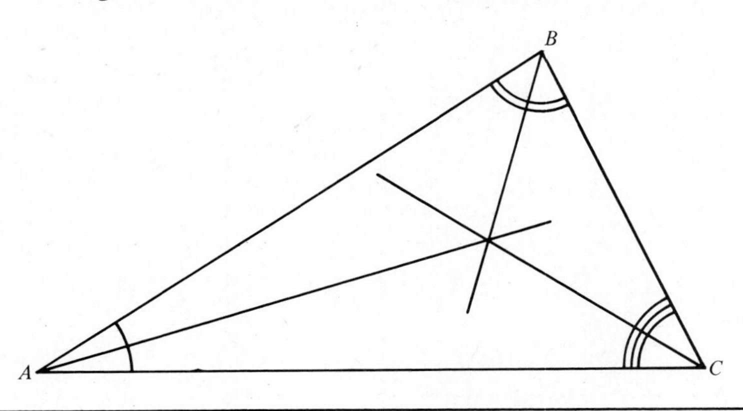

Q16 Verify that the angle bisectors of the triangle below meet in one point by constructing the bisectors of the angles with a compass and straightedge.

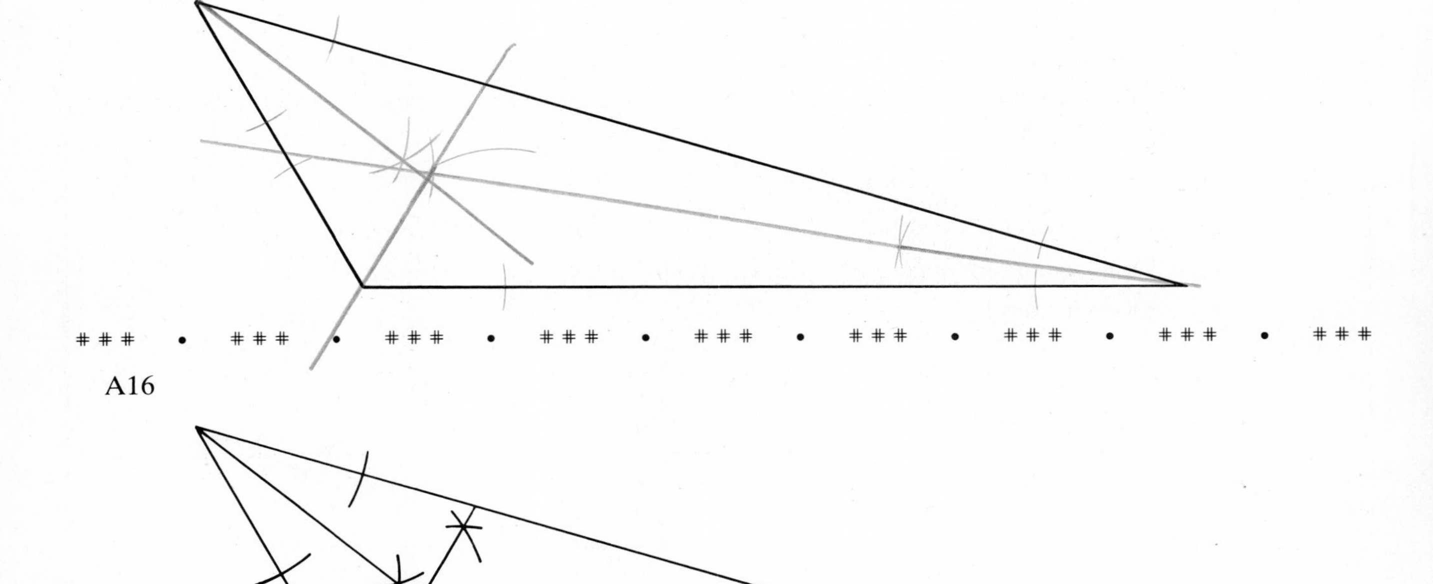

• # # # • # # # • # # # • # # # • # # # • # # # • # # # • # # # • # #

A16

*Q17 Many designs can be made by using only the tools of constructions and the techniques that have been presented in this section. Using larger circles, copy the same designs as shown.

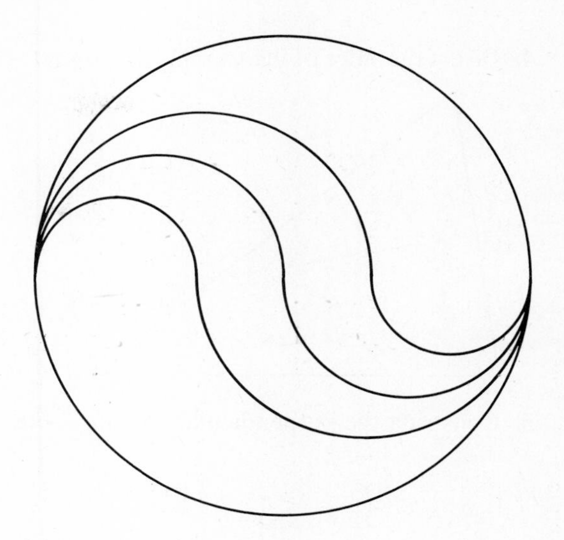

\# # # • # # # • # # # • # # # • # # # • # # # • # # # • # # # • # # #

*A17 Answers will vary.

This completes the instruction for this section.

4.6 EXERCISE

Perform the constructions of this exercise on your own paper.

1. What are the tools of geometric constructions?
2. Construct a regular hexagon with each side of the length given.

   ____

3. Construct an equilateral triangle by using the construction of problem 2 but connect only three of the intersection points with line segments.

4. Bisect the sides of the triangle below and verify that the bisectors of the sides all meet in a point.

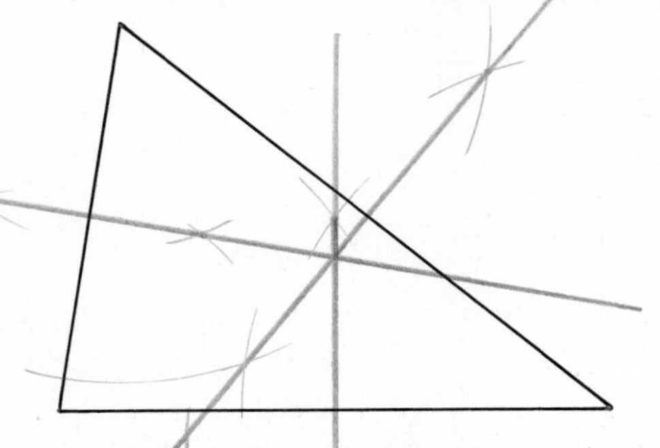

5. Construct the perpendiculars to line $\overleftrightarrow{AB}$ at point A and from point C.

6. What relationship do the two perpendiculars constructed in problem 5 have to each other?
7. Construct the altitudes of triangle ABC from vertex A and vertex B.

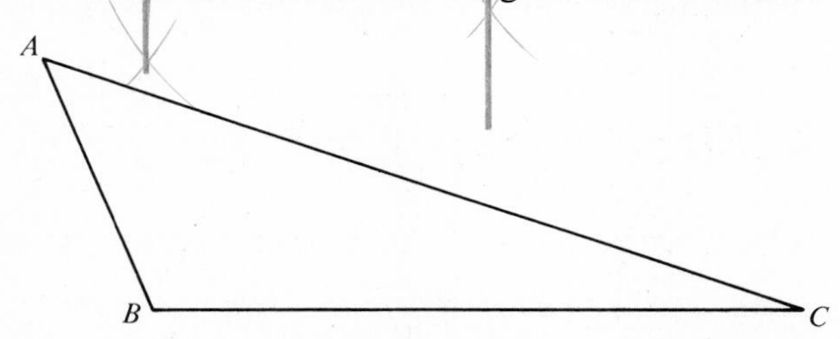

8. Copy $\measuredangle A$ at B. Measure the angles to verify your accuracy.

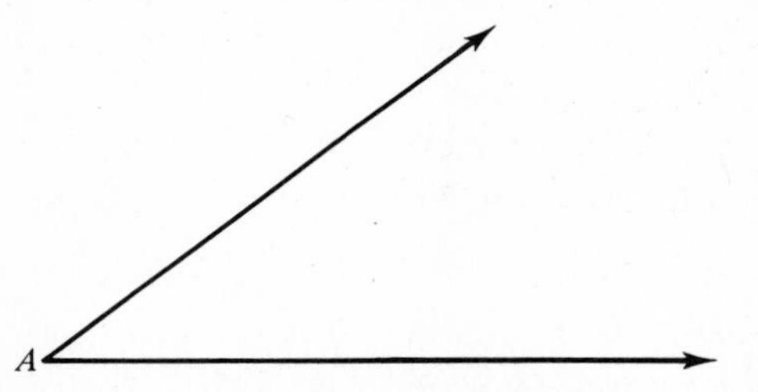

9. Construct a line parallel to $\overleftrightarrow{AB}$ through C.

C

A B

10. Divide the line segment $\overline{AB}$ into three equal lengths.

A B

11. Use the following figure to do the following:
   a. Bisect $\measuredangle A$ and $\measuredangle B$.
   b. Connect the intersection point of the bisectors with vertex C.

c. Verify by measuring the two angles formed at vertex C that $\measuredangle C$ is also bisected.

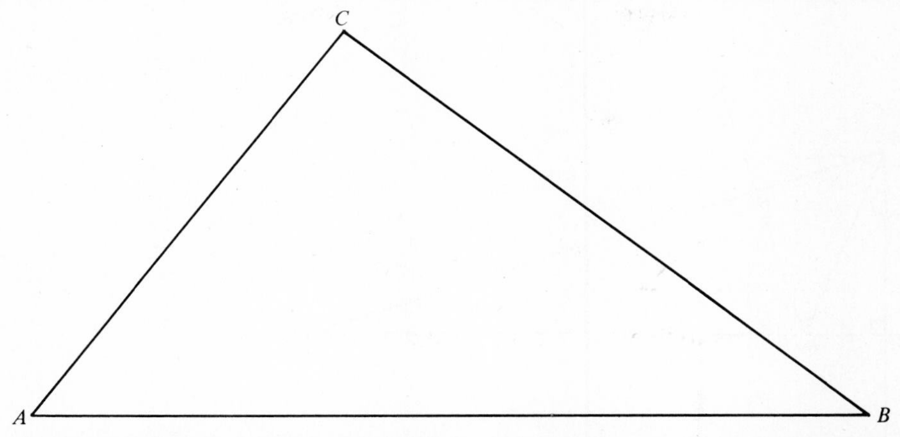

***12.** Construct designs such as those in Question 17 by using only the tools of geometric construction.

4.6 EXERCISE ANSWERS

1. Compass and straightedge.

2.

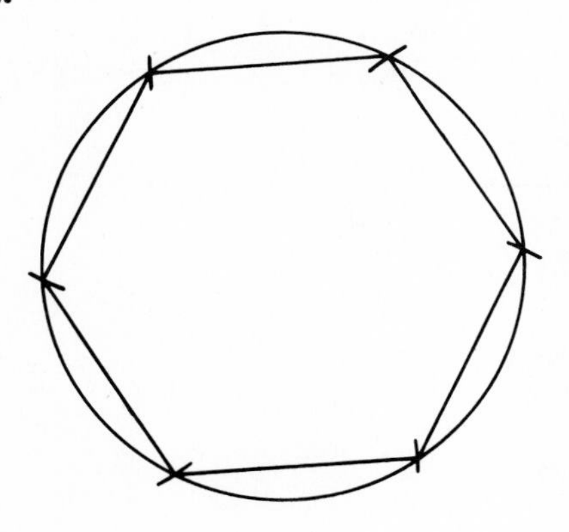

3.

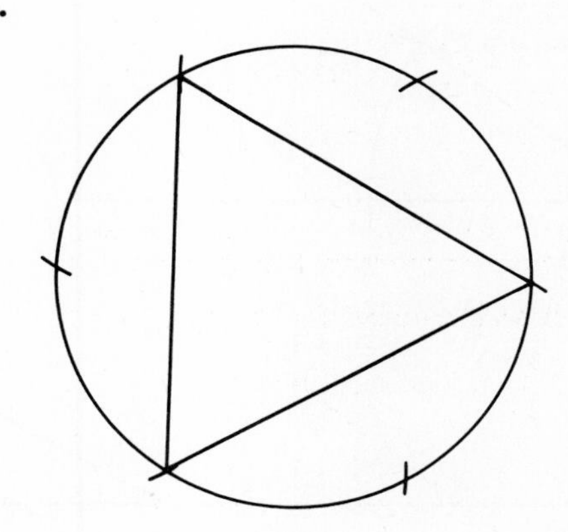

4.

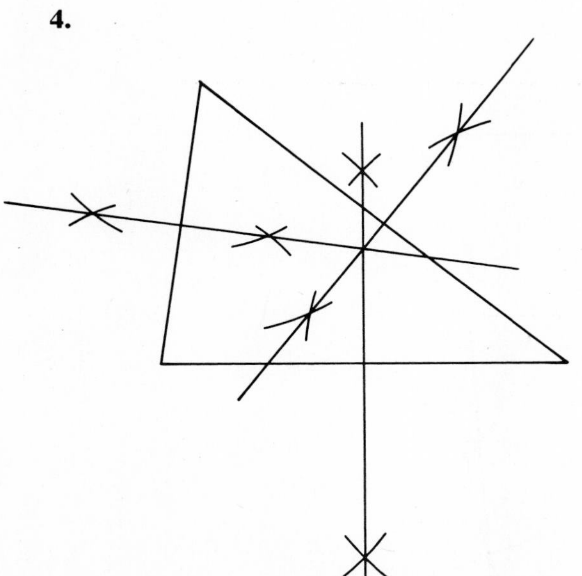

5.

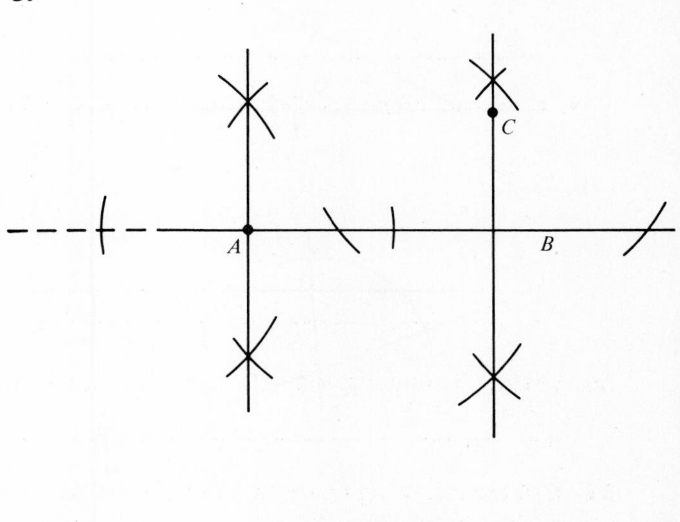

6. They are parallel.

7.

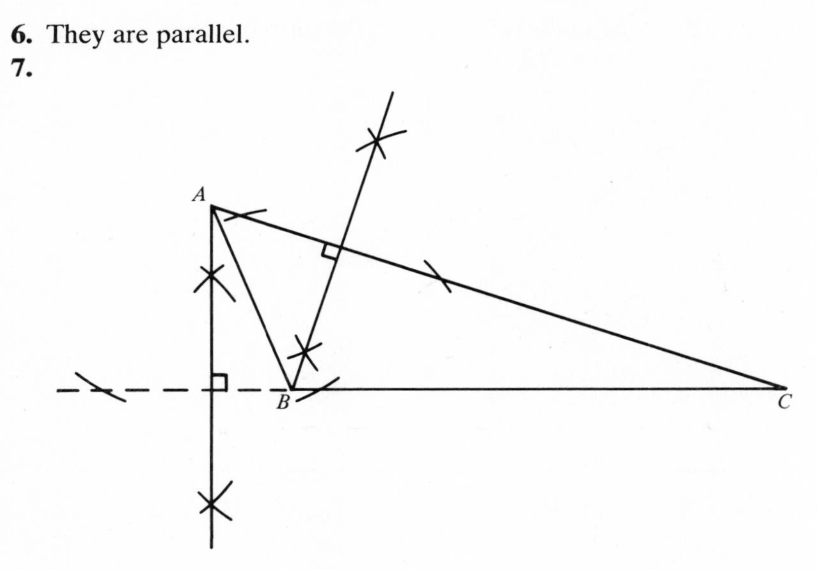

8.

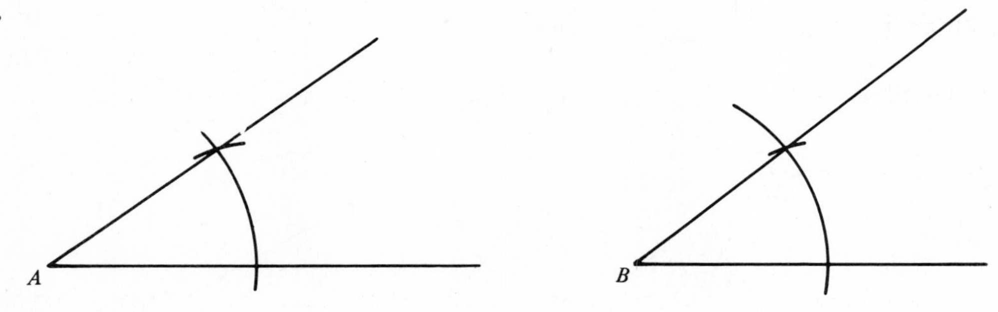

9.

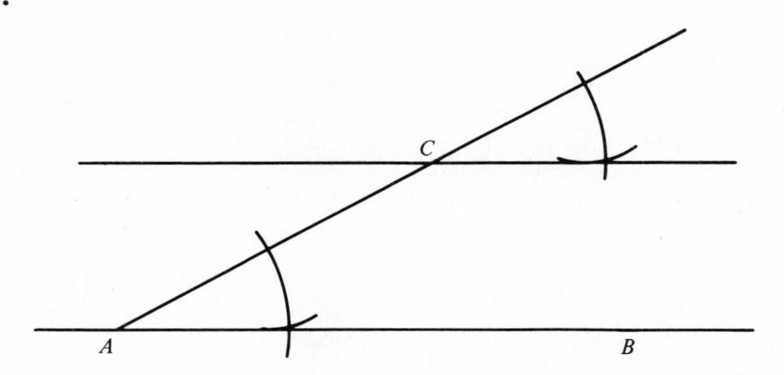

10.

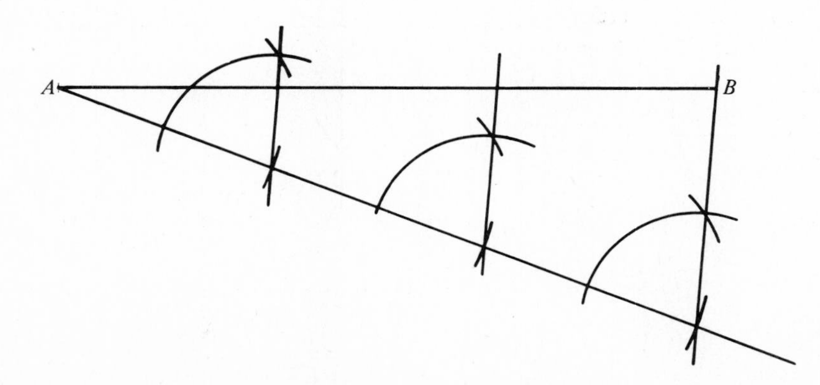

11.

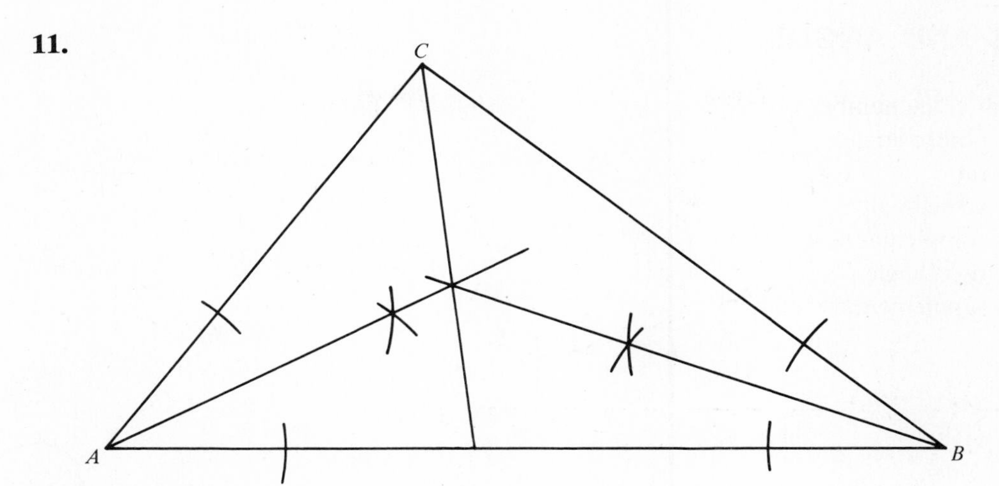

***12.** The designs will be of several varieties.

CHAPTER 4 SAMPLE TEST

At the completion of Chapter 4 you should be able to work the following problems.

4.1 POINTS, LINES, AND PLANES

1. The following suggest one of the three geometric objects: a point, a line, or a plane. Indicate which.

a. intersection of two walls in a room.

b. tabletop

c. stretched rubber band

d. grain of sand

e. stiff piece of cardboard

2. Answer true or false to the following statements. All lines referred to are on the figure below.

a. A point has no dimensions.

b. $l_3 \| l_4$.

c. $l_2 \perp l_3$.

d. $l_1 \| l_2$.

e. $l_2 \| l_4$.

f. Pictures are imperfect representations of points, lines, and planes.

g. l_2 and l_4 intersect in another point besides point B.

h. Many different planes contain all three of the points A, B, and C.

i. Other lines besides the one drawn pass through C and D.

j. Many other line segments different from $\overline{FA}$ contain point C.

k. $\measuredangle CAB$ is a right angle.

l. $\measuredangle EBA = \measuredangle DBA$.

m. $\measuredangle ACD$ is a right angle.

n. DB represents a line segment.

o. $\overleftrightarrow{CA}$ is a line.

p. E is on $\overline{DB}$.

q. F is on $\overleftrightarrow{CA}$.

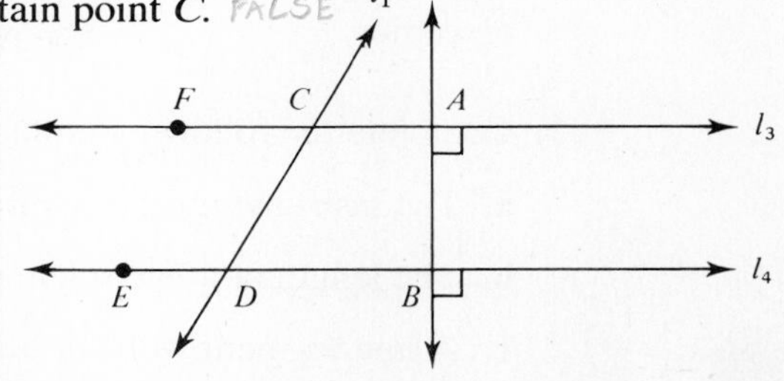

4.2 RAYS AND ANGLES

3. Match the number of the geometric figures with their names:

a. obtuse angle
b. ray
c. acute angle
d. complementary angles
e. right angle
f. supplementary angles

1. **2.**

3. **4.**

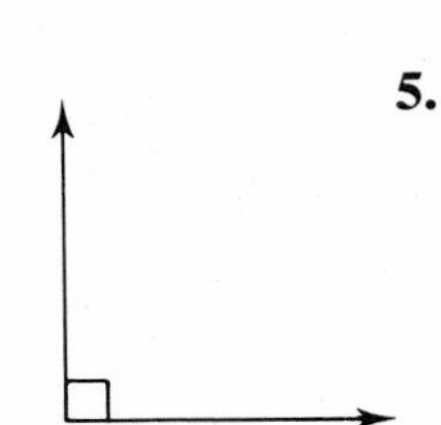

5.

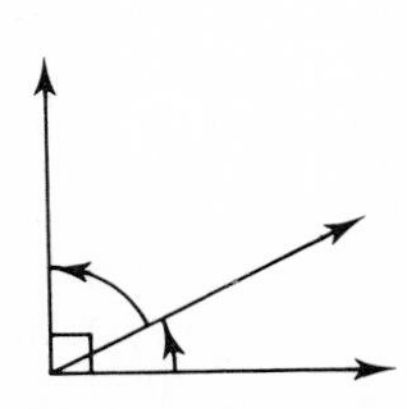

6.

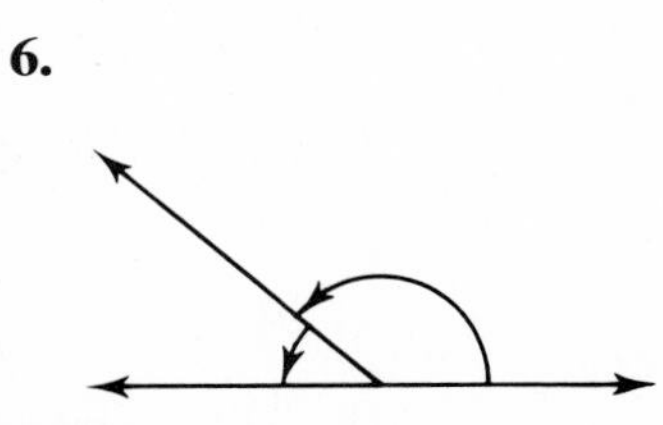

4. Answer true or false to the following statements. All angles and rays referred to are on the figure shown.

a. F is a point on $\measuredangle ABC$.
b. The vertex of $\measuredangle ACB$ is B.
c. $\overrightarrow{EC} = \overrightarrow{CE}$.
d. $\overrightarrow{EB} = \overrightarrow{EC}$.
e. $\overrightarrow{BF}$ is a side of $\measuredangle ABC$.
f. $\overrightarrow{BC}$ is a side of $\measuredangle ABC$.
g. $\measuredangle EBD$ has vertex B.
h. $\measuredangle EBF$ is obtuse.
i. $\measuredangle ACB$ is acute.
j. The sum of the measures of the angles of triangle ABC is 360°.
k. An angle is two line segments with a common endpoint.
l. A protractor measures the size of an angle.

5. Match the angles with their approximate measures:

a.

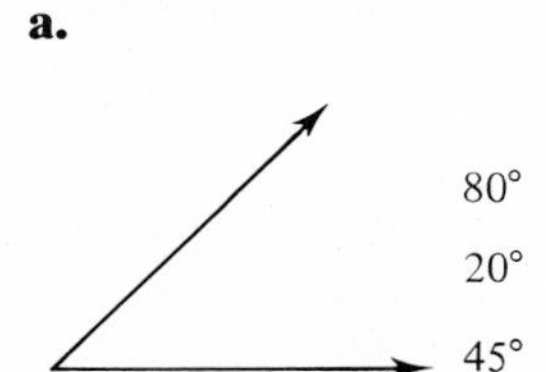

b.

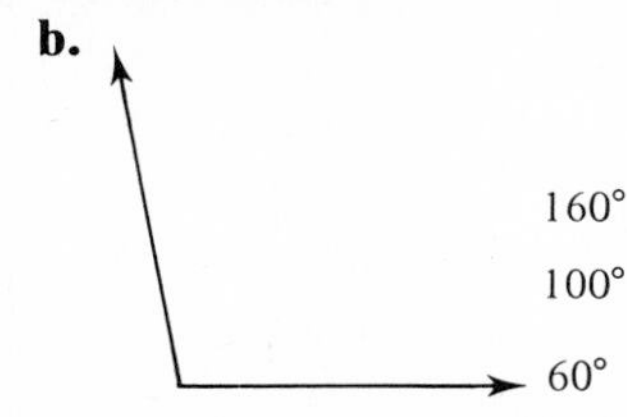

c.

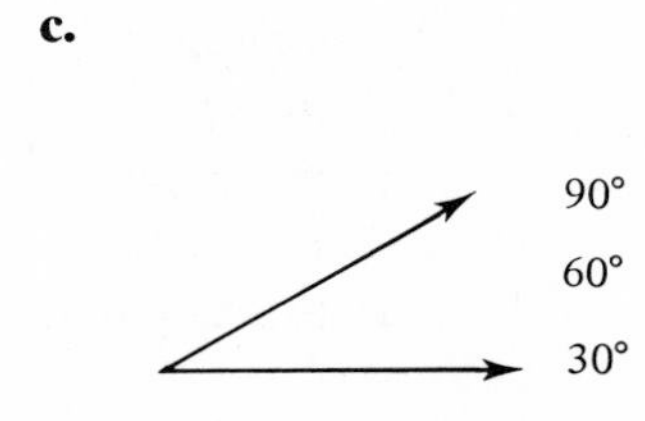

4.3 CIRCLES AND POLYGONS

6. Name an example of each geometric term from the figure:

a. diameter
b. radius
c. center

7. Complete the following sentences:

a. The instrument used to draw circles is a ____________.

b. The length of a line segment from the center to a point on the circle is the ____________.

c. A line segment with endpoints on the circle containing the center is a ____________.

8. Use the figure to name all lines or segments which are described by the term. O is the center of the circle.

a. radii
b. diameters
c. chords
d. tangents

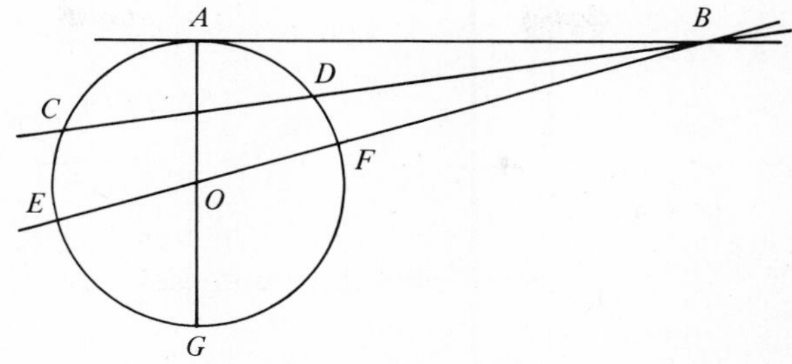

9. Use the figure to name three arcs which contain point D.

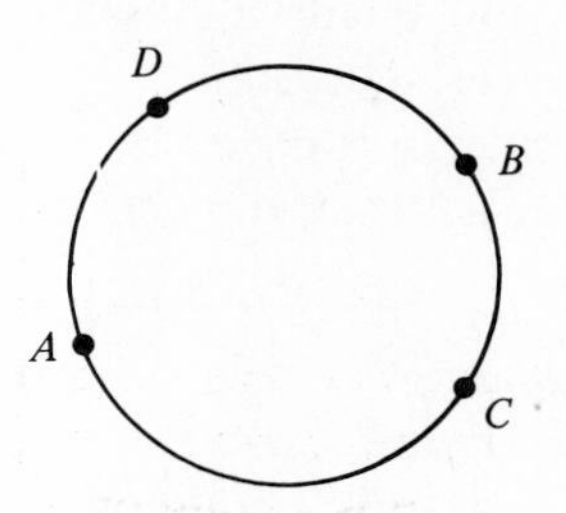

10. List the letters of the figures that are polygons:

a. 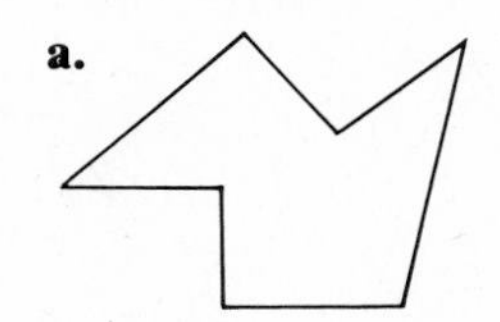

b. 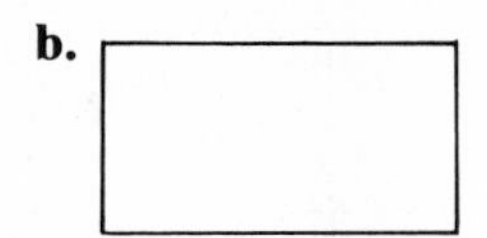

c. 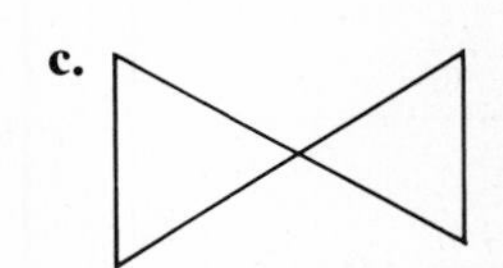

d. 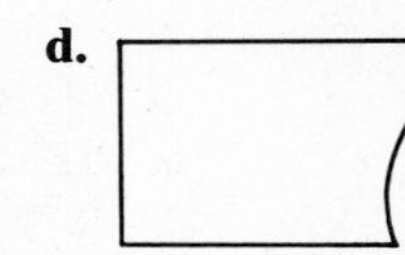

e. 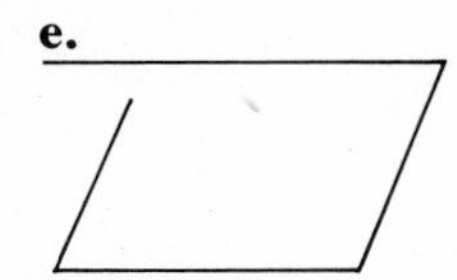

f. 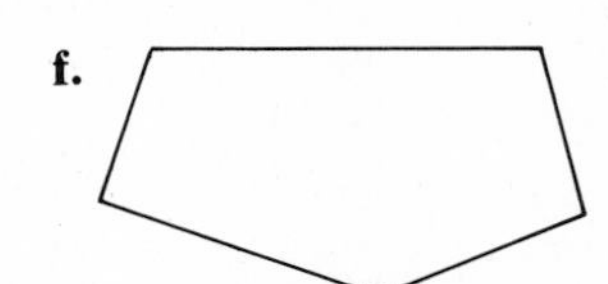

g. 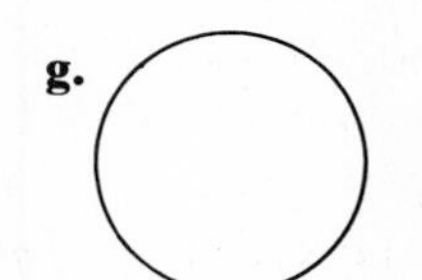

h. 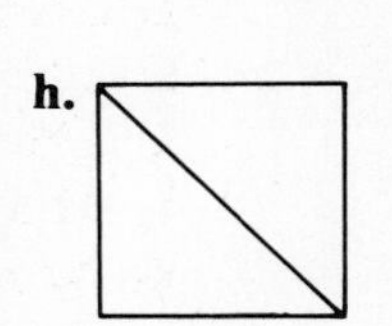

11. Match the number of the appropriate polygon with its name:

a. triangle
b. octagon
c. pentagon
d. hexagon
e. quadrilateral

1. 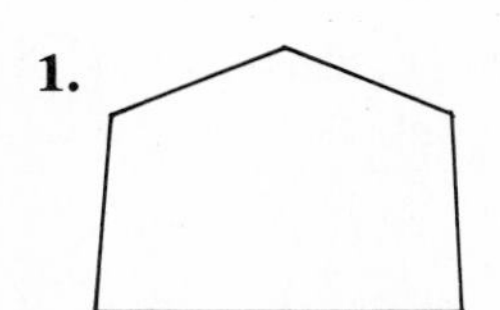

2. 

3. 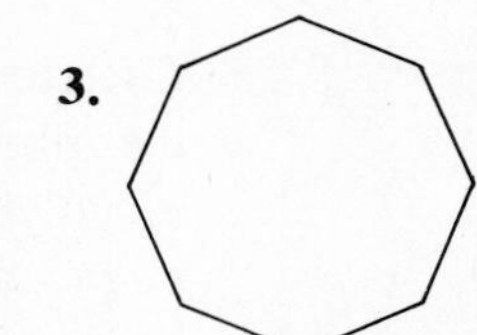

4. **5.** 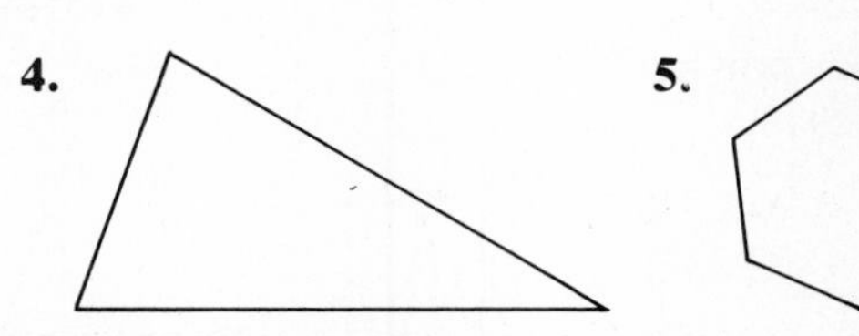

12. Explain how a triangular region differs from a triangle.

13. Each descriptive word applies to several of the triangles at the right. List *all* the numbers of triangles that can be described by each term. The number in parentheses indicates how many there are of each.

a. equilateral (1)
b. right (2)
c. scalene (2)
d. isosceles (4)
e. obtuse (2)
f. acute (2)

1. 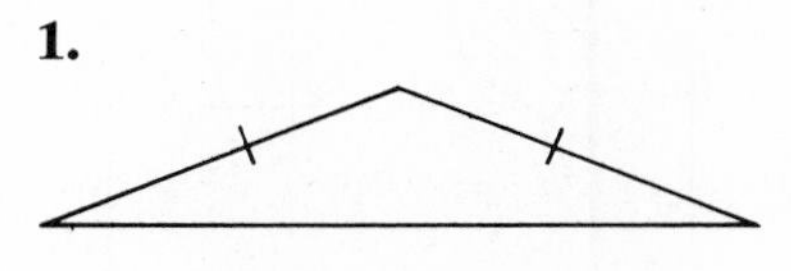

2. 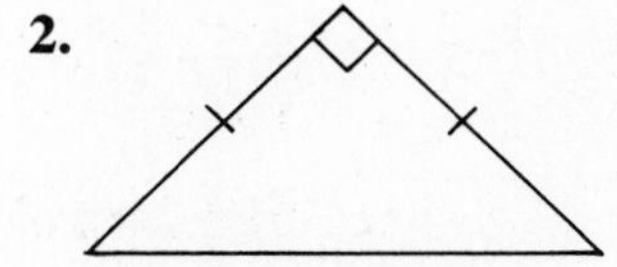

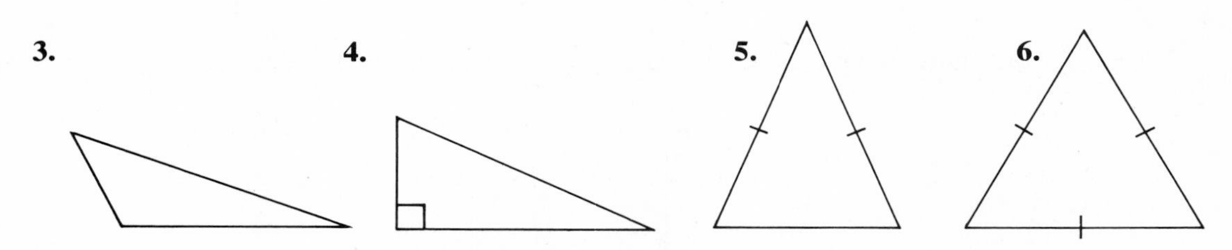

14. Each term applies to several of the following figures. List *all* the numbers of figures that can be described by each term. The number in parentheses indicates how many there are of each.

a. parallelogram (4)
b. quadrilateral (7)
c. square (1)
d. rectangle (2)
e. trapezoid (2)

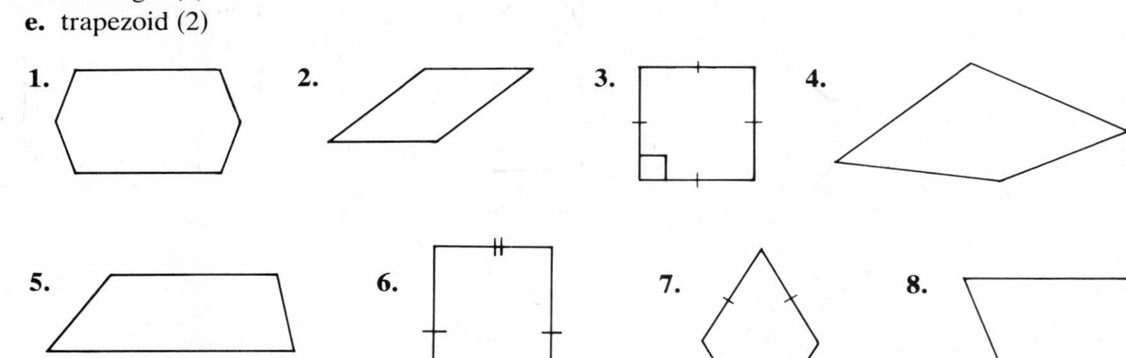

4.4 POLYHEDRA, PRISMS, AND PYRAMIDS

15. Each term applies to several of the following figures. List *all* the numbers of figures that can be described by each term. The number in parentheses indicates how many there are of each.

a. polyhedra (7)
b. prism (4)
c. pyramid (2)
d. regular polyhedra (3)
e. cube (1)

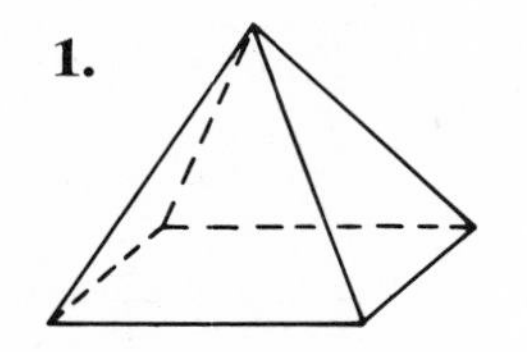

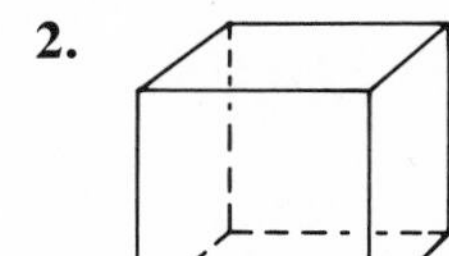

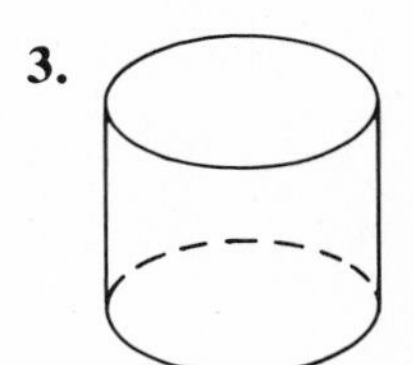

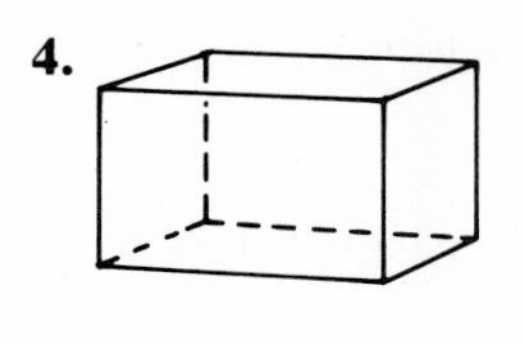

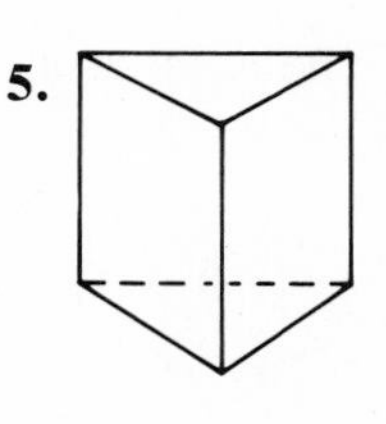

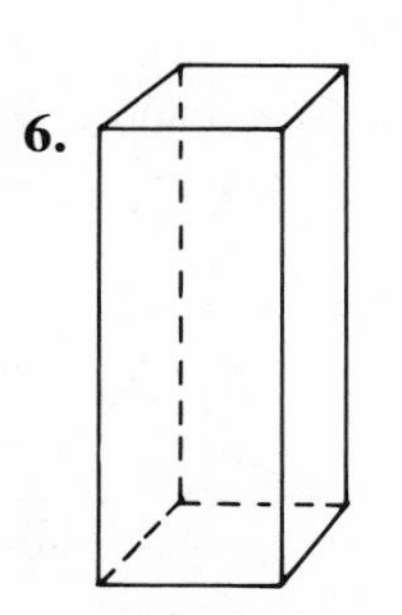

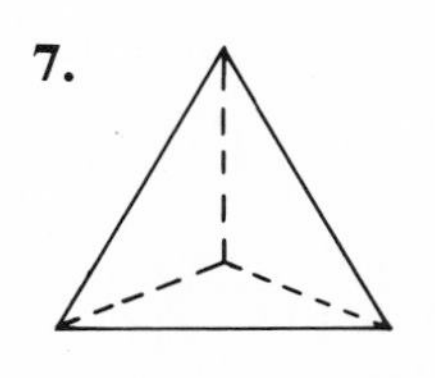

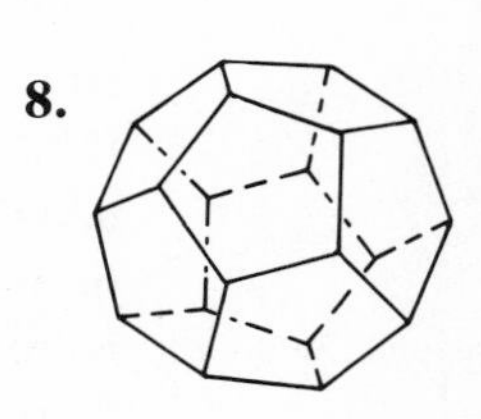

16. Match the name of the geometric term with the number of the arrow pointing to the object:

a. face
b. edge
c. vertex
d. apex
e. base
f. altitude

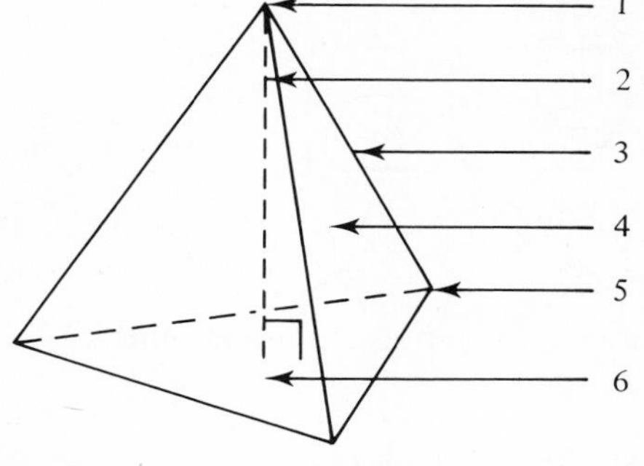

17. Indicate the letters of the figures that have the hidden lines drawn correctly:

a.

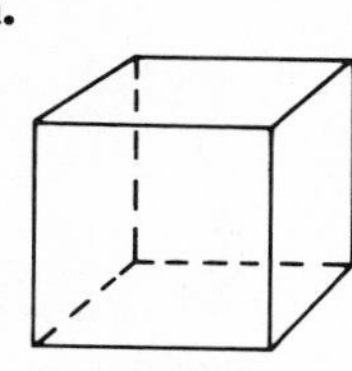

b.

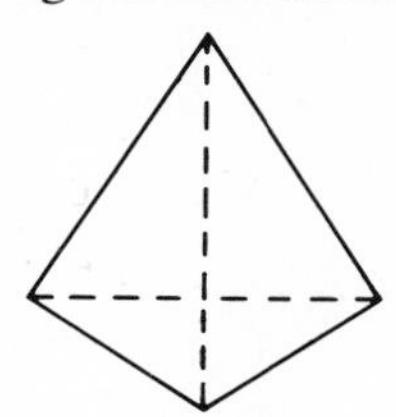

c.

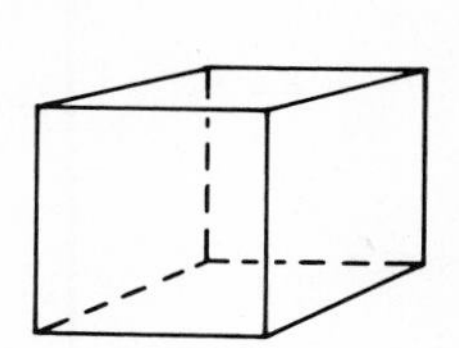

d.

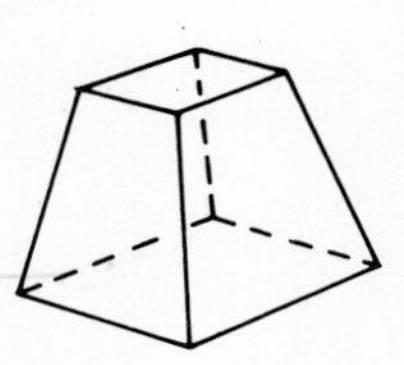

e.

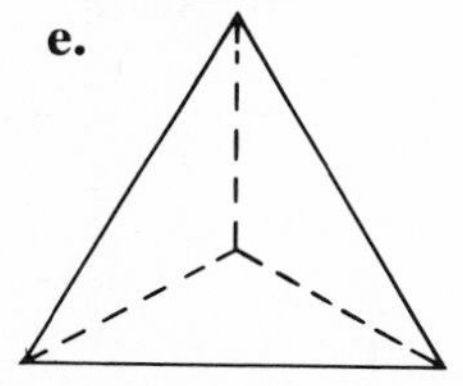

18. Sketch the hidden lines of the prisms:

a.

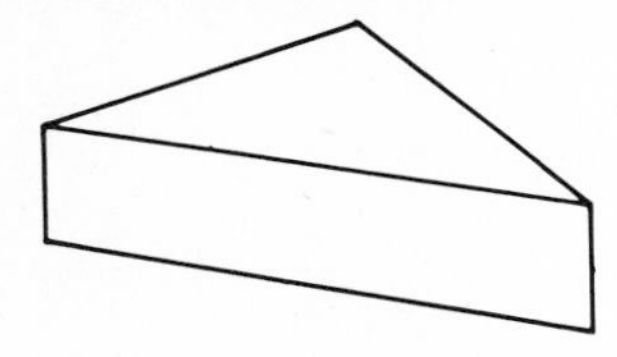

b.

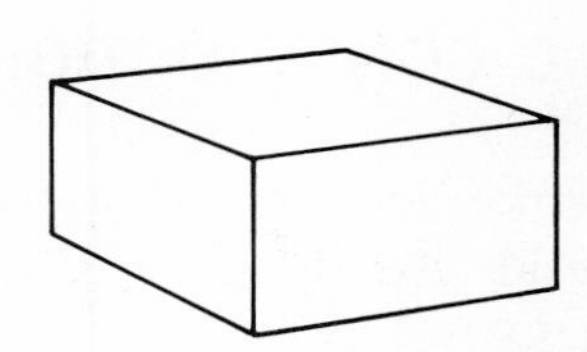

4.5 CYLINDERS, CONES, AND SPHERES

19. Match the name of the geometric term with the number of the arrow pointing to the object:

Cylinder
a. axis
b. base
c. lateral surface

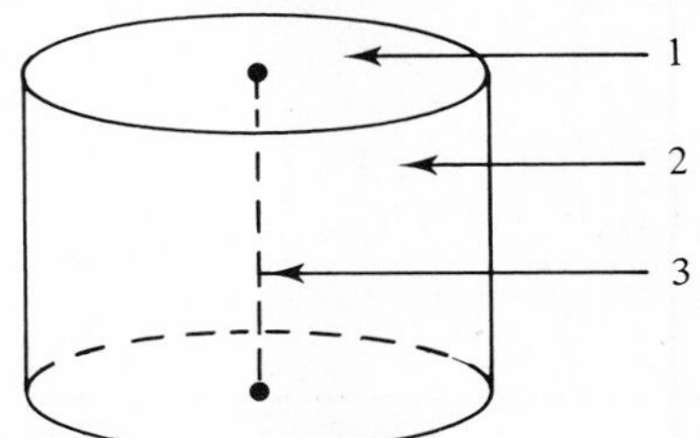

Cone
d. base
e. vertex
f. axis

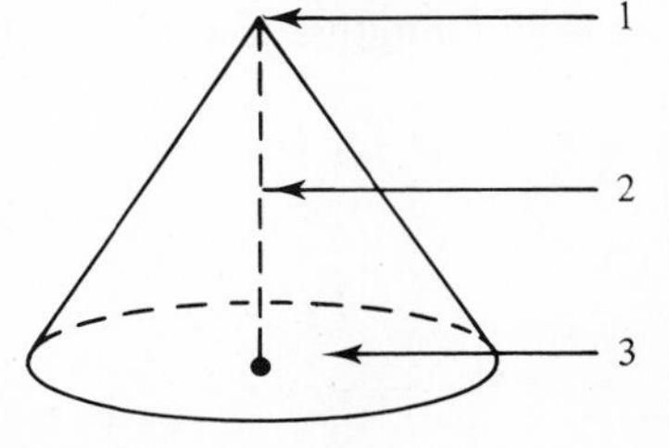

Sphere
g. center
h. radius
i. diameter
j. great circle

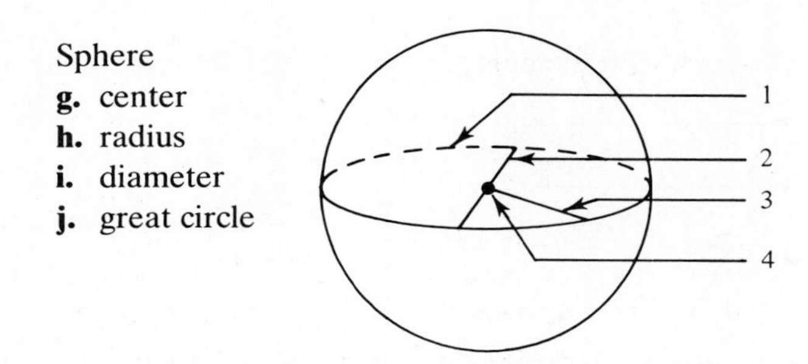

20. Complete the sketches of the figures, including the hidden lines.

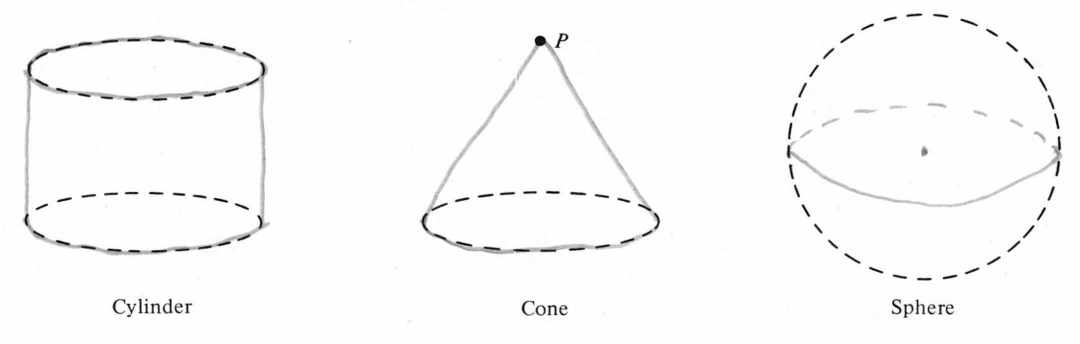

4.6 GEOMETRIC CONSTRUCTIONS

21. Consider triangle ABC and do the following constructions:
a. Bisect side $\overline{AB}$.
b. Bisect $\measuredangle ABC$.
c. Find the altitude from C.

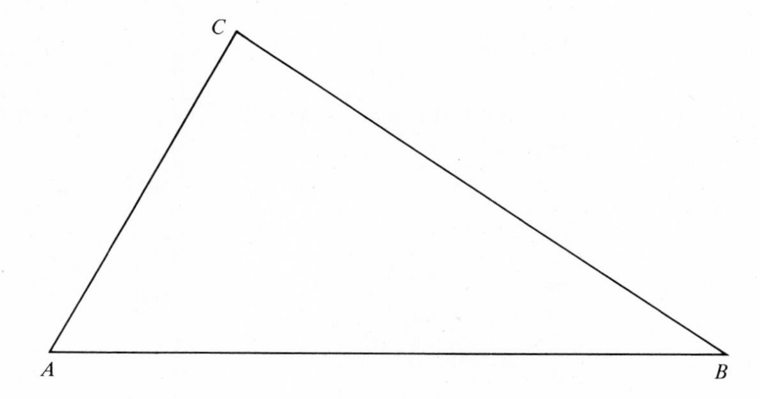

22. Divide the line segment into three equal lengths. Use the method of constructing parallel lines by copying an angle.

CHAPTER 4 SAMPLE TEST ANSWERS

1. **a.** line **b.** plane **c.** line **d.** point
e. plane

2. **a.** true **b.** true **c.** true **d.** false
e. false **f.** true **g.** false **h.** false
i. false **j.** true **k.** true **l.** true
m. false **n.** false **o.** true **p.** false
q. true

3. **a.** 3 **b.** 2 **c.** 1 **d.** 5
e. 4 **f.** 6

4. **a.** true **b.** false **c.** false **d.** true
e. true **f.** true **g.** true **h.** true
i. true **j.** false **k.** false **l.** true

5. **a.** 45° **b.** 100° **c.** 30°

6. **a.** $\overline{BC}$ **b.** $\overline{OB}$ or $\overline{OC}$ **c.** 0

7. **a.** compass **b.** radius **c.** diameter

8. **a.** $\overline{OA}$, $\overline{OF}$, $\overline{OG}$, $\overline{OE}$ **b.** $\overline{AG}$, $\overline{EF}$
c. $\overline{CD}$, $\overline{EF}$, $\overline{AG}$ **d.** $\overleftrightarrow{AB}$

9. arc *AB*, arc *CB*, and arc *AC*

10. a, b, and f

11. **a.** 4 **b.** 3 **c.** 1 **d.** 5
e. 2

12. A triangular region includes the points inside the triangle.

13. **a.** 6 **b.** 2 and 4 **c.** 3 and 4 **d.** 1, 2, 5, and 6
e. 1 and 3 **f.** 5 and 6

14. **a.** 2, 3, 6, and 7 **b.** 2, 3, 4, 5, 6, 7, and 8 **c.** 3
d. 3 and 6 **e.** 5 and 8

15. **a.** 1, 2, 4, 5, 6, 7, and 8 **b.** 2, 4, 5, and 6 **c.** 1 and 7
d. 2, 7, and 8 **e.** 2

16. **a.** 4 (or 6) **b.** 3 **c.** 5 (or 1) **d.** 1
e. 6 **f.** 2

17. a, c, d, and e

18. **a.** **b.**

19. **a.** 3 **b.** 1 **c.** 2 **d.** 3
e. 1 **f.** 2 **g.** 4 **h.** 3
i. 2 **j.** 1

20. **a.** **b.** **c.**

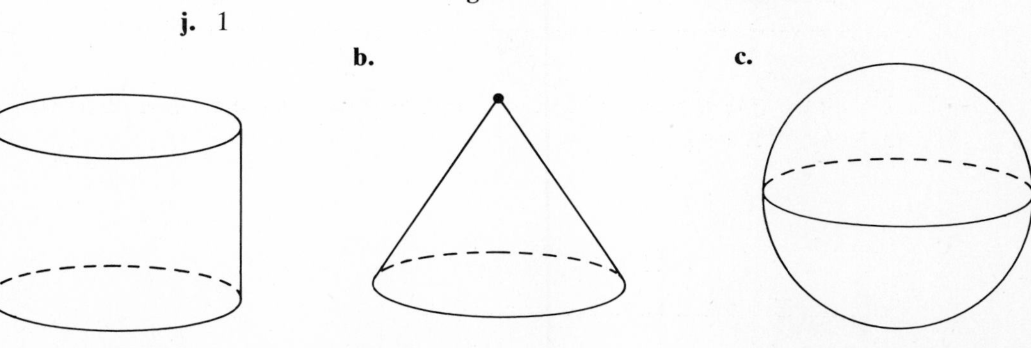

21.

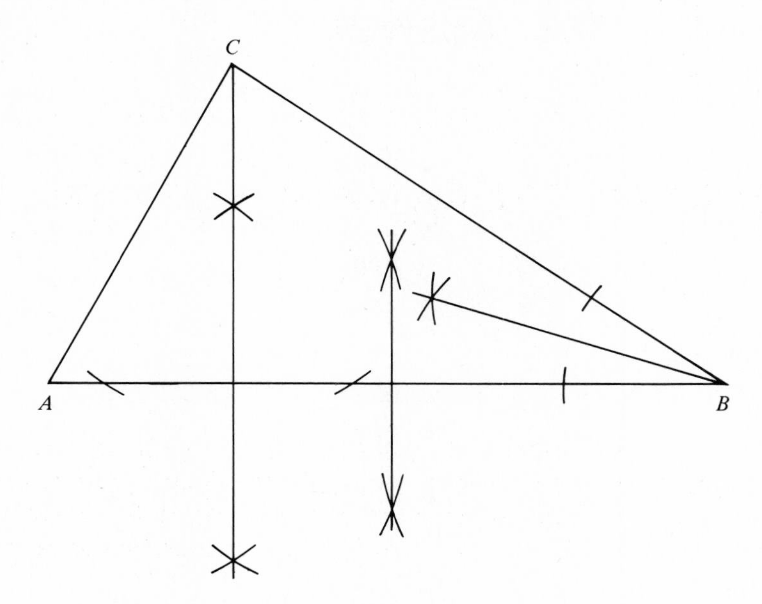

22.

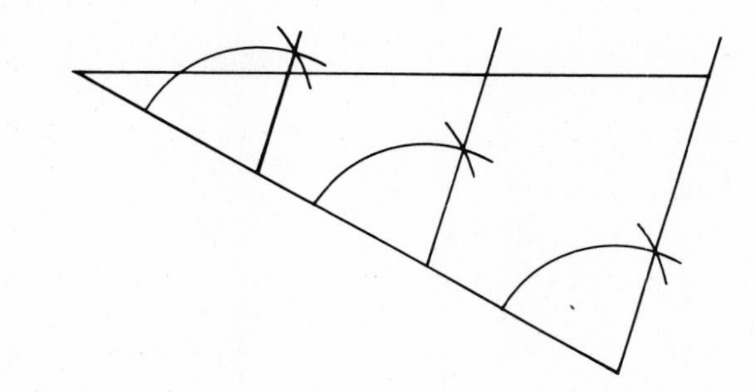

CHAPTER 5

MEASUREMENT AND GEOMETRY

In Chapter 4 certain geometric objects were defined. Relationships between these objects were stated as properties. At that time measurement of line segments was not required. In this chapter measurement is introduced and properties of geometry that make use of the idea of measurement are studied. This geometry is called *metric* geometry. Some of the basic notions of measurement will be discussed first.

An inch ruler and a centimetre ruler are required to study Chapter 5. Such a ruler is provided below. The ruler below may be cut from the text and used whenever a measurement is required. The English portion has two scales, one marked off each $\frac{1}{16}$ inch and the other each $\frac{1}{10}$ inch. The metric part is graduated in millimetres. On the metric scale, the numbers indicate centimetres.

5.1 MEASUREMENT OF LINE SEGMENTS

1 How many people are in the room in which you are located? How many rooms are there in the building? The answers to these questions can be obtained by counting and an *exact* result obtained.

Q1 How many words are there in the statement of this question? ______

• ### • ### • ### • ### • ### • ### • ### • ### •

A1 11

2 Measurement is the comparison of an observed quantity with a standard unit. The result of the measurement is a number times the standard unit. When counting a set of objects, the standard unit of measurement is one object.

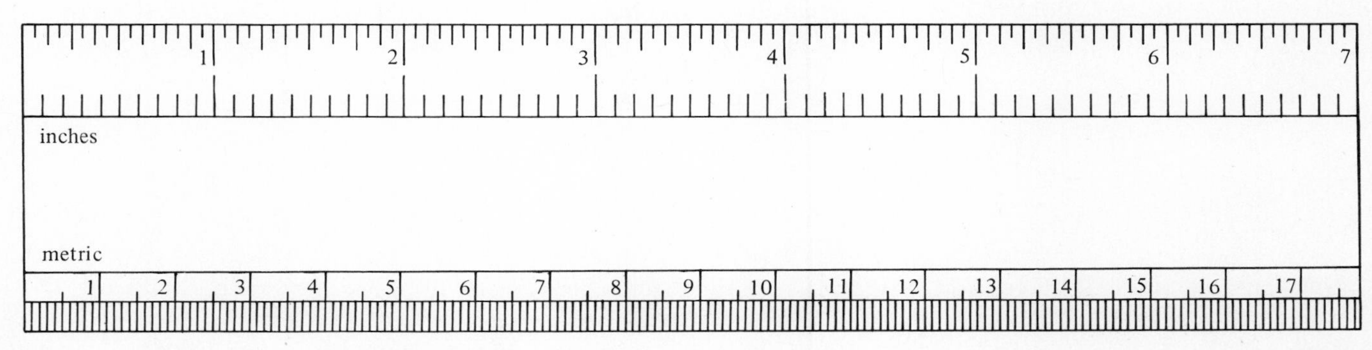

Q2 **a.** How many letters are in the following list?

a, b, c, d, e ______

b. What was the standard unit of measurement in part a? ____________

\# # # • # # # • # # # • # # # • # # # • # # # • # # # • # # # • # # #

A2 **a.** 5 **b.** one letter

3 A standard unit for measuring length is an inch. One inch is a length equal to the following line segment:

____

To measure a line segment in inches, a ruler with subdivisions showing inches is held next to the line segment, and the number of inches is read as carefully as possible. For greater accuracy, look down on the end of the segment from directly above the ruler. Although counting results in a measurement that is exact, measuring with a ruler gives an approximate result. You can use this method to tell the difference between an exact measurement and an approximate measurement. The only measurement that is exact is one that results from counting. Any measurement resulting from the use of a measuring instrument is approximate.

Q3 Decide whether the measurement would be approximate or exact:

a. length of your index finger ____________

b. number of letters in the word "bump" ____________

c. circumference of your head ____________

\# # # • # # # • # # # • # # # • # # # • # # # • # # # • # # # • # # #

A3 **a.** approximate **b.** exact **c.** approximate

4 The *precision* of a measurement is dependent upon the instrument used to measure. The precision of the instrument is determined by the smallest subdivision of the instrument. A ruler marked only at the inch marks is precise to a whole number of inches. A ruler marked at the quarter-inch positions

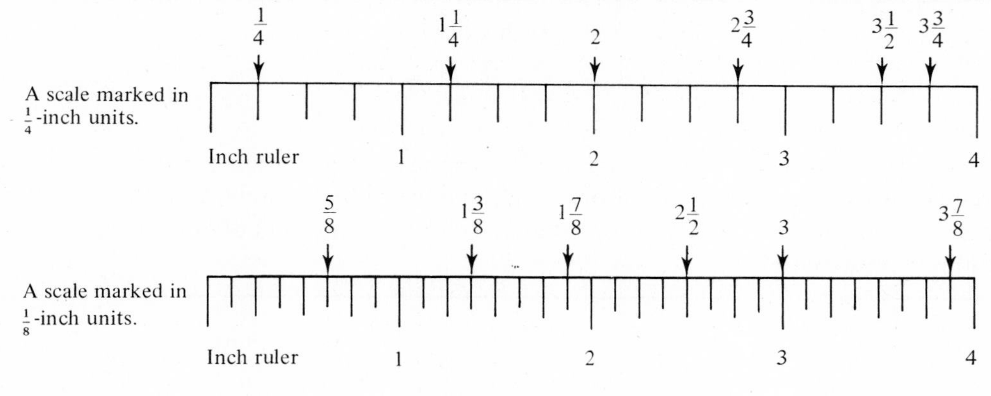

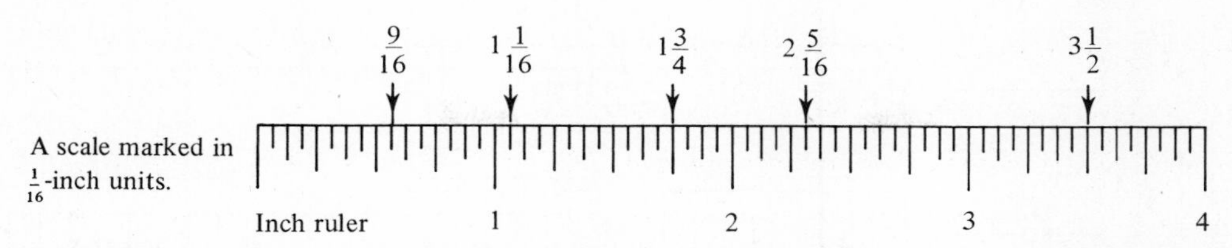

is precise to a quarter of an inch. A ruler marked at the one-sixteenth-inch positions is precise to sixteenths of an inch. The precision of a measurement is the same as the precision of the instrument used to obtain the measurement.

On the rulers above study each point to which the arrow is directed until you agree with the measurement given.

Q4 **a.** What is the precision of the measurements given in the first example in Frame 4? ________

b. What is the precision of the measurements given in the second example in Frame 4? ________

c. What is the precision of the measurements given in the third example in Frame 4? ________

\# # # • # # # • # # # • # # # • # # # • # # # • # # # • # # # • # # #

A4 **a.** $\frac{1}{4}$ inch **b.** $\frac{1}{8}$ inch **c.** $\frac{1}{16}$ inch

5 When measuring a line segment one must first decide how precise the resulting measurement needs to be. If only a measurement to the nearest inch is necessary, a ruler with only the inch marks is required. The bar below would have a measurement of three inches if it was measured to the *nearest* inch.

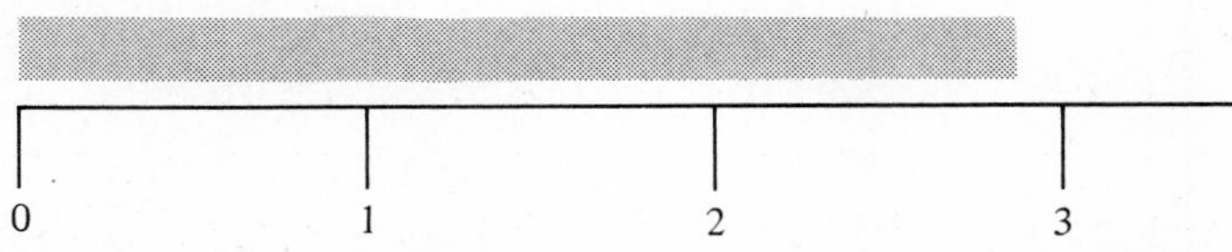

Q5 Use an inch ruler to measure the bars below to the nearest inch (use the ruler on the first page of this chapter if necessary):

a. ________

b. ________

c. ________

d. ________

e. ________

\# # # • # # # • # # # • # # # • # # # • # # # • # # # • # # # • # # #

A5 **a.** 3 inches **b.** 4 inches **c.** 1 inch **d.** 1 inch
e. 5 inches

Q6 Indicate the measurement of each line segment precise to the unit indicated:

a. $\frac{1}{4}$ unit ________

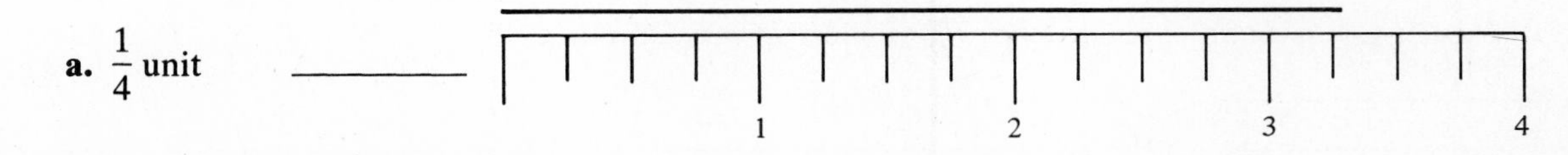

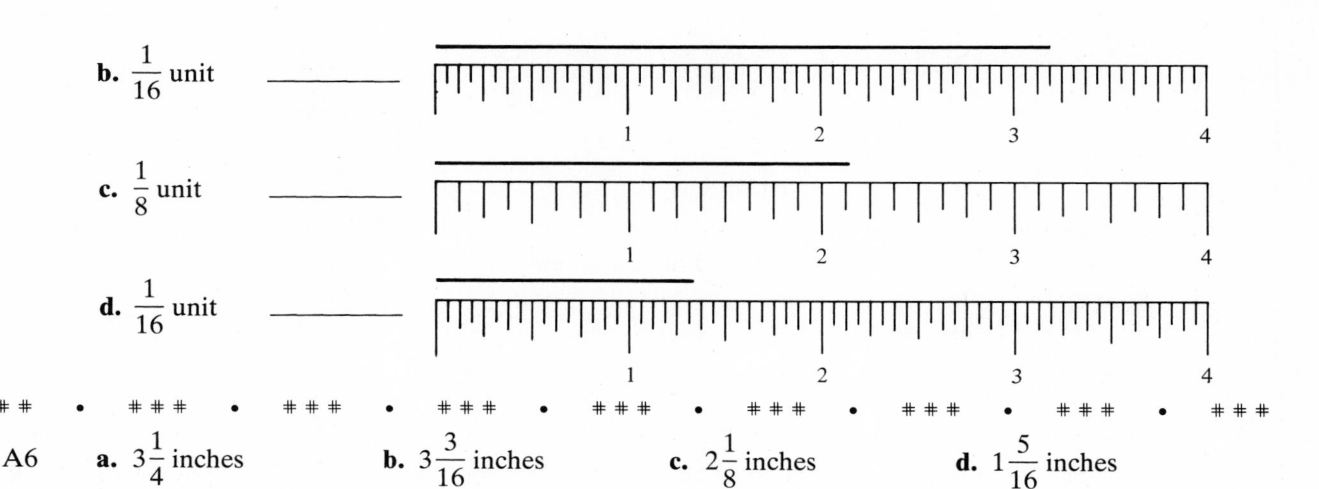

b. $\frac{1}{16}$ unit ______

c. $\frac{1}{8}$ unit ______

d. $\frac{1}{16}$ unit ______

\# \# \# • \# \# \# • \# \# \# • \# \# \# • \# \# \# • \# \# \# • \# \# \# • \# \# \# • \# \# \#

A6 **a.** $3\frac{1}{4}$ inches **b.** $3\frac{3}{16}$ inches **c.** $2\frac{1}{8}$ inches **d.** $1\frac{5}{16}$ inches

---

Q7 Read the scale at the places marked:

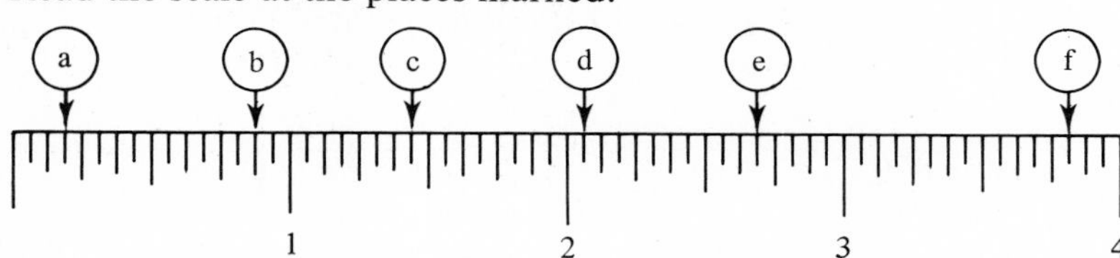

a. ______ **b.** ______ **c.** ______ **d.** ______

e. ______ **f.** ______

\# \# \# • \# \# \# • \# \# \# • \# \# \# • \# \# \# • \# \# \# • \# \# \# • \# \# \# • \# \# \#

A7 **a.** $\frac{3}{16}$ inches **b.** $\frac{7}{8}$ inches **c.** $1\frac{7}{16}$ inches **d.** $2\frac{1}{16}$ inches

e. $2\frac{11}{16}$ inches **f.** $3\frac{13}{16}$ inches

6 All measurements of line segments are approximate. The degree of precision of your reported measurement must be decided before you measure. The line segments below are all the same length. The measurement of their length is approximated to different degrees of precision.

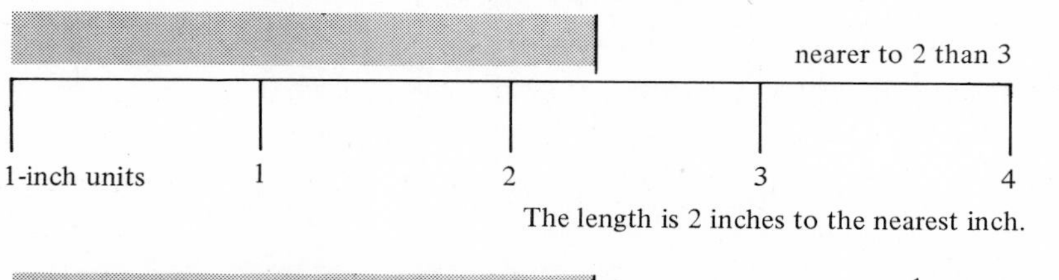

The length is 2 inches to the nearest inch.

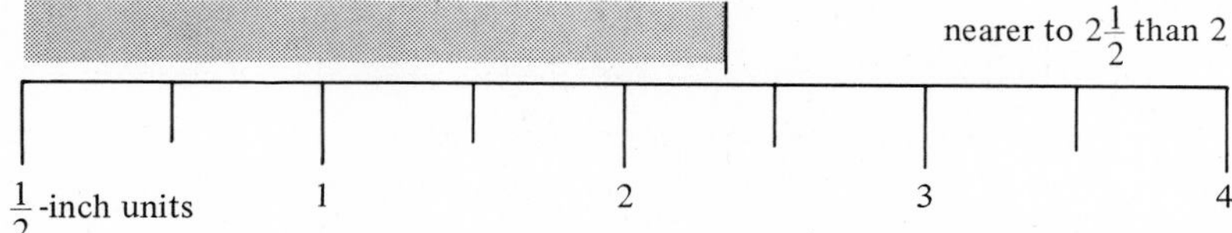

The length is $2\frac{1}{2}$ inches to the nearest $\frac{1}{2}$ inch.

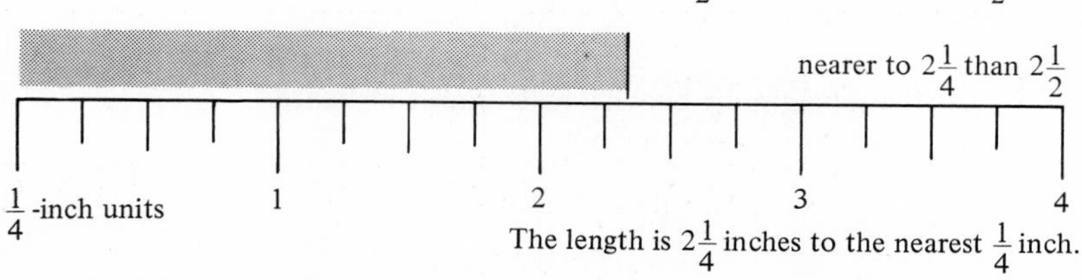

The length is $2\frac{1}{4}$ inches to the nearest $\frac{1}{4}$ inch.

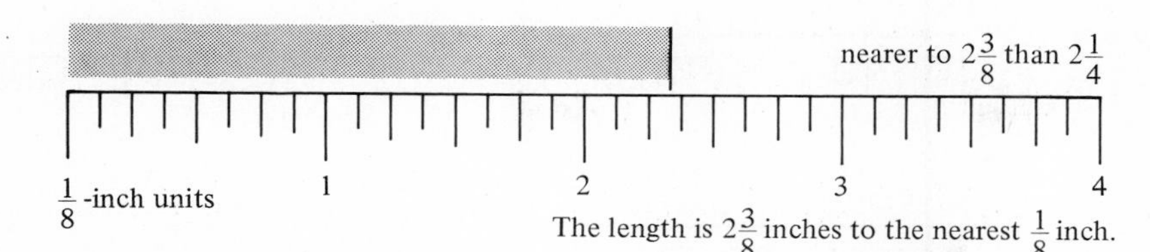

The length is $2\frac{3}{8}$ inches to the nearest $\frac{1}{8}$ inch.

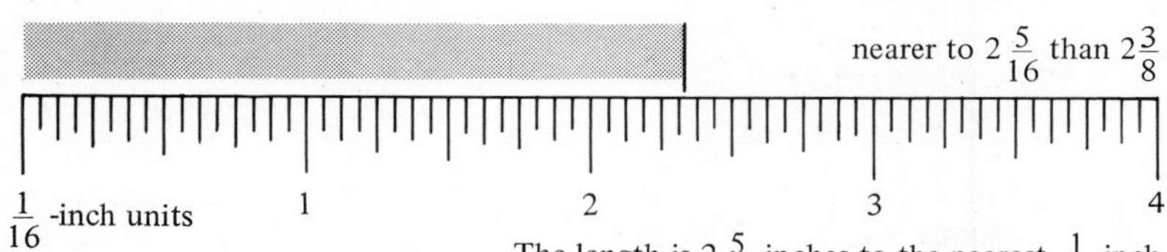

The length is $2\frac{5}{16}$ inches to the nearest $\frac{1}{16}$ inch.

Notice that the bars are the same length, but the measurements all differ from each other. Each is a correct measurement to the precision indicated for that measurement.

Q8 Approximate the length of the line segments to the degree of precision indicated:

a. 1 inch ________

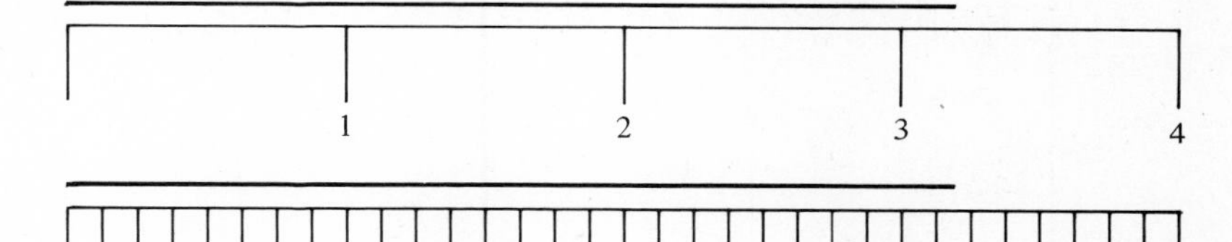

b. $\frac{1}{8}$ inch ________

c. $\frac{1}{16}$ inch ________

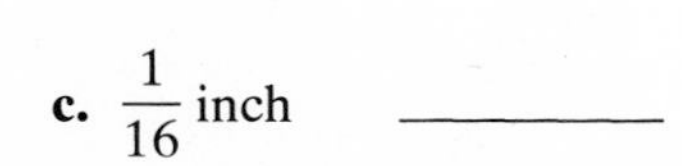

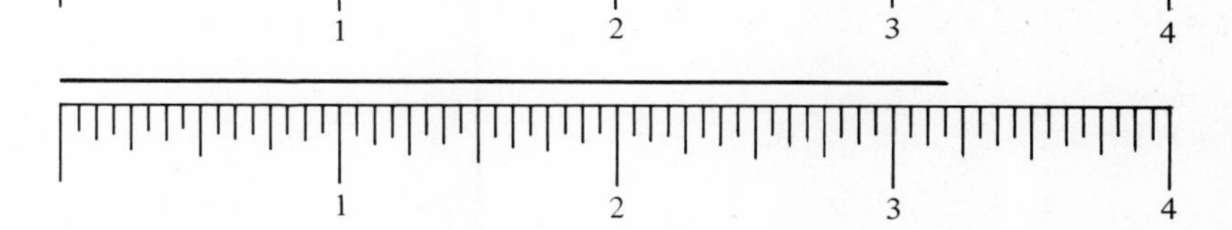

\# \# \# • \# \# \# • \# \# \# • \# \# \# • \# \# \# • \# \# \# • \# \# \# • \# \# \# • \# \# \#

A8 **a.** 3 inches **b.** $3\frac{1}{4}$ inches **c.** $3\frac{3}{16}$ inches

7 If precision to the nearest $\frac{1}{4}$ inch is required but only a ruler with smaller subdivisions is available, it can be used. All subdivisions smaller than the $\frac{1}{4}$ subdivisions are ignored. Thus, the measurement of the following line segment is $2\frac{1}{2}$ inches to the nearest $\frac{1}{2}$ inch and $2\frac{5}{8}$ inches to the nearest $\frac{1}{8}$ inch:

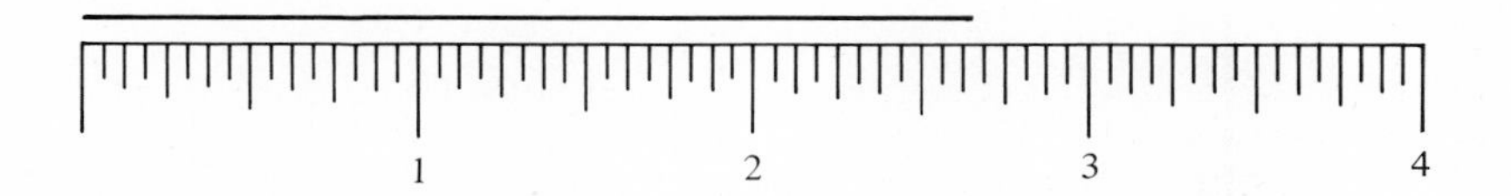

Q9 Approximate the length of the following line segments to the degree of precision indicated:

a. 1 inch ________

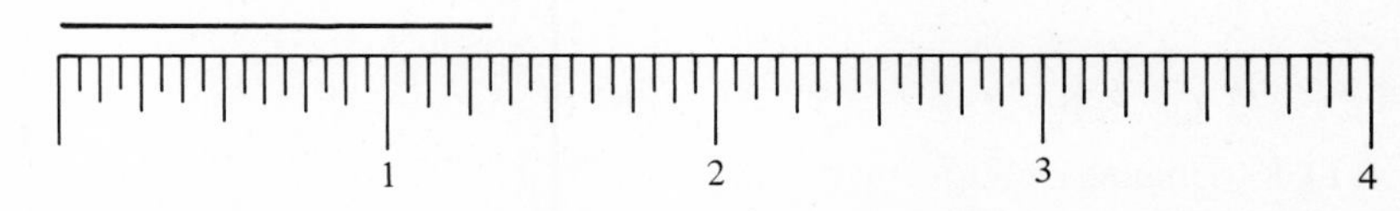

b. $\frac{1}{2}$ inch ________

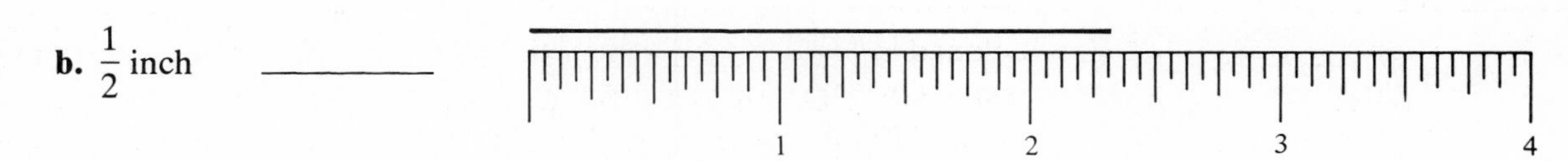

c. $\frac{1}{4}$ inch ________

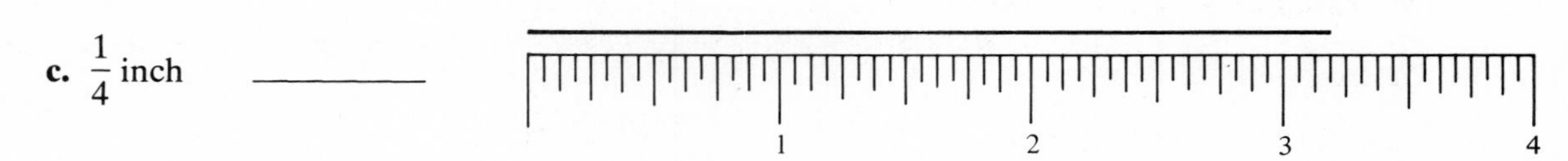

\# \# \# • \# \# \# • \# \# \# • \# \# \# • \# \# \# • \# \# \# • \# \# \# • \# \# \# • \# \# \#

A9 **a.** 1 inch **b.** $2\frac{1}{2}$ inches **c.** $3\frac{1}{4}$ inches

---

Q10 Measure the lengths below to the nearest $\frac{1}{8}$ inch.

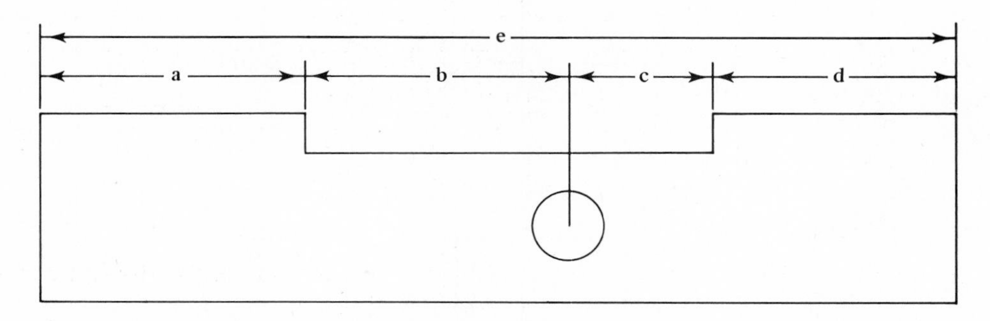

\# \# \# • \# \# \# • \# \# \# • \# \# \# • \# \# \# • \# \# \# • \# \# \# • \# \# \# • \# \# \#

A10 **a.** $1\frac{3}{8}$ inch **b.** $1\frac{3}{8}$ inch **c.** $\frac{3}{4}$ inch **d.** $1\frac{1}{4}$ inch **e.** $4\frac{3}{4}$ inch:

8 The inch is one of the units used in the English (U.S. customary) system of measurement.* Other units of length are the foot and the yard. The system of measurement that uses the inch, foot, and yard is called the English system, because King Edward I of England in the thirteenth century ordered that a permanent measuring stick made of iron serve as a master standard yardstick. At the same time he declared that a foot measurement would be $\frac{1}{3}$ of the standard yard and the inch would be $\frac{1}{36}$ of the standard yard. A mile became 5,280 feet. To summarize:

1 mile = 5,280 feet
1 yard = 36 inches
1 foot = 12 inches

*The system of measurement historically used for most everyday measurement in the United States is called the U.S. customary or English system of measurement.

Q11 What is the name of the system of measurement that uses the inch, foot, and yard? ____________

\# \# \# • \# \# \# • \# \# \# • \# \# \# • \# \# \# • \# \# \# • \# \# \# • \# \# \# • \# \# \#

A11 English or U.S. customary

9 The metric system was developed in France at the end of the eighteenth century as Napoleon rose to power. The *metre* was to be one ten-millionth part of the distance from the north pole to the equator on a great circle through Paris. The metre is slightly longer than the yard. The metric system is presently used in most countries of the world.

If the bar shown here were attached at the end of a standard yardstick, the result would be a metrestick (1 metre long):

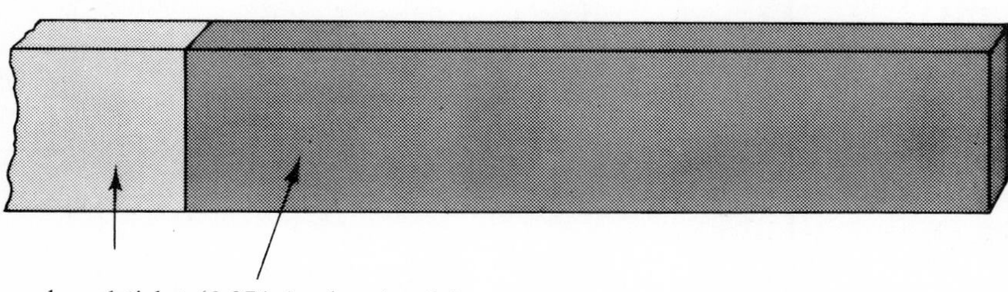

1 yardstick + (3.37 in.) = 1 metrestick

Q12 What system of measurement is used in most countries of the world? ____________

• ### • ### • ### • ### • ### • ### • ### •

A12 metric

Q13 Arrange the following units of measurement in order from shortest to longest: foot, metre, inch, mile, yard. ____________, ____________, ____________, ____________, ____________

• ### • ### • ### • ### • ### • ### • ### •

A13 inch, foot, yard, metre, mile

Q14 Would the 100-metre dash be a longer or shorter race than the 100-yard dash? ____________

• ### • ### • ### • ### • ### • ### • ### •

A14 longer

10 The metre that was the base unit of the metric system was multiplied and divided by powers of 10 to obtain other linear units of measurement. Greek prefixes to the term metre were used for multiples of the unit, and Latin prefixes were used for subdivisions. The results were as follows:

1 kilometre = 1000 metres | 1 decimetre = 0.1 metre = $\frac{1}{10}$ metre

1 hectometre = 100 metres | 1 centimetre = 0.01 metre = $\frac{1}{100}$ metre

1 dekametre = 10 metres | 1 millimetre = 0.001 metre = $\frac{1}{1000}$ metre

1 metre = 1 metre | 1 micrometre (micron) = 0.000 001 metre = $\frac{1}{1\,000\,000}$ metre

In the metric system when there are more than four digits to the right or left of the decimal point, the numerals are separated by a space into groups of three digits, starting at the decimal point. No commas are inserted.

Q15 Between each of the pairs of metric units place a $>$ (greater than) or $<$ (less than) symbol to make a true statement:

a. 1 kilometre ______ 1 metre

b. 1 centimetre ______ 1 metre

c. 1 dekametre ______ 1 decimetre

d. 1 millimetre ______ 1 centimetre

• ### • ### • ### • ### • ### • ### • ### •

A15 **a.** $>$ **b.** $<$ **c.** $>$ **d.** $<$

11 A decimetre, a centimetre, and a millimetre can be drawn on this page. Below is a decimetre and it is subdivided into 10 centimetres.

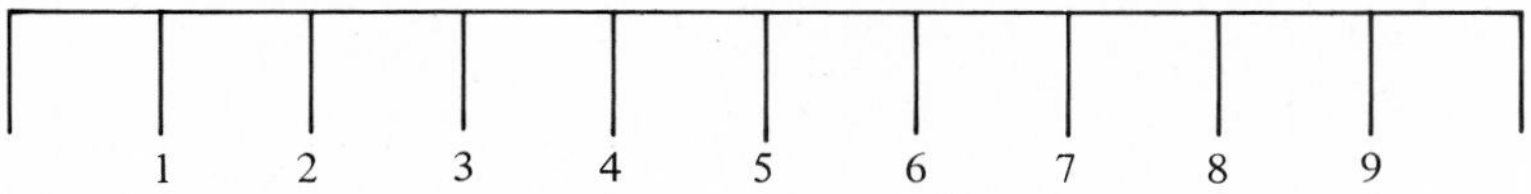

1 2 3 4 5 6 7 8 9

1 decimetre = 10 centimetres

If a centimetre were subdivided into millimetres, it would look like the following:

1 centimetre = 10 millimetres

A centimetre rule can also be used to measure the length of segments. The following bar is 7 centimetres long, measured to the nearest centimetre.

nearer 7 than 6

1 2 3 4 5 6 7 8 9

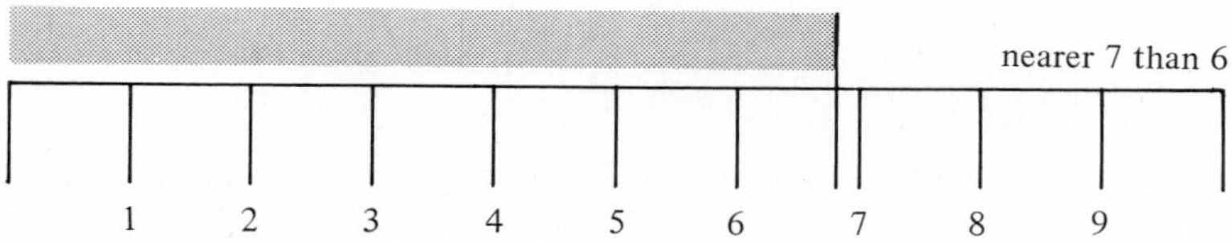

Q16 Use a centimetre ruler to measure the bars to the nearest centimetre:

a. ______

b. ______

c. ______

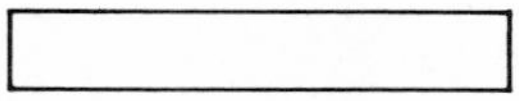

d. ______

e. ______

• ### • ### • ### • ### • ### • ### • ### •

A16 **a.** 6 centimetres **b.** 9 centimetres **c.** 3 centimetres **d.** 5 centimetres **e.** 10 centimetres

12 When a centimetre ruler is also marked in millimetres, a measurement can be read to the nearest millimetre. A millimetre is one tenth of a centimetre. The following measurements are reported two ways. The two results are equal.

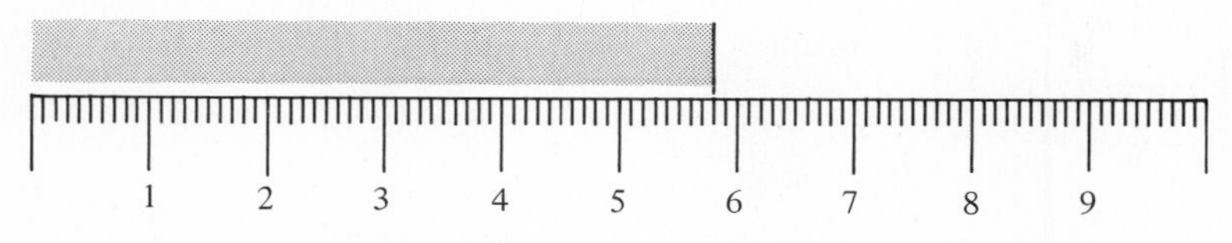

58 millimetres or 5.8 centimetres

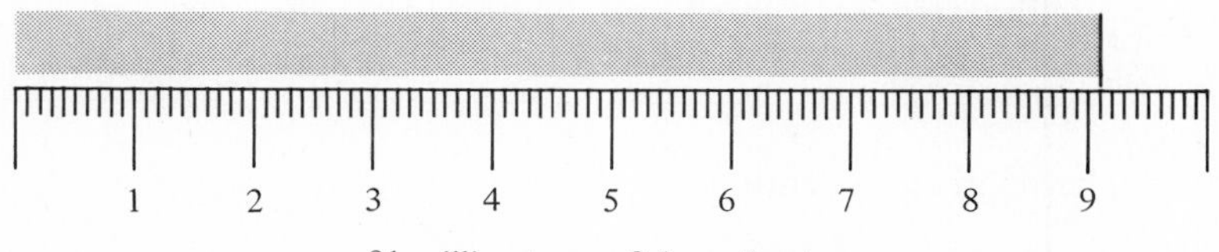

91 millimetres or 9.1 centimetres

Q17 Read the scales to the nearest millimetre:

a.

______

b.

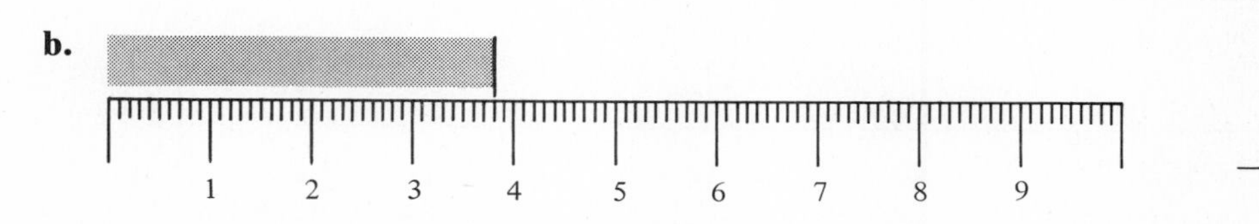

______

• ### • ### • ### • ### • ### • ### • ### •

A17 **a.** 72 millimetres **b.** 38 millimetres

Q18 Read the scales below to the nearest $\frac{1}{10}$ of a centimetre:

a.

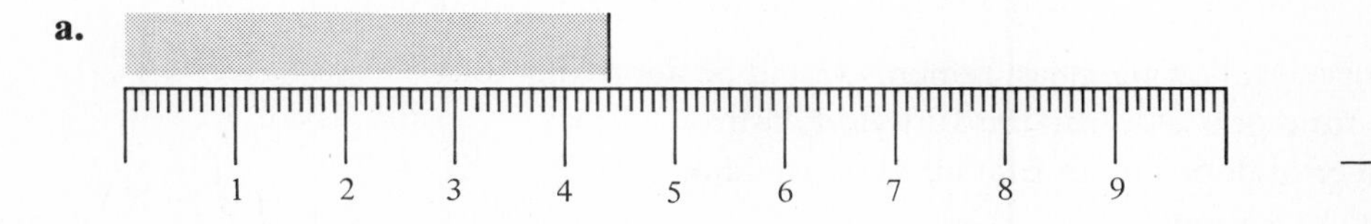

______

b.

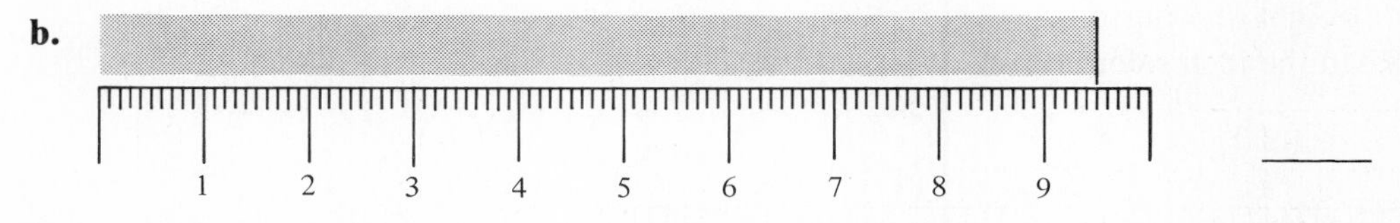

______

• ### • ### • ### • ### • ### • ### • ### •

A18 **a.** 4.4 centimetres **b.** 9.5 centimetres

Q19 Measure the bars to the precision indicated:

a. ▭ $\frac{1}{10}$ centimetre ______

b. 1 millimetre ______

c. 1 millimetre ______

d. 1 centimetre ______

e. $\frac{1}{10}$ centimetre ______

\# \# \# • \# \# \# • \# \# \# • \# \# \# • \# \# \# • \# \# \# • \# \# \# • \# \# \# • \# \# \#

A19 **a.** 4.0 centimetres **b.** 29 millimetres **c.** 43 millimetres **d.** 4 centimetres
e. 3.8 centimetres

Q20 Measure the lengths below to the nearest $\frac{1}{10}$ centimetre.

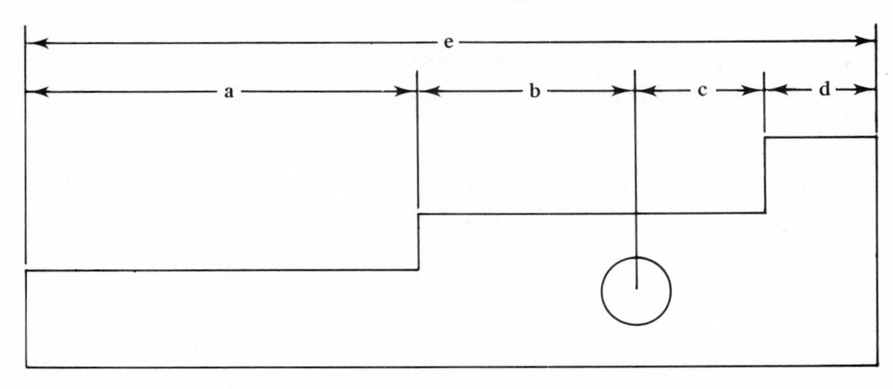

\# \# \# • \# \# \# • \# \# \# • \# \# \# • \# \# \# • \# \# \# • \# \# \# • \# \# \# • \# \# \#

A20 **a.** 5.2 centimetres **b.** 2.9 centimetres **c.** 1.7 centimetres **d.** 1.5 centimetres
e. 11.3 centimetres

This completes the instruction for this section.

5.1 EXERCISE

1. Which of the following measurements would be approximate?
   a. time for a person to run the 100-yard dash
   b. number of floors in the Empire State Building
   c. length of a room
   d. height of a person
2. Read the inch ruler at each of the arrows:

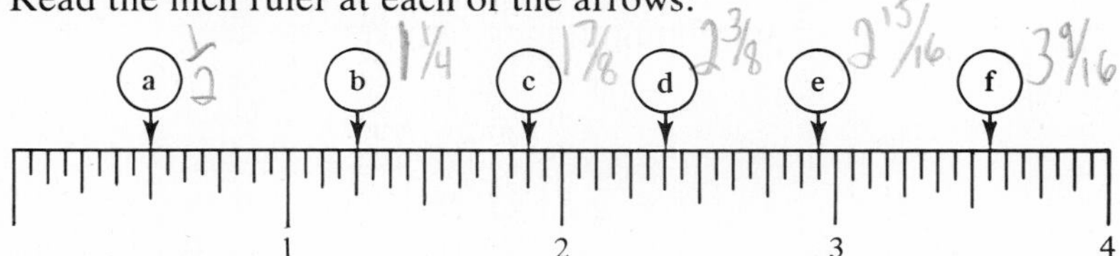

3. Approximate the length of the line segments to the precision indicated:

a. $\frac{1}{4}$ inch

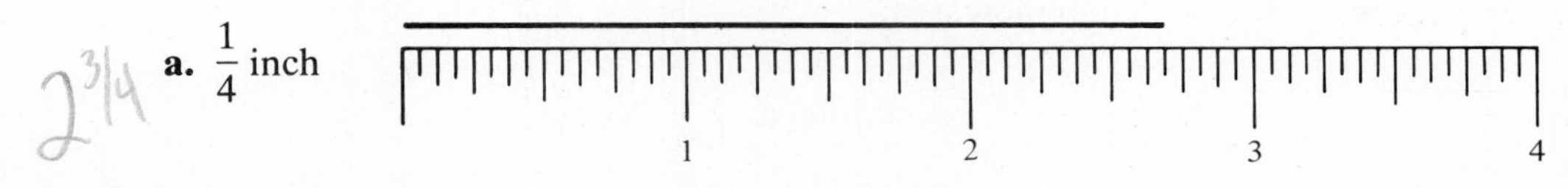

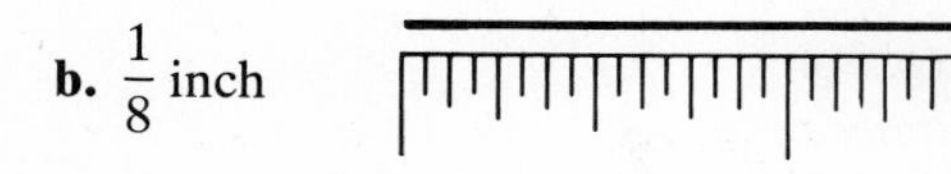

b. $\frac{1}{8}$ inch

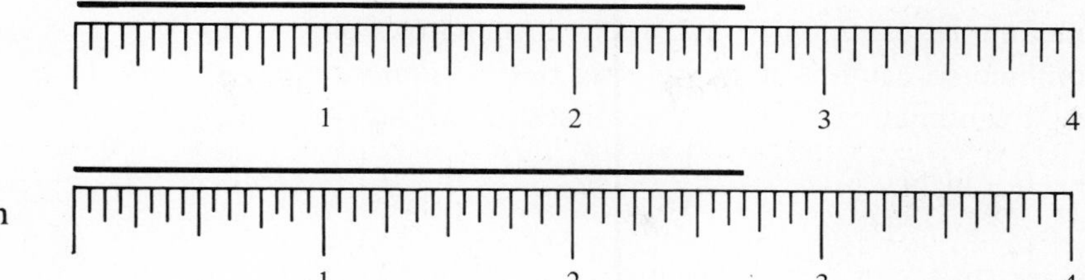

c. $\frac{1}{16}$ inch

4. Arrange the following units from shortest to longest: foot, metre, yard, inch, centimetre, mile, millimetre.
5. Arrange the following units from shortest to longest: metre, millimetre, kilometre, decimetre, centimetre.
6. Approximate the length of the bars to the precision indicated:

a. 1 centimetre

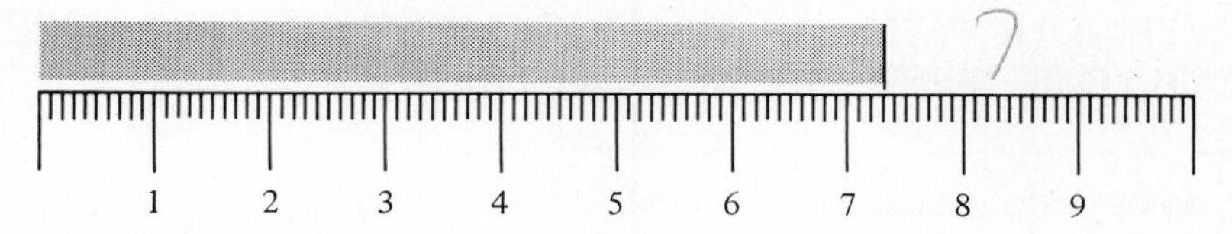

b. $\frac{1}{10}$ centimetre

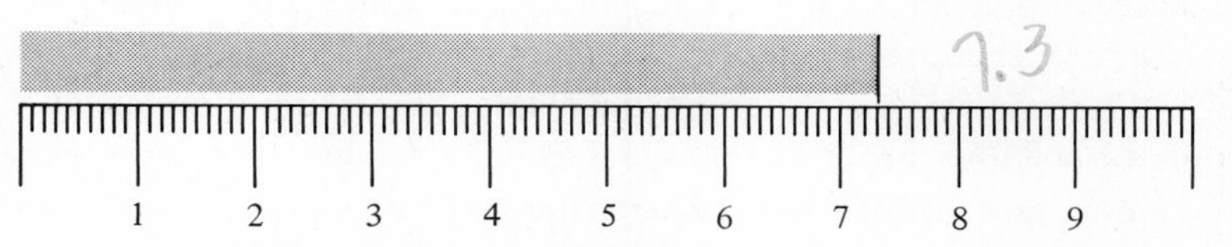

c. 1 millimetre

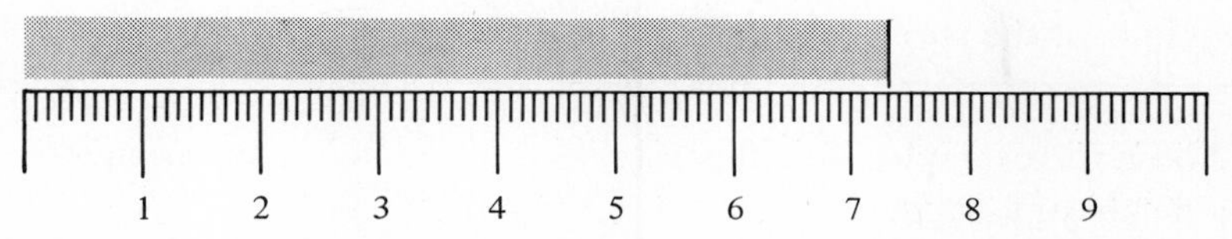

7. Measure the lengths on the figure below to the nearest $\frac{1}{16}$ inch.

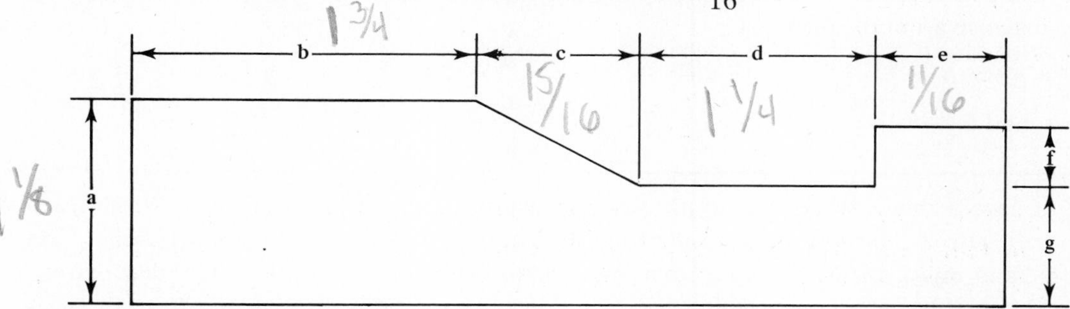

8. Measure the lengths on the figure of problem 7 to the nearest 0.1 centimetre.

5.1 EXERCISE ANSWERS

1. a, c, and d
2. **a.** $\frac{1}{2}$ inch **b.** $1\frac{1}{4}$ inches **c.** $1\frac{7}{8}$ inches **d.** $2\frac{3}{8}$ inches
   e. $2\frac{15}{16}$ inches **f.** $3\frac{9}{16}$ inches
3. **a.** $2\frac{3}{4}$ inches **b.** $2\frac{5}{8}$ inches **c.** $2\frac{11}{16}$ inches

4. millimetre, centimetre, inch, foot, yard, metre, mile
5. millimetre, centimetre, decimetre, metre, kilometre
6. a. 7 centimetres **b.** 7.3 centimetres **c.** 73 millimetres
7. a. $1\frac{1}{16}$ inches **b.** $1\frac{13}{16}$ inches **c.** $\frac{7}{8}$ inch **d.** $1\frac{1}{4}$ inches
e. $\frac{11}{16}$ inch **f.** $\frac{5}{16}$ inch **g.** $\frac{5}{8}$ inch
8. a. 2.8 centimetres **b.** 4.6 centimetres **c.** 2.2 centimetres **d.** 3.1 centimetres
e. 1.7 centimetres **f.** 0.8 centimetres **g.** 1.6 centimetres

5.2 INTRODUCTION TO MEASUREMENT OF DISTANCE, AREA, AND VOLUME (ENGLISH AND METRIC)

1 If an object is to be measured, the first task is to decide the unit of measurement to be used. One has a choice of units in some cases. However, the choice is not entirely arbitrary (one would not measure height in bushels, for example).

When measuring distances on a line, *linear* units are used. Distance is said to be a one-dimensional measure. Examples include: the height of a person, the width of a page, the distance to the moon, and the length of a kite string. Notice that the idea of distance on a line is extended to the length of a curve (kite string), which is not straight. The length of the curve is the same as the length would be if the string were laid out in a straight line.

Q1 Choose the examples that illustrate need for a linear measurement: ________
a. height of a flagpole
b. weight of a person
c. flying distance between New York City and Detroit
d. distance around your waist

• ### • ### • ### • ### • ### • ### • ### •

A1 **a**, **c**, and **d**

2 You have a choice of systems to use when measuring distance in linear units. Within each system there is also a choice of units. In the English system of measurement there are inches, feet, yards, rods, and miles. In the metric system, micrometres, millimetres, centimetres, decimetres, metres, and kilometres are all linear units.

Inches were used as one of the units of measurement in Section 5.1. If twelve of these units are placed end to end, the result is a line segment 1 foot long. This is the length of the standard foot ruler. Three of the foot rulers placed end to end form a yard, the length of a standard yardstick. These relationships are summarized below with the abbreviations for each unit:

1 foot (ft) = 12 inches (in.)

1 rod (rd) = $16\frac{1}{2}$ feet (ft)

1 yard (yd) = 3 feet (ft)

1 mile (mi) = 1,760 yards (yd)

Another symbol for 12 inches is 12″; 3 feet is 3′.

Q2 Choose the linear units from the following: ________

a. centimetre **b.** pound **c.** degree **d.** yard

• # # # • # # # • # # # • # # # • # # # • # # # • # # # • # #

A2 **a** and **d**: A pound measures weight and a degree measures an angle or a temperature.

Q3 If a distance is measured in feet, could it also be measured in centimetres? ________

• # # # • # # # • # # # • # # # • # # # • # # # • # # # • # #

A3 yes

Q4 If a distance is measured in inches, could it also be measured in pounds? ________

• # # # • # # # • # # # • # # # • # # # • # # # • # # # • # #

A4 no: Pounds do not represent a linear measurement.

---

3 Metric units are easier to work with mathematically, compared with English units, because they are related by multiples of 10. A centimetre is shown here. A decimetre which is 10 times as long as a centimetre is also shown.

1

1 centimetre

1 2 3 4 5 6 7 8 9

1 decimetre = 10 centimetres

Unfortunately, the size of the page does not allow a line segment 1 metre long. It is slightly longer than 1 yard. One thousand metres makes a kilometre that is over $\frac{1}{2}$ mile.

If a centimetre is divided into 10 parts, each is 1 millimetre long.

1

1 centimetre = 10 millimetres

A micrometre (micron) is the length of one of 1,000 equal parts of a millimetre. A micrometre cannot be shown on this page because of limitations on the accuracy of printing machines. A human red blood cell is about 7 micrometres across. Special measuring devices are used to measure to such accuracy. (One device is called a micrometre.) The diameter of a hair from your head is probably about 50 micrometres. These relationships are summarized below with symbols for each unit;

1 millimetre (mm) = 1000 micrometres (microns) (μm)

1 centimetre (cm) = 10 millimetres (mm)

1 decimetre (dm) = 10 centimetres (cm)

1 metre (m) = 10 decimetres (dm)

1 dekametre(dam) = 10 metres (m)

1 hectometre (hm) = 10 dekametres (dam)

1 kilometre (km) = 10 hectometres (hm)

Q5 Choose the most likely measurement for each of the following:

a. width of a finger ______
(1) 1 centimetre (2) 1 metre (3) 1 micrometre

b. thickness of a fingernail ______
(1) 300 micrometres (2) 300 decimetres (3) 300 centimetres

c. diameter of the wood part of a wooden pencil ______
(1) $\frac{3}{4}$ kilometre (2) $\frac{3}{4}$ centimetre (3) $\frac{3}{4}$ millimetre

d. diameter of a fine pencil lead for a mechanical pencil ______
(1) 1 metre (2) 1 millimetre (3) 1 centimetre

• ### • ### • ### • ### • ### • ### • ### •

A5 **a.** (1) **b.** (1) **c.** (2) **d.** (2)

Q6 Choose the most appropriate answer:

a. length of a horse racetrack ______
(1) 1 metre (2) 1 decimetre (3) 1 kilometre

b. one car length ______
(1) 5 metres (2) 5 centimetres (3) 5 kilometres

c. the height of a telephone pole ______
(1) 10 kilometres (2) 10 centimetres (3) 10 metres

d. diameter of the wheel of a 10-speed bicycle ______
(1) $\frac{3}{4}$ centimetre (2) $\frac{3}{4}$ metre (3) $\frac{3}{4}$ kilometre

• ### • ### • ### • ### • ### • ### • ### •

A6 **a.** (3) **b.** (1) **c.** (3) **d.** (2)

Q7 Match the symbol for a metric unit with the unit:

a. cm ______ (1) hectometre
b. μm ______ (2) decimetre
c. mm ______ (3) metre
d. m ______ (4) micrometre
e. km ______ (5) dekametre
f. dm ______ (6) kilometre
g. hm ______ (7) millimetre
h. dam ______ (8) centimetre

• ### • ### • ### • ### • ### • ### • ### •

A7 **a.** (8) **b.** (4) **c.** (7) **d.** (3) **e.** (6) **f.** (2) **g.** (1) **h.** (5)

4 *Area* is a measure of the region within a closed curve (including polygons) in a plane. Geometric figures that do not enclose a region do not have an area. The angle of Figure 1 and the object of Figure 2 do not have area.

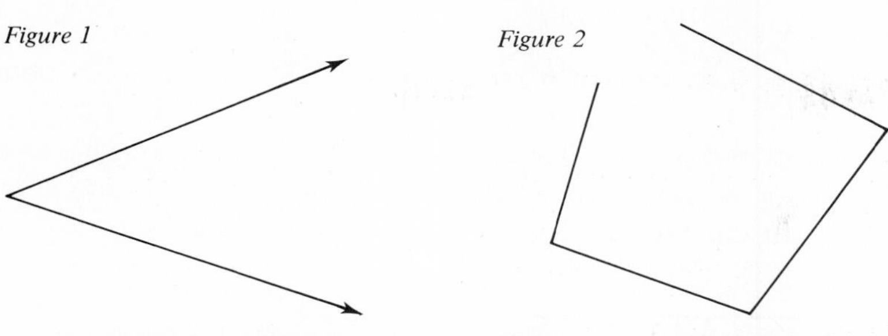

Irregular shapes that enclose a finite region have area, but the measurement of that area is difficult to obtain and usually is only roughly approximated. The following figure is of this nature. The areas of certain geometric regions, such as circular regions, triangular regions, and many others, are obtained from standard formulas that will be studied in this chapter.

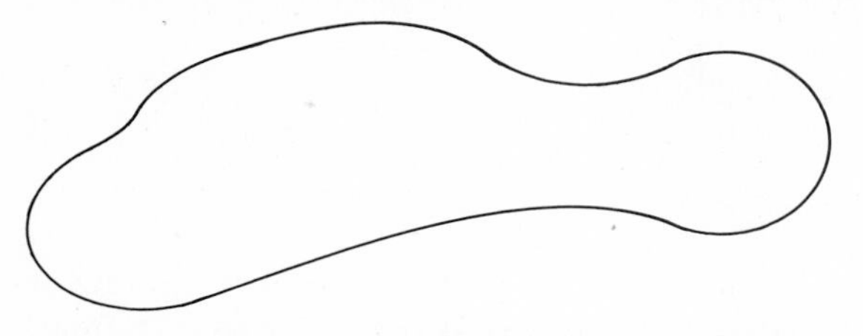

Q8 Choose the geometric regions that have an area. ______

a.

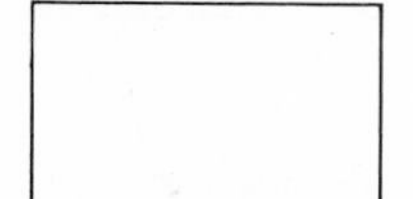

b.

c.

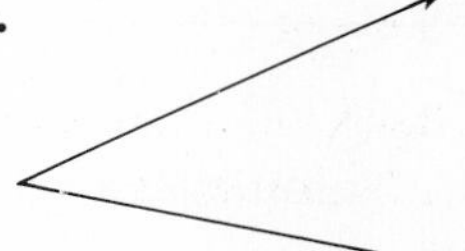

d.

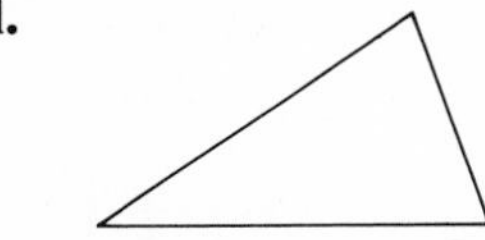

e.

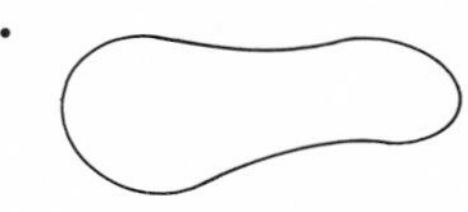

f.

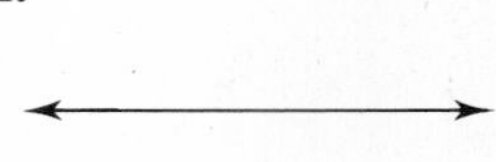

• # # # • # # # • # # # • # # # • # # # • # # # • # # # • # #

A8 **a**, **d**, and **e**

5 A square region with all four sides 1 inch long is said to have an area of 1 square inch. This area is considered to be a unit, and areas of other closed figures are given as some number of square inches. "Square inch" will be symbolized as in.^{2}. Other square units will be symbolized in a similar manner, that is, square feet as ft^2, square centimetres as cm^2, and so on. The figures below show a square inch

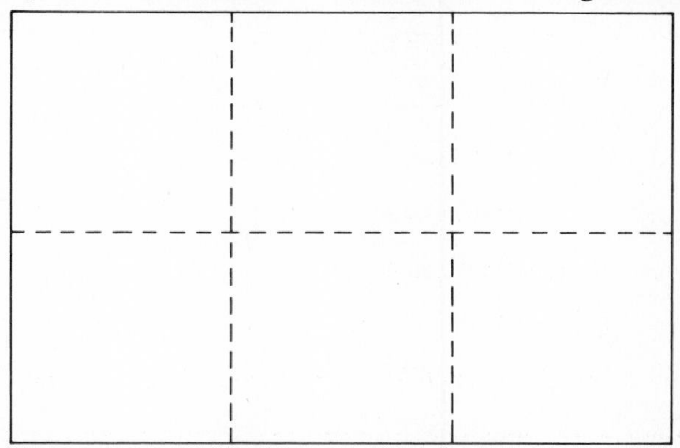

1 square inch
1 in.^{2}

6 square inches
6 in.^{2}

(Note: It is the publisher's convention to abbreviate inch or inches as in. (with the period) to avoid any confusion with the word *in*. Other English abbreviations such as ft, yd, and mi will be used without the period. m, cm, and mm are metric symbols, not abbreviations, and are written without the period.)

and also a rectangular region with an area of 6 square inches. This area may be obtained by counting.

In the metric system a square region with all four of its sides 1 centimetre long has an area of 1 square centimetre. Areas of other geometric figures are given as some number of square centimetres. A square centimetre and a rectangle of 12 square centimetres are shown.

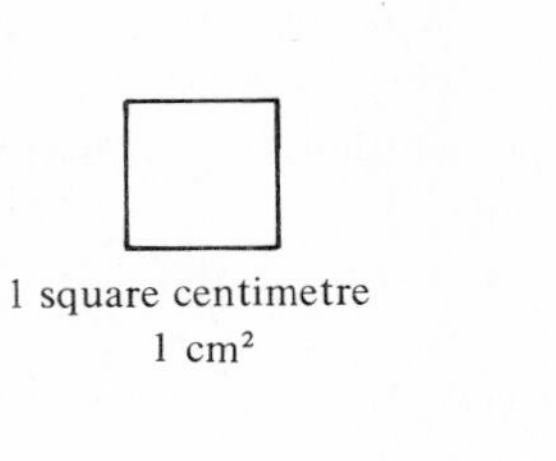
1 square centimetre
1 cm^2

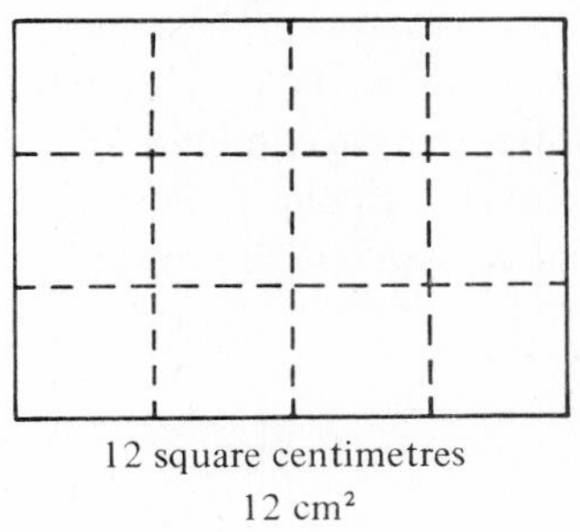
12 square centimetres
12 cm^2

Area is a measure of the polygonal region rather than the polygon, because the region includes all the points inside. It is the space inside that is being measured. (Other texts may use the less-precise phrase "area of a square." If you read such a phrase, interpret it as "area of a square region.")

Q9 Use the 1-centimetre grid to count the number of square centimetres in the following polygon regions:

a.

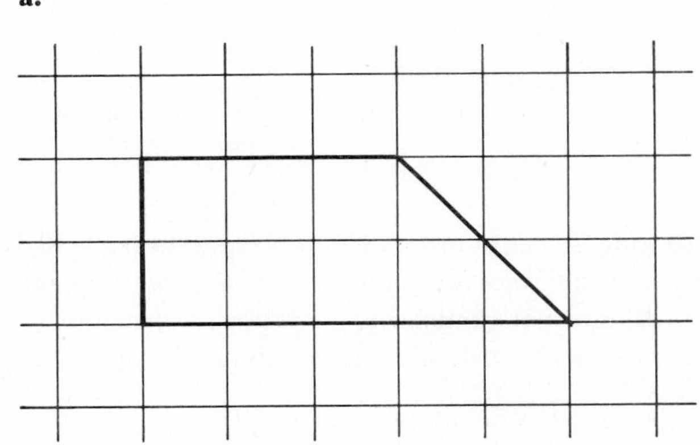

b.

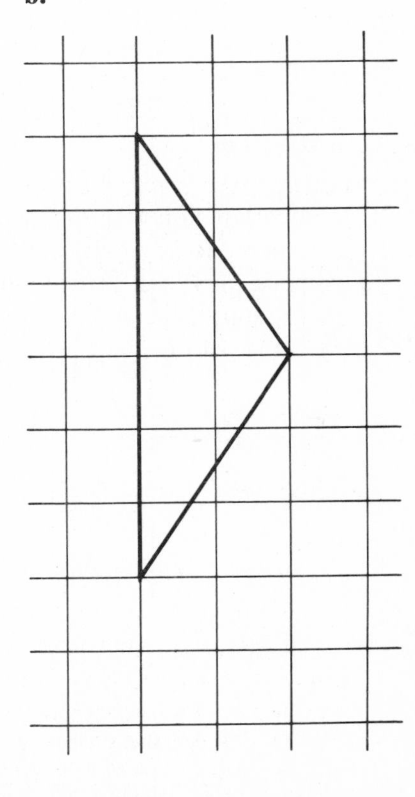

a. ______ cm^2

b. ______ cm^2

\# \# \# • \# \# \# • \# \# \# • \# \# \# • \# \# \# • \# \# \# • \# \# \# • \# \# \# • \# \# \#

A9 **a.** 8 cm^2

b. 6 cm^2: You may obtain this by mentally piecing triangles together to make whole squares, as follows:

a.

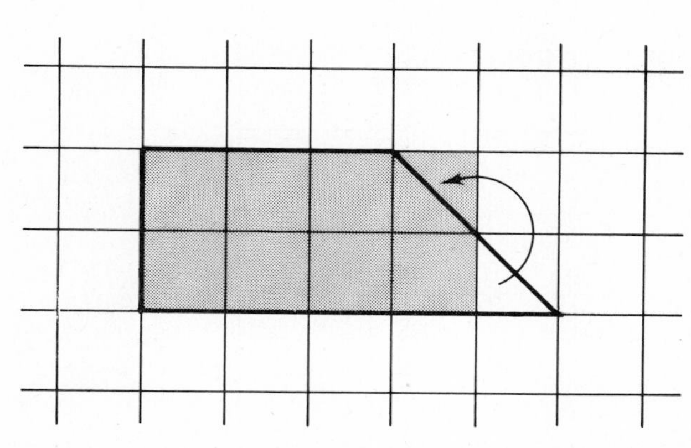

b.

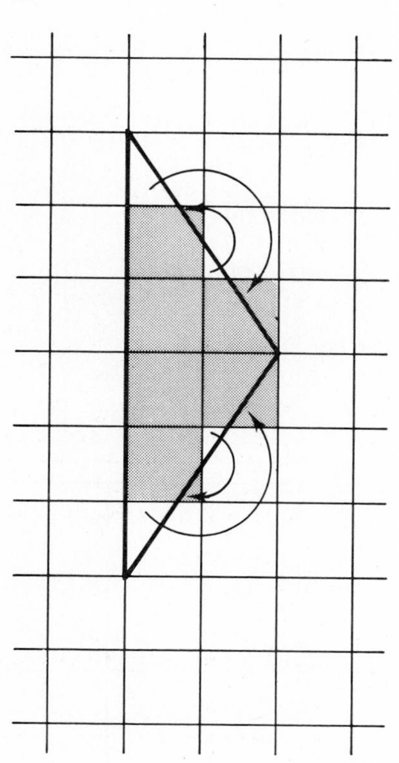

6 Other units that can be used in the metric system to measure area besides the square centimetre are: square millimetre, square decimetre, square metre, and square kilometre. Another unit used to measure area in the metric system is the *hectare*. It equals 10 000 square metres.

The relationship in size between a square centimetre and a square decimetre is possible to show with a drawing. It illustrates the problem for all conversions of square units. A decimetre contains 10 centimetres, but a square decimetre contains 100 square centimetres, as follows:

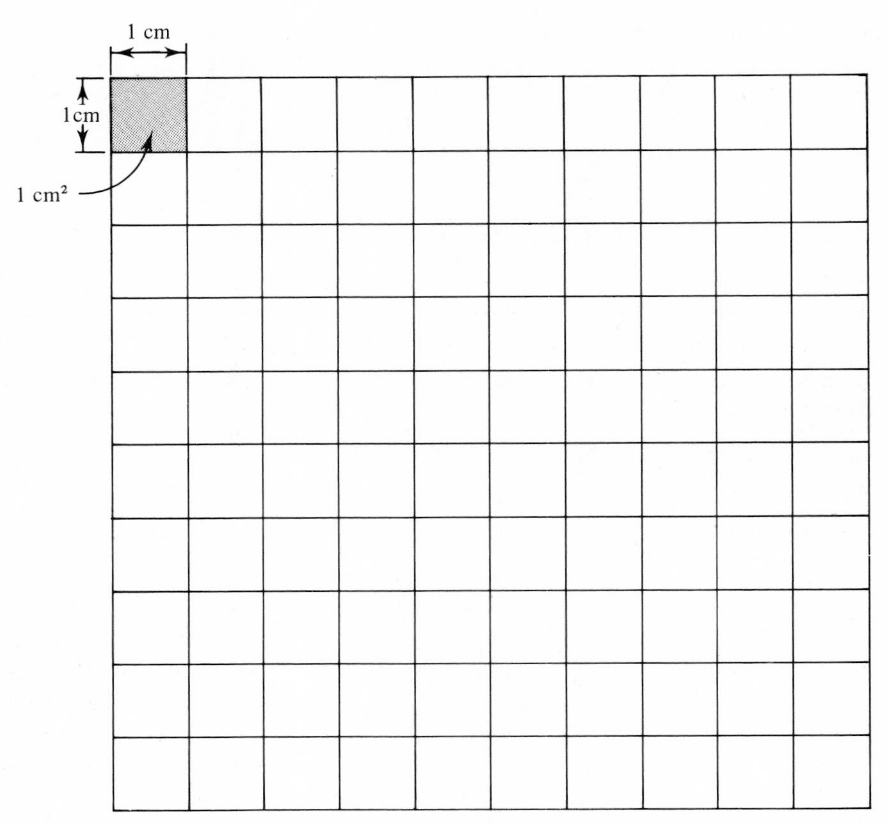

1 square decimetre = 100 square centimetres

Similar figures could be drawn to show the relationships contained in this chart:

1 square centimetre = 100 square millimetres
1 square decimetre = 100 square centimetres
1 square metre = 100 square decimetres = 10 000 square centimetres
1 square kilometre = 1 000 000 square metres
1 hectare = 10 000 square metres

Q10 Use the diagrams (drawn to scale) below to answer the questions:

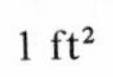

in.^{2}

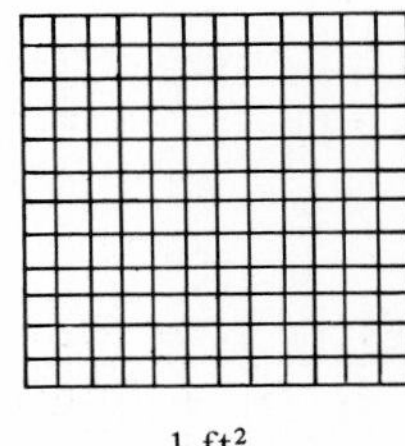

1 ft^2

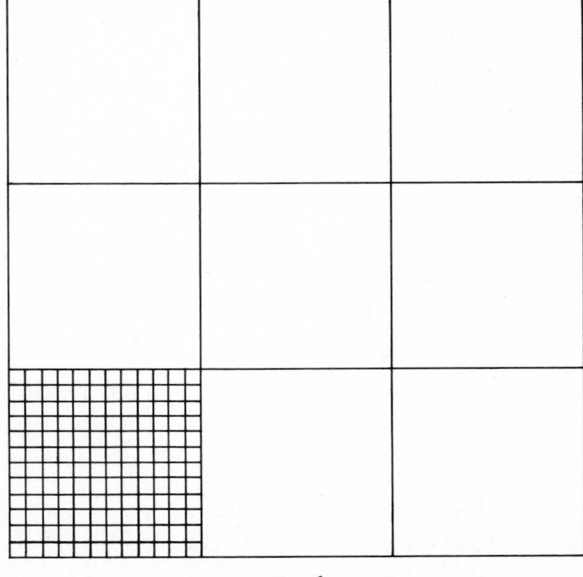

1 yd^2

a. 1 square foot = ______ square inches

b. 1 square yard = ______ square feet

c. 1 square yard = ______ square inches

• # # # • # # # • # # # • # # # • # # # • # # # • # # # • # #

A10 **a.** 144 **b.** 9 **c.** 1,296

7 A unit of square measure usually used for land measurement in the English system is the acre. One acre contains 43,560 square feet. One hectare in the metric system contains approximately 2.471 acres. The symbol for "approximately equals" is "$\doteq$."

Summarizing the answers to Q10 and the above remarks:

1 square foot = 144 square inches

1 square yard = 9 square feet

1 acre = 43,560 square feet

1 hectare $\doteq$ 2.471 acres

The number of the smaller square units in a larger square unit may be obtained by multiplying. For example, since there are 3 feet in a yard, there are 3×3, or 9, square feet in a square yard. Also, since there are 12 inches in a foot, there are 12×12, or 144, square inches in a square foot.

Q11 **a.** Since there are 36 inches in a yard, there are ______ × ______, or ______, square inches in a square yard.

b. Since there are 1,760 yards in a mile, there are ______ × ______, or ______, square yards in a square mile.

c. Since there are 1000 micrometres in a millimetre, there are ______ × ______, or ______, square micrometres in a square millimetre.

• # # # • # # # • # # # • # # # • # # # • # # # • # # # • # #

A11 **a.** 36, 36, 1,296
b. 1,760, 1,760, 3,097,600
c. 1000, 1000, 1 000 000

8 *Volume* is a measure of the space within a closed solid figure. Irregular-shaped solid figures have volume, but the measurement of these volumes is difficult to obtain. The volumes of certain solid figures, such as prisms, cones, and spheres, are obtained by using standard formulas that will be studied in this chapter.

One basic unit of volume is the cubic inch. This is a cube with each edge measuring 1 inch. To find the volume of a rectangular prism in cubic inches, think of the number of cubes, 1 inch on each edge, that could be stacked inside the prism.

The prism shown here would have 15 cubes on the bottom level. This is the same number as the area of the rectangular base. Above *each* of these bottom cubes would be three more, making a stack 4 cubes high. Therefore, the total volume would be 60 cubic inches. "Cubic inch" will be symbolized by "in.^{3}." Other cubic units will be symbolized in a similar manner, that is, cubic centimetres as cm^3, cubic yards as yd^3, and so on.

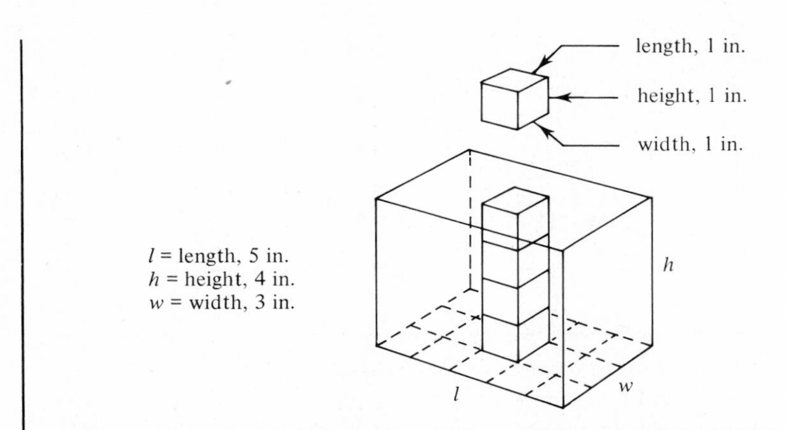

Q12 Use the method of Frame 8 to determine the volume of the prism at the right. ________

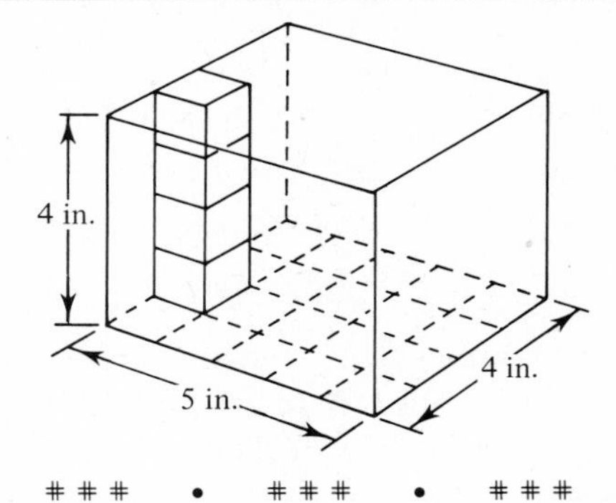

\# \# \# • \# \# \# • \# \# \# • \# \# \# • \# \# \# • \# \# \# • \# \# \# • \# \# \# • \# \# \#

A12 80 in.^{3}

9 The volume of a prism can be obtained from the dimensions: the length (l), width (w), and height (h). Just as the volume of the rectangular prism in Frame 8 could be found by multiplying the length times the width to get the area of the base and then multiplying that times the height, the volume of any rectangular prism may be found the same way. The formula would be $V = lwh$.

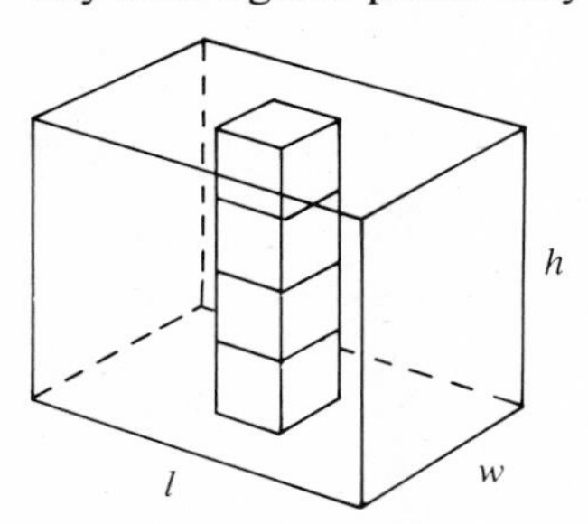

To find the number of cubic inches in a cubic foot, the same formula may be used. The dimensions would be $l = 12$ in., $w = 12$ in., $h = 12$ in.

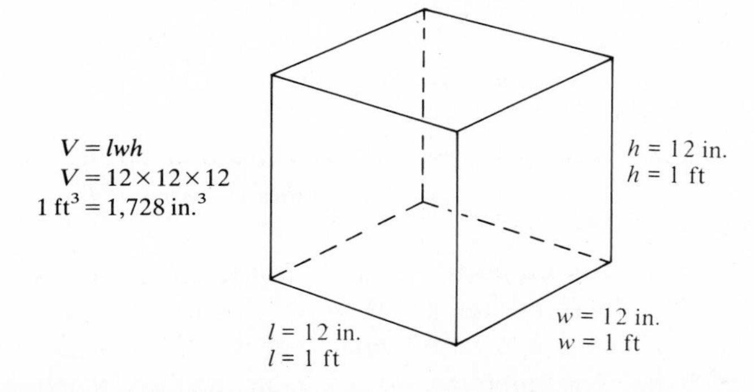

Q13 Find the number of cubic inches in a rectangular solid with dimensions: $l = 8.0$ in., $w = 10.0$ in., $h = 4.5$ in.

\# \# \# • \# \# \# • \# \# \# • \# \# \# • \# \# \# • \# \# \# • \# \# \# • \# \# \# • \# \# \#

A13 360 in.^{3}: $V = lwh$
$V = (8.0)(10.0)(4.5)$
$V = 360$

Q14 Use the formula for the volume of a prism to find the number of cubic feet in 1 cubic yard.

\# \# \# • \# \# \# • \# \# \# • \# \# \# • \# \# \# • \# \# \# • \# \# \# • \# \# \# • \# \# \#

A14 27 ft^3: $V = lwh$
$V = 3(3)(3)$
$V = 27$
$1 \text{ yd}^3 = 27 \text{ ft}^3$

Q15 Find the number of cubic centimetres in 1 cubic decimetre.

\# \# \# • \# \# \# • \# \# \# • \# \# \# • \# \# \# • \# \# \# • \# \# \# • \# \# \# • \# \# \#

A15 1000 cm^3: $V = lwh$
$V = 10(10)(10)$
$V = 1000$
$1 \text{ dm}^3 = 1000 \text{ cm}^3$

10 Summarizing the relationships between different units of volume:

1 cubic foot = 1,728 cubic inches
1 cubic yard = 27 cubic feet
1 cubic decimetre = 1000 cubic centimetres

Other units of liquid volume are the pint, quart, and gallon in the English system and the litre in the metric system. The relationships that exist between them and relate them to the cubic measurements follow:

1 quart (qt) = 2 pints (pt)
1 gallon (gal) = 4 quarts (qt) = 8 pints (pt)
1 quart (qt) = 57.75 cubic inches (in.^{3})
1 litre (l) $\doteq$ 1.05 quarts (qt)
1 litre (l) = 1000 cubic centimetres (cm^3)

Q16 Indicate whether each unit measures distance, area, or volume:

a. ft^2 __________ **b.** quart __________

c. cm^2 __________ **d.** cm^3 __________

e. liter __________ **f.** millimetre __________

g. $in.^3$ __________ **h.** pint __________

• ### • ### • ### • ### • ### • ### • ### •

A16 **a.** area **b.** volume **c.** area
d. volume **e.** volume **f.** distance
g. volume **h.** volume

Q17 Use the statement 1 litre $\doteq$ 1.05 quarts to tell which is larger, a litre or a quart. __________

• ### • ### • ### • ### • ### • ### • ### •

A17 litre: It takes more than 1 quart to equal 1 litre.

Q18 Indicate all units of measurement that could be used to report the measurement required (there can be more than one correct):

a. length of a room __________
(1) $in.^3$ (2) litre (3) feet (4) decimetres

b. volume of a barrel __________
(1) gallons (2) ft^2 (3) cm^3 (4) micrometres

c. area of a bulletin board __________
(1) quarts (2) $in.^2$ (3) feet (4) cm^2

d. surface area of a pyramid __________
(1) $in.^2$ (2) quart (3) feet (4) yd^2

• ### • ### • ### • ### • ### • ### • ### •

A18 **a.** (3) and (4) **b.** (1) and (3) **c.** (2) and (4) **d.** (1) and (4)

---

11 It helps to visualize some of the units presented in this section by comparing them with familiar objects which are approximately that size.

English

1 cubic inch	cube the size of a marshmallow
1 quart	quart milk carton
1 gallon	gallon milk carton or a gallon ice cream container
1 cubic foot	cube with each edge the length of a foot ruler such as a box of 12 in. × 12 in. floor tile 12 in. high

Metric

millimetre	diameter of a paper-clip wire
centimetre	width of a large paper clip
cubic centimetre	sugar cube
litre	slightly larger than 1 quart milk carton

Q19 Arrange the following units in the following list from smallest to largest: litre, cubic foot, cubic inch, quart. __________

• ### • ### • ### • ### • ### • ### • ### •

A19 cubic inch, quart, litre, cubic foot

This completes the instruction for this section.

5.2 EXERCISE

1. Label each of the following as (1) a distance, (2) an area, (3) a volume, or (4) none of these:
 a. length of a pencil
 b. floor space in a room
 c. amount you weigh
 d. amount of water in a bucket
 e. circumference of your head
 f. amount of window space in a house
 g. amount of gravel on a driveway
 h. amount of blood in your body
 i. depth of water in a lake

2. What are the names of the two systems of measurement used in the world today?

3. What is the quantity that is measured in linear units?

4. Choose the linear units from the following: centimetres, square miles, gallons, inches, metres, cubic yards, rods.

5. If a distance is measured in metres, could it also be measured in centimetres?

6. If an area is measured in square feet, could it also be measured in decimetres?

7. Choose the most likely object to have the given measurement:
 a. 7 in.: (1) length of an adult arm, (2) length of a pencil, (3) diameter of a light bulb
 b. $2\frac{1}{2}$ cm: (1) diameter of the cap of a pop bottle, (2) width of a tooth, (3) width of an adult hand
 c. 1 mm: (1) thickness of a human hair, (2) thickness of a fine pencil lead, (3) length of an adult's big toe
 d. 1 km: (1) length of an automobile, (2) length of a football field, (3) length of six city blocks

8. Choose the units that could be used to measure area from the following: centimetres, square feet, square metres, gallons, cubic inches, yards, hectares, acres.

9. Use the fact that there are 100 centimetres in a metre to find the number of square centimetres in a square metre.

10. Choose the most likely object to have the given measurement:
 a. 1 in.^{2}: (1) fingernail on an adult's little finger, (2) cross section of a wooden pencil, (3) postage stamp
 b. 1 ft^2: (1) surface of a brick, (2) surface of an asbestos floor tile, (3) surface of an adult's hand
 c. 20 ft^2: (1) surface of a door to a house, (2) area of a living room, (3) area of the windshield on a car
 d. 1 cm^2:

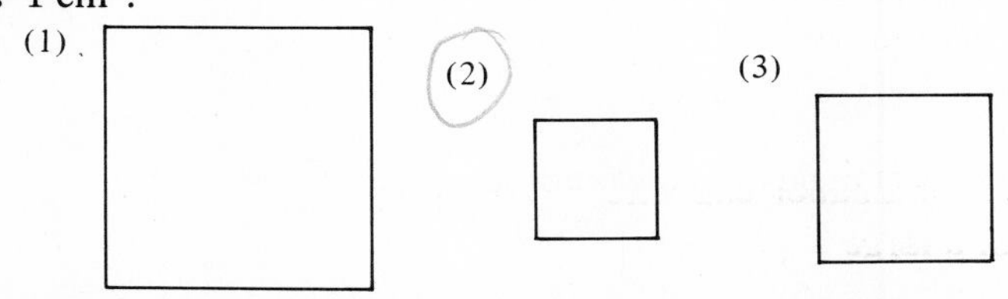

11. Choose the units of measure that could be used to measure volume: cubic inch, acre, gallon, pint, cubic centimetre, decimetre, square foot, litre, cubic yard, hectare.

12. Use the fact that there are 100 centimetres in a metre to find the number of cubic centimetres in a cubic metre.

13. Choose all units of measurement that could be used to report the measurement required:
 a. space in the trunk of an automobile: (1) square feet, (2) cubic metres, (3) feet, (4) cubic feet
 b. size of a field: (1) cubic yards, (2) hectares, (3) acres, (4) square yards
 c. milk in a bottle: (1) pints, (2) cubic centimetres, (3) centimetres, (4) gallons
 d. capacity of the lungs: (1) cubic inches, (2) square centimetres, (3) cubic centimetres, (4) hectares

5.2 EXERCISE ANSWERS

1. **a.** (1) **b.** (2) **c.** (4) **d.** (3) **e.** (1) **f.** (2) **g.** (3) **h.** (3) **i.** (1)
2. English and metric
3. distance
4. centimetres, inches, metres, rods
5. yes
6. no
7. **a.** (2) **b.** (1) **c.** (2) **d.** (3)
8. square feet, square metres, hectares, acres
9. 10 000
10. **a.** (3) **b.** (2) **c.** (1) **d.** (2)
11. cubic inch, gallon, pint, cubic centimetre, litre, cubic yard
12. 1 000 000
13. **a.** (2) and (4) **b.** (2), (3), and (4) **c.** (1), (2), and (4) **d.** (1) and (3)

5.3 PERIMETERS

1 The *perimeter* of a polygon is the total length of all its sides. For any geometric figure in a plane, the perimeter is the length of the boundary of the geometric figure. Notice that perimeter is a measure of length (distance).

The perimeter of a rectangle can be found by adding the lengths of the sides:

Example: Find the perimeter of the rectangle below.

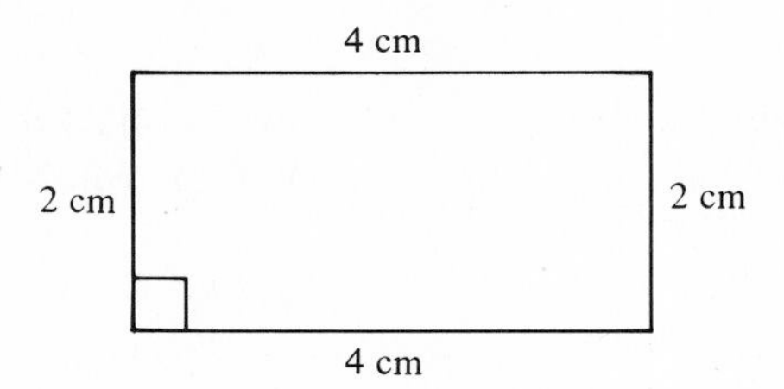

Solution: The perimeter of the rectangle above would be

$p = 2+4+2+4$
$p = 12$

Therefore, the perimeter is 12 cm.

Q1 Which would have a larger perimeter, a triangle with each side 2 inches or a square with each side 2 inches? ________

\# \# \# • \# \# \# • \# \# \# • \# \# \# • \# \# \# • \# \# \# • \# \# \# • \# \# \# • \# \# \#

A1 square

Q2 Find the perimeter of the figures:

a.

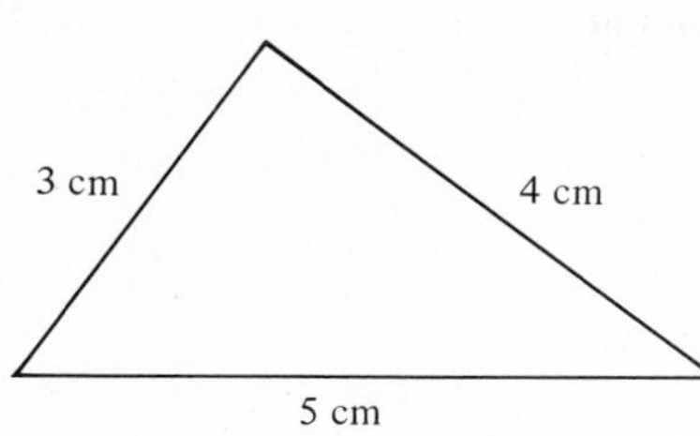

b.

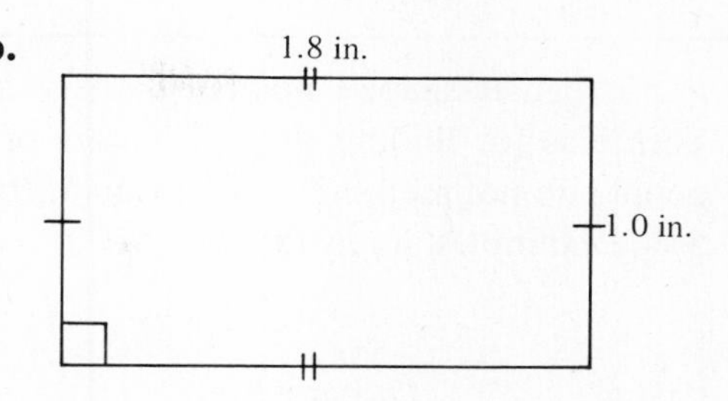

\# # # • # # # • # # # • # # # • # # # • # # # • # # # • # # # • # # #

A2 **a.** 12 cm: $p = 3 + 4 + 5$
$p = 12$

b. 5.6 in.: $p = 1.8 + 1.0 + 1.8 + 1.0$
$p = 5.6$

Q3 Sometimes the full perimeter is not needed for a particular purpose. A stair is to be carpeted both on the rise and on the run of the stair. How long should a piece of carpet be in order to cover the stair from A to B. Each rise is 8 inches and each run is 9 inches.

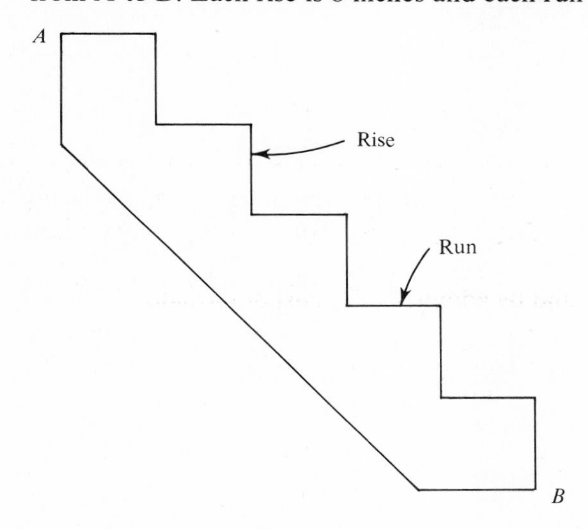

\# # # • # # # • # # # • # # # • # # # • # # # • # # # • # # # • # # #

A3 85 inches

2	Since perimeter is a length, each time a perimeter is found, the unit of measurement must be reported. Many units of length have been introduced. For example, in the English system there are inch, foot, yard, rod, and mile. In the metric system there are the micrometre, millimetre, centimetre, decimetre, metre, and kilometre. The perimeter of a polygon could be measured in any of these units or in other units of linear measure.

Q4 Indicate yes if the unit could be used to measure perimeter. Otherwise indicate no.

a. inch ______ **b.** square foot ______

c. mile ______ **d.** acre ______

e. centimetre ______ **f.** yard ______

g. cubic feet ______ **h.** rod ______

\# # # • # # # • # # # • # # # • # # # • # # # • # # # • # # # • # # #

A4 **a.** yes **b.** no **c.** yes **d.** no **e.** yes **f.** yes **g.** no **h.** yes

---

3 For irregular-shaped polygons, there is no other way to find perimeters but to add the sides. Formulas for finding the perimeters of these polygons are very similar to each other, but most people do not memorize the formula. Rather, they remember the idea of what perimeter means. Some examples are as follows:

triangle $p = a + b + c$

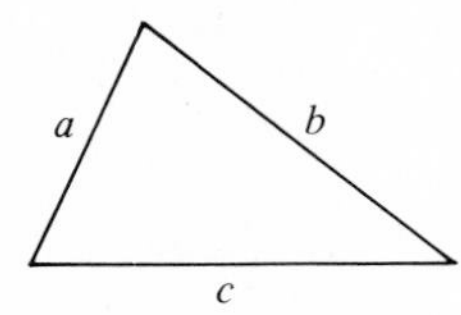

quadrilateral $p = a + b + c + d$

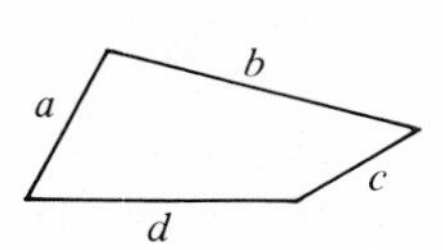

pentagon $p = a + b + c + d + e$

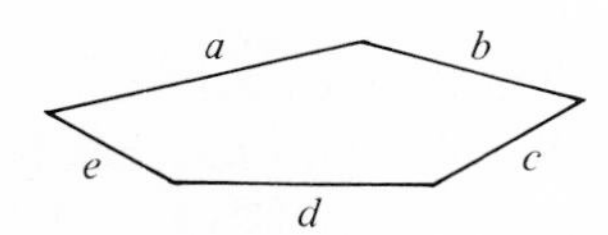

Q5 Find the perimeter of the polygons:

a.

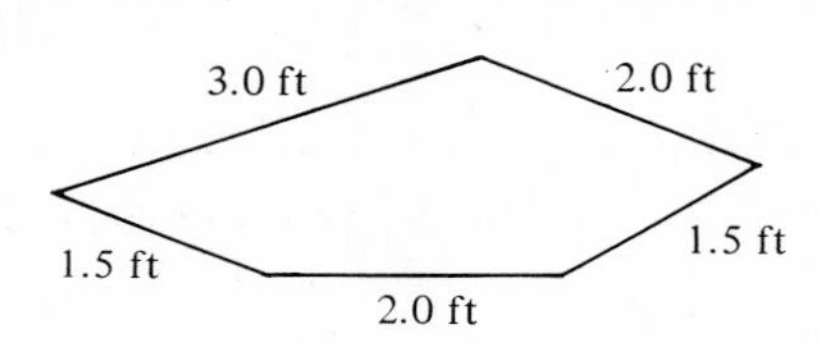

b.

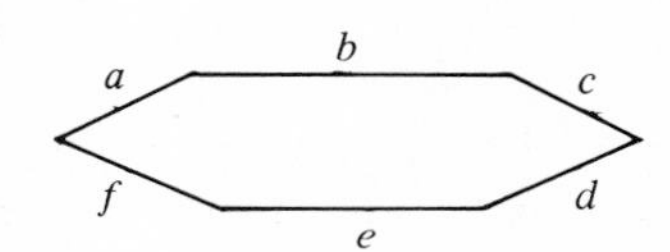

$a = 9$ yd $b = 18$ yd $c = 9$ yd
$d = 11$ yd $e = 16$ yd $f = 11$ yd

\# # # • # # # • # # # • # # # • # # # • # # # • # # # • # # # • # # #

A5 **a.** 10.0 ft: $p = a + b + c + d + e$
$p = 3.0 + 2.0 + 1.5 + 2.0 + 1.5$
$p = 10.0$

b. 74 yd: $p = a + b + c + d + e + f$
$p = 9 + 18 + 9 + 11 + 16 + 11$
$p = 74$

Q6 Find the perimeter of the template below. Measurements are in centimetres.

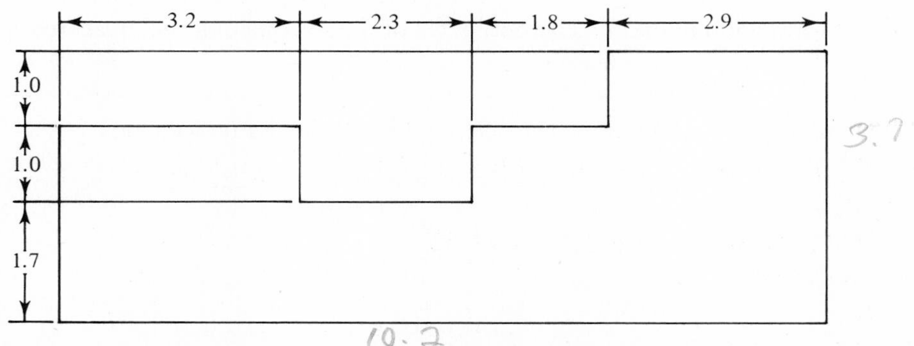

\# # # • # # # • # # # • # # # • # # # • # # # • # # # • # # # • # # #

A6 29.8 centimetres: The perimeter is the sum of the numbers on the figure below.

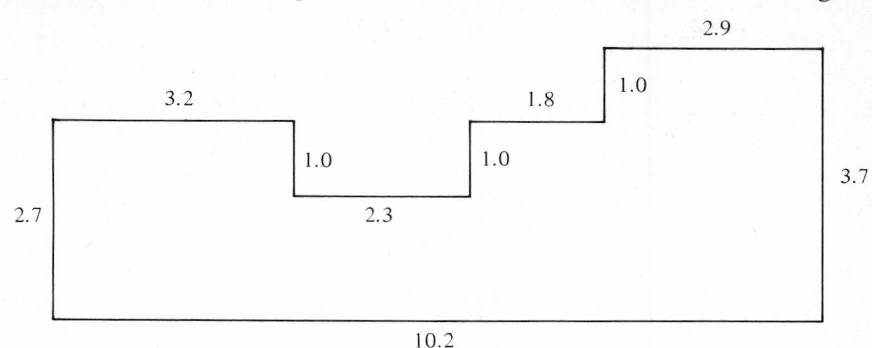

Q7 A length of stock steel of hexagonal cross section measures 0.87 inches from point A to point B. Find the perimeter of the cross section if all sides of the hexagon are equal.

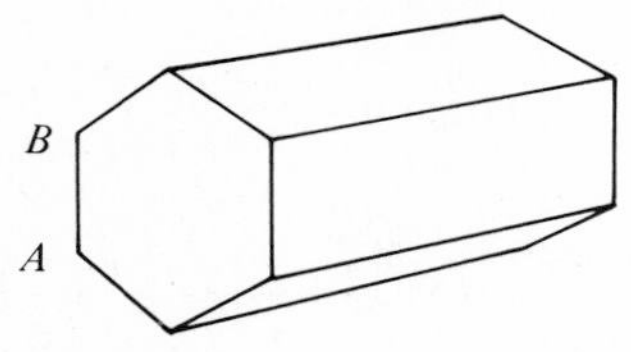

• ### • ### • ### • ### • ### • ### • ### •

A7 5.22 inches

4 For rectangles and squares, formulas can be written which shorten the work somewhat. If the length is l and the width is w,

$p = l + w + l + w$
$p = l + l + w + w$
$p = 2l + 2w$

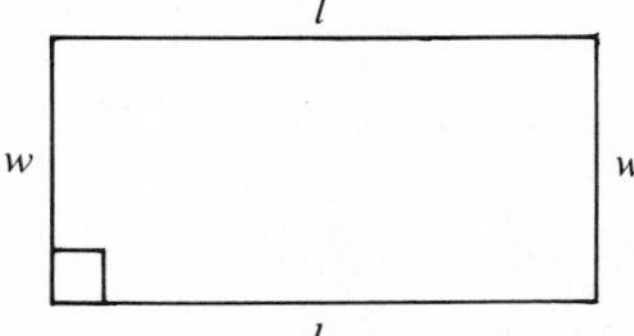

This gives a formula for the perimeter of a rectangle:

$p = 2l + 2w$

A formula for the perimeter of a square may be obtained from the formula for the perimeter of a rectangle where $l = s$ and $w = s$:

$p = 2l + 2w$
$p = 2s + 2s$
$p = 4s$

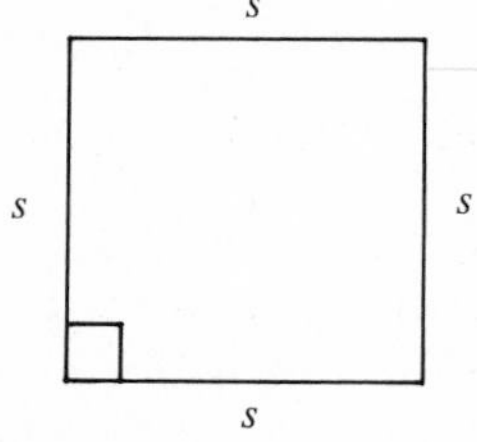

Therefore, $p = 4s$.

Example: Find the perimeter of a rectangle with width 7.0 in. and length 8.4 in.

Solution: $p = 2l + 2w$
$p = 2(8.4) + 2(7.0)$ [Note: 2(8.4) means 2×8.4]
$p = 16.8 + 14.0$
$p = 30.8$

Therefore, the perimeter is 30.8 in.

Q8 Find the perimeters of the following rectangles by substituting in the appropriate formula:

a.

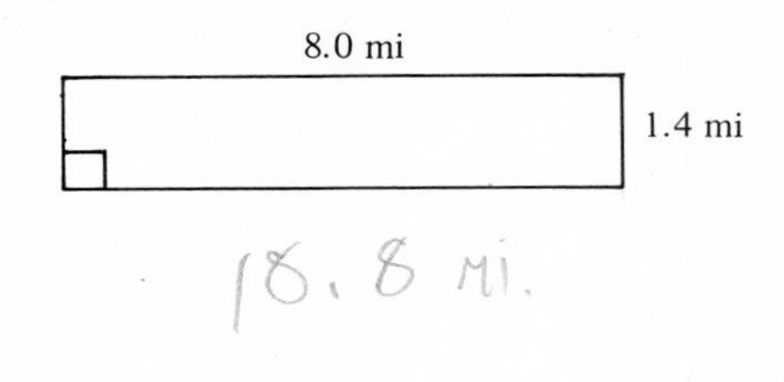

b.

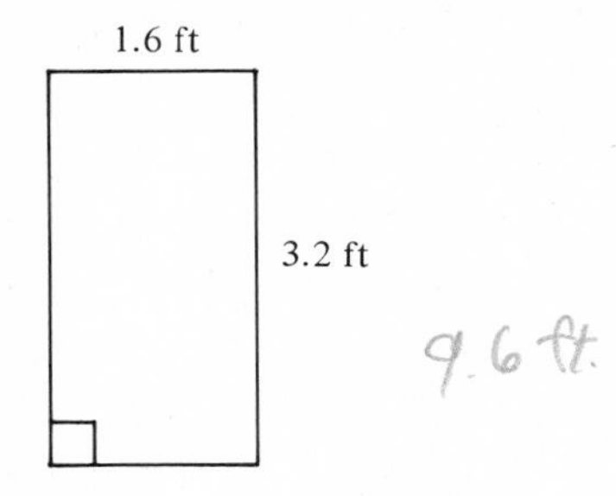

c.

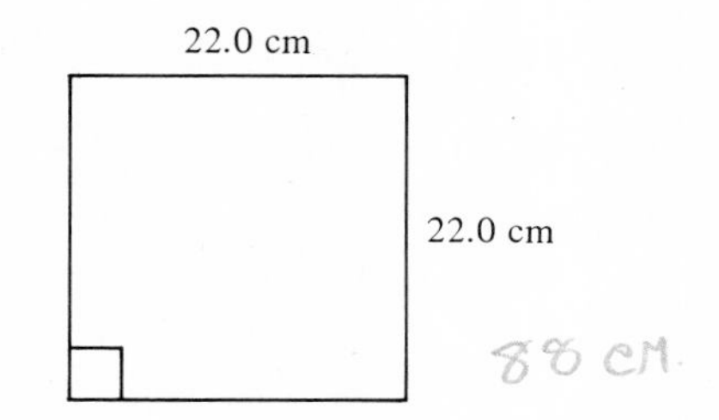

d.

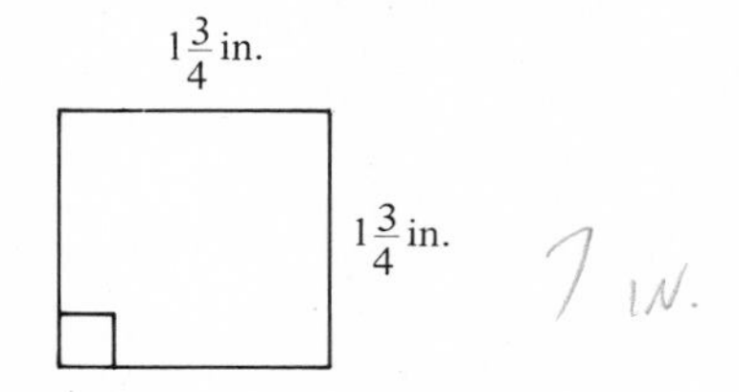

\# \# \# • \# \# \# • \# \# \# • \# \# \# • \# \# \# • \# \# \# • \# \# \# • \# \# \# • \# \# \#

A8 **a.** 18.8 mi: $p = 2l + 2w$
$p = 2(8.0) + 2(1.4)$
$p = 16.0 + 2.8$
$p = 18.8$

b. 9.6 ft: $p = 2l + 2w$
$p = 2(3.2) + 2(1.6)$
$p = 6.4 + 3.2$
$p = 9.6$

c. 88.0 cm: $p = 4s$
$p = 4(22.0)$
$p = 88.0$

d. 7 in.: $p = 4s$
$p = 4\left(1\frac{3}{4}\right)$
$p = 4\left(\frac{7}{4}\right)$
$p = 7$

5 The perimeter of a circle is called its *circumference*. The circumference of a circle is found with another formula. To obtain the circumference it is necessary to know the diameter of the circle and the value of the number π (pi). The number π is the ratio of the circumference to the diameter of a circle. The value of π is a constant; that is, it never changes. However, it is impossible to write a decimal or fraction that is exactly equal to π. One is therefore forced to use approximations of π rounded off to various numbers of decimal places. Since it is not possible to say that π *equals* another number, the symbol "$\doteq$" is used to mean approximately equal. The symbol "$\doteq$" says "approximately equal to." Various approximations follow:

1. $\pi \doteq 3.14159265$, precise to 8 decimal places.
2. $\pi \doteq 3.14$, precise to 2 decimal places. This approximation is acceptable for calculations in this book.
3. $\pi \doteq 3\frac{1}{7}$, also precise to 2 decimal places. This value is commonly used when calculating with fractions.

4. $\pi \doteq \frac{355}{113}$, precise to 6 decimal places. This is an easy approximation to remember (think 1, 1, 3, 3, 5, 5) for use with an electronic calculator.

Q9 Is π a variable or a constant? ______

\# # # • # # # • # # # • # # # • # # # • # # # • # # # • # # # • # # #

A9 constant

Q10 **a.** Give two approximations of π that are precise to two decimal places. ______ and ______

b. Give two approximations of π that are precise to six decimal places. ______ and ______

\# # # • # # # • # # # • # # # • # # # • # # # • # # # • # # # • # # #

Q10 **a.** 3.14 and $3\frac{1}{7}$ **b.** 3.141593 and $\frac{355}{113}$

Q11 Can an exact value for π be written as a decimal? ______

\# # # • # # # • # # # • # # # • # # # • # # # • # # # • # # # • # # #

A11 no

6 The circumference of a circle is found by using the formula $c = \pi d$, where d is the diameter. Since the diameter is twice the radius ($d = 2r$), a second formula for the circumference is $c = 2\pi r$, where r is the radius:

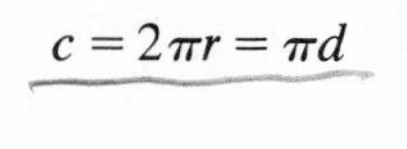

$c = 2\pi r = \pi d$

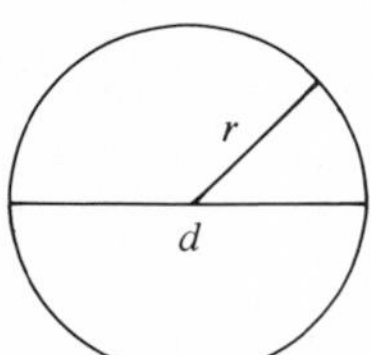

Example: If the diameter of a circle is 2.5 in., what is its circumference? Round off to tenths.

Solution: $c = \pi d$

$c = (3.14)(2.5)$ [Note: (3.14)(2.5) means 3.14×2.5]

$c = 7.850$

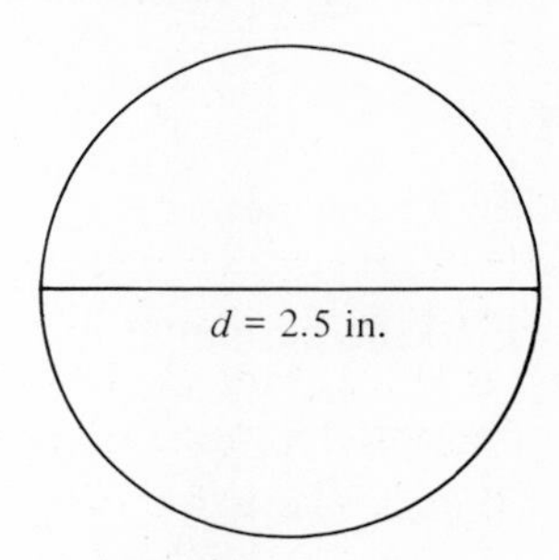

Therefore, the circumference is 7.9 in.

Use $\pi \doteq \frac{22}{7}$ when measurements are given as common fractions. Use $\pi \doteq 3.14$ when measurements are given as decimal fractions.

Q12 Find the circumference of a circle with diameter 8.5 cm (use $\pi \doteq 3.14$). Round off to ones.

\# # # • # # # • # # # • # # # • # # # • # # # • # # # • # # # • # # #

A12 27 cm: $c = \pi d$

$c = (3.14)(8.5)$

$c = 26.690$

Q13 What is the circumference of a circle whose *radius* is $4\frac{3}{4}$ ft? $\left(\text{Use } c = 2\pi r \text{ and } \pi \doteq 3\frac{1}{7}.\right)$

• # # # • # # # • # # # • # # # • # # # • # # # • # # # • # #

A13 $29\frac{6}{7}$ ft: $c = 2\pi r$

$$c = 2\left(3\frac{1}{7}\right)\left(4\frac{3}{4}\right)$$

$$c = \frac{2}{1}\left(\frac{22}{7}\right)\left(\frac{19}{4}\right)$$

$$c = 29\frac{6}{7}$$

Q14 A gear is to be cut from a gear blank with a diameter of 18.00 centimetres. What is its circumference? Round off to tenths.

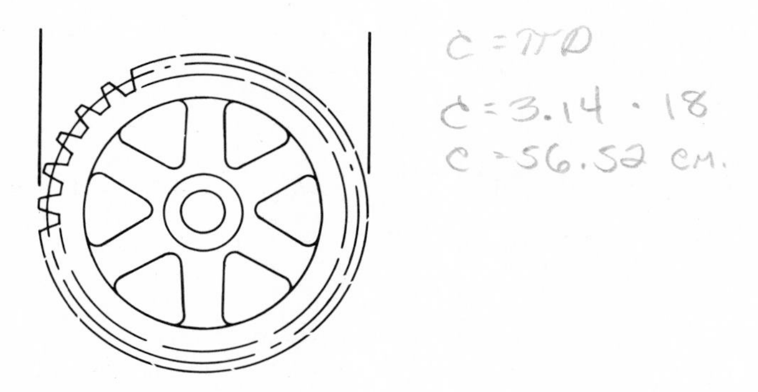

• # # # • # # # • # # # • # # # • # # # • # # # • # # # • # #

A14 56.5 centimetres

Q15 Find the perimeter of the hole in the plate. Round off to ones.

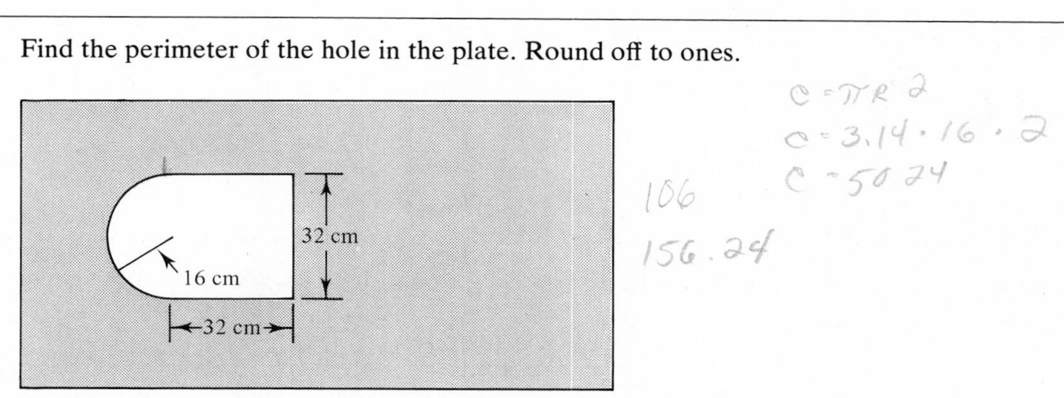

• # # # • # # # • # # # • # # # • # # # • # # # • # # # • # #

A15 146 centimetres

Q16 Two pulleys each with diameter 22 cm and whose centers are 38 cm apart are connected with a belt. Calculate the length of the belt. Round off to the nearest centimetre.

\# # # • # # # • # # # • # # # • # # # • # # # • # # # • # # # • # # #

A16 145 cm: $\text{length} = \pi d + 2(38) = 3.14(22) + 76 = 69.08 + 76$

7 A milling machine spindle with diameter d inches turns at a certain number of revolutions per minute (rpm). From the knowledge of the number of rpms and the diameter of the spindle we can calculate the cutting speed of the milling machine in feet per minute.

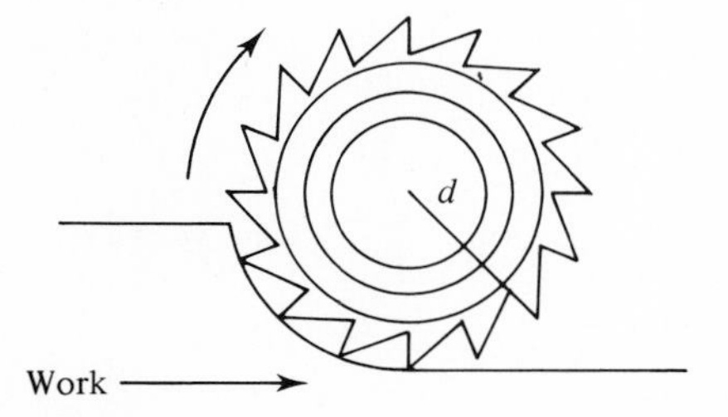

Step 1: Find the distance, D, in inches that a point on the spindle travels in one revolution.

$D = \pi d$ inches

Step 2: Convert this distance to feet.

$$D = \frac{\pi d}{12} \text{ feet}$$

Step 3: Find the distance traveled in one minute, and you have the cutting speed (cs) of the spindle in feet per minute.

$$\text{cs} = (\text{rpm})\left(\frac{\pi d}{12}\right)$$

Example: Find the cutting speed in feet per minute of a 4.0 inch diameter spindle moving at 300 rpm.

Solution:

$$\text{cs} = (\text{rpm})\left(\frac{\pi d}{12}\right)$$

$$\text{cs} = (300)\left(\frac{3.14 \times 4.00}{12}\right)$$

$$\text{cs} = 314$$

Therefore, the cutting speed is 314 feet per minute.

Q17 Find the cutting speed in feet per minute of the milling machines below:

a. The diameter of the spindle is 3.5 inches and it moves at 500 rpm. Round off to tens.

b. The diameter of the spindle is 6.0 inches and it moves at 350 rpm. Round off to tens.

\# # # • # # # • # # # • # # # • # # # • # # # • # # # • # # # • # # #

A17 **a.** 460 feet per minute **b.** 550 feet per minute

8 Problems have been given in which the dimensions of the polygons or circles have been supplied. One should also be able to measure the appropriate dimensions and compute the perimeter of a figure. The following problems will require that you measure with an inch or centimetre ruler and calculate the perimeters.

Q18 Compute the perimeter in inches $\left(\text{measure to the nearest } \frac{1}{8} \text{ inch}\right)$:

a.

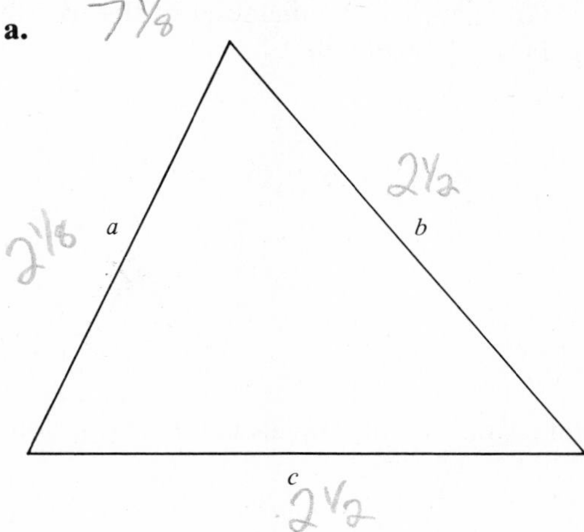

b.

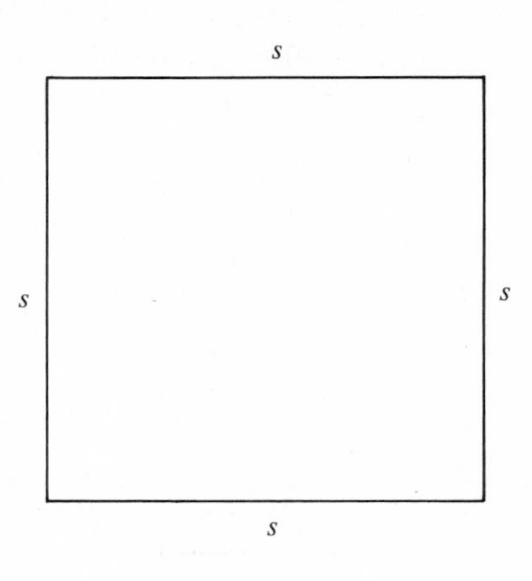

\# \# \# • \# \# \# • \# \# \# • \# \# \# • \# \# \# • \# \# \# • \# \# \# • \# \# \# • \# \# \#

A18 **a.** $7\frac{1}{8}$ in.: $p = a + b + c$

$$p = 2\frac{1}{8} + 2\frac{1}{2} + 2\frac{1}{2}$$

$$p = 7\frac{1}{8}$$

b. 7 in.: $p = 4s$

$$p = 4\left(1\frac{3}{4}\right)$$

$$p = 7$$

Q19 Compute the perimeter in centimetres $\left(\text{measure to the nearest } \frac{1}{10} \text{ centimetre}\right)$:

a.

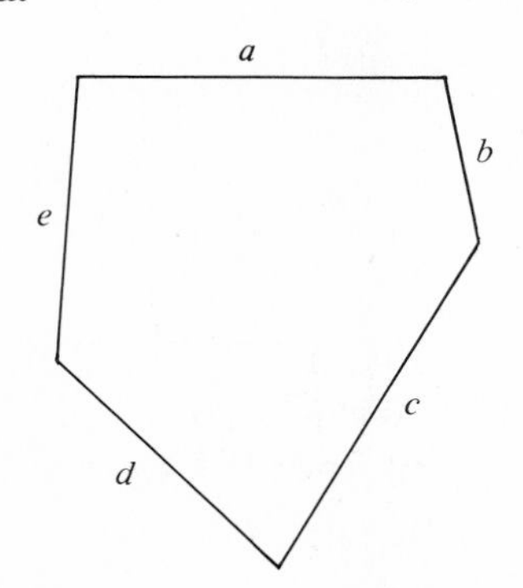

b.

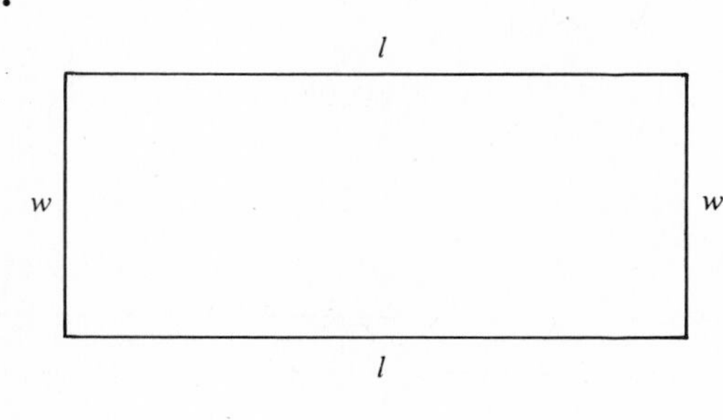

\# \# \# • \# \# \# • \# \# \# • \# \# \# • \# \# \# • \# \# \# • \# \# \# • \# \# \# • \# \# \#

A19 **a.** 12.5 cm: $p = a + b + c + d + e$

$p = 3.0 + 1.4 + 3.2 + 2.5 + 2.4$

$p = 12.5$

b. 16.4 cm: $p = 2l + 2w$

$p = 2(5.7) + 2(2.5)$

$p = 11.4 + 5.0$

$p = 16.4$

Q20 Compute the circumference of each circle:

a. Measure the diameter of circle 1 to the nearest $\frac{1}{4}$ inch.

b. Measure the diameter of circle 2 to the nearest $\frac{1}{2}$ centimetre.

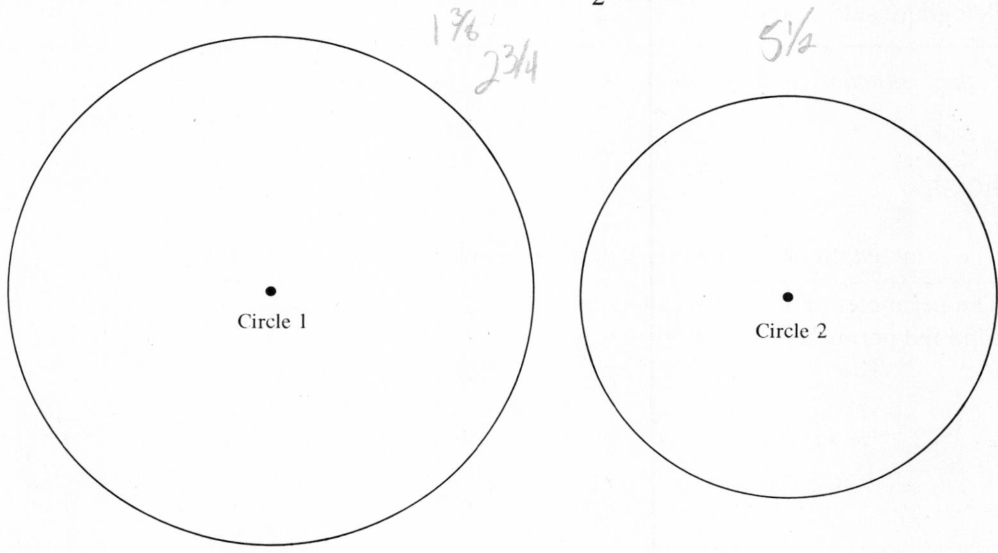

A20 **a.** $8\frac{9}{14}$ in.: $c = \pi d$

$$c = 3\frac{1}{7}\left(2\frac{3}{4}\right)$$

$$c = \frac{\overset{11}{\cancel{22}}}{7} \cdot \frac{11}{\underset{2}{\cancel{4}}}$$

$$c = \frac{121}{14}$$

$$c = 8\frac{9}{14}$$

b. 17.27 cm: $c = \pi d$

$c = (3.14)(5.5)$

$c = 17.27$

or

$17\frac{2}{7}$ cm: $c = \pi d$

$$c = \frac{22}{7}\left(5\frac{1}{2}\right)$$

$$c = \frac{22}{7} \cdot \frac{11}{2}$$

$$c = \frac{242}{14}$$

$$c = 17\frac{2}{7}$$

A summary of formulas for perimeters is provided. Some have been considered in the previous material. Others may be useful in further work.

Geometric figure	Formula for perimeter
Triangle: sides a, b, c	$p = a + b + c$
Quadrilateral: sides a, b, c, d	$p = a + b + c + d$
Pentagon: sides a, b, c, d, e	$p = a + b + c + d + e$
Rectangle: length l, width w	$p = 2l + 2w$
Square: side s	$p = 4s$
Circle: radius r	$c = 2\pi r$ or $c = \pi d$
Equilateral triangle: side s	$p = 3s$
Regular pentagon: side s	$p = 5s$
Parallelogram: sides a and b	$p = 2a + 2b$

This completes the instruction for this section.

5.3 EXERCISE

1. The total length of the boundary of a polygon is called its ________.
2. The perimeter of a circle is called its ________.
3. Find the perimeter of the polygons below:

a.

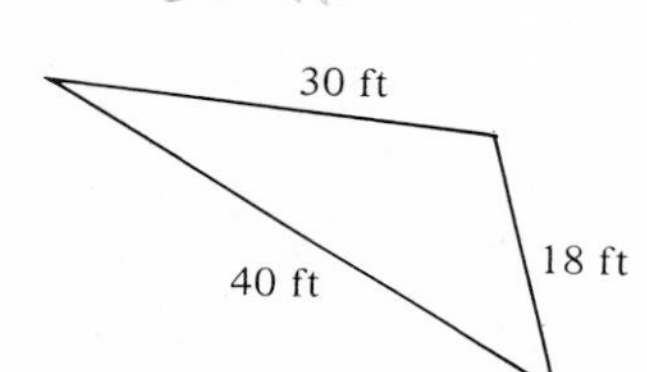

b.

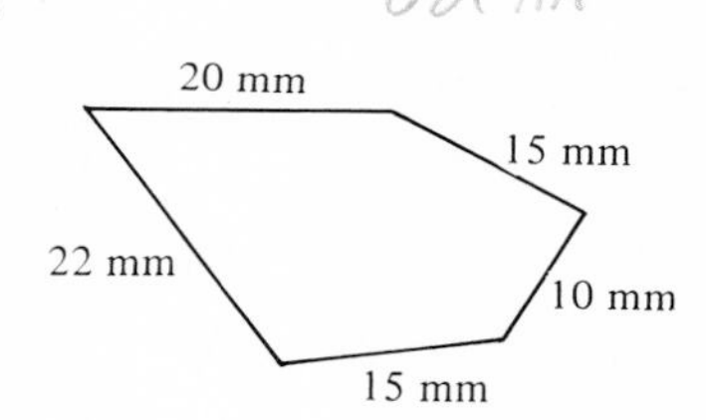

4. Find the perimeter of a pentagon with sides to the nearest $\frac{1}{8}$ inch of the following lengths: $1\frac{3}{4}$, $2\frac{1}{4}$, 3, $1\frac{7}{8}$, and $2\frac{1}{2}$ inches.
5. Use the formula $p = 2l + 2w$ to compute the perimeter of the rectangles:

a.

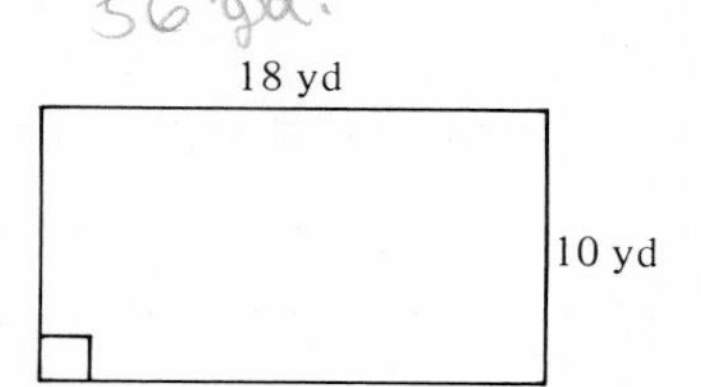

b.

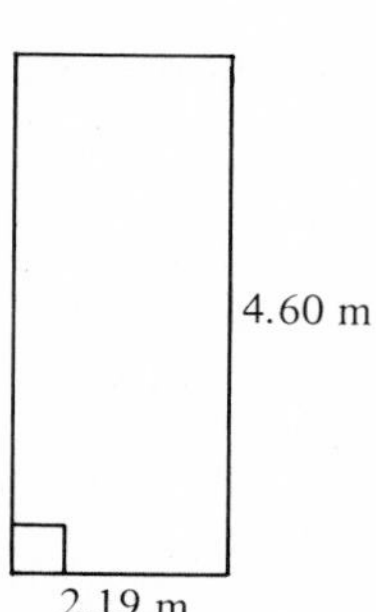

6. Use the formula $p = 4s$ to compute the perimeter of the following:
   a. square with a side of 4.7 centimetres
   b. square with a side of $1\frac{7}{8}$ inches

7. Use the formula $c = 2\pi r$ or $c = \pi d$ to compute the circumference of the circles:

a. 60.445 ft.

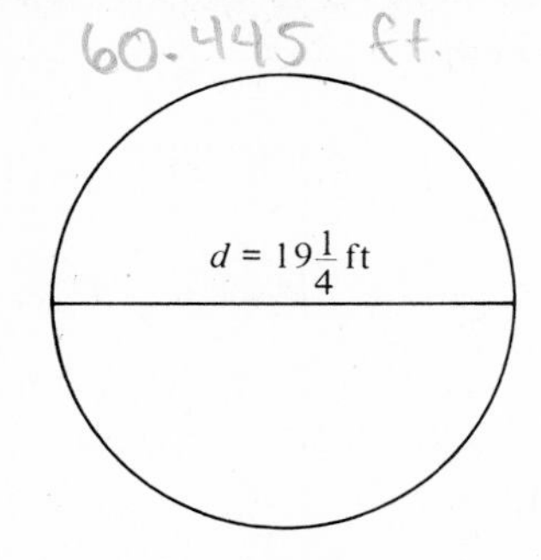

b. 10.048 in.

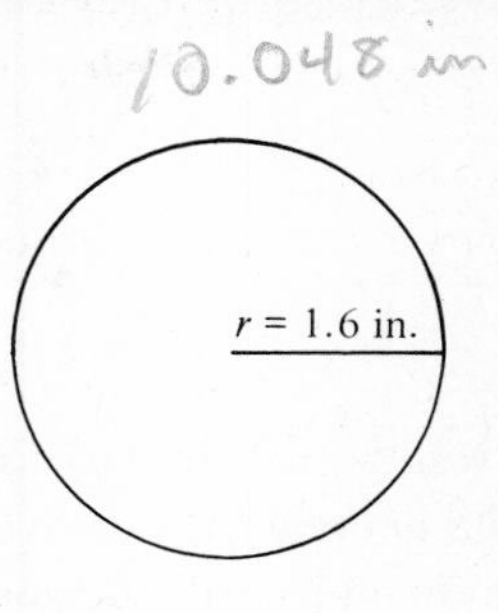

8. Use a centimetre rule to measure the figures to the nearest 0.1 centimetre and compute their perimeters in centimetres:

a.

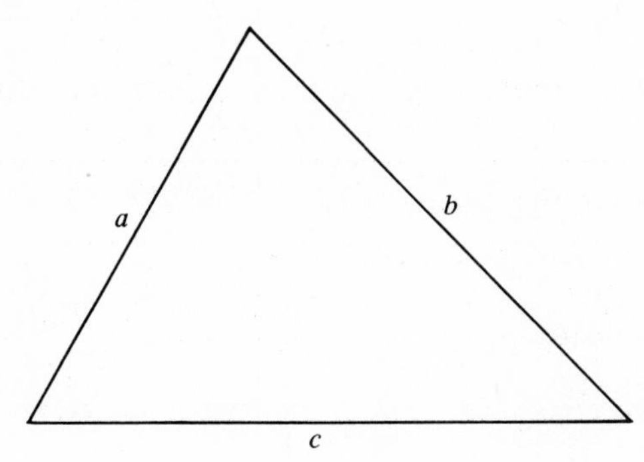

b.

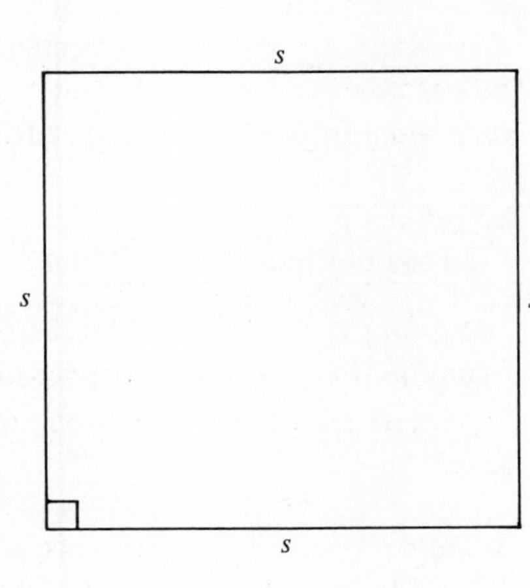

c.

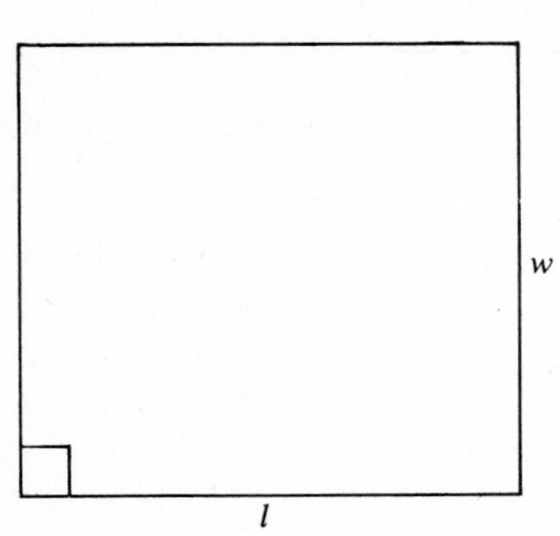

d.

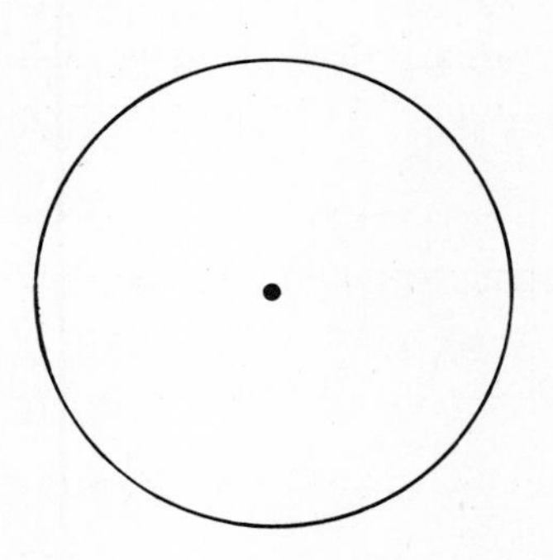

9. Use an inch ruler to measure the figures to the nearest $\frac{1}{8}$ inch, and compute their perimeters:

a.

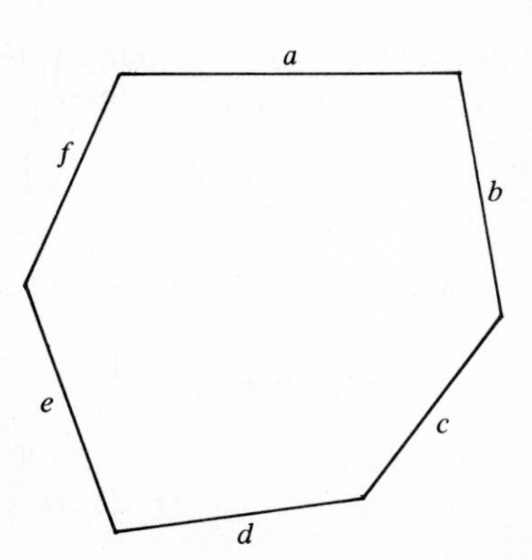

b.

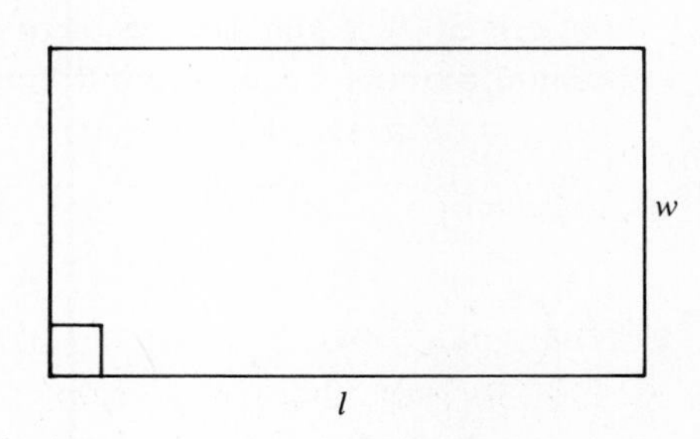

c.

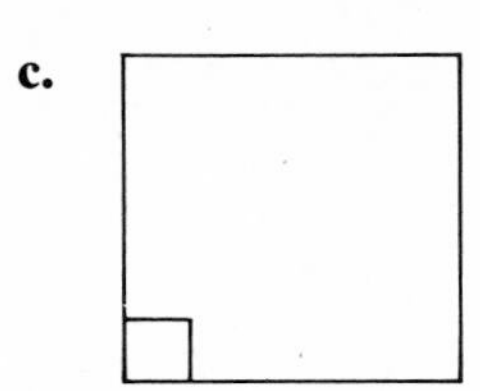

d.

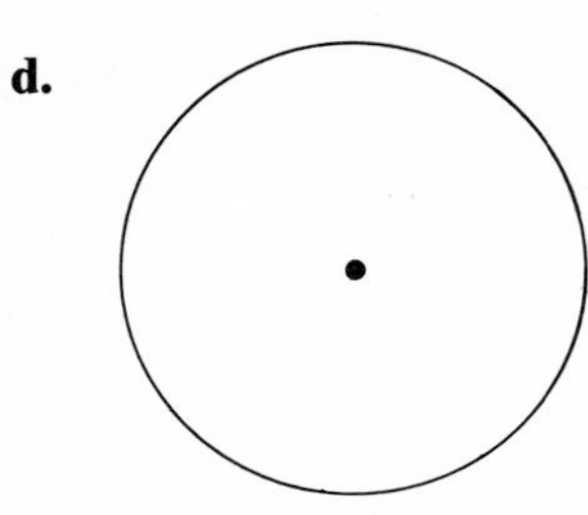

10. To order wallpaper it is necessary to know the perimeter of a room. If the dimensions of the room to the nearest inch were 10 ft 6 in. × 15 ft, what was its perimeter?

11. Match the formula with the geometric figure:

a. $p = 2l + 2w$	(1) regular pentagon
b. $c = \pi d$	(2) equilateral triangle
c. $p = 4s$	(3) circle
d. $p = 5s$	(4) rectangle
e. $c = 2\pi r$	(5) any triangle
f. $p = a + b + c$	(6) quadrilateral
g. $p = 3s$	(7) square
h. $p = a + b + c + d$	

12. Find the perimeter of the block.

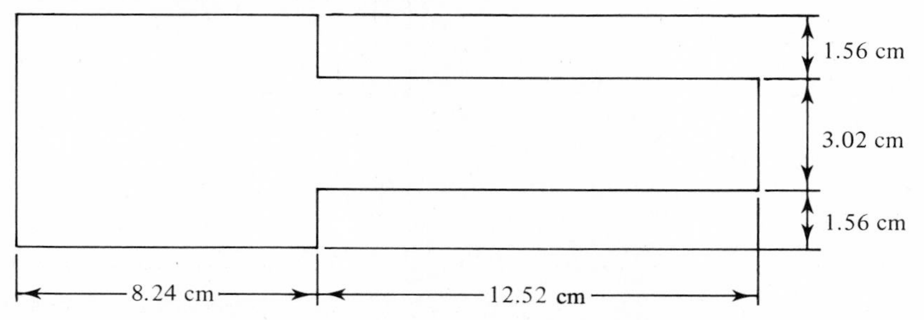

13. A step gauge is illustrated below. Find the perimeter, the dimensions are given in centimetres.

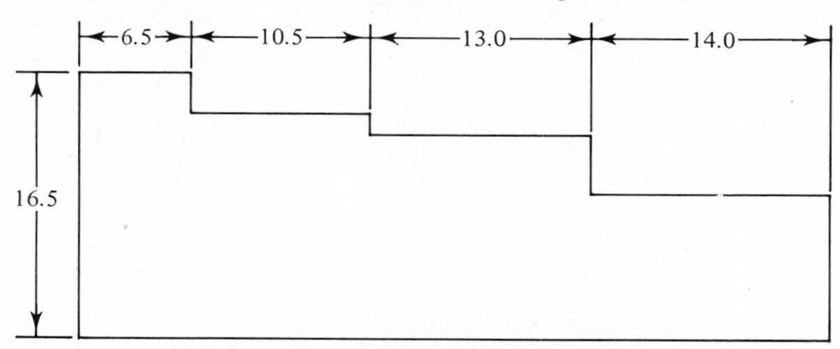

14. A track is laid out with a semicircle on each end of two straightaways. The radius of the inside lane is 50 feet and the radius of the second lane is 53 feet. How much farther does a runner travel running in the second lane compared to the first lane in one time around the track? Round off to the nearest foot.

5.3 EXERCISE ANSWERS

1. perimeter
2. circumference
3. a. 88 ft **b.** 82 mm
4. $11\frac{3}{8}$ in.
5. a. 56 yd **b.** 13.58 m
6. a. 18.8 cm **b.** $7\frac{1}{2}$ in.
7. a. $60\frac{1}{2}$ ft **b.** 10.048 in.
8. 16.1 cm **b.** 19.2 cm **c.** 16.6 cm **d.** 11.932 cm
9. a. 6 in. **b.** $6\frac{1}{4}$ in. **c.** $3\frac{1}{2}$ in. **d.** $3\frac{13}{14}$ in.
10. 51 ft
11. a. 4 **b.** 3 **c.** 7 **d.** 1
e. 3 **f.** 5 **g.** 2 **h.** 6
12. 53.80 cm
13. 121.0 cm
14. 19 ft

5.4 AREAS

1 The units that are used to report area within geometric figures were discussed in Section 5.2. Some examples are square inches, square feet, square centimetres, and square metres. The numerical value of the area is usually determined by a formula. For example, the area of a rectangular region is obtained with the formula $A = lw$, where l is the length and w is the width. The length and width are linear measurements. However, when inches are multiplied by inches, the result is a number with the unit square inches. Before the numbers are substituted into formulas, all measurements must be in the same units. For example,

$$A = lw$$
$$A = 25 \cdot 12^*$$
$$A = 300$$

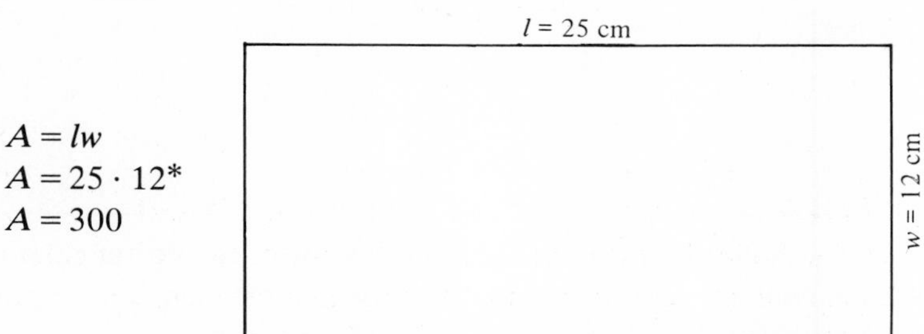

Therefore, the area is 300 cm^2.

*The raised dot also indicates multiplication. That is, $25 \cdot 12 = 25 \times 12$.

Q1 Find the area of a rectangular floor with one dimension 10.5 ft and the other dimension 12.0 ft.

\# \# \# • \# \# \# • \# \# \# • \# \# \# • \# \# \# • \# \# \# • \# \# \# • \# \# \# • \# \# \#

A1 126 ft^2: $A = lw$
$A = (12.0)(10.5)$
$A = 126.00$

Q2 Find the area in square yards of a lot with dimensions 30 yd 2 ft by 53 yd 1 ft. (Convert feet to fractions of a yard by dividing by 3.)

\# \# \# • \# \# \# • \# \# \# • \# \# \# • \# \# \# • \# \# \# • \# \# \# • \# \# \# • \# \# \#

A2 $1{,}635\frac{5}{9} \text{ yd}^2$: $30 \text{ yd } 2 \text{ ft} = 30\frac{2}{3} \text{ yd}$ $A = lw$
$53 \text{ yd } 1 \text{ ft} = 53\frac{1}{3} \text{ yd}$ $A = 53\frac{1}{3} \cdot 30\frac{2}{3}$
$A = 1{,}635\frac{5}{9}$

---

2 The formula for the area of a square region is a special case of the formula for the area of a rectangular region. If $l = w$ in the formula $A = lw$, then it becomes $A = ww = w^2$. Rather than using w for the width of a square, the variable s is usually used. Therefore, the area of a square region is computed with the formula

$A = s^2$ (Note: s^2 means $s \times s$. s^2 is read "s squared.")

Example: Find the area of a square card table that is 30 inches on a side.

Solution: $A = s^2$
$A = (30)^2$
$A = 900$

Therefore, the area is 900 in.^2.

Q3 Find the area of a field that is 45 yards square. Round off to hundreds.

\# \# \# • \# \# \# • \# \# \# • \# \# \# • \# \# \# • \# \# \# • \# \# \# • \# \# \# • \# \# \#

A3 $2{,}000 \text{ yd}^2$: $A = s^2$
$A = 45^2$
$A = 2{,}025$

Q4 Find the area of the plate by two methods. First use the sum of the areas of two rectangular regions, then use the difference of two rectangular regions. Round off to the nearest inch.

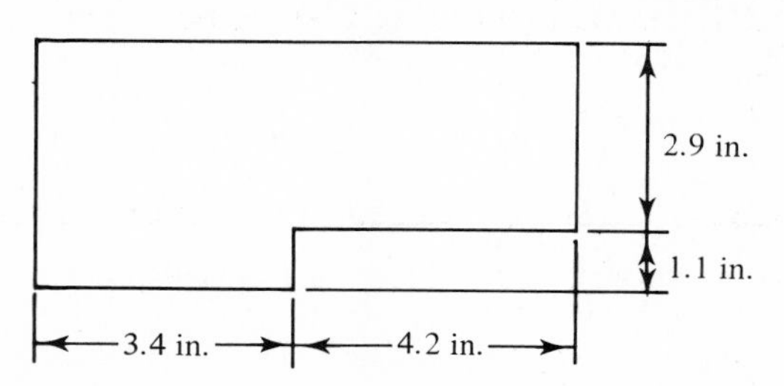

\# \# \# • \# \# \# • \# \# \# • \# \# \# • \# \# \# • \# \# \# • \# \# \# • \# \# \# • \# \# \#

A4 26 in.: Divide the region as is shown below.

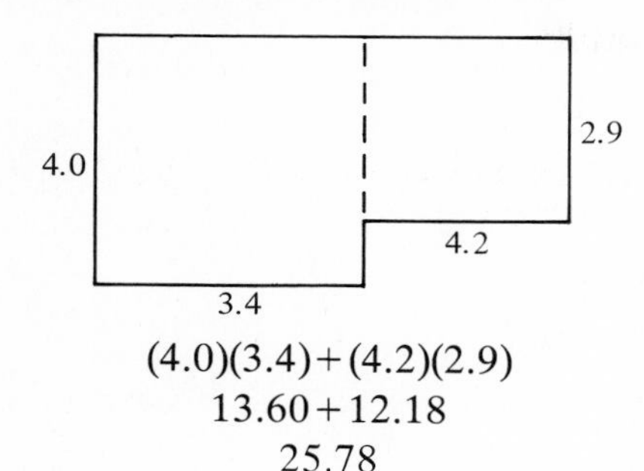

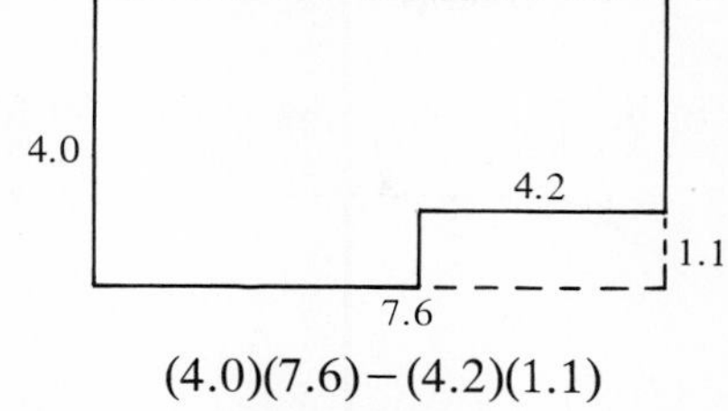

(4.0)(3.4) + (4.2)(2.9)
13.60 + 12.18
25.78

(4.0)(7.6) − (4.2)(1.1)
30.40 − 4.62
25.78

Q5 Find the area of the shaded region. Round off to ones.

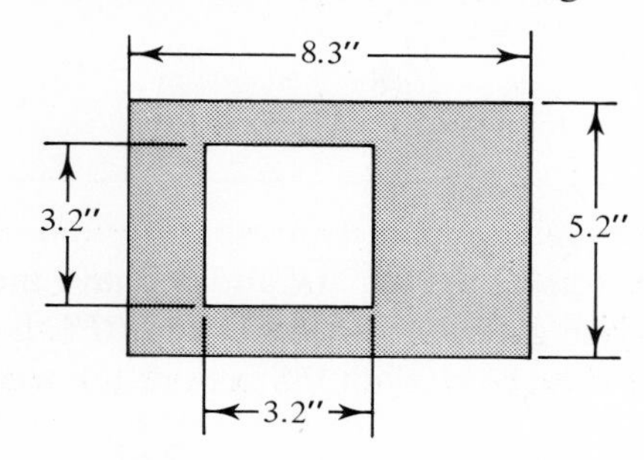

\# \# \# • \# \# \# • \# \# \# • \# \# \# • \# \# \# • \# \# \# • \# \# \# • \# \# \# • \# \# \#

A5 33 in.^{2}

3 The following figure is a parallelogram. The opposite sides are parallel. The bases of a parallelogram are any two opposite sides (usually the longest side is considered a base). The altitude of a parallelogram is the perpendicular distance between the bases.

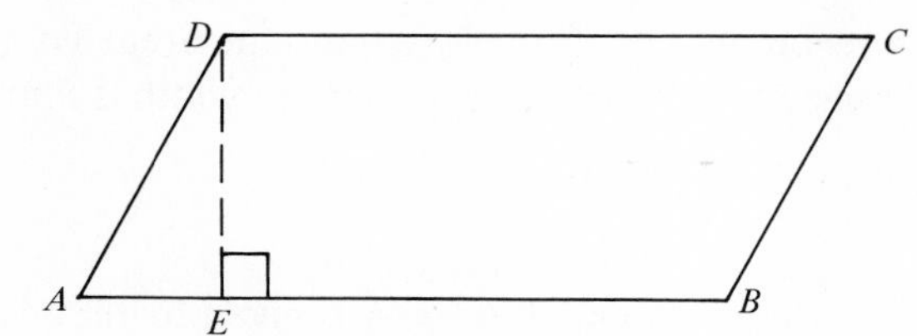

AREA of PARALLELOGRAM
A = BASE · HEIGHT
A = bh

Example 1: Name the bases of the parallelogram.

Solution: $\overline{AB}$ and $\overline{DC}$ ($\overline{AD}$ and $\overline{BC}$ could be considered bases, but it would not be as convenient to find the altitude)

Example 2: The altitude of parallelogram *ABCD* is the length of what line segment?

Solution: $\overline{DE}$

The area of a parallelogram is found by multiplying the length of a base, *b*, times the altitude (height), *h*. The formula is

$A = bh$

Example 3: Find the area of a parallelogram with base 22 cm and altitude 14 cm. Round off to tens.

Solution: $A = bh$
$A = 22(14)$
$A = 308$ Therefore, the area is 310 cm^2.

Q6 **a.** Find the area of a parallelogram with base 202 m and altitude 26.0 m. Round off to tens.

b. Find the area of the parallelogram. Round off to tens.

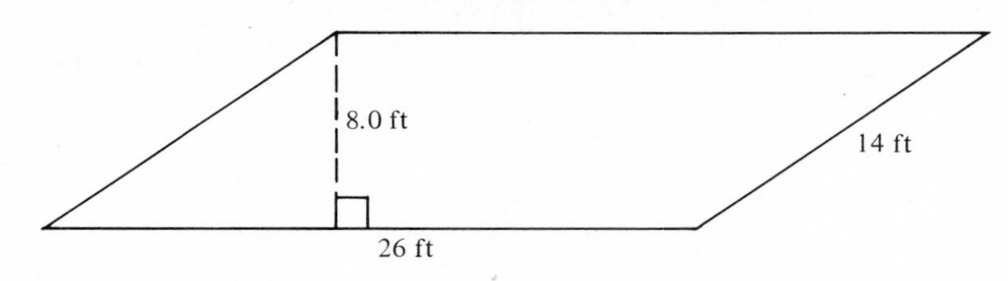

\# \# \# • \# \# \# • \# \# \# • \# \# \# • \# \# \# • \# \# \# • \# \# \# • \# \# \# • \# \# \#

A6 **a.** 5250 m^2: $A = bh$
$A = 202(26)$
$A = 5252$

b. 210 ft^2: $A = bh$
$A = 26(8)$
$A = 208$

4 To derive a formula for the area of a region within a parallelogram with a base b and altitude (height) h, a perpendicular is drawn from D to $\overleftrightarrow{AB}$, and from C to $\overleftrightarrow{AB}$. $\overline{AE}$ and $\overline{BF}$ have the same length, and $\overline{DE}$ and $\overline{CF}$ have the same length (h), so the triangular regions AED and BFC have the same area. Since the area of the triangular region AED is replaced with the area of the triangular

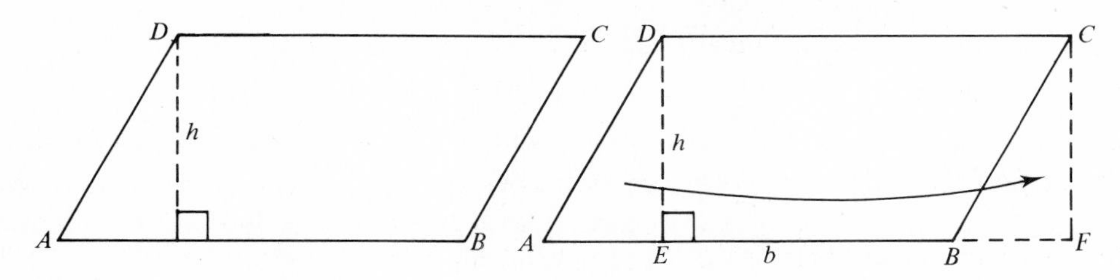

region BFC, the parallelogram $ABCD$ and the rectangle $EFCD$ enclose the same area. To find the area of region $EFCD$, the length and width are needed. The length is h and the width is b, because $\overline{EF}$ and $\overline{AB}$ have the same length.

area of region $ABCD$ = area of region $EFCD = bh$

Therefore, the area within a parallelogram with base b and altitude h is given by the formula

$A = bh$

Example: Find the area of a parallelogram with altitude 4.06 inches and base 8.05 inches. Round off to tenths.

Solution: $A = bh$
$A = (8.05)(4.06)$
$A = 32.6830$

Therefore, the area is 32.7 in.^{2}.

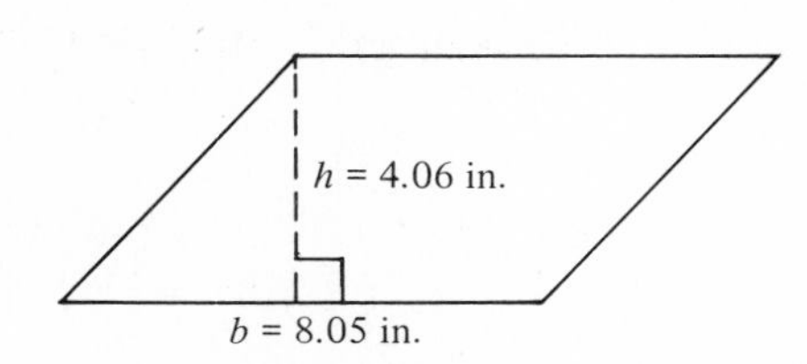

Q7 Find the area of the region within a parallelogram with base 12.5 inches and altitude 6.0 inches. Round off to ones.

\# \# \# • \# \# \# • \# \# \# • \# \# \# • \# \# \# • \# \# \# • \# \# \# • \# \# \# • \# \# \#

A7 75 in.^{2}: $A = bh$
$A = (12.5)(6.0)$
$A = 75.00$

Q8 The cross section of a prism is shown below. Find the area of the cross section. Round off to tenths.

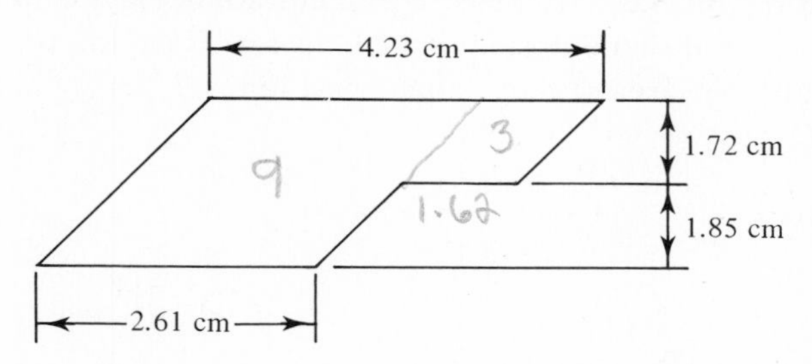

\# \# \# • \# \# \# • \# \# \# • \# \# \# • \# \# \# • \# \# \# • \# \# \# • \# \# \# • \# \# \#

A8 12.1 cm^2: The area may be obtained by finding the difference between the areas shown below. $(4.23)(3.57) - (1.62)(1.85)$

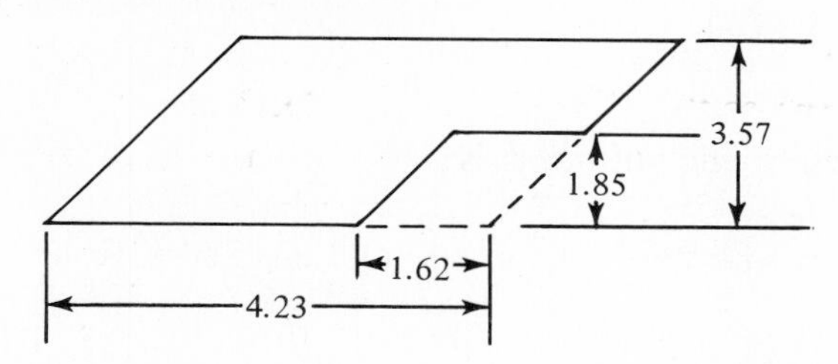

Q9 Find the area of the shaded region below. Round off to ones.

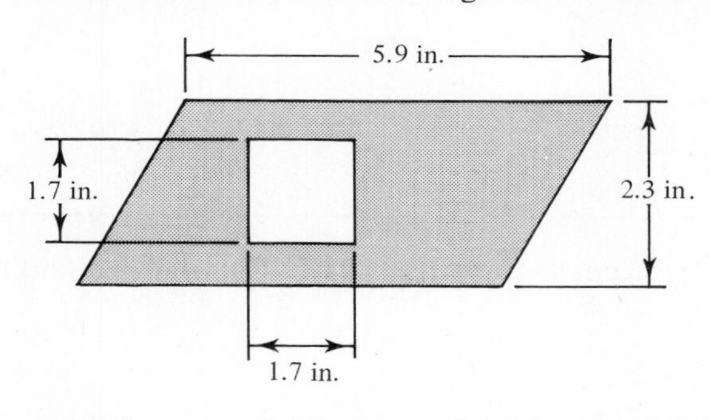

A= 5.9 · 2.3
A = 13.57
A = 10.68

A = 1.7^2
A = 2.89

\# \# \# • \# \# \# • \# \# \# • \# \# \# • \# \# \# • \# \# \# • \# \# \# • \# \# \# • \# \# \#

A9 12 in.^{2}

Q10 Find the total area of the space enclosed in the following floor plan:

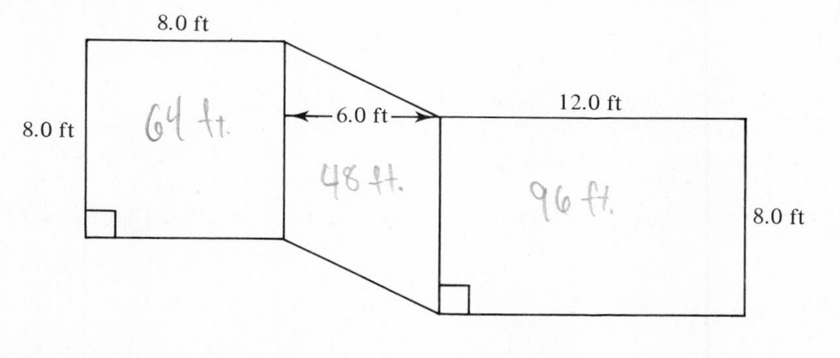

208 ft.

\# \# \# • \# \# \# • \# \# \# • \# \# \# • \# \# \# • \# \# \# • \# \# \# • \# \# \# • \# \# \#

A10 208 ft^2: $A = s^2$, $A = 8.0^2$, $A = 64$; $A = bh$, $A = (8.0)(6.0)$, $A = 48$; $A = lw$, $A = (12.0)(8.0)$, $A = 96$

5 The formula for the area of the triangular region ABC with base b and altitude h may be found by drawing another triangle with the same size and shape upside down above it. Since the triangular regions ABC and ACD have the same area, the area of the triangular region ABC is equal to one half the area within parallelogram $ABCD$.

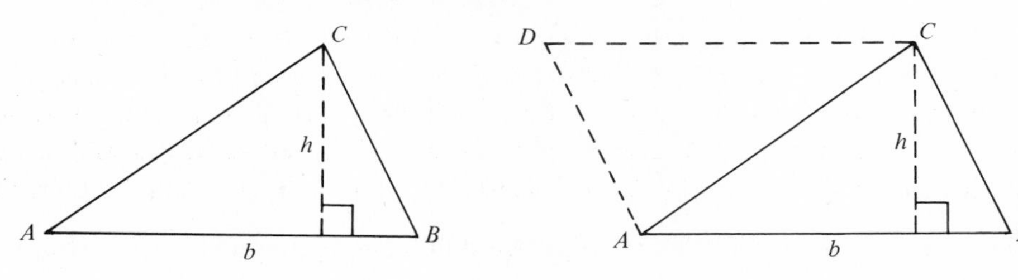

area of region $ABC = \frac{1}{2}$ area of region $ABCD = \frac{1}{2}bh$

The formula for the area of a triangle with base b and altitude h is $A = \frac{1}{2}bh$.

Example: Find the area of a triangular region with base 10.5 m and altitude 5.6 m. Round off to ones.

Solution: $A = \frac{1}{2}bh$

$A = \frac{1}{2}(10.5)(5.6)$

$A = 29.40$

Therefore, the area is 29 m^2.

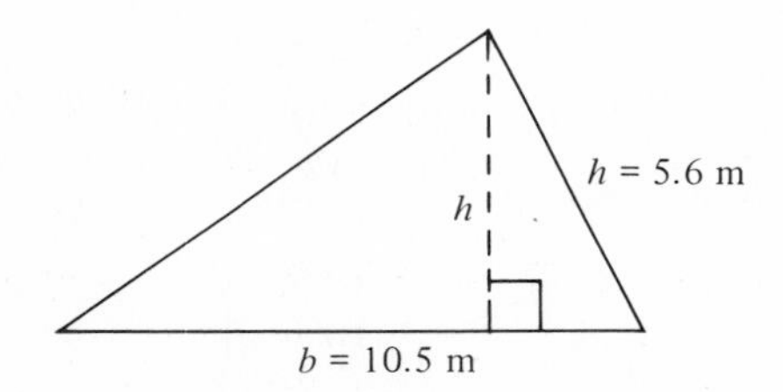

Q11 Find the area of a triangular region with base 4.6 cm and altitude 3.7 cm. Round off to tenths.

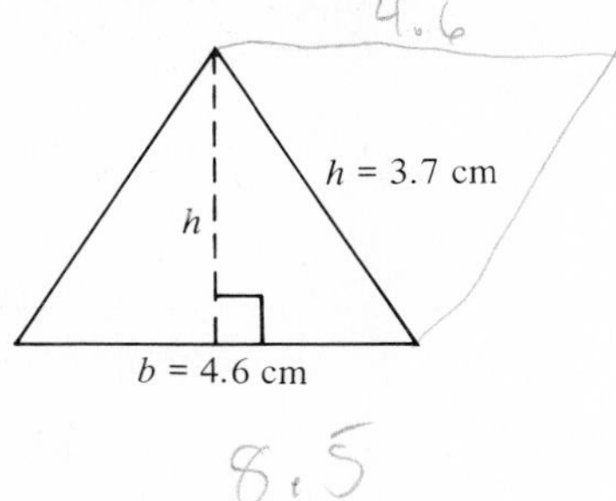

8.5

• ### • ### • ### • ### • ### • ### • ### •

A11 8.5 cm^2: $A = \frac{1}{2}bh$

$A = \frac{1}{2}(4.6)(3.7)$

$A = 8.51$

Q12 Find the area of a triangular region with base 4.2 ft and altitude 4.0 yd. Round off to ones.

\# # # • # # # • # # # • # # # • # # # • # # # • # # # • # # # • # # #

A12 25 ft²: $A = \frac{1}{2}bh$

$A = \frac{1}{2}(4.2)(12.0)$ (*Note*: 4.0 yd must be changed to feet.)

6 The formula for the area of a trapezoidal region with altitude h and bases b_1 (read "bee sub-one") and b_2 (read "bee sub-two") can be found by drawing another trapezoid upside down beside it which is equal in size and shape. (In b_1 and b_2, the 1 and 2 are called *subscripts* and are used to distinguish between the two bases.) A parallelogram $AEFD$ is formed which includes twice the area

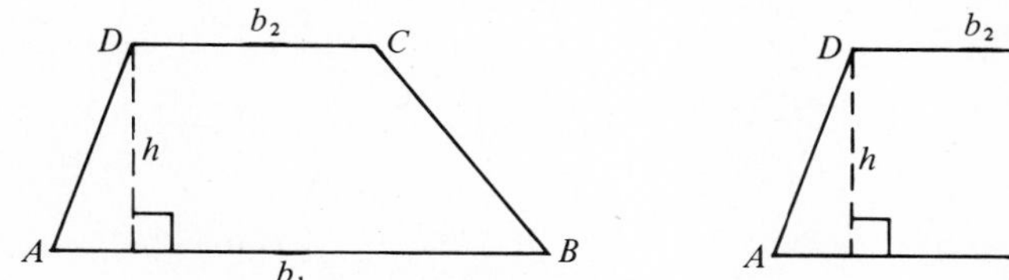

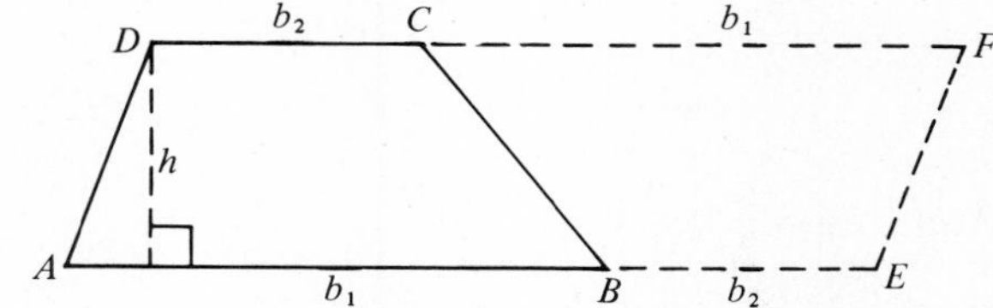

of the trapezoidal region $ABCD$. The base of the new parallelogram is $b_1 + b_2$ and its altitude is h. Therefore, the area of the trapezoid region $ABCD$ is $\frac{1}{2}$ the area of the region within parallelogram $AEFD$.

area of region $ABCD = \frac{1}{2}$ area of region $AEFD = \frac{1}{2}h(b_1 + b_2)$

Therefore, the area of a trapezoidal region with altitude h and bases b_1 and b_2 is given by the formula

$$A = \frac{1}{2}h(b_1 + b_2)$$

Example: Find the area of a trapezoidal region with altitude 7.2 cm and bases 5.3 cm and 4.7 cm. Round off to ones.

Solution: $A = \frac{1}{2}h(b_1 + b_2)$

$A = \frac{1}{2}(7.2)(5.3 + 4.7)$

$A = (3.6)(10.0)$

$A = 36.00$

Therefore, the area is 36 cm².

Q13 **a.** Find the area of the trapezoidal region shown. Round off to thousandths.

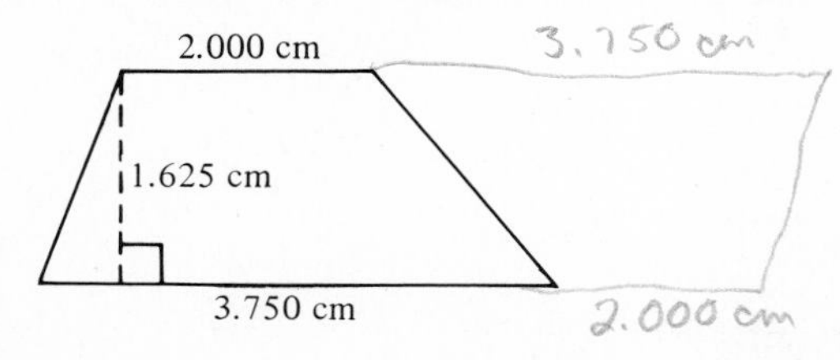

b. Find the area of a trapezoidal region if $b_1 = 7.8$ m, $b_2 = 4.3$ m and $h = 5.0$ m. Round off to ones.

\# \# \# • \# \# \# • \# \# \# • \# \# \# • \# \# \# • \# \# \# • \# \# \# • \# \# \# • \# \# \#

A13 **a.** 4.672 cm^2: $A = \frac{1}{2}h(b_1 + b_2)$

$A = \frac{1}{2}(1.625)(2.000 + 3.750)$

$A = 4.671\ 875$

b. 30 m^2: 30.25

Q14 Find the area of a trapezoidal region with altitude 20.0 metres and bases 19.7 and 14.3 metres. Round off to ones.

\# \# \# • \# \# \# • \# \# \# • \# \# \# • \# \# \# • \# \# \# • \# \# \# • \# \# \# • \# \# \#

A14 340 m^2: $A = \frac{1}{2}h(b_1 + b_2)$

$A = \frac{1}{2}(20.0)(19.7 + 14.3)$

$A = 340.0$

Q15 Find the cross sectional area of the V-block.

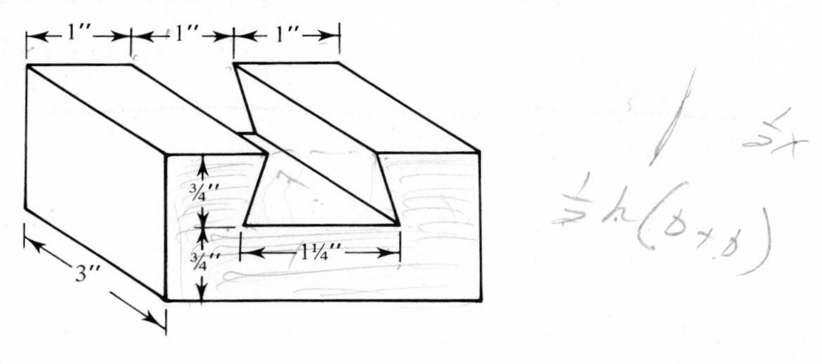

\# \# \# • \# \# \# • \# \# \# • \# \# \# • \# \# \# • \# \# \# • \# \# \# • \# \# \# • \# \# \#

A15 $3\frac{21}{32}$ in.^{2}: $A = (3)\left(1\frac{1}{2}\right) - \left(\frac{1}{2}\right)\left(\frac{3}{4}\right)\left(1\frac{1}{4} + 1\right)$

---

7 The area of a circular region is found with a formula that contains the constant π. The values of 3.14 or $3\frac{1}{7}$ may be used for the problems of this section.

The area of a circular region with radius r is given by

$A = \pi r^2$

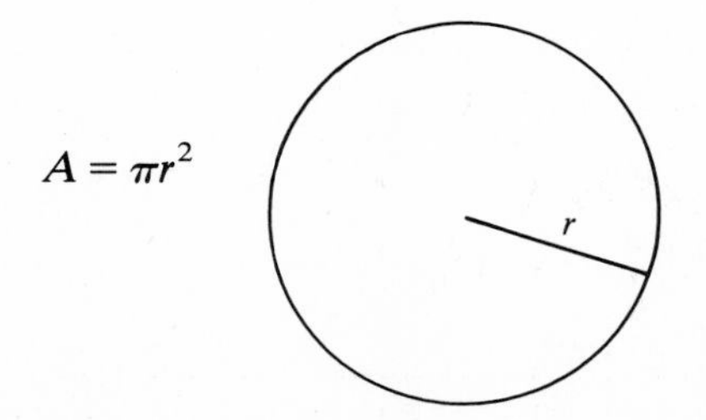

Example: If the radius of a circle is 3.5 cm, find the area of the region within it. Round off to ones.

Solution: $A = \pi r^2$
$A = (3.14)(3.5)^2$
$A = 38.4650$

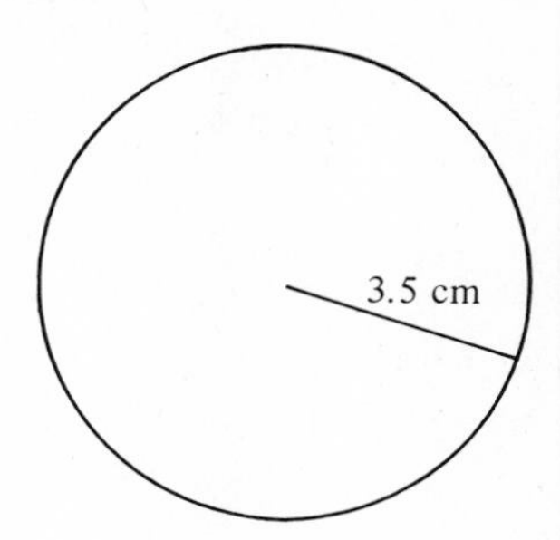

Therefore, the area is 38 cm^2.

Q16 Find the area of a circular region with radius 2.5 feet (use $\pi \doteq 3.14$ when you are given a measurement in decimal form). Round off to ones.

\# \# \# • \# \# \# • \# \# \# • \# \# \# • \# \# \# • \# \# \# • \# \# \# • \# \# \# • \# \# \#

A16 20 ft^2: $A = \pi r^2$
$A = (3.14)(2.5)(2.5)$
$A = 19.6250$

Q17 Find the area of the following region. (*Hint*: To find the area of a semicircular region, take $\frac{1}{2}$ the area of a circular region of the same radius.) Round off to a whole number of square feet.

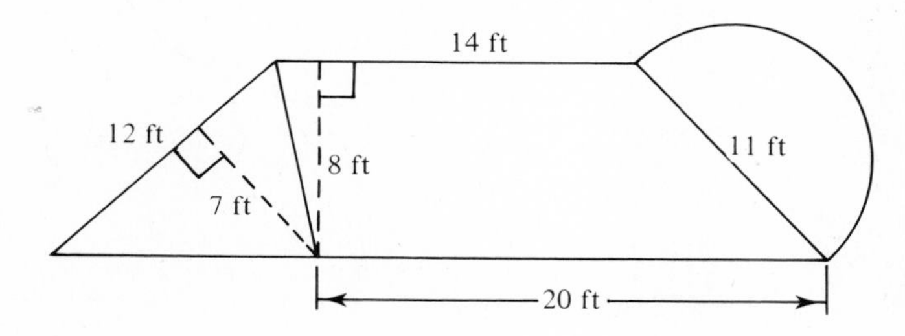

\# \# \# • \# \# \# • \# \# \# • \# \# \# • \# \# \# • \# \# \# • \# \# \# • \# \# \# • \# \# \#

A17 225 ft^2: $A = \frac{1}{2}bh$ $\quad A = \frac{1}{2}h(b_1 + b_2)$ $\quad A = \frac{1}{2}\pi r^2$

$A = \frac{1}{2}(12)(7.0)$ $\quad A = \frac{1}{2}(8.0)(20 + 14)$ $\quad A = \frac{1}{2}(3.14)(5.5)^2$

$A = 42$ $\quad A = 136$ $\quad A = 47.4925$

$A = 47$

total area $= 42 + 136 + 47 = 225$ ft^2

Q18 Measure the following geometric figures and determine their area in in.^{2}. Measure to the nearest $\frac{1}{8}$ inch.

a.

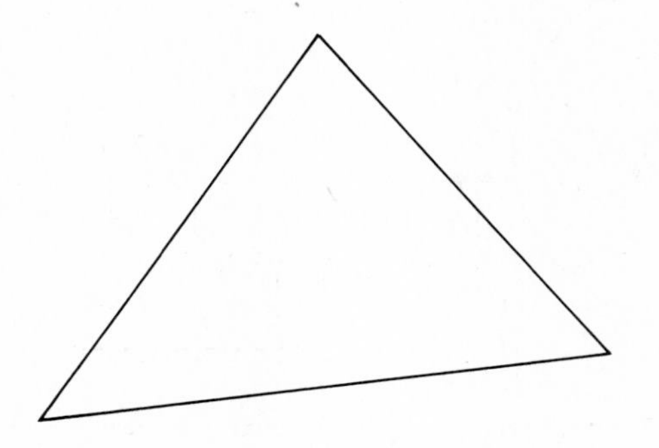

b.

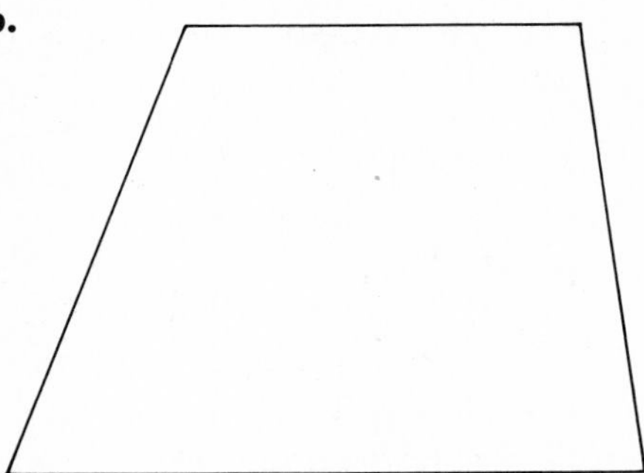

c.

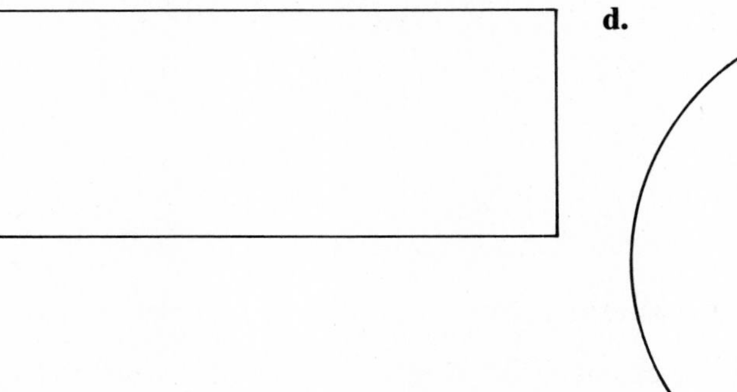

d.

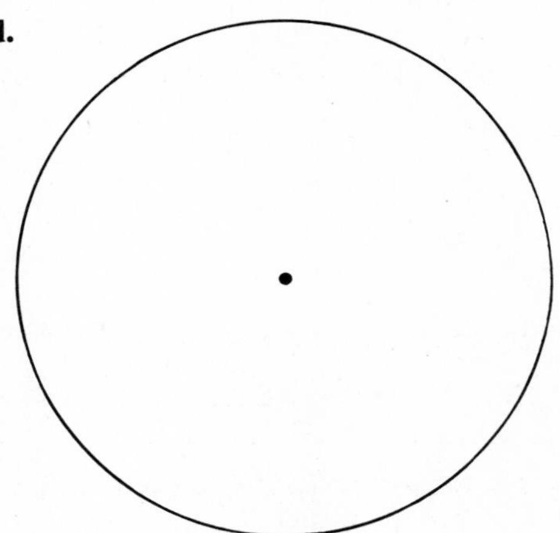

\# \# \# • \# \# \# • \# \# \# • \# \# \# • \# \# \# • \# \# \# • \# \# \# • \# \# \# • \# \# \#

A18 **a.** $1\frac{35}{64}$ in.^{2}: $A = \frac{1}{2}bh$

$$A = \frac{1}{2}\left(2\frac{1}{4}\right)\left(1\frac{3}{8}\right)$$

$$A = 1\frac{35}{64}$$

b. $2\frac{121}{128}$ in.^{2}: $A = \frac{1}{2}h(b_1 + b_2)$

$$A = \frac{1}{2}\left(1\frac{5}{8}\right)\left(2\frac{1}{4} + 1\frac{3}{8}\right)$$

$$A = 2\frac{121}{128}$$

c. $2\frac{1}{2}$ in.^{2}: $A = lw$

$$A = \left(2\frac{1}{2}\right)(1)$$

$$A = 2\frac{1}{2}$$

d. $3\frac{219}{224}$ in.^{2}: $A = \pi r^2$

$$A = 3\frac{1}{7}\left(1\frac{1}{8}\right)^2$$

$$A = 3\frac{219}{224}$$

Q19 Measure the geometric figures in Q18 with a centimetre ruler to the nearest 0.1 cm and compute the area. Round off your answer to part a to tenths and your answers to parts b, c and d to ones.

\# \# \# • \# \# \# • \# \# \# • \# \# \# • \# \# \# • \# \# \# • \# \# \# • \# \# \# • \# \# \#

A19 **a.** 10.0 cm^2: $A = \frac{1}{2}bh$

$A = \frac{1}{2}(5.7)(3.5)$

$A = 9.975$

b. 19 cm^2: $A = \frac{1}{2}h(b_1 + b_2)$

$A = \frac{1}{2}(4.1)(5.7 + 3.5)$

$A = 18.86$

c. 17 cm^2: $A = lw$

$A = (6.4)(2.6)$

$A = 16.64$

d. 26 cm^2: $A = \pi r^2$

$A = (3.14)(2.9)^2$

$A = 26.4074$

8 This section will be concluded with a summary of the formulas that are used to compute area.

Geometric regions	Formula for area
Triangle	$A = \frac{1}{2}bh$
Square	$A = s^2$
Rectangle	$A = lw$
Parallelogram	$A = bh$
Trapezoid	$A = \frac{1}{2}h(b_1 + b_2)$
Circle	$A = \pi r^2$

This completes the instruction for this section.

5.4 EXERCISE

1. Find the area of a rectangular region with length 25.7 feet and width 9.6 feet. Round off to tens.
2. Find the area of a square region with side 14.8 cm. Round off to ones.
3. Find the area of the region within a parallelogram with base 120 feet and altitude 82 feet. Round off to hundreds.
4. Find the area of a triangular region with base 79 metres and altitude 84 metres. Round off to hundreds.
5. Find the area of a trapezoidal region with altitude 1.42 cm and bases 0.36 cm and 0.48 cm. Round off to hundredths.
6. Find the area of a circular region with radius 1.8 cm. Round off to ones.
7. Find the areas of the following regions. Round off parts a and f to tens and parts b, c, d, and e to ones.

a.

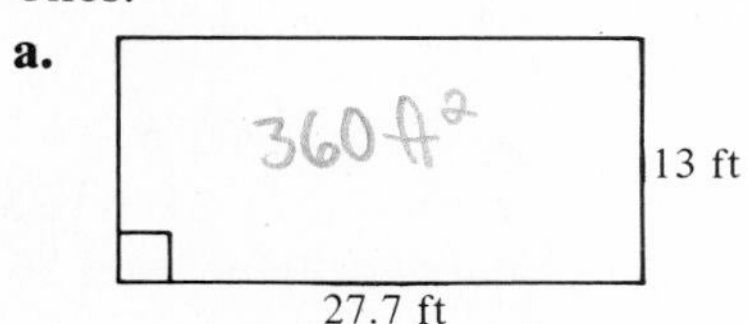

27.7 ft

b.

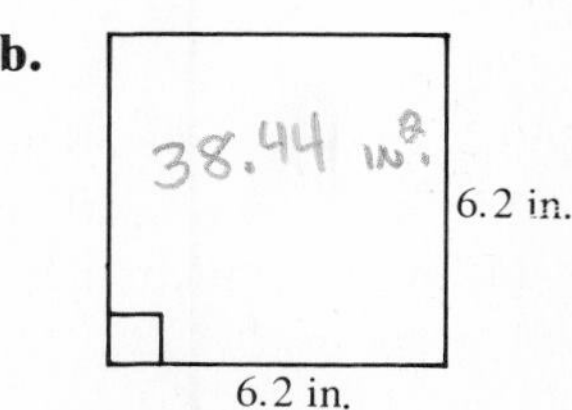

c.

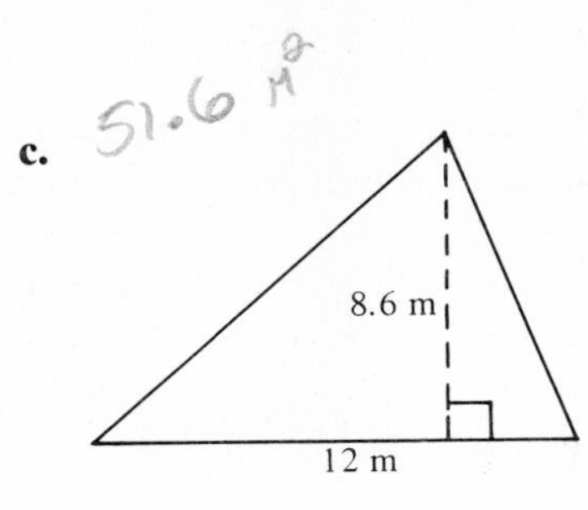

d.

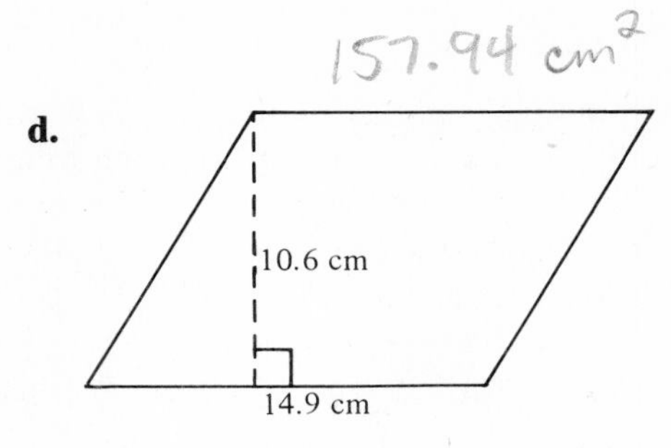

e.

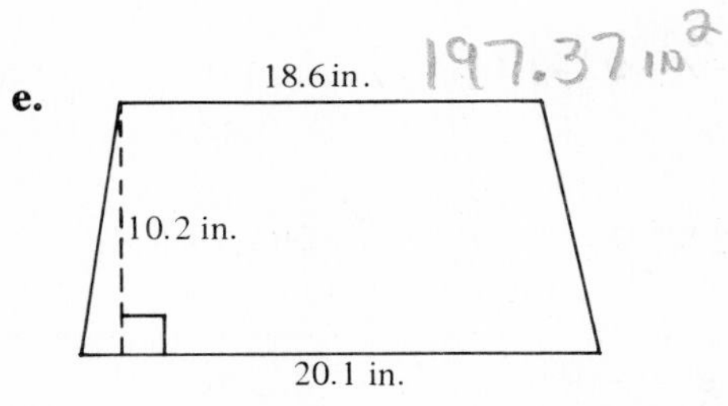

f.

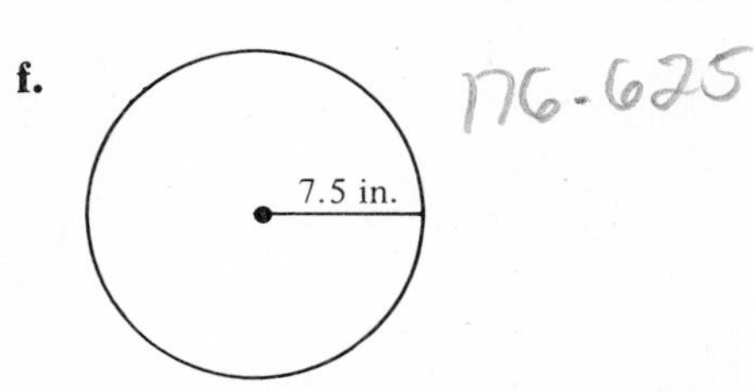

8. Find the area of the region. Round off to the nearest whole number.

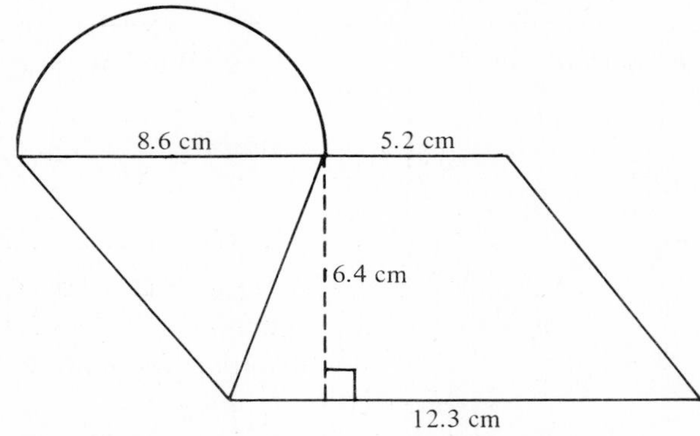

9. Match the geometric figure with the formula for the area of that figure:

a. Square	(1) $A = bh$
b. Rectangle	(2) $A = \frac{1}{2}bh$
c. Trapezoid	(3) $A = s^2$
d. Circle	(4) $A = \frac{1}{2}h(b_1 + b_2)$
e. Triangle	(5) $A = lw$
f. Parallelogram	(6) $A = \pi r^2$

5.4 EXERCISE ANSWERS

1. 250 ft^2: 246.72
2. 219 cm^2: 219.04
3. 9,800 ft^2: 9,840
4. 3300 m^2: 3318 (Note: Commas are not used in the metric system.)
5. 0.60 cm^2: 0.5964
6. 10 cm^2: 10.1736
7. **a.** 360 ft^2: 360.1 **b.** 38 in.^{2}: 38.44 **c.** 52 m^2: 51.6
d. 158 cm^2: 157.94 **e.** 197 in.^{2}: 197.37 **f.** 180 in.^{2}: 176.625
8. 113 cm^2: 29 cm^2 in the semicircular region, 28 cm^2 in the triangular region, and 56 cm^2 in the region within the trapezoid.
9. **a.** 3 **b.** 5 **c.** 4 **d.** 6 **e.** 2 **f.** 1

5.5 THE PYTHAGOREAN THEOREM

1 A relationship among the three sides of any right triangle was known to man far back into history. The relationship was used to lay out land after the floods in Egypt and continues to be useful for many purposes. The relationship is named for the ancient Greek mathematician Pythagoras.

First consider some vocabulary for right triangles.

A right triangle is a triangle with one right (90°) angle. In the right triangle shown here, side c is opposite $\measuredangle C$. Side b is opposite $\measuredangle B$. Side a is opposite $\measuredangle A$. Side c, the side opposite the right angle, is the *hypotenuse*. The other two are called the *legs* of the right triangle.

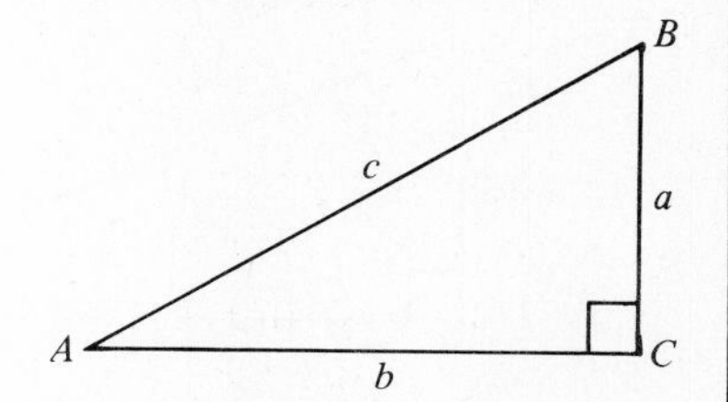

Q1 Given the right triangle shown here, identify:

a. the legs ______

b. the hypotenuse ______

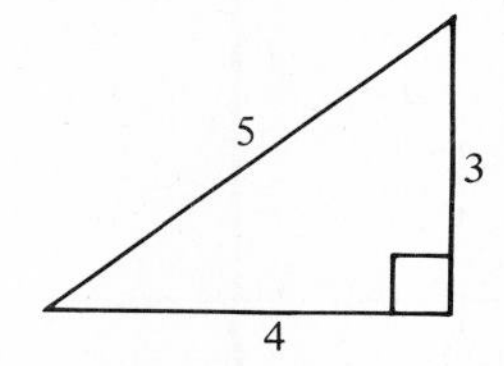

\# \# \# • \# \# \# • \# \# \# • \# \# \# • \# \# \# • \# \# \# • \# \# \# • \# \# \# • \# \# \#

A1 **a.** 3 and 4 (either order) **b.** 5

2 Consider the right triangle with legs of 3 cm and 4 cm and a hypotenuse of 5 cm. A square is constructed on each of the three sides of the right triangle. If the number of square centimetres in the square regions constructed on the legs are added, they exactly equal the number of square

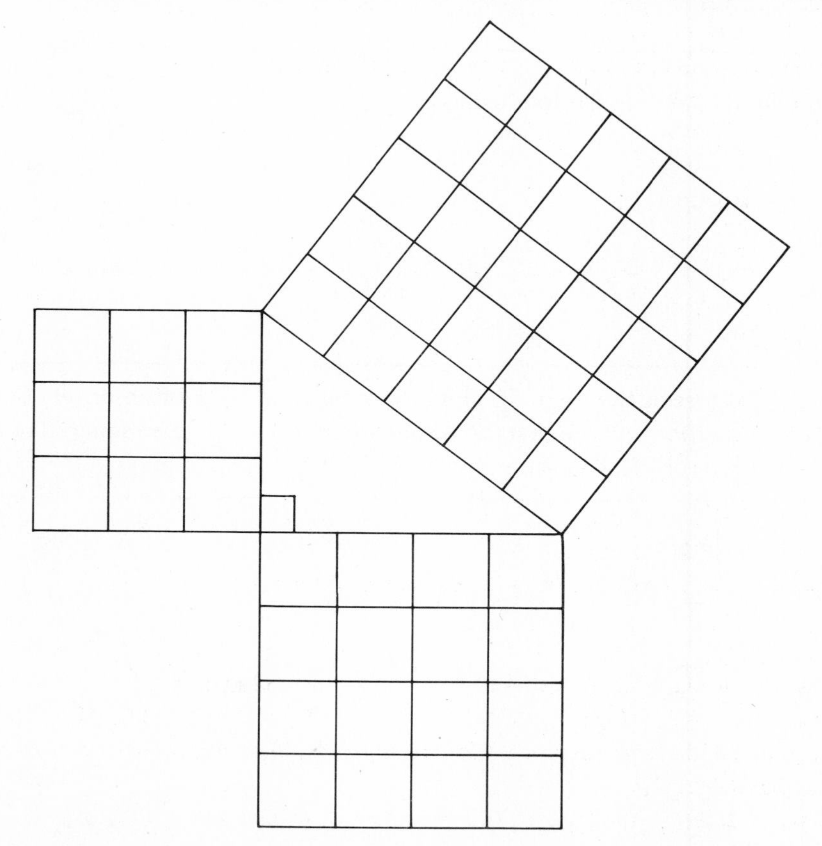

centimeters in the square region on the hypotenuse. The relationship can be stated mathematically as $3^2+4^2=5^2$. This triangle is an example of the *Pythagorean theorem,* which could be stated: *The sum of the squares of the legs of a right triangle is equal to the square of the hypotenuse.*

Q2 Given the right triangle shown, verify that the Pythagorean relationship holds.

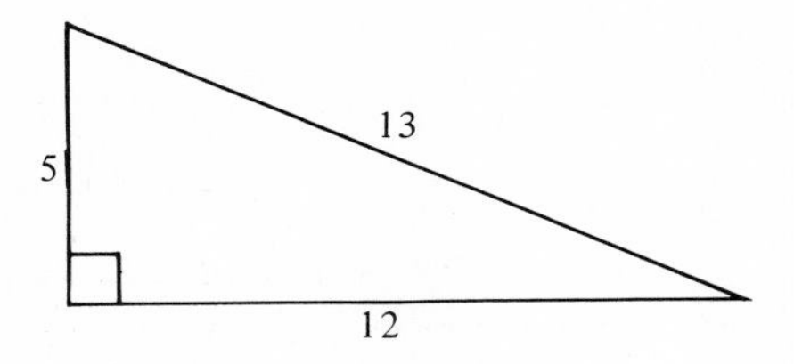

\# \# \# • \# \# \# • \# \# \# • \# \# \# • \# \# \# • \# \# \# • \# \# \# • \# \# \# • \# \# \#

A2 $5^2+12^2=13^2$
$25+144=169$

3 The Pythagorean theorem is true for any right triangle. In general, given any right triangle with legs a and b and hypotenuse c,

$a^2+b^2=c^2$

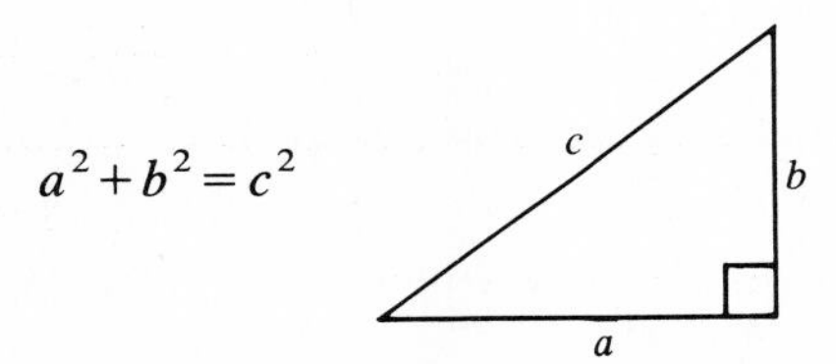

Q3 Given the triangle at the right, complete the following: $r^2=$ ________

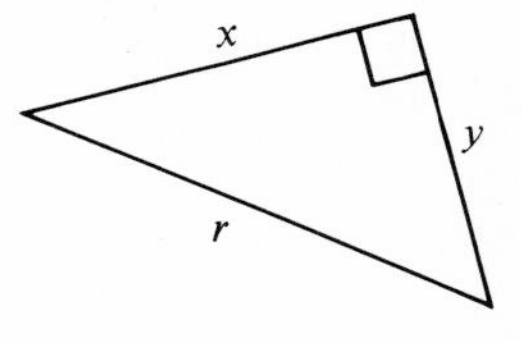

\# \# \# • \# \# \# • \# \# \# • \# \# \# • \# \# \# • \# \# \# • \# \# \# • \# \# \# • \# \# \#

A3 x^2+y^2 or y^2+x^2

4 When working with the Pythagorean theorem, an equation such as $x^2=25$ often arises. One can examine the meaning of the equation and see that the whole number that could be squared to get 25 is 5. Five is called the *square root* of 25. The square root of 25 is written $\sqrt{25}$.

Q4 Write the answers to the following:

a. The square root of 81 is ______.

b. $\sqrt{9}=$ ______.

c. The square root of 100 is ______.

d. $\sqrt{\frac{1}{4}}=$ ______.

\# \# \# • \# \# \# • \# \# \# • \# \# \# • \# \# \# • \# \# \# • \# \# \# • \# \# \# • \# \# \#

A4 **a.** 9 **b.** 3 **c.** 10 **d.** $\frac{1}{2}$

Q5 Answer the following:

a. Write the square root of 36 in two ways. ________

b. The square root of 53 is written ________.

c. $\sqrt{17}$ means ________________.

\# \# \# • \# \# \# • \# \# \# • \# \# \# • \# \# \# • \# \# \# • \# \# \# • \# \# \# • \# \# \#

A5 **a.** $\sqrt{36}$, 6 **b.** $\sqrt{53}$ **c.** the square root of 17

5 Determine the value of the hypotenuse of the right triangle shown.

$$a^2+b^2=c^2$$
$$9+16=c^2$$
$$25=c^2$$
$$\sqrt{25}=c$$
$$5=c$$

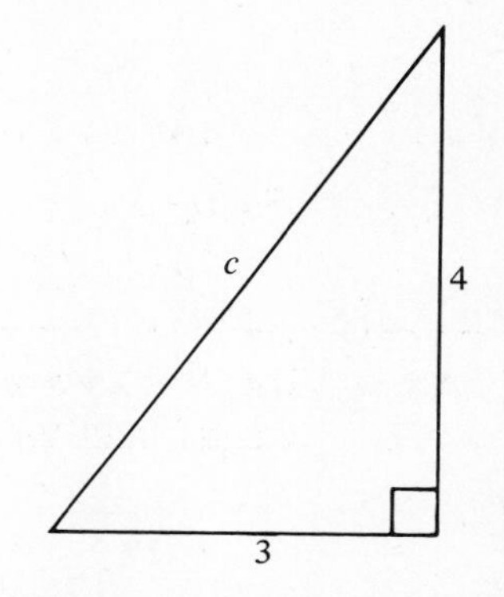

Q6 Determine the value of the hypotenuse of the right triangle shown.

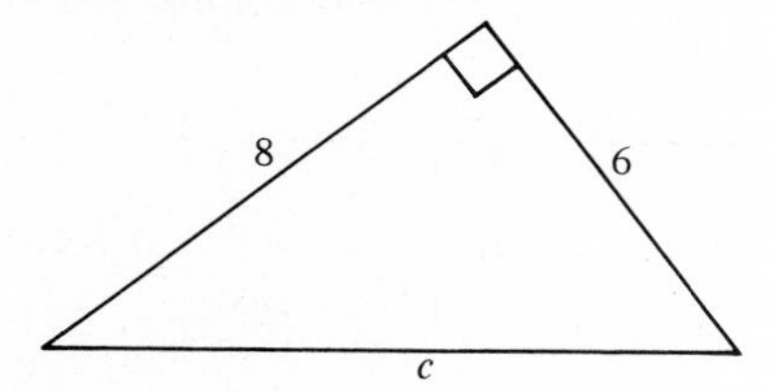

\# \# \# • \# \# \# • \# \# \# • \# \# \# • \# \# \# • \# \# \# • \# \# \# • \# \# \# • \# \# \#

A6 10: $8^2+6^2=c^2$

$$64+36=c^2$$
$$100=c^2$$
$$\sqrt{100}=c$$
$$10=c$$

6 The symbol $\sqrt{\ }$ is called a *radical sign*. When the number under the radical sign is so large that the square root is not recognized, a table can be used to find the square root. See Table IX in the Appendix. The numbers in the column with heading "Square" are called *perfect squares*. They are obtained by squaring the whole numbers in the column with heading "Number." The first few perfect squares are 1, 4, 9, 16, 25, ...

Example: $\sqrt{1,764}=?$

Solution: Find a number in the table which when squared gives 1,764. The number is 42. Therefore, $\sqrt{1,764}=42$.

Q7 Use the table to find the following square roots:

a. $\sqrt{361} =$ _____ **b.** $\sqrt{1{,}225} =$ _____ **c.** $\sqrt{3{,}969} =$ _____ **d.** $\sqrt{121} =$ _____

• # # # • # # # • # # # • # # # • # # # • # # # • # # # • # #

A7 **a.** 19 **b.** 35 **c.** 63 **d.** 11

Q8 Determine the value of the hypotenuse of the right triangle shown.

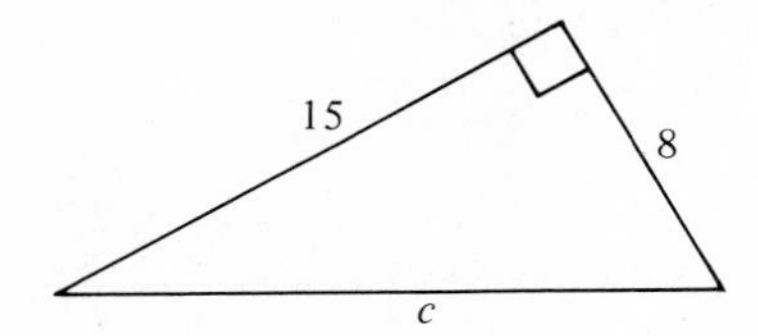

• # # # • # # # • # # # • # # # • # # # • # # # • # # # • # #

A8 17: $8^2 + 15^2 = c^2$

$$64 + 225 = c^2$$
$$289 = c^2$$
$$\sqrt{289} = c$$
$$17 = c$$

7 A different form of the Pythagorean theorem is used when you know the hypotenuse and one leg, and are finding the length of the other leg. The equations are:

$a^2 = c^2 - b^2$

or

$b^2 = c^2 - a^2$

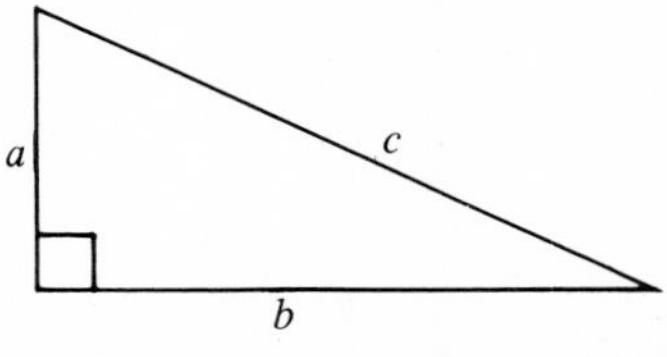

The following two examples illustrate the use of the Pythagorean theorem when solving for the value of a leg given the other leg and the hypotenuse.

Example 1: Find a.

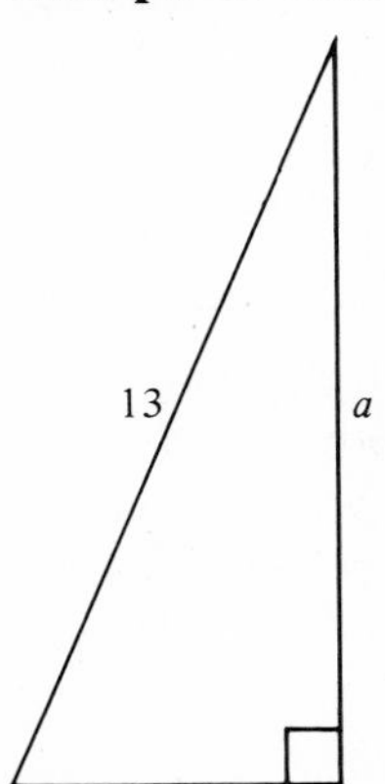

Solution: $a^2 = c^2 - b^2$

$$a^2 = 13^2 - 5^2$$
$$a^2 = 169 - 25$$
$$a^2 = 144$$
$$a = \sqrt{144}*$$
$$a = 12$$

Example 2: Find b.

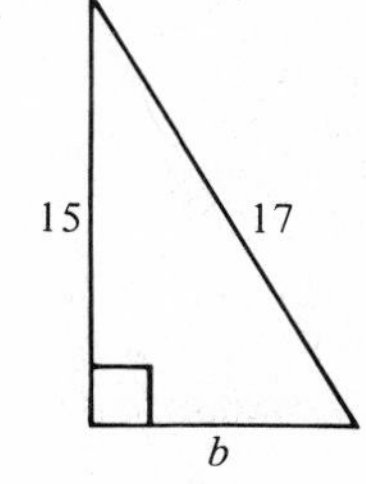

Solution: $b^2 = c^2 - a^2$

$$b^2 = 17^2 - 15^2$$
$$b^2 = 289 - 225$$
$$b^2 = 64$$
$$b = 8$$

*This step is usually done mentally when the number is a perfect square.

Q9 Use the technique of Frame 7 to determine a in the right triangle shown.

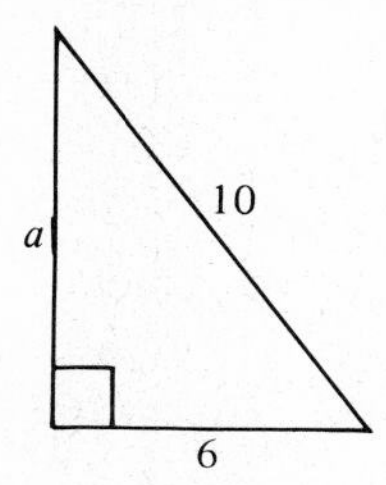

\# \# \# • \# \# \# • \# \# \# • \# \# \# • \# \# \# • \# \# \# • \# \# \# • \# \# \# • \# \# \#

A9 8: $a^2 = 10^2 - 6^2$

$a^2 = 100 - 36$

$a^2 = 64$

$a = 8$

---

8 If there is no whole number which, when squared, is equal to the number under the radical sign, the table can still be used to approximate the square root.

Example: $\sqrt{24} = ?$

Solution: Since no whole number squared is 24, look for 24 in the "Number" column and find a decimal approximation of $\sqrt{24}$ in the "Square Root" column. Therefore, $\sqrt{24} \doteq 4.899$.

The positive square root of 24 is approximately equal to 4.899 rounded off to three decimal places.

Q10 Find the following positive square roots to three decimal places:

a. $\sqrt{5} \doteq$ _____ **b.** $\sqrt{68} \doteq$ _____ **c.** $\sqrt{64} =$ _____ **d.** $\sqrt{14} \doteq$ _____

\# \# \# • \# \# \# • \# \# \# • \# \# \# • \# \# \# • \# \# \# • \# \# \# • \# \# \# • \# \# \#

A10 **a.** 2.236 **b.** 8.246 **c.** 8.000 **d.** 3.742

Q11 Find c in the figure shown.

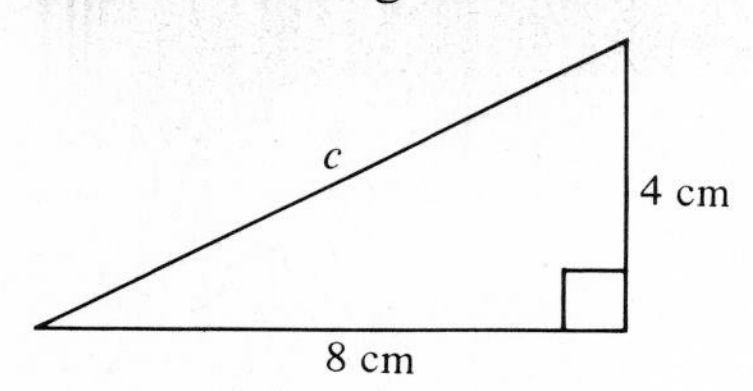

\# \# \# • \# \# \# • \# \# \# • \# \# \# • \# \# \# • \# \# \# • \# \# \# • \# \# \# • \# \# \#

A11 8.944 cm: $8^2 + 4^2 = c^2$

$64 + 16 = c^2$

$80 = c^2$

$\sqrt{80} = c$

$8.944 \doteq c$

Q12 Find a in the figure shown.

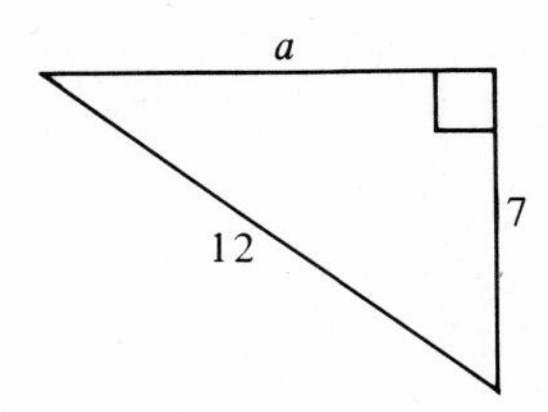

\# \# \# • \# \# \# • \# \# \# • \# \# \# • \# \# \# • \# \# \# • \# \# \# • \# \# \# • \# \# \#

A12 9.747: $a^2 = 12^2 - 7^2$
$a^2 = 144 - 49$
$a^2 = 95$
$a = \sqrt{95}$
$a \doteq 9.747$

Q13 An antenna is to be erected and steadied with guy wires attached 80 feet above the ground and 60 feet away from the center of the base of the antenna. How long will each guy wire be?

\# \# \# • \# \# \# • \# \# \# • \# \# \# • \# \# \# • \# \# \# • \# \# \# • \# \# \# • \# \# \#

A13 100 ft: $c^2 = 80^2 + 60^2$
$c^2 = 6{,}400 + 3{,}600$
$c^2 = 10{,}000$
$c = 100$

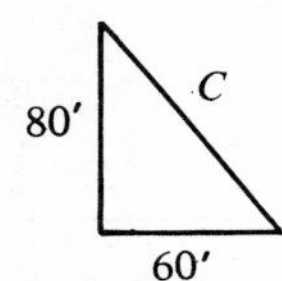

Q14 Find the area of the triangle shown. (*Hint*: First find the altitude h.) Round off to ones.

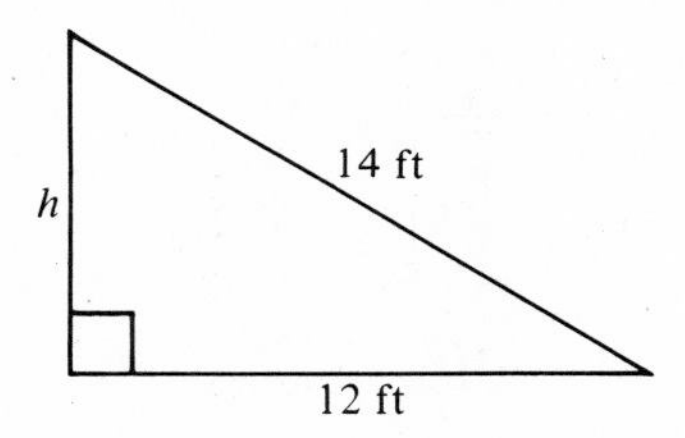

\# \# \# • \# \# \# • \# \# \# • \# \# \# • \# \# \# • \# \# \# • \# \# \# • \# \# \# • \# \# \#

A14 43 ft^2: $h^2 = 14^2 - 12^2$
$h^2 = 196 - 144$
$h^2 = 52$
$h = \sqrt{52}$
$h \doteq 7.211$

$A = \frac{1}{2}hb$

$A = \frac{1}{2}(7.211)(12)$

$A = 43.266$

Q15 Find the area of the triangle shown. (*Hint*: First find altitude h.)

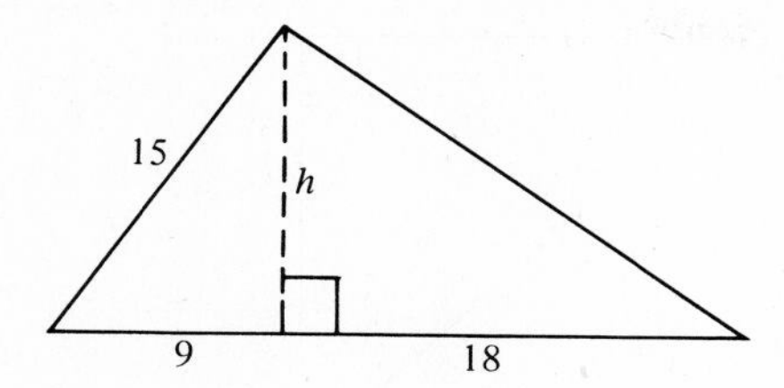

\# \# \# • \# \# \# • \# \# \# • \# \# \# • \# \# \# • \# \# \# • \# \# \# • \# \# \# • \# \# \#

A15 162 ft^2:

$$h^2+9^2=15^2$$
$$h^2+81=225$$
$$h^2=144$$
$$h^2=\sqrt{144}$$
$$h=12$$

$$A=\frac{1}{2}bh$$
$$A=\frac{1}{2}(27)(12)$$
$$A=162$$

This completes the instruction for this section.

5.5 EXERCISE

Find the unknown side on each of the following triangles:

1.

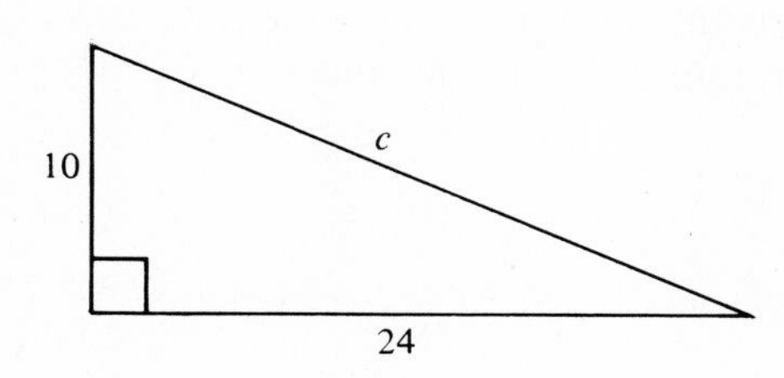

2.

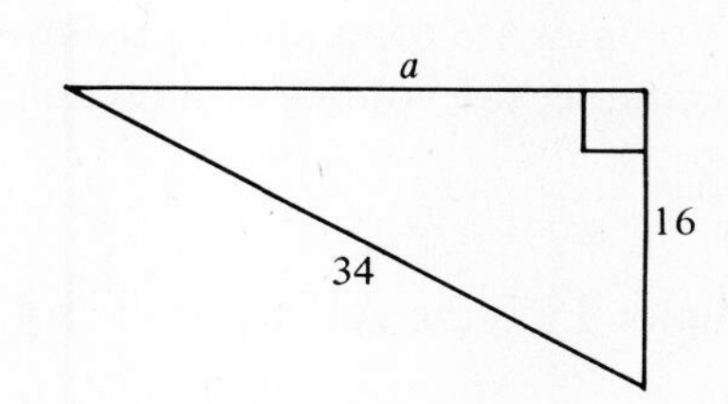

3.

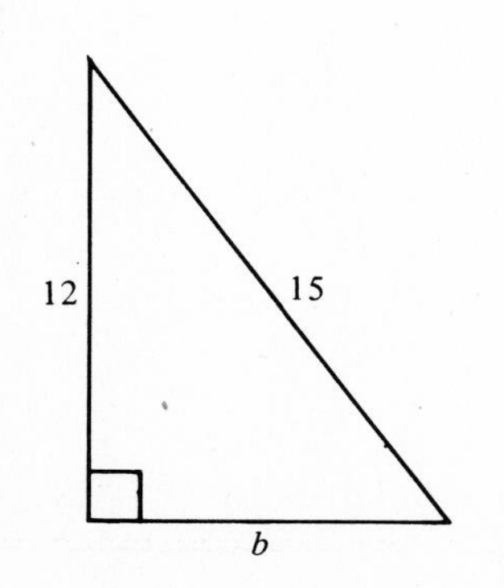

4.

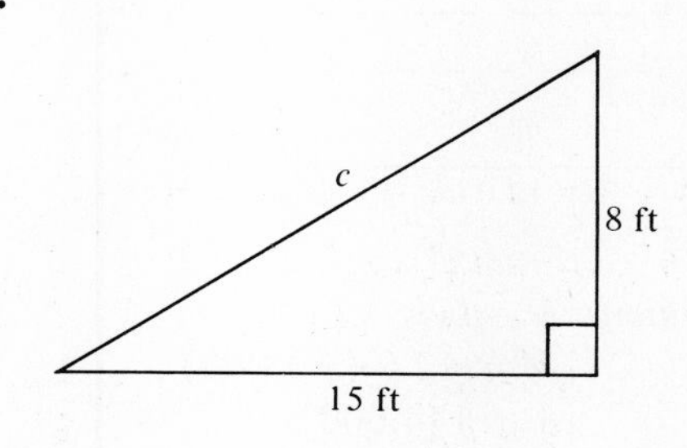

5.

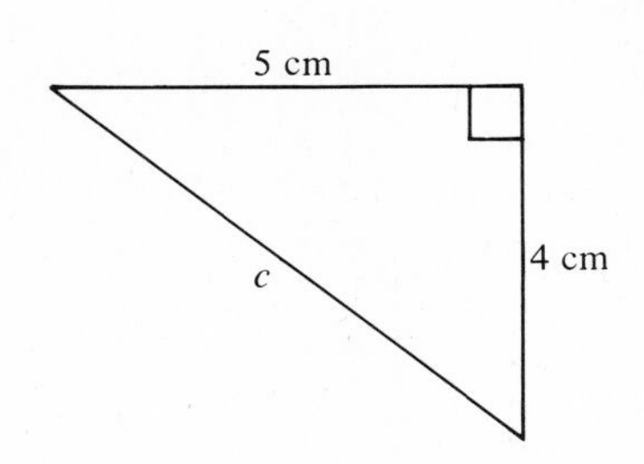

6.

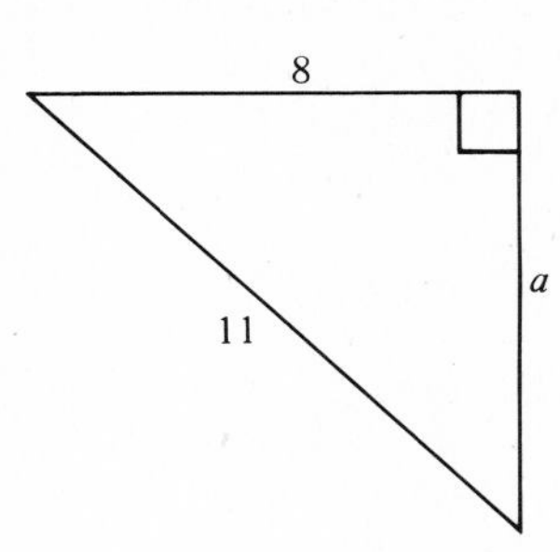

7. Find the area of the triangle. Round off to tens.

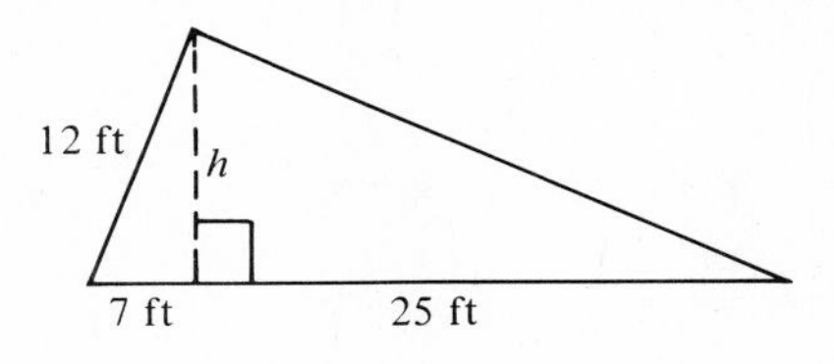

5.5 EXERCISE ANSWERS

1. 26 **2.** 30 **3.** 9 **4.** 17 ft
5. 6.403 cm **6.** 7.550 **7.** 160 ft^2: 155.952

5.6 VOLUMES

1 The *volume* of a solid object can be thought of as the amount that the object "holds." It is a measure of the enclosed space within the object. As was discussed in Section 5.2, the units of measurement for volume are cubic units. Examples are: cubic inches, cubic feet, cubic yards, cubic centimetres, and cubic metres.

Volumes are computed by using a formula to combine various linear measurements of the solid. For example, the volume of a rectangular prism is found with the formula

$$V = lwh$$

Example: Find the volume of the solid.

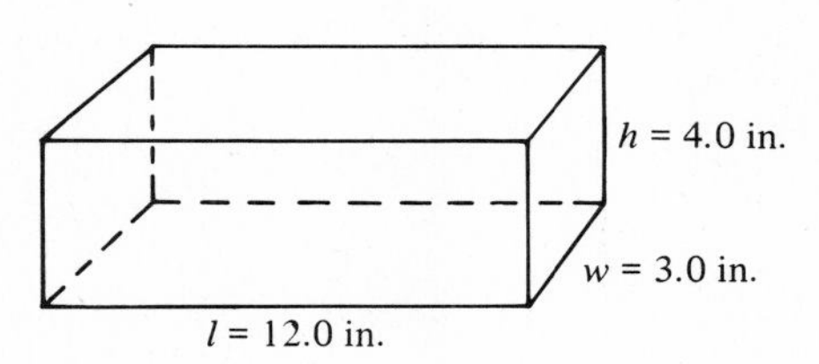

Solution: $V = lwh$
$V = (12.0)(3.0)(4.0)$
$V = 144.000$

Therefore, the volume is 144 cubic inches.

Q1 Find the volumes of the following prisms:

a.

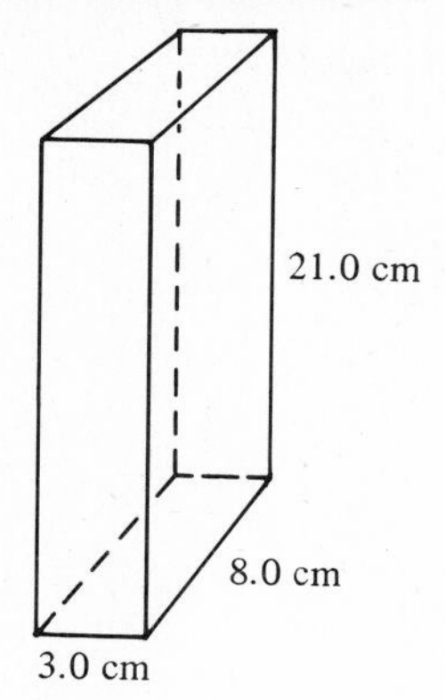

b.

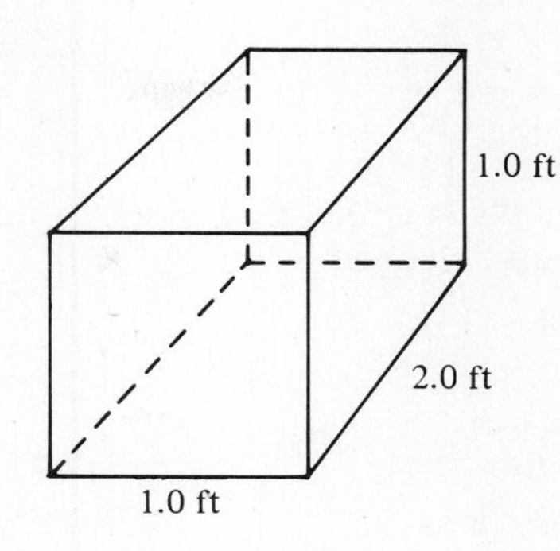

\# \# \# • \# \# \# • \# \# \# • \# \# \# • \# \# \# • \# \# \# • \# \# \# • \# \# \# • \# \# \#

A1 **a.** 504 cm^3: $V = lwh$
$V = (3.0)(8.0)(21.0)$
$V = 504.000$

b. 2 ft^3: $V = lwh$
$V = (1.0)(2.0)(1.0)$
$V = 2.000$

Q2 A quenching tank is to be built that is 2.5 ft by 3.0 ft and 2.0 ft deep.

a. How many cubic feet of coolant will it hold? Round off to ones.

b. If oil is used as a coolant and weighs 60 pounds per ft^3, how much would the contents for the tank weigh when full?

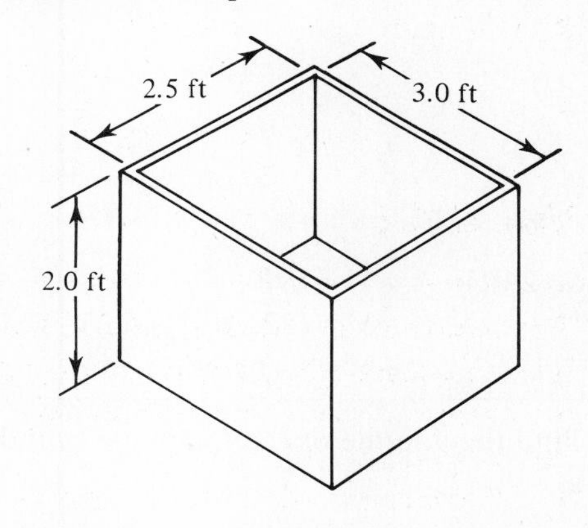

\# \# \# • \# \# \# • \# \# \# • \# \# \# • \# \# \# • \# \# \# • \# \# \# • \# \# \# • \# \# \#

A2 **a.** 15 ft^3 **b.** 900 pounds

2 The formula for the volume of a rectangular prism fits a general form that can also be used for all prisms and cylinders. The volume of each is found by multiplying the area of the base by the altitude (height perpendicular to the base):

$V = Bh$

More specific formulas follow:

Rectangular prism
$V = Bh$
$V = lwh$

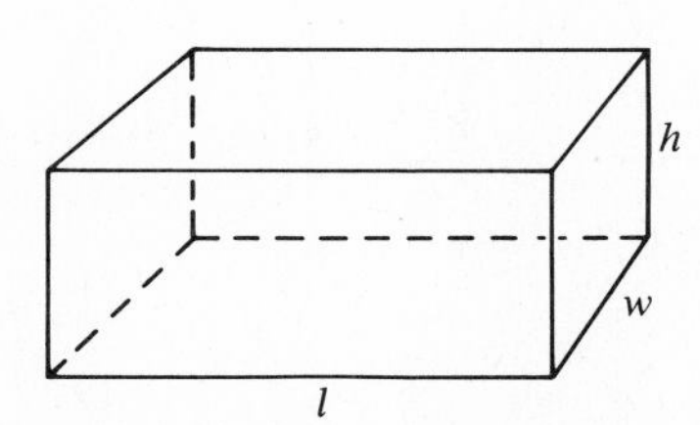

Area of the base
$B = lw$

Cube
$V = Be$
$V = e^2 e$
$V = e^3$

(Note: e^3 means $e \times e \times e$.)
e^3 is read "e cubed."

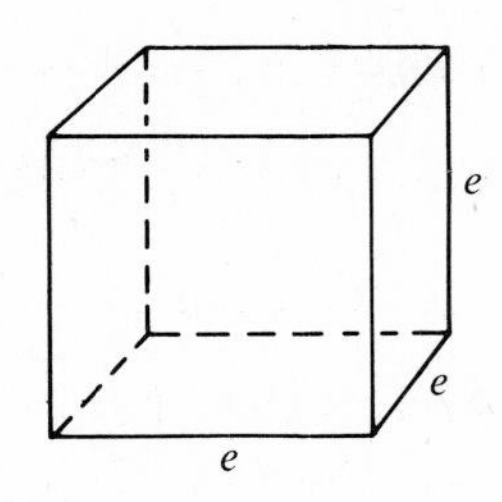

Area of the base
$B = e^2$

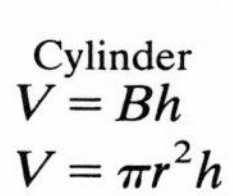

Cylinder
$V = Bh$
$V = \pi r^2 h$

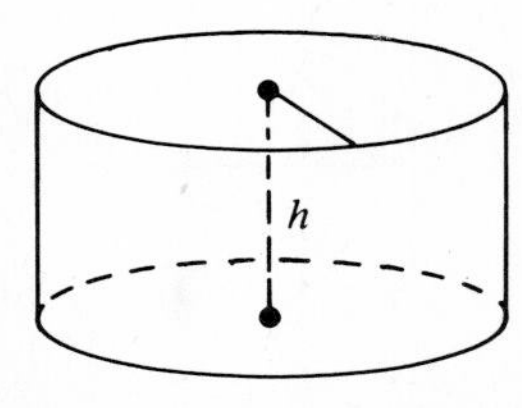

Area of the base
$B = \pi r^2$

Q3 Find the volume of each of the prisms (round answer off to nearest tens):

a.

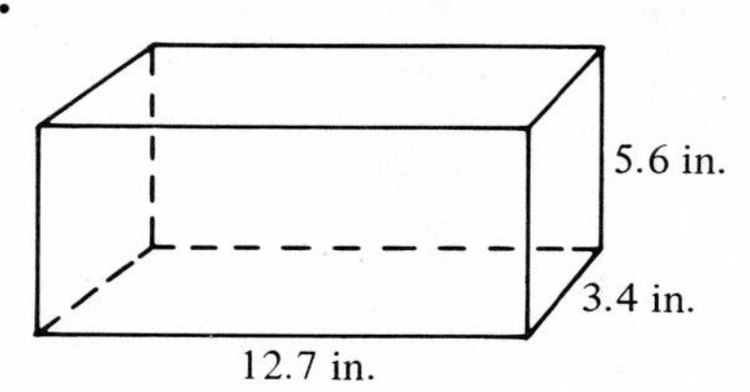

b.

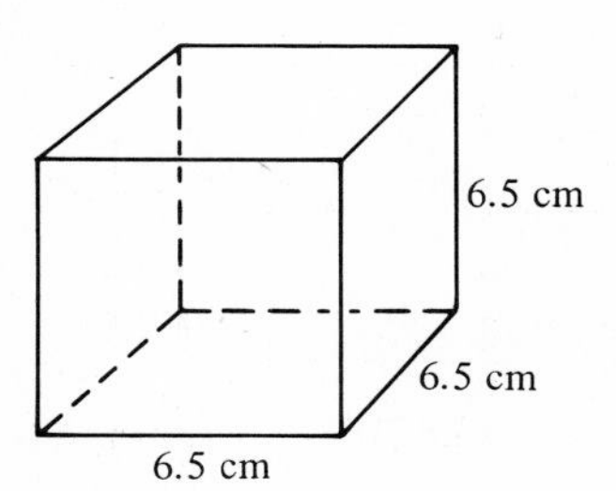

\# \# \# • \# \# \# • \# \# \# • \# \# \# • \# \# \# • \# \# \# • \# \# \# • \# \# \# • \# \# \#

A3 **a.** 240 in.^{3}: $V = lwh$
$V = (12.7)(3.4)(5.6)$
$V = 241.808$

b. 270 cm^3: $V = e^3$
$V = 6.5^3$
$V = 274.625$

Q4 Find the volume of the following cylinders (round off to whole numbers):

a.

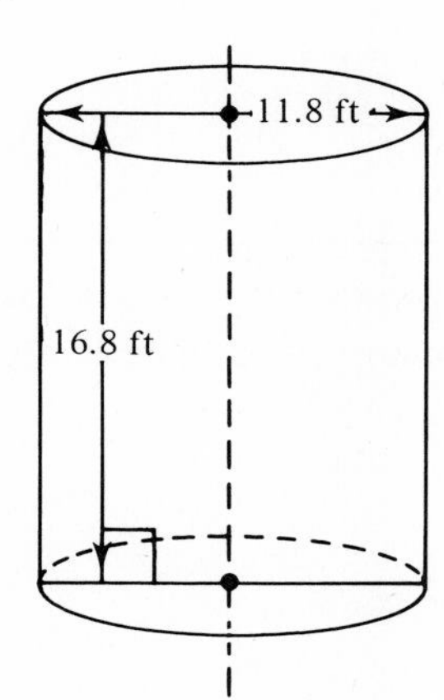

b.

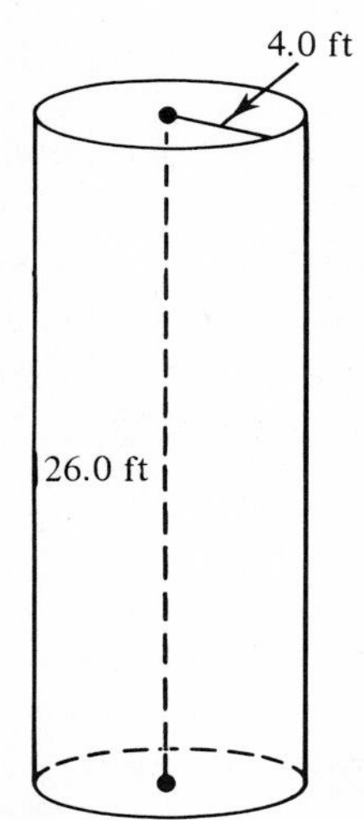

\# \# \# • \# \# \# • \# \# \# • \# \# \# • \# \# \# • \# \# \# • \# \# \# • \# \# \# • \# \# \#

A4 **a.** 1,836 ft^3: $V = \pi r^2 h = (3.14)(5.9)(5.9)(16.8) = 1{,}836.29712$

b. 1,306 ft^3: $V = \pi r^2 h = (3.14)(4.0)(4.0)(26.0) = 1{,}306.24$

3 Some prisms have bases that can be broken down into elementary shapes. These can be used to find the area of the base. The volume is then the area of the base times the height.

Example: Find the volume of the casting:

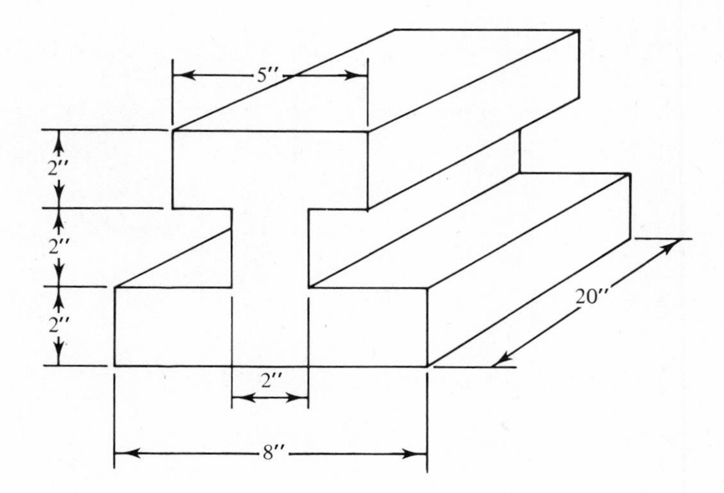

Solution: Consider the cross section to be the base. Use rectangles to find the area of the cross section.

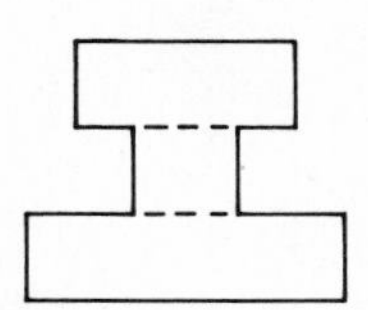

$B = (2)(5) + (2)(2) + (2)(8)$
$B = \quad 10 \quad + \quad 4 \quad + \quad 16$
$B = 30$

Hence, the area of the base is 30 $in.^2$

$V = Bh$
$V = (30)(20)$
$V = 600$

Therefore, the volume is 600 $in.^3$

Q5 Find the volume of concrete, in ft^3, needed to build a foundation with dimensions as shown at the top of the next page. (Convert all measurements to feet by dividing the number of inches by 12.)

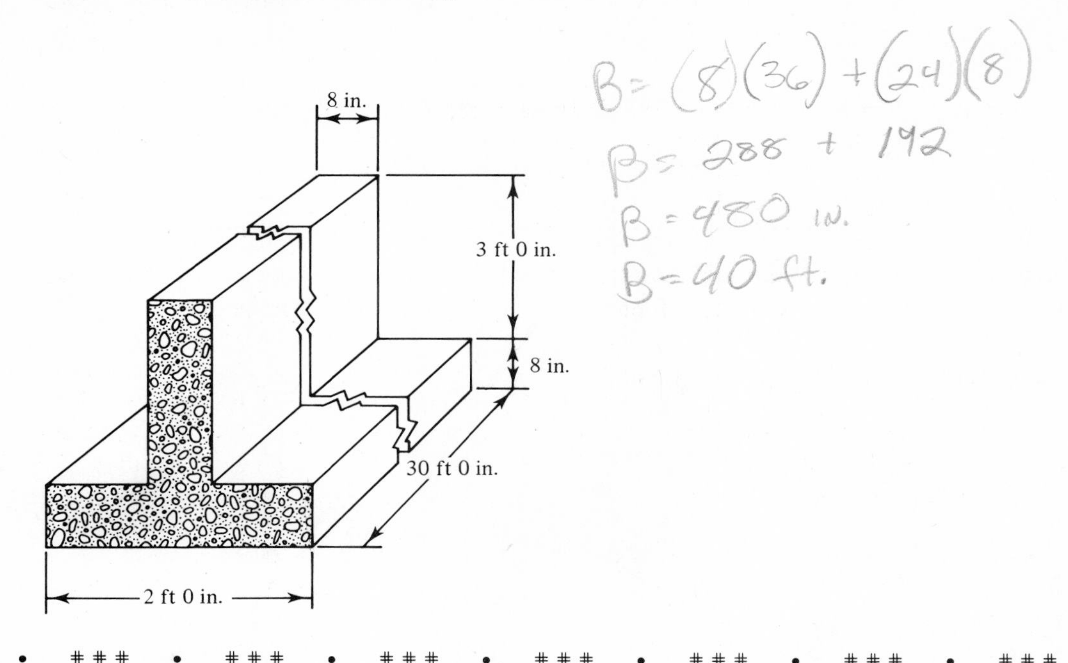

• ### • ### • ### • ### • ### • ### • ### •

A5 40 ft^3: $B = (2)\left(\frac{8}{12}\right) + (3)\left(\frac{8}{12}\right)$ $V = Bh$

$B = \frac{4}{3} + 2$ $V = \left(3\frac{1}{3}\right)(30)$

$B = 3\frac{1}{3}$ $V = 40$

Q6 Find the volume of the metal in the casting. (You may want to subtract volumes to get it.) use $\pi = \frac{22}{7}$. Round off to ones at the end of your solution.

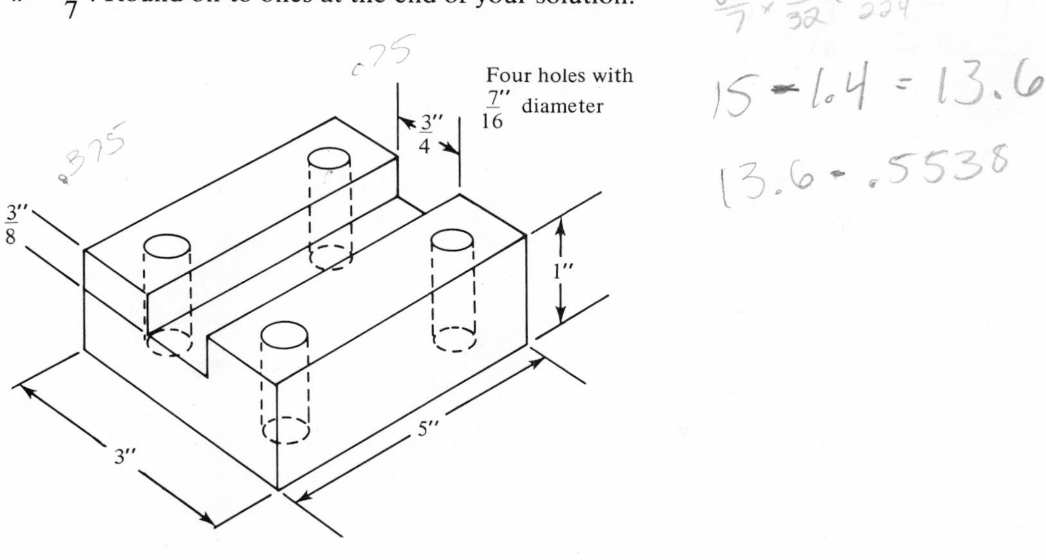

• ### • ### • ### • ### • ### • ### • ### •

A6 13 in.^{3}: Volume of the prism before subtracting the cylinders is $\frac{435}{32} = 13.59$ in.^{3}. Volume of the four cylinders is $\frac{4312}{7168} = 0.60$ in.^{3}. The final volume is $13.59 - 0.60 = 12.99$.

4 In a triangular prism the parallel faces of the prism are called bases. If the prism rests on one of these parallel faces, it is called *the* base. In the figure shown here the base of the prism is the triangular region ABC.

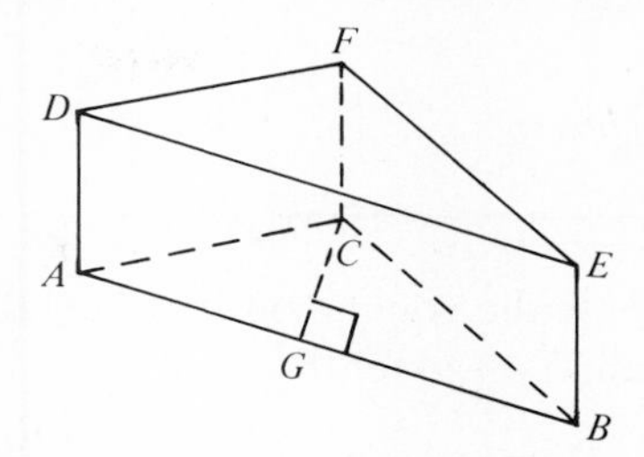

In a triangle the base of the triangle is arbitrary. Any of the sides of the triangle can be considered a base and the altitude to that base is a perpendicular segment from the opposite vertex to that base. The following three triangles have the same area. However, different numbers would be used to calculate that area, depending upon which side was chosen as the base to be placed in the formula $A = \frac{1}{2}bh$.

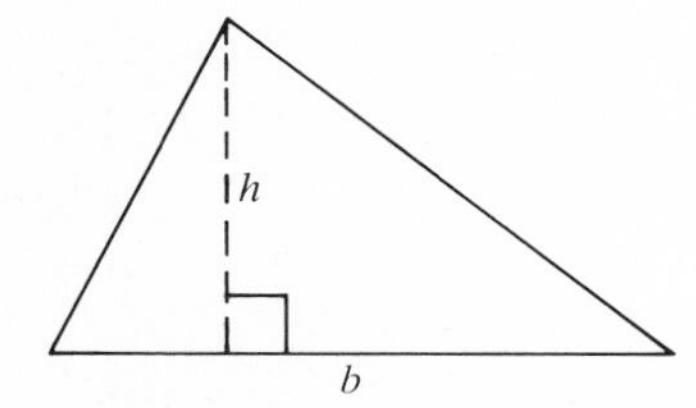

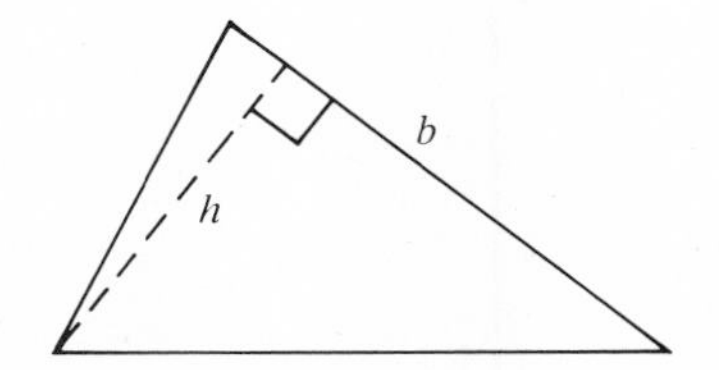

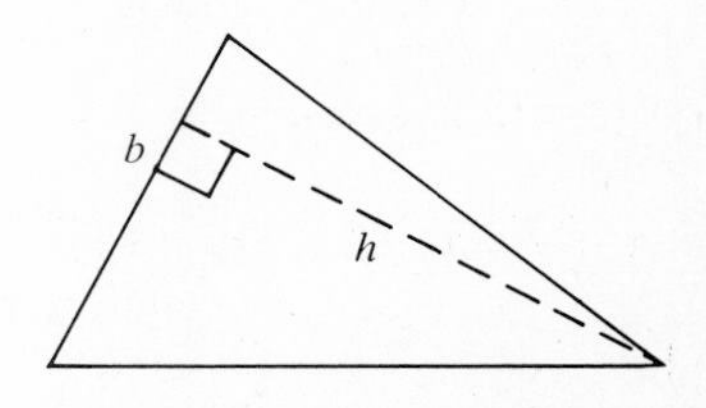

Q7 Use the triangular prism shown here to answer the following:

a. Name the bases of the prism: ABC and DEF.

b. Name *the* base of the prism ABC.

c. Name the sides of the base triangle that could be used as the base of the triangle. AB, AC, cb

d. Name three segments whose length is the altitude of the prism. AD, CF, BE

e. Name a segment that could be used as an altitude of the base triangle. CG

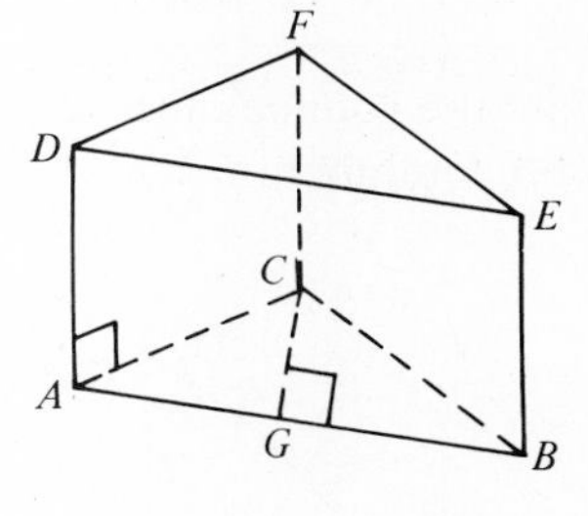

\# \# \# • \# \# \# • \# \# \# • \# \# \# • \# \# \# • \# \# \# • \# \# \# • \# \# \# • \# \# \# • \# \# \#

A7 **a.** ABC and DEF **b.** ABC **c.** $\overline{AB}$, $\overline{BC}$, or $\overline{AC}$

d. $\overline{AD}$, $\overline{BE}$, or $\overline{CF}$ **e.** $\overline{CG}$

5 The volume of a triangular prism is found by multiplying the area of the base times the height of the prism. Notice that the area of the base is found by multiplying the base of the triangular base by the altitude of the triangular base.

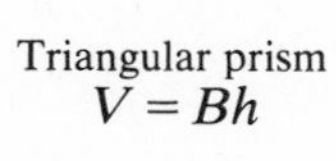
Triangular prism
$V = Bh$

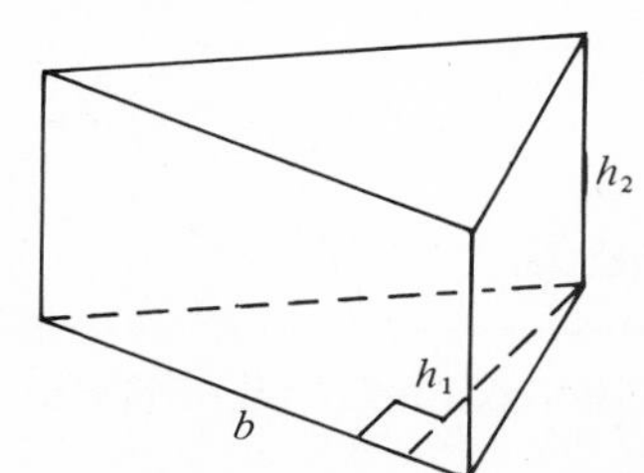

Area of the base
$B = \frac{1}{2}bh_1$

So,

$$V = \frac{1}{2}bh_1h_2$$

Q8 Compute the volume of the triangular prisms:

a. Round off to tenths.

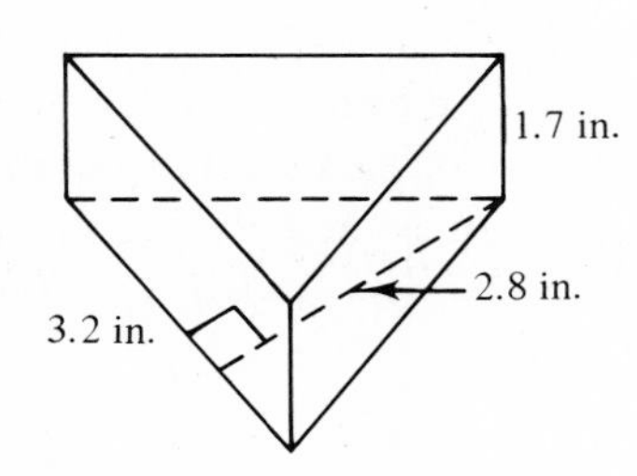

b. Round off to tens.

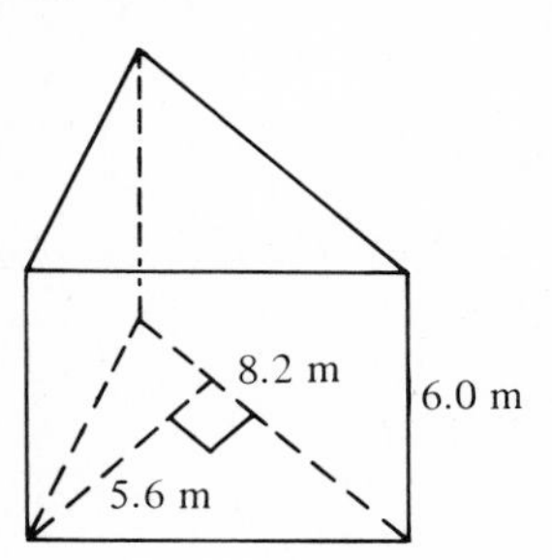

\# \# \# • \# \# \# • \# \# \# • \# \# \# • \# \# \# • \# \# \# • \# \# \# • \# \# \# • \# \# \#

A8 **a.** 7.6 in.^{3}: $V = \frac{1}{2}bh_1h_2$

$$V = \frac{1}{2}(3.2)(2.8)(1.7)$$

$$V = 7.616$$

b. 140 m^3: $V = \frac{1}{2}bh_1h_2$

$$V = \frac{1}{2}(8.2)(5.6)(6.0)$$

$$V = 137.760$$

Q9 Find the volume of the prism shown. (Notice that the prism is not resting on its base.) Round off to a whole number.

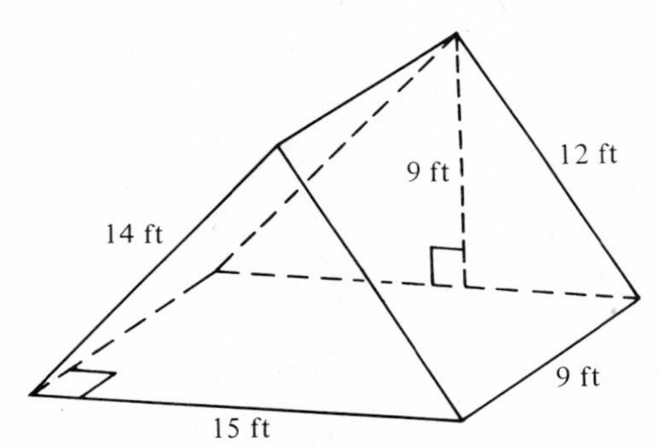

$V = ½ bh_1h_2$
$V = \frac{1}{2}(15)(9)(9)$
$V = \frac{1}{2}(1215)$
$V = 607.5$ ft^3

\# \# \# • \# \# \# • \# \# \# • \# \# \# • \# \# \# • \# \# \# • \# \# \# • \# \# \# • \# \# \#

A9 608 ft^3: $V = \frac{1}{2}bh_1h_2$

$$V = \frac{1}{2}(15)(9)(9)$$

$$V = 607.5$$

6 The volume of pyramids and cones are related to the volume of the prism or cylinder with the same base and altitude. Consider the following figures:

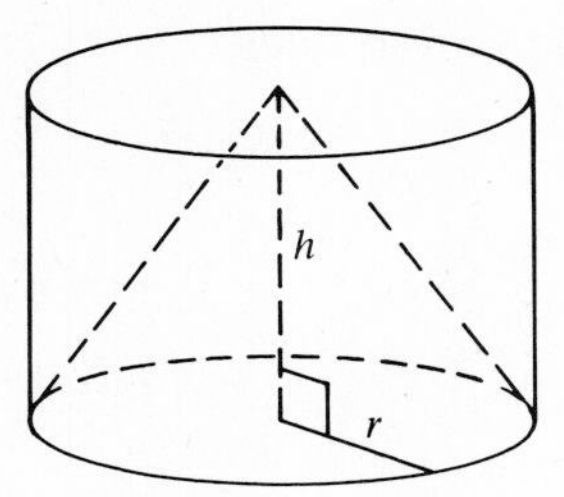

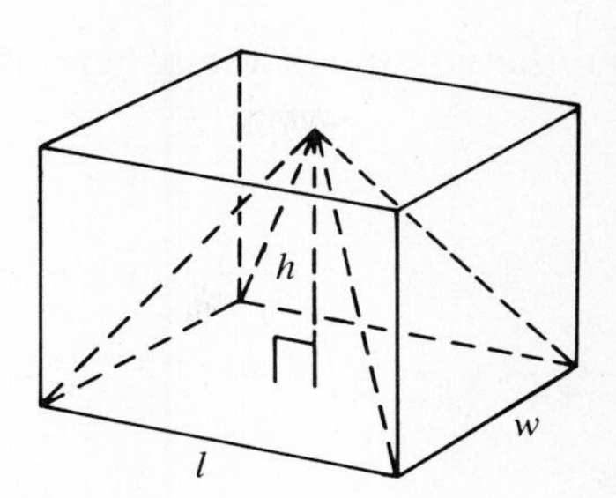

Volume of cylinder: $V = \pi r^2 h$ Volume of prism: $V = lwh$

Volume of cone: $V = \frac{1}{3}\pi r^2 h$ Volume of pyramid: $V = \frac{1}{3}lwh$

The volume of a pyramid or cone is always one-third of the area of the base times the altitude. This applies to all pyramids, no matter what the shape of the base.

$$V = \frac{1}{3}Bh$$

Q10 Find the volume of the following figures (round off to tens):

a.

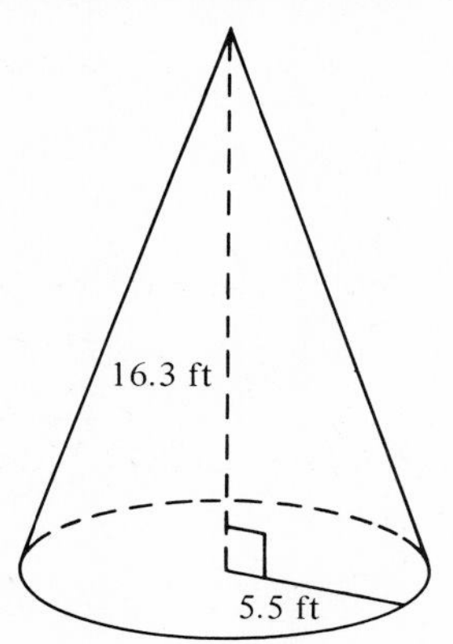

b.

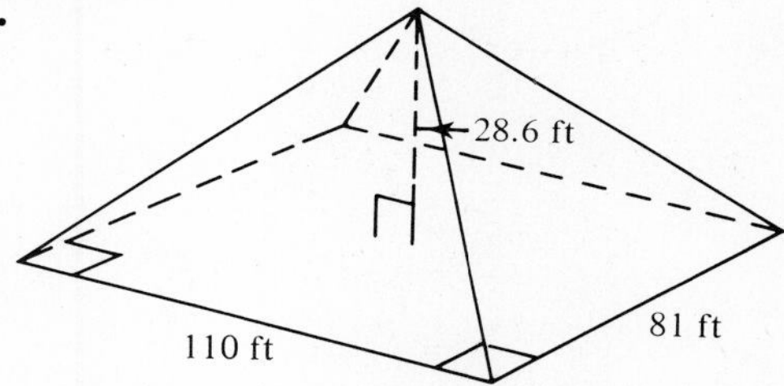

\# \# \# • \# \# \# • \# \# \# • \# \# \# • \# \# \# • \# \# \# • \# \# \# • \# \# \# • \# \# \#

A10 **a.** 520 ft^3: $V = \frac{1}{3}\pi r^2 h$

$V = \frac{1}{3}(3.14)(5.5)(5.5)(16.3)$

$V = 516.08517$

b. 84,940 ft^3: $V = \frac{1}{3}lwh$

$V = \frac{1}{3}(110)(81)(28.6)$

$V = 84{,}942$

7 The volume of a sphere is found with the formula

$$V = \frac{4}{3}\pi r^3$$

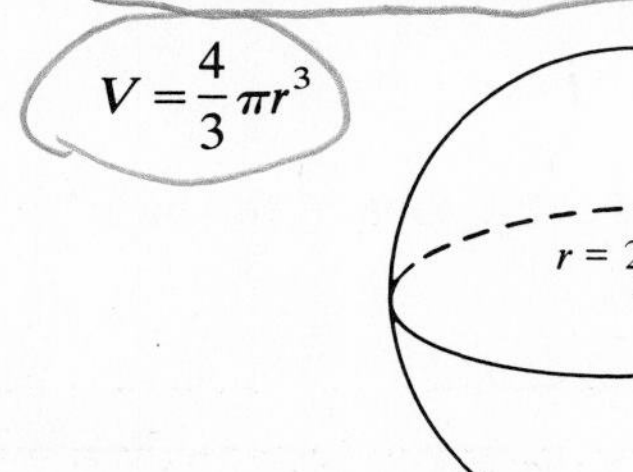

Example: Find the volume of the sphere above with $r = 2.4$ ft. Round off to ones.

Solution:

$$V = \frac{4}{3}\pi r^3$$

$$V = \frac{4}{3}(3.14)(2.4)(2.4)(2.4)$$

$$V = 57.87648$$

Therefore, the volume is 58 ft^3.

Q11 Find the volume of a sphere with radius 8.4 in. Round off to hundreds.

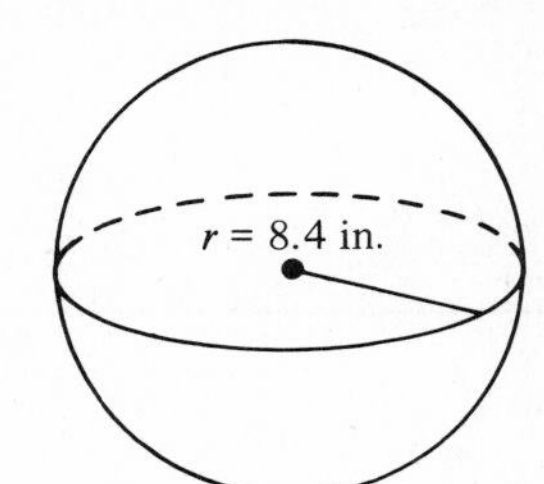

\# \# \# • \# \# \# • \# \# \# • \# \# \# • \# \# \# • \# \# \# • \# \# \# • \# \# \# • \# \# \#

A11 2,500 in.^{3}: $V = \frac{4}{3}\pi r^3$

$$V = \frac{4}{3}(3.14)(8.4)(8.4)(8.4)$$

$$V = 2{,}481.4540$$

8 This section will close with a summary of the formulas for volume.

Geometric figure	Formula for volume
Prism or cylinder	$V = Bh$, where B is the area of the base
Cube	$V = e^3$
Rectangular prism	$V = lwh$
Cylinder	$V = \pi r^2 h$
Triangular prism	$V = \frac{1}{2}bh_1h_2$
Cone	$V = \frac{1}{3}\pi r^2 h$
Pyramid	$V = \frac{1}{3}Bh$, where B is the area of the base
Sphere	$V = \frac{4}{3}\pi r^3$

This completes the instruction for this section.

5.6 EXERCISE

Find the volumes of the following figures:

1.

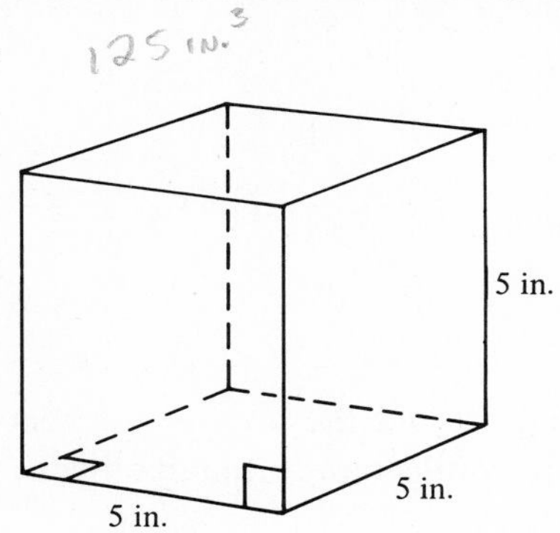

2. Round off to tenths.

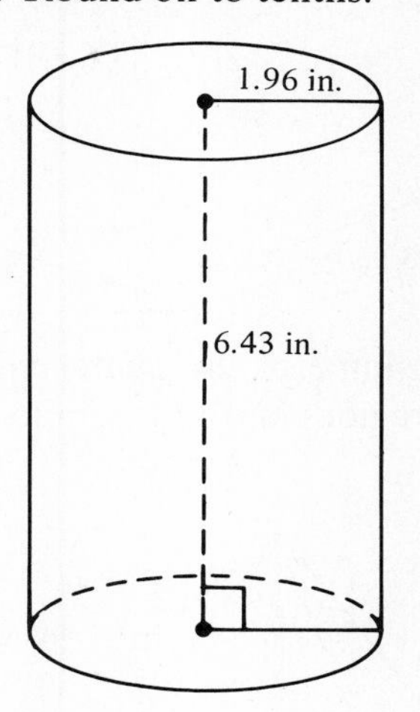

3.

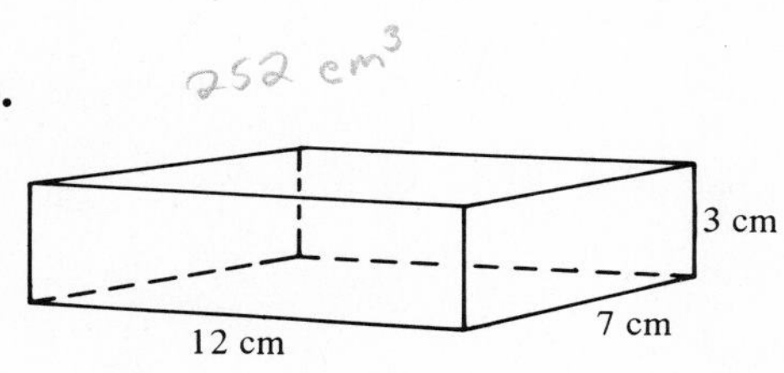

4. Round off to tens.

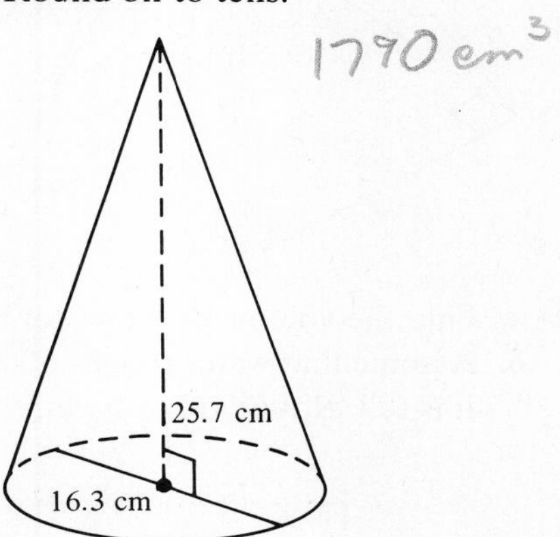

5. Round off to tens

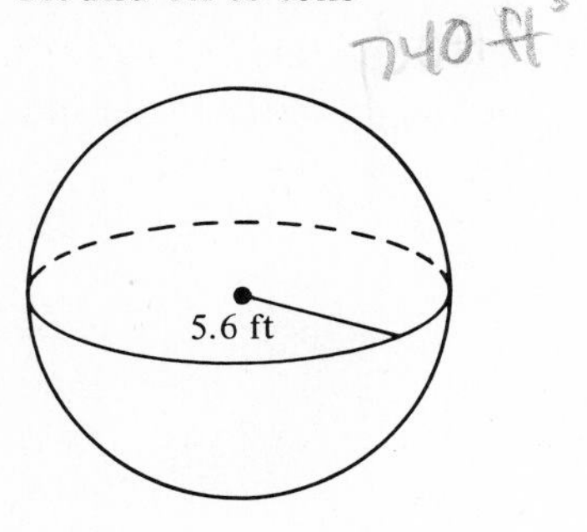

6. Round off to tens.

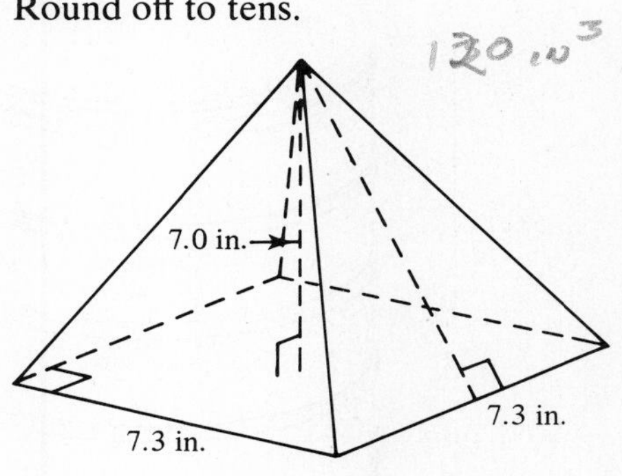

7. Round off to tens.

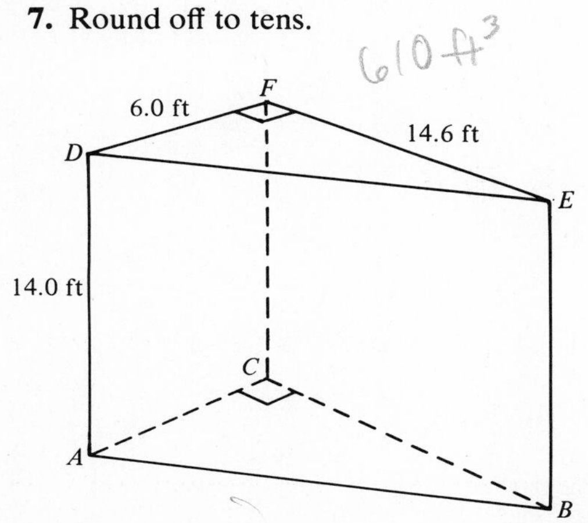

8. Find the total volume of air space in the building. Round off to thousands.

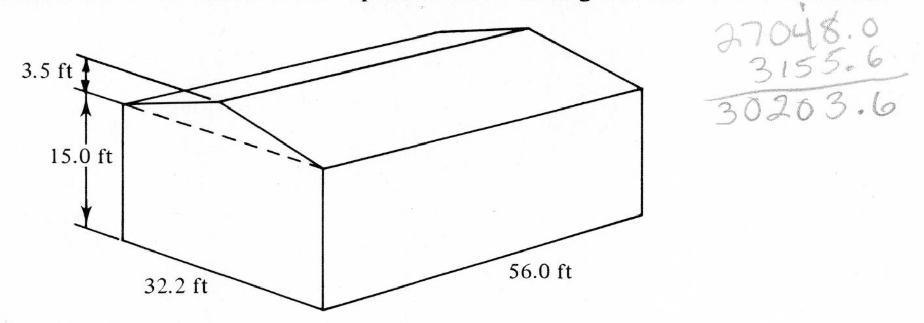

9. Find the volume of the pentagonal pyramid at right. (Find the area of the rectangular and triangular regions and add them to determine the area of the base.) Round off your final answer to ones.

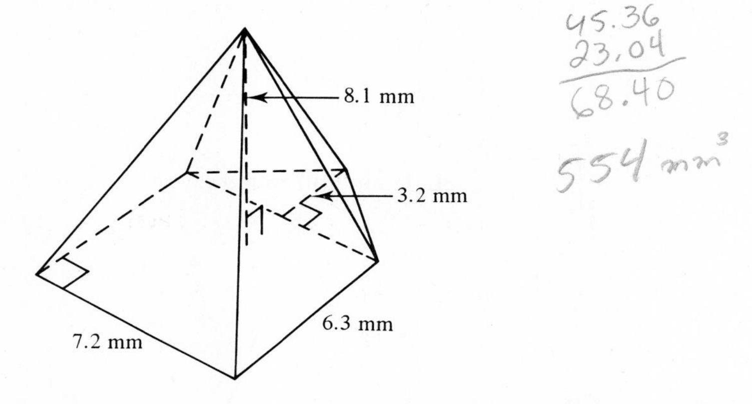

10. a. Find the volume of the water tank shown below. Round to ones.

b. Assume that water weighs 60.4 pounds per ft^3, find the weight of the water in the tank when it is full. Round off to hundreds.

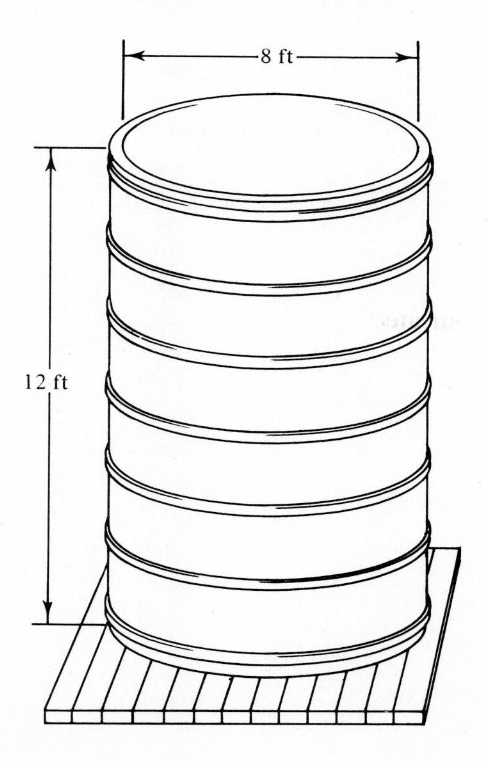

5.6 EXERCISE ANSWERS

1. 125 in.^{3}
2. 77.6 in.^{3}
3. 252 cm^3
4. 1790 cm^3
5. 740 ft^3
6. 120 in.^{3}
7. 610 ft^3
8. 30,000 ft^3
9. 154 m^3
10. a. 603 ft^3 **b.** 36,400 pounds

CHAPTER 5 SAMPLE TEST

At the completion of Chapter 5 you should be able to work the following problems.

5.1 MEASUREMENT OF LINE SEGMENTS

1. Which of the following measurements are approximate rather than exact?
a. height of the Empire State Building
b. number of brothers and sisters of George Washington
c. capacity of a soda bottle

2. Indicate the precision of each measurement:

a.

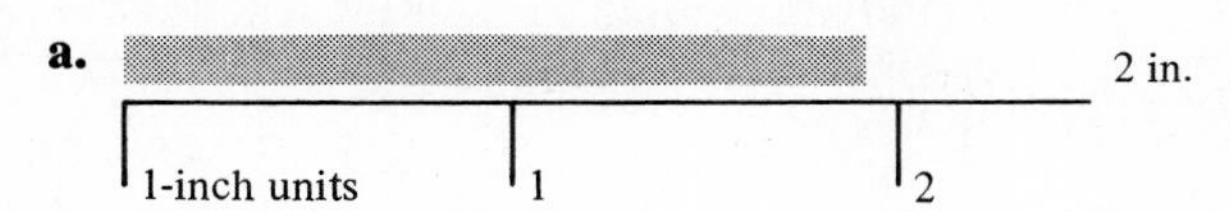

b.

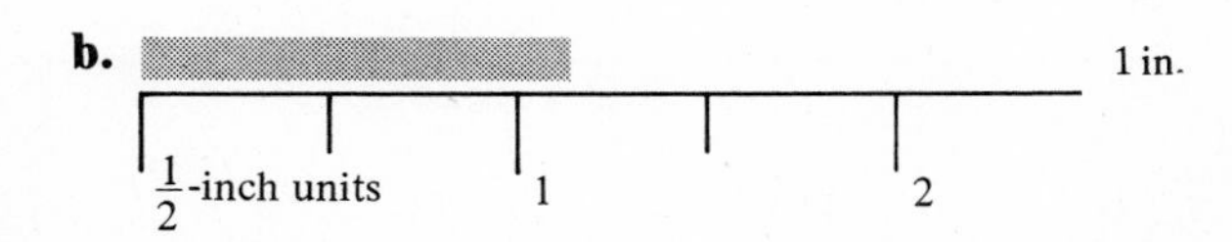

c.

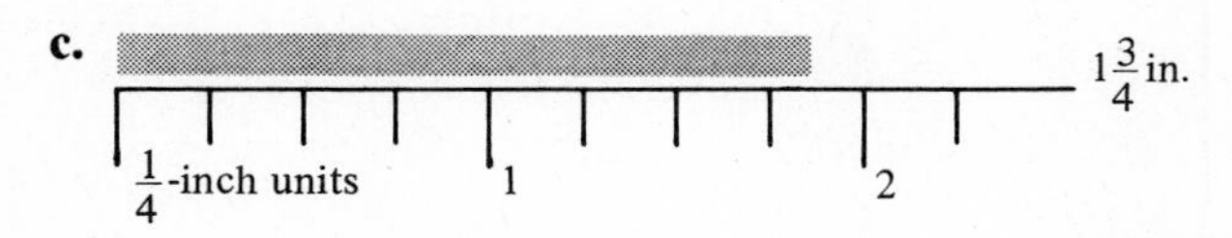

3. Measure each segment to the precision indicated with an inch or centimetre ruler:

a. $\frac{1}{4}$ in. ______________________________

b. 1 in. ________________

c. $\frac{1}{8}$ in. ______________________

d. $\frac{1}{10}$ cm ________________

e. 1 cm __________

f. 1 mm ______________________________

5.2 INTRODUCTION TO MEASUREMENT OF DISTANCE, AREA, AND VOLUME

4. Label each as being a distance, an area, or a volume:
- **a.** amount of wallpaper needed in a room
- **b.** length of a brick
- **c.** amount of water in a swimming pool
- **d.** capacity of a human lung

5. Label each unit as a unit of distance, area, or volume:
- **a.** cubic inch
- **b.** square centimetre
- **c.** decimetre
- **d.** kilometre
- **e.** square yard

6. Choose the most likely measurement for each object:
- **a.** diameter of a 60-watt light bulb:

  (1) $2\frac{1}{2}$ in. (2) $2\frac{1}{2}$ cm (3) $2\frac{1}{2}$ m
- **b.** area of the surface of a \$1 bill:

  (1) 15 cm^2 (2) 15 yd^2 (3) 15 $in.^2$

5.3 PERIMETERS

(Formulas for perimeters may be found immediately preceding the sample test answers.)

7. Find the perimeter of each of the following figures:

a.

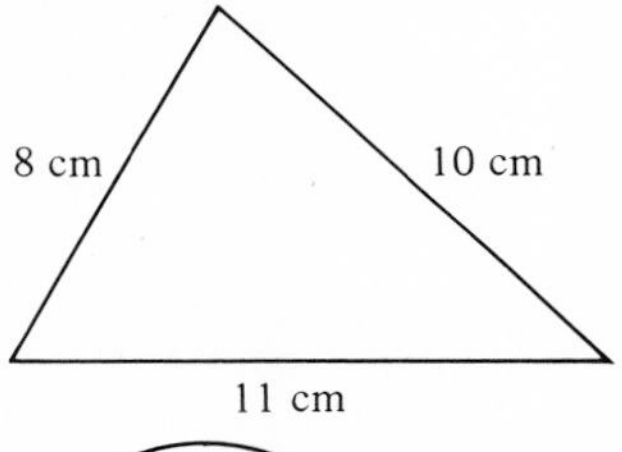

b.

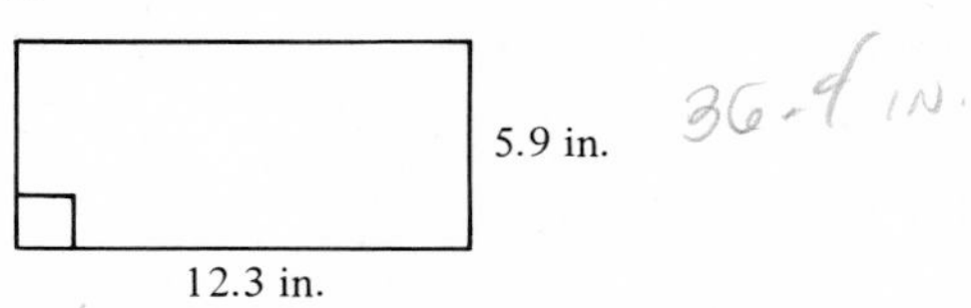

c.

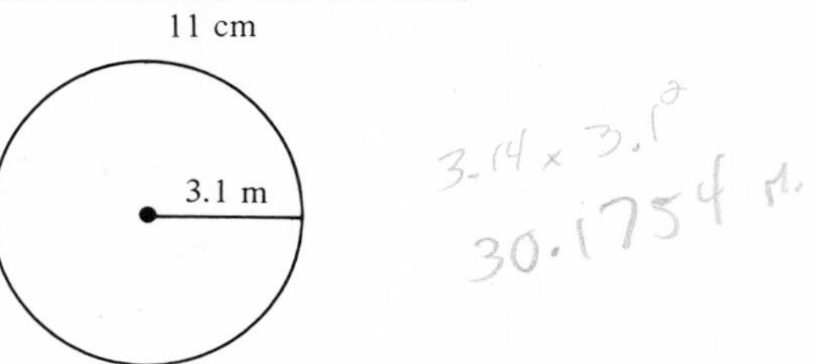

8. Measure to the precision indicated and compute the perimeter:

a. Nearest $\frac{1}{8}$ in.

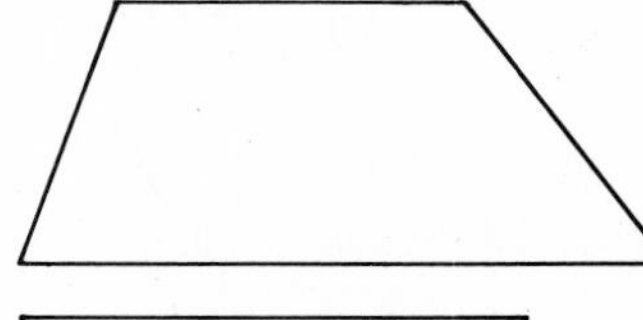

b. Nearest 0.1 cm

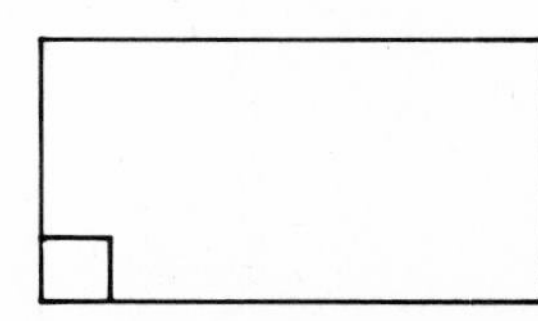

5.4 AREAS

(A summary of formulas is included with this test, immediately preceding the sample test answers.)

9. Find the area of each of the following figures:

a. Round off to ones.

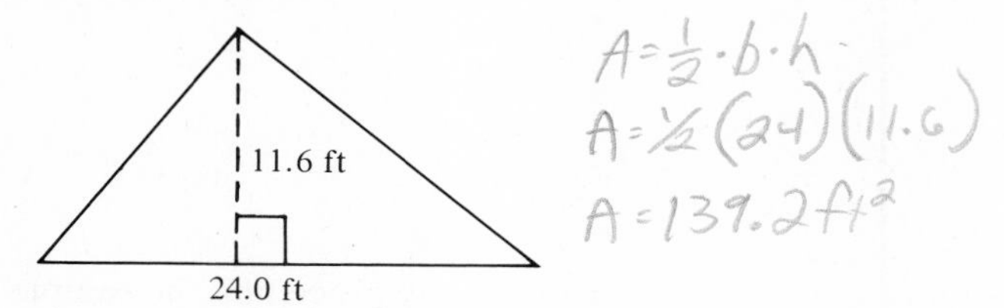

b. Round off to tenths.

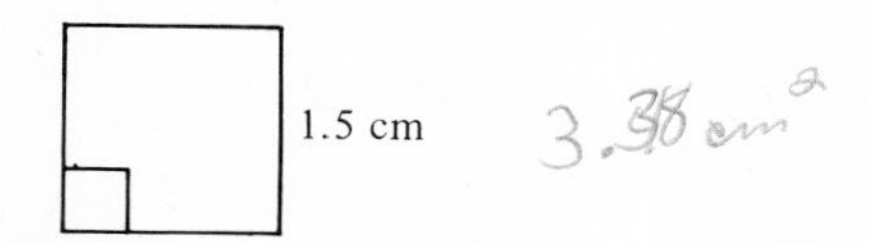

c. Round off to ones.

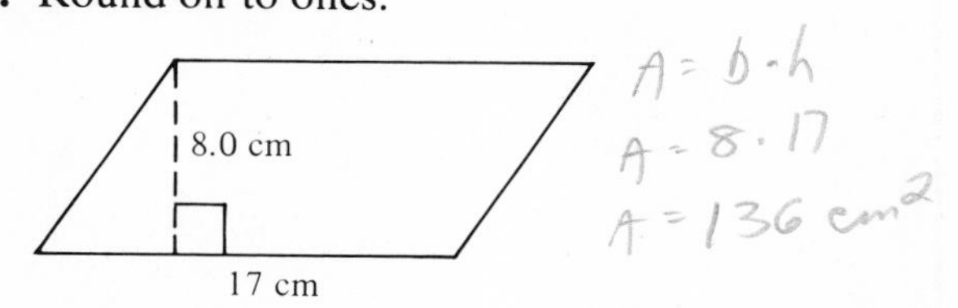

10. Measure to the precision indicated and compute the area.

a. Nearest $\frac{1}{8}$ in.

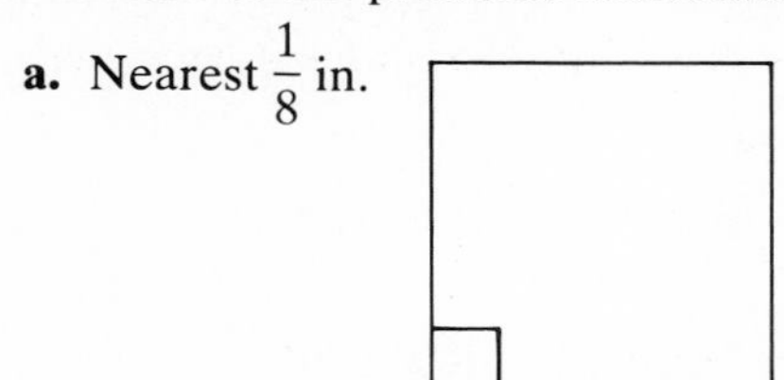

b. Nearest 0.1 cm (round off to tenths)

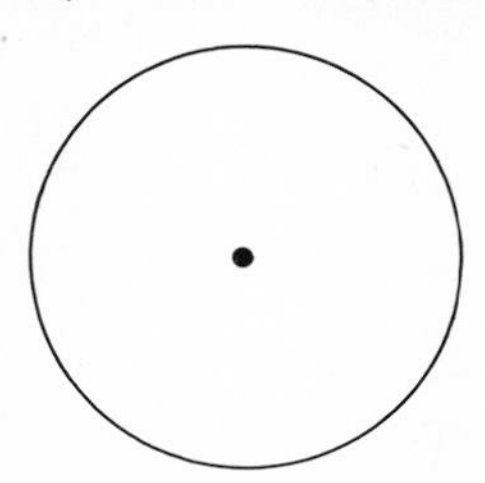

5.5 THE PYTHAGOREAN THEOREM

(Use Table IX in the Appendix.)

11. Find the missing side on each of the following triangles:

a.

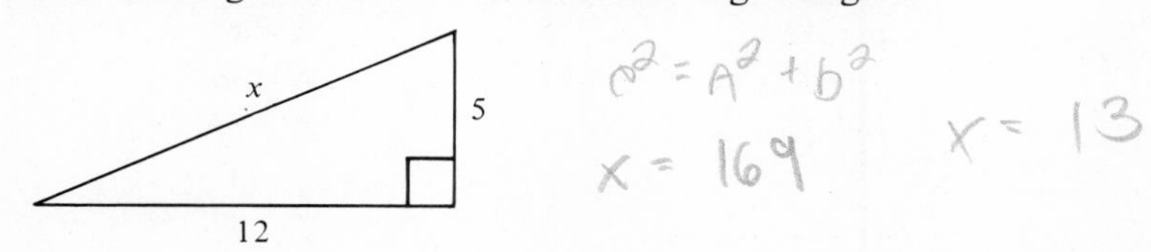

b.

x

6

10

c.

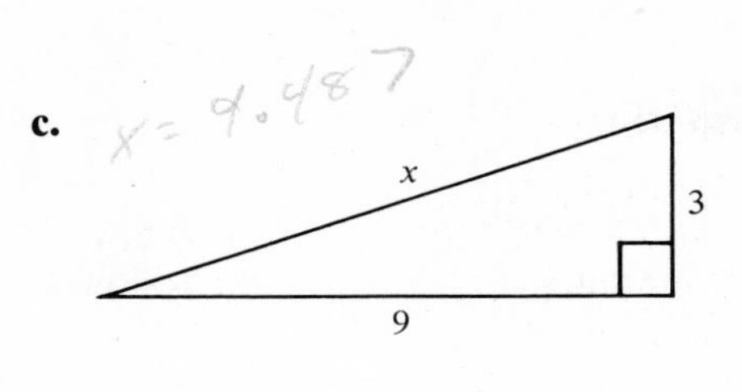

5.6 VOLUMES

(A summary of formulas is included with this test, immediately preceding the sample test answers.)

12. Find the volumes of each of the following figures:

a. Round off to tens.

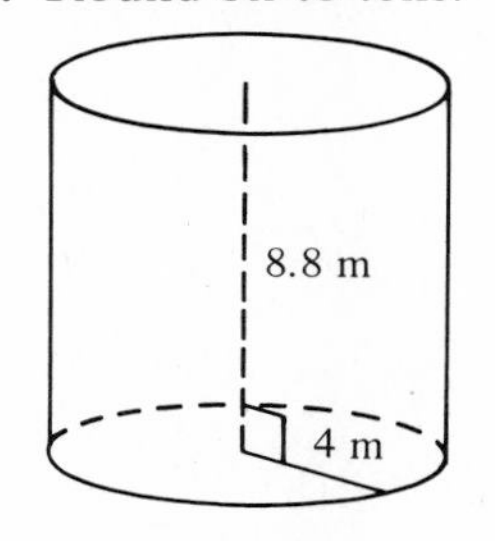

b. Round off to tenths.

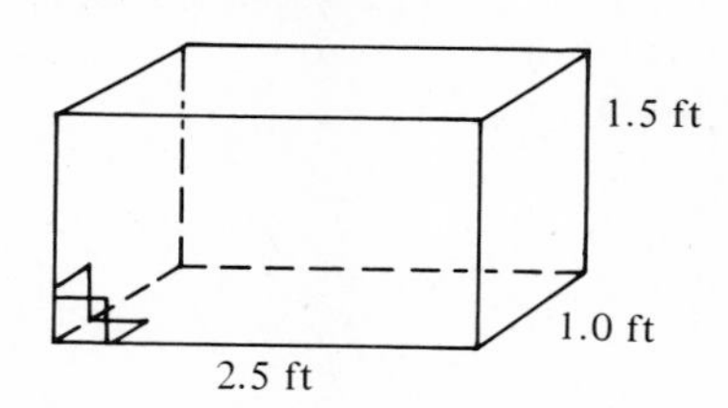

c. Round off to tens.

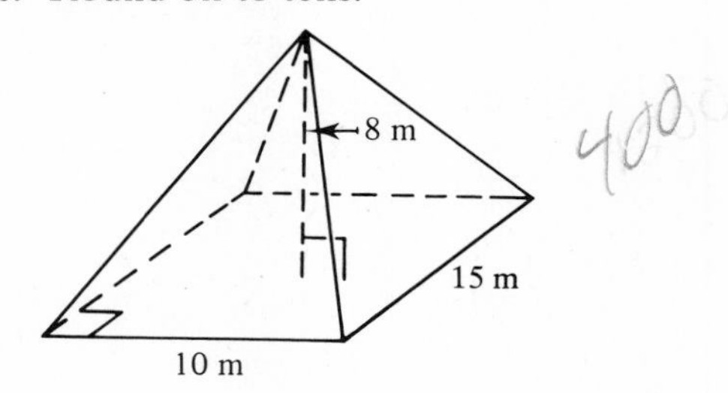

Summary of Formulas

Geometric figure	Perimeter	Area
Triangle	$p = a + b + c$	$A = \frac{1}{2}bh$
Quadrilateral	$p = a + b + c + d$	—
Rectangle	$p = 2l + 2w$	$A = lw$
Square	$p = 4s$	$A = s^2$
Circle	$c = 2\pi r = \pi d$	$A = \pi r^2$
Parallelogram	$p = 2a + 2b$	$A = bh$
Trapezoid	—	$A = \frac{1}{2}h(b_1 + b_2)$

Geometric figure	Volume
Prism or cylinder	$V = Bh$
Cube	$V = e^3$
Rectangular prism	$V = lwh$
Triangular prism	$V = \frac{1}{2}bh_1h_2$
Cylinder	$V = \pi r^2 h$
Cone	$V = \frac{1}{3}\pi r^2 h$
Pyramid	$V = \frac{1}{3}Bh$
Sphere	$V = \frac{4}{3}\pi r^3$

CHAPTER 5 SAMPLE TEST ANSWERS

1. a and c

2. a. 1 in. **b.** $\frac{1}{2}$ in. **c.** $\frac{1}{4}$ in.

3. a. $1\frac{3}{4}$ in. **b.** 1 in. **c.** $1\frac{3}{8}$ in. **d.** 2.8 cm
e. 2 cm **f.** 46 mm

4. a. area **b.** distance **c.** volume **d.** volume

5. a. volume **b.** area **c.** distance **d.** distance
e. area

6. a. 1 **b.** 3

7. a. 29 cm **b.** 36.4 in. **c.** 19.468 m

8. a. $4\frac{1}{4}$ in. **b.** 10.4 cm

9. a. 139 ft^2 **b.** 2.3 cm^2 **c.** 136 cm^2

10. a. 1 $in.^2$ **b.** 6.2 cm^2

11. a. 13 **b.** 8 **c.** 9.487

12. a. 440 m^3 **b.** 3.8 ft^3 **c.** 400 m^3

CHAPTER 6

THE METRIC SYSTEM

6.1 UNITS OF LENGTH AND AREA

1 The metric system of measurement is used today throughout the world. Because of its widespread use, it is important for the informed technical student to be familiar with the various units of the metric system.

The base unit of length in the metric system is the *metre*.* Originally defined as one ten-millionth of the distance from the North Pole to the Equator, the metre is today defined for greater accuracy in terms of the orange-red wavelength of radiating krypton gas (1 650 763.72† wavelengths in vacuum of the orange-red line of the spectrum of krypton 86). As shown in the following scale drawing of a metre stick and a yardstick, the length of a metre is slightly longer than 1 yard. More precisely, 1 metre is approximately equal to 1.1 yards or 39.37 inches.

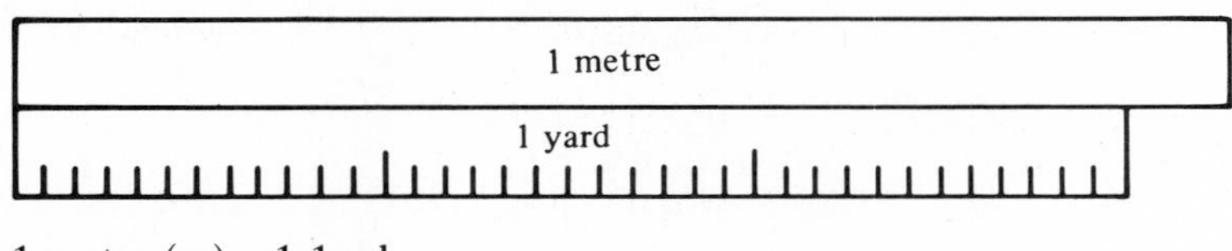

1 metre (m) = 1.1 yd
1 m = 39.37 in.

*Another spelling for metre is meter. Both spellings are accepted by the United States National Bureau of Standards.
†In the metric system numbers are written with digits in groups of three counting from the decimal point. A comma is not used. Instead, a space is left to avoid confusion, since many countries use a comma for the decimal point. In numbers of four digits, the space is not recommended unless four-digit numbers are grouped in a column with numbers of five digits or more. Examples: 2135 or 2 135 instead of 2,135. 982 125.013 25 instead of 982,125.01325.

Q1 Answer true or false:

a. The metre is the base unit of length in the metric system. ______

b. A yard is longer than a metre. ______

c. A metre is longer than a yard. ______

• ### • ### • ### • ### • ### • ### • ### •

A1 **a.** true **b.** false **c.** true

2 Metric measurement is based on a decimal system which means that its units are related by certain multiples and divisions of ten. The multiples of ten most commonly used are denoted by the Greek-derived prefixes:

kilo — 1000
hecto — 100
deka — 10

The divisions (fractions) of ten most commonly used are denoted by the Latin-derived prefixes:

deci — 0.1 or $\frac{1}{10}$

centi — 0.01 or $\frac{1}{100}$

milli — 0.001 or $\frac{1}{1000}$

Metric measurements of length are written using any of the designated prefixes in front of the base unit metre.

Examples: 1 dekametre = 10 metres
1 millimetre = 0.001 metre

Q2 Use the prefixes of Frame 2 to give the equivalents of each of the following in terms of a metre:

a. 1 hectometre = _______
b. 1 decimetre = _______
c. 1 centimetre = _______
d. 1 kilometre = _______
e. 1 millimetre = _______
f. 1 dekametre = _______

• ### • ### • ### • ### • ### • ### • ### •

A2 **a.** 100 metres
b. $\frac{1}{10}$ or 0.1 metre
c. $\frac{1}{100}$ or 0.01 metre
d. 1000 metres
e. $\frac{1}{1000}$ or 0.001 metre
f. 10 metres

3 For simplicity, symbols are used to stand for each of the units of length. These are:

kilometre — km
hectometre — hm
dekametre — dam
metre — m
decimetre — dm
centimetre — cm
millimetre — mm

The symbols are not considered abbreviations and thus are written without periods.

Q3 Write the symbol for each of the following:

a. centimetre ______
b. kilometre ______
c. millimetre ______
d. decimetre ______

• ### • ### • ### • ### • ### • ### • ### •

A3 **a.** cm **b.** km **c.** mm **d.** dm

Q4 Complete each equation by filling in the correct multiple or division of ten:

a. 1 cm = ________ m **b.** 1 km = ________ m

c. 1 dm = ________ m **d.** 1 mm = ________ m

• ### • ### • ### • ### • ### • ### • ### •

A4 **a.** 0.01 **b.** 1000 **c.** 0.1 **d.** 0.001

4 The metric units of length are summarized as follows:

1 kilometre (km) = 1000 metres
1 hectometre (hm) = 100 metres
1 dekametre (dam) = 10 metres
1 metre (m) = 1 metre
1 decimetre (dm) = 0.1 metre
1 centimetre (cm) = 0.01 metre
1 millimetre (mm) = 0.001 metre

Q5 Complete each of the following equations:

a. 1 m = ________ mm **b.** 1 cm = ________ m

c. 1 km = ________ m **d.** 1 dam = ________ m

• ### • ### • ### • ### • ### • ### • ### •

A5 **a.** 1000 **b.** 0.01 **c.** 1000 **d.** 10

5 From Frame 4 it can be seen that consecutive metric units differ by either a multiple or division of 10. The actual-size metric rule below demonstrates this relationship for the units decimetre, centimetre, and millimetre.

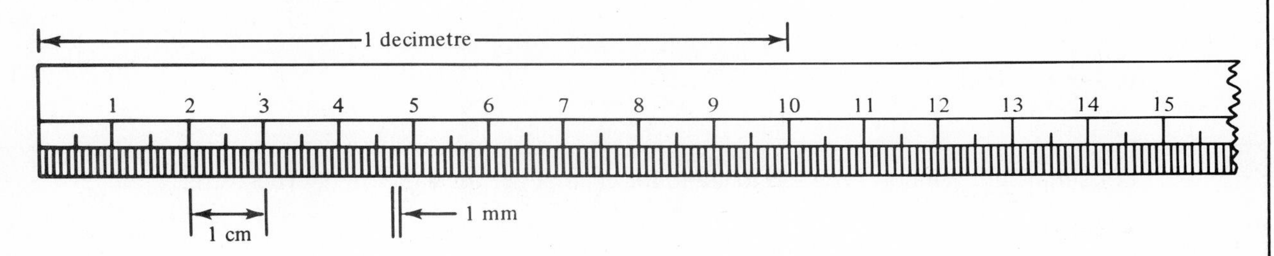

The relationships demonstrated by the metric rule are:

1 dm = 10 cm or 1 cm = 0.1 dm
1 cm = 10 mm or 1 mm = 0.1 cm

Ten decimetres placed end to end equal 1 metre. Thus, the decimetre and metre are related as follows:

1 m = 10 dm or 1 dm = 0.1 m

Notice that a centimetre is about the width of your little finger, while a decimetre is approximated by measuring across the knuckles of your clenched fist.

Q6 Estimate the length of each of the following lines in the unit shown:

a. (cm) ______________________ ______

b. (dm) ____ ______

c. (mm) ______ ______

d. (cm) ______ ______

\# # # • # # # • # # # • # # # • # # # • # # # • # # # • # # # • # # #

A6 **a.** 5 **b.** 1 **c.** 5 **d.** 1

Q7 Complete each of the following equations:

a. 1 mm = ______ cm **b.** 1 dm = ______ cm

c. 1 m = ______ cm **d.** 1 km = ______ m

\# # # • # # # • # # # • # # # • # # # • # # # • # # # • # # # • # # #

A7 **a.** 0.1 **b.** 10 **c.** 100 **d.** 1000

6 Conversion within the metric system is simplified if the similarity between metric units and our decimal number system is pointed out. Recall from Chapter 3 the following place-value diagram.

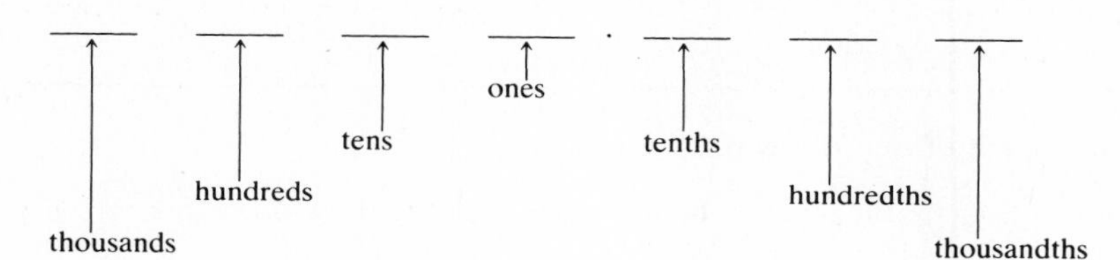

Using its prefix, each metric unit of length can be appropriately placed in the above diagram. For example, since kilometre has prefix "kilo" (1000), kilometre (km) is placed in the thousands place. Since centimetre has prefix "centi" $\left(\frac{1}{100}\right)$, centimetre (cm) is placed in the hundredths place.

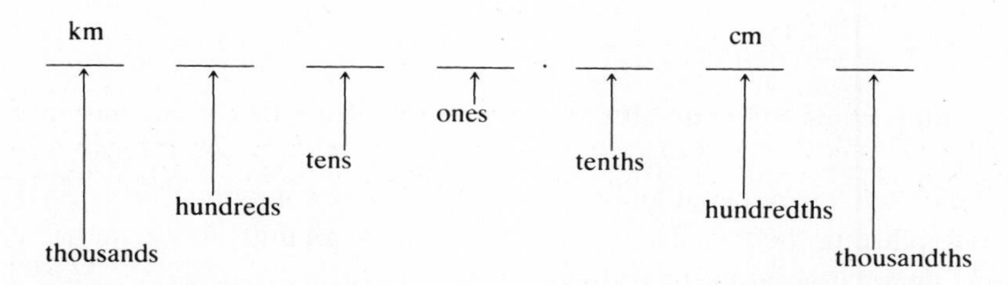

Q8 Complete the diagram below by filling in the appropriate metric unit of length symbols:

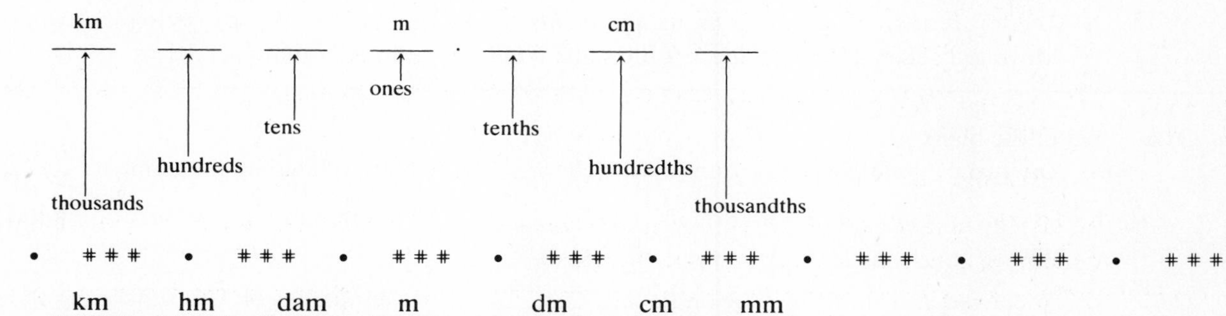

\# # # • # # # • # # # • # # # • # # # • # # # • # # # • # # # • # # #

A8 km hm dam m . dm cm mm

7 Notice that the place values and metric units of length in the diagram of Q8 are related by 10's in the following ways:

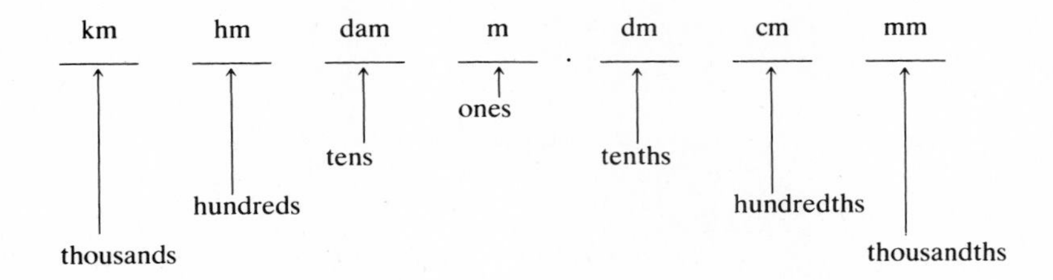

1. Each unit is equal to 10 times the unit to its right.

Examples: 1 km = 10 hm
1 hm = 10 dam
1 dam = 10 m

2. Each unit is equal to 0.1 times $\left(\frac{1}{10}\text{ of}\right)$ the unit to its left.

Examples: 1 mm = 0.1 cm
1 cm = 0.1 dm
1 dm = 0.1 m

Q9 Fill in 10 or 0.1 to complete each of the following:

a. 1 km = ______ hm
b. 1 cm = ______ dm
c. 1 dm = ______ cm
d. 1 dm = ______ m

\# # # • # # # • # # # • # # # • # # # • # # # • # # # • # # # • # # #

A9 **a.** 10 **b.** 0.1 **c.** 10 **d.** 0.1

8 The diagram of Frame 7 can be simplified in the following chart:

$$\text{km} \overset{10}{—} \text{hm} \overset{10}{—} \text{dam} \overset{10}{—} \text{m} \overset{10}{—} \text{dm} \overset{10}{—} \text{cm} \overset{10}{—} \text{mm}$$

The 10's indicate that any unit is equal to the unit on its right multiplied by 10 and the unit on its left divided by 10.

Multiplying or dividing by ten in a decimal number system is done by moving the decimal point one place to the right or one place to the left. Thus, conversions between units in the metric system can be made by moving the decimal point in the following ways:

1. To convert any unit to a unit on its *right, multiply* by the number of 10's separating the two units. (Move the decimal point to the right the same number of places as there are 10's.)
2. To convert any unit to a unit on its *left, divide* by the number of 10's separating the two units. (Move the decimal point to the left the same number of places as there are 10's.)

Q10 Fill in the blank:

a. Any two consecutive metric units on the chart in Frame 8 are related by the number ________.

b. To convert any unit to a unit on its right, ________ by the number of 10's separating the two units.

c. To convert any unit to a unit on its left, ________ by the number of 10's separating the two units.

\# # # • # # # • # # # • # # # • # # # • # # # • # # # • # # # • # # #

A10 **a.** 10 **b.** multiply **c.** divide

9 Consider the following examples:

Example 1: 45.8 dm = ______ m

km $\overset{10}{—}$ hm $\overset{10}{—}$ dam $\overset{10}{—}$ m $\overset{10}{—}$ dm $\overset{10}{—}$ cm $\overset{10}{—}$ mm

Solution: "m" is *one* 10 to the *left* of "dm" on the chart. Move the decimal point *one* place to the *left* (45.8). Thus, 45.8 dm = 4.58 m.

Example 2: 0.5 m = ______ mm

km $\overset{10}{—}$ hm $\overset{10}{—}$ dam $\overset{10}{—}$ m $\overset{10}{—}$ dm $\overset{10}{—}$ cm $\overset{10}{—}$ mm

Solution: "mm" is *three* 10's to the *right* of "m" on the chart. Move the decimal point *three* places to the *right* (0.500). Thus, 0.5 m = 500 mm.

Q11 168 cm = ______ m

• ### • ### • ### • ### • ### • ### • ### •

A11 1.68: km $\overset{10}{—}$ hm $\overset{10}{—}$ dam $\overset{10}{—}$ m $\overset{10}{—}$ dm $\overset{10}{—}$ cm $\overset{10}{—}$ mm "m" is *two* 10's to the *left* of "cm" on the chart. Move the decimal point *two* places to the *left* (168.).

Q12 15.5 cm = ______ mm

• ### • ### • ### • ### • ### • ### • ### •

A12 155: km $\overset{10}{—}$ hm $\overset{10}{—}$ dam $\overset{10}{—}$ m $\overset{10}{—}$ dm $\overset{10}{—}$ cm $\overset{10}{—}$ mm "mm" is *one* 10 to the *right* of "cm" on the chart. Move the decimal point *one* place to the *right* (15.5).

Q13 The width of an electrical switch plate is 70 mm. What is the width in centimetres?

• ### • ### • ### • ### • ### • ### • ### •

A13 7 cm: 70 mm = ____ cm
"cm" is *one* 10 to the *left* of "mm" on the chart. Move the decimal point *one* place to the *left* (70.).

Q14 A building has dimensions 120 m by 70 m. What are the dimensions in dekametres?

• ### • ### • ### • ### • ### • ### • ### •

A14 12 dam by 7 dam

Q15 Some tools are pictured at the top of page 296 with a certain dimension given in millimetres. Convert each dimension to the metric unit shown.

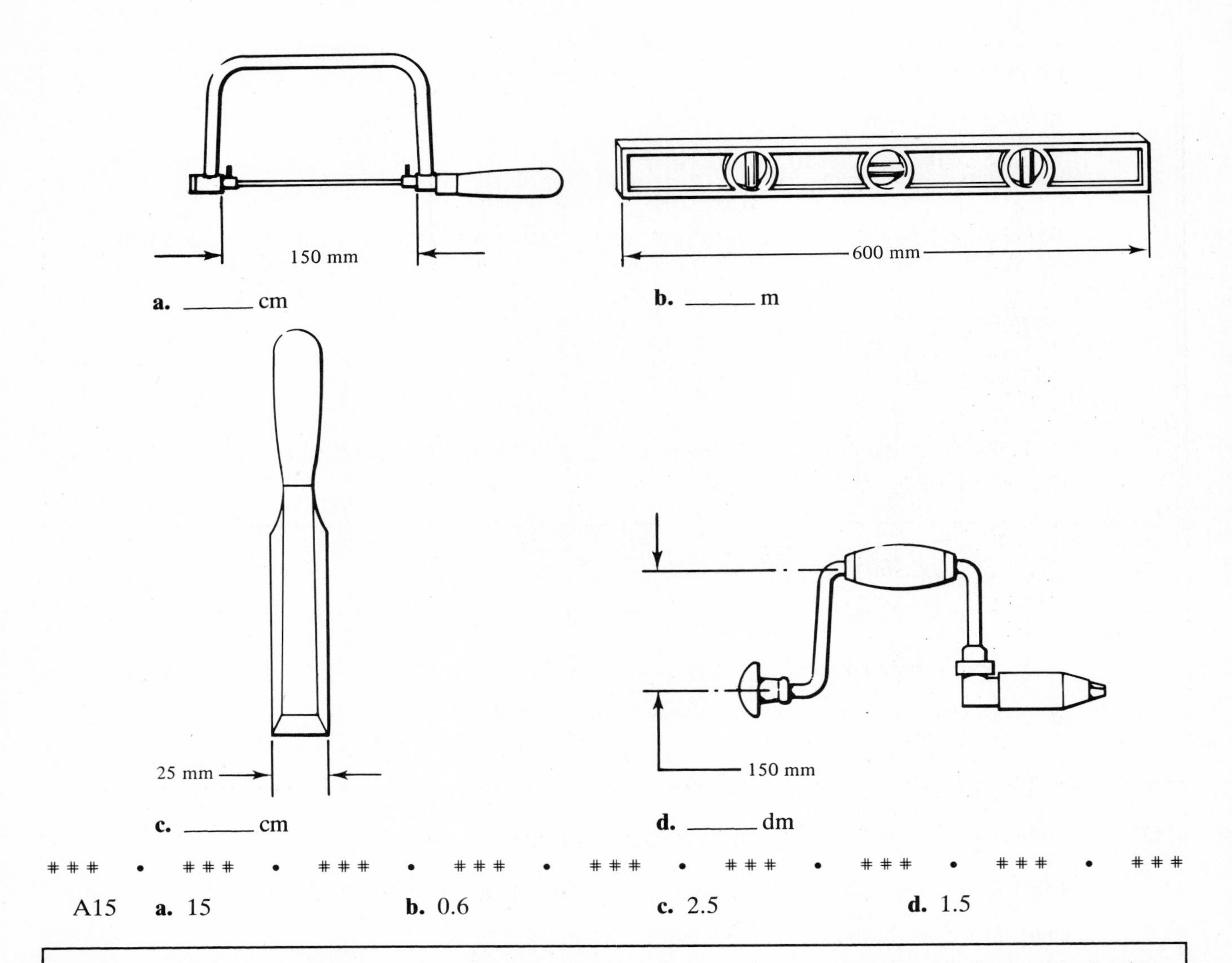

a. ______ cm **b.** ______ m

c. ______ cm **d.** ______ dm

• ### • ### • ### • ### • ### • ### • ### • ### •

A15 **a.** 15 **b.** 0.6 **c.** 2.5 **d.** 1.5

---

10 In the metric system, as in the U.S. customary system of measurement, area is measured using square units. Metric units of area are the square metre (m^2), square decimetre (dm^2), square centimetre (cm^2), square millimetre (mm^2), and for larger areas the square dekametre (dam^2), square hectometre (hm^2), and square kilometre (km^2). The relationship among dm^2, cm^2, and mm^2 can be seen in the actual size drawing of 1 square decimetre on page 297.

The relationships demonstrated by the drawing are:

$1\ dm^2 = 100\ cm^2$ or $1\ cm^2 = 0.01\ dm^2$
$1\ cm^2 = 100\ mm^2$ or $1\ mm^2 = 0.01\ cm^2$

The relationship between the square metre (m^2) and square decimetre (dm^2) is shown using the area formula and the equation $1\ m = 10\ dm$.

$$A = l \cdot w$$
$$A = 1\ m \cdot 1\ m = 10\ dm \cdot 10\ dm$$
$$A = 1\ m^2 \qquad = 100\ dm^2$$

Thus, m^2 and dm^2 are related:

$1\ m^2 = 100\ dm^2$ or $1\ dm^2 = 0.01\ m^2$

It is important to notice that consecutive units of area from mm^2 to m^2 are related by a multiple or division of 100.

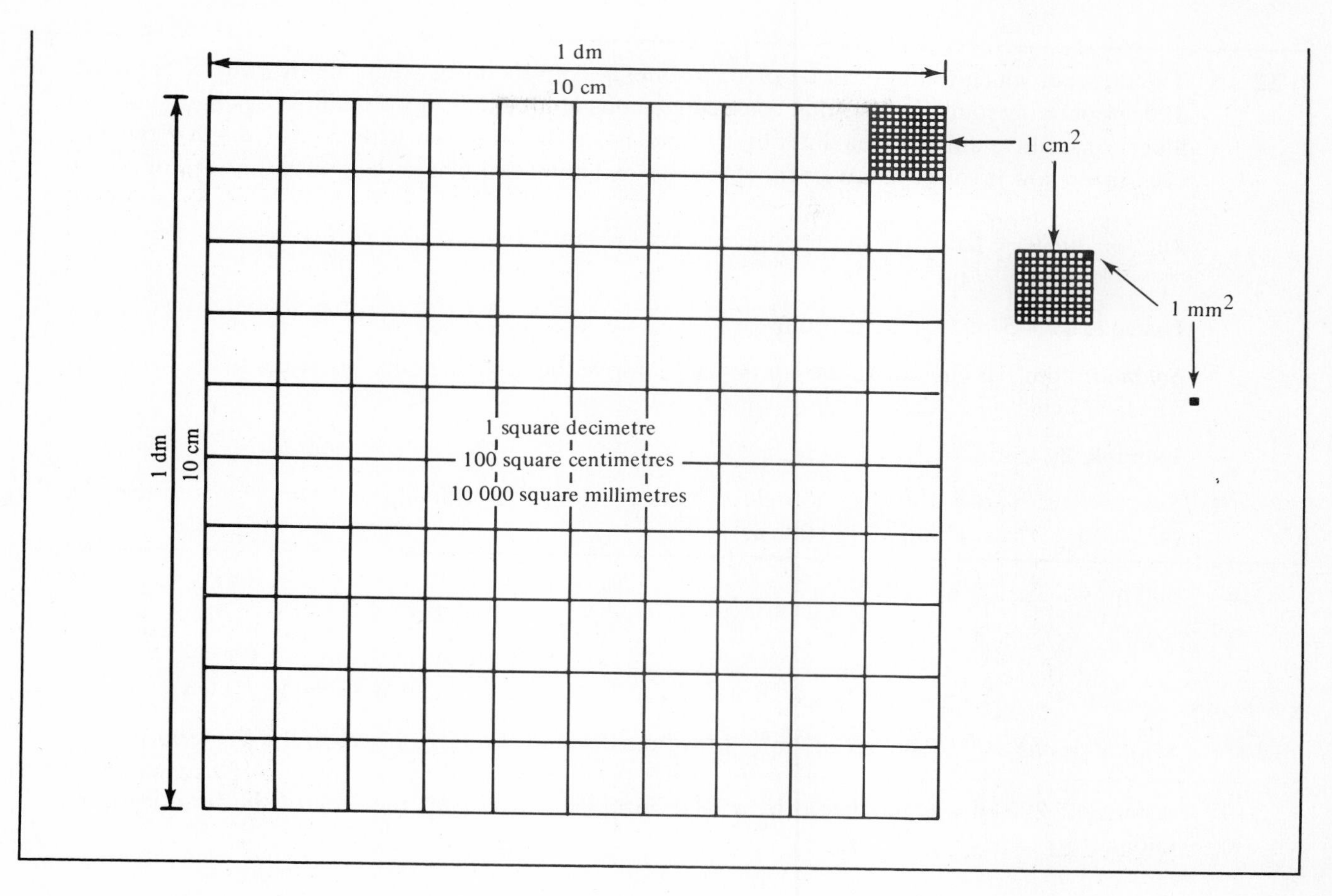

Q16 Complete each equation using the information of Frame 10.

a. 1 m^2 = ________ dm^2 **b.** 1 dm^2 = ________ cm^2
c. 1 mm^2 = ________ cm^2 **d.** 1 cm^2 = ________ dm^2

\# # # • # # # • # # # • # # # • # # # • # # # • # # # • # # # • # # #

A16 **a.** 100 **b.** 100 **c.** 0.01 **d.** 0.01

11 Large areas are measured using the square dekametre (dam^2), square hectometre (hm^2), and square kilometre (km^2). When dealing with land measurements the square dekametre is commonly called an *are* (a) and the square hectometre is called a *hectare* (ha). One hectare is approximately equal to 2.5 acres.

As in the case of the smaller units of area, the consecutive units from dam^2 to km^2 are related by a multiple or division of 100. These relationships are:

1 dam^2 (a) = 100 m^2 or 1 m^2 = 0.01 dam^2 (a)
1 hm^2 (ha) = 100 dam^2 (a) or 1 dam^2 (a) = 0.01 hm^2 (ha)
1 km^2 = 100 hm^2 (ha) or 1 hm^2 (ha) = 0.01 km^2

Q17 Write the common names for the following when land is to be measured:

a. hm^2 __________ **b.** dam^2 __________

\# # # • # # # • # # # • # # # • # # # • # # # • # # # • # # # • # # #

A17 **a.** hectare (ha) **b.** are (a)

12 The diagram which follows can be used to simplify conversion between metric units of area. The 100's separating units indicate that consecutive units of area differ by a multiple or division of 100. Since consecutive units of area differ by 100, and not 10 as is the case with units of length (Frame 8), each space now involves a movement of the decimal point *two* places right or left instead of one.

$$\text{km}^2 \overset{100}{\text{——}} \underset{\text{(ha)}}{\text{hm}^2} \overset{100}{\text{——}} \underset{\text{(a)}}{\text{dam}^2} \overset{100}{\text{——}} \text{m}^2 \overset{100}{\text{——}} \text{dm}^2 \overset{100}{\text{——}} \text{cm}^2 \overset{100}{\text{——}} \text{mm}^2$$

Example 1: 575 cm^2 = ________ dm^2

Solution: "dm^2" is *one* 100 to the *left* of "cm^2." Move the decimal point *two* places to the *left* (5.75.). Thus, 575 cm^2 = 5.75 dm^2.

Example 2: 45 km^2 = ____________ a

Solution: "a" is *two* 100's to the *right* of "km^2." Move the decimal point *four* places to the *right* (45.0000.). Thus, 45 km^2 = 450 000 a.

Q18 800 m^2 = ________ a

• ### • ### • ### • ### • ### • ### • ### •

A18 8: "a" is *one* 100 to the *left* of "m^2." Move the decimal point *two* places to the *left* (8.00.).

Q19 A farm of 200 ha was purchased by a land developer. What was the area of the farm in square kilometres?

• ### • ### • ### • ### • ### • ### • ### •

A19 2: "km^2" is *one* 100 to the *left* of "ha." Move the decimal point *two* places to the *left* (2.00.).

Q20 The area of a small garage floor which measures 6 m × 3.6 m is 21.6 m^2. What is the area of the floor in square decimetres?

• ### • ### • ### • ### • ### • ### • ### •

A20 2160: 21.6 m^2 = ________ dm^2
"dm^2" is *one* 100 to the *right* of "m^2." Move the decimal point *two* places to the *right* (21.60.).

This completes the instruction for this section.

6.1 EXERCISE

1. The base unit of length in the metric system is the ________.
2. The base metric unit of length is approximately equal to what U.S. customary unit?
3. Complete each equation by filling in the correct multiple or division of ten:
   a. 1 m = ________ mm   **b.** 1 dam = ________ m
   c. 1 m = ________ km   **d.** 1 cm = ________ m
4. Estimate the length of the following line in the units shown:
   a. (dm) ____________________________________

b. (cm) ____________

c. (mm) ____

5. Complete each of the following equations:

a. 48.5 m = ______ dm **b.** 7500 m = ______ km

c. 1.25 dm = ______ mm **d.** 0.25 km = ______ hm

6. The perimeter of a building site measures 50 hm. What is the perimeter in metres?

7. Convert the dimensions of the tools below to the metric unit shown:

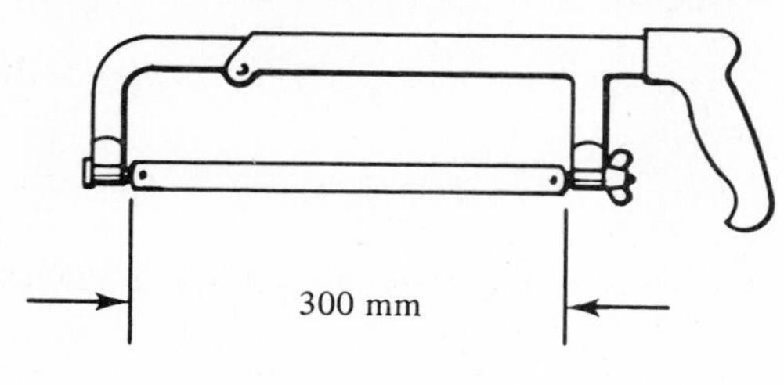

b. ______ cm

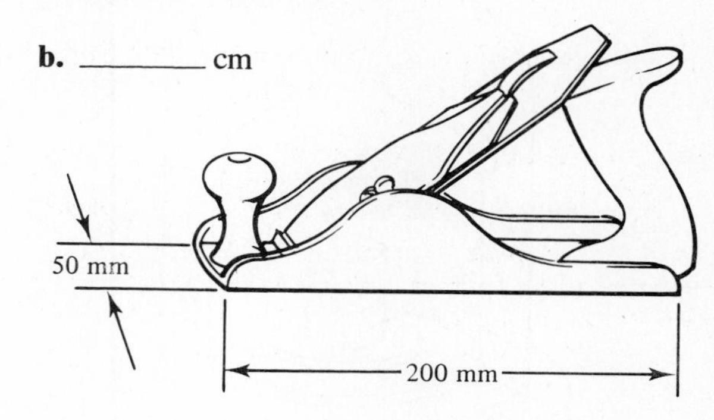

a. ______ dm

c. ______ cm

8. Complete each of the following equations:

a. 1 a = ______ m^2 **b.** 1 ha = ______ a

c. 8.5 cm^2 = ______ mm^2 **d.** 250 000 m^2 = ______ hm^2

9. A metal part with a surface area of 8.7 cm^2 is to be covered with an expensive metal alloy. What is the surface area of the part in square millimetres?

10. An apartment complex is to be constructed on a 150 square hectometre site.

a. What is the area of the site in hectares?

b. What is the area of the site in ares?

6.1 EXERCISE ANSWERS

1. metre

2. yard

3. a. 1000 **b.** 10 **c.** 0.001 **d.** 0.01

4. a. 1 dm **b.** 5 cm **c.** 10 mm

5. a. 485 **b.** 7.5 **c.** 125 **d.** 2.5

6. 5000 m

7. a. 3 **b.** 5 **c.** 20

8. a. 100 **b.** 100 **c.** 850 **d.** 25

9. 870 mm^2

10. a. 150 ha **b.** 15 000 a

6.2 UNITS OF MASS (WEIGHT) AND VOLUME

1 The base unit of mass (weight)* in the metric system is the *kilogram.* The only base unit still defined by a man-made object, the kilogram is defined as the weight of a cylinder of platinum-iridium alloy

*Mass is the quantity of matter that a body possesses; weight is force, the earth's attraction for a given mass. Technically, mass and weight are two different things. In commercial and everyday use, the term weight nearly always means mass.

kept by the International Bureau of Weights and Measures in Sèvres, France. A duplicate of this cylinder is present in the National Bureau of Standards in the United States. When compared with a unit of the U.S. customary system, the mass of 1 kilogram is approximately 2.2 pounds.

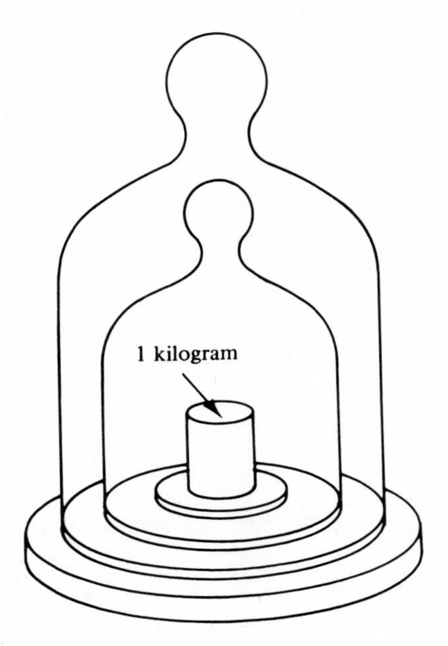

The original base unit was the gram, defined as the weight of a cubic centimetre of water at the temperature of melting ice. The change was later made to the kilogram as a base unit, owing to the difficulty involved in working with something as small as the gram. The units for mass are powers of 10 using the same prefixes as with the units of length. The units, their symbols, and equivalents are:

1 kilogram (kg) = 1000 grams
1 hectogram (hg) = 100 grams
1 dekagram (dag) = 10 grams
1 gram (g) = 1 gram
1 decigram (dg) = 0.1 gram
1 centigram (cg) = 0.01 gram
1 milligram (mg) = 0.001 gram

Of the units listed, the most commonly used are the kilogram, the milligram by pharmacists, and the gram. A unit not listed above is the metric ton. It is defined as:

1 metric ton (t) = 1000 kg

Q1 **a.** The base unit of mass in the metric system is the ____________.

b. The approximate mass of 1 kg is ______ pounds.

\# # # • # # # • # # # • # # # • # # # • # # # • # # # • # # # • # # #

A1 **a.** kilogram **b.** 2.2

Q2 Complete each equation using the information of Frame 1:

a. 1 cg = __________ g

b. __________ g = 1 kg

c. __________ mg = 1 g

d. 10 g = __________ dag

e. __________ kg = 1 t

\# # # • # # # • # # # • # # # • # # # • # # # • # # # • # # # • # # #

A2 **a.** 0.01 **b.** 1000 **c.** 1000 **d.** 1 **e.** 1000

2 Some common items and the approximate mass of each are shown below.

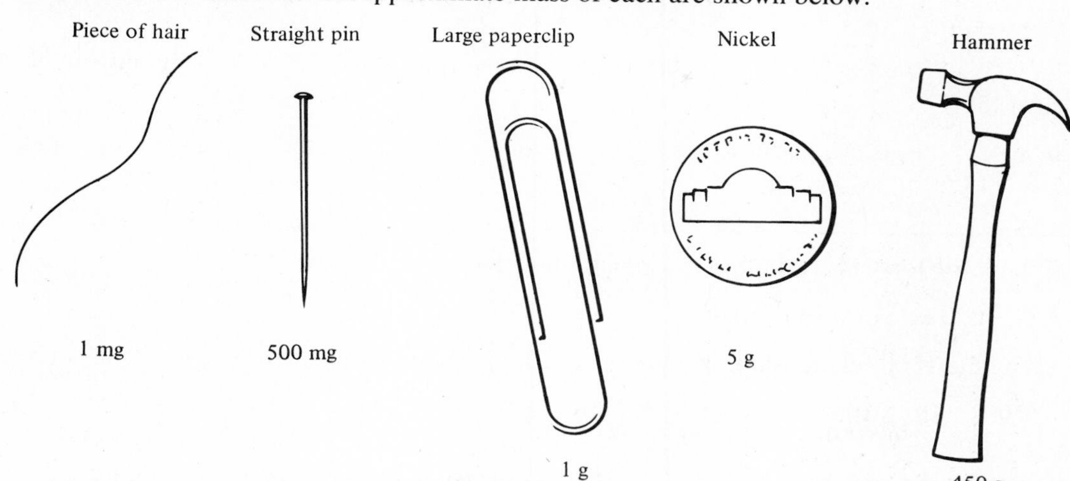

The mass of an adult male is approximately 80 kg, while that of a subcompact car is approximately 1 metric ton.

Q3 Write the following units in order from smallest to largest: kilogram, milligram, gram, ton.

____

\# \# \# • \# \# \# • \# \# \# • \# \# \# • \# \# \# • \# \# \# • \# \# \# • \# \# \# • \# \# \#

A3 milligram, gram, kilogram, ton

3 Conversions between metric units of mass are possible using the procedures of the previous section. The following chart lists the metric units of mass from largest (kg) to smallest (mg). *The 10's indicate that any unit is equal to the unit on its right multiplied by 10 and the unit on its left divided by 10.*

$$\text{kg} \overset{10}{—} \text{hg} \overset{10}{—} \text{dag} \overset{10}{—} \text{g} \overset{10}{—} \text{dg} \overset{10}{—} \text{cg} \overset{10}{—} \text{mg}$$

The procedure for performing conversions between metric units of mass is as follows:

1. To convert any unit to a unit on its *right, multiply* by the number of 10's separating the two units. (Move the decimal point to the right the same number of places as there are 10's.)
2. To convert any unit to a unit on its *left, divide* by the number of 10's separating the two units. (Move the decimal point to the left the same number of places as there are 10's.)

Example: 85 kg = __________ g

Solution: $\text{kg} \overset{10}{—} \text{hg} \overset{10}{—} \text{dag} \overset{10}{—} \text{g} \overset{10}{—} \text{dg} \overset{10}{—} \text{cg} \overset{10}{—} \text{mg}$

"g" is *three* 10's to the *right* of "kg" on the chart. Move the decimal point *three* places to the *right* (85.000). Thus, 85 kg = 85 000 g.

Q4 3500 mg = ______ g

\# \# \# • \# \# \# • \# \# \# • \# \# \# • \# \# \# • \# \# \# • \# \# \# • \# \# \# • \# \# \#

A4 3.5: "g" is *three* 10's to the *left* of "mg" on the chart. Move the decimal point *three* places to the *left* (3500.).

Q5 Perform each of the following conversions:

a. 50.7 dag = ______ g **b.** 45 cg = ______ g

• ### • ### • ### • ### • ### • ### • ### •

A5 **a.** 507: "g" is *one* 10 to the *right* of "dag" on the chart. Move the decimal point *one* place to the *right* (50.7.).

b. 0.45: "g" is *two* places to the *left* of "g" on the chart. Move the decimal point *two* places to the *left* (.45.).

---

4 The metric ton (t) is related to the kilogram by the following two equations:

1 t = 1000 kg or 1 kg = 0.001 t

Using the 1000 relationship, the metric ton can be added to the chart of metric units of mass.

$$t \overset{1000}{—} kg \overset{10}{—} hg \overset{10}{—} dag \overset{10}{—} g \overset{10}{—} dg \overset{10}{—} cg \overset{10}{—} mg$$

Since the ton and kg are related by 1000, conversions between the two units are performed by moving the decimal point three places to the *right* (multiplying by 1000) or three places to the *left* (dividing by 1000).

Example: 45 000 kg = ______ t

Solution: "t" is to the *left* of "kg" on the chart. They are related by 1000. Move the decimal point three places to the *left* (45.000.). Thus, 45 000 kg = 45 t.

Q6 The shipment mass (weight) was 250 kg. What is the mass in metric tons?

• ### • ### • ### • ### • ### • ### • ### •

A6 0.25 t: "t" is to the *left* of "kg" on the chart and related by 1000. Move the decimal point three places to the *left*.

Q7 What is the mass in kilograms of 400 metric tons?

• ### • ### • ### • ### • ### • ### • ### •

A7 400 000 kg: "kg" is to the *right* of "t" on the chart and related by 1000. Move the decimal point three places to the *right* (400.000.). Thus, 400 t = 400 000 kg.

Q8 Perform each conversion:

a. 5.5 t = ______ kg **b.** 1850 kg = ______ t

• ### • ### • ### • ### • ### • ### • ### •

A8 **a.** 5500 **b.** 1.85

---

5 The volume or capacity of an object can be thought of as the amount that the object holds. Volume can be expressed in units of liquid measure or in cubic units. A cube that has edges of length 1 cm is said to have a volume of 1 cubic centimetre (1 cm^3). A cubic centimetre is shown at the top of page 303. Its size is approximately that of a sugar cube. If a cube with edges of length 10 cm (1 dm) is formed, the result is 1 cubic decimetre (dm^3) or 1000 cubic centimetres.

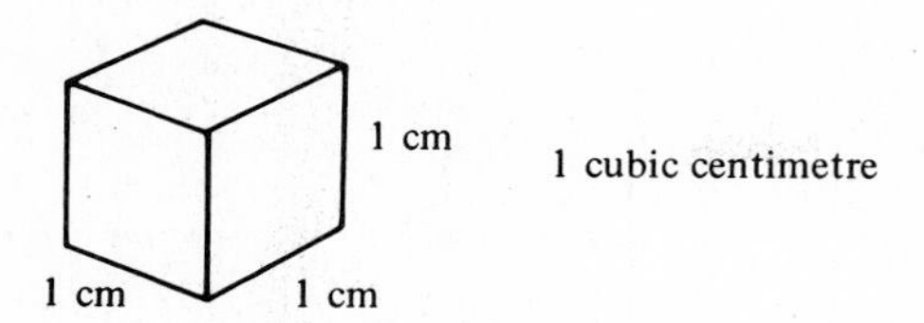

The unit that is generally used as the metric unit of volume or capacity is the litre (l). One litre is defined as 1000 cubic centimetres. The large cube shown has a volume of 1 litre. Compared with the U.S. customary system, the litre is slightly larger than a quart.

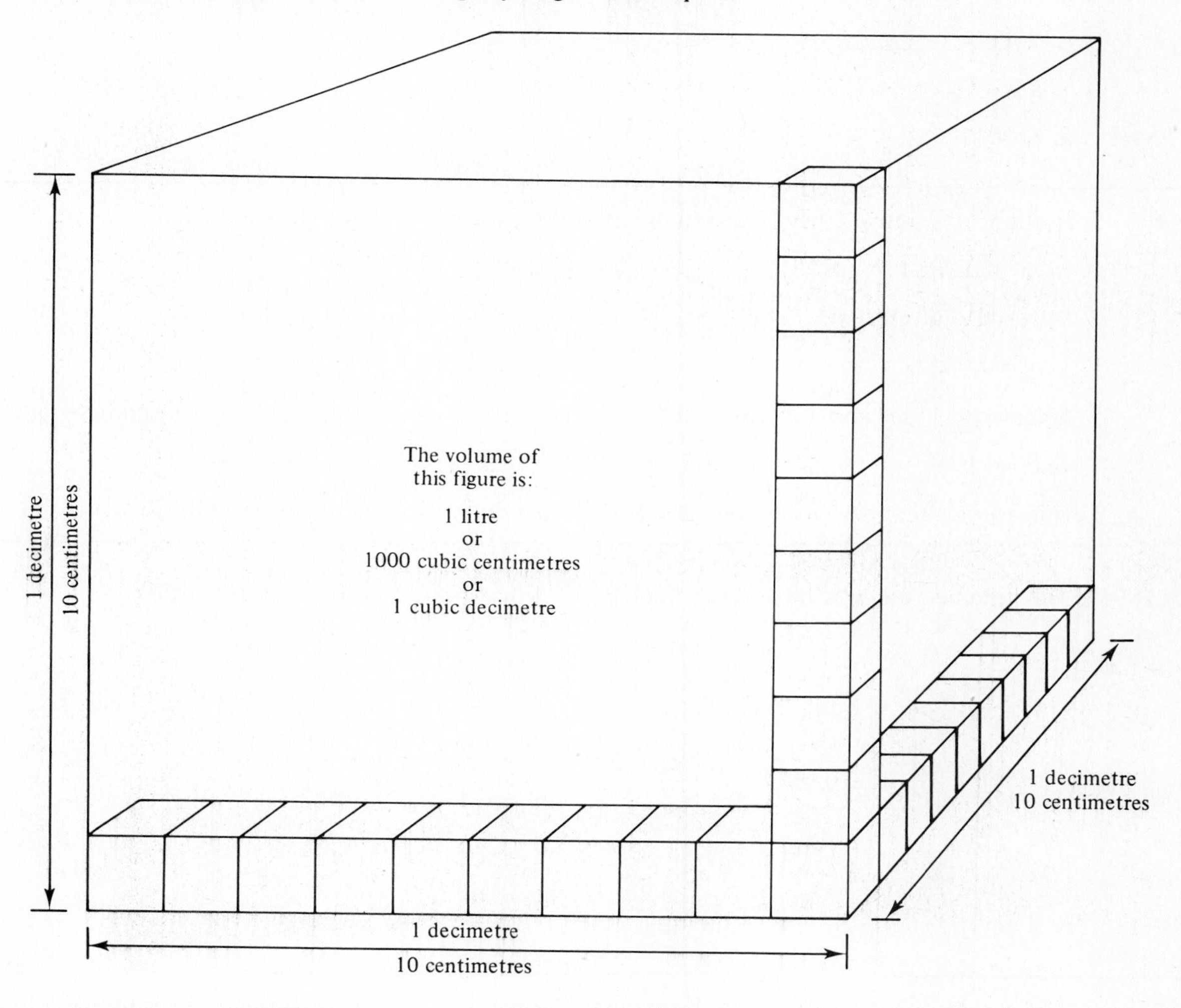

The metric units of liquid volume are defined as follows:

1 kilolitre (kl) = 1000 litres
1 hectolitre (hl) = 100 litres
1 dekalitre (dal) = 10 litres
1 litre (l) = 1 litre
1 decilitre (dl) = 0.1 litre
1 centilitre (cl) = 0.01 litre
1 millilitre (ml) = 0.001 litre

Q9 Answer true or false:

a. A litre is larger than a quart. ______

b. 1 l = 1000 cubic centimetres ________

c. 1 l = 1 cubic decimetre ________

• ### • ### • ### • ### • ### • ### • ### •

A9 **a.** true **b.** true **c.** true

Q10 Complete each equation using the information of Frame 5.

a. 1 ml = ________ l **b.** 1000 cm^3 = ________ l

c. 1 dal = ________ l **d.** 1 kl = ________ l

• ### • ### • ### • ### • ### • ### • ### •

A10 **a.** 0.001 **b.** 1 **c.** 10 **d.** 1000

6 In the frame above a litre was defined as 1000 cubic centimetres. Thus,

$1 \text{ cm}^3 = 0.001 \text{ l}$

It was also given that

1 ml = 0.001 l

Since both 1 cm^3 and 1 ml are equal to 0.001 l, the two units are equal to each other. That is,

$1 \text{ cm}^3 = 1 \text{ ml}$

It is, therefore, possible to interchange the units millilitre and cubic centimetre.

Q11 The volume of a cube measuring 1 cm on a side can be named in what two ways?

a. ________ **b.** ________

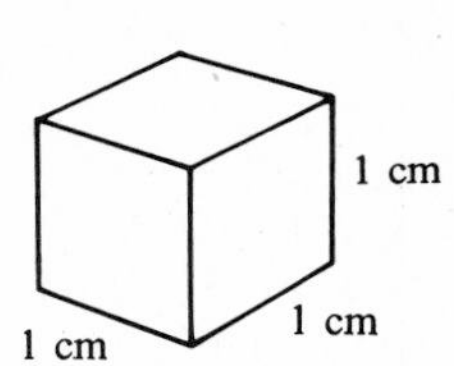

• ### • ### • ### • ### • ### • ### • ### •

A11 **a.** 1 cm^3 **b.** 1 ml (any order)

7 The metric units of volume can be arranged in order from largest to smallest for use in conversion.

$$\text{kl} \frac{10}{} \text{hl} \frac{10}{} \text{dal} \frac{10}{} \text{l} \frac{10}{} \text{dl} \frac{10}{} \text{cl} \frac{10}{} \text{ml (cm}^3)$$

Example: 3.8 l = ________ ml

Solution: $\text{kl} \frac{10}{} \text{hl} \frac{10}{} \text{dal} \frac{10}{} \text{l} \frac{10}{} \text{dl} \frac{10}{} \text{cl} \frac{10}{} \text{ml (cm}^3)$

"ml" is *three* 10's to the *right* of "l" on the chart. Move the decimal point *three* places to the *right* (3.800‸). Thus, 3.8 l = 3800 ml.

Q12 25 l = ______ dal

• ### • ### • ### • ### • ### • ### • ### •

A12 2.5: "dal" is *one* 10 to the *left* of "l" on the chart. Move the decimal point *one* place to the *left* (25.). Thus, 25 l = 2.5 dal.

Q13 200 ml = ______ cm^3

\### • ### • ### • ### • ### • ### • ### • ### • ###

A13 200: 1 ml = 1 cm^3, therefore 200 ml = 200 cm^3

Q14 Perform each conversion:

a. 750 ml = ______ l

b. 1.5 kl = ______ l

\### • ### • ### • ### • ### • ### • ### • ### • ###

A14 **a.** 0.75 **b.** 1500

---

8 For volume measured in cubic units, a common unit is the cubic metre (m^3). One cubic metre is the volume of a cube with edges equal to 1 metre. The cubic metre is the unit for selling such things as gravel and cement. Other units of volume are the cubic decimetre (dm^3) and the cubic centimetre (cm^3). The relationship between m^3 and dm^3 is seen using the volume formula and the relationship 1 m = 10 dm.

$$V = l \cdot w \cdot h$$
$$V = 1\text{ m} \cdot 1\text{ m} \cdot 1\text{ m} = 10\text{ dm} \cdot 10\text{ dm} \cdot 10\text{ dm}$$
$$V = 1\text{ m}^3 \qquad = 1000\text{ dm}^3$$

Thus, 1 m^3 = 1000 dm^3 or 1 dm^3 = 0.001 m^3. The relationship between cm^3 and dm^3 as pictured in Frame 5 is 1 dm^3 = 1000 cm^3 or 1 cm^3 = 0.001 dm^3.

Q15 **a.** For volume measured in cubic units, a common unit is the ______.

b. For liquid volume, a common unit is the ______.

\### • ### • ### • ### • ### • ### • ### • ### • ###

A15 **a.** m^3 **b.** litre

Q16 Complete each equation using the information of Frame 8:

a. 1 m^3 = ______ dm^3

b. 1 cm^3 = ______ dm^3

\### • ### • ### • ### • ### • ### • ### • ### • ###

A16 **a.** 1000 **b.** 0.001

---

9 Notice that any two units of volume from the list m^3, dm^3, cm^3, differ by a multiple of 1000. This fact is summarized for conversion purposes in the following diagram:

$$m^3 \overset{1000}{\text{———}} dm^3 \overset{1000}{\text{———}} cm^3$$

Since consecutive units of volume differ by 1000, conversion between any two consecutive units involves a movement of the decimal point *three* places to the right or *three* places to the left.

Example: 500 dm^3 = ______ m^3

Solution: "m^3" is one 1000 to the *left* of "dm^3." Move the decimal point three places to the *left* (.500.). Thus, 500 dm^3 = 0.5 m^3.

Q17 8.5 m^3 = ______ dm^3

\### • ### • ### • ### • ### • ### • ### • ### • ###

A17 8500: "dm^3" is *one* 1000 to the *right* of "m^3." Move the decimal point *three* places to the *right* (8.500.).

Q18 The volume of a storage tank is 25 cubic metres. What is the volume in cubic decimetres?

• ### • ### • ### • ### • ### • ### • ### •

A18 25 000 dm^3

Q19 A home excavation is 4 m wide, 8 m long, and 2 m high. The cubic metres of material to be removed is 64 m^3. What is the volume of material in cubic decimetres?

• ### • ### • ### • ### • ### • ### • ### •

A19 64 000 dm^3

---

10 One litre has a volume of 1000 cubic centimetres or 1 cubic decimetre. This fact can be used to determine the number of litres a given container will hold.

Example: A case has inside dimensions of 20 cm high by 70 cm long by 40 cm wide for a volume of 56 000 cm^3. What is the capacity of the case in litres?

Solution: 1000 cm^3 = 1 litre. To find the capacity in litres, divide the volume of the tank by 1000. Since $56\,000 \div 1000 = 56$, the capacity of the case is 56 litres.

Q20 A box has a volume of 8000 cubic centimetres. What is its capacity in litres?

• ### • ### • ### • ### • ### • ### • ### •

A20 8 litres: $8000 \div 1000 = 8$

Q21 A gasoline tank is 3 decimetres high, 2 decimetres wide, and 2 decimetres deep, for a volume of 12 cubic decimetres. What is the capacity of the tank in litres?

• ### • ### • ### • ### • ### • ### • ### •

A21 12 l: 1 cubic decimetre = 1 litre

This completes the instruction for this section.

6.2 EXERCISE

1. The base unit of mass in the metric system is the ________.

2. The approximate mass of 1 kg in the U.S. customary system is ________.

3. Complete each equation:

a. 1 kg = ______ g **b.** 1 mg = ______ g

c. 1 t = ______ kg **d.** 1 dag = ______ g

4. Arrange the units in order from the largest to smallest: ton, mg, g, kg.

5. Perform each of the following conversion:

a. 3.2 t = ______ kg **b.** 1750 g = ______ kg

c. 250 mg = ______ g **d.** 0.5 kg = ______ g

6. Find the mass in kilograms of 7.5 metric tons.

7. The mass of a shipment was recorded as 2000 kg. What is the mass in metric tons?

8. For liquid volume the commonly used metric unit is the ______.

9. For volume measured in cubic units the commonly used metric unit is the ______.

10. Complete each equation:

a. 1 kl = ______ l **b.** 1 ml = ______ l

c. 1 cm^3 = ______ ml **d.** 1 l = ______ dm^3

11. Perform each of the following conversions:

a. 7.5 kl = ______ l **b.** 1500 ml = ______ l

c. 500 m^3 = ______ cm^3 **d.** 35 m^3 = ______ dm^3

12. A milk tank has a volume of 150 000 cubic centimetres. What is the capacity of the tank in litres?

6.2 EXERCISE ANSWERS

1. kilogram	**2.** 2.2 pounds		
3. a. 1000	**b.** 0.001	**c.** 1000	**d.** 10
4. ton, kg, g, mg			
5. a. 3200	**b.** 1.75	**c.** 0.25	**d.** 500
6. 7500 kg	**7.** 2 t	**8.** litre	**9.** cubic metre (m^3)
10. a. 1000	**b.** 0.001	**c.** 1	**d.** 1
11. a. 7500	**b.** 1.5	**c.** 500 000 000	**d.** 35 000
12. 150 litres			

6.3 METRIC–ENGLISH CONVERSION

1 During the period of gradual change from the U.S. customary system there will be initially, for some, confusion as what a given measurement in one system means in terms of the other system. Although, ideally, everyone will quickly learn to "think metric," most likely this will not happen overnight. The purpose of this section, therefore, is to develop the ability to perform conversions between units of the two systems.

Recall from Section 6.1 the scale drawing comparing 1 metre and 1 yard.

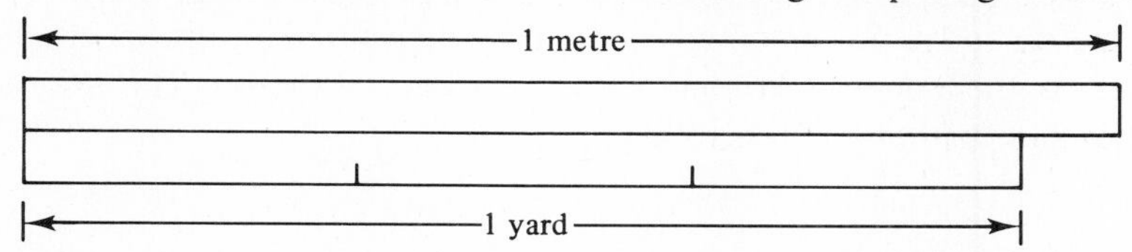

The approximate relationship between the metre and yard is:

1 m = 1.1 yd or 1 yd = 0.9 m*

The approximate relationship between a metre and inches is:

1 m = 39.37 in.

*In this text, the use of the "=" sign between a customary and a metric measurement will be understood to mean "is approximately equal to."

Q1 **a.** Which is longer, a metre or a yard? ________
b. If a person is 2 m tall, is he taller or shorter than 6 feet (2 yd)? ________
c. Write the following in order from smallest to largest: yard, inch, foot, metre. ________

____

• ### • ### • ### • ### • ### • ### • ### •

A1 **a.** a metre **b.** taller **c.** inch, foot, yard, metre

2 The relationship among the inch, centimetre, and millimetre can be seen in a comparison of the metric and customary rulers. Notice that a large paper clip approximates the metric units centimetre

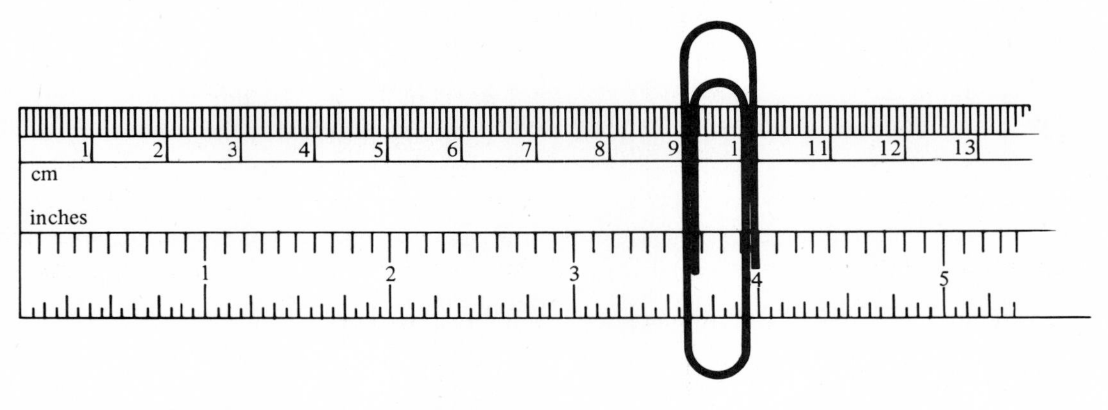

and millimetre quite closely; that is,

1 centimetre = width of a large paper clip
1 millimetre = diameter of a paper-clip wire

Q2 **a.** Use a large paper clip to estimate the length of the following line

____

b. Which is larger, 1 cm or 1 in.? ______
c. Write the following in order from smallest to largest: mm, in., ft, yd, m, cm. ________

• ### • ### • ### • ### • ### • ### • ### •

A2 **a.** 5 cm **b.** 1 in. **c.** mm, cm, in., ft, yd, m

3 Conversions between the metric and U.S. customary systems of measurement are most efficiently made using a chart similar to the one that follows and a pocket calculator. The approximate relationships between metric and customary units of length are:

To Convert	*Into*	*Multiply by*	*Conversion Fact*
millimetres (mm)	inches (in.)	0.04	1 mm = 0.04 in.
centimetres (cm)	inches (in.)	0.4	1 cm = 0.4 in.
metres (m)	feet (ft)	3.3	1 m = 3.3 ft
metres (m)	yards (yd)	1.1	1 m = 1.1 yd
inches (in.)	centimeters (cm)	2.5	1 in. = 2.5 cm
feet (ft)	centimetres (cm)	30	1 ft = 30 cm
yards (yd)	metres (m)	0.9	1 yd = 0.9 m

Example 1: 12 m = ______ yd

Solution: To convert "m" into "yd" multiply by 1.1. Thus, since 12(1.1) = 13.2, 12 m = 13.2 yd.

Example 2: 440 yd = ______ m

Solution: To convert "yd" to "m" multiply by 0.9. Thus, since 440(0.9) = 396, 440 yd = 396 m

Q3 Perform each conversion:

a. 6 in. = ______ cm **b.** 100 yd = ______ m

c. 15 mm = ______ in. **d.** 400 m = ______ yd

• ### • ### • ### • ### • ### • ### • ### •

A3 **a.** 15 **b.** 90 **c.** 0.6 **d.** 440

4 When approximate equivalents between the metric and customary system are used repeatedly, it is sometimes helpful to have conversion charts available. The accompanying chart lists the millimetre equivalents for fractions of an inch given in fraction and decimal form. The use of the chart is shown in the following example:

Example: Use the chart to find: $\frac{5}{8}$ in. = ________ mm

Solution: Locate $\frac{5}{8}$ in the inch column on the chart. Reading to the right of $\frac{5}{8}$, the chart gives the decimal equivalent of $\frac{5}{8}$ (0.625) and the millimetre equivalent (15.875). Thus, $\frac{5}{8}$ in. = 15.875 mm.

Fractions of an Inch to Millimetres

inch		*mm*	*inch*		*mm*	*inch*		*mm*	*inch*		*mm*
1/64	0.016	0.397	17/64	0.266	6.747	33/64	0.516	13.097	49/64	0.766	19.447
1/32	0.031	0.794	9/32	0.281	7.144	17/32	0.531	13.494	25/32	0.781	19.844
3/64	0.047	1.191	19/64	0.297	7.541	35/64	0.547	13.891	51/64	0.797	20.241
1/16	0.062	1.588	5/16	0.312	7.938	9/16	0.562	14.288	13/16	0.812	20.638
5/64	0.078	1.984	21/64	0.328	8.334	37/64	0.578	14.684	53/64	0.828	21.034
3/32	0.094	2.381	11/32	0.344	8.731	19/32	0.594	15.081	27/32	0.844	21.431
7/64	0.109	2.778	23/64	0.359	9.128	39/64	0.609	15.478	55/64	0.859	21.828
1/8	0.125	3.175	3/8	0.375	9.525	5/8	0.625	15.875	7/8	0.875	22.225
9/64	0.141	3.572	25/64	0.391	9.922	41/64	0.641	16.272	57/64	0.891	22.622
5/32	0.156	3.969	13/32	0.406	10.319	21/32	0.656	16.669	29/32	0.906	23.019
11/64	0.172	4.366	27/64	0.422	10.716	43/64	0.672	17.066	59/64	0.922	23.416
3/16	0.188	4.762	7/16	0.438	11.112	11/16	0.688	17.462	15/16	0.938	23.812
13/64	0.203	5.159	29/64	0.453	11.509	45/64	0.703	17.859	61/64	0.953	24.209
7/32	0.219	5.556	15/32	0.469	11.906	23/32	0.719	18.256	31/32	0.969	24.606
15/64	0.234	5.953	31/64	0.484	12.303	47/64	0.734	18.653	63/64	0.984	25.003
1/4	0.250	6.350	1/2	0.500	12.700	3/4	0.750	19.050	1	1.000	25.400

Q4 Use the chart to complete the following:

a. $\frac{15}{16}$ in. = ________ mm **b.** 0.125 in. = ________ mm

• ### • ### • ### • ### • ### • ### • ### •

A4 **a.** 0.938 **b.** 3.175: $0.125 \text{ in.} = \frac{1}{8} \text{ in.}$

Q5 Metric wrenches are stamped with the same size as the size of the head or nut. The M10 metric wrench shown represents a distance across the flats of 10 mm. Use the chart to find the fractional part of an inch closest to 10 mm.

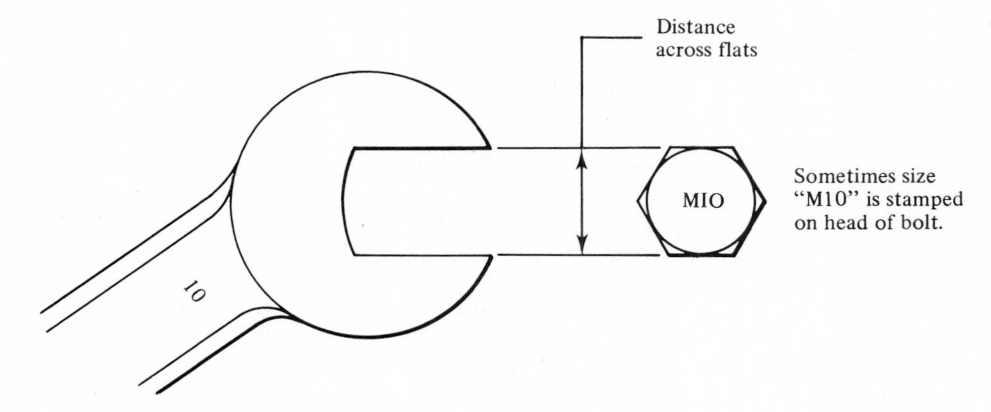

\# # # • # # # • # # # • # # # • # # # • # # # • # # # • # # # • # # #

A5 $\frac{13}{32}$: The chart reading closest to 10 mm is 10.319 mm. Its fractional equivalent is $\frac{13}{32}$ in.

Q6 The boring tools shown will not need replacement when converting to the metric system. They can be used until wearing out because their sizes convert to approximate metric equivalents, which is satisfactory in woodworking. Find millimetre equivalents for each of the fractional inches below and round each value to the nearest millimetre.

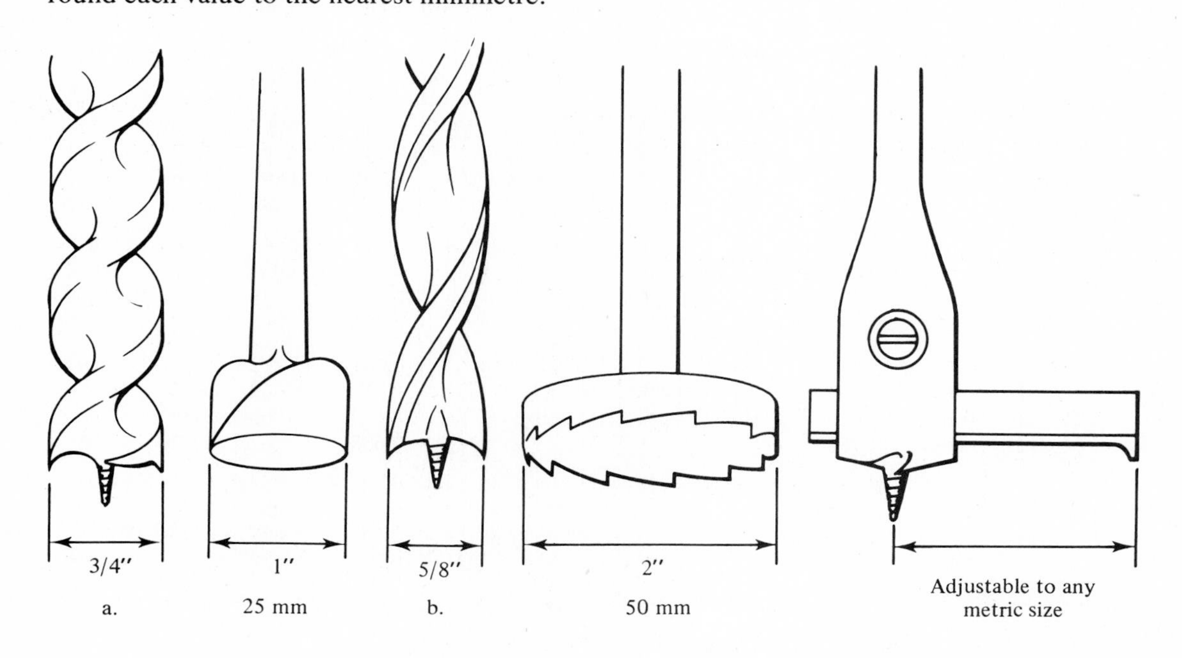

a. ________ **b.** ________

\# # # • # # # • # # # • # # # • # # # • # # # • # # # • # # # • # # #

A6 **a.** 19 mm **b.** 16 mm

5 The metric unit of measure for long distances is the kilometre. It is comparable to the U.S. customary unit, the mile. Since 1 metre is approximately 3 feet, 1 kilometre is 1000 metres or 1000 (3 feet) = 3000 ft. Recall that 5280 ft = 1 mile. Thus 1 kilometre is a little longer than $\frac{1}{2}$ mile. For conversion purposes, the mile and kilometre are approximately related as follows:

To Convert	*Into*	*Multiply by*	*Conversion Fact*
kilometres (km)	miles (mi)	0.6	1 km = 0.6 mi
miles (mi)	kilometres (km)	1.6	1 mi = 1.6 km

Q7 **a.** Which is larger, a kilometre or a mile? ________

b. Write the following in order from smallest to largest: in., cm, mm, yd, ft, mi, m, km.

____

\# # # • # # # • # # # • # # # • # # # • # # # • # # # • # # # • # # #

A7 **a.** mile **b.** mm, cm, in., ft, yd, m, km, mi

Q8 Perform each conversion:

a. 10 mi = ______ km **b.** 50 km = ______ mi

\# # # • # # # • # # # • # # # • # # # • # # # • # # # • # # # • # # #

A8 **a.** 16 **b.** 30

6 The following is a scale that gives approximate relationships between kilometres and miles. It can be used to check answers to conversion problems or to quickly estimate one reading in terms of the other.

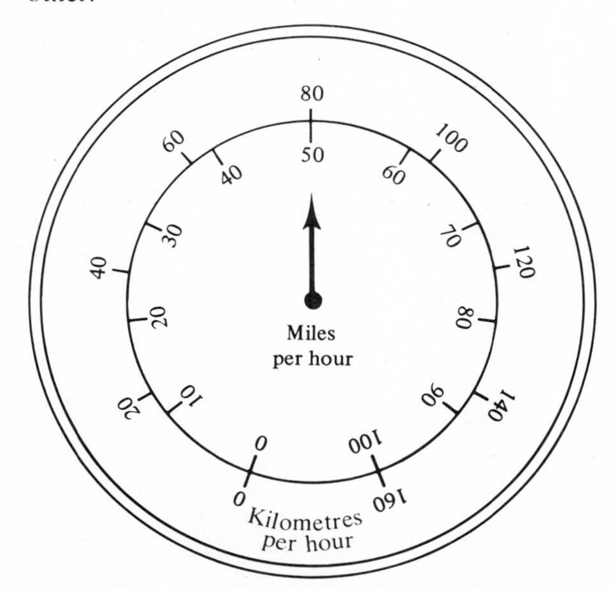

Q9 What is the approximate speed limit, in miles per hour, indicated by each of the metric speed-limit signs (kilometres per hour):

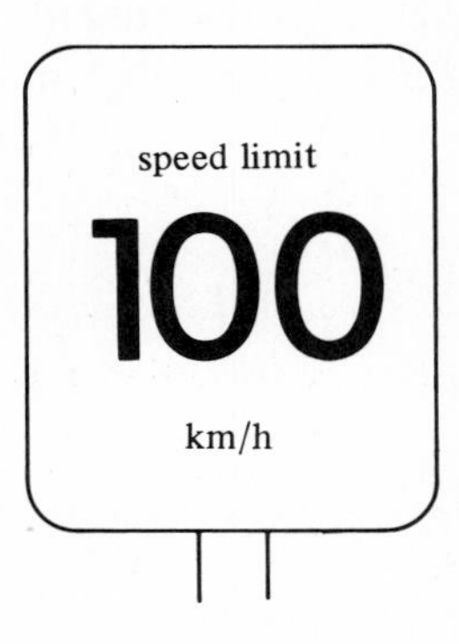

speed limit
40
km/h

a. ________ mph **b.** ________ mph **c.** ________ mph

\# \# \# • \# \# \# • \# \# \# • \# \# \# • \# \# \# • \# \# \# • \# \# \# • \# \# \# • \# \# \#

A9 **a.** 63 **b.** 25 **c.** 50

Q10 Approximate the distance to each of the following cities:

a. Detroit, 70 kilometres ________

b. Lansing, 105 kilometres ________

c. Mackinac Bridge, 450 kilometres ________

\# \# \# • \# \# \# • \# \# \# • \# \# \# • \# \# \# • \# \# \# • \# \# \# • \# \# \# • \# \# \#

A10 **a.** 42 mi **b.** 63 mi **c.** 270 mi

7 Units of area in the metric and U.S. customary systems are related in the same way as their corresponding linear (length) units. Thus, since 1 metre is larger than 1 yard, 1 square metre is larger than 1 square yard. Similarly, just as 1 inch is larger than 1 centimetre, 1 square inch is larger than 1 square centimetre.

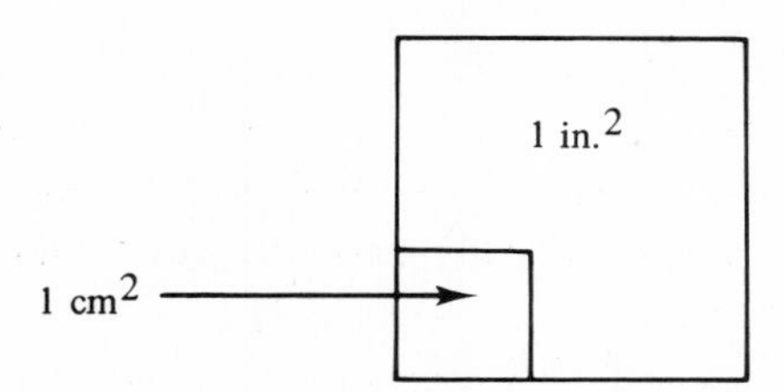

Some of the approximate relationships for units of area are:

To Convert	*Into*	*Multiply by*	*Conversion Fact*
square centimetres (cm^2)	square inches ($in.^2$)	0.16	$1\ cm^2 = 0.16\ in.^2$
square metres (m^2)	square yards (yd^2)	1.2	$1\ m^2 = 1.2\ yd^2$
square inches ($in.^2$)	square centimetres (cm^2)	6.5	$1\ in.^2 = 6.5\ cm^2$
square feet (ft^2)	square metres (m^2)	0.09	$1\ ft^2 = 0.09\ m^2$
square yards (yd^2)	square metres (m^2)	0.8	$1\ yd^2 = 0.8\ m^2$

Q11 A 9×12 yd rug covers an area of 108 yd^2. What is the area measure in square metres?

$108 \text{ yd}^2 = \_\_\_\_\_ \text{ m}^2$

• ### • ### • ### • ### • ### • ### • ### •

A11 86.4

Q12 The area of the bottom of a swimming pool is 1800 square feet. Find the area in square metres.

• ### • ### • ### • ### • ### • ### • ### •

A12 162 m^2

---

8 Large areas are measured in hectares or square kilometres in the metric system of measurement and in acres or square miles in the English system of measurement. These units are related as follows:

To Convert	*Into*	*Multiply by*	*Conversion Fact*
hectares (ha)	acres	2.5	1 ha = 2.5 acres
square kilometres (km^2)	square miles (mi^2)	0.4	1 km^2 = 0.4 mi^2
square miles (mi^2)	square kilometres (km^2)	2.6	1 mi^2 = 2.6 km^2
acres	hectares (ha)	0.4	1 acre = 0.4 ha

Q13 The area of a small farm is 2 hectares. What is the area of the farm in acres?

• ### • ### • ### • ### • ### • ### • ### •

A13 5 acres

Q14 A land development site has an area of 12 square miles. What is the area of the site in square kilometres?

• ### • ### • ### • ### • ### • ### • ### •

A14 31.2 km^2

---

9 The metric units gram and kilogram are used in the same way as the U.S. customary units ounce and pound. The approximate relationships between these units are:

To Convert	*Into*	*Multiply by*	*Conversion Fact*
grams (g)	ounces (oz)	0.035	1 g = 0.035 oz
kilograms (kg)	pounds (lb)	2.2	1 kg = 2.2 lb
ounces (oz)	grams (g)	28	1 oz = 28 g
pounds (lb)	grams (g)	454	1 lb = 454 g
pounds (lb)	kilograms (kg)	0.454	1 lb = 0.454 kg

To help yourself think metric, notice that a kilogram is more than twice as large as a pound, while a gram is only approximately $\frac{1}{28}$ of an ounce. Some common comparisons with the gram are the weight of a nickel (5 g), and the weight of a pound of butter (454 g).

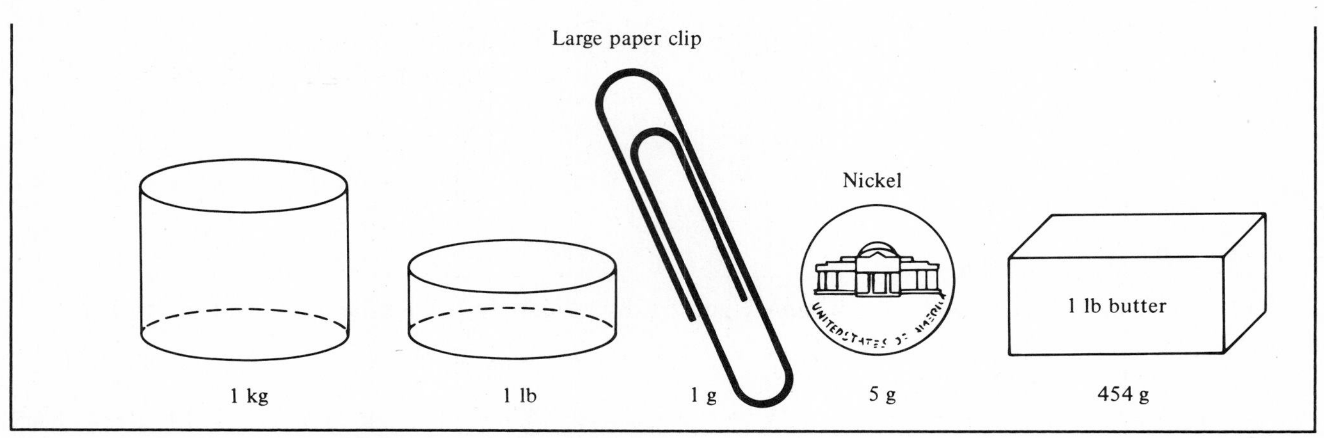

Q15 Write the following in order from smallest to largest: g, lb, kg, oz, mg. ________

• ### • ### • ### • ### • ### • ### • ### •

A15 mg, g, oz, lb, kg

Q16 A man's weight is 83 kg. What is his weight in pounds?

• ### • ### • ### • ### • ### • ### • ### •

A16 182.6 lb

Q17 A truck has a mass of 7700 pounds. What is the mass of the truck in kilograms?

• ### • ### • ### • ### • ### • ### • ### •

A17 3496 kg

Q18 A company markets its mushrooms throughout the world. What metric weight, in grams, should be placed on 8 ounce, 1 pound, and $1\frac{1}{2}$ pound cans of mushrooms? (Use 1 lb = 454 g.)

a. 8 oz ________ **b.** 1 pound ________ **c.** $1\frac{1}{2}$ lb ________

• ### • ### • ### • ### • ### • ### • ### •

A18 **a.** 227 g **b.** 454 g **c.** 681 g

---

10 Notice that the answers of Q15 could have been completed with two different conversion facts. For example, 8 ounces can be (1) converted directly to grams or (2) thought of as $\frac{1}{2}$ pound and then converted to grams.

(1) 8 oz = ______ g

To convert "oz" to "g" multiply by 28. Thus, 8 oz = 224 g.

(2) $\frac{1}{2}$ lb = ______ g

To convert "lb" to "g" multiply by 454. Thus, $\frac{1}{2}$ lb = 227 g.

The fact that the answers are different brings up an important point about performing conversions with *approximate* relationships between units. When using approximate relations between units, answers may vary slightly depending on the approximation that is used. Although the answers vary slightly, both are considered correct.

Q19 Perform the conversion 16 oz = ______ g in the two ways indicated:

a. 16 oz = 448 g (Use 1 oz = 28 g) **b.** 16 oz (1 lb) = 454 g (Use 1 lb = 454 g)

c. Do the answers to parts **a.** and **b.** differ slightly? ______

d. Which answer is considered correct? ______

\# # # • # # # • # # # • # # # • # # # • # # # • # # # • # # # • # # #

A19 **a.** 448 **b.** 454 **c.** yes **d.** both

11 Common units of liquid volume in the metric and U.S. customary systems are the litre and the quart. Recall from Section 6.2 that a litre is slightly larger than a quart. Because the litre and quart are nearly equal, the litre is related to the pint and gallon in much the same way as the quart.

The approximate relationships are:

To Convert	*Into*	*Multiply by*	*Conversion Fact*
litres (l)	pints (pt)	2.1	1 l = 2.1 pt
litres (l)	quarts (qt)	1.06	1 l = 1.06 qt
litres (l)	gallons (gal)	0.26	1 l = 0.26 gal
pints (pt)	litres (l)	0.47	1 pt = 0.47 l
quarts (qt)	litres (l)	0.95	1 qt = 0.95 l
gallons (gal)	litres (l)	3.8	1 gal = 3.8 l

Q20 **a.** Which is larger, a quart of milk or a litre of milk? ______

b. Which is larger, $\frac{1}{2}$ gal of milk (2 qt) or 2 l of milk? ______

c. Which is larger, 1 pt ($\frac{1}{2}$ qt) of cream or $\frac{1}{2}$ l? ______

\# # # • # # # • # # # • # # # • # # # • # # # • # # # • # # # • # # #

A20 **a.** litre **b.** 2 l **c.** $\frac{1}{2}$ l

Q21 A container holds $2\frac{1}{2}$ litres of wine. Find its volume in quarts.

2.5 · 1.06 = 2.65 qt.

• ### • ### • ### • ### • ### • ### • ### •

A21 2.65 qt

Q22 What is the capacity of a 20 gallon gasoline tank in litres?

• ### • ### • ### • ### • ### • ### • ### •

A22 76 l

12 Small amounts of liquid volume are measured in millilitres or ounces. Recall that a millilitre is equal in volume to a cubic centimetre, a cube with sides equal to 1 centimetre. The approximate size of a millilitre can be imagined from the fact that it takes 5 millilitres to equal 1 teaspoonful. The approximate relationships between millilitre and ounce are:

To Convert	*Into*	*Multiply by*	*Conversion Fact*
millilitres (ml)	fluid ounces (oz)	0.03	1 ml = 0.03 oz
fluid ounces (oz)	millilitres (ml)	30	1 oz = 30 ml

Q23 Coca-Cola markets its soft drink in a 10 fluid ounce can. What is the equivalent size in millilitres?

• ### • ### • ### • ### • ### • ### • ### •

A23 300 ml

Q24 A drinking glass is made to hold a volume of approximately 250 ml. What is its volume in fluid ounces?

• ### • ### • ### • ### • ### • ### • ### •

A24 7.5 ounces

13 Volume measures in cubic units are approximately related in the metric and U.S. customary systems as follows:

To Convert	*Into*	*Multiply by*	*Conversion Fact*
cubic centimetres (cm^3)	cubic inches ($in.^3$)	0.06	$1\ cm^3 = 0.06\ in.^3$
cubic metres (m^3)	cubic feet (ft^3)	35	$1\ m^3 = 35\ ft^3$
cubic metres (m^3)	cubic yards (yd^3)	1.3	$1\ m^3 = 1.3\ yd^3$
cubic inches ($in.^3$)	cubic centimetres (cm^3)	16	$1\ in.^3 = 16\ cm^3$
cubic feet (ft^3)	cubic metres (m^3)	0.03	$1\ ft^3 = 0.03\ m^3$
cubic yard (yd^3)	cubic metres (m^3)	0.76	$1\ yd^3 = 0.76\ m^3$

Q25 A man ordered 10 cubic yards of gravel for his driveway. What is the equivalent order in cubic metres?

• ### • ### • ### • ### • ### • ### • ### •

A25 7.6 m^3

Q26 The volume of a home tool box 56 cm by 22 cm × 25 cm is 30 800 cm^3. What is the volume in cubic inches?

• ### • ### • ### • ### • ### • ### • ### •

A26 1848 $in.^3$

14 In countries using the metric system two scales are used to measure temperature. The *Kelvin* scale is used for calculations involving gases while the *Celsius* scale is used for such things as air and body temperature measurement.

The Kelvin scale is named in honor of the British physicist and mathematician Lord Kelvin (William Thomson, 1824–1907). The Celsius scale is named after the Swedish astronomer Anders Celsius (1701–1744) who invented it.

The symbols for Kelvin (K) and degree Celsius (°C) are capitalized since they are derived from the names of persons. The Kelvin symbol is written without a degree symbol. Both symbols are written without a period and with one space separating them from the numerical reading.

Examples: 310.15 K not 310.15K
0 °C not 0°C
100 °C not 100° C

Q27 Circle the temperature readings below which are written correctly.

310.15 °K, 25 °C, 10°C, 200 K, 30 °C.

• ### • ### • ### • ### • ### • ### • ### •

A27 25 °C, 200 K: 310 °K (no degree symbol needed)
10°C (needs space between 10 and °C)
30 °C. (°C written without period)

15 Temperature in the U.S. customary system is measured using the Fahrenheit scale. The Fahrenheit scale was named in honor of Gabriel Fahrenheit (1686–1736), a German physicist. The three scales are compared below:

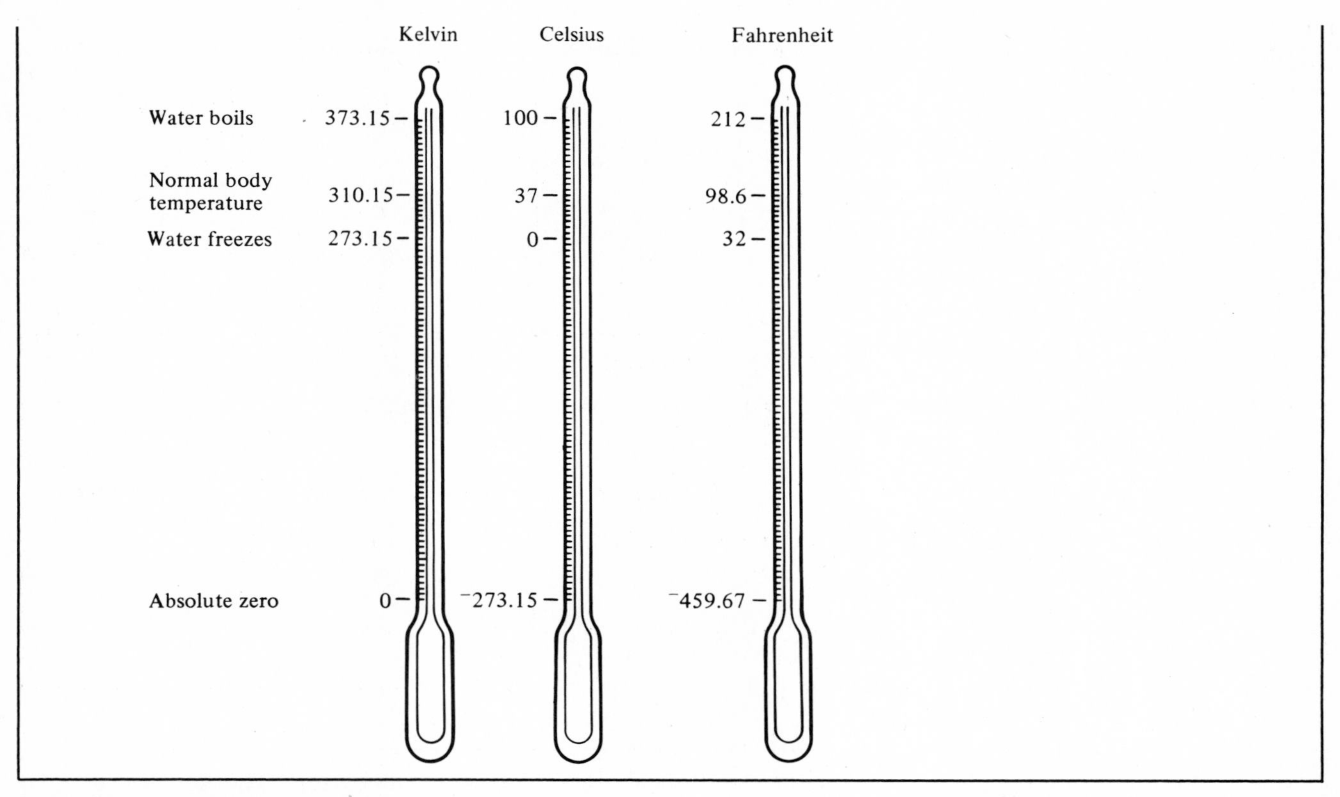

Q28 Complete the table:

	Normal Body Temperature	*Freezing Point*	*Boiling Point*
a. Celsius	_____	_____	_____
b. Fahrenheit	_____	_____	_____

\# # # • # # # • # # # • # # # • # # # • # # # • # # # • # # # • # # #

A28 **a.** 37 °C, 0 °C, 100 °C **b.** 98.6 °F, 32 °F, 212 °F

16 Notice that the difference between the freezing and boiling points is 100 on both the Celsius and Kelvin scales. In fact, the only difference between the two scales is that each Kelvin temperature is 273.15 units more than the corresponding Celsius temperature. That is, Celsius (C) and Kelvin (K) temperatures are related by the equations:

$K = C + 273.15$

$C = K - 273.15$

Example 1: What is the Kelvin temperature for 20 °C?

Solution: $K = C + 273.15$

$K = 20 + 273.15$

$K = 293.15$

Thus, 20 °C = 293.15 K.

Example 2: What is the Celsius temperature for 370.15 K?

Solution: $C = K - 273.15$

$C = 370.15 - 273.15$

$C = 97$

Thus, 370.15 K = 97 °C.

Q29 **a.** What is the Kelvin temperature for 50 °C? _______

b. What is the Celsius temperature for 310.15 K? _______

\# # # • # # # • # # # • # # # • # # # • # # # • # # # • # # # • # # #

A29 **a.** 323.15 K **b.** 37 °C

17 The conversion formulas for Celsius (C) and Fahrenheit (F) temperatures are as follows:

$C = \frac{F-32}{1.8}$ and $F = 1.8C + 32$

To change 75 °F to its corresponding Celsius temperature, we use the formula $C = \frac{F-32}{1.8}$ as follows:

$$C = \frac{F-32}{1.8} = \frac{75-32}{1.8} = \frac{43}{1.8} = 23.9$$

Hence, 75 °F = 23.9 °C.

All temperatures will be rounded off to the nearest tenth.

Q30 50 °F = _______ °C

\# # # • # # # • # # # • # # # • # # # • # # # • # # # • # # # • # # #

A30 10: $C = \frac{F-32}{1.8} = \frac{50-32}{1.8} = \frac{18}{1.8} = 10$

18 To change 25 °C to its corresponding Fahrenheit temperature, we use the formula F = 1.8C + 32 as follows:

$F = 1.8C + 32$

$F = 1.8(25) + 32$

$F = 45 + 32$

$F = 77$ Hence, 25 °C = 77 °F

Q31 80 °C = _______ °F

\# # # • # # # • # # # • # # # • # # # • # # # • # # # • # # # • # # #

A31 176: $F = 1.8C + 32 = 1.8(80) + 32 = 144 + 32 = 176$

Q32 Complete each of the following:

a. 95 °F = _____ °C **b.** 0 °C = _____ °F

c. 0 °F = _____ °C **d.** 10 °C = _____ °F

\# # # • # # # • # # # • # # # • # # # • # # # • # # # • # # # • # # #

A32 **a.** 35 **b.** 32 **c.** $^{-}17.8$ **d.** 50

19 The following chart is a comparison of the Celsius and Fahrenheit scales. It can be used for quick approximations between the two scales.

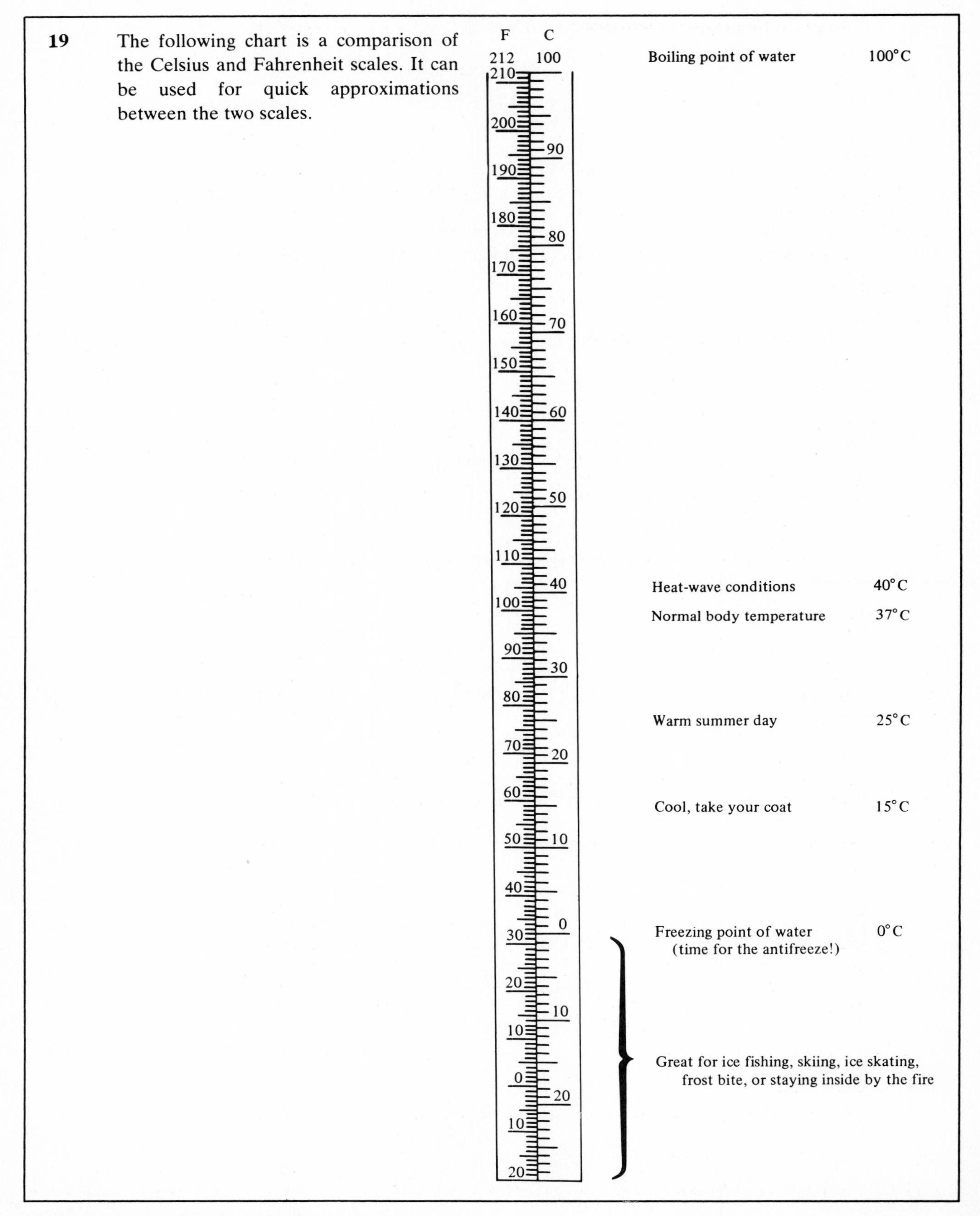

This completes the instruction for this section.

Summary of Approximate Conversion Factors—Metric to Customary Conversion

To Convert	*Into*	*Multiply by*	*Conversion Fact*
	Length		
millimetres (mm)	inches (in.)	0.04	1 mm = 0.04 in.
centimetres (cm)	inches (in.)	0.4	1 cm = 0.4 in.
metres (m)	feet (ft)	3.3	1 m = 3.3 ft
metres (m)	yards (yd)	1.1	1 m = 1.1 yd
kilometres (km)	miles (mi)	0.6	1 km = 0.6 mi
	Area		
square centimetres (cm^2)	square inches ($in.^2$)	0.16	1 cm^2 = 0.16 $in.^2$
square metres (m^2)	square yards (yd^2)	1.2	1 m^2 = 1.2 yd^2
	Mass (weight)		
grams (g)	ounces (oz)	0.035	1 g = 0.035 oz
kilograms (kg)	pounds (lb)	2.2	1 kg = 2.2 lb
	Volume		
millilitres (ml)	fluid ounces (oz)	0.03	1 ml = 0.03 oz
litres (l)	pints (pt)	2.1	1 l = 2.1 pt
litres (l)	quarts (qt)	1.06	1 l = 1.06 qt
litres (l)	gallons (gal)	0.26	1 l = 0.26 gal
cubic centimetres (cm^3)	cubic inches ($in.^3$)	0.06	1 cm^3 = 0.06 $in.^3$
cubic metres (m^3)	cubic feet (ft^3)	35	1 m^3 = 35 ft^3
cubic metres (m^3)	cubic yards (yd^3)	1.3	1 m^3 = 1.3 yd^2

Summary of Approximate Conversion Factors—Customary to Metric Conversion

To Convert	*Into*	*Multiply by*	*Conversion Fact*
	Length		
inches (in.)	centimetres (cm)	2.5	1 in. = 2.5 cm
feet (ft)	centimetres (cm)	30	1 ft = 30 cm
yards (yd)	metres (m)	0.9	1 yd = 0.9 m
miles (mi)	kilometres (km)	1.6	1 mi = 1.6 km
	Area		
square inches ($in.^2$)	square centimetres (cm^2)	6.5	1 $in.^2$ = 6.5 cm^2
square feet (ft^2)	square metres (m^2)	0.09	1 ft^2 = 0.09 m^2
square yards (yd^2)	square metres (m^2)	0.8	1 yd^2 = 0.8 m^2
	Mass (weight)		
ounces (oz)	grams (g)	28	1 oz = 28 g
pounds (lb)	kilograms (kg)	0.454	1 lb = 0.454 kg
	Volume		
ounces (oz)	millilitres (ml)	30	1 oz = 30 ml
pints (pt)	litres (l)	0.47	1 pt = 0.47 l
quarts (qt)	litres (l)	0.95	1 qt = 0.95 l
gallons (gal)	litres (l)	3.8	1 gal = 3.8 l
cubic inches ($in.^3$)	cubic centimetres (cm^3)	16.39	1 $in.^3$ = 16.39 cm^3
cubic feet (ft^3)	cubic metres (m^3)	0.03	1 ft^3 = 0.03 m^3
cubic yards (yd^3)	cubic metres (m^3)	0.76	1 yd^3 = 0.76 m^3

6.3 EXERCISE

1. Place each of the following sets of units in order from smaller to larger:

a. m, cm, in., yd, mm, ft, km, mi

b. oz, g, lb, kg, mg

c. ml, oz, qt, l, gal

2. Perform each of the following conversions using the summary of approximate conversion factor tables at the end of this section:

a. 80 mi = ________ km **b.** 67 in. = ________ cm

c. 2.5 mm = ________ in. **d.** 150 km = ________ mi

3. A draftsman recorded the following measurements for a small car garage: length 6000 mm, width 3600 mm, window width 600 mm, and overhead door dimensions 2300 mm × 2000 mm. Find the measurements of the garage in inches (rounded to the nearest inch).

4. Perform each of the following conversions:

a. 50 cm^3 = ________ $in.^3$ **b.** 144 yd^3 = ________ m^3

c. 81 yd^2 = ________ m^2 **d.** 85 acres = ________ ha

5. The area of a land development site is 25 hectares. What is the area of the site in acres?

6. Perform each of the following conversions:

a. 12 oz = ______ g **b.** 5 lb = ______ kg

7. A man weighs 180 pounds. What is his weight in kilograms?

8. Perform each of the following conversions:

a. 4 qt = ______ l **b.** 300 ml = ______ oz

9. A man's gas purchase was 20 litres. What is his purchase in gallons?

10. Perform the following conversions between the Celsius and Fahrenheit temperature scales:

a. 212 °F = ______ °C **b.** 45 °C = ______ °F

c. 0 °C = ______ °F **d.** 37 °C = ______ °F

6.3 EXERCISE ANSWERS

1. a. mm, cm, in., ft, yd, m, km, mi
b. mg, g, oz, lb, kg
c. ml, oz, qt, l, gal

2. a. 128 **b.** 167.5 **c.** 0.1 **d.** 90

3. length 240 in., width 144 in., window width 24 in., overhead door 92 × 80 in.

4. a. 3 **b.** 109.44 **c.** 64.8 **d.** 34

5. 62.5 acres **6. a.** 336 **b.** 2.27

7. 81 kg **8. a.** 3.8 **b.** 9 **9.** 5.2 gal

10. a. 100 **b.** 113 **c.** 32 **d.** 98.6

CHAPTER 6 SAMPLE TEST

At the completion of Chapter 6 you should be able to work the following problems. [*Note*: The two tables showing summaries of approximate conversion factors (on page 321) will be provided when you are taking a post-test on this chapter.]

6.1 UNITS OF LENGTH AND AREA

1. The base unit of length in the metric system is the ____________.

2. Write the symbol for each of the following:

a. centimetre **b.** kilometre

c. metre **d.** millimetre

3. Complete each of the following equations:

a. 1 m = ________ cm **b.** 1 mm = ________ m

c. 1 km = ________ m **d.** 1 dm = ________ cm

4. Write from memory the seven metric units of length arranged from largest to smallest.

5. A 10 separates each of the units in the list of Question 4. The 10's indicate that any unit is equal to the unit on its right ________ by 10 and the unit on its left ________ by 10.

6. Perform each of the following conversions:

a. 500 m = ________ km **b.** 39.5 m = ________ cm

c. 46 dm = ________ mm **d.** 75 mm = ________ m

7. Write the common names for the following when land is measured in:

a. dam^2 **b.** hm^2

8. A shopping center is to be built on a 5 hectare site. What is the area of the site in ares?

9. Complete each of the following equations:

a. 500 dm^2 = ________ m^2 **b.** 4200 cm^2 = ________ dm^2

c. 450 000 a = ________ ha **d.** 75 km^2 = ________ a

6.2 UNITS OF MASS AND VOLUME

10. The base unit of mass in the metric system is the ________.

11. Complete each of the following equations:

a. 1 g = ________ mg **b.** 1 kg = ________ g

c. 1 t = ________ kg **d.** 1 hg = ________ g

12. Write from memory the seven metric units of mass in order from largest to smallest.

13. Perform each of the following conversions:

a. 5 t = ________ kg **b.** 1575 g = ________ kg

c. 750 mg = ________ g **d.** 0.5 kg = ________ g

14. The mass of a truck load shipment was 2.5 t. What is the mass in kilograms?

15. The metric unit of volume which most closely approximates a quart is the ________.

16. A litre is the equivalent of ________ cubic decimetre.

17. Complete each equation:

a. 1 ml = ________ cm^3 **b.** 1 l = ________ ml

18. Perform each of the following conversions:

a. 5 m^3 = ________ dm^3 **b.** 25 kl = ________ l

c. 500 ml = ________ l **d.** 30 cm^3 = ________ ml

6.3 METRIC–ENGLISH CONVERSION

19. Rank the following sets of units in order from largest to smallest:

a. oz, g, lb, mg, kg

b. l, gal, qt, oz, ml

c. cm, mm, m, in., ft, yd, mi, km

20. Use the tables of approximate conversion factors to perform the following conversions:

a. 80 in. = 200 cm **b.** 55 mi = 88 km

c. 2 lb = .908 kg **d.** 25 l = 6.5 gal

21. Complete the following table:

	°C	°F
a. Normal body temperature	37	98.6
b. Freezing point of water	0	32
c. Boiling point of water	100	212

22. Perform the following conversions (round answers to the nearest tenth):

a. 20 °C = 68 °F **b.** 85 °F = 29.4 °C

c. 35 °C = 308.15 K **d.** 300.15 K = 27 °C

CHAPTER 6 SAMPLE TEST ANSWERS

1. metre **2. a.** cm **b.** km **c.** m **d.** mm
3. a. 100 **b.** 0.001 **c.** 1000 **d.** 10
4. km, hm, dam, m, dm, cm, mm
5. multiplied, divided
6. a. 0.5 **b.** 3950 **c.** 4600 **d.** 0.075
7. a. are **b.** hectare **8.** 500 a
9. a. 5 **b.** 42 **c.** 4500 **d.** 750 000
10. kilogram **11. a.** 1000 **b.** 1000 **c.** 1000 **d.** 100
12. kg, hg, dag, g, dg, cg, mg
13. a. 5000 **b.** 1.575 **c.** 0.75 **d.** 500
14. 2500 kg **15.** litre **16.** one **17. a.** 1 **b.** 1000
18. a. 5000 **b.** 25 000 **c.** 0.5 **d.** 30
19. a. kg, lb, oz, g, mg
b. gal, l, qt, oz, ml
c. mi, km, m, yd, ft, in., cm, mm
20. a. 200 **b.** 88 **c.** 0.908 **d.** 6.5
21. a. 37, 98.6 **b.** 0, 32 **c.** 100, 212
22. a. 68 **b.** 29.4 **c.** 308.15 **d.** 27

CHAPTER 7

SIGNED NUMBERS

7.1 INTRODUCTION

1 The whole numbers* can be pictured using what is called a *number line* as follows:

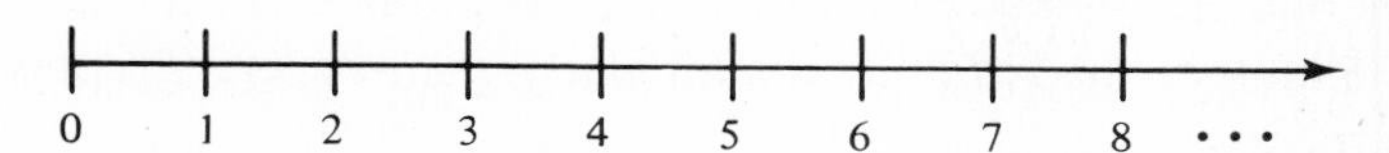

Each dot on the number line indicates the point where a specific number occurs and is called the *graph* of the number. The number that corresponds with each dot is called the *coordinate* of the point. The arrow on the end of the number line and the three dots that follow the last number shown both serve to indicate that the whole numbers are infinite or unending.

*The whole numbers are 0, 1, 2, 3, and so on. The next whole number is formed by adding one to the preceding whole number.

Q1 Graph the following whole numbers by placing dots at the appropriate points on the number line:
a. 7 **b.** 2 **c.** 0

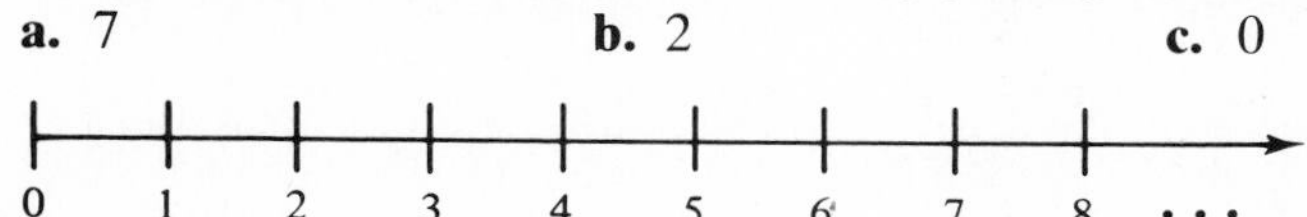

\# \# \# • \# \# \# • \# \# \# • \# \# \# • \# \# \# • \# \# \# • \# \# \# • \# \# \# • \# \# \#

A1

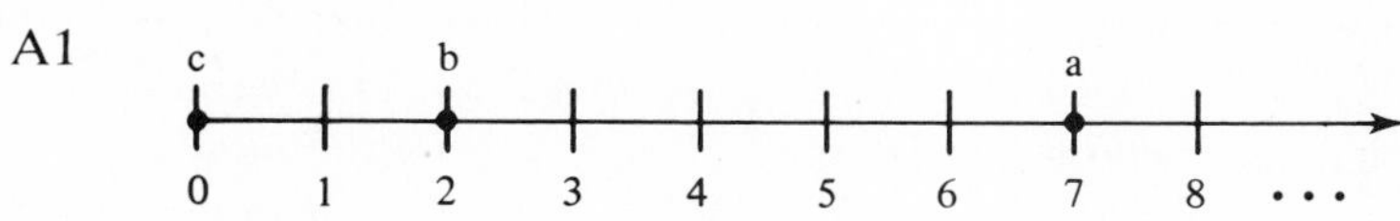

2 The number line of Frame 1 pictures the whole numbers spaced an equal distance apart, with each number indicating the *number of units that it is to the right of zero.* That is, the number 2 indicates the point that is 2 units to the right of zero. A new collection of numbers will now be defined by extending the number line to the *left* of zero and again marking off points an equal distance apart. This collection is named the *negative integers* and is written $^-1$, $^-2$, $^-3$, . . . (read "negative one," "negative two," "negative three," etc.). In the same way that 2 represents 2 units to the *right of zero,* the negative integer $^-2$ (negative two) will represent a distance of 2 units to the *left of zero.* To emphasize the difference in direction between the negative integers ($^-1$, $^-2$, $^-3$, . . .) and the natural numbers (1, 2, 3, . . .), the natural numbers are sometimes written $^+1$, $^+2$, $^+3$, . . . (read "positive one," "positive two," "positive three," etc.). The natural numbers are also correctly referred to as the *positive integers.* Thus, any positive integer can be written either with or without the positive sign. The positive integer $^+7$, for example, is also correctly represented as 7.

Q2 Graph each of the following numbers by placing dots at the appropriate points on the number line:

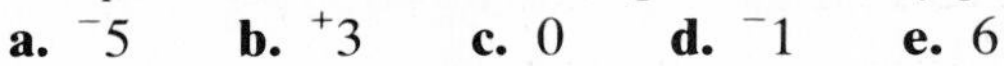

a. $^-5$ **b.** $^+3$ **c.** 0 **d.** $^-1$ **e.** 6

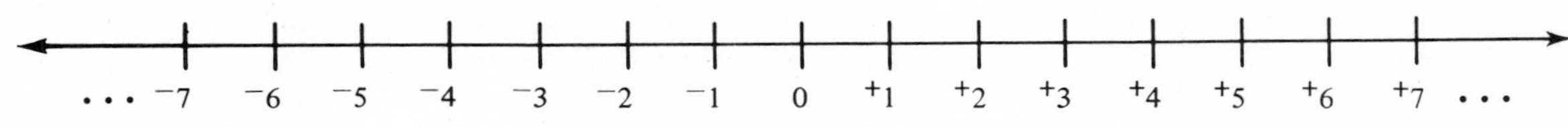

\# \# \# • \# \# \# • \# \# \# • \# \# \# • \# \# \# • \# \# \# • \# \# \# • \# \# \# • \# \# \#

A2

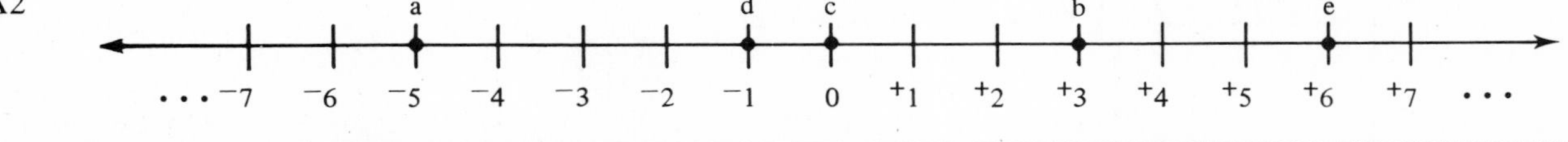

3 The collection that contains the negative integers, zero, and the positive integers is called the *integers*. The integers are pictured on the number line as follows:

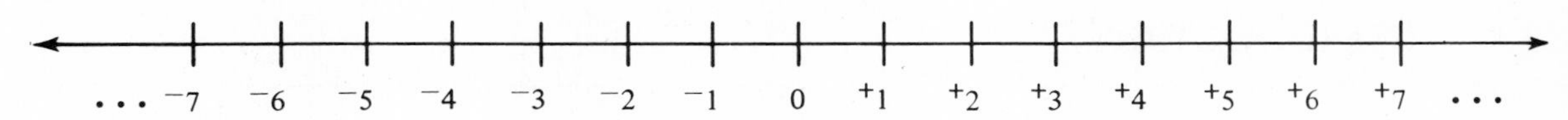

Each of the integers can be thought of as having two parts, *distance* and *direction*. The integer $^+5$, for example, represents a distance of 5 units to the *right* of zero. Similarly, the integer $^-9$ represents a distance of 9 units to the *left* of zero.

Q3 Write the distance and direction represented by each of the following integers:

	Distance	*Direction*
a. $^-6$	6 units	left of zero
b. $^+42$	________	________
c. $^-42$	________	________
d. 0	________	________

\# \# \# • \# \# \# • \# \# \# • \# \# \# • \# \# \# • \# \# \# • \# \# \# • \# \# \# • \# \# \#

A3 **b.** 42 units, right of zero
c. 42 units, left of zero
d. 0 units, neither direction

4 The integers, . . . , $^-3$, $^-2$, $^-1$, 0, $^+1$, $^+2$, $^+3$, . . . , are also referred to as *signed numbers* because each one (except zero) has a direction designated by either a "$^-$" or a "$^+$" sign. Zero is an integer but is considered neither positive nor negative since its coordinate is neither to the right nor to the left of the zero point. Notice that the positive and negative signs are raised so that they will not be confused with addition and subtraction signs.

Q4 What is the sign of each of the following integers?

a. $^+7$ ________ **b.** 6 ________

c. $^-3$ ________ **d.** 0 ________

\# \# \# • \# \# \# • \# \# \# • \# \# \# • \# \# \# • \# \# \# • \# \# \# • \# \# \# • \# \# \#

A4 **a.** positive **b.** positive **c.** negative
d. none: Zero has no sign.

5 As a result of the definition of the integers, each integer can be paired with a second integer that is the same distance from zero but in a different direction. The paired integers are called *opposites* and

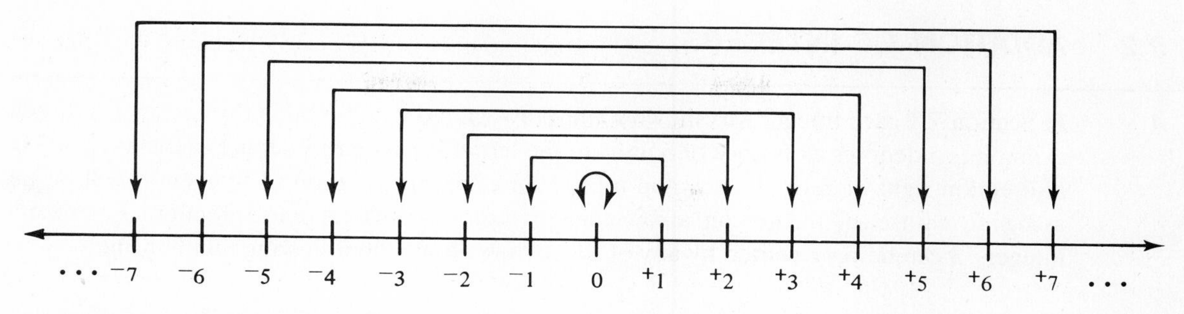

are shown in the accompanying drawing. Notice that each of the integers has a unique opposite. The opposite of $^{+}4$ is $^{-}4$. The opposite of $^{-}6$ is $^{+}6$. The opposite of 0 is 0. That is, zero is the only integer that is its own opposite.

Q5 Write the opposite of each of the following integers:

a. 5 -5 **b.** $^{-}7$ $+7$ **c.** 0 0

d. $^{+}1$ -1 **e.** $^{-}95$ $+95$ **f.** 125 -125

\# \# \# • \# \# \# • \# \# \# • \# \# \# • \# \# \# • \# \# \# • \# \# \# • \# \# \# • \# \# \#

A5 **a.** $^{-}5$ **b.** $^{+}7$ **c.** 0 **d.** $^{-}1$ **e.** $^{+}95$ **f.** $^{-}125$

This completes the instruction for this section.

7.1 EXERCISE

1. The collection . . . $^{-}2$, $^{-}1$, 0, 1, 2, . . . is called the ________.
2. Each integer is thought of as as having what two parts?
3. Write the distance and direction represented by each of the following integers:
   a. 15 **b.** $^{-}3$ **c.** 0 **d.** $^{+}7$
4. Graph each of the following integers: **a.** $^{-}2$ **b.** 4 **c.** 0 **d.** $^{+}3$

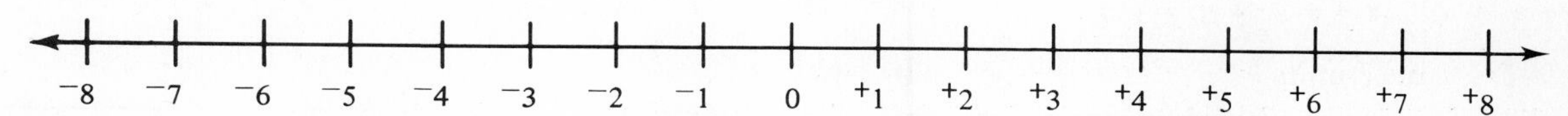

5. Write the opposite of each of the following integers:
   a. $^{+}2$ **b.** $^{-}3$ **c.** 11 **d.** 0

7.1 EXERCISE ANSWERS

1. integers
2. direction, distance
3. **a.** 15 units, right of zero
   b. 3 units, left of zero
   c. 0 units, neither right nor left
   d. 7 units, right of zero
4.

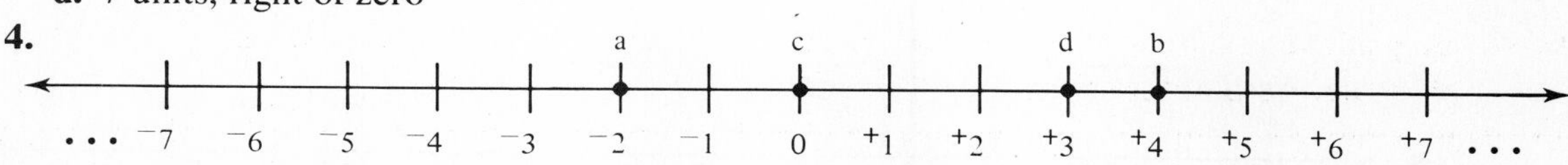

5. **a.** $^{-}2$ **b.** $^{+}3$ **c.** $^{-}11$ **d.** 0

7.2 ADDITION OF INTEGERS

1 In Section 7.1 each integer was shown to have associated with it both a distance and a direction. For example, $^{-}5$ denotes a distance of 5 units to the left of zero, and $^{+}3$ denotes a distance of 3 units and a direction right of zero. To develop a procedure for the addition of integers it will be helpful to associate with each integer an arrow that pictures its distance and direction. For example, the integer $^{+}3$ can be represented by any of the arrows above the following number line:

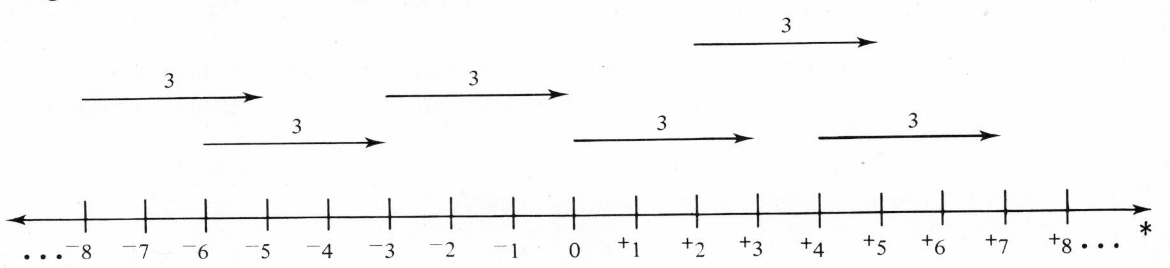

In this and the following sections the integer $^{+}3$ will be represented by any arrow that is *3 units long* and *points to the right.*

* When drawing the number line, it is common to omit the · · · 's. it is understood that the line extends infinitely in both directions.

Q1 **a.** What is the length of each arrow on the following number line? 2 units

b. What direction is indicated by each of the arrows? right

c. What integer is represented by each of the arrows? $^{+}2$

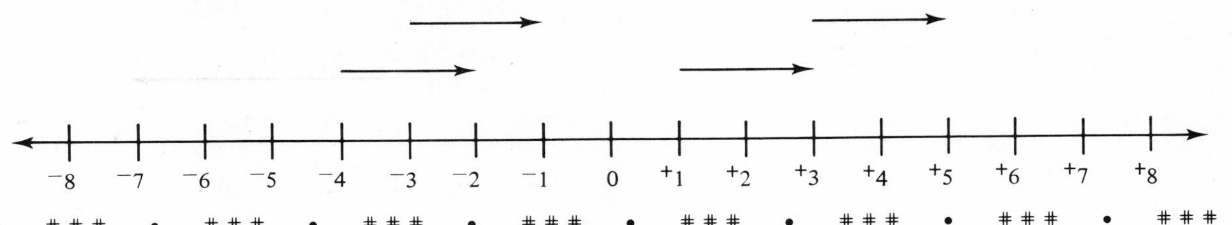

• # # # • # # # • # # # • # # # • # # # • # # # • # # # • # #

A1 **a.** 2 units **b.** right **c.** $^{+}2$

2 The integer $^{-}5$ is represented by each of the arrows above the following number line:

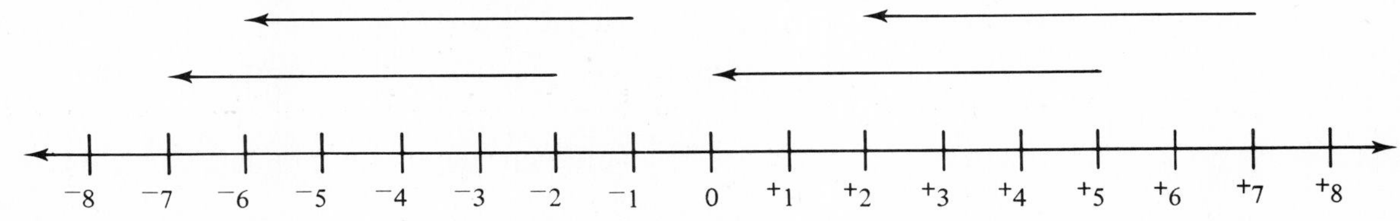

Each arrow represents $^{-}5$ since each is 5 *units in length* and each *points to the left.*

Q2 What integer is represented by each of the following arrows? $^{-}1$

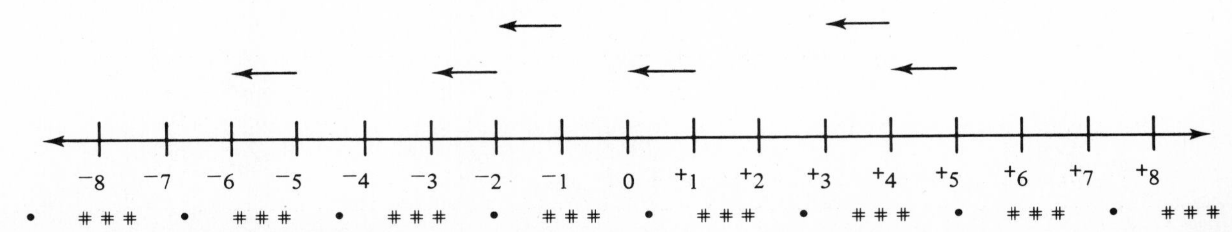

• # # # • # # # • # # # • # # # • # # # • # # # • # # # • # #

A2 $^{-}1$: Each is 1 unit long and each points to the left.

Q3 Draw three arrows which each represent the integer $^{-}4$.

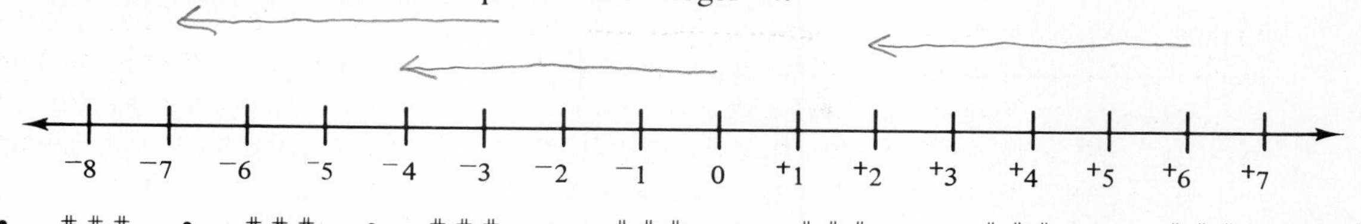

\# \# \# • \# \# \# • \# \# \# • \# \# \# • \# \# \# • \# \# \# • \# \# \# • \# \# \# • \# \# \#

A3 The arrows shown must each be 4 units in length and point to the left. For example,

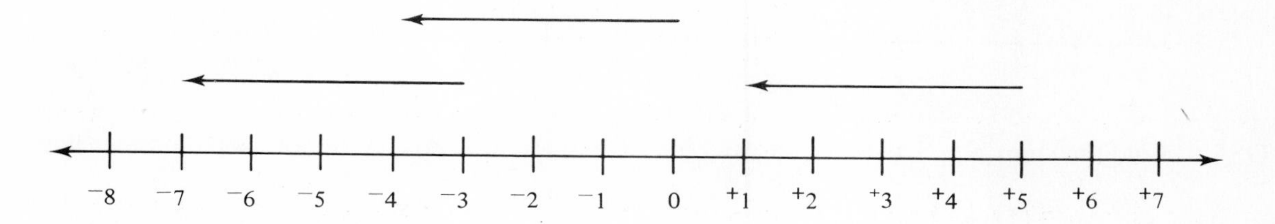

3 To add any two integers on the number line, use the following procedure:

Step 1: Represent the first addend* by an arrow that starts at zero.

Step 2: From the end of the first arrow, draw a second arrow to represent the second addend.

Step 3: Read the coordinate of the point at the end of the second arrow.

Examples:

1. $^{+}5 + {}^{-}3$

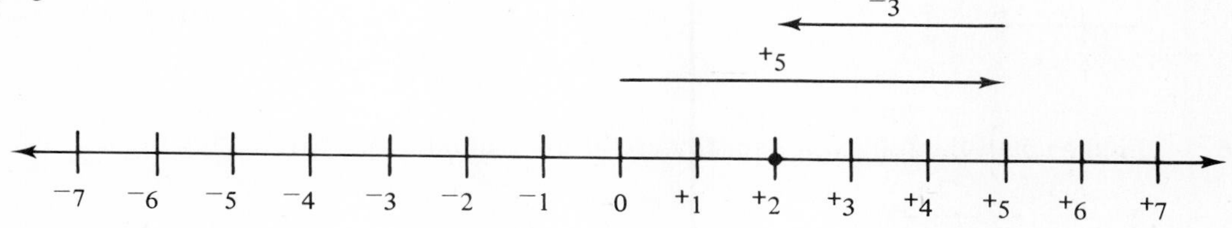

Therefore, $^{+}5 + {}^{-}3 = {}^{+}2$.

2. $^{-}4 + {}^{-}3$

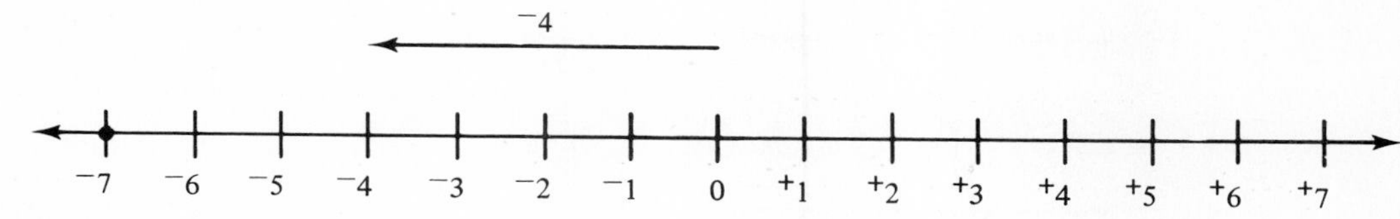

Therefore, $^{-}4 + {}^{-}3 = {}^{-}7$.

3. $^{+}2 + {}^{-}8$

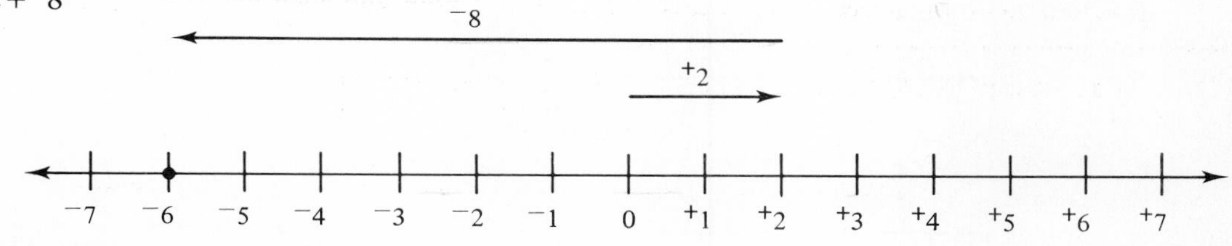

Therefore, $^{+}2 + {}^{-}8 = {}^{-}6$.

*An addend is a value that is being added to another value.

Q4 Find the sum $^{-}3+{}^{+}4$ by use of arrows for each of the addends on the following number line:

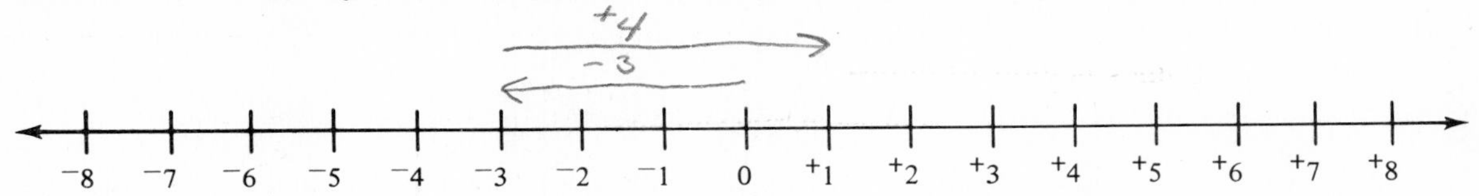

\# \# \# • \# \# \# • \# \# \# • \# \# \# • \# \# \# • \# \# \# • \# \# \# • \# \# \# • \# \# \#

A4 $^{+}1$:

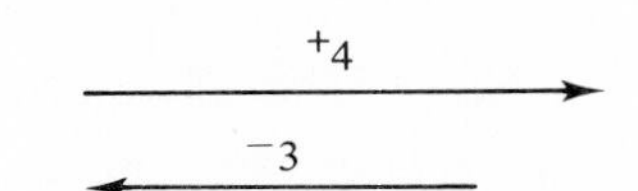

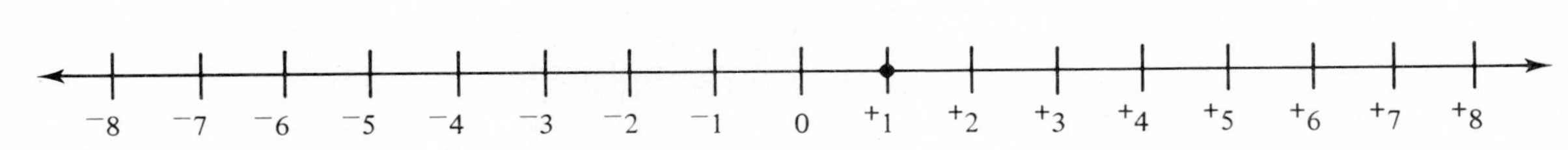

Q5 Find the sum $^{+}5+{}^{-}7$ by use of arrows for each of the addends on the following number line:

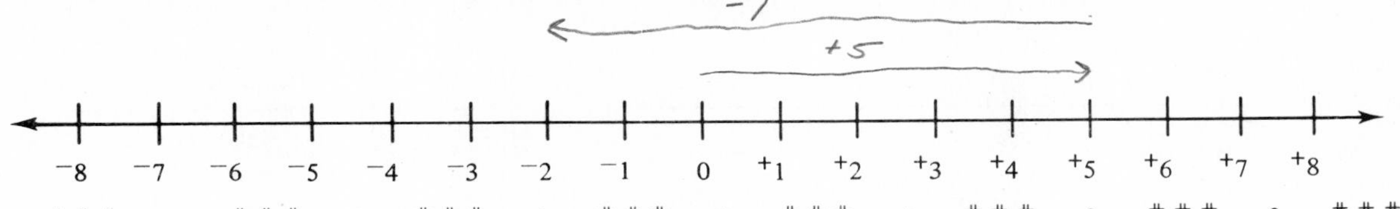

\# \# \# • \# \# \# • \# \# \# • \# \# \# • \# \# \# • \# \# \# • \# \# \# • \# \# \# • \# \# \#

A5 $^{-}2$:

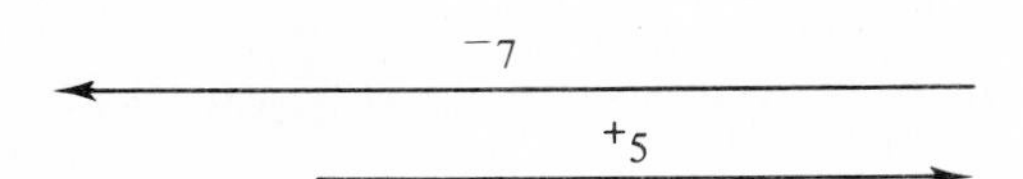

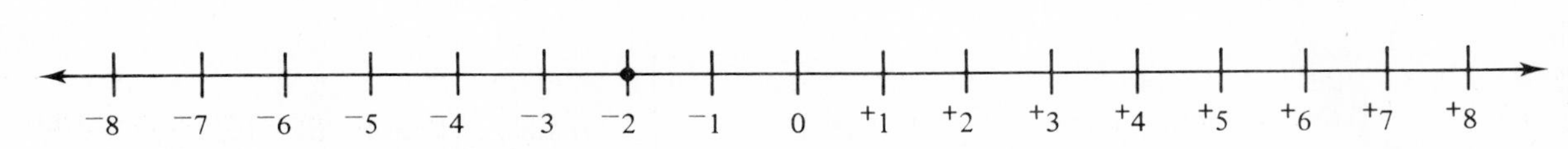

Q6 Find each of the following sums by use of the number line provided:

a. $^{-}4+{}^{-}5=$ $^{-}9$

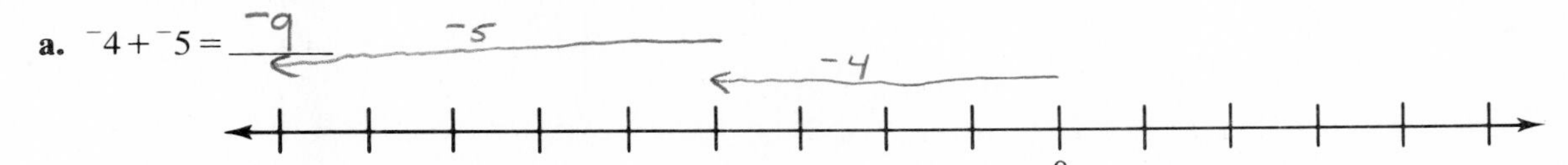

b. $^{+}3+{}^{+}2=$ $^{+}5$

0

c. $^{-}6+4=$ $^{-}2$

0

d. $^{-}5+{}^{+}9=$ $^{+}4$

0

e. $^{-}6+{}^{+}6=$ 0

0

\# \# \# • \# \# \# • \# \# \# • \# \# \# • \# \# \# • \# \# \# • \# \# \# • \# \# \# • \# \# \#

A6 **a.** $^{-}9$ **b.** $^{+}5$ **c.** $^{-}2$ **d.** $^{+}4$ **e.** 0

4 It is sometimes helpful to notice certain facts about the sum of two integers. Study each of the following three example sets to see if you can discover the three facts demonstrated.

Sum of two positives	*Sum of two negatives*	*Sum of a positive and a negative*
$^{+}5+^{+}3=^{+}8$	$^{-}5+^{-}3=^{-}8$	$^{-}5+^{+}3=^{-}2$
$^{+}7+^{+}6=^{+}13$	$^{-}7+^{-}6=^{-}13$	$^{+}7+^{-}6=^{+}1$
$^{+}2+^{+}9=^{+}11$	$^{-}2+^{-}9=^{-}11$	$^{+}2+^{-}2=0$

The three facts that correspond with the preceding examples are:

1. The sum of two positive integers is a positive integer.
2. The sum of two negative integers is a negative integer.
3. The sum of a positive integer and a negative integer is sometimes positive, sometimes negative, and sometimes zero.

Q7 Use the facts of Frame 4 or a number line to find each of the following sums:

a. $^{-}3+^{-}6=$ <u>$^{-}9$</u> **b.** $^{+}5+^{+}8=$ <u>13</u>

c. $^{+}4+^{-}1=$ <u>3</u> **d.** $^{+}2+^{-}7=$ <u>$^{-}5$</u>

e. $^{+}3+^{-}3=$ <u>0</u> **f.** $^{-}5+^{+}5=$ <u>0</u>

• # # # • # # # • # # # • # # # • # # # • # # # • # # # • # #

A7 **a.** $^{-}9$ (fact 2)

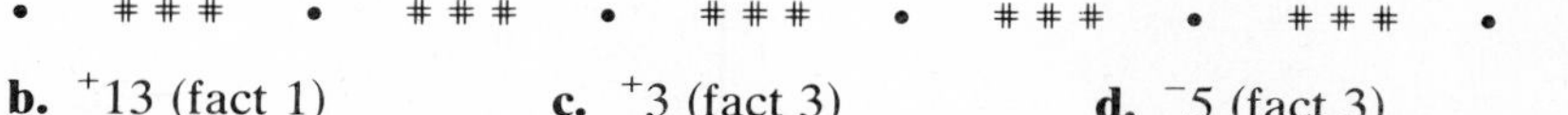

b. $^{+}13$ (fact 1) **c.** $^{+}3$ (fact 3) **d.** $^{-}5$ (fact 3)

e. 0 (fact 3) **f.** 0

5 Recall from Section 7.1 that every integer has a unique opposite, and that opposites represent the same distance but different directions. Consider the sum of $^{-}7$ and its opposite, $^{+}7$, shown on the following number line:

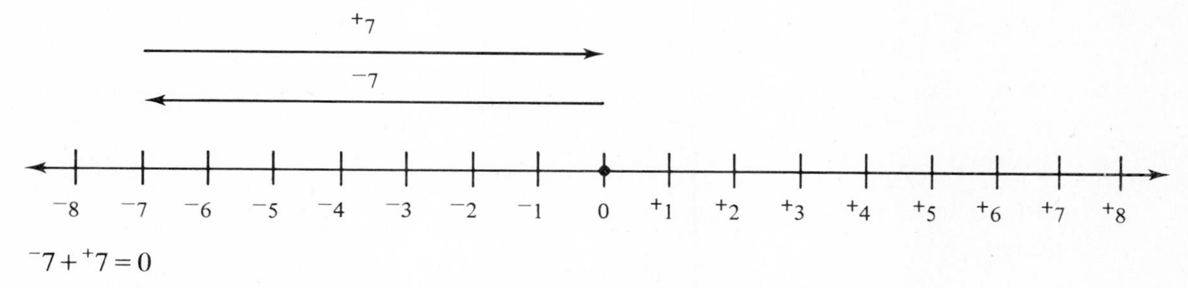

$^{-}7+^{+}7=0$

If the order of the addends is reversed, the sum remains unchanged:

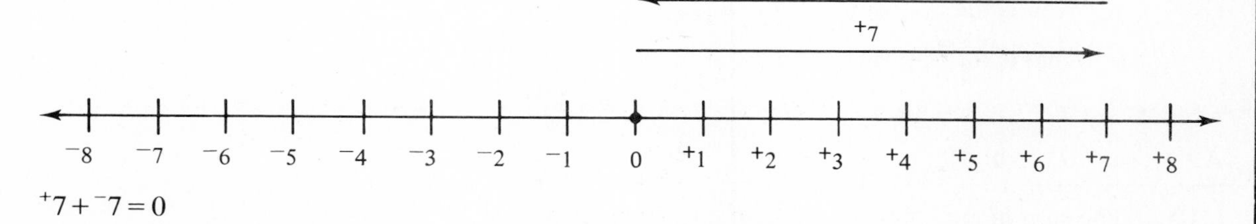

$^{+}7+^{-}7=0$

Thus, regardless of the order of the addends, the sum of $^{+}7$ and its opposite, $^{-}7$, is 0:

$^{+}7+^{-}7=^{-}7+^{+}7=0$

Q8 **a.** Find the sum $^{-}5 + {}^{+}5$.

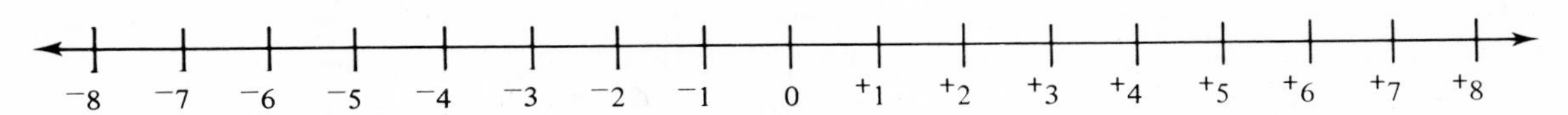

b. Find the sum $^{+}5 + {}^{-}5$.

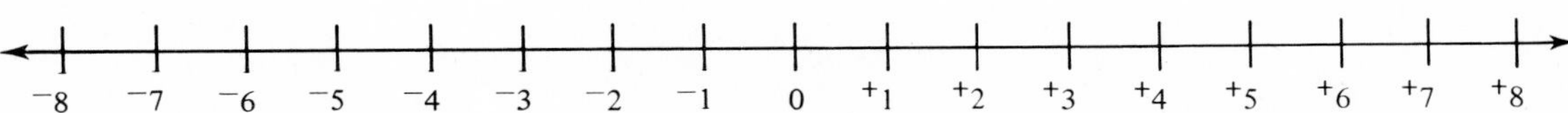

c. Is the sum in part **a** the same as the sum in part **b**? yes

\# # # • # # # • # # # • # # # • # # # • # # # • # # # • # # # • # # #

A8 **a.** 0 **b.** 0 **c.** yes: $^{-}5 + {}^{+}5 = {}^{+}5 + {}^{-}5 = 0$

6 Because of the distance and direction relationships between opposites, the sum of any integer and its opposite is zero. This fact is generalized $a + {}^{-}a = {}^{-}a + a = 0$.*

Examples:

Opposites	Sum
$^{+}4, {}^{-}4$	$^{+}4 + {}^{-}4 = 0$

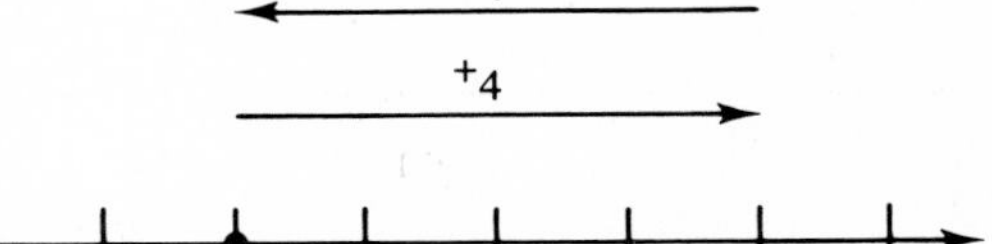

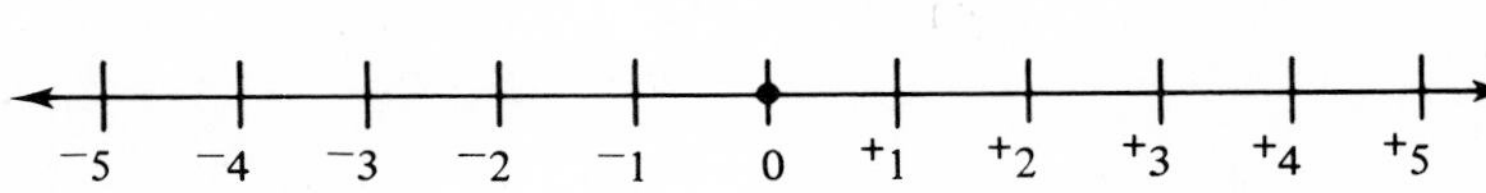

Opposites	Sum
$^{-}3, {}^{+}3$	$^{-}3 + {}^{+}3 = 0$

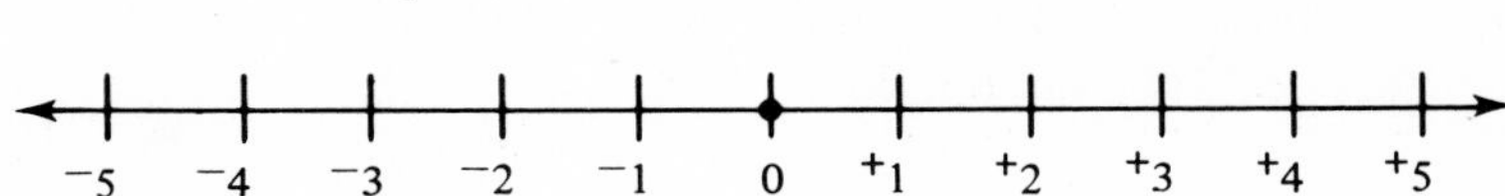

Opposites	Sum
0, 0	$0 + 0 = 0$

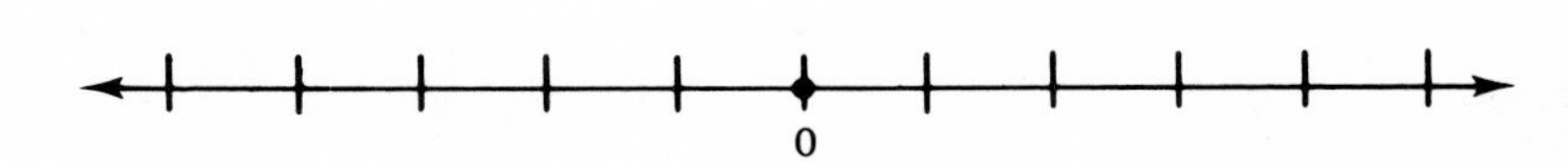

(Notice that the arrow representing zero is simply a dot at 0, because the integer zero has no length or direction.)

*The symbol ^{-}a means "the opposite of a."

Q9 **a.** What is the opposite of $^{+}6$? $^{-}6$

b. What is the opposite of $^{-}9$? $^{+}9$

\# # # • # # # • # # # • # # # • # # # • # # # • # # # • # # # • # # #

A9 **a.** $^{-}6$ **b.** $^{+}9$

Q10 The sum of any integer and its opposite is ______.

\# # # • # # # • # # # • # # # • # # # • # # # • # # # • # # # • # # #

A10 zero

Q11 Find each of the following sums:

a. $^{-}3 + ^{-}3 =$ ~~~~$^{-}6$

b. $^{+}5 + ^{-}5 =$ 0

c. $^{+}7 + ^{+}7 =$ 14

d. $^{-}6 + ^{+}6 =$ 0

e. $^{-}8 + ^{+}8 =$ 0

f. $0 + 0 =$ 0

• # # # • # # # • # # # • # # # • # # # • # # # • # # # • # #

A11 **a.** $^{-}6$: The sum of two negatives is a negative.
b. 0: The sum of two opposites is zero.
c. $^{+}14$: The sum of two positives is a positive.
d. 0 **e.** 0 **f.** 0

Q12 Write a number in the blank to make each of the following a true statement:

a. $^{+}2 + ^{-}2 =$ 0

b. $^{-}7$ $+ ^{+}7 = 0$

c. $^{-}15 +$ 15 $= 0$

d. $^{+}13 + ^{-}13 =$ 0

• # # # • # # # • # # # • # # # • # # # • # # # • # # # • # #

A12 **a.** 0 **b.** $^{-}7$ **c.** $^{+}15$ **d.** 0

7 The sum of any natural number and zero is the natural number: for example, $0+2=2$ and $8+0=8$. The same is true of any integer and zero.

Examples: $^{-}3 + 0 = ^{-}3$
$0 + ^{+}12 = ^{+}12$

This fact is called the *addition property of zero* and is generalized $a + 0 = 0 + a = a$ for any integer a.

Q13 Find each of the following sums:

a. $0 + ^{-}5 =$ $^{-}5$

b. $^{-}4 + ^{-}9 =$ $^{-}13$

c. $^{+}4 + ^{-}4 =$ 0

d. $^{+}11 + 0 =$ 11

e. $^{+}5 + ^{+}7 =$ 12

f. $^{-}7 + ^{+}3 =$ $^{-}4$

• # # # • # # # • # # # • # # # • # # # • # # # • # # # • # #

A13 **a.** $^{-}5$ **b.** $^{-}13$ **c.** 0 **d.** $^{+}11$ **e.** $^{+}12$ **f.** $^{-}4$

8 To find the sum of more than two integers, use the methods presented earlier to add the integers two at a time. For example, the sum $^{-}3 + ^{+}7 + ^{-}6$ is found:

$$\begin{aligned} ^{-}3 + ^{+}7 + ^{-}6 &= (^{-}3 + ^{+}7) + ^{-}6 \\ &= ^{+}4 + ^{-}6 \\ &= ^{-}2 \end{aligned}$$

[Note: $(^{-}3 + ^{+}7)$ is read "the quantity $^{-}3 + ^{+}7$."]

Notice that each sum of two integers can be found by starting at zero and using the number-line procedure if necessary.

Q14 Find the sum $^{+}3 + ^{-}2 + ^{+}5$.

$(3 + ^{-}2) + 5$
$1 + 5 = 6$

• # # # • # # # • # # # • # # # • # # # • # # # • # # # • # #

A14 $\quad {}^{+}6$: $\quad {}^{+}3+{}^{-}2+{}^{+}5=({}^{+}3+{}^{-}2)+{}^{+}5$

$$= {}^{+}1+{}^{+}5$$
$$= {}^{+}6$$

Q15 $\quad$ Find the sum ${}^{-}4+{}^{-}7+{}^{+}3$. $= (^{-}4+{}^{-}7)+3$

$= {}^{-}11+3$

$= -8$

\# \# \# • \# \# \# • \# \# \# • \# \# \# • \# \# \# • \# \# \# • \# \# \# • \# \# \# • \# \# \#

A15 $\quad {}^{-}8$: $\quad {}^{-}4+{}^{-}7+{}^{+}3=({}^{-}4+{}^{-}7)+{}^{+}3$

$$= {}^{-}11+{}^{+}3$$
$$= {}^{-}8$$

9 The sum ${}^{-}3+{}^{+}5+{}^{-}6+{}^{-}2$ is found as follows:

$$\begin{aligned} {}^{-}3+{}^{+}5+{}^{-}6+{}^{-}2 &= ({}^{-}3+{}^{+}5)+{}^{-}6+{}^{-}2 \\ &= {}^{+}2+{}^{-}6+{}^{-}2 \\ &= ({}^{+}2+{}^{-}6)+{}^{-}2 \\ &= {}^{-}4+{}^{-}2 \\ &= {}^{-}6 \end{aligned}$$

Q16 $\quad$ Find each of the following sums:

a. ${}^{-}3+{}^{-}5+{}^{+}8$ $\quad$ 0

b. ${}^{-}4+{}^{+}5+{}^{-}1+{}^{-}3$ $\quad$ -3

c. ${}^{-}2+{}^{-}3+{}^{-}6$ $\quad$ -11

d. ${}^{+}1+{}^{+}5+{}^{+}7$ $\quad$ 13

\# \# \# • \# \# \# • \# \# \# • \# \# \# • \# \# \# • \# \# \# • \# \# \# • \# \# \# • \# \# \#

A16 $\quad$ **a.** 0 $\quad$ **b.** ${}^{-}3$ $\quad$ **c.** ${}^{-}11$ $\quad$ **d.** ${}^{+}13$

10 An important property of mathematics is the *commutative property of addition*. The commutative property of addition states that regardless of the *order* in which two numbers are added, the sum is the same. Many examples can be used to demonstrate this fact for the integers.

Examples: ${}^{-}1+{}^{+}2={}^{+}2+{}^{-}1$ (both sums are ${}^{+}1$)
${}^{-}7+{}^{-}4={}^{-}4+{}^{-}7$ (both sums are ${}^{-}11$)
${}^{+}9+{}^{-}2={}^{-}2+{}^{+}9$ (both sums are ${}^{+}7$)

The commutative property of addition is generalized $a+b=b+a$ for any integers a and b.

Q17 $\quad$ Verify that $a+b=b+a$ is true for $a={}^{-}5$ and $b={}^{+}8$.

$-5+8=8+{}^{-}5$

\# \# \# • \# \# \# • \# \# \# • \# \# \# • \# \# \# • \# \# \# • \# \# \# • \# \# \# • \# \# \#

A17 $\quad {}^{-}5+{}^{+}8={}^{+}8+{}^{-}5$ (both sums are ${}^{+}3$)

11 A second important mathematical property is the *associative property of addition*. It states that regardless of the grouping of addends, when three or more numbers are being added the sums are the same. Two examples using integers are as follows:

1. $(^{-}3 + {}^{+}4) + {}^{-}5 = {}^{-}3 + ({}^{+}4 + {}^{-}5)$
$^{+}1 + {}^{-}5 = {}^{-}3 + {}^{-}1$
$^{-}4 = {}^{-}4$

2. $(^{+}7 + {}^{+}5) + {}^{-}10 = {}^{+}7 + ({}^{+}5 + {}^{-}10)$
$^{+}12 + {}^{-}10 = {}^{+}7 + {}^{-}5$
$^{+}2 = {}^{+}2$

The associative property of addition is generalized $(a+b)+c = a+(b+c)$ for any integers a, b, and c.

Q18 Verify that $(a+b)+c = a+(b+c)$ is true for $a = {}^{-}2$, $b = {}^{-}7$, and $c = {}^{+}3$. – 6

\# # # • # # # • # # # • # # # • # # # • # # # • # # # • # # # • # # #

A18 $(^{-}2 + {}^{-}7) + {}^{+}3 = {}^{-}2 + ({}^{-}7 + {}^{+}3)$
$^{-}9 + {}^{+}3 = {}^{-}2 + {}^{-}4$
$^{-}6 = {}^{-}6$

12 Signed numbers can be used to study motion in a hydraulic system. Frequently it is necessary to interconnect several fluid passages with a valve. A spool valve is often required. The spool valve is cylindrically shaped, with two or more lands and with annular grooves between the lands. The valve is closely fitted to a round bore and slides in the bore on a pressurized film or fluid. Fluid passages are open or closed to each other, depending on the valve-land positions.

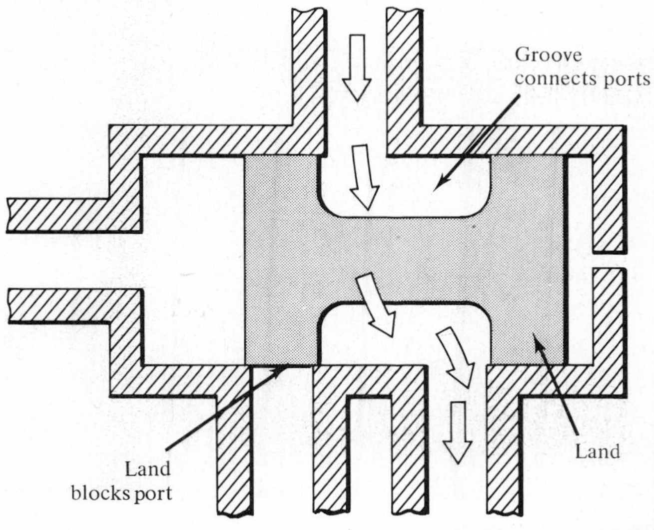

Spool valves can be positioned manually, but are more often positioned by springs or pressure. There will frequently be a spring force tending to move the valve in one direction and a reaction force tending to move it back. Sometimes there is an auxiliary force which acts in the same direction as the spring. The reaction and auxiliary forces are usually the result of hydraulic pressure on one end of the spool or the other.

A diagram of a typical spool valve system is shown on the next page. Notice that the spring in the chamber to the left of the spool valve tends to move the spool valve to the right. Often pressure is built up in the same chamber as the spring, which also tends to move the spool valve to the right. We call this an auxiliary force.

Let us refer to any forces which tend to move the spool valve to the right as positive forces and those which tend to move it to the left as negative forces. In the following diagram the spring force is positive, the reaction force is negative, and the auxiliary force is positive.

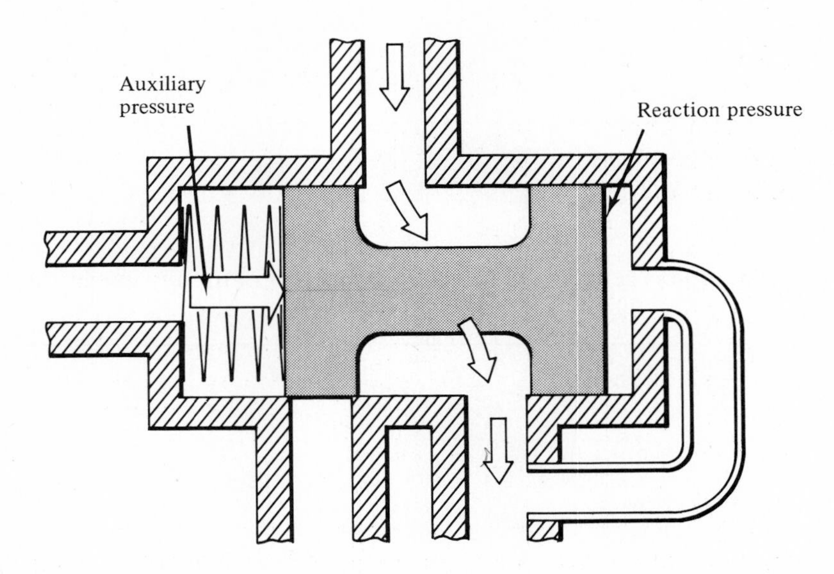

Example: In a power steering flow control valve there is a constant spring force on a spool valve of $^{+}5$ lb. Both the auxiliary force and the reaction force are variable. If at a given instant the auxiliary force is $^{+}2$ lb and the reaction force is $^{-}8$ lb:

a. What is the sum of the forces on the spool?
b. Which direction will the spool valve move?

Solution:

a. The sum of the forces on the spool is $^{+}5 + {}^{+}2 + {}^{-}8 = {}^{-}1$.
b. Since the sum of the forces is negative, the valve moves left.

Q19 If the spring force is $^{+}5$ lb, the auxiliary force $^{+}3$ lb, and the reaction force $^{-}4$ lb:

a. What is the sum of the forces on the spool?
b. Which direction will the spool valve move?

• ### • ### • ### • ### • ### • ### • ### •

A19 **a.** $^{+}5 + {}^{+}3 + {}^{-}4 = {}^{+}4$ **b.** The spool valve moves right.

Q20 Below is shown a chart where the spring force on the spool valve, the auxiliary force, and the reaction force are given. Find the sum of the forces and indicate whether the spool valve moves right or left. (The first one is done for you.)

	Spring Force	*Auxiliary Force*	*Reaction Force*	*Sum*	*Direction*
a.	$^{+}5$	$^{+}2$	$^{-}3$	$^{+}4$	right
b.	$^{+}5$	$^{+}3$	$^{-}6$		
c.	$^{+}5$	$^{+}1$	$^{-}7$		
d.	$^{+}5$	0	$^{-}6$		
e.	$^{+}5$	$^{+}3$	$^{-}8$		

• ### • ### • ### • ### • ### • ### • ### •

A20 **b.** $^{+}2$, right **c.** $^{-}1$, left **d.** $^{-}1$, left **e.** 0, no motion

Q21 The signs of voltage through resistors in a circuit are determined by the direction of current flow. Find the sum of the known voltages in this circuit. (This problem exemplifies Kirchhoff's law on the sum of voltages in a circuit.)

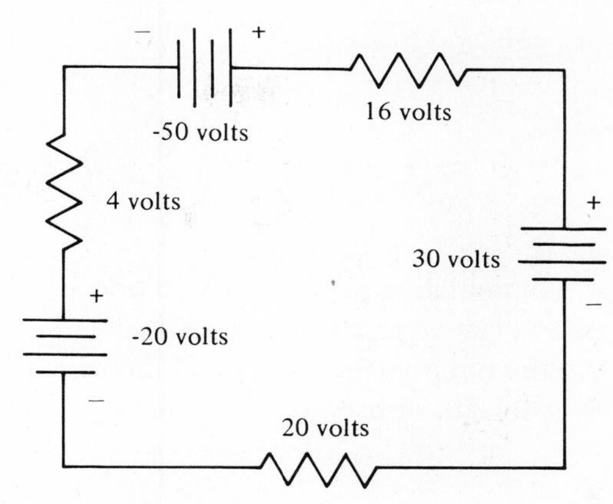

• ### • ### • ### • ### • ### • ### • ### •

A21 0

Q22 A metal producer specializing in the construction of gears uses a Brinell hardness test gauge for inspection purposes. For surface durability of hard steel bevel gears, the Brinell gauge reading should be between 210 and 245. This producer set as its standard a reading of 225. Quality control inspectors give readings above or below 225 by signed numbers. For example, a reading of $^{-}10$ means the hardness test shows the durability to be $225+({}^{-}10)=215$. Find the durability for each of the following inspection readings:

a. $^{-}8$ **b.** $^{+}11$ **c.** $^{-}13$ **d.** $^{+}6$

• ### • ### • ### • ### • ### • ### • ### •

A22 **a.** 217 **b.** 236 **c.** 212 **d.** 231

This completes the instruction for this section.

7.2 EXERCISE

1. Find each of the following sums:

a. $^{-}3+{}^{-}5$ **b.** $^{+}5+{}^{+}9$
c. $^{+}7+{}^{-}6$ **d.** $^{-}6+{}^{+}6$
e. $^{-}9+0$ **f.** $^{-}3+{}^{+}4+{}^{-}3$
g. $^{-}7+{}^{+}6+{}^{+}1$ **h.** $0+{}^{+}4$
i. $^{-}4+{}^{-}3+{}^{+}7+{}^{+}4$ **j.** $^{-}1+{}^{+}6+{}^{-}6+{}^{+}1$

2. Write a number in the blank to make each of the following a true statement:

a. _____ $+{}^{-}2=0$ **b.** $^{-}8+{}^{+}8=$ _____
c. $^{-}12+$ _____ $={}^{-}12$ **d.** $^{-}12+$ _____ $=0$

3. The commutative and associative properties of addition are true for all integers. Identify each of the following as demonstrating the commutative or the associative property of addition.

a. $4+{}^{-}5={}^{-}5+4$
b. $({}^{-}3+6)+{}^{-}4={}^{-}3+(6+{}^{-}4)$
c. $^{-}1+{}^{-}2={}^{-}2+{}^{-}1$
d. $({}^{-}5+0)+5={}^{-}5+(0+5)$

4. The known voltages in a circuit are: $^{-}15$, 10, $^{-}4$, 16, $^{-}7$. Find the sum of the voltages.

5. A metal producer specializing in the construction of gears uses a Brinell hardness test gauge for inspection purposes (see Q22). This producer set as its standard a reading of 220. Find the durability for each of the following inspection readings.

a. $^{-}3$ **b.** 6 **c.** $^{-}15$ **d.** 0

7.2 EXERCISE ANSWERS

1. a. $^{-}8$ **b.** $^{+}14$ **c.** $^{+}1$ **d.** 0
e. $^{-}9$ **f.** $^{-}2$ **g.** 0 **h.** $^{+}4$
i. $^{+}4$ **j.** 0
2. a. $^{+}2$ **b.** 0 **c.** 0 **d.** $^{+}12$
3. a. commutative property of addition
b. associative property of addition
c. commutative property of addition
d. associative property of addition
4. 0
5. a. 217 **b.** 226 **c.** 205 **d.** 220

7.3 SUBTRACTION OF INTEGERS

1 Subtraction and addition can be thought of as opposite operations.* Consider, for example, the effect on any number x of first adding 5 and then performing the opposite operation of subtracting 5.

x
$x+5$ add 5
$x+5-5$ subtract 5
x

Notice that the operation of subtraction undoes what the operation of addition does, and the result is again the number x. The operation of addition also undoes an equal subtraction.

x
$x-5$ subtract 5
$x-5+5$ add 5
x

Thus, regardless of the order in which they are done, the operations of addition and subtraction are opposites. One operation undoes the other operation.

* Some mathematicians refer to addition and subtraction as "inverse operations."

Q1 **a.** What is the opposite operation of addition? ____________

b. What is the opposite of adding 7 to any number? ____________

\# # # • # # # • # # # • # # # • # # # • # # # • # # # • # # # • # # #

A1 **a.** subtraction **b.** subtracting 7

Q2 **a.** What is the opposite operation of subtraction? ____________

b. What is the opposite of subtracting 3 from any number? ____________

\# # # • # # # • # # # • # # # • # # # • # # # • # # # • # # # • # # #

A2 **a.** addition **b.** adding 3

2 Using the idea of addition and subtraction as opposite operations, the procedure for adding integers on the number line is easily modified for subtracting any two integers. Recall the procedure for finding the sum of any two integers on the number line.

$^-5 + {}^+7$:

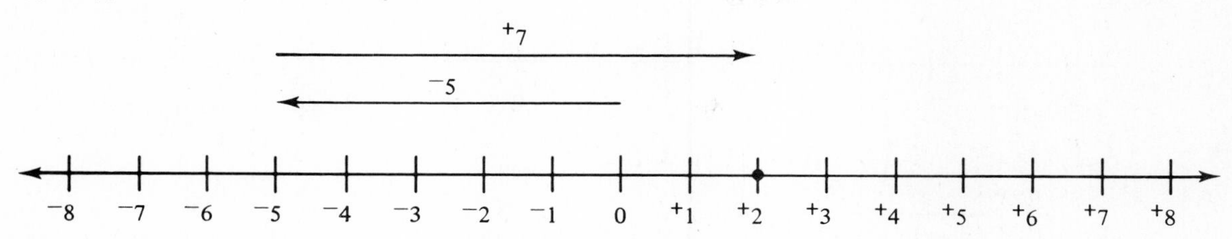

Step 1: Draw the arrow for the first addend ($^-5$) from zero.

Step 2: From the tip of the first arrow, draw the arrow for the second addend ($^+7$).

Step 3: Read the answer below the tip of the arrow for the second addend.

Thus, $^-5 + {}^+7 = {}^+2$.

Q3 Find the sum $^+3 + {}^-9$ using the number line given.

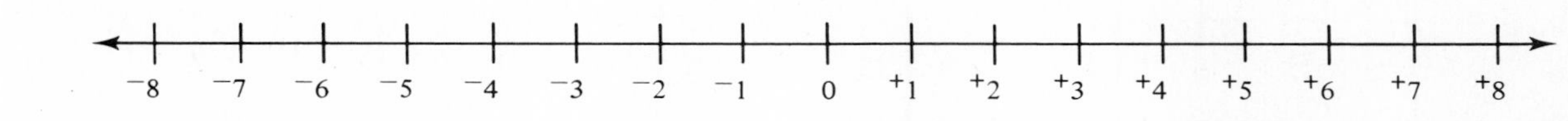

\# \# \# • \# \# \# • \# \# \# • \# \# \# • \# \# \# • \# \# \# • \# \# \# • \# \# \# • \# \# \# • \# \# \#

A3 $^-6$:

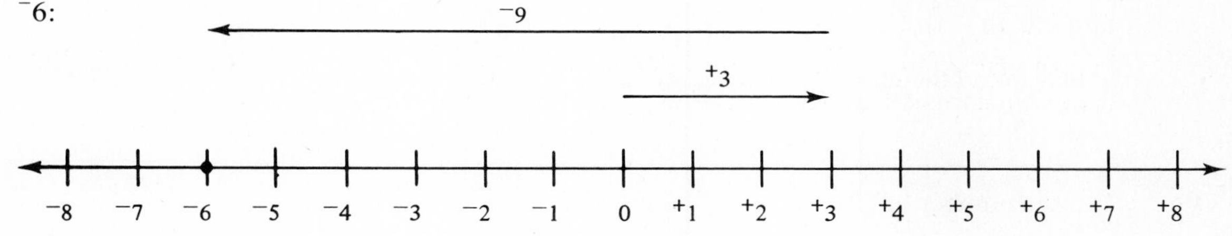

3 To find the difference $^-3 - {}^-5$, the following steps can be used:

Minuend Subtrahend

↘ ↙

$^-3 - {}^-5$

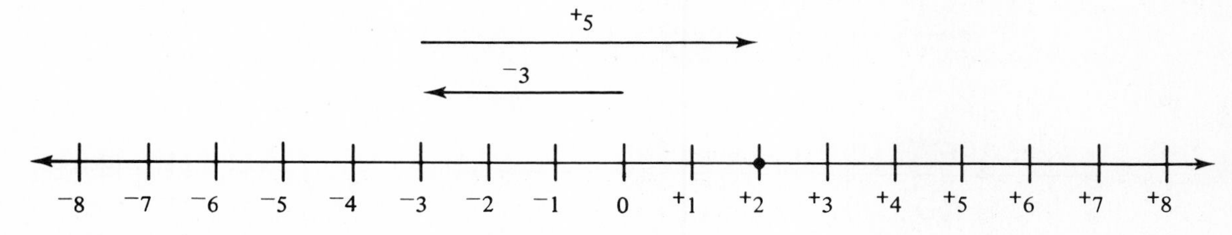

Step 1: Draw the arrow for the minuend ($^-3$).

Step 2: Since subtraction is the opposite of addition, draw the arrow for the *opposite* of the subtrahend ($^+5$) from the tip of the first arrow.

Step 3: Read the answer below the tip of the second arrow.

Thus, $^-3 - {}^-5 = {}^+2$.

Q4 Use the following number line to find the difference ${}^{-}7 - {}^{-}4$:

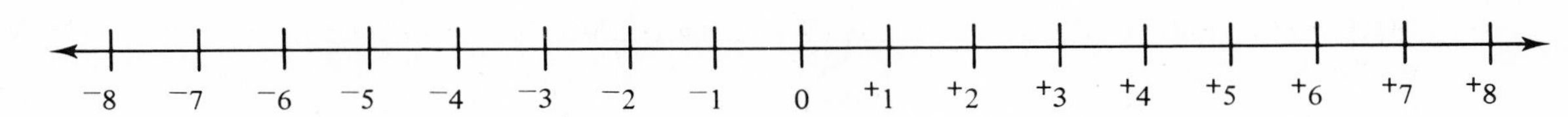

\# \# \# • \# \# \# • \# \# \# • \# \# \# • \# \# \# • \# \# \# • \# \# \# • \# \# \# • \# \# \#

A4 ${}^{-}3$:

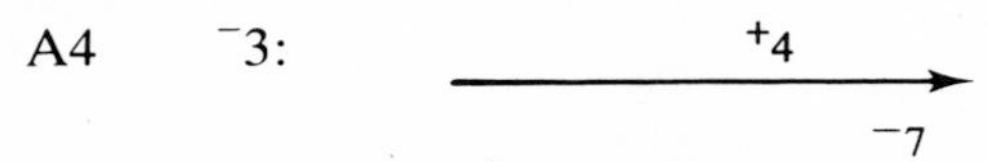

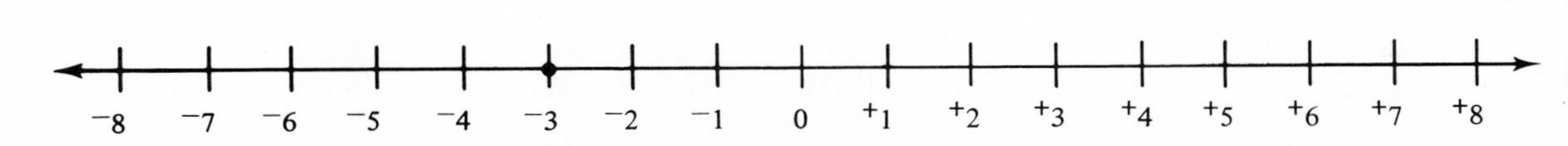

4 The difference ${}^{-}2 - {}^{+}4$ can be found as follows:

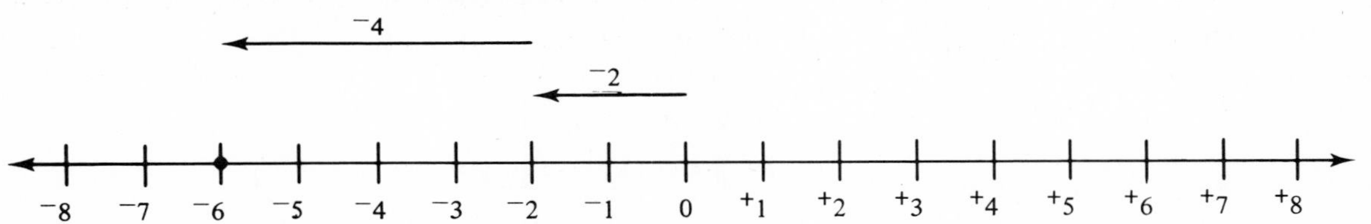

Step 1: Draw the arrow for the minuend (${}^{-}2$).

Step 2: Since subtraction is the opposite of addition, draw the arrow for the *opposite* of the subtrahend (${}^{-}4$).

Step 3: Read the answer below the tip of the second arrow.

Thus, ${}^{-}2 - {}^{+}4 = {}^{-}6$.

Q5 Use the number line provided to find the difference ${}^{+}3 - {}^{+}10$.

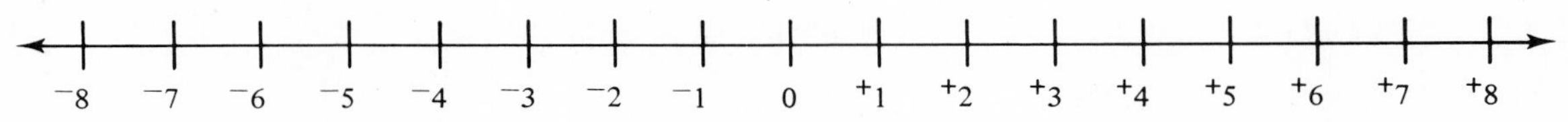

\# \# \# • \# \# \# • \# \# \# • \# \# \# • \# \# \# • \# \# \# • \# \# \# • \# \# \# • \# \# \#

A5 ${}^{-}7$:

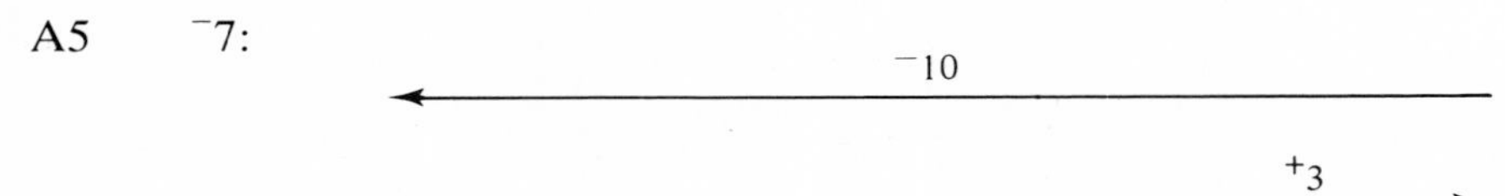

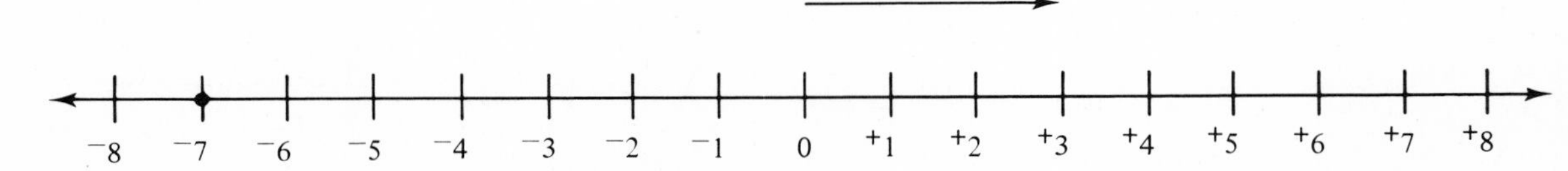

5 We repeat here the examples of Frames 3 and 4:

Minuend		*Subtrahend*		*Difference*
${}^{-}3$	$-$	${}^{-}5$	$=$	${}^{+}2$
${}^{-}2$	$-$	${}^{+}4$	$=$	${}^{-}6$

In each case the difference was found by first drawing the arrow for the minuend and then drawing the arrow for the *opposite* of the subtrahend.

This procedure is the basis for defining subtraction in terms of addition: *To find the difference of two integers, add the minuend to the opposite of the subtrahend.* That is, $a - b = a + {}^{-}b$ for any integers a and b.

Examples:

	Minuend		*Subtrahend*		*Opposite of Subtrahend*		*Difference*
1.	${}^{+}7$	$-$	$\boxed{{}^{+}4}$	$= {}^{+}7 +$	$\boxed{{}^{-}4}$	$=$	${}^{+}3$
2.	${}^{-}3$	$-$	$\boxed{{}^{+}5}$	$= {}^{-}3 +$	$\boxed{{}^{-}5}$	$=$	${}^{-}8$
3.	${}^{-}4$	$-$	$\boxed{{}^{-}6}$	$= {}^{-}4 +$	$\boxed{{}^{+}6}$	$=$	${}^{+}2$
4.	${}^{+}2$	$-$	$\boxed{{}^{+}6}$	$= {}^{+}2 +$	$\boxed{{}^{-}6}$	$=$	${}^{-}4$

Notice that in each example, two changes are involved: The operation sign for subtraction is changed to addition, and the subtrahend is changed to its opposite.

Q6 Find the difference ${}^{-}3 - {}^{+}1$ using the number line provided.

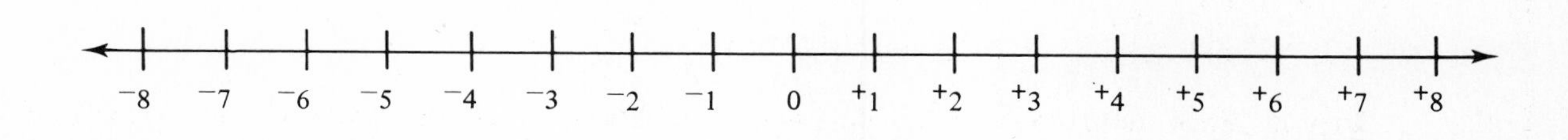

\# \# \# • \# \# \# • \# \# \# • \# \# \# • \# \# \# • \# \# \# • \# \# \# • \# \# \# • \# \# \#

A6 ${}^{-}4$: ${}^{-}3 - {}^{+}1 = {}^{-}3 + {}^{-}1$

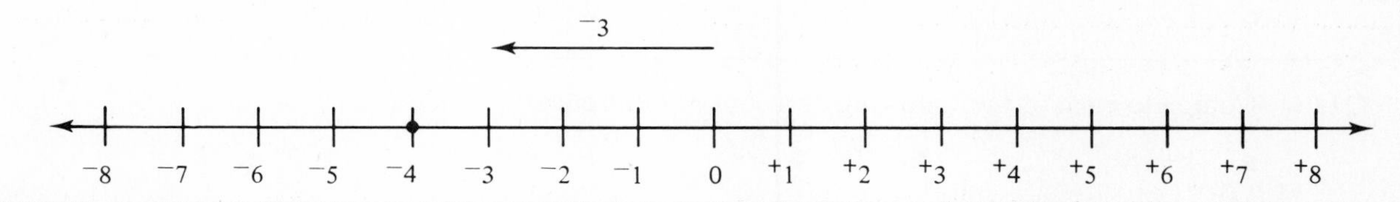

Q7 Find the difference by rewriting as a sum.

${}^{-}4 - {}^{+}7 = {}^{-}4 +$ ___ $=$ ___

\# \# \# • \# \# \# • \# \# \# • \# \# \# • \# \# \# • \# \# \# • \# \# \# • \# \# \# • \# \# \#

A7 ${}^{-}7$, ${}^{-}11$

Q8 Find each of the following differences:

a. ${}^{-}2 - {}^{+}5$ **b.** ${}^{+}7 - {}^{-}5$

c. ${}^{+}1 - {}^{+}9$ **d.** ${}^{-}5 - {}^{-}3$

\# \# \# • \# \# \# • \# \# \# • \# \# \# • \# \# \# • \# \# \# • \# \# \# • \# \# \# • \# \# \#

A8 **a.** ${}^{-}7$: ${}^{-}2 - {}^{+}5 = {}^{-}2 + {}^{-}5 = {}^{-}7$
b. ${}^{+}12$: ${}^{+}7 - {}^{-}5 = {}^{+}7 + {}^{+}5 = {}^{+}12$
c. ${}^{-}8$: ${}^{+}1 - {}^{+}9 = {}^{+}1 + {}^{-}9 = {}^{-}8$
d. ${}^{-}2$: ${}^{-}5 - {}^{-}3 = {}^{-}5 + {}^{+}3 = {}^{-}2$

6 It is important to realize that the procedure of "adding the opposite" is done only with *subtraction*.

1. To find a *sum*, follow the procedure for adding integers directly.
2. To find a *difference*, *rewrite the problem as a sum* (by adding the opposite of the subtrahend to the minuend) and follow the procedure for adding integers.

Examples:

$^{-}4 - {}^{+}6 = {}^{-}4 + {}^{-}6 = {}^{-}10$

$^{+}5 + {}^{-}6 = {}^{-}1$

$^{-}7 + {}^{-}4 = {}^{-}11$

$^{-}4 - {}^{-}3 = {}^{-}4 + {}^{+}3 = {}^{-}1$

Q9 Which of the following problems must be rewritten?

a. $^{-}4 - {}^{-}5$ **b.** $^{-}2 + {}^{-}1$

c. $^{+}6 + {}^{-}9$ **d.** $^{+}6 - {}^{+}15$ ________

• # # # • # # # • # # # • # # # • # # # • # # # • # # # • # #

A9 **a** and **d**: Subtraction problems must be rewritten (**b** and **c** are addition problems and thus can be answered directly).

Q10 Complete the problems of Q9.

• # # # • # # # • # # # • # # # • # # # • # # # • # # # • # #

A10 **a.** $^{-}4 - {}^{-}5 = {}^{-}4 + {}^{+}5 = {}^{+}1$ **b.** $^{-}3$

c. $^{-}3$ **d.** $^{+}6 - {}^{+}15 = {}^{+}6 + {}^{-}15 = {}^{-}9$

Q11 Complete each of the following as a sum or difference as indicated:

a. $^{-}4 + {}^{-}8$ **b.** $^{+}7 - {}^{+}4$ **c.** $^{-}6 - {}^{+}4$ **d.** $^{-}6 + {}^{+}7$

e. $^{+}2 - {}^{+}5$ **f.** $^{-}4 - {}^{+}8$ **g.** $^{+}4 + {}^{+}7$ **h.** $^{-}1 - {}^{-}1$

i. $^{+}11 + {}^{-}9$ **j.** $0 - {}^{-}5$ **k.** $^{-}4 + 0$ **l.** $^{+}1 - {}^{+}1$

m. $^{+}5 + {}^{-}5$ **n.** $^{-}2 + {}^{-}3$ **o.** $^{-}3 - {}^{-}3$ **p.** $^{+}5 + {}^{-}7$

q. $^{+}5 - {}^{+}9$ **r.** $^{-}1 + {}^{+}7$ **s.** $^{-}2 - 0$ **t.** $0 - {}^{+}1$

• # # # • # # # • # # # • # # # • # # # • # # # • # # # • # #

A11 **a.** $^{-}12$ **b.** $^{+}3$ **c.** $^{-}10$ **d.** $^{+}1$

e. $^{-}3$ **f.** $^{-}12$ **g.** $^{+}11$ **h.** 0

i. $^{+}2$ **j.** $^{+}5$ **k.** $^{-}4$ **l.** 0

m. 0 **n.** $^{-}5$ **o.** 0 **p.** $^{-}2$

q. $^{-}4$ **r.** $^{+}6$ **s.** $^{-}2$ **t.** $^{-}1$

7 When evaluating number sentences involving a combination of sums and differences, rewrite each of the differences as a sum and proceed as in Section 7.2, Frame 8. For example, the expression

$^{+}3 + {}^{-}7 - {}^{+}5$

sum

difference

involves a sum and a difference. First rewrite the difference as

$^{+}3 + {}^{-}7 + {}^{-}5$

The problem is now completed:

$^{-}4 + {}^{-}5$
$^{-}9$

Study the following examples:

1. $^{-}4 - {}^{-}2 + {}^{+}3$
$^{-}4 + {}^{+}2 + {}^{+}3$
$^{-}2 + {}^{+}3$
$^{+}1$

2. $^{-}3 - {}^{+}4 - {}^{+}6 + {}^{+}7$
$^{-}3 + {}^{-}4 + {}^{-}6 + 7$
$^{-}13 + 7$
$^{-}6$

Q12 Evaluate $^{-}3 + {}^{+}5 - {}^{+}7$.

• ### • ### • ### • ### • ### • ### • ### •

A12 $^{-}5$: $^{-}3 + {}^{+}5 - 7 = {}^{-}3 + {}^{+}5 + {}^{-}7 = {}^{+}2 + {}^{-}7 = {}^{-}5$

Q13 Evaluate $^{+}1 - {}^{+}5 - {}^{+}6$.

• ### • ### • ### • ### • ### • ### • ### •

A13 $^{-}10$: $^{+}1 - {}^{+}5 - {}^{+}6 = {}^{+}1 + {}^{-}5 + {}^{-}6 = {}^{-}4 + {}^{-}6 = {}^{-}10$

Q14 Evaluate $^{+}4 + {}^{-}3 - {}^{+}7 - {}^{-}6$.

• ### • ### • ### • ### • ### • ### • ### •

A14 0: $^{+}4 + {}^{-}3 - {}^{+}7 - {}^{-}6 = {}^{+}4 + {}^{-}3 + {}^{-}7 + {}^{+}6 = {}^{+}1 + {}^{-}7 + {}^{+}6 = {}^{-}6 + {}^{+}6 = 0$

Q15 Evaluate:

a. $^{-}2 - {}^{+}5 + {}^{+}3$

b. $^{+}3 + {}^{-}5 + {}^{-}4$

c. $^{+}4 - {}^{-}6 + {}^{+}7$

d. $^{-}6 - {}^{+}2 + {}^{-}7$

e. ${}^{-}2-{}^{+}3-{}^{+}4$

f. ${}^{+}7+{}^{-}2-{}^{+}8+{}^{+}3$

g. ${}^{+}4-0+{}^{-}4$

h. $0-{}^{+}4-{}^{+}3+{}^{+}2$

i. ${}^{+}6-{}^{+}3-{}^{+}5$

j. ${}^{-}6+{}^{-}3-{}^{+}10$

• ### • ### • ### • ### • ### • ### • ### •

A15 **a.** ${}^{-}4$: ${}^{-}2-{}^{+}5+{}^{+}3={}^{-}2+{}^{-}5+{}^{+}3={}^{-}7+{}^{+}3={}^{-}4$
b. ${}^{-}6$
c. ${}^{+}17$
d. ${}^{-}15$: ${}^{-}6-{}^{+}2+{}^{-}7={}^{-}6+{}^{-}2+{}^{-}7={}^{-}15$
e. ${}^{-}9$
f. 0: ${}^{+}7+{}^{-}2-{}^{+}8+{}^{+}3={}^{+}7+{}^{-}2+{}^{-}8+{}^{+}3=0$
g. 0
h. ${}^{-}5$: $0-{}^{+}4-{}^{+}3+{}^{+}2=0+{}^{-}4+{}^{-}3+{}^{+}2={}^{-}7+{}^{+}2={}^{-}5$
i. ${}^{-}2$: ${}^{+}6-{}^{+}3-{}^{+}5={}^{+}6+{}^{-}3+{}^{-}5={}^{+}3+{}^{-}5={}^{-}2$
j. ${}^{-}19$

8 Consider the following examples:

${}^{+}3-{}^{+}4={}^{+}3+{}^{-}4={}^{-}1$
${}^{+}4-{}^{+}3={}^{+}1$

The expressions ${}^{+}3-{}^{+}4$ and ${}^{+}4-{}^{+}3$ are not equivalent since they have different evaluations. Thus, ${}^{+}3-{}^{+}4={}^{+}4-{}^{+}3$ is a false statement. ${}^{+}3-{}^{+}4\neq{}^{+}4-{}^{+}3$ is a true statement. ("$\neq$" means "is not equal to.")

Q16 **a.** Are ${}^{-}2-{}^{+}3$ and ${}^{+}3-{}^{-}2$ equivalent expressions? ______

b. Is ${}^{-}2-{}^{+}3={}^{+}3-{}^{-}2$ a true statement? ______

• ### • ### • ### • ### • ### • ### • ### •

A16 **a.** no: ${}^{-}2-{}^{+}3={}^{-}5$, whereas ${}^{+}3-{}^{-}2={}^{+}5$
b. no

9 Frame 8 demonstrates that when the *order* of the numbers in a subtraction problem is changed the answers are not always the same. Thus, it is said that subtraction is *not* a commutative operation.

Q17 **a.** Are ${}^{-}5-4$ and $4-{}^{-}5$ equivalent expressions? ______

b. Is subtraction a commutative operation? ______

c. Is addition a commutative operation? ______

• ### • ### • ### • ### • ### • ### • ### •

A17 **a.** no: ${}^{-}5-4={}^{-}9$, whereas $4-{}^{-}5={}^{+}9$
b. no **c.** yes

Q18 Verify that $(^{+}3 - {}^{+}5) - {}^{+}7$ and $^{+}3 - (^{+}5 - {}^{+}7)$ do not have the same evaluation.

• ### • ### • ### • ### • ### • ### • ### •

A18 $(^{+}3 - {}^{+}5) - {}^{+}7 = (^{+}3 + {}^{-}5) + {}^{-}7 = {}^{-}2 + {}^{-}7 = {}^{-}9$
$^{+}3 - (^{+}5 - {}^{+}7) = {}^{+}3 - (^{+}5 + {}^{-}7) = {}^{+}3 - {}^{-}2 = {}^{+}3 + {}^{+}2 = {}^{+}5$

Q19 $(^{+}3 - {}^{+}5) - {}^{+}7 = {}^{+}3 - (^{+}5 - {}^{+}7)$ is a ________ statement.
true/false

• ### • ### • ### • ### • ### • ### • ### •

A19 false

10 Question 19 demonstrates that when the *grouping* of three or more numbers in a subtraction problem is changed the answers are not always the same. Thus, it is said that subtraction is *not* an associative operation.

Q20 **a.** Are $(^{+}5 - {}^{+}2) - {}^{+}1$ and $^{+}5 - (^{+}2 - {}^{+}1)$ equivalent expressions? ______

b. Is subtraction an associative operation? ______

c. Is addition an associative operation? ______

• ### • ### • ### • ### • ### • ### • ### •

A20 **a.** no: $(^{+}5 - {}^{+}2) - {}^{+}1 = {}^{+}2$ whereas $^{+}5 - (^{+}2 - {}^{+}1) = {}^{+}4$.
b. no **c.** yes

11 Technicians often use the operation of subtraction in situations where quantities are represented by integers. One such situation is the calculation of differences in temperature readings.

Example: A chemist recorded a morning temperature of a solution at $^{-}5$ °C and an afternoon temperature of the same solution at 12 °C. What was the change in temperature of the solution.

Solution: The change in temperature is found by subtracting the first reading from the second.

$12 - {}^{-}5 = 12 + 5 = 17$

Thus, the change in temperature is 17 °C.

Q21 The temperature of a chemical varies from $^{-}10$ °C to 15 °C over a given period of time. Find the change in temperature.

• ### • ### • ### • ### • ### • ### • ### •

A21 25 °C: $15 - {}^{-}10 = 15 + 10 = 25$

Q22 A laboratory technician observed a low temperature of $^{-}125$ °C and a high temperature of $^{-}103$ °C. How much did the temperature vary?

• ### • ### • ### • ### • ### • ### • ### •

A22 22 °C: $^{-}103 - ^{-}125 = ^{-}103 + 125 = 22$

Q23 The refrigeration equipment of a supermarket failed and the temperature of a freezer went from $^{-}7$ °C to 2 °C. Find the change in temperature.

• ### • ### • ### • ### • ### • ### • ### •

A23 9 °C: $2 - ^{-}7 = 2 + 7 = 9$

This completes the instruction for this section.

7.3 EXERCISE

1. Find the following sums:
a. $^{+}3 + ^{-}7$ **b.** $^{+}12 + ^{-}9$ **c.** $^{-}5 + ^{-}6$ **d.** $^{-}9 + ^{+}12$
e. $^{-}4 + ^{+}11$ **f.** $^{+}6 + ^{-}6$ **g.** $^{-}2 + ^{+}2$ **h.** $^{+}4 + 0$
i. $0 + ^{-}7$ **j.** $^{-}3 + ^{-}4$

2. Write the opposite for each of the following integers:
a. $^{-}5$ **b.** 0 **c.** $^{+}6$ **d.** $^{-}4$ **e.** 2 **f.** $^{-}1$

3. Find the following differences:
a. $^{-}2 - 4$ **b.** $^{-}5 - ^{-}2$ **c.** $^{-}7 - ^{-}5$ **d.** $^{-}3 - 0$
e. $^{+}6 - ^{+}9$ **f.** $^{-}4 - ^{-}4$ **g.** $^{+}3 - ^{+}2$ **h.** $0 - ^{-}1$
i. $0 - ^{+}3$ **j.** $^{-}2 - ^{-}8$

4. Complete each of the following:
a. $^{+}7 + ^{-}3$ **b.** $^{-}3 + 0$ **c.** $^{-}6 - ^{+}4$ **d.** $^{+}7 - ^{+}5$
e. $^{-}5 + ^{-}9$ **f.** $^{+}7 + ^{-}7$ **g.** $0 + ^{-}4$ **h.** $^{-}6 - ^{-}7$
i. $^{-}1 - ^{+}1$ **j.** $0 + ^{+}3$ **k.** $^{-}3 - ^{+}4$ **l.** $^{-}4 + ^{+}9$
m. $^{-}5 - 0$ **n.** $0 - ^{-}5$ **o.** $^{+}6 + ^{-}9$ **p.** $^{+}4 - ^{+}5$
q. $^{-}3 + ^{+}12$ **r.** $0 + ^{-}4$ **s.** $^{-}5 + ^{+}5$ **t.** $0 - ^{-}2$

5. Complete each of the following:
a. $^{+}4 - ^{+}3 + ^{-}7$ **b.** $^{-}2 - ^{-}5 - ^{-}6$ **c.** $^{-}6 + ^{+}7 + ^{-}3$
d. $^{-}4 + ^{+}6 - ^{-}4 - ^{+}6$ **e.** $^{+}2 + ^{-}2 - 0$ **f.** $^{+}3 - ^{+}2 - ^{+}4$
g. $^{+}4 + ^{-}7 - ^{+}3$ **h.** $0 - ^{+}4 + ^{+}7$ **i.** $^{-}2 - ^{+}5 - ^{+}6$
j. $^{+}6 - 0 + ^{-}3$

6. The temperature of a chemical varies from $^{-}7$ °C to 12 °C over a given period of time. Find the change in temperature.

7. The outdoor temperature dropped overnight from 15 °C to $^{-}13$ °C. What was the change in temperature?

8. The refrigeration equipment on a truck failed and the temperature changed from $^{-}25$ °C to $^{-}10$ °C. Find the change in temperature.

7.3 EXERCISE ANSWERS

1. a. $^{-}4$ **b.** $^{+}3$ **c.** $^{-}11$ **d.** $^{+}3$ **e.** $^{+}7$ **f.** 0 **g.** 0
h. $^{+}4$ **i.** $^{-}7$ **j.** $^{-}7$

2. a. $^{+}5$ **b.** 0 **c.** $^{-}6$ **d.** $^{+}4$ **e.** $^{-}2$ **f.** $^{+}1$

3. a. $^{-}6$ **b.** $^{-}3$ **c.** $^{-}2$ **d.** $^{-}3$ **e.** $^{-}3$ **f.** 0 **g.** $^{+}1$
h. $^{+}1$ **i.** $^{-}3$ **j.** $^{+}6$

4. a. $^{+}4$ **b.** $^{-}3$ **c.** $^{-}10$ **d.** $^{+}2$ **e.** $^{-}14$ **f.** 0 **g.** $^{-}4$
h. $^{+}1$ **i.** $^{-}2$ **j.** $^{+}3$ **k.** $^{-}7$ **l.** $^{+}5$ **m.** $^{-}5$ **n.** $^{+}5$
o. $^{-}3$ **p.** $^{-}1$ **q.** $^{+}9$ **r.** $^{-}4$ **s.** 0 **t.** $^{+}2$

5. a. $^-6$ **b.** $^+9$ **c.** $^-2$ **d.** 0 **e.** 0 **f.** $^-3$ **g.** $^-6$
h. $^+3$ **i.** $^-13$ **j.** $^+3$
6. 19 °C
7. $^-28$ °C
8. 15 °C

7.4 MULTIPLICATION OF INTEGERS

1 The operation of multiplication was developed as a shortcut procedure for addition. For example, the product $2 \cdot 3$ can be represented either as the sum of 2 threes or as the sum of 3 twos:

$$2 \cdot 3 = \underbrace{3 + 3}_{\text{2 addends of 3}} = 6$$

or

$$3 \cdot 2 = \underbrace{2 + 2 + 2}_{\text{3 addends of 2}} = 6$$

Similarly, the product $^+7 \cdot {}^+4$ can be represented as either 7 positive fours or 4 positive sevens.

$$^+7 \cdot {}^+4 = {}^+4 + {}^+4 + {}^+4 + {}^+4 + {}^+4 + {}^+4 + {}^+4 = {}^+28$$

or

$$^+4 \cdot {}^+7 = {}^+7 + {}^+7 + {}^+7 + {}^+7 = {}^+28$$

Q1 Write $^+5 \cdot {}^+6$ as a sum in two ways.

\### • ### • ### • ### • ### • ### • ### • ### • ###

A1 $^+5 \cdot {}^+6 = {}^+6 + {}^+6 + {}^+6 + {}^+6 + {}^+6 = {}^+30$

or

$^+6 \cdot {}^+5 = {}^+5 + {}^+5 + {}^+5 + {}^+5 + {}^+5 + {}^+5 = {}^+30$

Q2 Write $^+3 \cdot {}^+9$ as a sum in two ways.

\### • ### • ### • ### • ### • ### • ### • ### • ###

A2 $^+3 \cdot {}^+9 = {}^+9 + {}^+9 + {}^+9 = {}^+27$

or

$^+9 \cdot {}^+3 = {}^+3 + {}^+3 + {}^+3 + {}^+3 + {}^+3 + {}^+3 + {}^+3 + {}^+3 + {}^+3 = {}^+27$

2 The procedure of writing a product as a sum can also be used to find the product of two integers. For example, the product $^+4 \cdot {}^-2$ can be written as the sum of 4 negative twos:

$$^+4 \cdot {}^-2 = \underbrace{{}^-2 + {}^-2 + {}^-2 + {}^-2}_{\text{4 addends of }{}^-2}$$

Since the sum on the right is equal to $^-8$ the product $^+4 \cdot {}^-2$ is $^-8$:

$^+4 \cdot {}^-2 = {}^-8$

Similarly, the product $^+3 \cdot {}^-7$ is $^-21$, because

$$^+3 \cdot {}^-7 = {}^-7 + {}^-7 + {}^-7$$
$$= {}^-21$$

Q3 Write $^+2 \cdot {}^-6$ as a sum. ____________

• # # # • # # # • # # # • # # # • # # # • # # # • # # # • # #

A3 $^+2 \cdot {}^-6 = {}^-6 + {}^-6$

Q4 Find the product $^+2 \cdot {}^-6$

• # # # • # # # • # # # • # # # • # # # • # # # • # # # • # #

A4 $^-12$

Q5 Find the product $^+3 \cdot {}^-5$ by writing it as a sum.

• # # # • # # # • # # # • # # # • # # # • # # # • # # # • # #

A5 $^-15$: $^+3 \cdot {}^-5 = {}^-5 + {}^-5 + {}^-5 = {}^-15$

Q6 Find the product $^+7 \cdot {}^-1$ by writing it as a sum.

• # # # • # # # • # # # • # # # • # # # • # # # • # # # • # #

A6 $^-7$: $^+7 \cdot {}^-1 = {}^-1 + {}^-1 + {}^-1 + {}^-1 + {}^-1 + {}^-1 + {}^-1 = {}^-7$

Q7 Find each of the following products:

a. $^+5 \cdot {}^-2 =$ -10 **b.** $^+6 \cdot {}^-9 =$ -54 **c.** $^+1 \cdot {}^-4 =$ -4

d. $0 \cdot {}^+4 =$ 0 **e.** $^+3 \cdot {}^+3 =$ 9 **f.** $^+2 \cdot {}^-8 =$ -16

• # # # • # # # • # # # • # # # • # # # • # # # • # # # • # #

A7 **a.** $^-10$ **b.** $^-54$ **c.** $^-4$ **d.** 0 **e.** $^+9$ **f.** $^-16$

3 The product $^-5 \cdot {}^+4$ can be found by computing the sum of 4 negative fives:

$$^-5 \cdot {}^+4 = \underbrace{{}^-5 + {}^-5 + {}^-5 + {}^-5}_{\text{4 addends of } {}^-5}$$
$$= {}^-20$$

Q8 Find the product $^-7 \cdot {}^+5$ by writing it as a sum.

• # # # • # # # • # # # • # # # • # # # • # # # • # # # • # #

A8 $^-35$: $^-7 \cdot {}^+5 = {}^-7 + {}^-7 + {}^-7 + {}^-7 + {}^-7 = {}^-35$

Q9 Find the product $^{+}5 \cdot {}^{-}7$.

• ### • ### • ### • ### • ### • ### • ### •

A9 $^{-}35$: $^{+}5 \cdot {}^{-}7 = {}^{-}7 + {}^{-}7 + {}^{-}7 + {}^{-}7 + {}^{-}7 = {}^{-}35$

Q10 Find the product $^{-}4 \cdot {}^{+}6$.

• ### • ### • ### • ### • ### • ### • ### •

A10 $^{-}24$: $^{-}4 \cdot {}^{+}6 = {}^{-}4 + {}^{-}4 + {}^{-}4 + {}^{-}4 + {}^{-}4 + {}^{-}4 = {}^{-}24$

Q11 Find the product $^{-}7 \cdot {}^{+}9$.

• ### • ### • ### • ### • ### • ### • ### •

A11 $^{-}63$

4 In each of the preceding products where the two factors had different signs (one positive and one negative), the product was negative. For example,

$\underbrace{^{+}5 \cdot {}^{-}9}_{\text{different signs}} = \underbrace{^{-}45}_{\text{negative product}}$ $\underbrace{^{-}7 \cdot {}^{+}8}_{\text{different signs}} = \underbrace{^{-}56}_{\text{negative product}}$

These examples demonstrate the following rule for multiplying integers with different signs: *The product of two integers with different signs (one positive and one negative) is a negative integer.*

To make the multiplication of integers consistent with the multiplication of natural numbers, the rule for multiplying two positive integers is as follows: *The product of two positive integers is a positive integer.* For example,

$^{+}2 \cdot {}^{+}7 = {}^{+}14$ $5 \cdot 8 = 40$

Recall from Section 7.1 that the "+" sign is frequently omitted from positive numbers such as in the second example above. A number written without a sign is assumed to be positive.

Q12 Find the product in each of the following:

a. $^{-}2 \cdot {}^{+}3 =$ ______ **b.** $^{+}6 \cdot {}^{-}3 =$ ______ **c.** $4 \cdot 7 =$ ______

d. $^{+}1 \cdot {}^{-}5 =$ ______ **e.** $^{+}10 \cdot {}^{-}8 =$ ______ **f.** $^{+}5 \cdot {}^{+}3 =$ ______

g. $^{-}7 \cdot {}^{+}7 =$ ______ **h.** $^{-}8 \cdot {}^{+}6 =$ ______

• ### • ### • ### • ### • ### • ### • ### •

A12 **a.** $^{-}6$ **b.** $^{-}18$ **c.** 28 **d.** $^{-}5$
e. $^{-}80$ **f.** 15 **g.** $^{-}49$ **h.** $^{-}48$

5 Two important properties of whole numbers are also true for the integers. These are the *multiplication property of zero* and the *multiplication property of one*. The multiplication property of zero states that the product of any number and zero is zero.

Examples: $0 \cdot {}^{-}5 = 0$
$8 \cdot 0 = 0$

In general, $a \cdot 0 = 0 \cdot a = 0$ for any integer a.

The multiplication property of one states that the product of any number and one is the identical number.

Examples: $1 \cdot {}^{-}7 = {}^{-}7$
$2 \cdot 1 = 2$

In general, $a \cdot 1 = 1 \cdot a = a$ for any integer a.

Q13 Find each of the following products:

a. ${}^{-}4 \cdot 0 =$ ______ **b.** ${}^{+}8 \cdot 1 =$ ______

c. ${}^{-}3 \cdot {}^{+}4 =$ ______ **d.** $0 \cdot {}^{+}5 =$ ______

e. $6 \cdot 7 =$ ______ **f.** ${}^{+}9 \cdot {}^{-}9 =$ ______

• ### • ### • ### • ### • ### • ### • ### •

A13 **a.** 0 **b.** 8 **c.** ${}^{-}12$ **d.** 0 **e.** 42 **f.** ${}^{-}81$

6 The product of two integers with *different signs* is *negative*. The product of *two positive integers* is a *positive* integer. To discover the product of *two negative integers*, study the following series of products and notice the pattern that is present in the answers on the right.

${}^{-}2 \cdot {}^{+}4 = {}^{-}8$
${}^{-}2 \cdot {}^{+}3 = {}^{-}6$
${}^{-}2 \cdot {}^{+}2 = {}^{-}4$
${}^{-}2 \cdot {}^{+}1 = {}^{-}2$
${}^{-}2 \cdot 0 = 0$
${}^{-}2 \cdot {}^{-}1 = ?$
${}^{-}2 \cdot {}^{-}2 = ?$

The pattern in the products on the right is that each answer increases by 2. Hence, to complete the pattern, the products are:

${}^{-}2 \cdot {}^{-}1 = {}^{+}2$
${}^{-}2 \cdot {}^{-}2 = {}^{+}4$

It is, thus, appropriate to state the rule for the product of two negative integers as follows: *The product of two negative integers is a positive integer.* For example,

${}^{-}7 \cdot {}^{-}4 = {}^{+}28$ ${}^{-}11 \cdot {}^{-}5 = {}^{+}55$

Q14 Find the product ${}^{-}4 \cdot {}^{-}6$.

• ### • ### • ### • ### • ### • ### • ### •

A14 ${}^{+}24$ (or simply 24)

Q15 ${}^{-}3 \cdot 5 =$ ______

• ### • ### • ### • ### • ### • ### • ### •

A15 ${}^{-}15$

Q16 $^{-}9 \cdot {}^{-}6 =$ ______

• # # # • # # # • # # # • # # # • # # # • # # # • # # # • # #

A16 54

Q17 Find the following products:

a. $^{-}3 \cdot 4 =$ ______ **b.** $^{-}4 \cdot {}^{-}9 =$ ______

c. $^{-}2 \cdot {}^{-}5 =$ ______ **d.** $9 \cdot {}^{-}8 =$ ______

e. $0 \cdot {}^{-}6 =$ ______ **f.** $^{-}12 \cdot 0 =$ ______

g. $5 \cdot {}^{-}1 =$ ______ **h.** $^{-}7 \cdot {}^{-}5 =$ ______

• # # # • # # # • # # # • # # # • # # # • # # # • # # # • # #

A17 **a.** $^{-}12$ **b.** 36 **c.** 10 **d.** $^{-}72$
e. 0 **f.** 0 **g.** $^{-}5$ **h.** 35

7 Study the effect of multiplying any integer by $^{-}1$ in the following examples:

$^{-}1 \cdot 4 = {}^{-}4 \qquad {}^{-}1 \cdot {}^{-}5 = {}^{+}5$

In the first example, multiplying 4 by negative one changes it to $^{-}4$, its opposite. In the second example, the product of $^{-}5$ and negative one changes $^{-}5$ to $^{+}5$, its opposite. The fact that negative one times a number is the opposite of the number is often called the *multiplication property of* $^{-}1$. In general, $^{-}1 \cdot a = a \cdot {}^{-}1 = {}^{-}a$ for any integer a. (Note: ^{-}a is read "the opposite of a.")

Q18 Find each of the following products:

a. $^{-}1 \cdot {}^{-}7 =$ ______ **b.** $1 \cdot {}^{-}7 =$ ______

c. $^{-}1 \cdot 9 =$ ______ **d.** $^{-}1 \cdot 0 =$ ______

• # # # • # # # • # # # • # # # • # # # • # # # • # # # • # #

A18 **a.** 7 **b.** $^{-}7$ **c.** $^{-}9$ **d.** 0

8 It is important for later work that the student be able to read the multiplication property of $^{-}1$ both from left to right and from right to left. Reading from left to right it says that multiplying by $^{-}1$ gives the opposite of the integer involved:

	The integer		Its opposite
$^{-}1 \cdot$	$^{+}6$	=	$^{-}6$
$^{-}1 \cdot$	$^{-}9$	=	$^{+}9$

Reading from right to left it says that any integer can be expressed as a product of $^{-}1$ and its opposite.

The integer		Its opposite
$^{-}5$	$= {}^{-}1 \cdot$	$^{+}5$
$^{-}7$	$= {}^{-}1 \cdot$	$^{+}7$
$^{+}6$	$= {}^{-}1 \cdot$	$^{-}6$
$^{+}3$	$= {}^{-}1 \cdot$	$^{-}3$

Q19 Complete the following statements using the multiplication property of $^{-}1$:

a. $^{-}1 \cdot {}^{+}7 =$ ______ **b.** $^{-}1 \cdot$ ______ $= {}^{+}5$

c. $^{-}3 = {}^{-}1 \cdot$ ______ **d.** ______ $\cdot {}^{-}8 = {}^{+}8$

\# \# \# • \# \# \# • \# \# \# • \# \# \# • \# \# \# • \# \# \# • \# \# \# • \# \# \# • \# \# \#

A19 **a.** $^{-}7$ **b.** $^{-}5$ **c.** $^{+}3$ **d.** $^{-}1$

Q20 Find the product:

a. $^{-}3 \cdot {}^{+}2 =$ ______ **b.** $^{+}2 \cdot {}^{-}3 =$ ______

\# \# \# • \# \# \# • \# \# \# • \# \# \# • \# \# \# • \# \# \# • \# \# \# • \# \# \# • \# \# \#

A20 **a.** $^{-}6$ **b.** $^{-}6$

Q21 **a.** Evaluate $^{-}7 \cdot {}^{-}8$. ______ **b.** Evaluate $^{-}8 \cdot {}^{-}7$. ______

\# \# \# • \# \# \# • \# \# \# • \# \# \# • \# \# \# • \# \# \# • \# \# \# • \# \# \# • \# \# \#

A21 **a.** 56 **b.** 56

Q22 **a.** Evaluate $3 \cdot {}^{-}5$. ______ **b.** Evaluate $^{-}5 \cdot 3$. ______

\# \# \# • \# \# \# • \# \# \# • \# \# \# • \# \# \# • \# \# \# • \# \# \# • \# \# \# • \# \# \#

A22 **a.** $^{-}15$ **b.** $^{-}15$

9 Questions 21 and 22 demonstrate that the integers are commutative with respect to the operation of multiplication. The *commutative property of multiplication* states that regardless of the order in which two or more numbers are multiplied, the products are the same.

Examples: 1. $^{-}3 \cdot {}^{+}9 = {}^{+}9 \cdot {}^{-}3$
$^{-}27 = {}^{-}27$

2. $^{-}12 \cdot {}^{-}5 = {}^{-}5 \cdot {}^{-}12$
$^{+}60 = {}^{+}60$

In general, $a \cdot b = b \cdot a$ for any integers a and b.

Another important property of the integers is the *associative property of multiplication.* It states that the grouping of factors in a product does not affect the answer.

Example: $7({}^{-}3 \cdot 2) = (7 \cdot {}^{-}3)2$
$7({}^{-}6) = ({}^{-}21)2$
$^{-}42 = {}^{-}42$

In general, $a(b \cdot c) = (a \cdot b)c$ for any integers a, b, and c.

Q23 **a.** Is $^{-}253 \cdot 479 = 479 \cdot {}^{-}253$ a true statement? ______

b. Is $^{-}6(7 \cdot {}^{-}3) = ({}^{-}6 \cdot 7) \cdot {}^{-}3$ a true statement? ______

\# \# \# • \# \# \# • \# \# \# • \# \# \# • \# \# \# • \# \# \# • \# \# \# • \# \# \# • \# \# \#

A23 **a.** yes: The commutative property of multiplication is true for all integers.
b. yes: The associative property of multiplication is true for all integers.

10 Frame 9 makes it possible to state the following procedure for finding a product of more than two numbers. To find the product of more than two numbers:

Step 1: Do all work within parentheses first.

Step 2: Find the product of two numbers at a time in *any order desired.*

Examples:

1. $^{-}2 \cdot {}^{+}4 \cdot {}^{-}5$
$^{+}10 \cdot {}^{+}4$ (since there are no parentheses, follow step 2)
40

2. $^{+}3(^{-}6 \cdot {}^{-}5) \cdot {}^{+}9$
$^{+}3(^{+}30) \cdot {}^{+}9$ (since parentheses are involved,
$^{+}90 \cdot {}^{+}9$ use step 1 and then step 2)
810

Q24 Find the product $^{-}2 \cdot {}^{-}3 \cdot {}^{+}5$.

• ### • ### • ### • ### • ### • ### • ### •

A24 30

Q25 Find the product $(^{-}1 \cdot {}^{+}3)(^{-}4 \cdot {}^{-}7)$.

• ### • ### • ### • ### • ### • ### • ### •

A25 $^{-}84$: $(^{-}1 \cdot {}^{+}3)(^{-}4 \cdot {}^{-}7) = {}^{-}3 \cdot {}^{+}28 = {}^{-}84$

Q26 Find each of the following products:

a. $^{-}2 \cdot {}^{-}4 \cdot 0$

b. $(^{-}3 \cdot {}^{+}4) \cdot {}^{-}5$

c. $^{-}6 \cdot {}^{+}11 \cdot {}^{+}10$

d. $^{-}1(^{+}3 \cdot {}^{-}5)$

e. $(^{+}4 \cdot 0)(^{-}7 \cdot {}^{-}6)$

f. $(^{-}2 \cdot {}^{+}3)(^{-}4 \cdot {}^{-}6)$

• ### • ### • ### • ### • ### • ### • ### •

A26 **a.** 0 **b.** 60 **c.** $^{-}660$ **d.** 15 **e.** 0 **f.** $^{-}144$

11 It has been established that the integers are commutative and associative with respect to addition and multiplication. Two additional properties of the integers are the *distributive property of multiplication over addition* and the *distributive property of multiplication over subtraction.* Examples of each are shown below:

(Distributive property of multiplication over addition) $2(^{-}5+7) = 2 \cdot {}^{-}5 + 2 \cdot 7$

(Distributive property of multiplication over subtraction) $3(^{-}4-6) = 3 \cdot {}^{-}4 - 3 \cdot 6$

The distributive properties state that the evaluations of the expressions on opposite sides of the equal sign are the same.

Q27 **a.** Evaluate $2(^-5+7)$ by working within the parentheses first.

b. Evaluate $2 \cdot {^-5} + 2 \cdot 7$

• ### • ### • ### • ### • ### • ### • ### •

A27 **a.** 4: $2(^-5+7) = 2(2) = 4$

b. 4: $2 \cdot {^-5} + 2 \cdot 7 = {^-10} + 14 = 4$

Q28 **a.** Evaluate $3(^-4-6)$ by working within the parentheses first.

b. Evaluate $3 \cdot {^-4} - 3 \cdot 6$

• ### • ### • ### • ### • ### • ### • ### •

A28 **a.** $^-30$: $3(^-4-6) = 3(^-4+{^-6}) = 3(^-10) = {^-30}$

b. $^-30$: $3 \cdot {^-4} - 3 \cdot 6 = {^-12} - 18 = {^-12} + {^-18} = {^-30}$

Q29 Use the examples of Frame 11 to complete the following (without evaluating):

$8(^-2+5) = 8 \cdot$ __-2__ $+ 8 \cdot$ __5__

• ### • ### • ### • ### • ### • ### • ### •

A29 $8(^-2+5) = 8 \cdot {^-2} + 8 \cdot 5$

Q30 Complete the following (without evaluating):

$7(^-3-{^-4}) = 7 \cdot$ __-3__ $- 7 \cdot$ __-4__

• ### • ### • ### • ### • ### • ### • ### •

A30 $7(^-3-{^-4}) = 7 \cdot {^-3} - 7 \cdot {^-4}$

---

12 The distributive properties are often classified as right or left depending on which side of the parentheses the number lies.

Examples:

1. Left distributive property of multiplication over addition

   $\underline{5}(2+{^-6}) = \underline{5} \cdot 2 + \underline{5} \cdot {^-6}$

2. Right distributive property of multiplication over addition

   $(2+{^-6})\underline{5} = 2 \cdot \underline{5} + {^-6} \cdot \underline{5}$

3. Left distributive property of multiplication over subtraction

   $\underline{3}(2-7) = \underline{3} \cdot 2 - \underline{3} \cdot 7$

4. Right distributive property of multiplication over subtraction

   $(2-7)\underline{3} = 2 \cdot \underline{3} - 7 \cdot \underline{3}$

Q31 **a.** Evaluate $5(2+{}^{-}6)$ by working within parentheses first.

b. Evaluate $5 \cdot 2+5 \cdot {}^{-}6$.

\# \# \# • \# \# \# • \# \# \# • \# \# \# • \# \# \# • \# \# \# • \# \# \# • \# \# \# • \# \# \#

A31 **a.** ${}^{-}20$: $5(2+{}^{-}6)=5({}^{-}4)={}^{-}20$ **b.** ${}^{-}20$: $5 \cdot 2+5 \cdot {}^{-}6=10+{}^{-}30={}^{-}20$

Q32 $3(5+{}^{-}7)=3 \cdot 5+3 \cdot {}^{-}7$ is an example of the ______ distributive property of ________ over ________.

\# \# \# • \# \# \# • \# \# \# • \# \# \# • \# \# \# • \# \# \# • \# \# \# • \# \# \# • \# \# \#

A32 left, multiplication, addition

This completes the instruction for this section.

7.4 EXERCISE

1. The product of two integers with different signs is a ____________ integer.
2. The product of any integer and zero is ____________.
3. The product of any two integers with the same sign is a ____________ integer.
4. Fill in the blanks so that a true statement results:
   - **a.** $\_\_\_\_ \cdot 5=0$ **b.** ${}^{-}1 \cdot 7=\_\_\_\_$
   - **c.** ${}^{+}8=\_\_\_\_ \cdot {}^{-}8$ **d.** ${}^{-}1 \cdot {}^{-}5=\_\_\_\_$
5. Find each of the following products:
   - **a.** ${}^{+}4 \cdot {}^{-}9$ **b.** ${}^{-}3 \cdot {}^{-}4$
   - **c.** ${}^{-}2 \cdot 0$ **d.** ${}^{-}1 \cdot {}^{+}9$
   - **e.** ${}^{-}1 \cdot {}^{+}7$ **f.** $5 \cdot {}^{-}7$
   - **g.** $0 \cdot {}^{-}9$ **h.** ${}^{-}6 \cdot {}^{-}9$
   - **i.** $11 \cdot {}^{-}5$ **j.** $7 \cdot {}^{-}8$
6. Find each of the following products:
   - **a.** ${}^{+}2 \cdot {}^{-}3 \cdot {}^{-}4$ **b.** $({}^{+}7 \cdot {}^{-}1) \cdot {}^{-}7$
   - **c.** ${}^{-}1 \cdot {}^{-}7 \cdot {}^{-}8$ **d.** ${}^{-}1 \cdot {}^{-}1 \cdot 0$
   - **e.** ${}^{-}5 \cdot 0 \cdot {}^{-}6 \cdot {}^{+}3$ **f.** $4 \cdot 3 \cdot {}^{-}5$
   - **g.** $({}^{-}4 \cdot {}^{-}4)({}^{-}2 \cdot {}^{+}2)$ **h.** ${}^{-}7(9 \cdot {}^{-}3) \cdot {}^{+}2$
   - **i.** ${}^{+}6({}^{-}3 \cdot {}^{+}2)$ **j.** $({}^{-}2 \cdot 4)(3 \cdot {}^{-}3)$
   - **k.** $({}^{+}6 \cdot {}^{-}3) \cdot {}^{+}2$ **l.** ${}^{-}4 \cdot 5({}^{-}6 \cdot 0)$
   - **m.** ${}^{-}2({}^{-}3+{}^{-}5)$ **n.** $({}^{-}7-{}^{-}4)4$

7.4 EXERCISE ANSWERS

1. negative
2. zero
3. positive
4. **a.** 0 **b.** ${}^{-}7$ **c.** ${}^{-}1$ **d.** 5
5. **a.** ${}^{-}36$ **b.** 12 **c.** 0 **d.** ${}^{-}9$ **e.** ${}^{-}7$ **f.** ${}^{-}35$ **g.** 0 **h.** 54 **i.** ${}^{-}55$ **j.** ${}^{-}56$
6. **a.** 24 **b.** 49 **c.** ${}^{-}56$ **d.** 0 **e.** 0 **f.** ${}^{-}60$ **g.** ${}^{-}64$ **h.** 378 **i.** ${}^{-}36$ **j.** 72 **k.** ${}^{-}36$ **l.** 0 **m.** 16 **n.** ${}^{-}12$

7.5 DIVISION OF INTEGERS

1 The operation of division is closely related to that of multiplication. A division problem is often checked using the multiplication operation. As was the case with addition and subtraction, multiplication and division are also opposite or inverse operations; that is, one undoes the effect of the other. For example,

$10 \div 2 = 5$ because $5 \cdot 2 = 10$ or $8\overline{)56}$ with quotient 7 because $7 \cdot 8 = 56$

Q1 $12 \div 3 = 4$, because _____ · _____ = _____.

• ### • ### • ### • ### • ### • ### • ### •

A1 $4 \cdot 3 = 12$

Q2 $7\overline{)63}$ with quotient 9, because _____ · _____ = _____.

• ### • ### • ### • ### • ### • ### • ### •

A2 $9 \cdot 7 = 63$

Q3 Use multiplication to find the quotient and write the check: $45 \div 9 =$ _____, because __________.

• ### • ### • ### • ### • ### • ### • ### •

A3 $45 \div 9 = 5$, because $5 \cdot 9 = 45$

Q4 Use multiplication to find the quotient and check the result: $82 \div 2 =$ _____, because __________.

• ### • ### • ### • ### • ### • ### • ### •

A4 $82 \div 2 = 41$, because $41 \cdot 2 = 82$

---

2 Consider the following quotient and check:

$^{-}27 \div {}^{+}9 = \square$ because $\square \cdot {}^{+}9 = {}^{-}27$

Since $^{-}3$ makes the product statement true, it also satisfies the quotient statement:

$^{-}27 \div {}^{+}9 = \boxed{^{-}3}$ because $\boxed{^{-}3} \cdot {}^{+}9 = {}^{-}27$

Q5 Place an integer in the $\square$ to form a true statement: $^{+}12 \div {}^{-}3 = \square$ because $\square \cdot {}^{-}3 = {}^{+}12$

• ### • ### • ### • ### • ### • ### • ### •

A5 $^{+}12 \div {}^{-}3 = \boxed{^{-}4}$, because $\boxed{^{-}4} \cdot {}^{-}3 = {}^{+}12$

Q6 Place an integer in the $\square$ to form a true statement: $^{-}16 \div {}^{+}2 = \square$ because $\square \cdot {}^{+}2 = {}^{-}16$

• ### • ### • ### • ### • ### • ### • ### •

A6 $^{-}16 \div {}^{+}2 = \boxed{^{-}8}$, because $\boxed{^{-}8} \cdot {}^{+}2 = {}^{-}16$

Q7 Find the quotient and write the check as a multiplication problem: $^{+}56 \div {}^{-}8 =$ _____ because _____ $\cdot {}^{-}8 = {}^{+}56$

• ### • ### • ### • ### • ### • ### • ### •

A7 $^{+}56 \div {}^{-}8 = \underline{{}^{-}7}$, because $\underline{{}^{-}7} \cdot {}^{-}8 = {}^{+}56$

Q8 Find the quotient and write the check as a multiplication problem: $^{-}30 \div {}^{+}10 =$ ______ because ______________

\# \# \# • \# \# \# • \# \# \# • \# \# \# • \# \# \# • \# \# \# • \# \# \# • \# \# \# • \# \# \#

A8 $\underline{{}^{-}3}$, because $\underline{{}^{-}3 \cdot {}^{+}10 = {}^{-}30}$

Q9 $^{+}15 \div {}^{-}3 =$ ______

\# \# \# • \# \# \# • \# \# \# • \# \# \# • \# \# \# • \# \# \# • \# \# \# • \# \# \# • \# \# \#

A9 $^{-}5$

3 Each of the problems in Q5 through Q9 involved the quotient of two integers with different signs. Study the problems and their answers below:

$^{+}12 \div {}^{-}3 = {}^{-}4$
$^{-}16 \div {}^{+}2 = {}^{-}8$
$^{+}56 \div {}^{-}8 = {}^{-}7$
$^{-}30 \div {}^{+}10 = {}^{-}3$
$^{+}15 \div {}^{-}3 = {}^{-}5$

Q10 What is true of all the answers in Frame 3? ______________________________

\# \# \# • \# \# \# • \# \# \# • \# \# \# • \# \# \# • \# \# \# • \# \# \# • \# \# \# • \# \# \#

A10 Each answer is a negative integer.

4 The examples of Frame 3 demonstrate the following definition for the division of two integers with different signs: *The quotient of two integers with different signs is a negative integer.*

A division problem is also correctly written as a fraction. Thus, $20 \div 4 = 5$ can also be written $\frac{20}{4} = 5$. Study the following examples:

$\frac{^{-}36}{^{+}4} = {}^{-}9$ $\quad$ $^{+}27 \div {}^{+}9 = {}^{+}3$

$48 \div {}^{-}6 = {}^{-}8$ $\quad$ $\frac{^{+}14}{^{-}2} = {}^{-}7$

Q11 Find the following quotients:

a. $\frac{^{-}25}{^{+}5} =$ -5 $\quad$ **b.** $^{-}25 \div {}^{+}5 =$ -5

c. $\frac{^{-}90}{^{+}10} =$ -9 $\quad$ **d.** $^{+}12 \div {}^{-}2 =$ -6

e. $^{+}1 \div {}^{-}1 =$ -1 $\quad$ **f.** $\frac{^{+}45}{^{-}9} =$ -5

\# \# \# • \# \# \# • \# \# \# • \# \# \# • \# \# \# • \# \# \# • \# \# \# • \# \# \# • \# \# \#

A11 **a.** $^{-}5$ **b.** $^{-}5$ **c.** $^{-}9$ **d.** $^{-}6$ **e.** $^{-}1$ **f.** $^{-}5$

5 To determine the sign of the *quotient* of two *integers* with the *same sign*, consider the integer that converts the following open sentence into a true statement.

$^-20 \div {}^-5 = \square$ because $\square \cdot {}^-5 = {}^-20$

The correct integer replacement is $^+4$. That is, $^-20 \div {}^-5 = {}^+4$, because $^+4 \cdot {}^-5 = {}^-20$.

Q12 Place an integer in the $\square$ to form a true statement: $^-15 \div {}^-3 = \square$ because $\square \cdot {}^-3 = {}^-15$

\# \# \# • \# \# \# • \# \# \# • \# \# \# • \# \# \# • \# \# \# • \# \# \# • \# \# \# • \# \# \#

A12 $^+5$: $^-15 \div {}^-3 = \boxed{^+5}$, because $\boxed{^+5} \cdot {}^-3 = {}^-15$.

Q13 Find the quotient and write the check as a multiplication problem: $^-63 \div {}^-7 =$ ______ because ______ $\cdot {}^-7 = {}^-63$

\# \# \# • \# \# \# • \# \# \# • \# \# \# • \# \# \# • \# \# \# • \# \# \# • \# \# \# • \# \# \#

A13 $^+9$: $^-63 \div {}^-7 = \underline{^+9}$, because $\underline{^+9} \cdot {}^-7 = {}^-63$

Q14 $^-81 \div {}^-9 =$ ______

\# \# \# • \# \# \# • \# \# \# • \# \# \# • \# \# \# • \# \# \# • \# \# \# • \# \# \# • \# \# \#

A14 $^+9$ (or simply 9)

6 The quotient of Q12 through Q14 involved integers with the same sign. The answer in each case was a positive integer. The rule suggested for quotients of this type is as follows: *The quotient of two integers with the same sign is a positive integer.* Some examples of the above rule are:

1. $\dfrac{^-36}{^-4} = {}^+9$

2. $^-26 \div {}^-2 = {}^+13$

3. $^+42 \div {}^+21 = {}^+2$

4. $\dfrac{18}{3} = 6$

Q15 Find the quotient $^-24 \div {}^-6$.

\# \# \# • \# \# \# • \# \# \# • \# \# \# • \# \# \# • \# \# \# • \# \# \# • \# \# \# • \# \# \#

A15 4

Q16 $\dfrac{^-56}{^-7} =$ ______

\# \# \# • \# \# \# • \# \# \# • \# \# \# • \# \# \# • \# \# \# • \# \# \# • \# \# \# • \# \# \#

A16 8

Q17 Find each of the following quotients:

a. $\dfrac{^-12}{^-3} =$ ______ **b.** $^+72 \div {}^-6 =$ ______ **c.** $^+4 \div {}^-4 =$ ______

d. $^-93 \div {}^-3 =$ ______ **e.** $\dfrac{^-19}{^-19} =$ ______ **f.** $14 \div 7 =$ ______

g. $\dfrac{^+24}{^-6} =$ ______ **h.** $\dfrac{^-24}{^+6} =$ ______ **i.** $^-28 \div {}^-7 =$ ______

j. $^{-}1 \div {}^{+}1 =$ _____ **k.** $^{-}1 \div {}^{-}1 =$ _____ **l.** $\frac{^{-}2}{^{-}2} =$ _____

\### • ### • ### • ### • ### • ### • ### • ### • ###

A17 **a.** 4 **b.** $^{-}12$ **c.** $^{-}1$ **d.** 31 **e.** 1 **f.** 2 **g.** $^{-}4$ **h.** $^{-}4$ **i.** 4 **j.** $^{-}1$ **k.** 1 **l.** 1

---

7 The *quotient* of two integers with *different signs* is a *negative* integer: for example, $^{-}12 \div {}^{+}6 = {}^{-}2$. The *product* of two integers with *different signs* is also a *negative* integer: for example, $^{-}4 \cdot {}^{+}6 = {}^{-}24$.

Q18 The quotient of two integers with the *same sign* is a __________ integer.

\### • ### • ### • ### • ### • ### • ### • ### • ###

A18 positive

Q19 The product of two integers with the *same sign* is a __________ integer.

\### • ### • ### • ### • ### • ### • ### • ### • ###

A19 positive

Q20 **a.** The quotient of two integers with *different signs* is a __________ integer.

b. The product of two integers with *different signs* is a __________ integer.

\### • ### • ### • ### • ### • ### • ### • ### • ###

A20 **a.** negative **b.** negative

---

8 The rules for multiplication and division are the same. They may be summarized:

1. *The product or quotient of two integers with the same sign is positive.*
2. *The product or quotient of two integers with different signs is negative.*

(*Note*: The above rules are sometimes abbreviated.)

In multiplication or division of integers:

1. Same sign—positive.
2. Different signs—negative.

This gives a quick and easy means to remember the signs in a multiplication or division problem.

Q21 Find the following products and quotients:

a. $^{-}15 \div {}^{-}3 =$ _____ **b.** $\frac{^{+}18}{^{-}3} =$ _____ **c.** $^{-}4 \cdot {}^{-}8 =$ _____

d. $\frac{^{-}14}{^{-}7} =$ _____ **e.** $^{+}7 \cdot {}^{-}3 =$ _____ **f.** $^{-}5 \cdot {}^{-}9 =$ _____

g. $\frac{^{-}12}{^{+}3} =$ _____ **h.** $^{+}7 \cdot {}^{-}6 =$ _____ **i.** $^{-}8 \cdot 0 =$ _____

j. $\frac{^{+}27}{^{-}9} =$ _____ **k.** $^{+}48 \div {}^{-}12 =$ _____ **l.** $^{-}32 \div {}^{-}4 =$ _____

\### • ### • ### • ### • ### • ### • ### • ### • ###

A21 **a.** 5 **b.** $^{-}6$ **c.** 32 **d.** 2 **e.** $^{-}21$ **f.** 45 **g.** $^{-}4$ **h.** $^{-}42$ **i.** 0 **j.** $^{-}3$ **k.** $^{-}4$ **l.** 8

9 In Section 7.4 the multiplication property of $^{-}1$ was stated. According to this property, *negative one times a number is the opposite of the number.* For example, $^{-}1 \cdot {}^{+}5 = {}^{-}5$ and $^{-}1 \cdot {}^{-}3 = {}^{+}3$. Consider the result of dividing an integer by $^{-}1$ in the following examples:

$$\frac{^{+}3}{^{-}1} = {}^{-}3 \qquad \frac{^{-}5}{^{-}1} = {}^{+}5$$

Q22 When $^{+}3$ is divided by $^{-}1$, the quotient is __-3__.

\# \# \# • \# \# \# • \# \# \# • \# \# \# • \# \# \# • \# \# \# • \# \# \# • \# \# \# • \# \# \#

A22 $^{-}3$: the opposite of $^{+}3$

Q23 When $^{-}5$ is divided by $^{-}1$, the quotient is __5__.

\# \# \# • \# \# \# • \# \# \# • \# \# \# • \# \# \# • \# \# \# • \# \# \# • \# \# \# • \# \# \#

A23 $^{+}5$: the opposite of $^{-}5$

10 The preceding examples demonstrate that the result of dividing an integer by $^{-}1$ is the same as when multiplying an integer by $^{-}1$. In each case, the answer is the opposite of the integer being multiplied or divided. The *division property of* $^{-}1$ states that $\frac{a}{^{-}1} = {}^{-}a$ is true for any integer a.

Q24 Complete each of the following:

a. $^{-}1 \cdot {}^{+}7 =$ ______ **b.** $\frac{^{+}7}{^{-}1} =$ ______ **c.** $^{-}1 \cdot {}^{-}5 =$ ______

d. $\frac{^{-}5}{^{-}1} =$ ______ **e.** $^{-}x = {}^{-}1 \cdot$ ______ **f.** $\frac{x}{^{-}1} =$ ______

\# \# \# • \# \# \# • \# \# \# • \# \# \# • \# \# \# • \# \# \# • \# \# \# • \# \# \# • \# \# \#

A24 **a.** $^{-}7$ **b.** $^{-}7$ **c.** 5 **d.** 5 **e.** $^{-}x = {}^{-}1 \cdot x$ **f.** $\frac{x}{^{-}1} = {}^{-}x$

11 The number zero is frequently confusing when involved as a divisor or dividend in a division problem. Consider the open sentence

$0 \div 5 = \square$ because $\square \cdot 5 = 0$

The integer that converts the open sentence to a true statement is zero:

$0 \div 5 = \boxed{0}$ because $\boxed{0} \cdot 5 = 0$

Q25 What integer converts the open sentence to a true statement? $0 \div {}^{-}9 = \square$ because $\square \cdot {}^{-}9 = 0$

\# \# \# • \# \# \# • \# \# \# • \# \# \# • \# \# \# • \# \# \# • \# \# \# • \# \# \# • \# \# \#

A25 0: $0 \div {}^{-}9 = \boxed{0}$, because $\boxed{0} \cdot {}^{-}9 = 0$

Q26 $0 \div {}^{+}2 =$ ______, because ______________

\# \# \# • \# \# \# • \# \# \# • \# \# \# • \# \# \# • \# \# \# • \# \# \# • \# \# \# • \# \# \#

A26 0, because $0 \cdot {}^{+}2 = 0$

Q27 $0 \div {}^{-}8 =$ _____

\# \# \# • \# \# \# • \# \# \# • \# \# \# • \# \# \# • \# \# \# • \# \# \# • \# \# \# • \# \# \#

A27 0

12 When zero is the divisor, the quotient is said to be *undefined*. To understand why, consider the open sentence

$4 \div 0 = \square$ because $\square \cdot 0 = 4$

There is no answer to the product $\square \cdot 0 = 4$, so there is no answer to the corresponding quotient $4 \div 0 = \square$. That is, $4 \div 0$ is undefined, because *no number* $\cdot\ 0 = 4$.

Q28 $7 \div 0 =$ __________, because __________

\# \# \# • \# \# \# • \# \# \# • \# \# \# • \# \# \# • \# \# \# • \# \# \# • \# \# \# • \# \# \#

A28 undefined, because no number $\cdot\ 0 = 7$

Q29 ${}^{-}3 \div 0 =$ __________

\# \# \# • \# \# \# • \# \# \# • \# \# \# • \# \# \# • \# \# \# • \# \# \# • \# \# \# • \# \# \#

A29 undefined

13 *When zero is the divisor* in a division problem, *the quotient is* said to be *undefined.* That is, $\frac{a}{0}$ = undefined for any integer a. For example,

$\frac{{}^{-}3}{0} =$ undefined $\quad {}^{+}12 \div 0 =$ undefined

When zero is divided by any nonzero integer, the quotient is zero. That is, $\frac{0}{a} = 0$ for any nonzero integer a.

Examples:

1. $\frac{0}{7} = 0$

2. $0 \div 6 = 0$

Q30 Complete each of the following quotients:

a. $0 \div {}^{-}2 =$ __________

b. ${}^{-}2 \div 0 =$ __________

c. $\frac{0}{{}^{+}7} =$ __________

d. $\frac{7}{0} =$ __________

e. ${}^{-}3 \div 0 =$ __________

f. $10 \div {}^{-}1 =$ __________

g. $\frac{{}^{-}4}{{}^{-}2} =$ __________

h. $0 \div {}^{-}4 =$ __________

i. $\frac{42}{{}^{-}6} =$ __________

j. $\frac{8}{0} =$ __________

\# \# \# • \# \# \# • \# \# \# • \# \# \# • \# \# \# • \# \# \# • \# \# \# • \# \# \# • \# \# \#

A30 **a.** 0 **b.** undefined **c.** 0
d. undefined **e.** undefined **f.** $^{-}10$
g. 2 **h.** 0 **i.** $^{-}7$
j. undefined

Q31 Evaluate:

a. $^{+}4 \div {}^{-}2 =$ _____ **b.** $^{-}2 \div 4 =$ _____

\# \# \# • \# \# \# • \# \# \# • \# \# \# • \# \# \# • \# \# \# • \# \# \# • \# \# \# • \# \# \#

A31 **a.** $^{-}2$ **b.** $\frac{^{-}2}{4} = \frac{^{-}1}{2}$

Q32 Is $^{+}4 \div {}^{-}2 = {}^{-}2 \div {}^{+}4$ a true statement? _____

\# \# \# • \# \# \# • \# \# \# • \# \# \# • \# \# \# • \# \# \# • \# \# \# • \# \# \# • \# \# \#

A32 no

14	The statement $a \div b = b \div a$ is false for most integers a and b. Therefore, the integers are not commutative with respect to the operation of division.

Q33 Evaluate:
a. $(^{-}20 \div {}^{-}10) \div {}^{+}2$ **b.** $^{-}20 \div (^{-}10 \div {}^{+}2)$

\# \# \# • \# \# \# • \# \# \# • \# \# \# • \# \# \# • \# \# \# • \# \# \# • \# \# \# • \# \# \#

A33 **a.** 1: $(^{-}20 \div {}^{-}10) \div {}^{+}2 = {}^{+}2 \div {}^{+}2 = 1$
b. 4: $^{-}20 \div (^{-}10 \div {}^{+}2) = {}^{-}20 \div {}^{-}5 = 4$

Q34 Is $(^{-}20 \div {}^{-}10) \div {}^{+}2 = {}^{-}20 \div (^{-}10 \div {}^{+}2)$ a true statement? _____

\# \# \# • \# \# \# • \# \# \# • \# \# \# • \# \# \# • \# \# \# • \# \# \# • \# \# \# • \# \# \#

A34 no

15	The statement $(a \div b) \div c = a \div (b \div c)$ is false for most integers a, b, and c. Therefore, the integers are not associative with respect to the operation of division.

Q35 Complete the following:

a. The integers _____ (are/are not) commutative with respect to multiplication.

b. The integers _____ (are/are not) commutative with respect to division.

c. $(^{-}16 \div {}^{+}4) \div {}^{+}2 = {}^{-}16 \div (^{+}4 \div {}^{+}2)$ _____ (is/is not) a true statement.

\# \# \# • \# \# \# • \# \# \# • \# \# \# • \# \# \# • \# \# \# • \# \# \# • \# \# \# • \# \# \#

A35 **a.** are **b.** are not **c.** is not

This completes the instruction for this section.

7.5 EXERCISE

1. Find each of the following quotients:

a. $72 \div {}^-9$ **b.** $\frac{^-18}{2}$ **c.** ${}^-15 \div {}^-5$ **d.** ${}^-5 \div {}^-5$

e. $0 \div {}^-7$ **f.** $\frac{^-24}{^-6}$ **g.** $\frac{10}{0}$ **h.** $\frac{0}{^-2}$

i. ${}^-13 \div {}^-1$ **j.** $\frac{^-8}{^-8}$ **k.** $\frac{0}{0}$

2. Find each of the following products and quotients:

a. $\frac{^-51}{3}$ **b.** $\frac{0}{^-12}$ **c.** ${}^-12 \cdot {}^-5$ $\frac{^-45}{^-3}$

e. $\frac{5}{^-1}$ **f.** $\frac{4}{0}$ **g.** ${}^-5 \cdot 0$ **h.** ${}^-4 \cdot 0$

i. $0 \div 3$ **j.** ${}^-1 \cdot 3$ **k.** $\frac{0}{^-5}$ **l.** ${}^-6 \cdot {}^-7$

m. $\frac{^-75}{^-15}$ **n.** $7 \cdot {}^-9$ **o.** ${}^-3 \cdot 0$ **p.** $\frac{^-8}{^-2}$

q. ${}^-1 \cdot 6$ **r.** ${}^-32 \div 4$ **s.** ${}^-8 \cdot {}^-1$ **t.** $7 \cdot 0$

3. The product or quotient of two integers with the same sign is a ____________ integer.

4. The product or quotient of two integers with different signs is a ____________ integer.

5. Answer true or false: The integers are

a. associative for division

b. commutative for multiplication

c. commutative for division

d. associative for multiplication

7.5 EXERCISE ANSWERS

1. a. ${}^-8$ **b.** ${}^-9$ **c.** 3 **d.** 1

e. 0 **f.** 4 **g.** undefined **h.** 0

i. 13 **j.** 1 **k.** undefined

2. a. ${}^-17$ **b.** 0 **c.** 60 **d.** 15

e. ${}^-5$ **f.** undefined **g.** 0 **h.** 0

i. 0 **j.** ${}^-3$ **k.** 0 **l.** 42

m. 5 **n.** ${}^-63$ **o.** 0 **p.** 4

q. ${}^-6$ **r.** ${}^-8$ **s.** 8 **t.** 0

3. positive **4.** negative

5. a. false **b.** true **c.** false **d.** true

7.6 OPERATIONS WITH RATIONAL NUMBERS

1 The rational numbers are an extension of the integers which are formed by including with the integers the positive and negative fractions. A rational number is defined as follows: *A rational number is any number which can be written in the form* $\frac{p}{q}$, *where p and q are integers and* $q \neq 0$.*

* The symbol "$\neq$" means "is not equal to."

Some examples of rational numbers are:

$\frac{2}{3}$ $\quad \frac{^-5}{7}$ $\quad 4$ $\quad ^-2$ $\quad 0$ $\quad 3\frac{1}{7}$

$\frac{2}{3}$ and $\frac{^-5}{7}$ are rational numbers, because they are written in the form $\frac{p}{q}$ and their denominators (q) are not zero. Integers, such as 4, $^-2$, and 0, are rational numbers, because they can be written in the form $\frac{p}{q}$ by using 1 as their denominators. That is, $4 = \frac{4}{1}$, $^-2 = \frac{^-2}{1}$, and $0 = \frac{0}{1}$. $3\frac{1}{7}$ is a rational number, because it can be written in the form $\frac{p}{q}$ using its improper fraction equivalent. That is, $3\frac{1}{7} = \frac{22}{7}$.

Q1 Which of the following are rational numbers? $^-5, \frac{3}{0}, \frac{1}{2}, 2\frac{1}{6}$ ________

\# \# \# • \# \# \# • \# \# \# • \# \# \# • \# \# \# • \# \# \# • \# \# \# • \# \# \# • \# \# \#

A1 $^-5$, $\frac{1}{2}$, and $2\frac{1}{6}$: $\frac{3}{0}$ is not a rational number, because its denominator is zero and its value is thus undefined.

Q2 Write in the $\frac{p}{q}$ form:

a. $^-5 =$ ______ **b.** $\frac{1}{2} =$ ______ **c.** $2\frac{1}{6} =$ ______

\# \# \# • \# \# \# • \# \# \# • \# \# \# • \# \# \# • \# \# \# • \# \# \# • \# \# \# • \# \# \#

A2 **a.** $\frac{^-5}{1}$ **b.** $\frac{1}{2}$ **c.** $\frac{13}{6}$

2 By the definition, the rational numbers include many different types of numbers. They include positive and negative *fractions*, both proper and improper, because these are already in the required $\frac{p}{q}$ form. They include positive and negative *mixed numbers*, because any mixed number can be rewritten as an improper fraction in the required $\frac{p}{q}$ form. Finally, the rational numbers include the natural numbers, the whole numbers, and the integers, because each of these numbers can be written in the $\frac{p}{q}$ form using the number 1 as the denominator q. Just as the natural numbers, whole numbers, and integers, the rational numbers are an infinite or unending collection.

Q3 Write each rational number in the $\frac{p}{q}$ form:

a. $^-3 =$ ______ **b.** $4\frac{2}{7} =$ ______ **c.** $0 =$ ______

\# \# \# • \# \# \# • \# \# \# • \# \# \# • \# \# \# • \# \# \# • \# \# \# • \# \# \# • \# \# \#

a. $\frac{^-3}{1}$ **b.** $\frac{30}{7}$ **c.** $\frac{0}{1}$

3 A positive mixed number can be written as the sum of its whole-number and fraction parts. For example, $3\frac{2}{5} = 3 + \frac{2}{5}$. The improper-fraction form is found:

$$3\frac{2}{5} = \frac{3 \cdot 5 + 2}{5}$$
$$= \frac{15+2}{5}$$
$$= \frac{17}{5}$$

A similar procedure can be used with negative mixed numbers:

$$^-3\frac{1}{2} = {}^-\left(3\frac{1}{2}\right)$$
$$= {}^-\left(\frac{3 \cdot 2 + 1}{2}\right)$$
$$= {}^-\left(\frac{7}{2}\right) \quad \text{(read "the opposite of seven halves")}$$
$$= \frac{^-7}{2} \quad \text{(read "negative seven halves")}$$

Q4 Find the improper-fraction form:

a. $^-4\frac{1}{7}$ $\frac{-29}{7}$

b. $^-2\frac{2}{9}$ $\frac{-20}{9}$

\# # # • # # # • # # # • # # # • # # # • # # # • # # # • # # # • # # #

A4 **a.** $\frac{^-29}{7}$: $^-4\frac{1}{7} = {}^-\left(4\frac{1}{7}\right)$
$$= {}^-\left(\frac{4 \cdot 7 + 1}{7}\right)$$
$$= {}^-\left(\frac{29}{7}\right)$$
$$= \frac{^-29}{7}$$

b. $\frac{^-20}{9}$: $^-2\frac{2}{9} = {}^-\left(2\frac{2}{9}\right)$
$$= {}^-\left(\frac{2 \cdot 9 + 2}{9}\right)$$
$$= {}^-\left(\frac{20}{9}\right)$$
$$= \frac{^-20}{9}$$

Q5 Write each of the rational numbers in the $\frac{p}{q}$ form:

a. $7 = \frac{7}{1}$

b. $^-5\frac{1}{4} = \frac{-21}{4}$

c. $\frac{2}{3} = \frac{2}{3}$

d. $^-4 = \frac{-4}{1}$

e. $4\frac{1}{2} = \frac{9}{2}$

f. $^-2\frac{4}{5} = \frac{-14}{5}$

\# # # • # # # • # # # • # # # • # # # • # # # • # # # • # # # • # # #

A5 **a.** $\frac{7}{1}$ **b.** $\frac{^-21}{4}$ **c.** $\frac{2}{3}$

d. $\frac{^-4}{1}$ **e.** $\frac{9}{2}$ **f.** $\frac{^-14}{5}$

4 The number line used to graph rational numbers in a manner similar to that used with integers. The rational numbers $\frac{3}{4}$ and $^-4\frac{2}{3}$ and their opposites are graphed on the following number line:

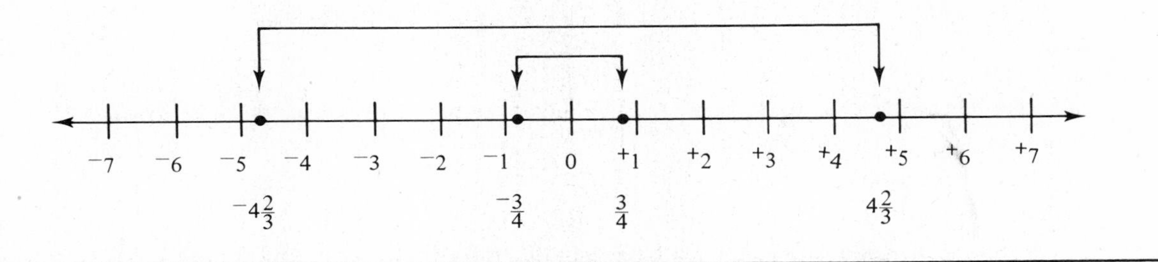

Q6 Graph the rational numbers and their opposites:

a. $^{-}2\frac{1}{2}$ **b.** $1\frac{2}{3}$ **c.** $^{-}4$ **d.** 0

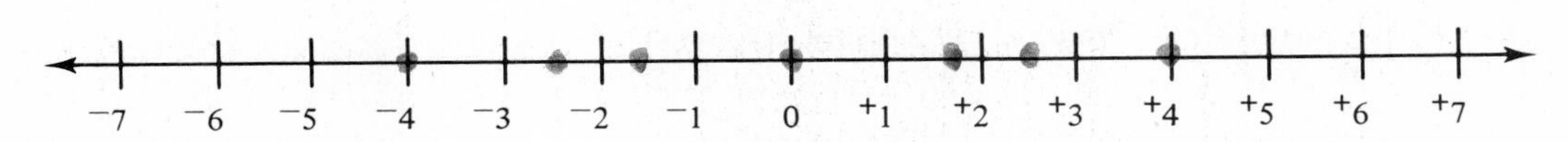

\# \# \# • \# \# \# • \# \# \# • \# \# \# • \# \# \# • \# \# \# • \# \# \# • \# \# \# • \# \# \#

A6

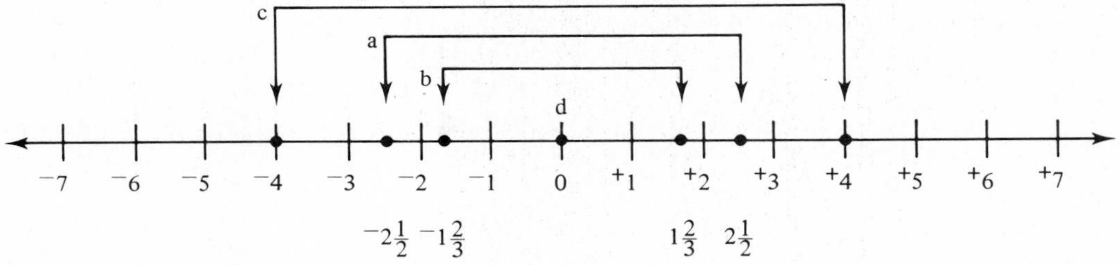

Note that zero is its own opposite.

5 In Section 2.5, procedures were developed for converting fractions into their decimal equivalents.* It is also possible to describe a *rational number* in terms of its decimal representation: *A rational number is any number whose decimal representation is either a terminating or a repeating decimal.*

Examples:

$\frac{1}{2} = 0.5$	terminating
$\frac{1}{3} = 0.\bar{3}$ (0.3333 · · ·)	repeating
$\frac{^{-}5}{8} = {}^{-}0.625$	terminating
$\frac{2}{7} = 0.\overline{285714}$ (0.285714285714 · · ·)	repeating
$^{-}4 = {}^{-}4.0$	terminating

The decimal representation of a rational number in $\frac{p}{q}$ form is found by dividing the numerator (p) by the denominator (q) and then applying the sign, if the rational number is negative.

Example 1: Find the decimal representation of $\frac{^{-}3}{4}$.

* If necessary, see Section 2.5 for a review of these procedures.

Solution:

$$\begin{array}{r} 0.75 \\ 4\overline{)3.00} \\ \underline{2\,8} \\ 20 \\ \underline{20} \\ 0 \end{array}$$

Thus, $\frac{^-3}{4} = {}^-0.75$. The decimal is *terminating* since its division comes out evenly (0 remainder).

Example 2: Find the decimal representation of $2\frac{5}{33}$.

Solution:

$$2\frac{5}{33} = \frac{71}{33}$$

$$\begin{array}{l} 2.1515 \cdots \text{ (repeating 15s)} \\ 33\overline{)71.0000} \\ \underline{66} \\ 50 \\ \underline{33} \\ 170 \\ \underline{165} \\ 50 \\ \underline{33} \\ 170 \\ \underline{165} \\ 5 \quad \text{etc.} \end{array}$$

Thus, $2\frac{5}{33} = 2.\overline{15}$. The decimal is infinite because its division does not come out evenly; it is *repeating* because it repeats itself in blocks of the digits 15. That is, $2.\overline{15} = 2.151515 \cdots$ (repeating 15s).

Q7 Find the decimal representation and state whether each is terminating or repeating:

a. $\frac{^-2}{3} =$ .666 R

b. $2\frac{1}{5} =$ 2.20 T

c. $5 =$ 5.0 T

d. $\frac{^-7}{9} =$ .7 R

e. $^-3\frac{1}{7} =$ 3.142857 R

f. $\frac{5}{8} =$ .625 T

g. $\frac{^{-}7}{11}=$ ________ ____________

\# # # • # # # • # # # • # # # • # # # • # # # • # # # • # # # • # # #

A7
a. $^{-}0.\overline{6}$, repeating
b. 2.2, terminating
c. 5.0, terminating
d. $^{-}0.\overline{7}$, repeating
e. $^{-}3.1\overline{42857}$, repeating
f. 0.625, terminating
g. $^{-}0.\overline{63}$, repeating

6 Since the fractions used in Chapter 1 are the same collection as the nonnegative rationals, the procedure used in the addition of rational numbers is much the same as that used for the addition of fractions.

To add two rational numbers:

Step 1: If the denominators are different, find a common denominator (see Section 1.4 for a review of LCDs).

Step 2: Write the sum of the numerators over the common denominator.

Step 3: Apply the rules for the addition of integers to the numerator.

Step 4: If possible, reduce the rational number to lowest terms.

Examples:

1. $\frac{3}{32}+\frac{^{-}11}{32}=\frac{3+{}^{-}11}{32}$ Step 2 (step 1 is not necessary)
$=\frac{^{-}8}{32}$ Step 3
$=\frac{^{-}1}{4}$ Step 4

2. $\frac{^{-}5}{8}+\frac{3}{7}=\frac{^{-}35}{56}+\frac{24}{56}$ Step 1 $\left(\text{because } \frac{^{-}5\times 7}{8\times 7}=\frac{^{-}35}{56} \text{ and } \frac{3\times 8}{7\times 8}=\frac{24}{56}\right)$
$=\frac{^{-}35+24}{56}$ Step 2
$=\frac{^{-}11}{56}$ Step 3

Q8 Find the sum:

a. $\frac{^{-}7}{8}+\frac{3}{8}$

b. $\frac{^{-}1}{6}+\frac{4}{5}$

\# # # • # # # • # # # • # # # • # # # • # # # • # # # • # # # • # # #

A8
a. $\frac{^{-}1}{2}$: $\frac{^{-}7}{8}+\frac{3}{8}=\frac{^{-}7+3}{8}$
$=\frac{^{-}4}{8}$
$=\frac{^{-}1}{2}$

b. $\frac{19}{30}$: $\frac{^{-}1}{6}+\frac{4}{5}$
$\frac{^{-}5}{30}+\frac{24}{30}$
$\frac{19}{30}$

Q9 Find the sum:

a. $\frac{^-2}{3}+\frac{^-5}{9}$ **b.** $\frac{5}{6}+\frac{^-7}{6}$

c. $\frac{^-2}{15}+\frac{19}{15}$ **d.** $\frac{7}{12}+\frac{4}{7}$

e. $\frac{5}{14}+\frac{^-3}{4}$ **f.** $\frac{^-7}{16}+\frac{^-1}{4}$

g. $\frac{7}{9}+\frac{^-4}{12}$ **h.** $\frac{^-1}{8}+\frac{^-4}{15}$

\# # # • # # # • # # # • # # # • # # # • # # # • # # # • # # # • # # #

A9 **a.** $^-1\frac{2}{9}$ **b.** $\frac{^-1}{3}$ **c.** $1\frac{2}{15}$ **d.** $1\frac{13}{84}$

e. $\frac{^-11}{28}$ **f.** $\frac{^-11}{16}$ **g.** $\frac{4}{9}$ **h.** $\frac{^-47}{120}$

7 To find a sum of two or more rational numbers where integers or mixed numbers are involved, the following procedures will be used:

1. Write all integers in $\frac{p}{q}$ form using a denominator of 1.

2. Write all mixed numbers as improper fractions in $\frac{p}{q}$ form.

3. Follow the rules for adding two rational numbers as previously stated, working from left to right.

Example 1: Find the sum $^-1\frac{1}{5}+3\frac{2}{5}$.

Solution:

$$^-1\frac{1}{5}+3\frac{2}{5}=\frac{^-6}{5}+\frac{17}{5}$$
$$=\frac{^-6+17}{5}$$
$$=\frac{11}{5}$$
$$=2\frac{1}{5}$$

Example 2: Find the sum $2\frac{4}{7}+{}^-5$.

Solution:

$$2\frac{4}{7}+{}^-5=\frac{18}{7}+\frac{^-5}{1}$$
$$=\frac{18}{7}+\frac{^-35}{7}$$
$$=\frac{18+{}^-35}{7}$$
$$=\frac{^-17}{7}$$
$$={}^-2\frac{3}{7}$$

Q10 Find the sum:

a. $^-2\frac{3}{5}+8$

b. $^-5\frac{2}{7}+\frac{3}{5}$

\# \# \# • \# \# \# • \# \# \# • \# \# \# • \# \# \# • \# \# \# • \# \# \# • \# \# \# • \# \# \# • \# \# \#

A10 **a.** $5\frac{2}{5}$: $^-2\frac{3}{5}+8=\frac{^-13}{5}+\frac{8}{1}$

$=\frac{^-13}{5}+\frac{40}{5}$

$=\frac{27}{5}$

$=5\frac{2}{5}$

b. $^-4\frac{24}{35}$: $^-5\frac{2}{7}+\frac{3}{5}=\frac{^-37}{7}+\frac{3}{5}$

$=\frac{^-185}{35}+\frac{21}{35}$

$=\frac{^-185+21}{35}$

$=\frac{^-164}{35}$

$=^-4\frac{24}{35}$

Q11 Find the sum:

a. $2\frac{7}{8}+^-6\frac{3}{8}$ $=\frac{23}{8}+\frac{^-51}{8}$

$=^-\frac{28}{8}$

$=^-3\frac{1}{2}$

b. $^-7\frac{2}{3}+4$ $=\frac{-23}{3}+\frac{4}{1}=\frac{-23}{3}+\frac{12}{3}$

$=\frac{-11}{3}=^-3\frac{2}{3}$

c. $6\frac{1}{5}+^-3\frac{4}{7}$ $=\frac{31}{5}+\frac{-25}{7}$

$=\frac{217}{35}+\frac{^-125}{35}$

$=\frac{92}{35}$

$=2\frac{22}{35}$

d. $^-1\frac{2}{3}+^-4\frac{5}{6}$ $=\frac{-5}{3}+\frac{-29}{6}$

$=\frac{-10}{6}+\frac{-29}{6}$

$=\frac{-39}{6}$

$=-6\frac{1}{2}$

e. $\frac{5}{9}+^-3\frac{1}{2}$

f. $\frac{^-5}{7}+^-3$

g. $^{-}8+\frac{3}{11}$

h. $^{-}4\frac{3}{8}+5$

i. $\frac{2}{3}+\frac{^{-}5}{8}+\frac{^{-}5}{6}$

j. $^{-}3\frac{1}{7}+^{-}2\frac{5}{8}$

\# \# \# • \# \# \# • \# \# \# • \# \# \# • \# \# \# • \# \# \# • \# \# \# • \# \# \# • \# \# \#

A11 a. $^{-}3\frac{1}{2}$ b. $^{-}3\frac{2}{3}$ c. $2\frac{22}{35}$ d. $^{-}6\frac{1}{2}$

e. $^{-}2\frac{17}{18}$ f. $^{-}3\frac{5}{7}$ g. $^{-}7\frac{8}{11}$ h. $\frac{5}{8}$

i. $\frac{^{-}19}{24}$ j. $^{-}5\frac{43}{56}$

Q12 Write the opposite for each rational number:

a. $\frac{2}{3}$ _____ b. $\frac{^{-}5}{8}$ _____ c. 7 _____

d. $^{-}4$ _____ e. $1\frac{2}{3}$ _____ f. $^{-}5\frac{4}{7}$ _____

\# \# \# • \# \# \# • \# \# \# • \# \# \# • \# \# \# • \# \# \# • \# \# \# • \# \# \# • \# \# \#

A12 a. $\frac{^{-}2}{3}$ b. $\frac{5}{8}$ c. $^{-}7$ d. 4

e. $^{-}1\frac{2}{3}$ f. $5\frac{4}{7}$

8 The definition of subtraction for integers is also true for the set of rational numbers. That is, $x-y=x+{}^{-}y$ *for all rational-number replacements of x and y.*

To subtract any two rational numbers:

Step 1: Use the definition of subtraction to rewrite the difference as a sum.

Step 2: Apply the procedures previously developed to find the sum.

Example 1: Find the difference $\frac{4}{5}-\frac{7}{12}$.

Solution:

$$\frac{4}{5}-\frac{7}{12}=\frac{4}{5}+\frac{^{-}7}{12}$$
$$=\frac{48}{60}+\frac{^{-}35}{60}$$
$$=\frac{13}{60}$$

Example 2: Find the difference $\frac{^{-}4}{7}-\frac{^{-}5}{6}$.

Solution:

$$\frac{^{-}4}{7}-\frac{^{-}5}{6}=\frac{^{-}4}{7}+\frac{5}{6}$$
$$=\frac{^{-}24}{42}+\frac{35}{42}$$
$$=\frac{11}{42}$$

Example 3: Find the difference $^{-}7\frac{1}{3}-{^{-}2}$.

Solution:

$$^{-}7\frac{1}{3}-{^{-}2}={^{-}7}\frac{1}{3}+2$$
$$=\frac{^{-}22}{3}+\frac{2}{1}$$
$$=\frac{^{-}22}{3}+\frac{6}{3}$$
$$=\frac{^{-}16}{3}$$
$$={^{-}5}\frac{1}{3}$$

Q13 Rewrite each difference as a sum (do not evaluate):

a. $\frac{1}{3}-\frac{4}{9}=$ $\frac{1}{3}+\frac{-4}{9}$

b. $\frac{^{-}3}{5}-\frac{^{-}2}{3}=$ $\frac{-3}{5}+\frac{2}{3}$

c. $^{-}4\frac{2}{3}-1\frac{2}{5}=$ $-4\frac{2}{3}+{^{-}1}\frac{2}{5}$

d. $3-{^{-}1}\frac{3}{10}=$ $\frac{3}{1}+1\frac{3}{10}$

\# \# \# • \# \# \# • \# \# \# • \# \# \# • \# \# \# • \# \# \# • \# \# \# • \# \# \# • \# \# \#

A13 **a.** $\frac{1}{3}+\frac{^{-}4}{9}$ **b.** $\frac{^{-}3}{5}+\frac{2}{3}$ **c.** $^{-}4\frac{2}{3}+{^{-}1}\frac{2}{5}$ **d.** $3+1\frac{3}{10}$

Q14 Find the difference:

a. $\frac{1}{3}-\frac{4}{9}$

b. $\frac{^{-}3}{5}-\frac{^{-}2}{3}$

c. $^{-}4\frac{2}{3} - 1\frac{2}{7}$

d. $3 - {}^{-}1\frac{3}{10}$

\# \# \# • \# \# \# • \# \# \# • \# \# \# • \# \# \# • \# \# \# • \# \# \# • \# \# \# • \# \# \#

A14 **a.** $\frac{^{-}1}{9}$: $\frac{1}{3} - \frac{4}{9} = \frac{1}{3} + \frac{^{-}4}{9}$
$= \frac{3}{9} + \frac{^{-}4}{9}$
$= \frac{^{-}1}{9}$

b. $\frac{1}{15}$: $\frac{^{-}3}{5} - \frac{^{-}2}{3} = \frac{^{-}3}{5} + \frac{2}{3}$
$= \frac{^{-}9}{15} + \frac{10}{15}$
$= \frac{1}{15}$

c. $^{-}5\frac{20}{21}$: $^{-}4\frac{2}{3} - 1\frac{2}{7} = {}^{-}4\frac{2}{3} + {}^{-}1\frac{2}{7}$
$= \frac{^{-}14}{3} + \frac{^{-}9}{7}$
$= \frac{^{-}98}{21} + \frac{^{-}27}{21}$
$= \frac{^{-}125}{21}$
$= {}^{-}5\frac{20}{21}$

d. $4\frac{3}{10}$: $3 - {}^{-}1\frac{3}{10} = 3 + 1\frac{3}{10}$
$= \frac{3}{1} + \frac{13}{10}$
$= \frac{30}{10} + \frac{13}{10}$
$= \frac{43}{10}$
$= 4\frac{3}{10}$

Q15 Find the difference:

a. $\frac{^{-}5}{24} - \frac{7}{24}$

b. $2\frac{3}{16} - 1\frac{5}{16}$

c. $\frac{^{-}7}{21} - \frac{^{-}5}{7}$

d. $^{-}2\frac{1}{9} - 3\frac{2}{5}$

e. $5\frac{3}{8} - 7\frac{4}{9}$

f. $\frac{^{-}3}{16} - 4$

g. $18 - \frac{5}{17}$

h. $\frac{^{-}6}{13} - 4\frac{1}{2}$

i. $16\frac{3}{10} - 5\frac{7}{8}$

j. $^{-}15\frac{1}{6} - {}^{-}2\frac{7}{15}$

\# \# \# • \# \# \# • \# \# \# • \# \# \# • \# \# \# • \# \# \# • \# \# \# • \# \# \# • \# \# \#

A15 **a.** $\frac{^-1}{2}$ **b.** $\frac{7}{8}$ **c.** $\frac{8}{21}$ **d.** $^-5\frac{23}{45}$

e. $^-2\frac{5}{72}$ **f.** $^-4\frac{3}{16}$ **g.** $17\frac{12}{17}$ **h.** $^-4\frac{25}{26}$

i. $10\frac{17}{40}$ **j.** $^-12\frac{7}{10}$

---

9 Let $\frac{a}{b}$ and $\frac{c}{d}$ stand for any two rational numbers. The product of $\frac{a}{b}$ and $\frac{c}{d}$ is defined as:

$$\frac{a}{b}\cdot\frac{c}{d}=\frac{ac}{bd}, \qquad bd \neq 0$$

where ac is the product of the numerators and bd is the product of the denominators.

Example 1: Find the product $\frac{^-3}{7}\cdot\frac{^-5}{4}$.

Solution:

$$\frac{^-3}{7}\cdot\frac{^-5}{4}=\frac{^-3\cdot{}^-5}{7\cdot 4}=\frac{15}{28}$$

As with fractions, a common procedure when multiplying rational numbers is to divide any numerator and any denominator by a common factor. This procedure is called *reducing* (sometimes referred to as "canceling").

Example 2: Find the product $\frac{^-3}{5}\cdot\frac{5}{12}$.

Solution:

$$\frac{\overset{-1}{\cancel{^-3}}}{\underset{1}{\cancel{5}}}\cdot\frac{\overset{1}{\cancel{5}}}{\underset{4}{\cancel{12}}}=\frac{^-1}{4}$$

Example 3: Find the product $\frac{^-9}{12}\cdot\frac{^-5}{7}$.

Solution:

$$\frac{\overset{^-3}{\cancel{^-9}}}{\underset{4}{\cancel{12}}}\cdot\frac{^-5}{7}=\frac{15}{28}$$

As is the case with the sum and difference of two rational numbers, the product of two rational numbers is always reduced to lowest terms.

Q16 Find the product:

a. $\frac{^-6}{5}\cdot\frac{^-2}{3}$ **b.** $\frac{15}{16}\cdot\frac{^-4}{25}$

A16 **a.** $\frac{4}{5}$: $\frac{\overset{^-2}{\cancel{^-6}}}{5}\cdot\frac{^-2}{\underset{1}{\cancel{3}}}=\frac{4}{5}$ **b.** $\frac{^-3}{20}$: $\frac{\overset{3}{\cancel{15}}}{\underset{4}{\cancel{16}}}\cdot\frac{\overset{^-1}{\cancel{^-4}}}{\underset{5}{\cancel{25}}}=\frac{^-3}{20}$

---

10 $\frac{^-2}{3}$ and $\frac{2}{^-3}$ are equivalent forms for the same rational number. Both represent a negative rational number since they involve a quotiént of integers with unlike signs. When writing a negative rational number, it is customary to place the negative sign on the number in the numerator. For example, $\frac{5}{^-8}$ is usually written $\frac{^-5}{8}$.

Q17 Find the product:

a. $\frac{^-12}{15}\cdot\frac{3}{6}$ **b.** $\frac{^-4}{6}\cdot\frac{^-13}{18}$

c. $\frac{7}{12}\cdot\frac{9}{^-11}$ **d.** $\frac{^-5}{7}\cdot\frac{6}{^-13}$

e. $\frac{5}{^-28}\cdot\frac{^-16}{17}$ **f.** $\frac{4}{^-7}\cdot\frac{7}{4}$

\# # # • # # # • # # # • # # # • # # # • # # # • # # # • # # # • # # #

A17 **a.** $\frac{^-2}{5}$ **b.** $\frac{13}{27}$ **c.** $\frac{^-21}{44}$ **d.** $\frac{30}{91}$ **e.** $\frac{20}{119}$ **f.** $^-1$

---

11 When finding a product of two or more rational numbers where integers or mixed numbers are involved:

1. Write all integers in $\frac{p}{q}$ form with a denominator of 1.

2. Write all mixed numbers as improper fractions in $\frac{p}{q}$ form.

3. Use the rule of Frame 9 to find the product.

Example 1: Find the product $5\frac{2}{9}\cdot{}^-3$.

Solution:

$$5\frac{2}{9}\cdot{}^-3=\frac{47}{\underset{3}{\cancel{9}}}\cdot\frac{\overset{^-1}{\cancel{^-3}}}{1}=\frac{^-47}{3}={}^-15\frac{2}{3}$$

Example 2: Find the product $^-5\frac{1}{3}\cdot{}^-3\frac{9}{16}$.

Solution:

$$^{-}5\frac{1}{3}\cdot{}^{-}3\frac{9}{16}=\frac{\overset{^{-}1}{\cancel{^{-}16}}}{\underset{1}{\cancel{3}}}\cdot\frac{\overset{^{-}19}{\cancel{^{-}57}}}{\underset{1}{\cancel{16}}}=19$$

Q18 Find the product:

a. $\frac{^{-}3}{11}\cdot 3$

b. $\frac{4}{^{-}5}\cdot\frac{^{-}10}{12}$

c. $^{-}5\frac{2}{9}\cdot{}^{-}12$

d. $2\frac{1}{3}\cdot{}^{-}1\frac{3}{4}$

e. $^{-}4\cdot\frac{^{-}1}{4}$

f. $3\frac{1}{5}\cdot{}^{-}1\frac{1}{2}$

• ### • ### • ### • ### • ### • ### • ### •

A18 **a.** $\frac{^{-}9}{11}$ **b.** $\frac{2}{3}$ **c.** $62\frac{2}{3}$ **d.** $^{-}4\frac{1}{12}$ **e.** 1 **f.** $^{-}4\frac{4}{5}$

12 Two rational numbers whose product is 1 are called *reciprocals*. Thus, 5 and $\frac{1}{5}$ are reciprocals, because $5\cdot\frac{1}{5}=\frac{\overset{1}{\cancel{5}}}{1}\cdot\frac{1}{\underset{1}{\cancel{5}}}=\frac{1}{1}=1$. Similarly, $\frac{^{-}5}{8}$ and $\frac{8}{^{-}5}$ are reciprocals, because $\frac{\overset{1}{\cancel{^{-}5}}}{\underset{1}{\cancel{8}}}\cdot\frac{\overset{1}{\cancel{8}}}{\underset{1}{\cancel{^{-}5}}}=\frac{1}{1}=1$

In general, to find the reciprocal of any nonzero rational number, write the number in $\frac{p}{q}$ form and interchange the numerator and denominator to form $\frac{q}{p}$.

Example 1: Write the reciprocal for $^{-}2\frac{1}{3}$.

Solution:

$^{-}2\frac{1}{3}=\frac{^{-}7}{3}$; thus, the reciprocal of $^{-}2\frac{1}{3}$ is $\frac{3}{^{-}7}$, which is written $\frac{^{-}3}{7}$

Example 2: Write the reciprocal for 3.

Solution:

$3=\frac{3}{1}$: thus, the reciprocal of 3 is $\frac{1}{3}$

Zero has no reciprocal, because $0=\frac{0}{1}$ and $\frac{1}{0}$ is undefined.

Q19 Write the reciprocal:

a. $\frac{2}{3}$ ______ b. $^{-}1\frac{5}{9}$ ______ c. 12 ______ d. $\frac{1}{13}$ ______

e. $7\frac{3}{8}$ ______ f. $^{-}2$ ______

• ### • ### • ### • ### • ### • ### • ### •

A19 a. $\frac{3}{2}$ b. $\frac{^{-}9}{14}$ c. $\frac{1}{12}$ c. 13 e. $\frac{8}{59}$ f. $\frac{^{-}1}{2}$

13 Let $\frac{a}{b}$ and $\frac{c}{d}$ stand for any two rational numbers with $\frac{c}{d} \neq 0$. Consider the simplification of the following quotient:

$$\frac{a}{b} \div \frac{c}{d} = \frac{\frac{a}{b}}{\frac{c}{d}} = \frac{\frac{a}{b} \cdot \frac{d}{c}}{\frac{c}{d} \cdot \frac{d}{c}} = \frac{\frac{a}{b} \cdot \frac{d}{c}}{1} = \frac{a}{b} \cdot \frac{d}{c}$$

Thus, the *definition of division for two rational numbers* (with a nonzero divisor) is stated:

$$\frac{a}{b} \div \frac{c}{d} = \frac{a}{b} \cdot \frac{d}{c} = \frac{ad}{bc}, \qquad \frac{c}{d} \neq 0$$

In words, the quotient of two rational numbers with a nonzero divisor is equal to the product of the first (dividend) times the reciprocal of the second (divisor).

Example 1: Find the quotient $\frac{2}{3} \div \frac{4}{9}$.

Solution:

$$\frac{2}{3} \div \frac{4}{9} = \frac{\overset{1}{\cancel{2}}}{\underset{1}{\cancel{3}}} \cdot \frac{\overset{3}{\cancel{9}}}{\underset{2}{\cancel{4}}} \qquad \left(\text{the reciprocal of } \frac{4}{9} \text{ is } \frac{9}{4}\right)$$

$$= \frac{3}{2}$$

$$= 1\frac{1}{2}$$

Example 2: Find the quotient $\frac{4}{5} \div \frac{^{-}6}{7}$.

Solution:

$$\frac{4}{5} \div \frac{^{-}6}{7} = \frac{\overset{2}{\cancel{4}}}{5} \cdot \frac{^{-}7}{\underset{3}{\cancel{6}}} \qquad \left(\text{the reciprocal of } \frac{^{-}6}{7} \text{ is } \frac{^{-}7}{6}\right)$$

$$= \frac{^{-}14}{15}$$

Q20 Find the quotient:

a. $\frac{4}{5} \div \frac{2}{3}$ b. $\frac{5}{6} \div \frac{^{-}15}{16}$

• ### • ### • ### • ### • ### • ### • ### •

A20 **a.** $1\frac{1}{5}$: $\frac{4}{5}\div\frac{2}{3}=\frac{\overset{2}{\cancel{4}}}{5}\cdot\frac{3}{\underset{1}{\cancel{2}}}=\frac{6}{5}$ **b.** $\frac{{}^{-}8}{9}$: $\frac{5}{6}\div\frac{{}^{-}15}{16}=\frac{\overset{1}{\cancel{5}}}{\underset{3}{\cancel{6}}}\cdot\frac{\overset{{}^{-}8}{\cancel{{}^{-}16}}}{\underset{3}{\cancel{15}}}=\frac{{}^{-}8}{9}$

---

14 When finding a quotient of two rational numbers where integers or mixed numbers are involved:

1. Write all integers in $\frac{p}{q}$ form with a denominator of 1.
2. Write all mixed numbers as improper fractions in $\frac{p}{q}$ form.
3. Use the rule of the previous frame to find the quotient.

Example 1: Find the quotient $\frac{{}^{-}5}{14}\div 3\frac{2}{7}$.

Solution:

$$\frac{{}^{-}5}{14}\div 3\frac{2}{7}=\frac{{}^{-}5}{14}\div\frac{23}{7}=\frac{{}^{-}5}{\underset{2}{\cancel{14}}}\cdot\frac{\overset{1}{\cancel{7}}}{23}=\frac{{}^{-}5}{46}$$

Example 2: Find the quotient ${}^{-}2\frac{5}{8}\div 12$.

Solution:

$$ {}^{-}2\frac{5}{8}\div 12=\frac{{}^{-}21}{8}\div\frac{12}{1}=\frac{\overset{{}^{-}7}{\cancel{{}^{-}21}}}{8}\cdot\frac{1}{\underset{4}{\cancel{12}}}=\frac{{}^{-}7}{32}$$

Q21 Find the quotient:

a. $\frac{{}^{-}11}{15}\div{}^{-}4\frac{2}{5}$ **b.** ${}^{-}8\div 2\frac{2}{3}$

\# \# \# • \# \# \# • \# \# \# • \# \# \# • \# \# \# • \# \# \# • \# \# \# • \# \# \# • \# \# \#

A21 **a.** $\frac{1}{6}$: $\frac{{}^{-}11}{15}\div{}^{-}4\frac{2}{5}=\frac{{}^{-}11}{15}\div\frac{{}^{-}22}{5}=\frac{\overset{{}^{-}1}{\cancel{{}^{-}11}}}{\underset{3}{\cancel{15}}}\cdot\frac{\overset{{}^{-}1}{\cancel{{}^{-}5}}}{\underset{2}{\cancel{22}}}=\frac{1}{6}$ **b.** ${}^{-}3$: ${}^{-}8\div 2\frac{2}{3}=\frac{{}^{-}8}{1}\div\frac{8}{3}=\frac{\overset{{}^{-}1}{\cancel{{}^{-}8}}}{1}\cdot\frac{3}{\underset{1}{\cancel{8}}}={}^{-}3$

Q22 Find the quotient:

a. $\frac{{}^{-}29}{50}\div 3\frac{1}{10}$ **b.** ${}^{-}4\frac{1}{5}\div{}^{-}3\frac{1}{3}$

c. ${}^{-}4\frac{1}{5}\div 3$ **d.** $\frac{1}{2}\div{}^{-}2$

e. $4 \div 4\frac{5}{8}$

f. $^{-}8 \div \frac{^{-}1}{8}$

g. $0 \div 3\frac{4}{7}$

h. $^{-}15 \div {}^{-}2\frac{5}{8}$

\# # # • # # # • # # # • # # # • # # # • # # # • # # # • # # # • # # #

A22 **a.** $\frac{^{-}29}{155}$ **b.** $1\frac{13}{50}$ **c.** $^{-}1\frac{2}{5}$ **d.** $\frac{^{-}1}{4}$

e. $\frac{32}{37}$ **f.** 64 **g.** 0 **h.** $5\frac{5}{7}$

15 Certain properties of integers have been established in previous sections. The integers are commutative and associative with respect to the operations of addition and multiplication. The distributive properties hold for multiplication over addition and subtraction. These properties are also true for the rational numbers. In addition, the following statements are true for any rational number a.

1. Addition property of zero:

$a + 0 = 0 + a = a$

2. Multiplication property of zero:

$a \cdot 0 = 0a = 0$

3. Multiplication property of one:

$a \cdot 1 = 1a = a$

Q23 The associative property of addition states that $(a+b)+c = a+(b+c)$. Verify that $(a+b)+c$ and $a+(b+c)$ are equivalent when $a = \frac{3}{5}$, $b = \frac{7}{10}$, and $c = \frac{9}{20}$.

\# # # • # # # • # # # • # # # • # # # • # # # • # # # • # # # • # # #

A23 $(a+b)+c=\left(\frac{3}{5}+\frac{7}{10}\right)+\frac{9}{20}$
$=\left(\frac{6}{10}+\frac{7}{10}\right)+\frac{9}{20}$
$=\frac{13}{10}+\frac{9}{20}$
$=\frac{26}{20}+\frac{9}{20}$
$=\frac{35}{20}$
$=1\frac{3}{4}$

$a+(b+c)=\frac{3}{5}+\left(\frac{7}{10}+\frac{9}{20}\right)$
$=\frac{3}{5}+\left(\frac{14}{20}+\frac{9}{20}\right)$
$=\frac{3}{5}+\frac{23}{20}$
$=\frac{12}{20}+\frac{23}{20}$
$=\frac{35}{20}$
$=1\frac{3}{4}$

Q24 The left distributive property of multiplication over subtraction states that $a(b-c)=ab-ac$. Verify that $a(b-c)$ and $ab-ac$ are equivalent when $a=\frac{2}{3}$, $b=\frac{4}{5}$, and $c=\frac{3}{7}$.

\# \# \# • \# \# \# • \# \# \# • \# \# \# • \# \# \# • \# \# \# • \# \# \# • \# \# \# • \# \# \#

A24 $a(b-c)=\frac{2}{3}\left(\frac{4}{5}-\frac{3}{7}\right)$
$=\frac{2}{3}\left(\frac{28}{35}-\frac{15}{35}\right)$
$=\frac{2}{3}\left(\frac{13}{35}\right)$
$=\frac{26}{105}$

$ab-ac=\frac{2}{3}\cdot\frac{4}{5}-\frac{2}{3}\cdot\frac{3}{7}$
$=\frac{8}{15}-\frac{2}{7}$
$=\frac{56}{105}-\frac{30}{105}$
$=\frac{26}{105}$

16 When a technician is working on proper steering alignment on an automobile, one of the adjustments that must be correct is the camber of the wheels. Camber is designed into the vehicle to

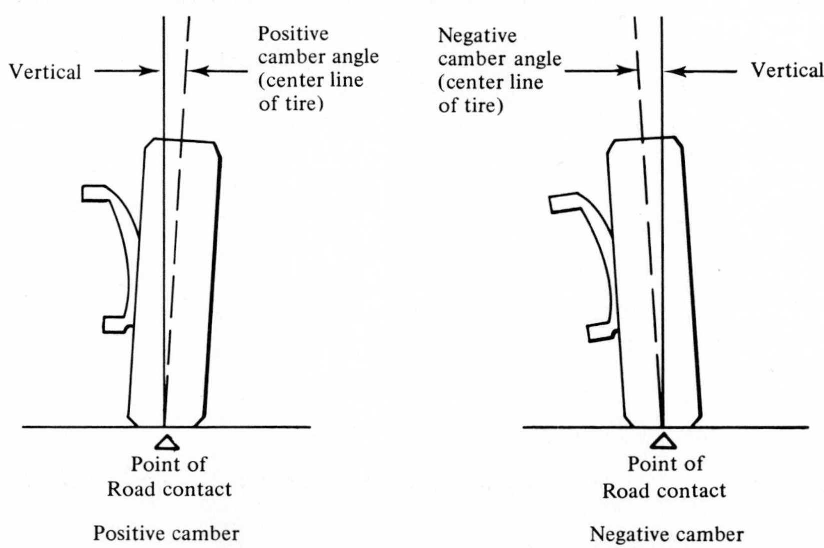

compensate for variations in loading, to increase steering stability, and to minimize tire wear. *Positive camber* means that the front wheels are inclined outward at the top. *Negative camber* means that the front wheels are inclined inward at the top. The proper adjustment of the wheels will result in a slight positive camber. Any wheel with incorrect camber tends to "lead" in a direction away from the direction the vehicle is being steered.

Q24 **a.** In the illustration is the wheel showing a positive camber or a negative camber?
b. Is this a preferred way of running the wheel or should it be adjusted?

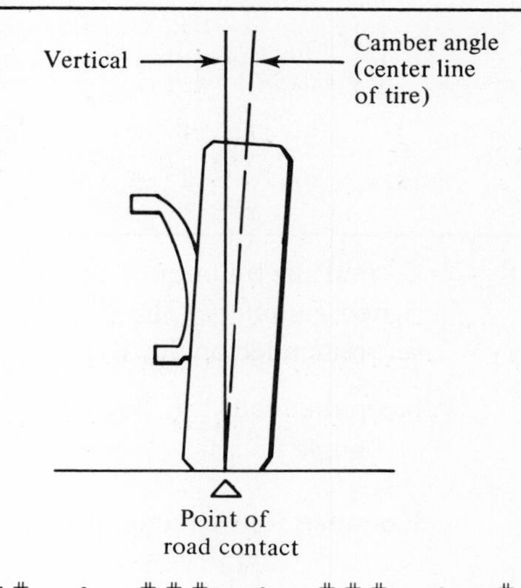

• ### • ### • ### • ### • ### • ### • ### •

A24 **a.** positive camber
b. It is preferable to leave it with a positive camber.

17 Consider the following application of the addition of rational numbers to camber adjustment in wheel alignment.

Example: A wheel with an initial camber angle of $\frac{^+3}{4}°$ has a change of $\frac{^-1}{2}°$ because of worn parts. A heavy piece of equipment is then mounted on one side of the car which changes it $^+2°$. What is the new camber angle?

Solution: $\frac{^+3}{4} + \frac{^-1}{2} + {}^+2 = \frac{3}{4} + \frac{^-2}{4} + \frac{8}{4} = \frac{^+9}{4} = {}^+2\frac{1}{4}$

Thus, the new camber angle is $^+2\frac{1}{4}°$.

Q25 The camber angle set by the factory is $\frac{^+1}{2}°$. After 20,000 miles the parts are worn to the point where the suspension is $\frac{^-3}{4}°$ out of adjustment. What is the new camber angle?

• ### • ### • ### • ### • ### • ### • ### •

A25 $\frac{^-1}{4}°$: $\frac{^+1}{2} + \frac{^-3}{4} = \frac{2}{4} + \frac{^-3}{4} = \frac{^-1}{4}$

Q26 A wheel with initial camber angle $\frac{^+3}{4}°$ suffers a change of camber angle of $^-1°$ because of worn parts and a change of camber angle of $\frac{^-1}{2}^\circ$ when the driver sits in the car. What is its new camber angle?

• ### • ### • ### • ### • ### • ### • ### •

A26 $\frac{^-3}{4}^\circ$: $\frac{^+3}{4} + {^-1} + \frac{^-1}{2} = \frac{3}{4} + \frac{^-4}{4} + \frac{^-2}{4} = \frac{^-3}{4}$

18 To find the amount of needed adjustment of a camber angle, a technician can use subtraction of signed numbers. The needed adjustment is found by subtracting the present angle from the recommended angle. That is,

recommended angle − present angle = needed adjustment

Example: The camber angle is measured and found to be $\frac{^-3}{4}^\circ$. How much adjustment is needed to bring it to $\frac{^+1}{2}^\circ$?

Solution: $\frac{^+1}{2} - \frac{^-3}{4} = \frac{1}{2} + \frac{^+3}{4} = \frac{2}{4} + \frac{3}{4} = \frac{5}{4} = 1\frac{1}{4}$

Thus, the adjustment needed is $1\frac{1}{4}^\circ$.

Q27 A wheel is measured and found to have a camber angle of $^-1\frac{1}{4}^\circ$. What adjustment is needed to bring it to $\frac{^+1}{2}^\circ$?

• ### • ### • ### • ### • ### • ### • ### •

A27 $1\frac{3}{4}^\circ$: $\frac{1}{2} - {^-1}\frac{1}{4} = \frac{2}{4} + 1\frac{1}{4} = 1\frac{3}{4}$

19 Multiplication of signed numbers is useful in finding the average change in camber angle.

Example: Suppose the average change in camber adjustment due to worn parts is $\frac{^-1}{10}^\circ$ per 1,000 miles. How far out of adjustment would a typical car be after 30,000 miles?

Solution: $\frac{^-1}{10} \times 30 = {^-3}$

Thus, a typical car would be out of adjustment by $^-3^\circ$.

Q28 Suppose the average change in camber adjustment due to worn parrts is $\frac{^-1}{10}^\circ$ per 1,000 miles.

a. What would be the measured camber angle after 45,000 miles?

b. What would be the measured camber angle after 45,000 miles if it was originally $\frac{1}{2}^\circ$?

\# \# \# • \# \# \# • \# \# \# • \# \# \# • \# \# \# • \# \# \# • \# \# \# • \# \# \# • \# \# \#

A28 **a.** ${^-4}\frac{1}{2}^\circ$: $\frac{^-1}{10}\times 45=\frac{^-45}{10}={^-4}\frac{1}{2}$ **b.** $\frac{1}{2}+{^-4}\frac{1}{2}={^-4}$

This completes the instruction for this section.

7.6 EXERCISE

1. Find the sum:

a. $\frac{2}{3}+\frac{^-5}{3}$ **b.** $\frac{^-4}{5}+\frac{3}{8}$ **c.** $\frac{^-1}{6}+\frac{^-4}{18}$

d. $\frac{^-6}{7}+0$ **e.** ${^-3}\frac{3}{4}+1\frac{1}{4}$ **f.** $15+{^-2}\frac{4}{11}$

g. $3\frac{2}{5}+\frac{^-17}{5}$ **h.** ${^-1}\frac{2}{3}+{^-7}\frac{3}{5}$ **i.** $\frac{3}{10}+\frac{^-7}{6}+\frac{5}{12}$

j. $\frac{^-5}{24}+\frac{7}{8}+\frac{^-7}{12}$

2. Find the difference:

a. $\frac{2}{3}-\frac{5}{3}$ **b.** $\frac{3}{7}-\frac{7}{9}$ **c.** $4-2\frac{4}{7}$

d. $2\frac{1}{3}-5\frac{1}{4}$ **e.** ${^-3}-2\frac{3}{7}$ **f.** $3\frac{2}{5}-4\frac{1}{7}$

g. $\frac{^-5}{8}-\frac{^-1}{8}$ **h.** ${^-6}\frac{1}{5}-{^-1}\frac{3}{7}$ **i.** ${^-2}\frac{1}{3}-{^-2}\frac{1}{3}$

j. $12\frac{4}{9}-3$

3. Find the product:

a. $\frac{^-4}{15}\cdot\frac{9}{16}$ **b.** $\frac{^-4}{5}\cdot\frac{^-2}{7}$ **c.** $\frac{^-3}{4}\cdot 1\frac{1}{3}$

d. $0\cdot 5\frac{7}{17}$ **e.** $\frac{^-6}{17}\cdot\frac{^-17}{6}$ **f.** ${^-1}\frac{2}{5}\cdot\frac{^-5}{7}$

g. $5\frac{1}{4}\cdot 11\frac{1}{3}$ **h.** $\frac{1}{4}\cdot\frac{^-2}{3}\cdot\frac{6}{7}$ **i.** ${^-2}\frac{1}{3}\cdot\frac{6}{7}\cdot\frac{^-1}{2}$

j. ${^-4}\frac{2}{3}\cdot\frac{2}{5}\cdot\frac{^-3}{14}$

4. Find the quotient:

a. $\frac{^-4}{15}\div\frac{3}{2}$ **b.** $\frac{^-7}{8}\div\frac{^-3}{4}$ **c.** ${^-2}\frac{1}{2}\div 5$

d. ${^-2}\div\frac{^-6}{11}$ **e.** ${^-3}\frac{2}{3}\div{^-3}\frac{1}{3}$ **f.** $\frac{2}{5}\div\frac{5}{2}$

g. $\frac{3}{17} \div {}^{-}1$ **h.** $0 \div 3$ **i.** $3 \div 0$

j. $\frac{^{-}5}{11} \div \frac{0}{1}$

5. Perform the indicated operation:

a. $1\frac{3}{7} + {}^{-}2\frac{4}{7}$ **b.** $\frac{^{-}5}{8} - \frac{4}{5}$ **c.** $\frac{8}{45} \cdot \frac{^{-}5}{16}$

d. ${}^{-}3\frac{1}{3} \div \frac{^{-}9}{10}$ **e.** $1\frac{2}{11} + 3\frac{15}{11}$ **f.** $\frac{^{-}8}{15} \cdot 0 \cdot 8\frac{1}{4}$

g. ${}^{-}5 \div \frac{^{-}7}{5}$ **h.** $4\frac{2}{3} \div {}^{-}8$ **i.** $2\frac{1}{3} + {}^{-}4 + 1\frac{2}{5}$

j. $\left(\frac{3}{5} \div \frac{^{-}2}{7}\right) \div \frac{^{-}3}{5}$

6. The camber angle set by the factory is $\frac{1}{2}^{\circ}$. After 35,000 miles the parts are worn to the point where the suspension is ${}^{-}1\frac{1}{4}^{\circ}$. What is the new camber angle?

7. The camber angle is measured and found to be ${}^{-}1\frac{1}{2}^{\circ}$. How much adjustment is needed to bring it to $\frac{1}{2}^{\circ}$?

8. Suppose the average change in camber adjustment due to worn parts is $\frac{^{-}1}{10}^{\circ}$ per 1,000 miles. How far out of adjustment would a typical car be after 40,000 miles?

7.6 EXERCISE ANSWERS

1. a. ${}^{-}1$ **b.** $\frac{^{-}17}{40}$ **c.** $\frac{^{-}7}{18}$ **d.** $\frac{^{-}6}{7}$

e. ${}^{-}2\frac{1}{2}$ **f.** $12\frac{7}{11}$ **g.** 0 **h.** ${}^{-}9\frac{4}{15}$

i. $\frac{^{-}9}{20}$ **j.** $\frac{1}{12}$

2. a. ${}^{-}1$ **b.** $\frac{^{-}22}{63}$ **c.** $1\frac{3}{7}$ **d.** ${}^{-}2\frac{11}{12}$

e. ${}^{-}5\frac{3}{7}$ **f.** $\frac{^{-}26}{35}$ **g.** $\frac{^{-}1}{2}$ **h.** ${}^{-}4\frac{27}{35}$

i. 0 **j.** $9\frac{4}{9}$

3. a. $\frac{^{-}3}{20}$ **b.** $\frac{8}{35}$ **c.** ${}^{-}1$ **d.** 0

e. 1 **f.** 1 **g.** $59\frac{1}{2}$ **h.** $\frac{^{-}1}{7}$

i. 1 **j.** $\frac{2}{5}$

4. a. $\frac{^{-}8}{45}$ **b.** $1\frac{1}{6}$ **c.** $\frac{^{-}1}{2}$ **d.** $3\frac{2}{3}$

e. $1\frac{1}{10}$ **f.** $\frac{4}{25}$ **g.** $\frac{^{-}3}{17}$ **h.** 0

i. undefined **j.** undefined

5. a. ${}^-1\frac{1}{7}$ **b.** ${}^-1\frac{17}{40}$ **c.** $\frac{{}^-1}{18}$ **d.** $3\frac{19}{27}$

e. $5\frac{6}{11}$ **f.** 0 **g.** $3\frac{4}{7}$ **h.** $\frac{{}^-7}{12}$

i. $\frac{{}^-4}{15}$ **j.** $3\frac{1}{2}$

6. $\frac{{}^-3°}{4}$ **7.** 2° **8.** ${}^-4°$

CHAPTER 7 SAMPLE TEST

At the completion of Chapter 7 you should be able to work the following problems.

7.1. INTRODUCTION

1. The collection of numbers . . . ${}^-2$, ${}^-1$, 0, 1, 2, . . . is called the ________.

2. Each integer is thought of as having what two parts? DIRECTION - VALUE

3. Graph each of the following integers on the number line:

a. 2 **b.** ${}^-7$ **c.** 5 **d.** 0

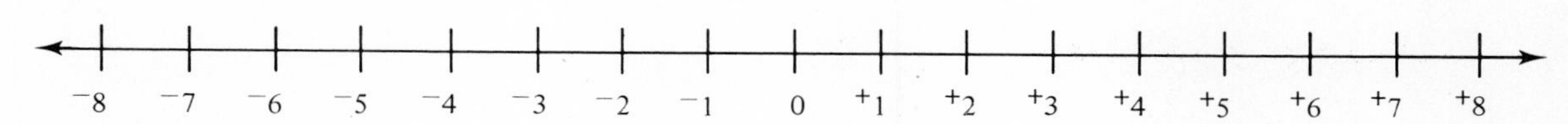

4. Write the opposite of each of the following integers:

a. ${}^-5$ **b.** 3 **c.** 0 **d.** 7

7.2 ADDITION OF INTEGERS

5. Find the sum:

a. ${}^-3+{}^-6$ **b.** ${}^-5+8$ **c.** $3+{}^-11$

d. $0+{}^-2$ **e.** $5+{}^-5$ **f.** ${}^-2+7+{}^-5$

g. ${}^-3+{}^-5+6$ **h.** $9+{}^-3+6$

6. The spool valve of a hydraulic system has a spring force of ${}^+5$ lb, an auxiliary force of ${}^+2$ lb, and a reaction force of ${}^-7$ lb.

a. What is the sum of the forces on the spool?

b. Which direction will the spool valve move?

7.3 SUBTRACTION OF INTEGERS

7. Find the difference:

a. ${}^-2-5$ **b.** $3-7$ **c.** $6-{}^-3$ **d.** ${}^-1-{}^-2$

e. $11-9$ **f.** $5-7-{}^-3$

8. Complete each of the following:

a. $^{-}5+7$ **b.** $^{-}5-7$ **c.** $0-5$ **d.** $^{-}3+0$

e. $^{-}4+^{-}9$ **f.** $6-^{-}9$ **g.** $^{-}4+7-6$ **h.** $2-1-^{-}1$

7.4 MULTIPLICATION OF INTEGERS

9. Find the product:

a. $^{-}2\cdot 5$ **b.** $^{-}4\cdot ^{-}6$ **c.** $^{-}3\cdot 0$ **d.** $5\cdot ^{-}9$

e. $^{-}1\cdot 13$ **f.** $0\cdot ^{-}1$ **g.** $^{-}2\cdot 4\cdot ^{-}5$ **h.** $(6\cdot ^{-}2)(^{-}3\cdot ^{-}2)$

i. $^{-}3(4+^{-}7)$ **j.** $2(^{-}4+4)$

7.5 DIVISION OF INTEGERS

10. Find the quotient:

a. $^{-}36\div 9$ **b.** $^{-}18\div ^{-}6$ **c.** $\frac{^{-}12}{4}$

d. $\frac{^{-}25}{^{-}5}$ **e.** $10\div 0$ **f.** $\frac{0}{^{-}6}$

g. $0\div ^{-}2$ **h.** $\frac{4}{0}$

7.6 OPERATIONS WITH RATIONAL NUMBERS

11. Find the sum:

a. $\frac{^{-}4}{7}+\frac{7}{12}$ **b.** $\frac{^{-}2}{3}+\frac{^{-}5}{8}$ **c.** $^{-}3\frac{1}{3}+2\frac{7}{9}$ **d.** $\frac{^{-}3}{16}+3$

12. Find the difference:

a. $\frac{5}{16}-\frac{3}{16}$ **b.** $1\frac{4}{5}-\frac{^{-}2}{3}$ **c.** $4\frac{1}{3}-5\frac{2}{5}$ **d.** $\frac{^{-}7}{9}-2$

13. Find the product:

a. $\frac{^{-}4}{15}\cdot\frac{^{-}5}{16}$ **b.** $3\cdot ^{-}4\frac{2}{3}$ **c.** $^{-}2\frac{1}{3}\cdot 0$ **d.** $^{-}5\frac{1}{3}\cdot ^{-}1\frac{2}{5}$

14. Find the quotient:

a. $\frac{^{-}9}{16}\div\frac{3}{8}$ **b.** $\frac{5}{12}\div 7$ **c.** $^{-}2\frac{1}{3}\div ^{-}4\frac{5}{16}$ **d.** $^{-}2\div\frac{^{-}8}{9}$

15. Perform the indicated operations:

a. $\frac{^{-}2}{3}-\frac{4}{7}$ **b.** $^{-}1\cdot\frac{3}{4}$ **c.** $\frac{^{-}2}{7}\div ^{-}1\frac{4}{21}$ **d.** $3\frac{1}{5}\cdot ^{-}4\frac{2}{5}$

e. $\frac{^{-}5}{12}-\frac{^{-}5}{12}$ **f.** $\frac{2}{7}\div\frac{24}{35}$ **g.** $\frac{^{-}6}{51}\cdot\frac{17}{18}\cdot\frac{^{-}1}{3}$ **h.** $^{-}1\frac{7}{8}+9\frac{3}{5}$

16. A wheel with an initial camber angle of $\frac{1^\circ}{2}$ has a change of $\frac{^{-}3^\circ}{4}$ because of worn parts. What is the new camber angle?

17. A wheel is measured and found to have a camber angle of $1\frac{1}{6}^\circ$. What adjustment is needed to bring it to $\frac{1}{2}^\circ$?

18. The average change in camber adjustment due to worn parts is $\frac{^-1}{10}^\circ$ per 1,000 miles. What is the camber angle after 15,000 miles?

CHAPTER 7 SAMPLE TEST ANSWERS

1. integers

2. distance, direction

3.

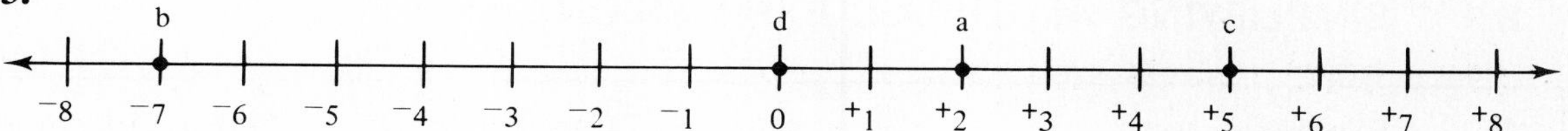

4. a. 5 **b.** $^-3$ **c.** 0 **d.** $^-7$

5. a. $^-9$ **b.** 3 **c.** $^-8$ **d.** $^-2$ **e.** 0 **f.** 0 **g.** $^-2$ **h.** 12

6. a. 0 **b.** no motion

7. a. $^-7$ **b.** $^-4$ **c.** 9 **d.** 1 **e.** 2 **f.** 1

8. a. 2 **b.** $^-12$ **c.** $^-5$ **d.** $^-3$ **e.** $^-13$ **f.** 15 **g.** $^-3$ **h.** 2

9. a. $^-10$ **b.** 24 **c.** 0 **d.** $^-45$ **e.** $^-13$ **f.** 0 **g.** 40 **h.** $^-72$
i. 9 **j.** 0

10. a. $^-4$ **b.** 3 **c.** $^-3$ **d.** 5 **e.** undefined **f.** 0 **g.** 0
h. undefined

11. a. $\frac{1}{84}$ **b.** $^-1\frac{7}{24}$ **c.** $\frac{^-5}{9}$ **d.** $2\frac{13}{16}$

12. a. $\frac{1}{8}$ **b.** $2\frac{7}{15}$ **c.** $^-1\frac{1}{15}$ **d.** $^-2\frac{7}{9}$

13. a. $\frac{1}{12}$ **b.** $^-14$ **c.** 0 **d.** $7\frac{7}{15}$

14. a. $^-1\frac{1}{2}$ **b.** $\frac{5}{84}$ **c.** $\frac{112}{207}$ **d.** $2\frac{1}{4}$

15. a. $^-1\frac{5}{21}$ **b.** $\frac{^-3}{4}$ **c.** $\frac{6}{25}$ **d.** $^-14\frac{2}{25}$
e. 0 **f.** $\frac{5}{12}$ **g.** $\frac{1}{27}$ **h.** $7\frac{29}{40}$

16. $\frac{^-1}{4}^\circ$ **17.** $\frac{^-2}{3}^\circ$ **18.** $^-1\frac{1}{2}^\circ$

CHAPTER 8

ALGEBRAIC EXPRESSIONS AND EQUATION SOLVING

8.1 SIMPLIFYING ALGEBRAIC EXPRESSIONS

1 Expressions such as $a+3$, $5y-zx$, and $9(b+7)$ are referred to as "open expressions." These expressions are also correctly called algebraic expressions. Since letters (variables) in an algebraic expression can be replaced by rational numbers, all properties valid for the rational numbers can be used when simplifying algebraic expressions. The purpose of this section is to develop skill in the use of these properties for the simplification of algebraic expressions.

The building blocks or components of algebraic expressions are called terms. When an algebraic expression shows only additions, the *terms* are the parts separated by plus signs. Subtraction signs appearing in an algebraic expression can be converted to addition signs using the definition of subtraction.

Examples:	*Terms*
$8y+3$	$8y$ and 3
$\frac{1}{2}x-4+2y=\frac{1}{2}x+{}^{-}4+2y$	$\frac{1}{2}x$, ${}^{-}4$, and $2y$
$x-3y-6=x+{}^{-}3y+{}^{-}6$	x, ${}^{-}3y$, and ${}^{-}6$

Q1 Identify the terms in the algebraic expression $4a+7ab-12$. ________

\# # # • # # # • # # # • # # # • # # # • # # # • # # # • # # # • # # #

A1 $4a$, $7ab$, and ${}^{-}12$

2 The *like terms* of an algebraic expression are terms that have exactly the same literal coefficients (letter factors). The like terms of the algebraic expression $7x-3y-\frac{2}{3}x+6$ are $\frac{{}^{-}2}{3}x$ and $7x$, because each has the literal coefficient x.

In the expression $\frac{4}{5}y-3xy+\frac{2}{3}x-2y+8xy$ there are two sets of like terms. They are $\frac{4}{5}y$ and ${}^{-}2y$, with the letter factor y, and ${}^{-}3xy$ and $8xy$, with the letter factors x and y.

Numbers without a literal factor are called *constants* and are considered to be like terms. That is, 5, ${}^{-}7$, $\frac{7}{8}$, 3, and so on, are constants and like terms.

Q2 Identify the like terms in each of the following algebraic expressions:

a. $4x-2y+3-6x$ ________

b. $6-2y+\frac{5}{9}$ ______

c. $5r-6s+2rs-\frac{1}{2}s$ ______

d. $6a-2b+3ab$ ______

\# \# \# • \# \# \# • \# \# \# • \# \# \# • \# \# \# • \# \# \# • \# \# \# • \# \# \# • \# \# \#

A2 **a.** $4x$, ^-6x **b.** 6, $\frac{5}{9}$ **c.** ^-6s, $\frac{^-1}{2}s$

d. There are no like terms in part **d**.

3 Algebraic expressions are simplified by combining (by addition or subtraction) like terms. The justification for combining like terms is the distributive property of multiplication over addition or the distributive property of multiplication over subtraction. For example, $5x-7x$ is simplified:

	Justification
$5x-7x$	
$(5-7)x$	right distributive property of multiplication over subtraction
^-2x	number fact: $5-7={}^-2$

Therefore, $5x-7x$ simplifies to ^-2x.

Q3 Use a distributive property to simplify the following expressions:

a. $4a-7a$ **b.** $\frac{2}{3}x+\frac{1}{4}x$

\# \# \# • \# \# \# • \# \# \# • \# \# \# • \# \# \# • \# \# \# • \# \# \# • \# \# \# • \# \# \#

A3 **a.** ^-3a: $4a-7a=(4-7)a={}^-3a$ **b.** $\frac{11}{12}x$: $\frac{2}{3}x+\frac{1}{4}x=\left(\frac{2}{3}+\frac{1}{4}\right)x=\frac{11}{12}x$

4 Consider the following simplification:

	Justification
$4x-3x$	
$(4-3)x$	right distributive property of multiplication over subtraction
$1x$	number fact: $4-3=1$
x	multiplication property of one

Therefore, $4x-3x=x$.

Q4 Simplify $9y-8y$.

\# \# \# • \# \# \# • \# \# \# • \# \# \# • \# \# \# • \# \# \# • \# \# \# • \# \# \# • \# \# \#

A4 y: $9y-8y=(9-8)y=1y=y$

Q5 Simplify $7b-6b$.

\# \# \# • \# \# \# • \# \# \# • \# \# \# • \# \# \# • \# \# \# • \# \# \# • \# \# \# • \# \# \#

A5 b

5 By the multiplication property of one, x also equals $1x$. This idea is used to simplify expressions such as $4x + x$ or $3b - b$.

Example 1: Simplify $4x + x$.

Solution:

	Justification
$4x + x$	
$4x + 1x$	multiplication property of one
$(4 + 1)x$	right distributive property of multiplication over addition
$5x$	number fact: $4 + 1 = 5$

Therefore, $4x + x = 5x$.

Example 2: Simplify $3b - b$.

Solution:

$3b - b$	
$3b - 1b$	multiplication property of one
$(3 - 1)b$	right distributive property of multiplication over subtraction
$2b$	number fact: $3 - 1 = 2$

Therefore, $3b - b = 2b$.

Q6 Simplify:

a. $7y + y$ **b.** $x + 2x$ **c.** $b - 5b$ **d.** $a - \frac{2}{3}a$

\# # # • # # # • # # # • # # # • # # # • # # # • # # # • # # # • # # #

A6 **a.** $8y$: $7y + y$; $7y + 1y$; $(7 + 1)y$; $8y$

b. $3x$: $x + 2x$; $1x + 2x$; $(1 + 2)x$; $3x$

c. $^{-}4b$: $b - 5b$; $1b - 5b$; $(1 - 5)b$; $^{-}4b$

d. $\frac{1}{3}a$: $a - \frac{2}{3}a$; $1a - \frac{2}{3}a$; $\left(1 - \frac{2}{3}\right)a$; $\frac{1}{3}a$

6 By the multiplication property of negative one, $^{-}1x = {}^{-}x$. Consider its use in the simplification of $x - 2x$.

	Justification
$x - 2x$	
$1x - 2x$	multiplication property of one
$(1 - 2)x$	right distributive property of multiplication over subtraction
$^{-}1x$	number fact: $1 - 2 = {}^{-}1$
^{-}x	multiplication property of negative one

Therefore, $x - 2x = {}^{-}x$.

Q7 Simplify:

a. $7x - 8x$ **b.** $12y - 13y$

• ### • ### • ### • ### • ### • ### • ### •

A7 **a.** ^{-}x: $7x - 8x = (7-8)x = {}^{-}1x = {}^{-}x$ **b.** ^{-}y

---

7 Notice that using a distributive property to simplify algebraic expressions is merely the process of combining the numerical coefficients of the like terms. Thus, the simplification process can be shortened as follows:

$^{-}5b + 3b = {}^{-}2b$ because $^{-}5 + 3 = {}^{-}2$

$\frac{^{-}1}{2}z - \frac{2}{3}z = \frac{^{-}7}{6}z$ because $\frac{^{-}1}{2} - \frac{2}{3} = \frac{^{-}7}{6}$

In algebra, numbers are usually written in $\frac{p}{q}$ form rather than as mixed numbers. Thus, in the preceding example, the answer is written as $\frac{^{-}7}{6}z$ rather than as $^{-}1\frac{1}{6}z$.

Q8 Simplify the following algebraic expressions by combining the numerical coefficients of the like terms:

a. $7y - 12y =$ ________ **b.** $4b - b =$ ________

c. $\frac{^{-}2}{3}x + \frac{4}{5}x =$ ________ **d.** $y - y =$ ________

• ### • ### • ### • ### • ### • ### • ### •

A8 **a.** $^{-}5y$: Because $7 - 12 = {}^{-}5$ **b.** $3b$: Because $4 - 1 = 3$

c. $\frac{2}{15}x$: Because $\frac{^{-}2}{3} + \frac{4}{5} = \frac{2}{15}$ **d.** 0: Because $1 - 1 = 0$

---

8 To simplify expressions involving more than two terms, a similar procedure is followed. Study the following examples:

$3y - 12y + 7$ $\quad$ $^{-}8x - 3x + 9 - 12$

$^{-}9y + 7$ $\quad$ $^{-}11x - 3$

Q9 Simplify each of the following expressions:

a. $^{-}3m - 7m + 9$ **b.** $11 - 1 + 7x - 3x$

c. $5y + 7y - 6 + 3$ **d.** $4 - 10 + 8y - \frac{3}{2}y$

• ### • ### • ### • ### • ### • ### • ### •

A9 **a.** $^{-}10m + 9$ **b.** $10 + 4x$ **c.** $12y - 3$ **d.** $^{-}6 + \frac{13}{2}y$

9 The expression $5x - 7$ may also be written $^{-}7 + 5x$. This fact is verified below:

	Justification
$5x - 7$	
$5x + ^{-}7$	definition of subtraction
$^{-}7 + 5x$	commutative property of addition

Q10 Verify that $3x - 4 = {}^{-}4 + 3x$ by completing the following:

	Justification
$3x - 4$	
a. ________	definition of subtraction
b. ________	commutative property of addition

• # # # • # # # • # # # • # # # • # # # • # # # • # # # • # #

A10 **a.** $3x + ^{-}4$ **b.** $^{-}4 + 3x$

10 The expression $^{-}8 - 7x$ may also be written $^{-}7x - 8$. The verification is as follows:

	Justification
$^{-}7x - 8$	
$^{-}7x + ^{-}8$	definition of subtraction
$^{-}8 + ^{-}7x$	commutative property of addition
$^{-}8 - 7x$	definition of subtraction

Q11 Verify that $^{-}2x - 9 = {}^{-}9 - 2x$ by completing the following:

	Justification
$^{-}2x - 9$	
a. ________	definition of subtraction
b. ________	commutative property of addition
c. ________	definition of subtraction

• # # # • # # # • # # # • # # # • # # # • # # # • # # # • # #

A11 **a.** $^{-}2x + ^{-}9$ **b.** $^{-}9 + ^{-}2x$ **c.** $^{-}9 - 2x$

11 Frame 9 showed that $5x - 7$ could be written $^{-}7 + 5x$. Frame 10 showed that $^{-}8 - 7x = {}^{-}7x - 8$. These frames demonstrate the following useful procedure: *In an algebraic expression, the terms may be rearranged in any order as long as the original sign of each term is left unchanged.*

Examples:

1. $5 - 2x = {}^{-}2x + 5$

2. $3x - \frac{4}{5} = \frac{^{-}4}{5} + 3x$

3. $\frac{^{-}3}{4}x - 5 = {}^{-}5 - \frac{3}{4}x$

4. $^{-}2x + 3 + 5x - 7 = {}^{-}2x + 5x + 3 - 7$

Notice in Example 4 that the terms are arranged so that the like terms are together.

Q12 Write equivalent expressions for each of the following:

a. $3x+2=$ __________ **b.** $^{-}5x-4=$ __________

c. $\frac{^{-}5}{8}x+12=$ __________ **d.** $^{-}4x+3-2x=$ __________

e. $^{-}2x-7-5x+6=$ ________________

• ### • ### • ### • ### • ### • ### • ### •

A12 **a.** $2+3x$ **b.** $^{-}4-5x$ **c.** $12-\frac{5}{8}x$

d. $^{-}4x-2x+3$ **e.** $^{-}2x-5x-7+6$

---

12 The expression $5x-2-3x-9$ may be simplified by rearranging and combining the like terms as follows:

$5x-2-3x-9$
$5x-3x-2-9$
$2x-11$

Q13 Simplify $7-3t+9-6t$ by rearranging and combining the like terms.

• ### • ### • ### • ### • ### • ### • ### •

A13 $^{-}9t+16$ or $16-9t$: $7-3t+9-6t$
$^{-}3t-6t+7+9$
$^{-}9t+16$

Q14 Simplify $4x-3-5x+6-x$.

• ### • ### • ### • ### • ### • ### • ### •

A14 $^{-}2x+3$ or $3-2x$: $4x-3-5x+6-x$
$4x-5x-x-3+6$
$^{-}2x+3$

Q15 Simplify the following algebraic expressions:

a. $^{-}3+5x+7$ **b.** $5y-6+2y$ **c.** $^{-}3x-4-x$

d. $x+3x+7-9$ **e.** $^{-}2z+3-5z-1$ **f.** $6t+3-7t-6+t$

• ### • ### • ### • ### • ### • ### • ### •

A15 **a.** $5x+4$ or $4+5x$ **b.** $7y-6$ or $^{-}6+7y$ **c.** $^{-}4x-4$ or $^{-}4-4x$

d. $4x-2$ or $^{-}2+4x$ **e.** $^{-}7z+2$ or $2-7z$ **f.** $^{-}3$

13 The associative and commutative properties of multiplication are used whenever there is a need to change the grouping or order of multiplication as an aid in the simplification of algebraic expressions. The associative property of multiplication is used to simplify $3(4x)$ as follows:

	Justification
$3(4x)$	
$(3 \cdot 4)x$	associative property of multiplication
$12x$	number fact

Thus, $3(4x) = 12x$.

Q16 Use the associative property of multiplication to simplify $^{-}5(7y)$.

\# \# \# • \# \# \# • \# \# \# • \# \# \# • \# \# \# • \# \# \# • \# \# \# • \# \# \# • \# \# \#

A16 $^{-}35y$: $^{-}5(7y) = (^{-}5 \cdot 7)y = {}^{-}35y$

14 The expression $7\left(\frac{^{-}3}{7}t\right)$ is simplified:

	Justification
$7\left(\frac{^{-}3}{7}t\right)$	
$\left(7 \cdot \frac{^{-}3}{7}\right)t$	associative property of multiplication
$^{-}3t$	number fact: $\frac{7}{1} \cdot \frac{^{-}3}{7} = {}^{-}3$

Thus, $7\left(\frac{^{-}3}{7}t\right) = {}^{-}3t$.

Q17 Simplify $^{-}9\left(\frac{^{-}1}{9}x\right)$.

\# \# \# • \# \# \# • \# \# \# • \# \# \# • \# \# \# • \# \# \# • \# \# \# • \# \# \# • \# \# \#

A17 x: $^{-}9\left(\frac{^{-}1}{9}x\right) = \left(^{-}9 \cdot \frac{^{-}1}{9}\right)x = 1x = x$

15 The expression $\frac{3}{5}\left(\frac{^{-}4}{9}y\right)$ is simplified:

	Justification
$\frac{3}{5}\left(\frac{^{-}4}{9}y\right)$	
$\left(\frac{3}{5} \cdot \frac{^{-}4}{9}\right)y$	associative property of multiplication
$\frac{^{-}4}{15}y$	number fact

Thus, $\frac{3}{5}\left(\frac{^{-}4}{9}y\right) = \frac{^{-}4}{15}y$.

Q18 Simplify $\frac{^-3}{4}\left(\frac{^-2}{7}z\right)$.

\# \# \# • \# \# \# • \# \# \# • \# \# \# • \# \# \# • \# \# \# • \# \# \# • \# \# \# • \# \# \#

A18 $\frac{3}{14}z$: $\frac{^-3}{4}\left(\frac{^-2}{7}z\right) = \left(\frac{^-3}{4}\cdot\frac{^-2}{7}\right)z = \frac{3}{14}z$

Q19 Simplify each of the following algebraic expressions:

a. $4\left(\frac{3}{4}x\right)$ **b.** $\frac{2}{3}\left(\frac{3}{2}y\right)$ **c.** $^-5\left(\frac{^-2}{15}z\right)$

d. $\frac{^-6}{7}\left(\frac{5}{12}t\right)$ **e.** $\frac{^-5}{9}\left(\frac{^-9}{5}m\right)$ **f.** $\frac{1}{7}(7x)$

\# \# \# • \# \# \# • \# \# \# • \# \# \# • \# \# \# • \# \# \# • \# \# \# • \# \# \# • \# \# \#

A19 **a.** $3x$ **b.** y **c.** $\frac{2}{3}z$ **d.** $\frac{^-5}{14}t$ **e.** m **f.** x

16 Both the commutative and associative properties of multiplication are used to simplify $\left(\frac{3}{4}x\right)\frac{2}{3}$ as follows:

	Justification
$\left(\frac{3}{4}x\right)\frac{2}{3}$	
$\frac{2}{3}\left(\frac{3}{4}x\right)$	commutative property of multiplication
$\left(\frac{2}{3}\cdot\frac{3}{4}\right)x$	associative property of multiplication
$\frac{1}{2}x$	number fact

Q20 Use the commutative and associative properties of multiplication to simplify $\left(\frac{2}{5}x\right)5$.

\# \# \# • \# \# \# • \# \# \# • \# \# \# • \# \# \# • \# \# \# • \# \# \# • \# \# \# • \# \# \#

A20 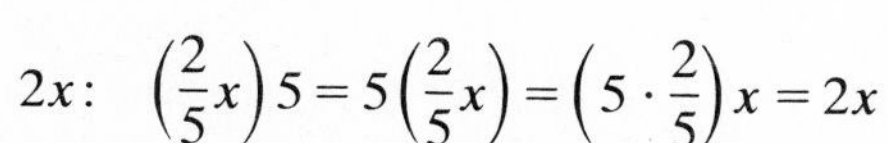 $2x$: $\left(\frac{2}{5}x\right)5 = 5\left(\frac{2}{5}x\right) = \left(5\cdot\frac{2}{5}\right)x = 2x$

Q21 Simplify $\left(\frac{^-3}{5}x\right)\cdot\frac{^-5}{3}$ by use of the commutative and associative properties.

\# \# \# • \# \# \# • \# \# \# • \# \# \# • \# \# \# • \# \# \# • \# \# \# • \# \# \# • \# \# \#

A21 x: $\left(\frac{^-3}{5}x\right) \cdot \frac{^-5}{3} = \frac{^-5}{3}\left(\frac{^-3}{5}x\right) = \left(\frac{^-5}{3} \cdot \frac{^-3}{5}\right)x = 1x = x$

17 The left distributive property of multiplication over addition and the left distributive property of multiplication over subtraction state that

$a(b+c) = ab + ac$

and

$a(b-c) = ab - ac$

for all rational-number replacements of a, b, and c.

Examples:

1. $3(7+5) = 3 \cdot 7 + 3 \cdot 5$
2. $2(1-9) = 2 \cdot 1 - 2 \cdot 9$

Q22 Use the left distributive property of multiplication over addition to fill in the blanks.

$5(2+9) =$ __________ + __________

• # # # • # # # • # # # • # # # • # # # • # # # • # # # • # #

A22 $5(2+9) = 5 \cdot 2 + 5 \cdot 9$

18 When parentheses are removed from an algebraic expression, the same procedure is used. For example, to remove the parentheses from $3(x+2)$, proceed as follows:

$3(x+2) = 3 \cdot x + 3 \cdot 2$
$= 3x + 6$

Q23 Use the left distributive property of multiplication over addition to remove the parentheses from $7(y+3)$.

• # # # • # # # • # # # • # # # • # # # • # # # • # # # • # #

A23 $7y + 21$: $7(y+3) = 7 \cdot y + 7 \cdot 3 = 7y + 21$

Q24 Use the left distributive property of multiplication over subtraction to remove the parentheses from $7(a-5)$.

• # # # • # # # • # # # • # # # • # # # • # # # • # # # • # #

A24 $7a - 35$: $7(a-5) = 7 \cdot a - 7 \cdot 5 = 7a - 35$

Q25 Remove the parentheses

a. $2(x-9)$

b. $5(x+1)$

c. $6(2-y)$

d. $8(2+c)$

• # # # • # # # • # # # • # # # • # # # • # # # • # # # • # #

A25 **a.** $2x-18$ **b.** $5x+5$ **c.** $12-6y$ **d.** $16+8c$

19 To remove the parentheses from $(a+3)$, recall that by the multiplication property of one:

$(a+3)=1\cdot(a+3)$

Thus,

$$\begin{aligned}(a+3)&=1\cdot(a+3)\\&=1\cdot a+1\cdot 3\\&=a+3\end{aligned}$$

Notice that the result is exactly the expression within the parentheses.

Examples:

$(b+2)=b+2$
$(7-y)=7-y$
$(2x-3)=2x-3$

Q26 Remove the parentheses

a. $(x-9)=$________ **b.** $(3-x)=$________ **c.** $(5x+6)=$________

• # # # • # # # • # # # • # # # • # # # • # # # • # # # • # #

A26 **a.** $x-9$ **b.** $3-x$ **c.** $5x+6$

20 The right distributive property of multiplication over addition and the right distributive property of multiplication over subtraction state that

$(a+b)c=ac+bc$

and

$(a-b)c=ac-bc$

for all rational numbers a, b, and c.

Examples:

1. $(3+5)2=3\cdot 2+5\cdot 2$
2. $(7-4)\frac{3}{8}=7\cdot\frac{3}{8}-4\cdot\frac{3}{8}$

Q27 Use the right distributive property of multiplication over addition to fill in the blanks.

$(4+9)7=$________$+$________

• # # # • # # # • # # # • # # # • # # # • # # # • # # # • # #

A27 $(4+9)7=4\cdot 7+9\cdot 7$

Q28 Use the right distributive property of multiplication over subtraction to fill in the blanks.

$\left(\frac{5}{6}-1\right)6=$________$-$________

• # # # • # # # • # # # • # # # • # # # • # # # • # # # • # #

A28 $\left(\frac{5}{6}-1\right)6=\frac{5}{6}\cdot 6-1\cdot 6$

21 The right distributive properties are also used to simplify algebraic expressions.

Examples:

1. $(x-2)3 = x \cdot 3 - 2 \cdot 3$
$= 3x - 6$ (it is customary to rewrite $x \cdot 3$ as $3x$)

2. $(5+r)3 = 5 \cdot 3 + r \cdot 3$
$= 15 + 3r$

Q29 Use the right distributive property of multiplication over addition to remove the parentheses from $(2+x)7$.

• # # # • # # # • # # # • # # # • # # # • # # # • # # # • # #

A29 $14+7x$: $(2+x)7 = 2 \cdot 7 + x \cdot 7$
$= 14 + 7x$

Q30 Use the right distributive property of multiplication over subtraction to remove the parentheses from $(y-4)5$.

• # # # • # # # • # # # • # # # • # # # • # # # • # # # • # #

A30 $5y-20$: $(y-4)5 = y \cdot 5 - 4 \cdot 5$
$= 5y - 20$

22 To remove the parentheses in the algebraic expression $3(4x-7)$, the following steps are used:

$3(4x-7) = 3 \cdot 4x - 3 \cdot 7$
$= 12x - 21$

Q31 Remove the parentheses from $7(6x-4)$.

• # # # • # # # • # # # • # # # • # # # • # # # • # # # • # #

A31 $42x-28$: $7(6x-4) = 7 \cdot 6x - 7 \cdot 4$
$= 42x - 28$

Q32 Remove the parentheses from $4\left(12-\frac{3}{4}x\right)$.

• # # # • # # # • # # # • # # # • # # # • # # # • # # # • # #

A32 $48-3x$

23 To remove the parentheses from $(3x-2)5$, the procedure is as follows:

$(3x-2)5 = 3x \cdot 5 - 2 \cdot 5$
$= 5 \cdot 3x - 2 \cdot 5$
$= 15x - 10$

Q33 Remove the parentheses from $(7x-4)9$.

• ### • ### • ### • ### • ### • ### • ### •

A33 $63x-36$: $(7x-4)9 = 7x \cdot 9 - 4 \cdot 9$
$= 9 \cdot 7x - 4 \cdot 9$
$= 63x - 36$

Q34 Remove the parentheses from $(5a+7)2$.

• ### • ### • ### • ### • ### • ### • ### •

A34 $10a+14$

24 To remove the parentheses from $4(^{-}x+7)$, recall that ^{-}x means $^{-}1x$ (multiplication property of negative one). Thus,

$$4(^{-}x+7) = 4(^{-}1x+7)$$
$$= 4 \cdot {}^{-}1x + 4 \cdot 7$$
$$= {}^{-}4x+28 \quad \text{or} \quad 28-4x$$

Q35 Remove the parentheses from $3(^{-}x-5)$

• ### • ### • ### • ### • ### • ### • ### •

A35 $^{-}3x-15$: $3(^{-}x-5) = 3(^{-}1x-5)$
$= 3 \cdot {}^{-}1x - 3 \cdot 5$
$= {}^{-}3x - 15$

25 When removing parentheses by use of the distributive properties, it is convenient to be able to find the result mentally (without showing work). For example, to remove parentheses from $5(3x-6)$, write only $5(3x-6) = 15x-30$.

Q36 Mentally remove the parentheses from $2(8-5t)$.

• ### • ### • ### • ### • ### • ### • ### •

A36 $16-10t$

Q37 Mentally remove the parentheses from $3\left(7-\frac{4}{3}y\right)$.

• ### • ### • ### • ### • ### • ### • ### •

A37 $21-4y$

Q38 Remove the parentheses from $(6z+1)9$.

• ### • ### • ### • ### • ### • ### • ### •

A38 $54z+9$

26 To remove the parentheses from $^{-}5(3x-4)$, the following steps can be used:

$$\begin{aligned} ^{-}5(3x-4) &= {}^{-}5\cdot 3x - {}^{-}5\cdot 4 \\ &= {}^{-}15x - {}^{-}20 \\ &= {}^{-}15x + 20 \end{aligned}$$

You should notice that $^{-}15x - {}^{-}20$ is equivalent to $^{-}15x+20$ because of the definition of subtraction. The expression $^{-}15x+20$ is considered to be in simplest form. The expression $20-15x$ may also be written.

Q39 Remove the parentheses from $^{-}4(7x-9)$ and write in simplest form.

• ### • ### • ### • ### • ### • ### • ### •

A39 $^{-}28x+36$ or $36-28x$: $\begin{aligned} ^{-}4(7x-9) &= {}^{-}4\cdot 7x - {}^{-}4\cdot 9 \\ &= {}^{-}28x - {}^{-}36 \\ &= {}^{-}28x+36 \end{aligned}$

Q40 Remove the parentheses from $^{-}5({}^{-}3x+4)$ and write in simplest form.

• ### • ### • ### • ### • ### • ### • ### •

A40 $15x-20$: $\begin{aligned} ^{-}5({}^{-}3x+4) &= {}^{-}5\cdot {}^{-}3x + {}^{-}5\cdot 4 \\ &= 15x + {}^{-}20 \\ &= 15x-20 \end{aligned}$

27 The expression $({}^{-}4x-7)\cdot {}^{-}8$ is simplified:

$$\begin{aligned} ({}^{-}4x-7)\cdot {}^{-}8 &= {}^{-}4x\cdot {}^{-}8 - 7\cdot {}^{-}8 \\ &= 32x - {}^{-}56 \\ &= 32x+56 \end{aligned}$$

Q41 Simplify $(3y+9)\cdot {}^{-}6$.

• ### • ### • ### • ### • ### • ### • ### •

A41 $^{-}18y-54$: $\begin{aligned} (3y+9)\cdot {}^{-}6 &= 3y\cdot {}^{-}6 + 9\cdot {}^{-}6 \\ &= {}^{-}18y + {}^{-}54 \\ &= {}^{-}18y-54 \end{aligned}$

Q42 Simplify $({}^{-}a-5)\cdot {}^{-}2$.

• ### • ### • ### • ### • ### • ### • ### •

A42 $2a+10$: $\begin{aligned} ({}^{-}a-5)\cdot {}^{-}2 &= {}^{-}a\cdot {}^{-}2 - 5\cdot {}^{-}2 \\ &= 2a - {}^{-}10 \\ &= 2a+10 \end{aligned}$

28 By the multiplication property of negative one, the expression $^{-}(3x+7)$ is equivalent to $^{-}1(3x+7)$. Thus $^{-}(3x+7)$ is simplified:

$$\begin{aligned} ^{-}(3x+7) &= {}^{-}1(3x+7) \\ &= {}^{-}1 \cdot 3x + {}^{-}1 \cdot 7 \\ &= {}^{-}3x + {}^{-}7 \\ &= {}^{-}3x - 7 \end{aligned}$$

An alternative method to the above is to notice that $^{-}(3x+7)$ means the *opposite of* $(3x+7)$, which can be found by taking the opposite of each term within the parentheses. Thus,

$$\begin{aligned} ^{-}(3x+7) &= {}^{-}3x + {}^{-}7 \\ &= {}^{-}3x - 7 \end{aligned}$$

Q43 Write an equivalent expression for $^{-}(5+2y)$ by forming the opposite of each term within the parentheses.

• ### • ### • ### • ### • ### • ### • ### •

A43 $^{-}5-2y$: $^{-}(5+2y) = {}^{-}5 + {}^{-}2y$

Q44 Simplify $^{-}1(5+2y)$ by removing the parentheses.

• ### • ### • ### • ### • ### • ### • ### •

A44 $^{-}5-2y$: $\begin{aligned} ^{-}1(5+2y) &= {}^{-}1 \cdot 5 + {}^{-}1 \cdot 2y \\ &= {}^{-}5 + {}^{-}2y \\ &= {}^{-}5 - 2y \end{aligned}$

Q45 Simplify $^{-}(4b-5)$

• ### • ### • ### • ### • ### • ### • ### •

A45 $^{-}4b+5$ or $5-4b$: $\begin{aligned} ^{-}(4b-5) &= {}^{-}1(4b-5) \\ &= {}^{-}1 \cdot 4b - {}^{-}1 \cdot 5 \\ &= {}^{-}4b - {}^{-}5 \\ &= {}^{-}4b + 5 \text{ or } 5-4b \end{aligned}$

or $\begin{aligned} ^{-}(4b-5) &= {}^{-}4b - {}^{-}5 \\ &= {}^{-}4b + 5 \text{ or } 5-4b \end{aligned}$

Q46 Simplify $^{-}(x-7)$.

• ### • ### • ### • ### • ### • ### • ### •

A46 $^{-}x+7$ or $7-x$

29 An algebraic expression is said to be in "simplest" form when:

1. All parentheses have been removed.
2. All like terms have been combined.

3. The definition of subtraction has been applied to remove all of the "raised" negative signs that can be removed.

Examples:	*Simplest Form?*
$3x+5-2x$	no, like terms not combined
$7x-3+2y$	yes
$4(x-2)+3x$	no, parentheses not removed and like terms not combined
$3+{}^-5y$	no, raised negative sign can be removed by writing $3-5y$
$9t-{}^-4$	no, raised negative sign can be removed by writing $9t+4$
${}^-5x-2$	yes

Q47 Indicate whether each of the following expressions are in simplest form. If the expression is not in simplest form, briefly state why.

a. $4x-2y+9$ ______ ______________________

b. $3(2-4y)-1$ ______ ______________________

c. $5t-6+9t$ ______ ______________________

d. $8x-5+{}^-y$ ______ ______________________

e. $5x+7y-3xy$ ______ ______________________

• # # # • # # # • # # # • # # # • # # # • # # # • # # # • # #

A47 **a.** yes
b. no, parentheses not removed and like terms not combined
c. no, like terms not combined
d. no, raised negative sign can be removed by writing $8x-5-y$
e. yes

Q48 Simplify $5-6t+3-9t$ by rearranging and combining the like terms.

• # # # • # # # • # # # • # # # • # # # • # # # • # # # • # #

A48 ${}^-15t+8$ or $8-15t$: $5-6t+3-9t$
${}^-6t-9t+5+3$
${}^-15t+8$ or $8-15t$

Q49 Simplify ${}^-3x+7x-5+x$.

• # # # • # # # • # # # • # # # • # # # • # # # • # # # • # #

A49 $5x-5$: ${}^-3x+7x-5+x$
${}^-3x+7x+x-5$
$5x-5$

30 To simplify expressions that involve parentheses:

1. Use the distributive properties to remove parentheses.
2. Rearrange and combine the like terms.

Examples:

1. $2(3y-7)+4$
$6y-14+4$
$6y-10$

2. $^{-}2x+\frac{4}{9}(x-4)$
$^{-}2x+\frac{4}{9}x-\frac{16}{9}$
$\frac{^{-}14}{9}x-\frac{16}{9}$ $\left(\text{Note: } ^{-}2+\frac{4}{9}=\frac{^{-}14}{9}\right)$

Q50 Simplify $5(2x-1)+7$.

• ### • ### • ### • ### • ### • ### • ### •

A50 $10x+2$: $5(2x-1)+7$
$10x-5+7$
$10x+2$

Q51 Simplify $3x+7(^{-}x+4)$.

• ### • ### • ### • ### • ### • ### • ### •

A51 $^{-}4x+28$ or $28-4x$: $3x+7(^{-}x+4)$
$3x+{^{-}7x}+7\cdot 4$
$^{-}4x+28$ or $28-4x$

Q52 Simplify each of the following:

a. $(7x+8)-3$ **b.** $3(2t-4)+7$ **c.** $8b+2(b-3)$

d. $(^{-}5+2x)-3x$ **e.** $(^{-}3x+4)-9$ **f.** $^{-}7(4-x)+3x$

• ### • ### • ### • ### • ### • ### • ### •

A52 **a.** $7x+5$ **b.** $6t-5$ **c.** $10b-6$
d. $^{-}x-5$ **e.** $^{-}3x-5$ **f.** $10x-28$

31 Recall that the expression $^{-}(x+3)$ could be simplified in either of two ways:

$^{-}(x+3)={^{-}1}(x+3)$
$={^{-}1}\cdot x+{^{-}1}\cdot 3$
$={^{-}x}+{^{-}3}$
$={^{-}x}-3$

or

$^{-}(x+3)={^{-}x}+{^{-}3}$
$={^{-}x}-3$

[$^{-}(x+3)$ means the opposite of $(x+3)$, which is the same as the opposite of each term within the parentheses]

Q53 Simplify:

a. $^{-}(y+9)=$ ________ **b.** $^{-}(2x+7)=$ ________

c. $^-\left(4z - \frac{4}{3}\right) =$ ________ **d.** $^-(b-12) =$ ________

e. $^-(5y+7) =$ ________

• ### • ### • ### • ### • ### • ### • ### •

A53 **a.** $^-y - 9$ **b.** $^-2x - 7$

c. $^-4z + \frac{4}{3}$ or $\frac{4}{3} - 4z$ **d.** $^-b + 12$ or $12 - b$

e. $^-5y - 7$

---

32 To simplify $4-(x+7)$ the following steps are used:

$$\begin{aligned} 4-(x+7) &= 4 + {}^-(x+7) \\ &= 4 + {}^-x + {}^-7 \\ &= {}^-x + {}^-3 \\ &= {}^-x - 3 \end{aligned}$$

Q54 Simplify $15-(2x+7)$.

• ### • ### • ### • ### • ### • ### • ### •

A54 $^-2x + 8$ or $8 - 2x$: $\begin{aligned} 15-(2x+7) &= 15 + {}^-(2x+7) \\ &= 15 + {}^-2x + {}^-7 \\ &= {}^-2x + 8 \text{ or } 8 - 2x \end{aligned}$

Q55 Simplify $^-4-(2b-3)$.

• ### • ### • ### • ### • ### • ### • ### •

A55 $^-2b - 1$: $\begin{aligned} 4-(2b-3) &= {}^-4 + {}^-(2b-3) \\ &= {}^-4 + {}^-2b - {}^-3 \\ &= {}^-4 + {}^-2b + 3 \\ &= {}^-2b - 1 \end{aligned}$

Q56 Simplify each of the following:

a. $4-(y+3)$ **b.** $2z-(z-5)$

c. $9-(2a-4)$ **d.** $6y-(2y+7)$

• ### • ### • ### • ### • ### • ### • ### •

A56 **a.** $^-y + 1$ or $1 - y$ **b.** $z + 5$

c. $^-2a + 13$ or $13 - 2a$ **d.** $4y - 7$

---

33 To simplify $(2x+7)-(3x-4)$, remove the parentheses and combine like terms as follows:

$$\begin{aligned} (2x+7)-(3x-4) &= (2x+7) + {}^-(3x-4) \\ &= (2x+7) + {}^-3x - {}^-4 \\ &= 2x + 7 + {}^-3x + 4 \\ &= {}^-x + 11 \end{aligned}$$

Notice that parentheses preceded by no sign (or a "+" sign) are removed by just dropping them, whereas parentheses preceded by a "−" sign are removed by rewriting the subtraction problem as an addition problem and taking the opposite of the terms within the parentheses.

Q57 Remove the parentheses only (do not simplify): $(x-5)-(2x+7)$

• ### • ### • ### • ### • ### • ### • ### •

A57 $x-5+{}^{-}2x+{}^{-}7$: $(x-5)-(2x+7)=(x-5)+{}^{-}(2x+7)$
$=x-5+{}^{-}2x+{}^{-}7$

Q58 Simplify the result in A57.

• ### • ### • ### • ### • ### • ### • ### •

A58 ${}^{-}x-12$

Q59 Remove the parentheses only (do not simplify): $(1-3b)-(4b-7)$

• ### • ### • ### • ### • ### • ### • ### •

A59 $1-3b+{}^{-}4b-{}^{-}7$

Q60 Simplify the result in A59.

• ### • ### • ### • ### • ### • ### • ### •

A60 ${}^{-}7b+8$ or $8-7b$

Q61 Simplify each of the following expressions:

a. $(5-4x)-(x+3)$ **b.** $(2x-5)-(3-5x)$ **c.** $({}^{-}5+b)-(b-5)$

d. ${}^{-}(4y+3)-(7y+3)$ **e.** $(z-9)-({}^{-}z-5)$ **f.** ${}^{-}(x+5)-(x-5)$

• ### • ### • ### • ### • ### • ### • ### •

A61 **a.** ${}^{-}5x+2$ or $2-5x$ **b.** $7x-8$ **c.** 0
d. ${}^{-}11y-6$ **e.** $2z-4$ **f.** ${}^{-}2x$

34 To simplify expressions such as $3(x-2)-4(x+3)$, the following steps are used:

$$\begin{aligned} 3(x-2)-4(x+3) &= 3(x-2)+{}^{-}4(x+3) \\ &= 3x-6+{}^{-}4x+{}^{-}12 \\ &= 3x+{}^{-}6+{}^{-}4x+{}^{-}12 \\ &= {}^{-}x+{}^{-}18 \\ &= {}^{-}x-18 \end{aligned}$$

Note: The work is usually shortened to:

$$3(x-2)-4(x+3)=3x-6-4x-12$$
$$={}^{-}x-18$$

Q62 Remove the parentheses only (do not simplify): $5(y+7)-2(y-4)$

• ### • ### • ### • ### • ### • ### • ### •

A62 $5y+35-2y+8$

Q63 Complete the simplification of A62.

• ### • ### • ### • ### • ### • ### • ### •

A63 $3y+43$

Q64 Remove the parentheses only (do not simplify): $2(2x-3)-4(x+5)$

• ### • ### • ### • ### • ### • ### • ### •

A64 $4x-6-4x-20$

Q65 Complete the simplification of A64.

• ### • ### • ### • ### • ### • ### • ### •

A65 ${}^{-}26$

Q66 Simplify each of the following algebraic expressions:

a. $(3x-4)-2(2x+5)$ **b.** ${}^{-}2(b+7)-(3b-5)$

c. $2(y+3)+3({}^{-}2y-7)$ **d.** $(y+7)-2(y+5)$

• ### • ### • ### • ### • ### • ### • ### •

A66 **a.** ${}^{-}x-14$ **b.** ${}^{-}5b-9$ **c.** ${}^{-}4y-15$ **d.** ${}^{-}y-3$

This completes the instruction for this section.

8.1 EXERCISE

1. Identify the like terms in each of the following algebraic expressions:

a. $3x-5y-6x+4$ **b.** $3r-2+7s-6$

2. Simplify by combining like terms:

a. $4x-2x$ **b.** $m-3-7m$

c. $7-4y-6-3y$ **d.** $\frac{3}{5}x-\frac{4}{7}x-2$

3. Simplify by use of the commutative and/or associative properties of addition:

a. $(3x-2)+5x$ **b.** $^{-}15+(7-8a)$

c. $^{-}x+\left(7x-\frac{2}{5}x\right)$ **d.** $\left(\frac{3}{5}b-\frac{2}{3}b\right)+b$

4. Simplify:

a. $4y+{}^{-}9$ **b.** $5-2r+2s-6r$

c. $\frac{1}{3}x+y-6xy-4y$ **d.** $4m+7-3m+{}^{-}7-m$

5. Simplify:

a. $2\left(\frac{3}{4}x\right)$ **b.** $\frac{^{-}1}{5}\left(\frac{^{-}4}{9}y\right)$ **c.** $\frac{^{-}1}{7}(7m)$

d. $\frac{5}{7}\left(\frac{7}{5}x\right)$ **e.** $\left(\frac{^{-}1}{9}z\right)\cdot{}^{-}9$ **f.** $\left(\frac{^{-}3}{5}x\right)\frac{2}{3}$

g. $(4y)\frac{5}{4}$ **h.** $\left(\frac{^{-}3}{5}x\right)20$

6. Simplify:

a. $2(x-3)$ **b.** $(y+7)\cdot{}^{-}3$ **c.** $^{-}4(2x-3)$

d. $(4-2x)5$ **e.** $(7x-9)$ **f.** $^{-}(7x-9)$

g. $^{-}(^{-}t+2)$ **h.** $^{-}5(^{-}7-3x)$ **i.** $(6x-5)\cdot{}^{-}8$

j. $^{-}7(3y+{}^{-}9)$

7. Simplify:

a. $^{-}4y+9+2y+5$ **b.** $\frac{4}{5}x-\frac{2}{3}x+2-9$ **c.** $(7x+8)-8$

d. $^{-}2(x+7)-3$ **e.** $\frac{^{-}4}{5}(y+10)+7$ **f.** $4-(z+7)$

g. $3(x+2)+7(x-5)$ **h.** $(y-6)-(4y+7)$ **i.** $^{-}7(x+2)-2(3x-7)$

j. $^{-}\left(\frac{1}{3}x+5\right)-(5-x)$

8.1 EXERCISE ANSWERS

1. a. $3x$ and $^{-}6x$ **b.** $^{-}2$ and $^{-}6$

2. a. $2x$ **b.** $^{-}6m-3$ **c.** $^{-}7y+1$ or $1-7y$ **d.** $\frac{1}{35}x-2$

3. a. $8x-2$ **b.** $^{-}8a-8$ **c.** $\frac{28}{5}x$ **d.** $\frac{14}{15}b$

4. a. $4y-9$ **b.** $^{-}8r+2s+5$ **c.** $\frac{1}{3}x-3y-6xy$ **d.** 0

5. a. $\frac{3}{2}x$ **b.** $\frac{4}{45}y$ **c.** ^{-}m **d.** x

e. z **f.** $\frac{^{-}2}{5}x$ **g.** $5y$ **h.** $^{-}12x$

6. a. $2x-6$ **b.** $^{-}3y-21$

c. $^{-}8x+12$ or $12-8x$ **d.** $20-10x$

e. $7x-9$ **f.** $^{-}7x+9$ or $9-7x$

g. $t-2$ **h.** $15x+35$

i. $^{-}48x+40$ or $40-48x$ **j.** $^{-}21y+63$ or $63-21y$

7. a. $^{-}2y+14$ or $14-2y$ **b.** $\frac{2}{15}x-7$ **c.** $7x$

d. $^{-}2x-17$ **e.** $\frac{^{-}4}{5}y-1$ **f.** $^{-}z-3$

g. $10x-29$ **h.** $^{-}3y-13$ **i.** $^{-}13x$

j. $\frac{2}{3}x-10$

8.2 EQUATIONS, FUNDAMENTAL PRINCIPLES OF EQUALITY

1 Section 8.1 dealt with the procedures involved in simplifying algebraic expressions. These skills will now be utilized in the study of equations. An equation is a statement in which the expressions on opposite sides of an equal sign represent the same number. Some examples of equations are:

$3+4=9-2$
$15=y+4$
$x-2=6$
$2x-3=5x+9$

The expressions on opposite sides of the equals sign are referred to as the left and right *sides* of the equation. For example,

$$\underbrace{4x-7}_{\text{left side}}=\underbrace{2-6x}_{\text{right side}}$$

Q1 Identify the left and right sides of the following equations:

a. $14-9=5$ left side ________; right side ________

b. $3=y+1$ left side ________; right side ________

c. $3x-7=8x+13$ left side ________; right side ________

• ### • ### • ### • ### • ### • ### • ### • ### •

A1 **a.** left side, $14-9$; right side, 5
b. left side, 3; right side, $y+1$
c. left side, $3x-7$; right side, $8x+13$

2 Equations that do not contain enough information to be judged as either true or false are often referred to as *open sentences*. Thus,

$18-y=14$ and $x+6=9$

are examples of open sentences, because they cannot be judged true or false until numbers are replaced for the unknown quantities y and x. If y is replaced by 10 in the open sentence $18-y=14$, the resulting statement $18-10=14$ is false. If x is replaced by 3 in the open sentence $x+6=9$, the resulting statement $3+6=9$ is true.

Q2 **a.** If x is replaced by 7 in the open sentence $x-2=5$, is the resulting statement true or false? TRUE

b. If y is replaced by 4 in the open sentence $12+y=15$, is the resulting statement true or false? FALSE

• ### • ### • ### • ### • ### • ### • ### • ### •

A2 **a.** true: $7-2=5$
b. false: $12+4 \neq 15$

3 Since it is not always possible to guess the solution to an equation, it is necessary to study some basic procedures that can be used to solve equations. Recall that an equation is a statement in which the expressions on opposite sides of the equal sign represent the same number. Since it is necessary to maintain this equality between sides, a basic rule in solving equations is that *whatever operation is performed on one side of an equation must also be performed on the other side of the equation.*

In general, solving an equation is like untying a knot, in that you always do the opposite of what has been done to form the equation. The equation has been solved when it has been changed to the form "a variable = a number" or "a number = a variable." If the variable is x, the form is "$x =$ a number" or "a number $= x$."

In the equation $x-2=10$, for example, 2 has been subtracted from x to equal 10. Since *the opposite of subtracting 2 is adding 2,* the equation can be solved by adding 2 to both sides as follows:

$$x-2=10 \quad \text{(add 2 to both sides)}$$
$$x-2+2=10+2$$
$$x=12$$

The solution can be checked by seeing if it converts the original open sentence into a true statement.

Check: $x-2=10$
$12-2 \stackrel{?}{=} 10$
$10=10$

So 12 is the correct solution, because 12 converts $x-2=10$ into the true statement $10=10$.

The solution of the preceding equation demonstrates the *addition principle of equality: If the same number is added to both sides of an equation, the result is another equation with the same solution.* In general, if $a=b$, then $a+c=b+c$ for any numbers a, b, and c.

Q3 Solve the following equations using the addition principle of equality and check the solutions:
a. $x-3=5$ **b.** $y-12={}^-7$ **c.** $3=x-6$

\# # # • # # # • # # # • # # # • # # # • # # # • # # # • # # # • # # #

A3 **a.** $x-3=5$ (3 was subtracted from x, so add 3 to both sides)
$x-3+3=5+3$
$x=8$

Check: $x-3=5$
$8-3 \stackrel{?}{=} 5$
$5=5$

b. $y-12={}^-7$ (12 was subtracted from x, so add 12 to both sides)
$y-12+12={}^-7+12$
$y=5$

Check: $y-12={}^-7$
$5-12 \stackrel{?}{=} {}^-7$
${}^-7={}^-7$

c. $3 = x - 6$ (6 was subtracted from x, so add 6 to both sides)

$3 + 6 = x - 6 + 6$

$9 = x$

Check: $3 = x - 6$

$3 \overset{?}{=} 9 - 6$

$3 = 3$

---

4 Observe that in the equation $x + 4 = 13$, 4 has been added to x to equal 13. The opposite of adding 4 is subtracting 4, so the equation can be solved by subtracting 4 from both sides as follows:

$x + 4 = 13$

$x + 4 - 4 = 13 - 4$

$x = 9$

Check: $x + 4 = 13$

$9 + 4 \overset{?}{=} 13$

$13 = 13$

The solution of the preceding equation demonstrates a second principle useful in solving equations, the *subtraction principle of equality: If the same number is subtracted from both sides of an equation, the result is another equation with the same solution.* In general, if $a = b$, then $a - c = b - c$ for any numbers a, b, and c.

Q4 Solve the following equations using the subtraction principle of equality and check each of the solutions:

a. $y + 1 = 16$

$y + 1 - 1 = 16 - 1$

$y = 15$

b. $x + 14 = {}^{-}29$

$x + 14 - 15 = {}^{-}29 - 14$

$x = {}^{-}43$

c. $34 = x + 11$

$34 - 11 = x + 11 - 11$

$23 = x$

\# \# \# • \# \# \# • \# \# \# • \# \# \# • \# \# \# • \# \# \# • \# \# \# • \# \# \# • \# \# \#

A4 **a.** $y + 1 = 16$ (1 was added to y, so subtract 1 from both sides)

$y + 1 - 1 = 16 - 1$

$y = 15$

Check: $y + 1 = 16$

$15 + 1 \overset{?}{=} 16$

$16 = 16$

b. $x + 14 = {}^{-}29$ (14 was added to x, so subtract 14 from both sides)

$x + 14 - 14 = {}^{-}29 - 14$

$x = {}^{-}43$

Check: $x + 14 = {}^{-}29$

${}^{-}43 + 14 \overset{?}{=} {}^{-}29$

${}^{-}29 = {}^{-}29$

c. $34 = x + 11$ (11 was added to x, so subtract 11 from both sides)

$34 - 11 = x + 11 - 11$

$23 = x$

Check: $34 = x + 11$

$34 \overset{?}{=} 23 + 11$

$34 = 34$

---

5 It is often necessary to simplify one or both sides of an equation by combining like terms before proceeding with the solution. For example,

$$x-5=15-2 \quad \text{(simplify by combining like terms)}$$
$$x-5=13$$
$$x-5+5=13+5$$
$$x=18$$

Q5 Solve the equation by first combining like terms: $y-7+2=13$

\# \# \# • \# \# \# • \# \# \# • \# \# \# • \# \# \# • \# \# \# • \# \# \# • \# \# \# • \# \# \#

A5
$$y-7+2=13$$
$$y-5=13$$
$$y-5+5=13+5$$
$$y=18$$

Q6 Use the addition and subtraction principles of equality to solve the following equations and check each of the solutions:

a. $x-2=5$	**b.** $y+7=11$	**c.** $1=x-7$
d. $x+3=3$	**e.** $12=x+8$	**f.** $x-7={}^{-}5$
g. $11+y={}^{-}7$	**h.** $x+6=5$	**i.** ${}^{-}3=y+9$
j. $7+x=2$	**k.** $y+7=2-11$	**l.** ${}^{-}4+x=7-14$

\# \# \# • \# \# \# • \# \# \# • \# \# \# • \# \# \# • \# \# \# • \# \# \# • \# \# \# • \# \# \#

A6

a. $x=7$	**b.** $y=4$	**c.** $8=x$ $(x=8)$	**d.** $x=0$
e. $4=x$ $(x=4)$	**f.** $x=2$	**g.** $y={}^{-}18$	**h.** $x={}^{-}1$
i. ${}^{-}12=y$ $(y={}^{-}12)$	**j.** $x={}^{-}5$	**k.** $y={}^{-}16$	**l.** $x={}^{-}3$

6 In each of the equations solved so far the understood coefficient of the variable is 1. That is, x is understood to be the same as $1x$. We now turn to equations in which the coefficient of the variable is a number other than 1. Some examples of this type are ${}^{-}5x=20$ and $\frac{3}{7}y=27$. Consider, first, the equation

$$4x=24$$

Recall that the term $4x$ means 4 times x. Since the variable x has been multiplied by 4 and the opposite of multiplying by 4 is dividing by 4, the equation can be solved by dividing both sides of the equation by 4 as follows:

$$4x=24 \qquad \text{Check:} \quad 4x=24$$
$$\frac{4x}{4}=\frac{24}{4} \qquad 4(6)\stackrel{?}{=}24$$
$$1x=6 \qquad 24=24$$
$$x=6$$

The procedure used in the preceding equation demonstrates the *division principle of equality: If both sides of an equation are divided by the same nonzero number, the result is another equation with the same solution.* (Zero is excluded since division by zero is impossible.) In general, if $a = b$, then $\frac{a}{c} = \frac{b}{c}$ for any numbers a, b, and c, $c \neq 0$.

Examples:

1. $^{-}3x = 75$ (x was multiplied by $^{-}3$, so divide both sides by $^{-}3$)

$$\frac{^{-}3x}{^{-}3} = \frac{75}{^{-}3}$$

$$1x = {}^{-}25$$

$$x = {}^{-}25$$

Check:

$$^{-}3x = 75$$

$$^{-}3(^{-}25) \stackrel{?}{=} 75$$

$$75 = 75$$

2. $^{-}17y = {}^{-}29$ (y was multiplied by $^{-}17$, so divide both sides by $^{-}17$)

$$\frac{^{-}17y}{^{-}17} = \frac{^{-}29}{^{-}17}$$

$$1y = \frac{29}{17}$$

$$y = \frac{29}{17}$$

Check:

$$^{-}17y = {}^{-}29$$

$$^{-}17\left(\frac{29}{17}\right) \stackrel{?}{=} {}^{-}29$$

$$^{-}29 = {}^{-}29$$

Notice that the number used to divide both sides is exactly the same as the coefficient of the variable.

Q7 Solve the following equations using the division principle of equality and check each of the solutions:

a. $2x = 10$ **b.** $^{-}4y = 12$ **c.** $4y = {}^{-}8$ **d.** $^{-}3x = {}^{-}7$

$\frac{2x}{2} = \frac{10}{2}$, $x = 5$ $\frac{-4y}{-4} = \frac{12}{-4}$, $y = -3$ $\frac{4y}{4} = \frac{-8}{4}$, $y = -2$ $\frac{-3x}{-3} = \frac{-7}{-3}$, $x = \frac{7}{3}$

\# \# \# • \# \# \# • \# \# \# • \# \# \# • \# \# \# • \# \# \# • \# \# \# • \# \# \# • \# \# \#

A7 **a.** $2x = 10$

$$\frac{2x}{2} = \frac{10}{2}$$

$$1x = 5$$

$$x = 5$$

Check: $2x = 10$, $2(5) \stackrel{?}{=} 10$, $10 = 10$

b. $^{-}4y = 12$

$$\frac{^{-}4y}{^{-}4} = \frac{12}{^{-}4}$$

$$1y = {}^{-}3$$

$$y = {}^{-}3$$

Check: $^{-}4y = 12$, $^{-}4(^{-}3) \stackrel{?}{=} 12$, $12 = 12$

c. $4y = {}^{-}8$

$$\frac{4y}{4} = \frac{^{-}8}{4}$$

$$1y = {}^{-}2$$

$$y = {}^{-}2$$

Check: $4y = {}^{-}8$, $4(^{-}2) \stackrel{?}{=} {}^{-}8$, $^{-}8 = {}^{-}8$

d. $^{-}3x = {}^{-}7$

$$\frac{^{-}3x}{^{-}3} = \frac{^{-}7}{^{-}3}$$

$$1x = \frac{7}{3}$$

$$x = \frac{7}{3}$$

Check: $^{-}3x = {}^{-}7$, $\frac{^{-}3}{1}\left(\frac{7}{3}\right) \stackrel{?}{=} {}^{-}7$, $^{-}7 = {}^{-}7$

7 When the coefficient of the variable in an equation is a fraction, a similar procedure can be followed. For example, in the equation $\frac{3}{4}x = 12$, because the variable x has been multiplied by $\frac{3}{4}$, the equation can be solved by dividing both sides by $\frac{3}{4}$.

$$\frac{3}{4}x = 12 \qquad \left(\text{divide by } \frac{3}{4}\right)$$

$$\frac{\frac{3}{4}x}{\frac{3}{4}} = \frac{12}{\frac{3}{4}}$$

$$1x = 12 \div \frac{3}{4}$$

$$x = 12 \cdot \frac{4}{3}$$

$$x = 16$$

The above solution can be simplified if it is recalled by dividing by $\frac{3}{4}$ is the same as multiplying by its reciprocal, $\frac{4}{3}$. Thus, the value of $1x$ (or x) can be found by multiplying both sides of the equation by $\frac{4}{3}$ as follows:

$$\frac{3}{4}x = 12$$

$$\frac{4}{3}\left(\frac{3}{4}x\right) = \frac{4}{3}(12)$$

$$1x = \frac{4}{3}\left(\frac{12}{1}\right)$$

$$x = 16$$

Check:

$$\frac{3}{4}x = 12$$

$$\frac{3}{4}(16) \stackrel{?}{=} 12$$

$$12 = 12$$

The procedure used to solve the preceding equation demonstrates the fourth principle useful in solving equations, the *multiplication principle of equality: If both sides of an equation are multiplied by the same nonzero number, the result is another equation with the same solution.* In general, if $a = b$, then $ac = bc$ for any numbers a, b, and c, $c \neq 0$.

Study the following examples of the multiplication principle of equality before proceeding to the problems of Q8.

Examples:

1.

$$\frac{5}{7}y = 10$$

$$\frac{7}{5}\left(\frac{5}{7}y\right) = \frac{7}{5}(10)$$

$$1y = \frac{7}{5} \cdot \frac{10}{1}$$

$$y = 14$$

Check:

$$\frac{5}{7}y = 10$$

$$\frac{5}{7}(14) \stackrel{?}{=} 10$$

$$10 = 10$$

2.

$$\frac{^-2}{3}x = \frac{4}{5}$$

$$\frac{^-3}{2}\left(\frac{^-2}{3}x\right) = \frac{^-3}{2}\left(\frac{4}{5}\right)$$

$$1x = \frac{^-6}{5}$$

$$x = \frac{^-6}{5} \text{ or } {}^-1\frac{1}{5}$$

Check:

$$\frac{^-2}{3}x = \frac{4}{5}$$

$$\frac{^-2}{3}\left(\frac{^-6}{5}\right) \stackrel{?}{=} \frac{4}{5}$$

$$\frac{4}{5} = \frac{4}{5}$$

Q8 Use the multiplication principle of equality to solve the following equations and check each of the solutions:

a. $\frac{1}{2}x = 12$ **b.** $\frac{4}{5}y = {}^{-}40$ **c.** $\frac{{}^{-}3}{7}x = \frac{5}{12}$ **d.** $\frac{{}^{-}6}{7}y = {}^{-}3$

$\frac{2}{1}\cdot\frac{1}{2}x = \frac{12}{1}\cdot\frac{2}{1}$ $\frac{5}{4}\cdot\frac{4}{5}y = \frac{-40}{1}\cdot\frac{5}{4}$ $\frac{-7}{3}\cdot\frac{-3}{7}x = \frac{5}{12}\cdot\frac{-7}{3}$ $\frac{-7}{6}\cdot\frac{-6}{7}y = \frac{-3}{1}\cdot\frac{-7}{6}$

$x = 24$ $y = -50$ $x = \frac{-35}{36}$ $y = \frac{21}{6} = \frac{7}{2}$

\# \# \# • \# \# \# • \# \# \# • \# \# \# • \# \# \# • \# \# \# • \# \# \# • \# \# \# • \# \# \#

A8 **a.**

$$\frac{1}{2}x = 12$$
$$\frac{2}{1}\left(\frac{1}{2}x\right) = \frac{2}{1}(12)$$
$$1x = \frac{2}{1}\left(\frac{12}{1}\right)$$
$$x = 24$$

Check:
$$\frac{1}{2}x = 12$$
$$\frac{1}{2}(24) \stackrel{?}{=} 12$$
$$12 = 12$$

b.

$$\frac{4}{5}y = {}^{-}40$$
$$\frac{5}{4}\left(\frac{4}{5}y\right) = \frac{5}{4}({}^{-}40)$$
$$1y = \frac{5}{4}\left(\frac{{}^{-}40}{1}\right)$$
$$y = {}^{-}50$$

Check:
$$\frac{4}{5}y = {}^{-}40$$
$$\frac{4}{5}({}^{-}50) \stackrel{?}{=} {}^{-}40$$
$${}^{-}40 = {}^{-}40$$

c.

$$\frac{{}^{-}3}{7}x = \frac{5}{12}$$
$$\frac{{}^{-}7}{3}\left(\frac{{}^{-}3}{7}x\right) = \frac{{}^{-}7}{3}\left(\frac{5}{12}\right)$$
$$1x = \frac{{}^{-}35}{36}$$
$$x = \frac{{}^{-}35}{36}$$

Check:
$$\frac{{}^{-}3}{7}x = \frac{5}{12}$$
$$\frac{{}^{-}3}{7}\left(\frac{{}^{-}35}{36}\right) \stackrel{?}{=} \frac{5}{12}$$
$$\frac{5}{12} = \frac{5}{12}$$

d.

$$\frac{{}^{-}6}{7}y = {}^{-}3$$
$$\frac{{}^{-}7}{6}\left(\frac{{}^{-}6}{7}y\right) = \frac{{}^{-}7}{6}({}^{-}3)$$
$$1y = \frac{7}{2}$$
$$y = \frac{7}{2}$$

Check:
$$\frac{{}^{-}6}{7}y = {}^{-}3$$
$$\frac{{}^{-}6}{7}\left(\frac{7}{2}\right) \stackrel{?}{=} {}^{-}3$$
$${}^{-}3 = {}^{-}3$$

8 The equation $\frac{x}{7} = 3$ can be solved in a manner similar to that used with the preceding equations. Using the understood coefficient 1 for x, the solution proceeds as follows:

$$\frac{x}{7} = 3$$
$$\frac{1x}{7} = 3$$
$$\frac{1}{7}x = 3$$
$$\frac{7}{1}\left(\frac{1}{7}x\right) = \frac{7}{1}(3)$$

$1x = 21$ (this step is usually omitted)

$$x = 21$$

Q9 Solve the following equations using the understood coefficient 1 for the variable:

a. $\frac{x}{5} = 2$ **b.** $\frac{^-y}{8} = 2$

\# \# \# • \# \# \# • \# \# \# • \# \# \# • \# \# \# • \# \# \# • \# \# \# • \# \# \# • \# \# \#

A9 **a.**

$$\frac{x}{5} = 2$$
$$\frac{1x}{5} = 2$$
$$\frac{1}{5}x = 2$$
$$\frac{5}{1} \cdot \frac{1}{5}x = \frac{5}{1}(2)$$
$$x = 10$$

b.

$$\frac{^-y}{8} = 2$$
$$\frac{^-1y}{8} = 2$$
$$\frac{^-1}{8}y = 2$$
$$\frac{^-8}{1} \cdot \frac{^-1}{8}y = \frac{^-8}{1}(2)$$
$$y = {}^-16$$

9 It is often necessary to solve equations of the form $^-x = a$ for some number a, that is, equations with a coefficient of $^-1$ on the variable. These can be solved using the multiplication principle of equality. For example,

$$^-x = 9$$
$$^-1x = 9$$
$$(^-1)(^-1x) = (^-1)(9)$$
$$x = {}^-9$$

Q10 Use the procedure of Frame 9 to solve each of the following equations:

a. $^-x = 5$ **b.** $^-y = {}^-7$

\# \# \# • \# \# \# • \# \# \# • \# \# \# • \# \# \# • \# \# \# • \# \# \# • \# \# \# • \# \# \#

A10 **a.**

$$^-x = 5$$
$$^-1x = 5$$
$$(^-1)(^-1x) = (^-1)(5)$$
$$x = {}^-5$$

b.

$$^-y = {}^-7$$
$$^-1y = {}^-7$$
$$(^-1)(^-1y) = {}^-1(^-7)$$
$$y = 7$$

Q11 Solve the following equations (do step 2 mentally):

a. $^-x = \frac{^-3}{5}$ **b.** $^-y = 0$

\# \# \# • \# \# \# • \# \# \# • \# \# \# • \# \# \# • \# \# \# • \# \# \# • \# \# \# • \# \# \#

A11 **a.**

$$^-x = \frac{^-3}{5}$$
$$(^-1)(^-x) = (^-1)\left(\frac{^-3}{5}\right)$$
$$x = \frac{3}{5}$$

b.

$$^-y = 0$$
$$(^-1)(^-y) = (^-1)(0)$$
$$y = 0$$

This completes the instruction for this section.

8.2 EXERCISE

1. Use the addition and subtraction principles of equality to solve the following equations, and check each of the solutions:

a. $x-3={}^-5$ **b.** $y+9=4$ **c.** $7=y+8$
d. $x-9={}^-9$ **e.** ${}^-2=5+y$ **f.** $3+y={}^-11$
g. $x-3=14-23$ **h.** ${}^-5-3=x+8$

2. Use the multiplication and division principles of equality to solve the following equations, and check each of the solutions:

a. $2x=10$ **b.** $\frac{2}{5}x=20$ **c.** ${}^-4y=12$ **d.** $\frac{3}{8}x=24$
e. $\frac{x}{5}={}^-3$ **f.** $5y={}^-15$ **g.** ${}^-4x={}^-8$ **h.** ${}^-y=12$
i. $\frac{x}{4}={}^-4$ **j.** $\frac{{}^-5}{7}y=10$ **k.** $5y={}^-6$ **l.** $\frac{{}^-3}{11}x=\frac{2}{3}$
m. ${}^-x={}^-1$ **n.** $13x={}^-26$ **o.** $\frac{x}{4}=\frac{3}{4}$ **p.** ${}^-7y={}^-5$
q. $\frac{9}{16}=\frac{{}^-3}{4}x$ **r.** $12=\frac{x}{2}$ **s.** $32y={}^-4$ **t.** $\frac{7}{8}y=0$
u. $\frac{x}{5}=\frac{1}{5}$ **v.** $\frac{{}^-5}{6}x=\frac{{}^-2}{3}$ **w.** $8={}^-y$ **x.** $3y={}^-4$
y. ${}^-7x=\frac{3}{5}$ **z.** $\frac{{}^-4}{9}={}^-3y$

3. Use the addition, subtraction, multiplication, and division principles of equality to solve the following equations, and check each of the solutions:

a. $x-3=7$ **b.** $4y={}^-12$ **c.** $\frac{{}^-2}{3}x=\frac{{}^-4}{5}$ **d.** $y+9=2$
e. $3+x=5$ **f.** ${}^-5y=25$ **g.** $\frac{1}{2}x=7$ **h.** $\frac{3}{4}y={}^-12$
i. $15=x-7$ **j.** $42={}^-6y$ **k.** $10=\frac{{}^-2}{5}y$ **l.** $0=\frac{4}{7}x$
m. $x-3={}^-5$ **n.** ${}^-12+x=5$ **o.** $x+7=11-5$ **p.** $\frac{{}^-4}{3}x=2$
q. $5y=\frac{3}{7}$ **r.** $13=4+x$ **s.** $5+x={}^-3$ **t.** ${}^-7y=\frac{14}{15}$

8.2 EXERCISE ANSWERS

1. a. $x={}^-2$ **b.** $y={}^-5$ **c.** ${}^-1=y$ **d.** $x=0$
e. ${}^-7=y$ **f.** $y={}^-14$ **g.** $x={}^-6$ **h.** ${}^-16=x$

2. a. $x=5$ **b.** $x=50$ **c.** $y={}^-3$ **d.** $x=64$
e. $x={}^-15$ **f.** $y={}^-3$ **g.** $x=2$ **h.** $y={}^-12$
i. $x={}^-16$ **j.** $y={}^-14$ **k.** $y=\frac{{}^-6}{5}$ **l.** $x=\frac{{}^-22}{9}$
m. $x=1$ **n.** $y={}^-2$ **o.** $x=3$ **p.** $y=\frac{5}{7}$
q. $x=\frac{{}^-3}{4}$ **r.** $x=24$ **s.** $y=\frac{{}^-1}{8}$ **t.** $y=0$

u. $x = 1$ **v.** $x = \frac{4}{5}$ **w.** $y = {}^{-}8$ **x.** $y = \frac{{}^{-}4}{3}$

y. $x = \frac{{}^{-}3}{35}$ **z.** $y = \frac{4}{27}$

3. a. $x = 10$ **b.** $y = {}^{-}3$ **c.** $x = \frac{6}{5}$ **d.** $y = {}^{-}7$

e. $x = 2$ **f.** $x = {}^{-}5$ **g.** $x = 14$ **h.** $y = {}^{-}16$

i. $x = 22$ **j.** $y = {}^{-}7$ **k.** $y = {}^{-}25$ **l.** $x = 0$

m. $x = {}^{-}2$ **n.** $x = 17$ **o.** $x = {}^{-}1$ **p.** $x = \frac{{}^{-}3}{2}$

q. $y = \frac{3}{35}$ **r.** $x = 9$ **s.** $x = {}^{-}8$ **t.** $y = \frac{{}^{-}2}{15}$

8.3 SOLVING EQUATIONS BY THE USE OF TWO OR MORE STEPS

1 Frequently it is necessary to use more than one step in solving an equation. Recall that if the variable is x, an equation is solved when it is changed to the form "$x =$ a number" or "a number $= x$." Thus the aim is to *isolate all terms involving variables on one side of the equation and all numbers on the opposite side*. By "isolate" it is meant that *only* terms involving variables are alone on one side of the equation and *only* number terms are alone on the other side of the equation.

In the equation $3x - 4 = 11$ the objective is to isolate the x term on the left side. Since 4 has been subtracted from $3x$, add 4 (the opposite of subtracting 4) to both sides of the equation.

$$3x - 4 = 11$$
$$3x - 4 + 4 = 11 + 4$$
$$3x = 15$$

With the number and variable terms isolated on opposite sides, reduce the $3x$ to $1x$ by use of the division principle of equality.

$$3x = 15 \qquad \text{Check: } 3x - 4 = 11$$
$$\frac{3x}{3} = \frac{15}{3} \qquad 3(5) - 4 \stackrel{?}{=} 11$$
$$x = 5 \qquad 11 = 11$$

Q1 **a.** Solve the equation $5x - 3 = 7$ by first adding 3 to both sides.

b. Check the solution to the equation.

\# # # • # # # • # # # • # # # • # # # • # # # • # # # • # # # • # # #

A1 **a.** $5x - 3 = 7$ (3 was subtracted, so add 3 to both sides)

$$5x - 3 + 3 = 7 + 3$$
$5x = 10$ (x was multiplied by 5, so divide both sides by 5)
$$\frac{5x}{5} = \frac{10}{5}$$
$$x = 2$$

b. Check: $5x - 3 = 7$
$$5(2) - 3 \stackrel{?}{=} 7$$
$$7 = 7$$

2 In the equation $^{-}9 = 5x + 6$, since 6 has been added to $5x$, do the opposite and subtract 6 from both sides.

$$^{-}9 = 5x + 6$$
$$^{-}9 - 6 = 5x + 6 - 6$$
$$^{-}15 = 5x$$

With the number and variable terms isolated on opposite sides, reduce the $5x$ to $1x$ by use of the division principle of equality.

$$^{-}15 = 5x \qquad \text{Check: } ^{-}9 = 5x + 6$$
$$\frac{^{-}15}{5} = \frac{5x}{5} \qquad ^{-}9 \stackrel{?}{=} 5(^{-}3) + 6$$
$$\qquad ^{-}9 \stackrel{?}{=} {^{-}15} + 6$$
$$^{-}3 = x \qquad ^{-}9 = {^{-}9}$$

Q2 **a.** Solve $12 = {^{-}4x} + 8$. **b.** Check the solution to the equation.

\# \# \# • \# \# \# • \# \# \# • \# \# \# • \# \# \# • \# \# \# • \# \# \# • \# \# \# • \# \# \#

A2 **a.**

$$12 = {^{-}4x} + 8$$
$$12 - 8 = {^{-}4x} + 8 - 8$$
$$4 = {^{-}4x}$$
$$\frac{4}{^{-}4} = \frac{^{-}4x}{^{-}4}$$
$$^{-}1 = x$$

b. Check:

$$12 = {^{-}4x} + 8$$
$$12 \stackrel{?}{=} {^{-}4}(^{-}1) + 8$$
$$12 \stackrel{?}{=} 4 + 8$$
$$12 = 12$$

3 In the equation $3 + 2x = 7$, the variable term will be isolated on the left side if 3 is subtracted from both sides:

$$3 + 2x = 7$$
$$3 + 2x - 3 = 7 - 3$$
$$2x = 4$$

The solution can now be completed by use of the division principle of equality:

$$\frac{2x}{2} = \frac{4}{2}$$
$$x = 2$$

Q3 Solve $7 - 5x = 12$ by first isolating the variable term.

\# \# \# • \# \# \# • \# \# \# • \# \# \# • \# \# \# • \# \# \# • \# \# \# • \# \# \# • \# \# \#

A3

$$7 - 5x = 12$$
$$7 - 5x - 7 = 12 - 7$$
$$^{-}5x = 5$$
$$\frac{^{-}5x}{^{-}5} = \frac{5}{^{-}5}$$
$$x = {^{-}1}$$

4 In the equation ${}^{-}17 = {}^{-}8 - 3x$, the variable term will be isolated on the right side if 8 is added to both sides:

$$
\begin{aligned}
{}^{-}17 &= {}^{-}8 - 3x \\
{}^{-}17 + 8 &= {}^{-}8 - 3x + 8 \\
{}^{-}9 &= {}^{-}3x
\end{aligned}
$$

The solution can now be completed by use of the division principle of equality:

$$
\begin{aligned}
\frac{{}^{-}9}{{}^{-}3} &= \frac{{}^{-}3x}{{}^{-}3} \\
3 &= x
\end{aligned}
$$

Q4 Solve $0 = {}^{-}5 + 7x$ by first isolating the variable term.

\# \# \# • \# \# \# • \# \# \# • \# \# \# • \# \# \# • \# \# \# • \# \# \# • \# \# \# • \# \# \#

A4

$$
\begin{aligned}
0 &= {}^{-}5 + 7x \\
0 + 5 &= {}^{-}5 + 7x + 5 \\
5 &= 7x \\
\frac{5}{7} &= \frac{7x}{7} \\
\frac{5}{7} &= x
\end{aligned}
$$

5 In the preceding equations, it is important to notice that the *addition or subtraction principles of equality* are used *first* with the *multiplication or division principles of equality* used in the *final step* of the problem. Study the following two examples before proceeding to Q5.

Examples:

1.

$$
\begin{aligned}
\frac{{}^{-}2}{3}x + 7 &= 1 \\
\frac{{}^{-}2}{3}x + 7 - 7 &= 1 - 7 \\
\frac{{}^{-}2}{3}x &= {}^{-}6 \\
\frac{{}^{-}3}{2} \cdot \frac{{}^{-}2}{3}x &= \frac{{}^{-}3}{2}({}^{-}6) \\
x &= 9
\end{aligned}
$$

Check:

$$
\begin{aligned}
\frac{{}^{-}2}{3}x + 7 &= 1 \\
\frac{{}^{-}2}{3}(9) + 7 &\stackrel{?}{=} 1 \\
{}^{-}6 + 7 &= 1 \\
1 &= 1
\end{aligned}
$$

2.

$$
\begin{aligned}
{}^{-}3 &= {}^{-}3 + 4x \\
{}^{-}3 + 3 &= {}^{-}3 + 4x + 3 \\
0 &= 4x \\
\frac{0}{4} &= \frac{4x}{4} \\
0 &= x
\end{aligned}
$$

Check:

$$
\begin{aligned}
{}^{-}3 &= {}^{-}3 + 4x \\
{}^{-}3 &\stackrel{?}{=} {}^{-}3 + 4(0) \\
{}^{-}3 &\stackrel{?}{=} {}^{-}3 + 0 \\
{}^{-}3 &= {}^{-}3
\end{aligned}
$$

Q5 Solve the following equations by first applying the addition or subtraction principles of equality and then the multiplication or division principles of equality:

a. $3x-7={}^{-}19$

b. $5=\frac{4}{5}y-3$

c. $0={}^{-}6+12x$

d. $25=7-2x$

\# \# \# • \# \# \# • \# \# \# • \# \# \# • \# \# \# • \# \# \# • \# \# \# • \# \# \# • \# \# \#

A5 **a.**
$$\begin{aligned}3x-7&={}^{-}19\\3x-7+7&={}^{-}19+7\\3x&={}^{-}12\\\frac{3x}{3}&=\frac{{}^{-}12}{3}\\x&={}^{-}4\end{aligned}$$

b.
$$\begin{aligned}5&=\frac{4}{5}y-3\\5+3&=\frac{4}{5}y-3+3\\8&=\frac{4}{5}y\\\frac{5}{4}(8)&=\frac{5}{4}\cdot\frac{4}{5}y\\10&=y\end{aligned}$$

c.
$$\begin{aligned}0&={}^{-}6+12x\\0+6&={}^{-}6+12x+6\\6&=12x\\\frac{6}{12}&=\frac{12x}{12}\\\frac{1}{2}&=x\end{aligned}$$

d.
$$\begin{aligned}25&=7-2x\\25-7&=7-2x-7\\18&={}^{-}2x\\\frac{18}{{}^{-}2}&=\frac{{}^{-}2x}{{}^{-}2}\\{}^{-}9&=x\end{aligned}$$

6 In the preceding equations a variable term occurred on only one side of the equation. If variable terms occur on both sides of the equation, the general procedure used in solving the equation is the same. That is, *isolate all terms involving variables on one side of the equation and all numbers on the opposite side.*

For example, consider the equation

$$3x-2=x+6$$

To isolate the variable terms on the left side, subtract x from both sides:

$$\begin{aligned}3x-2-x&=x+6-x\\2x-2&=6\end{aligned}$$

To isolate the numbers on the right, add 2 to both sides:

$$\begin{aligned}2x-2+2&=6+2\\2x&=8\end{aligned}$$

The final step involves the division principle of equality:

$$\begin{aligned}\frac{2x}{2}&=\frac{8}{2}\\x&=4\end{aligned}$$

Q6 **a.** Solve $5x-4=2x+5$ by isolating the terms that involve the variables on the left side and the numbers on the right side.

b. Check the solution.

\# \# \# • \# \# \# • \# \# \# • \# \# \# • \# \# \# • \# \# \# • \# \# \# • \# \# \# • \# \# \#

A6 **a.**

$$\begin{aligned} 5x-4&=2x+5 \\ 5x-4-2x&=2x+5-2x \\ 3x-4&=5 \\ 3x-4+4&=5+4 \\ 3x&=9 \\ \frac{3x}{3}&=\frac{9}{3} \\ x&=3 \end{aligned}$$

b.

$$\begin{aligned} 5x-4&=2x+5 \\ 5(3)-4&\stackrel{?}{=}2(3)+5 \\ 15-4&\stackrel{?}{=}6+5 \\ 11&=11 \end{aligned}$$

---

7 Consider the equation $x+3=15-5x$. If you decide to isolate the terms involving variables on the left side and the numbers on the right side, you must add $5x$ and subtract 3 from both sides.

$$\begin{aligned} x+3&=15-5x \\ x+3+5x&=15-5x+5x \\ 6x+3&=15 \\ 6x+3-3&=15-3 \\ 6x&=12 \\ \frac{6x}{6}&=\frac{12}{6} \\ x&=2 \end{aligned}$$

Check:

$$\begin{aligned} x+3&=15-5x \\ 2+3&\stackrel{?}{=}15-5(2) \\ 5&=5 \end{aligned}$$

Q7 **a.** Solve $4x+7=2x+8$ by isolating the terms involving variables on the left and the numbers on the right.

b. Check the solution.

\# \# \# • \# \# \# • \# \# \# • \# \# \# • \# \# \# • \# \# \# • \# \# \# • \# \# \# • \# \# \#

A7 **a.**

$$\begin{aligned} 4x+7&=2x+8 \\ 4x+7-2x&=2x+8-2x \\ 2x+7&=8 \\ 2x+7-7&=8-7 \\ 2x&=1 \\ \frac{2x}{2}&=\frac{1}{2} \\ x&=\frac{1}{2} \end{aligned}$$

b.

$$\begin{aligned} 4x+7&=2x+8 \\ 4\cdot\frac{1}{2}+7&\stackrel{?}{=}2\cdot\frac{1}{2}+8 \\ 2+7&\stackrel{?}{=}1+8 \\ 9&=9 \end{aligned}$$

8 The preceding problems have been solved by isolating the terms that involve variables on the left side. However, equations can be solved by isolating the variable on either side. You may wish to isolate the variable on the side that makes the coefficient of the variable positive.

Example 1: Solve $3x-1=x+5$ by isolating the variable on the left side.

Solution:
Subtract x from and add 1 to both sides:

$$\begin{aligned} 3x-1&=x+5\\ 3x-1-x&=x+5-x\\ 2x-1&=5\\ 2x-1+1&=5+1\\ 2x&=6\\ x&=3 \end{aligned}$$

Example 2: Solve $3x-1=x+5$ by isolating the variable on the right side.

Solution:
Subtract $3x$ and 5 from both sides:

$$\begin{aligned} 3x-1&=x+5\\ 3x-1-3x&=x+5-3x\\ {}^{-}1&={}^{-}2x+5\\ {}^{-}1-5&={}^{-}2x+5-5\\ {}^{-}6&={}^{-}2x\\ 3&=x \end{aligned}$$

The same solution was obtained by isolating the variable on either side.

Q8 Solve $5x-3=6x+9$ by isolating the variable on the left side.

• ### • ### • ### • ### • ### • ### • ### •

A8

$$\begin{aligned} 5x-3&=6x+9\\ 5x-3-6x&=6x+9-6x\\ {}^{-}x-3&=9\\ {}^{-}x-3+3&=9+3\\ {}^{-}x&=12\\ x&={}^{-}12 \end{aligned}$$

Q9 Solve $5x-3=6x+9$ by isolating the variable on the right side.

• ### • ### • ### • ### • ### • ### • ### •

A9

$$5x - 3 = 6x + 9$$
$$5x - 3 - 5x = 6x + 9 - 5x$$
$$^{-}3 = x + 9$$
$$^{-}3 - 9 = x + 9 - 9$$
$$^{-}12 = x$$

Q10 Solve by isolating the variable on the side that makes the coefficient of the variable positive:

a. $2x + 1 = {}^{-}x - 2$ **b.** ${}^{-}4x + 2 = 7x + 3$

\# \# \# • \# \# \# • \# \# \# • \# \# \# • \# \# \# • \# \# \# • \# \# \# • \# \# \# • \# \# \#

A10 **a.**

$$2x + 1 = {}^{-}x - 2$$
$$2x + 1 + x = {}^{-}x - 2 + x$$
$$3x + 1 = {}^{-}2$$
$$3x + 1 - 1 = {}^{-}2 - 1$$
$$3x = {}^{-}3$$
$$x = {}^{-}1$$

b.

$$^{-}4x + 2 = 7x + 3$$
$$^{-}4x + 2 + 4x = 7x + 3 + 4x$$
$$2 = 11x + 3$$
$$2 - 3 = 11x + 3 - 3$$
$$^{-}1 = 11x$$
$$\frac{^{-}1}{11} = x$$

9 To solve equations that involve parentheses, first simplify both sides of the equation wherever possible, and then proceed as before.

Examples:

1. $2x - (x + 2) = 7$

$2x - x - 2 = 7$ (remove parentheses)

$x - 2 = 7$ (combine like terms)

$x - 2 + 2 = 7 + 2$

$x = 9$

2. $x + 5 = x + (2x - 3)$

$x + 5 = x + 2x - 3$ (remove parentheses)

$x + 5 = 3x - 3$ (combine like terms)

$x + 5 - x = 3x - 3 - x$

$5 = 2x - 3$

$5 + 3 = 2x - 3 + 3$

$8 = 2x$

$4 = x$

Q11 Solve by first simplifying both sides of the equation:

a. $5x - (2x + 7) = 8$ **b.** $6x = 8 + (2x - 4)$

\# \# \# • \# \# \# • \# \# \# • \# \# \# • \# \# \# • \# \# \# • \# \# \# • \# \# \# • \# \# \#

A11 **a.**
$$5x-(2x+7)=8$$
$$5x-2x-7=8$$
$$3x-7=8$$
$$3x-7+7=8+7$$
$$3x=15$$
$$\frac{3x}{3}=\frac{15}{3}$$
$$x=5$$

b.
$$6x=8+(2x-4)$$
$$6x=8+2x-4$$
$$6x=4+2x$$
$$6x-2x=4+2x-2x$$
$$4x=4$$
$$\frac{4x}{4}=\frac{4}{4}$$
$$x=1$$

10 To solve $2(5x+3)-3(x-5)=7$, the following steps are used:

$$2(5x+3)-3(x-5)=7$$
$$10x+6-3x+15=7$$
$$7x+21=7$$
$$7x+21-21=7-21$$
$$7x={}^{-}14$$
$$\frac{7x}{7}=\frac{{}^{-}14}{7}$$
$$x={}^{-}2$$

Q12 Solve:

a. $5(x+6)=45$

b. $3(x-2)=x-2(3x-1)$

\# \# \# • \# \# \# • \# \# \# • \# \# \# • \# \# \# • \# \# \# • \# \# \# • \# \# \# • \# \# \#

A12 **a.**
$$5(x+6)=45$$
$$5x+30=45$$
$$5x+30-30=45-30$$
$$5x=15$$
$$\frac{5x}{5}=\frac{15}{5}$$
$$x=3$$

b.
$$3(x-2)=x-2(3x-1)$$
$$3x-6=x-6x+2$$
$$3x-6={}^{-}5x+2$$
$$3x-6+5x={}^{-}5x+2+5x$$
$$8x-6=2$$
$$8x-6+6=2+6$$
$$8x=8$$
$$\frac{8x}{8}=\frac{8}{8}$$
$$x=1$$

11 When solving equations such as $5-4(x+3)=1$, it is again important to first simplify both sides of the equation by removing parentheses and combining like terms.

Example:

$$5-4(x+3)=1$$
$$5-4x-12=1$$
$${}^{-}4x-7=1$$
$${}^{-}4x-7+7=1+7$$
$${}^{-}4x=8$$
$$\frac{{}^{-}4x}{{}^{-}4}=\frac{8}{{}^{-}4}$$
$$x={}^{-}2$$

Q13 Solve:

a. $5-3(x-2)=14$

b. $x+8={}^{-}10-6(2x-3)$

\# # # • # # # • # # # • # # # • # # # • # # # • # # # • # # # • # # #

A13 **a.**

$$5-3(x-2)=14$$
$$5-3x+6=14$$
$${}^{-}3x+11=14$$
$${}^{-}3x+11-11=14-11$$
$$\frac{{}^{-}3x}{{}^{-}3}=\frac{3}{{}^{-}3}$$
$$x={}^{-}1$$

b.

$$x+8={}^{-}10-6(2x-3)$$
$$x+8={}^{-}10-12x+18$$
$$x+8=8-12x$$
$$x+8+12x=8-12x+12x$$
$$13x+8=8$$
$$13x+8-8=8-8$$
$$13x=0$$
$$\frac{13x}{13}=\frac{0}{13}$$
$$x=0$$

12 Many times it is necessary to solve equations that involve several rational numbers (fractions). This is usually done by eliminating the fractions from the equation by use of the multiplication property of equality, and solving the resulting equation using the procedures of previous sections.

The fractions can be eliminated from the equation $\frac{x}{2}-4=\frac{x}{3}$ by multiplying both sides of the equation by each of the denominators, 2 and 3.

Example:

$$\frac{x}{2}-4=\frac{x}{3}$$
$$2\left(\frac{x}{2}-4\right)=2\left(\frac{x}{3}\right)$$
$$2\left(\frac{x}{2}\right)-2(4)=2\left(\frac{x}{3}\right)$$
$$x-8=\frac{2x}{3}$$

The first fraction has now been eliminated.

$$3(x-8)=3\left(\frac{2x}{3}\right)$$
$$3(x)-3(8)=3\left(\frac{2x}{3}\right)$$
$$3x-24=2x$$

All fractions have now been eliminated and the solution can be completed using the procedures already studied.

$$3x-24=2x$$
$$3x-24-2x=2x-2x$$
$$x-24=0$$
$$x-24+24=0+24$$
$$x=24$$

Check:

$$\frac{x}{2}-4=\frac{x}{3}$$
$$\frac{24}{2}-4\stackrel{?}{=}\frac{24}{3}$$
$$12-4\stackrel{?}{=}8$$
$$8=8$$

Q14 **a.** Eliminate the fractions from the equation $\frac{x}{2} = 5 + \frac{x}{3}$ by first multiplying both sides of the equation by 2 and then by 3.

b. Solve the resulting equation

c. Check the solution.

\# \# \# • \# \# \# • \# \# \# • \# \# \# • \# \# \# • \# \# \# • \# \# \# • \# \# \# • \# \# \#

A14 **a.**

$$\frac{x}{2} = 5 + \frac{x}{3}$$
$$2\left(\frac{x}{2}\right) = 2\left(5 + \frac{x}{3}\right)$$
$$2\left(\frac{x}{2}\right) = 2(5) + 2\left(\frac{x}{3}\right)$$
$$x = 10 + \frac{2x}{3}$$
$$3(x) = 3\left(10 + \frac{2x}{3}\right)$$
$$3(x) = 3(10) + 3 \cdot \frac{2x}{3}$$
$$3x = 30 + 2x$$

b.

$$3x - 2x = 30 + 2x - 2x$$
$$x = 30$$

c.

$$\frac{x}{2} = 5 + \frac{x}{3}$$
$$\frac{30}{2} \stackrel{?}{=} 5 + \frac{30}{3}$$
$$15 \stackrel{?}{=} 5 + 10$$
$$15 = 15$$

13 A much shorter procedure for eliminating several fractions from an equation is to multiply both sides by just *one* number. For example, rather than to multiply the equation $\frac{x}{4} - 3 = \frac{x}{5}$ by the two numbers 4 and 5, the fractions can be eliminated by multiplying both sides by *one* number which has both factors 4 and 5. The smallest number with both factors 4 and 5 is 20. Therefore, multiply both sides of the equation by 20.

$$20\left(\frac{x}{4} - 3\right) = 20\left(\frac{x}{5}\right)$$
$$20\left(\frac{x}{4}\right) - 20(3) = 20\left(\frac{x}{5}\right)$$
$$5x - 60 = 4x$$

With the fractions eliminated, the solution of the equation can now be completed:

$$5x - 60 = 4x$$
$$5x - 60 - 4x = 4x - 4x$$
$$x - 60 = 0$$
$$x - 60 + 60 = 0 + 60$$
$$x = 60$$

Check:

$$\frac{x}{4} - 3 = \frac{x}{5}$$
$$\frac{60}{4} - 3 \stackrel{?}{=} \frac{60}{5}$$
$$15 - 3 \stackrel{?}{=} 12$$
$$12 = 12$$

Q15 **a.** Eliminate the fractions from the equation $\frac{x}{3}-4=\frac{x}{5}$ by multiplying both sides by the smallest number with the factors 3 and 5.

b. Solve the resulting equation.

c. Check the solution.

• ### • ### • ### • ### • ### • ### • ### •

A15 **a.** The smallest number is 15.

$$15\left(\frac{x}{3}-4\right)=15\left(\frac{x}{5}\right)$$
$$15\left(\frac{x}{3}\right)-15(4)=15\left(\frac{x}{5}\right)$$
$$5x-60=3x$$

b.
$$5x-60-3x=3x-3x$$
$$2x-60=0$$
$$2x-60+60=0+60$$
$$2x=60$$
$$\frac{2x}{2}=\frac{60}{2}$$
$$x=30$$

c. Check:
$$\frac{x}{3}-4=\frac{x}{5}$$
$$\frac{30}{3}-4\stackrel{?}{=}\frac{30}{5}$$
$$10-4\stackrel{?}{=}6$$
$$6=6$$

14	The procedure of eliminating the fractions from an equation is called "clearing an equation of fractions." An equation can be cleared of fractions by multiplying both sides by the smallest number that all the denominators in the equation will divide into evenly. This number is called the least common denominator (LCD) for the fractions in the equation.

Q16 Solve the following equations by first clearing them of fractions using the LCD.

a. $\frac{2x}{3}-5=\frac{x}{4}$ **b.** $\frac{2}{3}x-\frac{2}{5}=\frac{2}{5}x$ **c.** $\frac{x}{2}+\frac{5}{2}=\frac{2x}{3}$

• ### • ### • ### • ### • ### • ### • ### •

A16 **a.** The LCD is 12.

$$12\left(\frac{2x}{3}-5\right)=12\left(\frac{x}{4}\right)$$
$$12\left(\frac{2x}{3}\right)-12(5)=12\left(\frac{x}{4}\right)$$
$$8x-60=3x$$
$$8x-60-3x=3x-3x$$
$$5x-60=0$$
$$5x-60+60=0+60$$
$$5x=60$$
$$\frac{5x}{5}=\frac{60}{5}$$
$$x=12$$

b. The LCD is 15.

$$15\left(\frac{2}{3}x-\frac{2}{5}\right)=15\left(\frac{2}{5}x\right)$$
$$15\left(\frac{2}{3}x\right)-15\left(\frac{2}{5}\right)=15\left(\frac{2}{5}x\right)$$
$$10x-6=6x$$
$$10x-6-6x=6x-6x$$
$$4x-6=0$$
$$4x-6+6=0+6$$
$$4x=6$$
$$\frac{4x}{4}=\frac{6}{4}$$
$$x=\frac{3}{2}$$

c. The LCD is 6. The solution is $x=15$.

15 The steps used when solving an equation with fractions are:

Step 1: Determine the LCD.

Step 2: Multiply both sides of the equation by the LCD. This step "clears" the equation of all fractions.

Step 3: Solve the resulting equation.

Step 4: Check the solution.

Q17 Solve each of the following equations using the preceding four-step procedure:

a. $\frac{x}{2} = 7 - \frac{2x}{3}$

b. $\frac{7x}{8} + \frac{5}{6} = \frac{1}{12}$

c. $\frac{9}{10} = \frac{^-3}{4}x + \frac{2}{5}$

d. $\frac{3y}{5} + \frac{5}{2} = \frac{^-y}{5} - \frac{3}{2}$

\# \# \# • \# \# \# • \# \# \# • \# \# \# • \# \# \# • \# \# \# • \# \# \# • \# \# \# • \# \# \#

A17 **a.** The LCD is 6.

$$6\left(\frac{x}{2}\right) = 6\left(7 - \frac{2x}{3}\right)$$
$$6\left(\frac{x}{2}\right) = 6(7) - 6\left(\frac{2x}{3}\right)$$
$$3x = 42 - 4x$$
$$3x + 4x = 42 - 4x + 4x$$
$$7x = 42$$
$$\frac{7x}{7} = \frac{42}{7}$$
$$x = 6$$

Check:
$$\frac{x}{2} = 7 - \frac{2x}{3}$$
$$\frac{6}{2} \overset{?}{=} 7 - \frac{2(6)}{3}$$
$$3 \overset{?}{=} 7 - \frac{12}{3}$$
$$3 = 3$$

b. The LCD is 24. The solution is $x = \frac{^-6}{7}$.

c. The LCD is 20. The solution is $x = \frac{^-2}{3}$.

d. The LCD is 10.

$$10\left(\frac{3y}{5} + \frac{5}{2}\right) = 10\left(\frac{^-y}{5} - \frac{3}{2}\right)$$
$$10\left(\frac{3y}{5}\right) + 10\left(\frac{5}{2}\right) = 10\left(\frac{^-y}{5}\right) - 10\left(\frac{3}{2}\right)$$
$$6y + 25 = {}^-2y - 15$$
$$6y + 25 + 2y = {}^-2y - 15 + 2y$$
$$8y + 25 = {}^-15$$
$$8y + 25 - 25 = {}^-15 - 25$$
$$8y = {}^-40$$
$$\frac{8y}{8} = \frac{^-40}{8}$$
$$y = {}^-5$$

Check:
$$\frac{3y}{5} + \frac{5}{2} = \frac{^-y}{5} - \frac{3}{2}$$
$$\frac{3(^-5)}{5} + \frac{5}{2} \overset{?}{=} \frac{^-(^-5)}{5} - \frac{3}{2}$$
$$^-3 + \frac{5}{2} \overset{?}{=} 1 - \frac{3}{2}$$
$$\frac{^-1}{2} = \frac{^-1}{2}$$

This completes the instruction for this section.

8.3 EXERCISE

1. Solve and check each of the following:

a. $2x - 3 = 9$ **b.** $15 = 4y + 7$ **c.** $6x + 11 = {}^-7$

d. ${}^-5y + 1 = {}^-4$ **e.** $\frac{2}{5}x - 3 = 11$ **f.** $6 = 6 - 7y$

g. $5 - y = 12$ **h.** $6 = 2x + 5$ **i.** $7 - \frac{1}{2}x = 3$

j. ${}^-3y = {}^-5$ **k.** $4 + 3y = 0$ **l.** $2 = {}^-3 + \frac{5}{7}y$

m. $6 = 9 - x$ **n.** $\frac{{}^-3}{4}x - 3 = 5$

2. Solve the following equations by isolating the terms that involve variables on one side and the numbers on the opposite side:

a. $2x + 3 = x + 4$ **b.** $7 + 3x = 4x - 2$

c. $12 + x = 2x + 3$ **d.** $5x + 3 = x - 1$

e. $4y = y + 9$ **f.** $2 - 3y = 2y + 2$

g. $x + 11 = {}^-x + 5$ **h.** ${}^-2y + 3 = {}^-5y + 5$

i. ${}^-7x + 3 - 5x = 0$ **j.** $9 = 3x - 15$

k. $4x - 11 = 2 - x$ **l.** $3 + 4y = {}^-2y + 1$

3. Solve:

a. $2(x - 4) = 10$ **b.** $2x + 3 = 4 + (x - 6)$

c. $6x - (x - 7) = 22$ **d.** $3(x - 5) = {}^-7 + 2(x - 4)$

e. $2(x - 3) - 3(2x + 5) = {}^-25$ **f.** $4 - 2(x + 1) = 2x + 4$

g. $3x + 10 = 7 - 5(2x + 15)$ **h.** $2(3x - 6) + 4 = 22 - 2(x - 1)$

4. Solve each of the following equations:

a. $\frac{4x}{5} - 2 = 10$ **b.** $\frac{2x}{3} + 3 = \frac{4}{5}$

c. $\frac{3y}{4} - \frac{2}{3} = \frac{5}{12}$ **d.** $\frac{x}{2} - 6 = \frac{x}{4}$

e. $\frac{1}{2}x - 7 = \frac{2}{3}x$ **f.** $\frac{3}{2} - \frac{x}{3} = 5 + \frac{x}{6}$

g. $\frac{3x}{4} - \frac{1}{2} = \frac{x}{4} + \frac{11}{2}$ **h.** $\frac{3}{4}x + \frac{5}{3} = \frac{1}{2}x - \frac{1}{3}$

8.3 EXERCISE ANSWERS

1. a. $x = 6$ **b.** $y = 2$ **c.** $x = {}^-3$ **d.** $y = 1$

e. $x = 35$ **f.** $y = 0$ **g.** $y = {}^-7$ **h.** $x = \frac{1}{2}$

i. $x = 8$ **j.** $y = \frac{5}{3}$ **k.** $y = \frac{{}^-4}{3}$ **l.** $y = 7$

m. $x = 3$ **n.** $x = \frac{{}^-32}{3}$

2. a. $x = 1$ **b.** $x = 9$ **c.** $x = 9$ **d.** $x = {}^-1$

e. $y = 3$ **f.** $y = 0$ **g.** $x = {}^-3$ **h.** $y = \frac{2}{3}$

i. $x = \frac{1}{4}$ **j.** $x = 8$ **k.** $x = \frac{13}{5}$ **l.** $y = \frac{{}^-1}{3}$

3. a. $x = 9$ **b.** $x = {}^-5$ **c.** $x = 3$ **d.** $x = 0$

e. $x = 1$ **f.** $x = \frac{^-1}{2}$ **g.** $x = {}^-6$ **h.** $x = 4$

4. a. $x = 15$ **b.** $x = \frac{^-33}{10}$ **c.** $y = \frac{13}{9}$ **d.** $x = 24$

e. $x = {}^-42$ **f.** $x = {}^-7$ **g.** $x = 12$ **h.** $x = {}^-8$

8.4 LITERAL EQUATIONS

1 An equation having more than one letter (variable) is sometimes called a *literal* equation. The literal equations most commonly used by technicians are *formulas*. Formulas are a mathematical way of representing the laws of the physical world. For example, $I = \frac{E}{R}$ is the formula for the amount of current in an electrical circuit. The formula uses variables to represent that the current (I) is equal to the voltage (E) divided by the resistance (R).

Q1 **a.** The equation $I = \frac{E}{R}$ is called a ________ equation.

b. The literal equations most commonly used by technicians are called ________.

• ### • ### • ### • ### • ### • ### • ### • ### •

A1 **a.** literal **b.** formulas

2 Occasionally we wish to solve a literal equation for one variable in terms of the others. This formula rearrangement is done in order to make formula evaluations easier, or to rewrite a given formula to emphasize a different variable.

Example: Solve the literal equation $A = lw$ for l.

Solution: The equation can be solved using the *division principle of equality*.

$$A = lw$$

$$\frac{A}{w} = \frac{lw}{w} \quad \text{(divide both sides by } w\text{)}$$

$$\frac{A}{w} = l$$

Thus, $l = \frac{A}{w}$.

Q2 Solve the equation $A = lw$ for w.

• ### • ### • ### • ### • ### • ### • ### • ### •

A2 $w=\frac{A}{l}$: $A=lw$

$$\frac{A}{l}=\frac{lw}{l} \quad \text{(divide both sides by } l\text{)}$$

$$\frac{A}{l}=w$$

Q3 The distance an object travels equals the average speed times the time traveled. This is expressed in the formula $d=rt$. Solve the equation $d=rt$ for r.

• ### • ### • ### • ### • ### • ### • ### • ### •

A3 $r=\frac{d}{t}$: $d=rt$

$$\frac{d}{t}=\frac{rt}{t} \quad \text{(divide both sides by } t\text{)}$$

$$\frac{d}{t}=r$$

---

3 Sometimes subscripts (smaller numbers or letters written to the right and below other letters) are used instead of using the same capital and small letters to distinguish *two different* variables. For example, in the formula $R_t=R_1+R_2$* the subscripts are t, 1, and 2.

The formula represents the total resistance in a series circuit, R_t as the sum of the other two resistances. The formula $R_t=R_1+R_2$ can be solved for R_2 using the *subtraction principle of equality*.

$$R_t=R_1+R_2$$

$$R_t-R_1=R_1+R_2-R_1 \quad \text{(subtract } R_1 \text{ from both sides)}$$

$$R_t-R_1=R_2$$

Thus, $R_2=R_t-R_1$.

*Read R_t, R_1, R_2 as R-sub t, R-sub 1, and R-sub 2.

Q4 Solve the equation $R_t=R_1+R_2$ for R_1.

• ### • ### • ### • ### • ### • ### • ### • ### •

A4 $R_1=R_t-R_2$

Q5 In a parallel circuit the current supplied to the circuit is equal to the sum of the currents flowing through all parts. This is represented by the formula $I=I_1+I_2+I_3$.

a. The subscripts in the formula are ________.

b. Solve the formula for I_2.

• ### • ### • ### • ### • ### • ### • ### • ### •

A5 **a.** 1, 2, 3 **b.** $I_2=I-I_1-I_3$ or $I_2=I-I_3-I_1$

4 The formula $X = X_L - X_C$ can be solved for X_L using the *addition principle of equality.*

$$X = X_L - X_C$$
$$X + X_C = X_L - X_C + X_C \quad \text{(add } X_C \text{ to both sides)}$$
$$X + X_C = X_L$$

Thus, $X_L = X + X_C$

Q6 Solve $I = E_1 - E_0$ for E_1.

• ### • ### • ### • ### • ### • ### • ### •

A6 $E_1 = I + E_0$:

$$I = E_1 - E_0$$
$$I + E_0 = E_1 - E_0 + E_0 \quad \text{(add } E_0 \text{ to both sides)}$$
$$I + E_0 = E_1$$

5 It is sometimes necessary to apply two principles of equality in order to solve for a given variable.

Example: Solve the equation $e = E + Ir$ for I.

Solution: In the equation $e = E + Ir$, the term containing I will be isolated on the right-hand side of the equation if E is subtracted from both sides.

$$e = E + Ir$$
$$e - E = Ir - E \quad \text{(subtract } E \text{ from both sides)}$$
$$e - E = Ir$$

The variable I can now be isolated by dividing both sides of the equation by r.

$$e - E = Ir$$
$$\frac{e - E}{r} = \frac{Ir}{r} \quad \text{(divide both sides by } r\text{)}$$
$$\frac{e - E}{r} = I$$

Thus, $I = \dfrac{e - E}{r}$.

Q7 Solve the equation $e = E - Ir$ for E.

• ### • ### • ### • ### • ### • ### • ### •

A7 $E = e + Ir$:

$$e = E - Ir$$
$$e + Ir = E - Ir + Ir \quad \text{(add } Ir \text{ to both sides)}$$
$$e + Ir = E$$

Q8 The formula for finding the perimeter of a rectangle with length l and width w is $p = 2l + 2w$. Solve the formula for w.

• # # # • # # # • # # # • # # # • # # # • # # # • # # # • # #

A8 $w = \frac{p-2l}{2}$:

$$p = 2l + 2w$$
$$p - 2l = 2l + 2w - 2l \quad \text{(subtract } 2l \text{ from both sides)}$$
$$p - 2l = 2w$$
$$\frac{p-2l}{2} = \frac{2w}{2} \quad \text{(divide both sides by 2)}$$
$$\frac{p-2l}{2} = w$$

Q9 The velocity of an object is given by the equation $V = V_0 + at$ where V is velocity, V_0 is initial velocity, a is acceleration, and t is time. Solve the equation for t.

• # # # • # # # • # # # • # # # • # # # • # # # • # # # • # #

A9 $t = \frac{V - V_0}{a}$:

$$V = V_0 + at$$
$$V - V_0 = V_0 + at - V_0 \quad \text{(subtract } V_0 \text{ from both sides)}$$
$$V - V_0 = at$$
$$\frac{V - V_0}{a} = \frac{at}{a} \quad \text{(divide both sides by } a\text{)}$$
$$\frac{V - V_0}{a} = t$$

---

6 The formula used to represent Boyle's law for gases is $\frac{P_1}{P_2} = \frac{V_2}{V_1}$ where P is pressure and V is volume. The subscripts 1 and 2 are used to distinguish between the initial and final pressure or volume. Since the equation is fractional in nature, it is necessary to first clear the equation of fractions using the *multiplication principle of equality*.

Example: Solve the equation $\frac{P_1}{P_2} = \frac{V_2}{V_1}$ for P_2.

Solution: The LCD is P_2V_1.

$$P_2V_1\left(\frac{P_1}{P_2}\right) = P_2V_1\left(\frac{V_2}{V_1}\right) \quad \text{(multiply both sides by } P_2V_1\text{)}$$
$$V_1P_1 = P_2V_2$$

With the fractions eliminated, the solution of the equation can now be completed using the division principle of equality.

$$\frac{V_1P_1}{V_2}=\frac{P_2V_2}{V_2} \quad \text{(divide both sides by } V_2\text{)}$$

$$\frac{V_1P_1}{V_2}=P_2$$

Thus, $P_2=\dfrac{V_1P_1}{V_2}$

Q10 Solve the equation $\dfrac{P_1}{P_2}=\dfrac{V_2}{V_1}$ for V_1.

\# \# \# • \# \# \# • \# \# \# • \# \# \# • \# \# \# • \# \# \# • \# \# \# • \# \# \# • \# \# \#

A10 $V_1=\dfrac{P_2V_2}{P_1}$: The LCD is P_2V_1.

$$P_2V_1\left(\frac{P_1}{P_2}\right)=P_2V_1\left(\frac{V_2}{V_1}\right)$$

$$V_1P_1=P_2V_2$$

$$\frac{V_1P_1}{P_1}=\frac{P_2V_2}{P_1}$$

$$V_1=\frac{P_2V_2}{P_1}$$

Q11 Solve the equation $\dfrac{P_1}{P_2}=\dfrac{V_2}{V_1}$ for P_1.

\# \# \# • \# \# \# • \# \# \# • \# \# \# • \# \# \# • \# \# \# • \# \# \# • \# \# \# • \# \# \#

A11 $P_1=\dfrac{P_2V_2}{V_1}$:

$$\frac{P_1}{P_2}=\frac{V_2}{V_1}$$

$$P_2V_1\left(\frac{P_1}{P_2}\right)=P_2V_1\left(\frac{V_2}{V_1}\right)$$

$$V_1P_1=P_2V_2$$

$$\frac{V_1P_1}{V_1}=\frac{P_2V_2}{V_1}$$

$$P_1=\frac{P_2V_2}{V_1}$$

Q12 $P_c = \frac{\pi D}{N}$ is the formula for finding the circular pitch (distance between centers of adjoining teeth on a gear). P_c is the circular pitch, D is the diameter of the pitch circle of the gear, and N is the number of teeth in the gear. Solve the formula for D.

\# \# \# • \# \# \# • \# \# \# • \# \# \# • \# \# \# • \# \# \# • \# \# \# • \# \# \# • \# \# \#

A12 $D = \frac{NP_c}{\pi}$:

$$P_c = \frac{\pi D}{N}$$
$$N(P_c) = N\left(\frac{\pi D}{N}\right)$$
$$NP_c = \pi D$$
$$\frac{NP_c}{\pi} = \frac{\pi D}{\pi}$$
$$\frac{NP_c}{\pi} = D$$

---

7 When variables are used in formulas to represent physical quantities, both capital and small letters may be used. The use of capital or small letters is usually arbitrary, but once a formula has gained acceptance among scientists or technicians in a certain form it is necessary to follow it exactly. The formula $P = \frac{Fs}{t}$, for example, has two small letters and two capital letters.

Q13 Newton's law of gravitational attraction between two bodies is given as $F = \frac{gMm}{r^2}$. It has been rewritten below in a number of ways. Which one is correct? ______

a. $f = \frac{GMM}{r^2}$ **b.** $F = \frac{gMm}{R^2}$ **c.** $F = \frac{gMm}{r^2}$

\# \# \# • \# \# \# • \# \# \# • \# \# \# • \# \# \# • \# \# \# • \# \# \# • \# \# \# • \# \# \#

A13 **c**

Q14 Solve $F = \frac{gMm}{r^2}$ for m.

\# \# \# • \# \# \# • \# \# \# • \# \# \# • \# \# \# • \# \# \# • \# \# \# • \# \# \# • \# \# \#

A14 $m = \frac{r^2F}{gM}$:

$$F = \frac{gMm}{r^2}$$
$$r^2(F) = r^2\left(\frac{gMm}{r^2}\right)$$
$$r^2F = gMm$$
$$\frac{r^2F}{gM} = \frac{gMm}{gM}$$
$$\frac{r^2F}{gM} = m$$

Q15 Solve $P = \frac{Fs}{t}$ for F.

• ### • ### • ### • ### • ### • ### • ### •

A15 $F = \frac{tP}{s}$:

$$P = \frac{Fs}{t}$$
$$t(P) = t\left(\frac{Fs}{t}\right)$$
$$tP = Fs$$
$$\frac{tP}{s} = \frac{Fs}{s}$$
$$\frac{tP}{s} = F$$

This completes the instruction for this section.

8.4 EXERCISE

1. Use the addition, subtraction, multiplication, or division principle of equality to solve each of the following literal equations for the variable indicated:

a. $at = V_2 - V_1$ for V_2
b. $d = rt$, for t
c. $E = IR$, for R
d. $s = vt$, for v
e. $MS = R - QS$, for R
f. $D = A + T$, for T
g. $at = V_2 - V_1$, for V_1
h. $I_E = I_C + I_B$, for I_C
i. $I = \frac{E}{R}$, for E
j. $l = \frac{A}{w}$, for A

2. Solve each literal equation for the variable indicated:

a. $H = E + PV$, for P
b. $e = E - Ir$, for I
c. $MS = R - QS$, for Q
d. $I = \frac{E}{R}$, for R
e. $\frac{P_1}{T_1} = \frac{P_2}{T_2}$, for P_1
f. $W = \frac{EI}{S}$, for I
g. $P = \frac{Fs}{t}$, for s
h. $\frac{F_1}{F_2} = \frac{A_1}{A_2}$, for A_2

8.4 EXERCISE ANSWERS

1. a. $V_2 = at + V_1$

b. $t = \dfrac{d}{r}$

c. $R = \dfrac{E}{I}$

d. $v = \dfrac{s}{t}$

e. $R = MS + QS$

f. $T = D - A$

g. $V_1 = V_2 - at$

h. $I_C = I_E - I_B$

i. $E = IR$

j. $A = lw$

2. a. $P = \dfrac{H - E}{V}$

b. $I = \dfrac{e - E}{r}$

c. $Q = \dfrac{MS - R}{^{-}S}$ or $\dfrac{R - MS}{S}$

d. $R = \dfrac{E}{I}$

e. $P_1 = \dfrac{T_1 P_2}{T_2}$

f. $I = \dfrac{WS}{E}$

g. $s = \dfrac{Pt}{F}$

h. $A_2 = \dfrac{F_2 A_1}{F_1}$

CHAPTER 8 SAMPLE TEST

At the completion of Chapter 8 you should be able to work the following problems.

8.1 SIMPLIFYING ALGEBRAIC EXPRESSIONS

1. Simplify by combining like terms:

a. $7x - 2y - 11x - 5$

b. $7 + 6r - 5s + 5s$

c. $2x + {^{-}11}$

d. $m - 5 + 7m + {^{-}4} - 9m$

2. Simplify using the associative, commutative, and distributive properties.

a. $2\left(\dfrac{3}{4}x\right)$

b. $\dfrac{^{-}5}{3}\left(\dfrac{^{-}3}{5}x\right)$

c. $2(x - 3)$

d. $^{-}4(2x - 8)$

e. $^{-}({^{-}t} + 2)$

f. $(4 - 2x)5$

3. Simplify:

a. $(5x + 6) - 4$

b. $^{-}3(x + 7) - 10$

c. $8 - (z - 5)$

d. $4(x - 5) + 3(x + 3)$

e. $(y - 3) - (3y + 8)$

f. $^{-}7(x + 9) - 5(2x + 1)$

8.2 EQUATIONS, FUNDAMENTAL PRINCIPLES OF EQUALITY

4. Solve using the addition and subtraction principles of equality.

a. $x + 7 = {^{-}11}$

b. $x - 7 = {^{-}7}$

c. $12 = x - 3$

d. $5 - 7 = x + 9$

5. Solve using the multiplication and division principles of equality.

a. $5y = {^{-}20}$

b. $\dfrac{2}{3}x = 12$

c. $\dfrac{4}{5} = \dfrac{^{-}3}{5}x$

d. $\dfrac{5}{9}y = 0$

8.3 SOLVING EQUATIONS BY THE USE OF TWO OR MORE STEPS

6. Solve each of the following equations:

a. $3x - 5 = 7$
b. $8 - y = 3$
c. $7 + 5y = 0$
d. $3x + 5 = x - 7$
e. $4 - 7y = 9y + 4$
f. $6x - (x - 7) = 22$
g. $\frac{1}{2}x - 7 = \frac{2}{3}x$
h. $\frac{5y}{7} - \frac{3}{2} = \frac{^{-}2y}{7} - \frac{5}{2}$

8.4 LITERAL EQUATIONS

7. Solve each literal equation for the variable indicated:

a. $E = IR$, for I
b. $X = X_L - X_C$, for X_L
c. $H = E + PV$, for E
d. $W = \frac{EI}{S}$, for S
e. $MS = R - QS$, for Q
f. $P = \frac{Fs}{t}$, for F
g. $\frac{P_1}{T_1} = \frac{P_2}{T_2}$, for T_2
h. $\frac{F_1}{F_2} = \frac{A_1}{A_2}$, for A_1

CHAPTER 8 SAMPLE TEST ANSWERS

1. a. $^{-}4x - 2y - 5$ **b.** $6r + 7$ **c.** $2x - 11$ **d.** $^{-}m - 9$

2. a. $\frac{3}{2}x$ **b.** x **c.** $2x - 6$ **d.** $^{-}8x + 32$
e. $t - 2$ **f.** $20 - 10x$

3. a. $5x + 2$ **b.** $^{-}3x - 31$ **c.** $13 - z$ **d.** $7x - 11$
e. $^{-}2y - 11$ **f.** $^{-}17x - 68$

4. a. $x = {}^{-}18$ **b.** $x = 0$ **c.** $x = 15$ **d.** $x = {}^{-}11$

5. a. $y = {}^{-}4$ **b.** $x = 18$ **c.** $x = \frac{^{-}4}{3}$ **d.** $y = 0$

6. a. $x = 4$ **b.** $y = 5$ **c.** $y = \frac{^{-}7}{5}$ **d.** $x = {}^{-}6$
e. $y = 0$ **f.** $x = 3$ **g.** $x = {}^{-}42$ **h.** $y = {}^{-}1$

7. a. $I = \frac{E}{R}$ **b.** $X_L = X + X_C$
c. $E = H - PV$ **d.** $S = \frac{EI}{W}$
e. $Q = \frac{MS - R}{^{-}S}$ or $\frac{R - MS}{S}$ **f.** $F = \frac{Pt}{s}$
g. $T_2 = \frac{T_1 P_2}{P_1}$ **h.** $A_1 = \frac{F_1 A_2}{F_2}$

CHAPTER 9

RATIO, PROPORTION, AND FORMULAS

9.1 INTRODUCTION TO RATIO AND PROPORTION

1 A fraction can be used to compare two quantities or numbers. Suppose that Sally has \$13 and Linda has \$39. Sally then has $\frac{13}{39}$ or $\frac{1}{3}$ as much money as Linda. When a fraction is used in this way to compare two numbers, the fraction is called a ratio. A *ratio* is the quotient of two quantities or numbers. The ratio $\frac{1}{3}$ is read "one to three" and can be written using a colon as 1 : 3.

Q1 **a.** 4 of 7 square regions are shaded. Write the ratio of the shaded regions to the total region. ______

b. Of the 12 rectangular regions, 5 are shaded. Write the ratio of shaded regions to the total region.

______

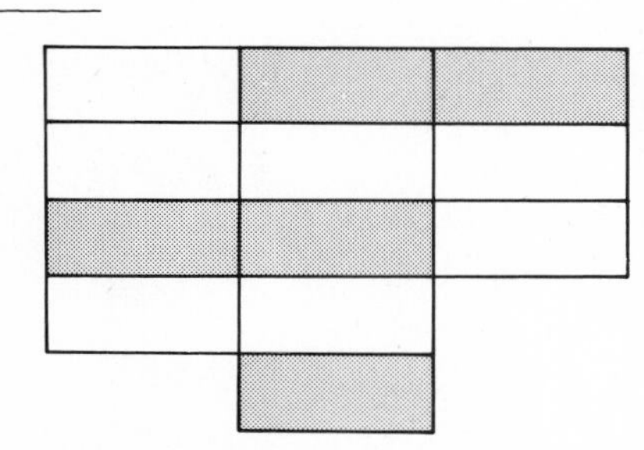

c. 6 of the 11 hexagonal regions are shaded. Write the indicated ratio. ______

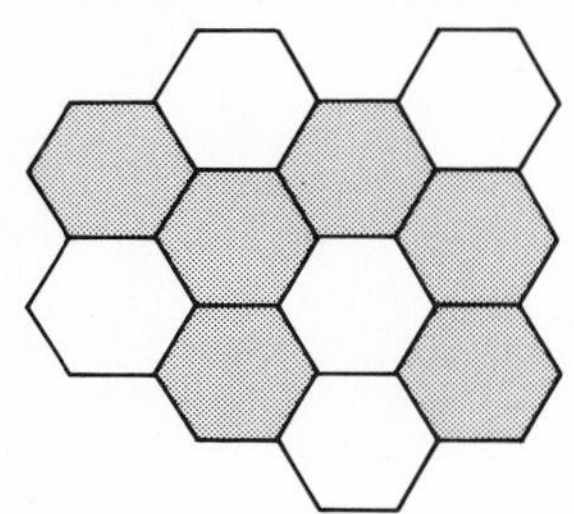

d. 5 of the 11 hexagonal regions are *not* shaded. What ratio represents this statement? ______

e. ______ of 6 circular regions are shaded.

f. Write the ratio. ______

g. 4 of ______ squares are circled.

h. Write the ratio. ______

i. ______ of ______ triangular regions are shaded.

j. Write the indicated ratio. ______

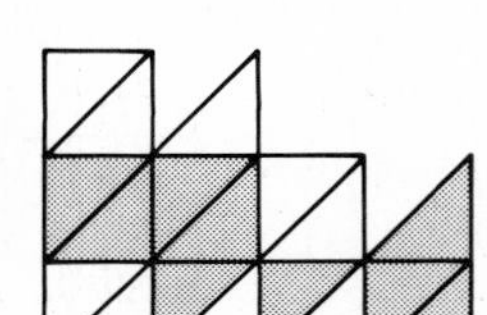

• ### • ### • ### • ### • ### • ### • ### •

A1 **a.** $\frac{4}{7}$ or 4 : 7 **b.** $\frac{5}{12}$ or 5 : 12 **c.** $\frac{6}{11}$ or 6 : 11 **d.** $\frac{5}{11}$ or 5 : 11

e. 5 **f.** $\frac{5}{6}$ or 5 : 6 **g.** 9 **h.** $\frac{4}{9}$ or 4 : 9

i. 9, 18 **j.** $\frac{9}{18}$ or 9 : 18 $\left(\frac{1}{2} \text{ or } 1:2\right)$

Q2 The comparison of one quantity to another by division is called a ________.

• ### • ### • ### • ### • ### • ### • ### •

A2 ratio

Q3 **a.** What is the ratio of 24 years to 9 years? ______

b. What is the ratio of 9 years to 24 years? ______

• ### • ### • ### • ### • ### • ### • ### •

A3 **a.** $\frac{8}{3}$: $\frac{24}{9} = \frac{8}{3}$ **b.** $\frac{3}{8}$

Q4 Find the following ratios as fractions reduced to the lowest terms:

a. 25 to 125 **b.** 25 to 10 **c.** 25 to 5 **d.** 25 to $\frac{1}{5}$

e. $\frac{1}{4}$ to $\frac{1}{2}$ **f.** $\frac{1}{4}$ to 2

• ### • ### • ### • ### • ### • ### • ### •

A4 **a.** $\frac{1}{5}$ **b.** $\frac{5}{2}$ **c.** $\frac{5}{1}$ **d.** $\frac{125}{1}$

e. $\frac{1}{2}$ **f.** $\frac{1}{8}$

Q5 The ratio $\frac{1}{2}$ is read "________."

• ### • ### • ### • ### • ### • ### • ### •

A5 one to two

2 Writing the ratio of 2 feet to 15 inches as $\frac{2}{15}$ would be incorrect, because 2 feet is not $\frac{2}{15}$ of 15 inches. two feet is actually 24 inches, so the ratio is really $\frac{24}{15}$ or $\frac{8}{5}$.

When a comparison between two quantities is desired, the quantities should be expressed in the same unit of measurement. However, the ratio is an abstract number (written without a unit of measurement).

Q6 Determine the ratio of $5 to 40 cents.

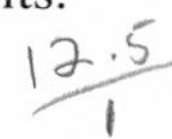

\# # # • # # # • # # # • # # # • # # # • # # # • # # # • # # # • # # #

A6 $\frac{25}{2}$: $5 = 500 cents, hence, $\frac{500}{40}$

Q7 Express as a ratio:

a. 15 inches to 2 feet. 3/8

b. 10 ounces to 1 pound. 5/8

c. 45 minutes to 1 hour. 3/4

d. a nickel to a dime. 1/2

\# # # • # # # • # # # • # # # • # # # • # # # • # # # • # # # • # # #

A7 **a.** $\frac{5}{8}$: 2 feet = 24 inches

b. $\frac{5}{8}$: 1 pound = 16 ounces

c. $\frac{3}{4}$: 1 hour = 60 minutes

d. $\frac{1}{2}$

Q8 Two gears have 45 and 15 teeth, respectively. What is the ratio of the numbers of teeth?

\# # # • # # # • # # # • # # # • # # # • # # # • # # # • # # # • # # #

A8 $\frac{3}{1}$: $\frac{45 \text{ teeth}}{15 \text{ teeth}}$

3 The relationship between gears connected together and their revolutions per minute (rpm) can be described using ratios. Consider the following illustration which shows a simplified version of the transmission of a car. The power would be applied by the engine to the clutch shaft. Because of the relative sizes of the gears, the clutch shaft gear is turning faster; that is, the clutch gear is turning more rpm than the drive-shaft gear. The ratio of the rpms of different size gears can be determine

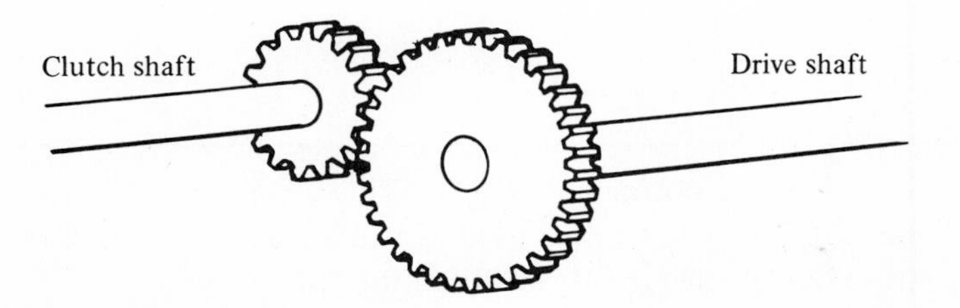

by the number of teeth in each gear. For the illustration,

$$\text{gear ratio} = \frac{\text{drive teeth}}{\text{clutch teeth}}$$

Example: Suppose the gear on the clutch shaft contained 10 teeth and the gear on the drive shaft contained 26 teeth, as would be the case in a typical low gear of a transmission. What is the gear ratio?

Solution: $\text{gear ratio} = \frac{\text{drive teeth}}{\text{clutch teeth}} = \frac{26}{10} = \frac{13}{5}$

The ratio $\frac{13}{5}$ means that for every 5 times the drive-shaft gear turns, the clutch-shaft gear turns 13 times.

Q9 If in second gear a clutch shaft has 5 teeth and the drive shaft has 8 teeth, what is the gear ratio?

• ### • ### • ### • ### • ### • ### • ### •

A9 $\frac{8}{5}$

4 In shop manuals, most gear ratios are given as some number compared (divided by) to 1. This is accomplished by dividing numerator and denominator of the ratio by the denominator.

Example: Express the ratio $\frac{13}{5}$ as some number compared to 1.

Solution: $\frac{13}{5} = \frac{13 \div 5}{5 \div 5} = \frac{2.6}{1}$

The ratio $\frac{2.6}{1}$ means that for each rpm of the drive shaft, the clutch shaft has 2.6 rpm.

Q10 Express the following ratios as some number compared to 1:

a. $\frac{3}{2}$ **b.** $\frac{8}{5}$ **c.** $\frac{5}{8}$ **d.** $\frac{14}{20}$

• ### • ### • ### • ### • ### • ### • ### •

A10 **a.** $\frac{1.5}{1}$ **b.** $\frac{1.6}{1}$ **c.** $\frac{0.625}{1}$ **d.** $\frac{0.7}{1}$

Q11 In second gear (Q9 above) for each rpm of the drive shaft, the clutch shaft has ______ rpm.

• ### • ### • ### • ### • ### • ### • ### •

A11 1.6

5 As you know from driving a car, you have more power in low gear. This means that the drive shaft produces more power than the clutch shaft. This continues to hold true in second gear. Of course, you are sacrificing speed since the drive shaft is turning slower than the clutch shaft.

Most cars have a gear ratio of 1 to 1 in high gear. If the clutch-shaft gear has 12 teeth on it, then the drive-shaft gear would have 12 teeth. The clutch shaft and drive shaft are now turning at the same rate.

Q12 Some cars have an overdrive gear. Find the gear ratio if the clutch-shaft gear has 20 teeth and the drive-shaft gear has 14 teeth.

• ### • ### • ### • ### • ### • ### • ### •

A12 $\frac{0.7}{1}$: gear ratio $= \frac{14}{20}$

As you can tell by the gear ratio, the drive shaft is turning faster than the clutch shaft. There is little power available so the overdrive is most useful on flat stretches and down hills.

6 Many ratios are often expressed using the word "per"—miles per gallon, cents per ounce, grams per cm^3, etc. To compute the value of the ratio when the word per is used, it may be helpful to interpret per as "divided by." For example, to find miles per gallon, divide the miles driven by gallons of gasoline consumed.

Example 1: Determine the miles per gallon if you traveled 140 miles and used 10 gallons of gasoline.

Solution: Miles per gallon means miles divided by gallons. $\frac{140 \text{ miles}}{10 \text{ gallons}} = 14$ miles per gallon.

Example 2: A can of tuna fish costs $0.45 for 6.5 ounces. Find the cost per ounce.

Solution: Cost per ounce means cost divided by ounces. $\frac{\$0.45}{6.5} = \$0.07$ (nearest cent).

It is important for all consumers to be able to determine the efficiency of their automobile, by figuring miles per gallon, and to do comparative shopping, by figuring cost per weight.

Q13 Determine the miles per gallon if you travel 220 miles and used 15.8 gallons of gasoline (nearest tenth).

• ### • ### • ### • ### • ### • ### • ### •

A13 13.9 miles per gallon: 220 miles ÷ 15.8 gallons

Q14 To determine cost per ounce you would divide ________ by ________.

• ### • ### • ### • ### • ### • ### • ### •

A14 cost by ounces

Q15 A can of corn weighing 12 ounces sells for $0.35. Determine the cost per ounce (nearest cent).

• ### • ### • ### • ### • ### • ### • ### •

A15 $0.03: $0.35 ÷ 12 ounces

Q16 Two dozen gaskets cost $59.04. Find the cost per gasket.

• # # # • # # # • # # # • # # # • # # # • # # # • # # # • # #

A16 $2.46: $\$59.04 \div 24$

Q17 Eight hundred forty-nine gallons of industrial chemical sold for $441.48. What was the price per gallon?

• # # # • # # # • # # # • # # # • # # # • # # # • # # # • # #

A17 $0.52: $\$441.48 \div 849$

Q18 A certain beverage costs $3.19 per 12-pack. If each bottle contains 12 ounces, find the cost of the beverage per ounce.

• # # # • # # # • # # # • # # # • # # # • # # # • # # # • # #

A18 $0.02: $\$3.19 \div 144$ ounces

7 A statement of equality between two ratios (or fractions) is called a *proportion*. A proportion is actually a special type of equation. The statement $\frac{1}{3} = \frac{13}{39}$ is a proportion and could be written

$1:3 = 13:39$

A proportion consists of four terms labeled as follows:

1st term : 2nd term = 3rd term : 4th term

In the proportion $\frac{1}{3} = \frac{13}{39}$, the first term is 1, the second term is 3, the third term is 13, and the fourth term is 39.

Q19 **a.** A statement of equality between two ratios is called a ____________.

b. The four parts of a proportion are called ____________.

c. In the proportion $\frac{2}{3} = \frac{10}{15}$, what is the third term? ____________

d. In the proportion $1:8 = 2:16$, what is the fourth term? ____________

• # # # • # # # • # # # • # # # • # # # • # # # • # # # • # #

A19 **a.** proportion **b.** terms **c.** 10 **d.** 16

Q20 Use colons to write the proportion $\frac{7}{2} = \frac{21}{6}$.

• # # # • # # # • # # # • # # # • # # # • # # # • # # # • # #

A20 $7:2 = 21:6$

Q21 Write the proportion $25:5=5:1$ in fraction form.

\# # # • # # # • # # # • # # # • # # # • # # # • # # # • # # # • # # #

A21 $\frac{25}{5}=\frac{5}{1}$

8 In a proportion, the first and fourth terms are called the *extremes* and the second and third terms are called the *means.* In the proportion

$$\frac{2}{5}=\frac{14}{35}$$

2 and 35 are the extremes and 5 and 14 are the means.

$2:5=14:35$

(2 and 35: extremes; 5 and 14: means)

An important property of proportions is that *the product of the means is equal to the product of the extremes.* In the proportion $\frac{2}{5}=\frac{14}{35}$ the product of the means is $5 \cdot 14=70$ and the product of the extremes is $2 \cdot 35=70$.

Q22 **a.** In a proportion the first and fourth terms are called the ____________.

b. In a proportion the product of the ____________ is equal to the product of the ____________.

c. Name the means of $\frac{7}{21}=\frac{1}{3}$. __________

d. What is the product of the extremes in $\frac{2}{5}=\frac{8}{20}$? __________

\# # # • # # # • # # # • # # # • # # # • # # # • # # # • # # # • # # #

A22 **a.** extremes **b.** means, extremes **c.** 21 and 1 **d.** 40: $2 \cdot 20$

9 The truth of a proportion can be determined by using the property that the product of the means is equal to the product of the extremes. The statement of equality between the ratios will be true only if the above property is satisfied. The proportion $\frac{7}{2}=\frac{14}{4}$ is true because

$$\underbrace{2 \cdot 14}_{\text{product of means}}=\underbrace{7 \cdot 4}_{\text{product of extremes}}$$

However, the proportion $\frac{6}{7}=\frac{5}{8}$ is false because

$7 \cdot 5 \neq 6 \cdot 8$

Q23 **a.** In the proportion $\frac{7}{8}=\frac{8}{9}$ the product of the means is __________ and

b. the product of the extremes is ________.

c. Is the proportion true or false? ________.

• ### • ### • ### • ### • ### • ### • ### •

A23 **a.** $8 \cdot 8 = 64$ **b.** $7 \cdot 9 = 63$ **c.** false

---

Q24 Express $\frac{7}{3} = \frac{21}{9}$ as a product of its means and extremes.

• ### • ### • ### • ### • ### • ### • ### •

A24 $3 \cdot 21 = 7 \cdot 9$

---

10 The property that the product of the means is equal to the product of the extremes, in a proportion, can be developed and generalized using the principles of equality. That is, for the proportion

$$\frac{a}{b} = \frac{c}{d}$$

the means are b and c and the extremes are a and d. The proportion can be rewritten as follows:

$$\frac{a}{b} = \frac{c}{d}$$

$$bd\left(\frac{a}{b}\right) = bd\left(\frac{c}{d}\right) \quad \text{(multiplying both sides by the LCD, } bd\text{)}$$

Hence

$$ad = bc$$

which states that the product of the extremes is equal to the product of the means. The last equation could also be written

$$bc = ad$$

Q25 Is $\frac{7}{3} = \frac{21}{9}$ true or false? ________

• ### • ### • ### • ### • ### • ### • ### •

A25 true: $3 \cdot 21 = 7 \cdot 9$
$63 = 63$

Q26 Is each of the following a true proportion?

a. $\frac{7}{14} = \frac{21}{42}$ ________

b. $\frac{4}{6} = \frac{20}{24}$ ________

c. $3:8 = 15:40$ ________

d. $6:9 = 48:72$ ________

e. $\frac{15}{20} = \frac{5}{4}$ ________

f. $\frac{9}{15} = \frac{72}{120}$ ________

• ### • ### • ### • ### • ### • ### • ### •

A26 **a.** yes: $14 \cdot 21 = 7 \cdot 42$

b. no: $6 \cdot 20 \neq 4 \cdot 24$

c. yes: $8 \cdot 15 = 3 \cdot 40$

d. yes: $9 \cdot 48 = 6 \cdot 72$

e. no: $20 \cdot 5 \neq 15 \cdot 4$

f. yes: $15 \cdot 72 = 9 \cdot 120$

Q27 Write the proportion $\frac{m}{n} = \frac{x}{y}$ so that the product of the means is equal to the product of the extremes.

• # # # • # # # • # # # • # # # • # # # • # # # • # # # • # # # • # #

A27 $nx = my$

This completes the instruction for this section.

9.1 EXERCISE

1. The comparison of one quantity to another by division is called a ________.
2. Find the following ratios in fraction form:

   a. 16 to 8 **b.** $\frac{9}{27}$ **c.** 25 to 10 **d.** 6 to 3

   e. $\frac{1}{4}$ to $\frac{1}{12}$ **f.** 0.2 to 2 **g.** 16 to 64 **h.** 9 to $\frac{1}{3}$

   i. 25 to 5 **j.** 6 to 2 **k.** $\frac{1}{4}$ to 2 **l.** 0.2 to $\frac{1}{10}$
3. Find the following ratios in fraction form:

   a. a quart to a pint **b.** a quart to a gallon

   c. a yard to two inches **d.** an inch to a foot

   e. five minutes to an hour **f.** a day to an hour
4. A transmission contains a drive gear with 20 teeth and a clutch gear with 12 teeth in second gear. What is the gear ratio in second gear (round off the numerator to the nearest hundredth)?
5. The differential assembly contains a pinion gear which is on the end of the drive shaft and a ring gear which is connected to the axle. If the gear ratio $= \frac{\text{ring teeth}}{\text{pinion teeth}}$, find the gear ratio if the ring gear has 36 teeth and the pinion gear has 9 teeth.
6. A pinion gear has 9 teeth and a ring gear, in the same assembly, has 34 teeth. What is the gear ratio (round off the numerator to the nearest hundredth)?
7. If a ring gear contains 38 teeth and the pinion gear 9 teeth, what is the rear axle gear ratio?
8. A small pulley turns 380 rpm and is belted to a larger one making 75 rpm. What is the ratio of their speeds?
9. If you traveled 320 miles and used 22.9 gallons of gasoline, how many miles per gallon is your car getting (nearest tenth)?
10. A can of washers weighing 12 ounces costs $0.44. What is the cost per ounce?
11. Five cubic centimetres of chemical weigh 85 grams. Find the weight per cubic centimetre.
12. Which is the better buy, Giant Economy Size $0.98 for 36 ounces or the Jumbo $1.44 for 48 ounces?
13. The statement of equality between two ratios is called a ________.
14. Name the means in the proportion $\frac{6}{13} = \frac{12}{26}$.
15. Name the extremes in the proportion $1:5 = 7:35$.
16. In a proportion, the product of the ________ is equal to the product of the ________.
17. Which of the following represent true proportions?

    a. $\frac{7}{2} = \frac{35}{10}$ **b.** $\frac{2}{7} = \frac{3}{11}$

    c. $6:2 = 23:8$ **d.** $9:11 = 63:77$

    e. $\frac{7}{15} = \frac{8}{14}$ **f.** $\frac{3}{18} = \frac{1}{6}$

9.1 EXERCISE ANSWERS

1. ratio

2. **a.** $\frac{2}{1}$ **b.** $\frac{1}{3}$ **c.** $\frac{5}{2}$ **d.** $\frac{2}{1}$ **e.** $\frac{3}{1}$ **f.** $\frac{1}{10}$

g. $\frac{1}{4}$ **h.** $\frac{27}{1}$ **i.** $\frac{5}{1}$ **j.** $\frac{3}{1}$ **k.** $\frac{1}{8}$ **l.** $\frac{2}{1}$

3. **a.** $\frac{2}{1}$ **b.** $\frac{1}{4}$ **c.** $\frac{18}{1}$ **d.** $\frac{1}{12}$ **e.** $\frac{1}{12}$ **f.** $\frac{24}{1}$

4. $\frac{1.67}{1}$

5. $\frac{4}{1}$

6. $\frac{3.78}{1}$

7. $\frac{4.22}{1}$

8. $\frac{76}{15}$

9. 14 miles per gallon

10. $0.04

11. 17 grams per cubic centimetre

12. Giant Economy Size

13. proportion

14. 13 and 12

15. 1 and 35

16. means, extremes

17. a, d, and f

9.2 SOLUTION OF A PROPORTION

1 Since a proportion is a special type of equation, it can be solved using the principles established for solving equations. For example, the proportion

$$\frac{n}{3}=\frac{4}{21}$$

may be solved as follows:

$$\frac{n}{3}=\frac{4}{21}$$

$$3\left(\frac{n}{3}\right)=3\left(\frac{4}{21}\right) \quad \text{(multiplying both sides by 3)}$$

$$n=\frac{4}{7}$$

Q1 Solve the proportion $\frac{n}{14}=\frac{51}{7}$.

\# # # • # # # • # # # • # # # • # # # • # # # • # # # • # # # • # # #

A1 102: $14\left(\frac{n}{14}\right)=14\left(\frac{51}{7}\right)$

$$n=2\cdot 51$$

Q2 Solve the following proportions:

a. $\frac{4}{7}=\frac{n}{28}$ **b.** $\frac{5}{14}=\frac{n}{42}$

\# # # • # # # • # # # • # # # • # # # • # # # • # # # • # # # • # # #

A2 **a.** 16: $28\left(\frac{4}{7}\right)=28\left(\frac{n}{28}\right)$ **b.** 15

Q3 Solve the following proportions:

a. $\frac{4}{17}=\frac{n}{34}$ **b.** $\frac{n}{30}=\frac{70}{175}$

\# # # • # # # • # # # • # # # • # # # • # # # • # # # • # # # • # # #

A3 **a.** 8 **b.** 12

---

2 The missing term in a proportion can be any of the four terms of the proportion. For example, in the proportion

$$\frac{15}{540}=\frac{2}{n}$$

the missing term is the fourth term. This proportion can be solved using principles of equality; however, the solution is simplified by using the property that the product of the means is equal to the product of the extremes. That is,

$$540 \cdot 2=15 \cdot n$$
$$\frac{540 \cdot 2}{15}=n$$
$$72=n$$

Q4 Solve the proportion $\frac{12}{n}=\frac{72}{30}$.

\# # # • # # # • # # # • # # # • # # # • # # # • # # # • # # # • # # #

A4 5: $n \cdot 72=12 \cdot 30$

$$n=\frac{12 \cdot 30}{72}$$

Q5 Solve the proportion $\frac{2}{n}=\frac{6}{15}$.

\# # # • # # # • # # # • # # # • # # # • # # # • # # # • # # # • # # #

A5 5: $n \cdot 6=2 \cdot 15$

$$n=\frac{2 \cdot 15}{6}$$

Q6 Solve the following proportions:

a. $\frac{18}{63}=\frac{440}{n}$ **b.** $\frac{108}{27}=\frac{52}{n}$

\# # # • # # # • # # # • # # # • # # # • # # # • # # # • # # # • # # #

A6 **a.** 1,540: $63 \cdot 440 = 18 \cdot n$

$$\frac{63 \cdot 440}{18} = n$$

b. 13

Q7 Solve the following proportions:

a. $\frac{n}{12} = \frac{2.5}{5}$ **b.** $\frac{0.75}{n} = \frac{19}{114}$

• ### • ### • ### • ### • ### • ### • ### •

A7 **a.** 6 **b.** 4.5

3 Various types of problems can be solved by writing the conditions of the problem as a proportion and solving for the missing term. Two general statements will aid in properly placing the conditions of a problem in the proportion.

1. Each ratio should be a comparison of similar things.
2. In every proportion both ratios must be written in the same order of value; that is, small to large or large to small.

Example: If 6 pencils cost 25 cents, how much will 12 pencils cost?

Solution: Let c represent the cost of 12 pencils and substitute the conditions of the problem in the following proportion:

$$\frac{\text{small number of pencils}}{\text{large number of pencils}} = \frac{\text{small cost}}{\text{large cost}}$$

Each ratio is a comparison of similar things. Each ratio is written in the same order of value. Hence,

$$\frac{6 \text{ pencils}}{12 \text{ pencils}} = \frac{\$.25}{c}$$

$$12 \cdot 25 = 6 \cdot c$$

$$50 = c$$

Therefore, 12 pencils will cost 50 cents. The quantities in the above proportion are *directly* related because an increase in the number of pencils causes a corresponding increase in the cost.

Q8 If 3 cans of cola cost 25 cents, how much will 15 cans cost?

$$\frac{\text{large number}}{\text{small number}} = \frac{\text{large cost}}{\text{small cost}}$$

• ### • ### • ### • ### • ### • ### • ### •

A8 \$1.25: $\frac{15 \text{ cans}}{3 \text{ cans}} = \frac{c}{\$0.25}$ ← large cost because 15 cans will cost more

---

Q9 In a proportion, if an increase in one quantity causes a corresponding increase in another quantity, the quantities are ________ related.

• ### • ### • ### • ### • ### • ### • ### •

A9 directly

Q10 An increase in the height of an object will cause a corresponding ______ in the length of its shadow.
increase/decrease

\# \# \# • \# \# \# • \# \# \# • \# \# \# • \# \# \# • \# \# \# • \# \# \# • \# \# \# • \# \# \#

A10 increase

Q11 If an 18-foot pole casts a 15-foot shadow, what will be the length of the shadow cast by a 50-foot pole?

\# \# \# • \# \# \# • \# \# \# • \# \# \# • \# \# \# • \# \# \# • \# \# \# • \# \# \# • \# \# \#

A11 $41\frac{2}{3}$ feet: $\frac{18 \text{ ft pole}}{50 \text{ ft pole}} = \frac{15 \text{ ft shadow}}{S}$ (S is the unknown length and will be larger than 15 ft)

Q12 A car can travel 47 miles on 3 gallons of gas. How far can it go on 18 gallons?

\# \# \# • \# \# \# • \# \# \# • \# \# \# • \# \# \# • \# \# \# • \# \# \# • \# \# \# • \# \# \#

A12 282 miles: $\frac{18 \text{ gal}}{3 \text{ gal}} = \frac{N}{47 \text{ mi}}$ or $\frac{3 \text{ gal}}{18 \text{ gal}} = \frac{47 \text{ mi}}{N}$

Q13 If \$1,000 worth of insurance cost \$20.80, what will \$15,000 of insurance cost (round off to the nearest cent)?

\# \# \# • \# \# \# • \# \# \# • \# \# \# • \# \# \# • \# \# \# • \# \# \# • \# \# \# • \# \# \#

A13 \$312: $\frac{\$1{,}000}{\$15{,}000} = \frac{\$20.80}{N}$

Q14 If valve stems cost 3 for 10 cents, what is the cost of 11 valve stems (round off to the nearest cent)?

\# \# \# • \# \# \# • \# \# \# • \# \# \# • \# \# \# • \# \# \# • \# \# \# • \# \# \# • \# \# \#

A14 \$0.37: $\frac{3 \text{ stems}}{11 \text{ stems}} = \frac{\$0.10}{c}$

Q15 If 3.5 tons of top soil cost \$20, what will 12.5 tons of top soil cost (round off to the nearest cent)?

\# \# \# • \# \# \# • \# \# \# • \# \# \# • \# \# \# • \# \# \# • \# \# \# • \# \# \# • \# \# \#

A15 $71.43

4 In a proportion, if an increase in one quantity causes a corresponding decrease in another quantity or a decrease in one quantity causes a corresponding increase in the other, the quantities are *indirectly* or inversely related. For example, if the number of workers on a job is increased, the time required to do the work is decreased. Problems involving indirect relationships can be solved using the principles established earlier.

Example: If 2 workers can build a garage in 5 days, how long will it take 6 workers, assuming that they all work at the same rate?

Solution: Since the number of days required to build the garage becomes smaller as the number of workers becomes larger, the quantities vary indirectly. Thus,

$$\frac{2 \text{ (smaller no. workers)}}{6 \text{ (larger no. workers)}} = \frac{N \text{ (smaller no. days)}}{5 \text{ (larger no. days)}}$$

$$6N = 10$$

$$N = 1\frac{2}{3} \text{ days}$$

Q16 Three workers can build a house in 10 days. Will it take 8 workers working at the same rate more or less time to build the house? ________

• ### • ### • ### • ### • ### • ### • ### •

A16 less

Q17 A plane flies from Detroit to Philadelphia in 1 hour and 40 minutes at 440 miles per hour. Will it take more or less time to make the trip flying at 380 miles per hour? ________

• ### • ### • ### • ### • ### • ### • ### •

A17 more

Q18 Eight students sell 3,800 tickets to a football game. Will it take more or fewer students to sell 5,200 tickets selling at the same rate? ________

• ### • ### • ### • ### • ### • ### • ### •

A18 more

Q19 If 4 men can clear the snow near a school in 6 hours, how long will it take 12 men working at the same rate to do the job?

$$\frac{4 \text{ men}}{12 \text{ men}} = \frac{(\quad)}{(\quad)}$$

(Let N equal the unknown time and ask yourself, "Does the time get smaller or larger?" The smaller time must be placed in the numerator and the larger time in the denominator.)

• ### • ### • ### • ### • ### • ### • ### •

A19 2 hours: smaller number of men → $\frac{4}{12} = \frac{N}{6}$ ← smaller time

Q20 A plane takes 3 hours at a speed of 320 miles per hour to go from Chicago to New York. How fast must the plane fly to make the trip in 2.5 hours?

• ### • ### • ### • ### • ### • ### • ### •

A20 384 mph: greatest time → $\frac{3 \text{ hr}}{2.5 \text{ hr}} = \frac{N}{320 \text{ mi}}$ ← greatest speed

(The problem could have been solved by placing the smaller numbers on top; that is, $\frac{2.5}{3} = \frac{320}{N}$.)

5 In the remainder of this section, specific applications of proportions will be presented from various technical areas. One example using inverse proportions is belted pulleys. When two pulleys are belted together as shown, the *revolutions per minute* (rpm) vary inversely to the size of the pulleys. That is, the smaller pulley, when belted to a larger one, will revolve more times per minute. The size of a pulley is usually given as the diameter (the distance across the center).

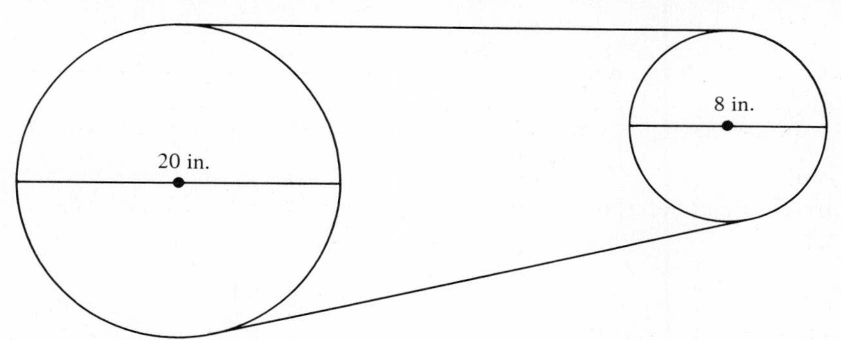

Example 1: A 20-inch pulley turning at 180 rpm drives an 8-inch pulley. Find the rpm of the 8-inch pulley.

Solution: The proportion would be set up as follows:

$$\frac{\text{diameter small pulley}}{\text{diameter large pulley}} = \frac{\text{small number rpm}}{\text{large number rpm}}$$

Since the smaller pulley will have the larger number of rpm, the larger number of rpm is the unknown quantity in the above proportion. Thus,

$$\frac{8 \text{ in.}}{20 \text{ in.}} = \frac{180 \text{ rpm}}{N}$$

$$N = 450 \text{ rpm}$$

Gears turning together have the same relationship as belted pulleys. The revolutions per minute of rotating gears vary inversely to the number of teeth in the gears.

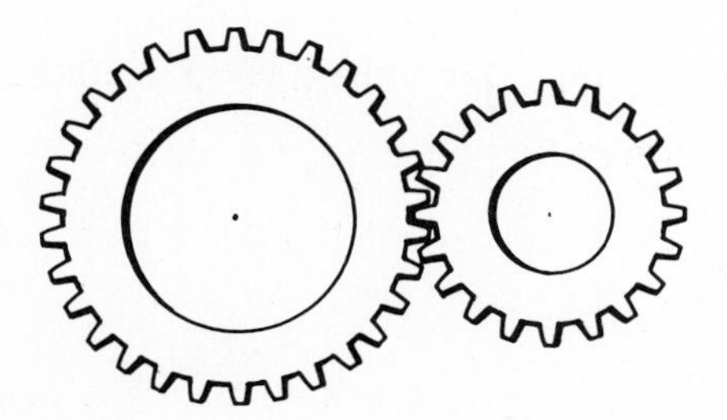

Example 2: A gear with 20 teeth revolving at 52 rpm meshes with a larger gear with 30 teeth. Find the rpm of the large gear.

Solution:

$$\frac{\text{teeth large gear}}{\text{teeth small gear}} = \frac{\text{large number rpm}}{\text{small number rpm}}$$

Since the large gear will revolve at a slower rate than the small gear, the unknown in the above proportion is the small number of rpm.

$$\frac{30 \text{ teeth}}{20 \text{ teeth}} = \frac{52 \text{ rpm}}{N}$$

$N = 34.7$ rpm (rounded off to the nearest tenth)

Q21 A 20-inch pulley turning 175 rpm drives a 5-inch pulley. How many rpm is the 5-inch pulley turning?

\# # # • # # # • # # # • # # # • # # # • # # # • # # # • # # # • # # #

A21 700 rpm: $\frac{20 \text{ in.}}{5 \text{ in.}} = \frac{N}{175 \text{ rpm}}$

Q22 A gear with 270 teeth turning at 600 rpm is connected to a gear with 180 teeth. How many rpm does the 180-tooth gear make?

\# # # • # # # • # # # • # # # • # # # • # # # • # # # • # # # • # # #

A22 900 rpm: $\frac{180 \text{ teeth}}{270 \text{ teeth}} = \frac{600}{N}$

Q23 How many teeth are required on a gear if it is to run at 270 rpm when driven by a gear with 42 teeth running at 180 rpm?

\# # # • # # # • # # # • # # # • # # # • # # # • # # # • # # # • # # #

A23 28 teeth: $\frac{270 \text{ rpm}}{180 \text{ rpm}} = \frac{42 \text{ teeth}}{N}$

If the gear is to turn at a faster rate, it must have fewer teeth.

Q24 A pinion gear with 20 teeth turning 18.6 rpm meshes with a ring gear with 42 teeth. Determine the rpm for the ring gear (round off to the nearest tenth).

\# # # • # # # • # # # • # # # • # # # • # # # • # # # • # # # • # # #

A24 8.9 rpm: $\dfrac{20 \text{ teeth}}{42 \text{ teeth}} = \dfrac{N}{18.6 \text{ rpm}}$

Q25 If a car going 20 mph turns 800 rpm, what would the same car going 55 mph be turning? (Hint: Does an increase in mph cause an increase or decrease in rpm?)

• ### • ### • ### • ### • ### • ### • ### •

A25 2200 rpm: $\dfrac{20 \text{ mph}}{55 \text{ mph}} = \dfrac{800 \text{ rpm}}{N}$

Q26 If 15 gallons of gas will drive a car 310 miles, how many gallons will be required to drive the same car at the same rate of consumption 200 miles (round off to the nearest tenth)?

• ### • ### • ### • ### • ### • ### • ### •

A26 9.7 gallons: $\dfrac{200 \text{ miles}}{310 \text{ miles}} = \dfrac{N}{15 \text{ gallons}}$

Q27 If the upkeep on 8 machines for 1 month is $87, what is the upkeep on 15 such machines for 1 month (round off to the nearest dollar)?

• ### • ### • ### • ### • ### • ### • ### •

A27 $163: $\dfrac{8 \text{ machines}}{15 \text{ machines}} = \dfrac{\$87}{N}$

Q28 If a gear on the clutch shaft with 10 teeth turns 120 rpm, how many rpm would the drive shaft turn if it has 26 teeth (round off to the nearest tenth)?

• ### • ### • ### • ### • ### • ### • ### •

A28 46.2 rpm

6 A *scale drawing* is a drawing or diagram representing a much larger area or figure. The scale drawing is proportional in every detail to the figure it represents. The contractor must interpret dimensions from the scale drawing to actual size. A standard scale is $\frac{1}{4}$ inch on the scale drawing representing one foot in actual size.

Example: On a scale drawing, a contractor measures a vent to be 2 inches from the floor. What will be the actual distance?

Solution: $\dfrac{\frac{1}{4}\text{ inch}}{\text{measured inches}} = \dfrac{1\text{ foot}}{\text{actual distance in feet}}$

$$\frac{\frac{1}{4}\text{ inch}}{2\text{ inch}} = \frac{1\text{ foot}}{N}$$

$$\frac{1}{4}\cdot N = 2$$

$$N = 8\text{ feet}$$

Q29 A scale drawing shows a wall to be $6\frac{3}{4}$ inches long. What is the actual length of the wall?

• ### • ### • ### • ### • ### • ### • ### •

A29 27 feet: $\dfrac{\frac{1}{4}\text{ inch}}{6\frac{3}{4}\text{ inch}} = \dfrac{1\text{ foot}}{N}$

Q30 A draftsman wishes to reduce an 18-foot length to a scale drawing. What would be the length on the drawing?

• ### • ### • ### • ### • ### • ### • ### •

A30 $4\frac{1}{2}$ inches: $\dfrac{\frac{1}{4}\text{ inch}}{N} = \dfrac{1\text{ foot}}{18\text{ foot}}$

Q31 If it takes 7 men 38 hours to partition an office area, how long will it take 11 men to do the same job (round off to the nearest hour)?

• ### • ### • ### • ### • ### • ### • ### •

A31 24 hours: $\dfrac{7\text{ men}}{11\text{ men}} = \dfrac{N}{38}$

7 A *lever* is a rigid bar pivoted to turn on a point called the *fulcrum*. An example of a lever and fulcrum is the teeter-totter shown below.

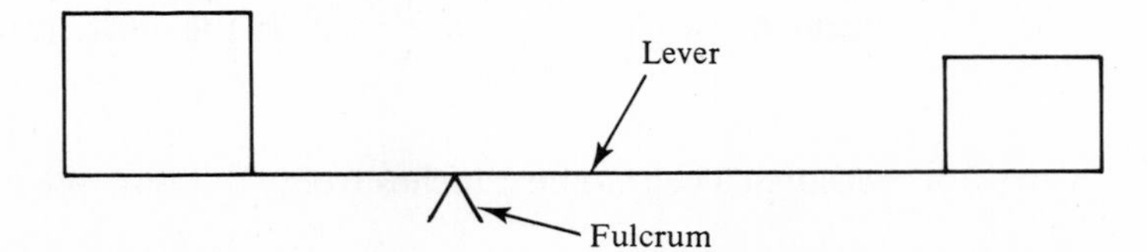

You may recall that a larger force (weight) on one end is balanced by placing the smaller force a greater distance away from the fulcrum. This idea can be illustrated using the following proportion:

$$\frac{\text{small force}}{\text{large force}} = \frac{\text{small distance from fulcrum}}{\text{large distance from fulcrum}}$$

A particular example of a lever is the brake pedal of a car. The foot pedal is actually a lever, with the bolt around which it pivots as the fulcrum. The force applied to the brake is increased by the action of the pedal in a proportion determined by the lengths of the pedal on each side of the fulcrum bolt.

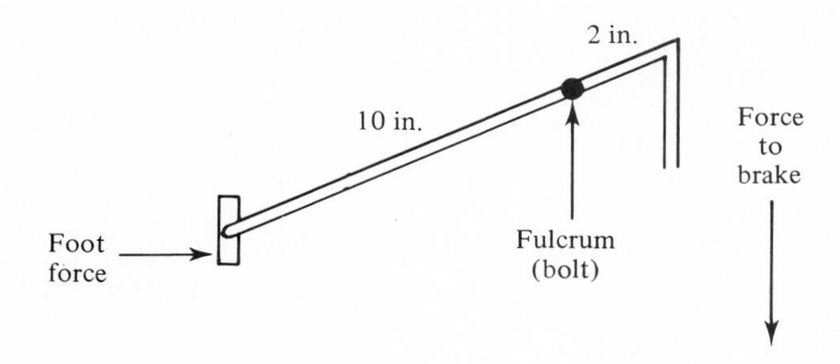

Example: In the figure above, find the force that would be transmitted to the brake when a force of 15 pounds is applied to the foot pedal.

Solution: $\frac{15 \text{ pounds}}{N} = \frac{2 \text{ in.}}{10 \text{ in.}}$

$N = 75$ pounds

Q32 A brake pedal measures 8 inches from the footrest to the fulcrum and 2.5 inches beyond the fulcrum. What force is transmitted to the brake when the foot pressure is 25 pounds?

\# \# \# • \# \# \# • \# \# \# • \# \# \# • \# \# \# • \# \# \# • \# \# \# • \# \# \# • \# \# \#

A32 80 pounds: $\frac{25 \text{ pounds}}{N} = \frac{2.5 \text{ inches}}{8 \text{ inches}}$

Q33 A force of 60 pounds is applied to one end of a lever 5.5 feet from the fulcrum. What force is transmitted to the other end if it is 1 foot from the fulcrum?

\# \# \# • \# \# \# • \# \# \# • \# \# \# • \# \# \# • \# \# \# • \# \# \# • \# \# \# • \# \# \#

A33 330 pounds: $\frac{1 \text{ foot}}{5.5 \text{ feet}} = \frac{60 \text{ pounds}}{N}$

This completes the instruction for this section.

9.2 EXERCISE

1. In a proportion, if an increase in one quantity causes a decrease in another quantity, the quantities are ______________ related.
2. In a proportion, if a decrease in one quantity causes a decrease in another quantity, the quantities are ______________ related.
3. Solve the following proportions:

   a. $\frac{5}{7}=\frac{N}{49}$ b. $\frac{49}{98}=\frac{7}{N}$ c. $\frac{6}{18}=\frac{N}{21}$ d. $\frac{5}{2}=\frac{15}{N}$

   e. $\frac{4}{3}=\frac{N}{21}$ f. $\frac{25}{X}=\frac{15}{24}$ g. $\frac{2}{N}=\frac{4}{8}$ h. $\frac{3}{N}=\frac{18}{27}$

   i. $\frac{9}{2}=\frac{5}{X}$ j. $\frac{\frac{2}{3}}{9}=\frac{6}{X}$ k. $\frac{\frac{1}{2}}{\frac{1}{4}}=\frac{\frac{X}{3}}{4}$ l. $\frac{0.3}{0.8}=\frac{N}{\frac{1}{8}}$

4. If 30 women do a job in 12 days, how long will it take 20 women to do the same work?
5. If the wing area of a model airplane is 3 square feet and it can lift 10 pounds, how much can a wing whose area is 149 square feet lift (round off to the nearest tenth)?
6. A swimming pool large enough for 6 people must hold 5000 gallons of water. How much water must a pool large enough for 15 people hold?
7. If a car goes 52 miles on 3 gallons of gas, how far will it go on 11 gallons (round off to the nearest tenth)?
8. If a certain machine screw is purchased at 3 boxes for $3.60, how much will 14 boxes cost.
9. A train takes 24 hours at a rate of 17 miles per hour to travel a certain distance. How fast would a plane have to fly in order to cover the same distance in 3 hours?
10. If electricity cost $0.10 for 4 kilowatt-hours, how many kilowatt-hours could be purchased for $7.20?
11. How many teeth are in a gear turning 125 rpm if it is driven by a gear with 50 teeth turning 400 rpm?
12. If a car going 70 mph turns 3000 rpm, how many rpm will it turn at 90 mph (round off to the nearest tenth)?
13. A pinion gear with 9 teeth meshes with a ring gear with 32 teeth. If the ring gear turns 100 rpm, how fast is the pinion gear turning (round off to the nearest tenth)?
14. A 30-inch pulley is belted to a 6-inch pulley. If the 30-inch pulley makes 250 rpm, how many rpm does the 6-inch pulley make?
15. A pulley with diameter of 8 inches turning 450 rpm is belted to a 6-inch pulley. How many rpm is the 6-inch pulley turning?
16. If 3 men can paint a house in an 8-hour day, how long will it take 5 men?
17. If 2 axes cost $7.70, what is the cost of 5 axes?
18. A gear turning 2100 rpm has 10 teeth and meshes with a 15-tooth gear in the transmission. How fast is the gear with 15 teeth turning?
19. A brake pedal measures 7.5 inches from the footrest to the fulcrum bolt and 2.5 inches beyond the fulcrum. What force is transmitted to the brake when the foot pressure is 8.2 pounds (round off to the nearest tenth)?
20. If 30 gallons of oil flow through an intake pipe in 20 minutes, how long will it take to fill a tank of 1250-gallon capacity?
21. A 12-inch pulley turning 325 rpm is belted to a pulley turning 1300 rpm. What is the size of the second pulley?
22. A truck was driven 230 miles on 40 gallons of fuel. How much fuel would be required to drive 600 miles (round off to the nearest tenth)?

23. If in a scale drawing $\frac{1}{4}$ inch equals 1 foot, $2\frac{1}{8}$ inches on the drawing would represent what distance?

24. If 15 machine parts cost $18.33, what would 6 parts cost (round off to the nearest cent)?

25. At one end of a lever is a force of 72 pounds 6.2 feet from the fulcrum. What force would be necessary, at the other end, 10.8 feet from the fulcrum if the lever is to balance (round off to the nearest tenth)?

9.2 EXERCISE ANSWERS

1. indirectly
2. directly
3. **a.** 35 **b.** 14 **c.** 7 **d.** 6
e. 28 **f.** 40 **g.** 4 **h.** $4\frac{1}{2}$
i. $1\frac{1}{9}$ **j.** 81 **k.** $1\frac{1}{2}$ **l.** 0.046875 $\left(\frac{3}{64}\right)$
4. 18 days
5. 496.7 pounds
6. 12,500 gallons
7. 190.7 miles
8. $16.80
9. 136 mph
10. 288 kilowatt-hours
11. 160 teeth
12. 3,857.1 rpm
13. 355.6 rpm
14. 1,250 rpm
15. 600 rpm
16. 4.8 hr or 4 hr 48 min
17. $19.25
18. 1,400 rpm
19. 24.6 pounds
20. $833\frac{1}{3}$ min or 13 hr 53 min 20 sec
21. 3 in.
22. 104.3 gal
23. 8.5 ft
24. $7.33
25. 41.3 pounds

9.3 SOLVING WORD PROBLEMS THAT INVOLVE FORMULAS

1 Formulas are useful in many situations in which certain information is needed.

Example: Determine the perimeter of a rectangle given the length and width to be 17.0 centimetres and 2.1 centimetres, respectively.

Solution: The formula $p = 2l + 2w$ for the perimeter of a rectangle can be used where p = perimeter, l = length, and w = width.

$p = 2l + 2w$
$p = 2(17.0) + 2(2.1)$
$p = 34.0 + 4.2$
$p = 38.2$

Therefore, the perimeter of the desired rectangle is 38.2 centimetres.

Q1 Determine the area of a trapezoid if the parallel sides (bases) are 12 inches and $7\frac{1}{2}$ inches and the altitude (height) is $3\frac{3}{4}$ inches. (Do not forget to record the final answer in square inches.) Use the formula $A=\frac{1}{2}h(b_1+b_2)$ for the area of a trapezoid where A = area, h = altitude (height), and b_1 and b_2 = parallel sides (bases).

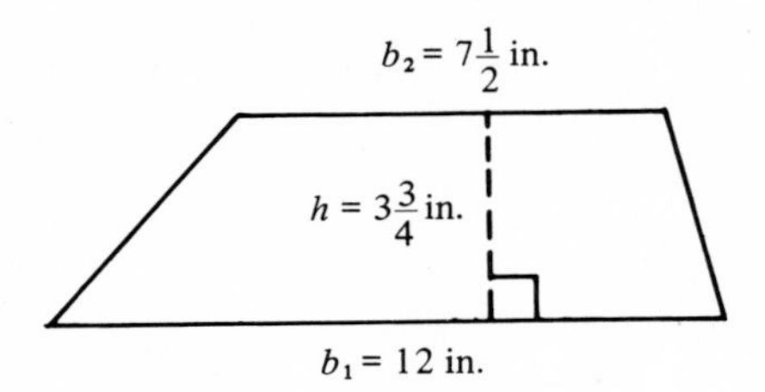

• ### • ### • ### • ### • ### • ### • ### •

A1 $36\frac{9}{16}$ square inches or 36.5625 square inches:

$$A=\frac{1}{2}h(b_1+b_2)$$

$$A=\frac{1}{2}\left(3\frac{3}{4}\right)\left(12+7\frac{1}{2}\right) \quad \text{or} \quad A=0.5(3.75)(12+7.5)$$

2 In the formula, $p=2l+2w$, p is called the *subject* of the formula. In this formula, the perimeter, p, is expressed in terms of the length, l, and the width, w. The subject of any formula is expressed in terms of the other variables present.

Q2 What is the subject of each of the following formulas?

a. $A=\frac{1}{2}bh$ ______ **b.** $c=2\pi r$ ______

c. $V=\frac{1}{3}\pi r^2 h$ ______ **d.** $a+b+c=p$ ______

• ### • ### • ### • ### • ### • ### • ### •

A2 **a.** A **b.** c **c.** V **d.** p

3 Often it is necessary to find the value of a variable that is not the subject of the formula. This can be done as long as values for all other variables in the formula are known.

Example: Find the length of a rectangle if the perimeter of the rectangle is 52 metres and the width of the rectangle is 12 metres.

Solution:

Step 1: Determine the appropriate formula. The formula for the perimeter of a rectangle is $p=2l+2w$, where p = perimeter, l = length, and w = width.

Step 2: Identify the given and unknown information from the statement of the problem.

$p = 52$ metres
l = unknown
$w = 12$ metres

Step 3: Substitute the values of the known information into the formula.

$p = 2l + 2w$
$52 = 2l + 2(12)$

Step 4: Solve the resulting equation.

$p = 2l + 2w$
$52 = 2l + 2(12)$
$52 = 2l + 24$
$28 = 2l$
$14 = l$

Step 5: Check the solution.

$2(14) + 2(12) \stackrel{?}{=} 52$
$28 + 24 \stackrel{?}{=} 52$
$52 = 52$

Therefore, the required length of the rectangle is 14 metres.

Q3 Find the height of a rectangular prism (box) if its volume is 315 cubic inches, the length of its base is 25 inches, and the width of its base is 3 inches. The formula for the volume of a rectangular prism is $V = lwh$, where V = volume, l = length of the base, w = width of the base, and h = height of the prism.

a. Identify the given and unknown information.

$V =$ __________ $l =$ __________

$w =$ __________ $h =$ __________

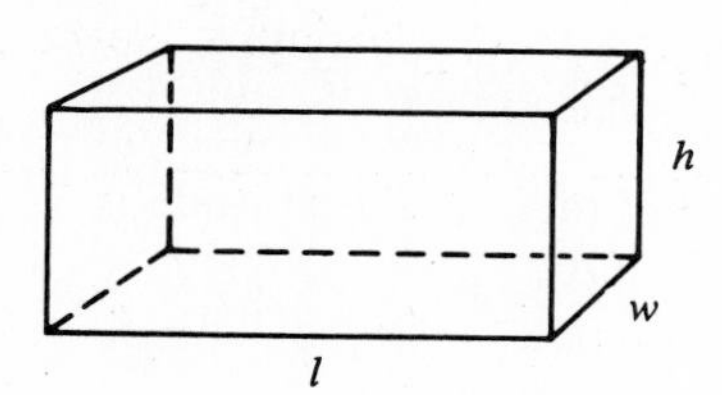

b. Substitute the values of the known information into the formula $V = lwh$.

c. Solve the resulting equation.

d. Check the solution.

e. Therefore, the required height is __________.

• # # # • # # # • # # # • # # # • # # # • # # # • # # # • # #

A3 **a.** $V = 315$ cubic inches
$l = 25$ inches
$w = 3$ inches
$h =$ unknown

b. $V = lwh$
$315 = 25(3)h$

c. $315 = 75h$
$4.2 = h$

d. $315 \stackrel{?}{=} 25(3)(4.2)$
$315 = 315$

e. 4.2 inches

Q4 Determine the required height for a cylindrical tank if the volume is to be 660 cubic feet and the diameter of the tank is to be 7 feet. $\left(\textit{Hint}: V=\pi r^2 h, \text{ where } r=\frac{d}{2}. \text{ Use } \pi \doteq \frac{22}{7}.\right)$

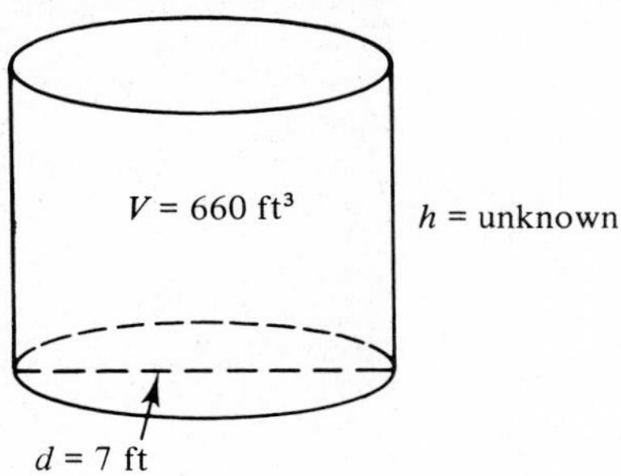

• ### • ### • ### • ### • ### • ### • ### •

A4 $17\frac{1}{7}$ feet:

$$V=\pi r^2 h$$
$$660=\frac{22}{7}\left(\frac{7}{2}\right)\left(\frac{7}{2}\right)h$$
$$660=\frac{77}{2}h$$
$$\frac{2}{77}(660)=\frac{2}{77}\left(\frac{77}{2}h\right)$$
$$17\frac{1}{7}=h$$

Therefore, the required height is $17\frac{1}{7}$ feet $\left(17 \text{ ft } 1\frac{5}{7} \text{ in.}\right)$.

Q5 Find the other parallel side of a trapezoid if the area is 24 square metres, the altitude is 3 metres, and the smaller parallel side is 2 metres. Use $A=\frac{1}{2}h(b_1+b_2)$.

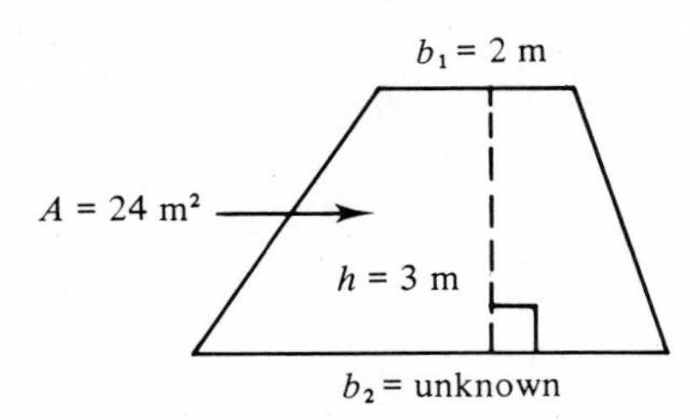

• ### • ### • ### • ### • ### • ### • ### •

A5 14 metres:

$$A=\frac{1}{2}h(b_1+b_2)$$
$$24=\frac{1}{2}(3)(2+b_2)$$
$$24=\frac{3}{2}(2+b_2)$$
$$\frac{2}{3}(24)=\frac{2}{3}\cdot\frac{3}{2}(2+b_2)$$
$$16=2+b_2$$
$$14=b_2$$

5 Whenever a formula can be used to determine missing information, the technique of substituting the known values and solving the resulting equation is often convenient. The examples used thus far have all concerned geometric objects. However, the technique being discussed can be applied to any situation in which a formula is available.

Example: Find the time required to accumulate \$300 interest if \$20,000 is invested at 6% simple interest.

Solution: The formula $i = prt$, where $i =$ interest, $p =$ principal (amount invested or borrowed), $r =$ annual rate of interest, and $t =$ time in years, could be used.

As before, it is important to identify the given and unknown information. That is, $i = \$300$, $p = \$20{,}000$, $r = 6\%$, and $t =$ unknown. Hence,

$$i = prt$$
$$300 = 20{,}000(6\%)t$$
$$300 = 20{,}000\left(\frac{6}{100}\right)t^*$$
$$300 = 1{,}200t$$
$$\frac{1}{4} = t$$

Therefore, the time required is $\frac{1}{4}$ of a year, or 3 months.

*Six percent could be expressed as 0.06, or $\frac{3}{50}$.

Q6 Given $d = rt$, where $d =$ distance, $r =$ rate, and $t =$ time, determine the rate at which a car must travel to cover a distance of 135 kilometres in 3 hours. Answer should be expressed in kilometres per hour (km/hr).

• ### • ### • ### • ### • ### • ### • ### • ### •

A6 45 km/hr:
$$d = rt$$
$$135 = r \cdot 3$$
$$135 = 3r$$
$$45 = r$$

5 If it is necessary to find the time it would take to make a trip when you know the distance to be traveled and you know how fast (rate or speed) you will be traveling, you could solve the formula $d = rt$ for t; that is, make t the subject of the formula. For example,

$$d = rt$$
$$\frac{d}{r} = \frac{rt}{r} \quad \text{Dividing both sides by } r$$
$$\frac{d}{r} = t \quad \text{Simplifying the right side}$$
$$t = \frac{d}{r}^*$$

*Review Section 8.4 if necessary.

Example: Find the time it takes to go 48 miles on a bicycle traveling 10 miles per hour (mph). Use the formula derived above.

Solution: $d = 48$ miles, $r = 10$ mph, and $t =$ unknown

$$t = \frac{d}{r} = \frac{48}{10} = 4.8$$

Hence, the time required is 4.8 hours.

Q7 (1) Solve the formula $d = rt$ for r and (2) find the rate you would have to travel in order to go 127 miles in 4 hours.

(1)

(2) ______ = 127 miles, ______ = 4 hours

______ = unknown

\### • ### • ### • ### • ### • ### • ### • ### • ###

A7 (1) $r = \frac{d}{t}$

(2) $31\frac{3}{4}$ mph or 31.75 mph:

$d = 127$ miles, $t = 4$ hours, $r =$ unknown

$$r = \frac{d}{t} = \frac{127}{4} = 31\frac{3}{4} \text{ or } 31.75$$

6 Since a person working in a technical area will be exposed to many formulas, skill in the use of formulas is essential to continued success on the job. The remaining portion of this section will illustrate numerous formulas taken directly from the world of work. Naturally, each individual will not come in contact with them all. However, exposure to them should give you some appreciation for the mathematics required of people working in related technical fields. (Note: It is the authors' hope that you will gain confidence in formula manipulation so that in future work in your field, the mathematics will not interfere with the comprehension of the application.)

A formula that is used by automotive technicians is

$$\frac{\text{length of piston stroke} \times \text{revolutions per minute}}{6} = \text{piston speed}$$

or

$$\frac{sn}{6} = p$$

where $s =$ length of piston stroke measured in inches,
$n =$ number of revolutions per minute (rpm), and
$p =$ speed of piston measured in feet per minute (ft/min).

Q8 (1) Solve the formula $\frac{sn}{6} = p$ for n and (2) find the number of rpm a piston would have to travel to move at 3200 ft/min if the piston has a 3-inch stroke.

(1)

(2) $p =$ ______, $s =$ ______

$n =$ ______

\### • ### • ### • ### • ### • ### • ### • ### • ###

A8 (1) $\frac{sn}{6} = p$

$\frac{6}{s} \cdot \frac{sn}{6} = p \cdot \frac{6}{s}$

$n = \frac{6p}{s}$

(2) 6400 rpm: $p = 3200$ ft/min
$s = 3$ in.
$n =$ unknown
$n = \frac{6p}{s} = \frac{6 \times 3200}{3} = 6400$

7 A formula used to calculate horsepower is

$$\frac{\text{torque} \times \text{number of rpm}}{5252} = \text{horsepower}$$

This formula can be written

$$\frac{Tn}{5252} = h$$

where $T =$ torque measured in foot-pounds (ft-lbs),
$n =$ number of revolutions per minute (rpm), and
$h =$ horsepower (hp).

Example: Find the horsepower of a car with 40 ft-lbs of torque which is turning at 2500 rpm.

Solution: $T = 40$ ft-lbs, $n = 2500$ rpm, and $h =$ unknown

$\frac{Tn}{5252} = h$ or $h = \frac{Tn}{5252}$

$h = \frac{40 \times 2500}{5252}$

$h = 19.0$ (rounded off to nearest tenth)

Hence, the car is operating at 19.0 hp.

Q9 (1) Solve the $\frac{Tn}{5252} = h$ for T and (2) find the torque for an engine with 60 horsepower which is turning 3000 rpm. Round off final answer to the nearest tenth.

(1)

(2) $h =$ ______, $n =$ ______

$T =$ ______

\### • ### • ### • ### • ### • ### • ### • ### • ###

A9 (1) $\frac{Tn}{5252} = h$

$5252 \frac{Tn}{5252} = 5252h$

$\frac{Tn}{n} = \frac{5252h}{n}$

$T = \frac{5252h}{n}$

(2) 105.0 ft-lbs: $h = 60$ hp, $n = 3000$ rpm
$T =$ unknown
$T = \frac{5252h}{n} = \frac{5252 \times 60}{3000}$

8 In order to convert the revolutions per minute (rpm) that an engine turns into miles per hour (mph) that an automobile travels, it is necessary to work with the circumference of the tire. There is a common formula used to calculate mph. We will show how that formula is obtained and how the formula for the circumference of a circle enters into it (Frame 10) but first we will practice using the formula. The formula can be stated as follows:

$$\text{mph} = \frac{\text{rpm} \times W}{R \times 168},$$

where mph = car speed in miles per hour,
rpm = engine crankshaft speed,
R = rear axle gear ratio, and
W = tire size (radius measured in inches).

Example: An engine is turning at 5000 rpm. It has a rear axle gear ratio of 4 (actually 4 to 1). The tires have a radius of 13 inches. Find the speed of the car.

Solution: rpm = 5000, $W = 13$ in., $R = 4$, mph = unknown

$$\text{mph} = \frac{\text{rpm} \times W}{R \times 168} = \frac{5000 \times 13}{4 \times 168} = 96.7 \quad \text{(rounded off)}$$

Hence, the speed of the car is 96.7 mph.

Q10 Find the rate the car is traveling (Frame 8) if the tires are changed to ones with a 14-inch radius but the rpm is left at 5000 and the gear ratio is left at 4. Round off the answer to nearest tenth.

• ### • ### • ### • ### • ### • ### • ### •

A10 104.2 mph: rpm = 5000, $W = 14$ in., $R = 4$, mph = unknown

$$\text{mph} = \frac{\text{rpm} \times W}{R \times 168} = \frac{5000 \times 14}{4 \times 168} = 104.2$$

9 The formula $\text{mph} = \frac{\text{rpm} \times W}{R \times 168}$ can be used to find the rpm of an engine when a car is traveling at a given speed. To do this the subject of the formula can be changed to rpm; that is, the above formula can be solved for rpm.

Example: Solve $\text{mph} = \frac{\text{rpm} \times W}{R \times 168}$ for rpm.

Solution:

$$\text{mph} = \frac{\text{rpm} \times W}{R \times 168}$$

$$\frac{R \times 168}{W} \cdot \text{mph} = \frac{\text{rpm} \times W}{R \times 168} \cdot \frac{R \times 168}{W}$$

$$\frac{168 \times R \times \text{mph}}{W} = \text{rpm}$$

$$\text{rpm} = \frac{\text{mph} \times R \times 168}{W}$$

Q11 Use the formula for rpm (Frame 9) to find the number of rpm an engine turns at the speeds given below. Suppose the tire radius is 12 inches and the gear ratio is 4.
a. 40 mph **b.** 60 mph **c.** 80 mph

• ### • ### • ### • ### • ### • ### • ### •

A11 **a.** 2240 rpm: $\text{rpm} = \dfrac{40 \times 4 \times 168}{12}$
b. 3360 rpm
c. 4480 rpm
(Note: It is convenient to have the formula solved for the unknown variable when repeated calculations are required.)

---

*Q12 How many *more* rpm would have to be turned to maintain 80 mph after $\frac{1}{4}$ inch has been worn off the tires in Q11**c**. Gear ratio is still 4 and the tires were originally 12 inches. (Hint: $\frac{1}{4}$ in. = 0.25 in.; therefore, $W = 11.75$ in.)

• ### • ### • ### • ### • ### • ### • ### •

*A12 95 rpm: $\text{rpm} = \dfrac{80 \times 4 \times 168}{11.75} = 4575$

$$\begin{array}{r} 4575 \\ -4480 \\ \hline 95 \end{array}$$ rpm more than before

10 Often you work with a formula that is provided to you. A criticism of training in technical fields is that you rarely see the derivation ("where it came from") of the formula. *The following discussion is optimal but highly recommended.* Its purpose is to derive the formula

$$\text{mph} = \frac{\text{rpm} \times W}{R \times 168}.$$

To arrive at this formula requires numerous steps, but basically two things are found:

1. the number of revolutions the tire makes in an hour, and
2. the number of revolutions the tire must turn in a mile.

If you know these two, the miles per hour (mph) is obtained by dividing the number of revolutions per hour (rph) by the revolutions per mile (rpm). This can be shown by using the units of measurements without any numbers. $\left(\text{Note: rph} = \dfrac{\text{rev}}{\text{hr}} \text{ and rpm} = \dfrac{\text{rev}}{\text{mi}}\right)$

$$\frac{\text{rev}}{\text{hr}} \div \frac{\text{rev}}{\text{mi}} = \frac{\text{rev}}{\text{hr}} \times \frac{\text{mi}}{\text{rev}} = \frac{1}{\text{hr}} \times \frac{\text{mi}}{1} = \frac{\text{mi}}{\text{hr}}$$

First, we will find the number of revolutions the tire makes in an hour. If the crankshaft turns a certain number of revolutions in a minute, it would turn 60 times that in an hour. The revolutions per hour of the crankshaft is then $60 \times \text{rpm}$. Since the axle turns slower than the crankshaft in a ratio determined by the gear ratio of the ring gear to the pinion gear, we must reduce the revolutions per hour by dividing by the gear ratio. Therefore, the number of revolutions per hour of the axle (which would be the same as the tire) is $\frac{60 \times \text{rpm}}{R}$ rev/hr.

Then we find the number of revolutions the tire must turn to travel a mile. We first find the circumference of the tire. This is obtained by measuring the distance in inches from the surface on which the tire is resting to the center of the rear axle shaft. We use this as the radius (W). The circumference is then $2\pi W$ inches. We change this to feet by dividing by 12. Therefore, the circumference is $\frac{2\pi W}{12}$ feet. To obtain the number of revolutions it turns in a mile we divide the circumference into 5280 (the number of feet in a mile).

$$\frac{5280}{\frac{2\pi W}{12}} = \frac{5280}{\frac{\pi W}{6}} = \frac{5280 \times 6}{\pi W} \text{ rev/mi}$$

Now we can divide the rev/hr by the rev/mi as was shown earlier and we will be done.

$$\frac{60 \times \text{rpm}}{R} \div \frac{5280 \times 6}{\pi W} = \frac{60 \times \text{rpm}}{R} \times \frac{\pi W}{5280 \times 6} = \frac{\text{rpm} \times \pi W}{R \times 528}$$

To get the formula mph $= \frac{\text{rpm} \times W}{R \times 168}$, we must replace π with 3.14 and divide that into 528. If we didn't simplify the formula like we did above, you would be working with much larger numbers every time you were to use the formula.

No questions are given with respect to Frame 10!

11 Often formulas are given already solved for the different variables present. For example,

$$d = rt \qquad r = \frac{d}{t} \qquad t = \frac{d}{r}$$

Technical areas have handbooks which supply the necessary formulas in all their equivalent forms. Naturally, having the formulas supplied in this manner saves the technician much time and energy (sometimes heartache as well).

It is the purpose of the following material to deal with a few of the basic formulas of electrical circuits from a mathematical viewpoint. More lengthy discussion of the underlying electrical concepts will be left to the specific application courses where these formulas are needed.

There are many forms of energy: mechanical, heat, light, electrical, atomic, molecular, and nuclear. They can be changed from one to another. That is, an electric motor uses electrical energy to produce mechanical energy. In the process some loss of energy is observed due to friction and heat. Naturally, the electrical technician will be studying many formulas that assist him in transforming these different forms of energy and that relate the different factors of the system. (Note: Einstein's famous formula, $E = mc^2$, in which E is energy, m is mass, and c is the speed of light in a vacuum*, transforms matter to energy.)

Ohm's law, $I = \frac{V}{R}$,

where I = current in amperes (amps),
V = potential difference in volts, and
R = resistance in ohms,

*$c = 3 \times 10^{10}$ cm/sec.

states that the value of the current in a resistance circuit supplied by steady direct current is equal to the potential difference divided by the resistance. If any two values are given the third may be determined.

$I = \frac{V}{R}$ may be written $IR = V$.

The latter formula means that the voltage drop, V, across a given resistance equals the current, I, flowing through it multiplied by the resistance, R.

Q13 Use $V = IR$ to find the voltage drop in a circuit when the current is 6 amps and the resistance is 2 ohms.

$I =$ ______ $R =$ ______ $V =$ ______

\# # # • # # # • # # # • # # # • # # # • # # # • # # # • # # # • # # #

A13 12 volts: $I = 6$ amps $R = 2$ ohms $V =$ unknown

$V = IR = 6 \times 2$

Q14 Find the current in a circuit if the voltage drop is 12 volts and the resistance is 6 ohms by (1) substituting the known values in $V = IR$ and solving the equation for I, and (2) solving $V = IR$ for I first and then substituting and evaluating.

(1) (2)

\# # # • # # # • # # # • # # # • # # # • # # # • # # # • # # # • # # #

A14 (1) 2 amps:

$$V = IR$$
$$12 = I \times 6$$
$$2 = I$$

(2) 2 amps:

$$V = IR$$
$$\frac{V}{R} = I$$
$$\frac{12}{6} = I$$
$$2 = I$$

Q15 Find the resistance of a circuit if the current is 100 amps and the voltage drop is 12.5 volts.

\# # # • # # # • # # # • # # # • # # # • # # # • # # # • # # # • # # #

A15 0.125 ohms or $\frac{1}{8}$ ohms:

$$V = IR$$
$$12.5 = 100 \cdot R$$
$$\frac{12.5}{100} = R$$
$$0.125 = R$$

or

$$V = IR$$
$$\frac{V}{I} = R$$
$$\frac{12.5}{100} = R$$
$$0.125 = R$$

12 You know that a lightbulb gets hot as electrical current passes through the filament. The amount of electrical energy transformed to heat energy as a current flows through a wire is given by the formula $W = RI^2t$

where W = heat energy (joules),
R = resistance of wire (ohms),
I = current (amps), and
t = time (sec).

Q16 Use $W = RI^2t$ to find the time in seconds when the heat energy is 791.4375 joules, the resistance is 33.5 ohms, and the current is 1.5 amps. Solve $W = RI^2t$ for t and then substitute.

• # # # • # # # • # # # • # # # • # # # • # # # • # # # • # #

A16 10.5 sec:

$$W = RI^2t$$
$$\frac{W}{RI^2} = t$$
$$t = \frac{W}{RI^2} = \frac{791.4375}{33.5 \times 1.5^2} = \frac{791.4375}{33.5 \times 2.25} = 10.5$$

Q17 Find the current passing through a wire to generate 360,000 joules of heat energy if the total resistance is 50 ohms during a period of 2 seconds. Substitute the values given into the formula $W = RI^2t$ and solve the resulting equation for I.

• # # # • # # # • # # # • # # # • # # # • # # # • # # # • # #

A17 60 amps:

$$W = RI^2t$$
$$360{,}000 = 50 \times I^2 \times 2$$
$$360{,}000 = 100I^2$$
$$3600 = I^2$$
$$\pm 60 = I$$

(Note: $^{-}60$ would have no practical meaning.)

13 The total resistance, R_T, in ohms, of a parallel circuit with resistances R_1 and R_2, in ohms, is given by the formula

$$R_T = \frac{1}{\frac{1}{R_1} + \frac{1}{R_2}}$$

Q18 Find the total resistance if $R_1 = 5$ ohms and $R_2 = 10$ ohms.

• # # # • # # # • # # # • # # # • # # # • # # # • # # # • # #

A18 $3\frac{1}{3}$ ohms: $R_T = \dfrac{1}{\frac{1}{R_1}+\frac{1}{R_2}} = \dfrac{1}{\frac{1}{5}+\frac{1}{10}} = \dfrac{1}{0.2+0.1} = \dfrac{1}{0.3} = \dfrac{10}{3} = 3\frac{1}{3}$

14 An equivalent form of $R_T = \dfrac{1}{\frac{1}{R_1}+\frac{1}{R_2}}$ may be obtained in the following manner.

1. $R_T = \dfrac{1}{\frac{1}{R_1}+\frac{1}{R_2}}$

2. $R_T = \dfrac{1 \cdot R_1 \cdot R_2}{\left(\frac{1}{R_1}+\frac{1}{R_2}\right)R_1 \cdot R_2}$

Multiplying numerator and denominator of the right side of (1) by R_1R_2.

3. $R_T = \dfrac{R_1R_2}{\frac{1}{R_1} \cdot R_1R_2 + \frac{1}{R_2} \cdot R_1 \cdot R_2}$

Applying the right distributive property of multiplication over addition to the denominator of (2).

4. $R_T = \dfrac{R_1R_2}{R_2+R_1}$

Simplifying the denominator of the right side of (3).

[Note: Solving 4 for R_2 will be an optional problem (Q20a).]

Q19 Use formula 4 above to find R_T when R_1 is 5 ohms and R_2 is 10 ohms.

• ### • ### • ### • ### • ### • ### • ### •

A19 $3\frac{1}{3}$ ohms: $R_T = \dfrac{R_1R_2}{R_1+R_2} = \dfrac{5 \cdot 10}{5+10} = \dfrac{50}{15} = 3\frac{1}{3}$

Compare this solution with that of Q18. Do you see the advantage of using the form of the formula from Frame 14 over the one from Frame 13?

15 An electrical technician must frequently find R_1 or R_2 given the other values. In such cases, it is convenient to have the formula expressed in terms of the correct subject.

Example: Find R_1 when $R_T = \frac{1}{5}$ ohm and $R_2 = \frac{1}{3}$ ohm. Use the formula $R_1 = \dfrac{R_TR_2}{R_2-R_T}$.*

*The derivation of the formula is shown below:

1. $R_T = \dfrac{R_1R_2}{R_1+R_2}$
2. $R_T(R_1+R_2) = R_1R_2$ — Multiplying both sides of 1 by R_1+R_2
3. $R_TR_1 + R_TR_2 = R_1R_2$ — Applying the left distributive property of multiplication over addition to the left side of 2
4. $R_TR_2 = R_1R_2 - R_TR_1$ — Subtracting R_TR_1 from both sides of 3
5. $R_TR_2 = R_1(R_2 - R_T)$ — Applying the left distributive property of mult. over subtraction to the right side of 4
6. $\dfrac{R_TR_2}{R_2-R_T} = R_1$ — Dividing both sides of 5 by R_2-R_T

$R_1 = \dfrac{R_TR_2}{R_2-R_T}$

Solution:

$$R_1=\frac{R_TR_2}{R_2-R_T}=\frac{\frac{1}{5}\cdot\frac{1}{3}}{\frac{1}{3}-\frac{1}{5}}=\frac{\frac{1}{15}}{\frac{2}{15}}=\frac{1}{2}$$

Therefore, $R_1=\frac{1}{2}$ ohm.

Q20 ***a.** Solve $R_T=\dfrac{R_1R_2}{R_1+R_2}$ for R_2.

b. Use the results of Q20a to find the value of R_2 when $R_T=\frac{1}{5}$ ohm and $R_1=\frac{1}{2}$ ohm.

• ### • ### • ### • ### • ### • ### • ### •

A20 ***a.**

$$R_T=\frac{R_1R_2}{R_1+R_2}$$
$$R_TR_1+R_TR_2=R_1R_2$$
$$R_TR_1=R_1R_2-R_TR_2$$
$$R_TR_1=R_2(R_1-R_T)$$
$$\frac{R_TR_1}{R_1-R_T}=R_2$$
$$R_2=\frac{R_TR_1}{R_1-R_T}$$

b. $\frac{1}{3}$ ohm: $R_2=\dfrac{R_TR_1}{R_1-R_T}=\dfrac{\frac{1}{5}\cdot\frac{1}{2}}{\frac{1}{2}-\frac{1}{5}}=\dfrac{\frac{1}{10}}{\frac{3}{10}}=\dfrac{1}{3}$

16 This section has illustrated a variety of uses of formulas. It is important to note that whenever a formula can be used to assist you in solving a particular problem, either

1. the formula has the correct subject (which means that following substitution of the known values, completion of the arithmetic will yield the desired solution—see Frame 7), or
2. it doesn't.

If it doesn't,

1. you can immediately substitute the known values and solve the resulting equation for the desired variable [see Q14(1)], or
2. you can change the subject of the formula to the desired variable and then substitute the known values and evaluate [see Q14(2)]. [Note: This choice is usually made when repeated calculations are required (see Q11).]

With the experience provided in this section, you should be able to cope with the wide variety of formulas with which you may be confronted. The Exercise which follows this section contains many additional formulas taken from a variety of technical fields. Your ability to solve them will greatly enhance your chances of success in your chosen area of interest.

This completes the instruction for this section.

9.3 EXERCISE

1. Given $p = a + b + c$, find a when $p = 32$, $b = 9$, and $c = 5$.
2. Given $p = 2l + 2w$, find w when $p = 29$ and $l = 4.5$.
3. Fiven $A = \frac{1}{2}bh$, find b when $A = 54$ and $h = 9$.
4. Given $c = 2\pi r$, find r when $c = 37.68$. Use $\pi \doteq 3.14$.
5. Given $V = \frac{1}{3}Bh$, find B when $V = 2{,}500$ and $h = 75$.
6. Given $A = p(1 + rt)$, find t when $A = 1{,}000$, $p = 750$, and $r = 0.06$.
7. Given $V_1 = V_0 + at$, find a when $V_1 = 50$, $V_0 = 42$, and $t = 2$.
8. Given $\frac{P_1V_1}{T_1} = \frac{P_2V_2}{T_2}$, find P_1 when $V_1 = 30$, $T_1 = 75$, $P_2 = 5$, $V_2 = 30$, and $T_2 = 60$.
9. Given $V = lwh$, find h when $V = 25$, $l = 10$, and $w = 10$.
10. Given $i = prt$, find p when $i = \$1{,}500$, $r = 5\%$, and $t = 3$ years.
11. Given $p = 2l + 2w$, determine the length of a rectangle if the perimeter is 35 centimetres and the width is 5.5 centimetres.
12. Given $A = bh$, where A = the area of a parallelogram, b = the base of the parallelogram, and h = the altitude of the parallelogram, determine the base of a parallelogram whose area is 51.8 square yards and whose height (altitude) is 14 yards.
13. Given $V = \pi r^2 h$, where V = the volume of a cylinder, r = the radius of the cylinder, and h = the height of the cylinder, find the height of a cylinder whose volume is 1,884 cubic centimetres when the radius of the cylinder is 10 centimetres. Use $\pi \doteq 3.14$.
14. Given $A = \frac{1}{2}h(b_1 + b_2)$, find the height of a trapezoid if the area of the trapezoid is 168 square metres when the two parallel bases are 15 and 17 metres.
15. Given $i = prt$, determine the time (in years) required to accumulate \$300 interest if \$1,000 is invested at $7\frac{1}{2}\%$ simple interest.
16. When working the following problems, select the appropriate formula from those given below:

$$d = rt, \quad p = \frac{sn}{6}, \quad h = \frac{Tn}{5252}, \quad \text{mph} = \frac{\text{rpm} \times W}{R \times 168}$$

   a. Find the time required to drive 2250 miles at an average speed of 55 mph. Round off to the nearest tenth.
   b. Find the torque (round off to nearest tenth) for an engine with 75 horsepower which is turning 3400 rpm.
   c. Find the number of revolutions per minute (round off to nearest tenth) the engine turns at a speed of 50 mph. The tire radius is 13 inches and the rear axle gear ratio is 4.
   d. Find the number of rpm an engine would turn in order for a piston in it to move at 2000 ft/min. if it has a 3-inch stroke.

17. To find the number of revolutions made by a driven gear in a given time you may use the formula

$$R = \frac{rt}{T}$$

where R = revolutions of the driven gear in a given time,
t = number of teeth of the driving gear,
r = revolutions driving gear makes in given time, and
T = teeth of driven gear.

a. A gear with 16 teeth, making 30 revolutions per minute, is driving a gear with 32 teeth. Find the revolutions per minute of the driven gear.

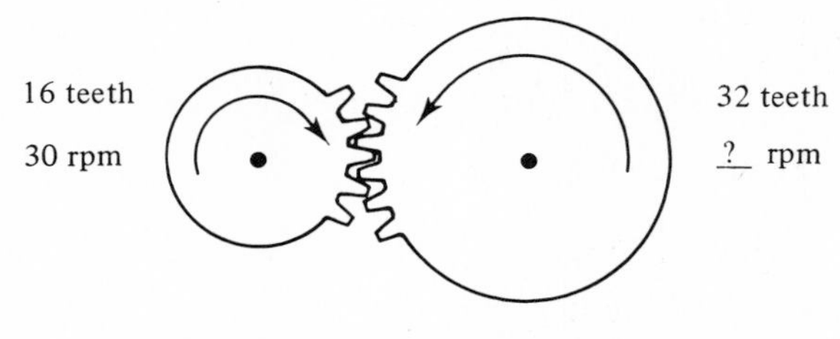

***b.** An axle should revolve at 360 rpm. The only engine available to run this axle is designed to run at 1800 rpm. If the gear on the axle has 120 teeth, how many teeth should there be on the gear used on the engine?

***18.** Find L_1 when $L_T = 3$ and $L_2 = 12$ if $L_T = \dfrac{L_1 L_2}{L_1 + L_2}$.

19. Find E when $P = 36$ and $R = 4$ if $P = \dfrac{E^2}{R}$.

20. The total resistance R_T in ohms in a parallel circuit with two resistances R_1, and R_2, in ohms, is given by the formula

$$R_T = \frac{1}{\frac{1}{R_1} + \frac{1}{R_2}} \quad \text{or} \quad R_T = \frac{R_1 R_2}{R_2 + R_1}$$

The electromotive force E, measured in volts, is given by $E = IR_T$ where I is current (amps).

The electrical power W, measured in watts, is given by the formula $W = IE$.

a. Find the total resistance, R_T, given the resistances R_1 and R_2 to be 10 ohms and 15 ohms, respectively.

b. Find the current, I, if the electromotive force, E, is 220 volts, given the total resistance, R_T, from problem 20a. Round off to the nearest tenth.

c. Find the electrical power, W, in watts if the current, I, is given from problem 20b and the electromotive force E is 220 volts.

21. The compression ratio, r, in an engine is given by the formula $r = \dfrac{c + d}{c}$

where c = volume of gases in final combustion chamber and
d = displacement of the cylinder.

The cubic centimetre displacement of a cylinder is 442 cubic centimetres (442 cc) and the final combustion chamber volume is 30 cc. What is the compression ratio (round off to nearest tenth)?

22. A formula which shows the relationship between the torque shown on a wrench dial and actual torque on the fastener is $T = D \times \left(\frac{L+A}{L}\right)$

where T = torque on fastener (ft-lbs),
D = dial reading on wrench (ft-lbs),
L = length of original wrench, and
A = length of adapter.

a. Find the torque on a fastener if a 19-inch wrench has a 19-inch adapter and the dial reading on the wrench is 60 ft-lbs.

b. You have to apply 90 ft-lbs of torque to a fastener. The torque wrench will not fit on the fastener unless you use a 9-inch adapter. If the torque wrench is 18 inches long, what should the dial value be in order to obtain the 90 ft-lbs?

23. Work, W, measured in ft-lbs, is determined by $W = fd$

where f = force (weight in lbs) and
d = distance, measured in feet.

An automotive technician moves a diesel engine weighing 1100 pounds from the floor to a workbench 4-feet high. How much work was done?

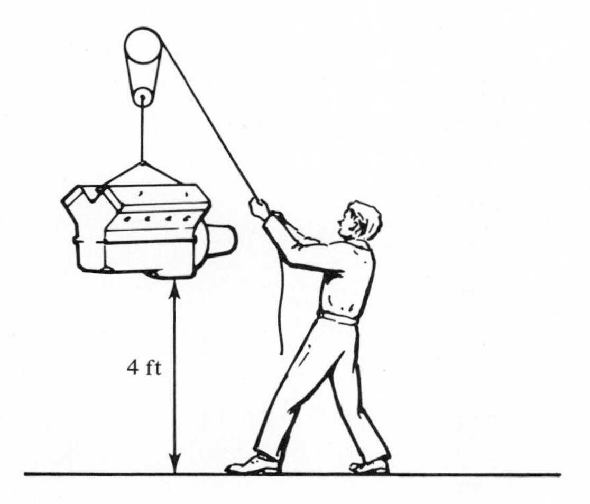

24. The thrust of a jet engine, F, measured in pounds, is given by the formula $F = \frac{M}{g}(v_2 - v_1)$

where M = mass rate of flow (lbs/sec),
g = acceleration of gravity (32.2 ft/sec/sec),
v_1 = final velocity (ft/sec), and
v_2 = initial velocity (ft/sec).

Compute the thrust of a jet engine in an airplane flying at 900 feet per second (v_1). The engine's mass rate of flow is 204 pounds per second. The velocity of the jet exhaust is 2100 feet per second (v_2). Round off to nearest tenth.

25. A cement company uses the formula $C = PN + d$ to determine the total charge, C

where P = price per brick,
N = number of bricks, and
d = delivery charge.

A mason orders 1000 bricks at $0.08 each from the above cement company. The total charge was $90. What was the delivery charge?

9.3 EXERCISE ANSWERS

1. 18 **2.** 10 **3.** 12 **4.** 6

5. 100 **6.** $\frac{50}{9}$ or $5\frac{5}{9}$ **7.** 4 **8.** $\frac{25}{6}$ or $4\frac{1}{6}$

9. $\frac{1}{4}$ or 0.25 **10.** \$10,000 **11.** 12 centimetres **12.** 3.7 yards

13. 6 centimetres **14.** 10.5 metres **15.** 4 years

16. a. 40.9 hrs **b.** 115.9 ft-lbs **c.** 2584.6 rpm **d.** 4000 rpm

17. a. 15 rpm ***b.** 24 teeth ***18.** $L_1 = 4$ **19.** $E = 12$

20. a. 6 ohms **b.** 36.7 amps **c.** 8074 watts

21. 15.7 **22. a.** 120 ft-lbs **b.** 60 ft-lbs

23. 4400 ft-lbs **24.** 7602.5 lbs **25.** \$10

CHAPTER 9 SAMPLE TEST

At the completion of Chapter 9 you should be able to work the following problems.

9.1 INTRODUCTION TO RATIO AND PROPORTION

1. Determine the ratio of (write in fraction form):
- **a.** 15 years to 10 years
- **b.** \$5.00 to 60 cents
- **c.** 1 foot to 4 yards
- **d.** $2\frac{1}{2}$ to $3\frac{1}{4}$

2. a. Name the means in the proportion $\frac{3}{8} = \frac{12}{32}$.
- **b.** What is the product of the extremes in the proportion $7:12 = 14:24$?
- **c.** Is $\frac{2}{3} = \frac{4}{5}$ a true proportion?
- **d.** Why?

3. a. A small pulley turns 420 rpm and is belted to a larger one turning 80 rpm. What is the ratio of their speeds?
- **b.** One hundred eighteen machine screws cost \$3.27. What is the cost per screw (round off to the nearest cent)?

9.2 SOLUTION OF A PROPORTION

4. Solve the following proportions:
- **a.** $\frac{16}{52} = \frac{M}{36}$
- **b.** $\frac{5}{12} = \frac{M}{36}$
- **c.** $\frac{M}{12} = \frac{2.5}{5}$
- **d.** $\frac{4}{5} = \frac{7}{M}$

5. Solve the following problems:
- **a.** Four bricklayers can brick a building in 15 hours. How long would it take with 6 men working at the same rate?
- **b.** If rain is falling at the rate of 0.5 inch per hour, how many inches will fall in 15 minutes?

c. A well produces 4,800 gallons of water in 120 minutes. How long will it take to produce 5,700 gallons of water?
d. An airplane travels from one city to another in 160 minutes at the rate of 270 miles per hour. How long will it take to make the return trip, traveling at 240 miles per hour?

6. A gear with 45 teeth turning 440 rpm meshes with a gear with 220 teeth. How many rpm does the 220-tooth gear turn?

7. An object, placed on a lever, is 6 feet from the fulcrum and balances a second object 8 feet from the fulcrum. If the first object weighs 180 pounds, how much does the second object weigh?

8. A measurement on a scale drawing $\left(\frac{1}{4}\text{ in.} = 1\text{ ft}\right)$ is $6\frac{3}{8}$ inches. What is the actual distance represented?

9. A pulley 15 inches in diameter turning 700 rpm is belted to a pulley turning 420 rpm. What is the diameter of the second pulley (round off to the nearest tenth)?

10. A gear with 18 teeth drives a gear with 42 teeth. If the smaller gear turns 25 rpm, how many rpm does the larger gear turn (round off to the nearest tenth)?

9.3 SOLVING WORD PROBLEMS THAT INVOLVE FOMULAS

11. When solving these problems, substitute the given information into the appropriate formula, and solve the resulting equation for the desired value:
a. Given $V_1 = V_0 + at$, find t when $V_1 = 18.5$, $V_0 = 3.5$, and $a = 5$.
b. Given $\frac{P_1V_1}{T_1} = \frac{P_2V_2}{T_2}$, find V_2 when $P_1 = 70$, $V_1 = 30$, $T_1 = 90$, $P_2 = 85$, and $T_2 = 102$.
c. Given the simple interest formula $i = prt$, where i = interest, p = principal, r = annual rate of interest, and t = time in years, determine the annual rate of interest required to accumulate \$165 interest if \$1,500 is invested for 2 years at simple interest.
d. Given $p = 2l + 2w$, where p = perimeter of a rectangle, l = length of the rectangle, w = width of the rectangle, find the width of a rectangle if the perimeter is 41 centimetres and the length is 13 centimetres.
e. Given $T = 2\pi r^2 + 2\pi rh$, where T = total surface of a cylinder, r = radius of the circular base, and h = height (altitude) of the cylinder, determine the height of a cylinder if the total surface of the cylinder is 748 square millimetres and the radius of the base is 7 millimetres. Use $\pi \doteq \frac{22}{7}$.
f. Given the formula $\text{mph} = \frac{\text{rpm} \times W}{R \times 168}$

where mph = car speed in miles per hour,
rpm = engine speed in revolutions per minute,
R = axle gear ratio, and
W = tire radius in inches.

Find the mph of a car with tire radius 12 inches, an axle gear ratio of 4, and an engine speed of 3000 rpm. Round off answer to nearest tenth.

12. a. Solve the formula in problem 11f for rpm.
b. Use the formula obtained in 12a to find the rpm of an engine driving a car at 52 mph if the axle gear ratio is 4 and the car has 14-inch tires.

13. A cement company uses the formula $C = PN + d$ to determine the total charge, C

where P = price per brick,
N = number of bricks, and
d = delivery charge.

a. A construction company orders 4000 bricks at $0.075 per brick. The total charge was $315. What was the delivery charge?

b. If the above construction company were to have ordered only 2000 bricks, each brick would have been $0.08 each. The delivery charge nevertheless would have been the same. In this case, what would be the total charge?

CHAPTER 9 SAMPLE TEST ANSWERS

1. a. $\frac{3}{2}$ **b.** $\frac{25}{3}$ **c.** $\frac{1}{12}$ **d.** $\frac{10}{13}$

2. a. 8 and 12 **b.** 168 **c.** no **d.** $3 \times 4 \neq 2 \times 5$

3. a. $\frac{21}{4}$ **b.** $0.03

4. a. $\frac{144}{13}$ or $11\frac{1}{13}$ **b.** 15 **c.** 6 **d.** $\frac{35}{4}$ or $8\frac{3}{4}$

5. a. 10 hours **b.** $\frac{1}{8}$ inch (0.125)

c. 142 minutes 30 seconds, or $142\frac{1}{2}$ minutes

d. 180 minutes

6. 90 rpm **7.** 135 pounds **8.** 25 ft 6 in. **9.** 25.1 in.

10. 10.7 rpm

11. a. 3 **b.** 28 **c.** 5.5% **d.** 7.5 centimetres

e. 10 millimetres **f.** 53.6 mph

12. a. $\text{rpm} = \frac{\text{mph} \times R \times 168}{W}$ **b.** 2496 rpm

13. a. $15 **b.** $175

CHAPTER 10

INSTRUMENTATION

The student studying in a technical field will be exposed to hundreds of measuring devices. Accuracy in reading the measurement is extremely important. This chapter will present numerous dials, gauges, and scales which may be encountered.

10.1 UNIFORM AND NONUNIFORM SCALES

1 There are two types of scales common to most measuring devices: uniform and nonuniform.

In a *uniform scale*, each graduation (space) is of the same length and represents the same value as all other graduations on the scale.

Example:

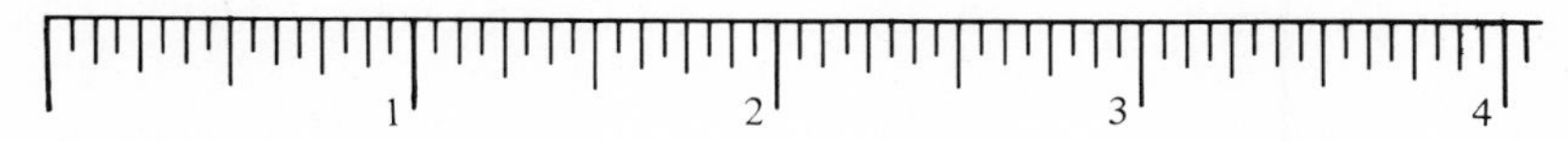

The portion of the ruler above is an example of a uniform scale in which each graduation represents $\frac{1}{16}$ of an inch.

A *nonuniform scale* is one in which the graduations vary in length and/or value.

Examples:

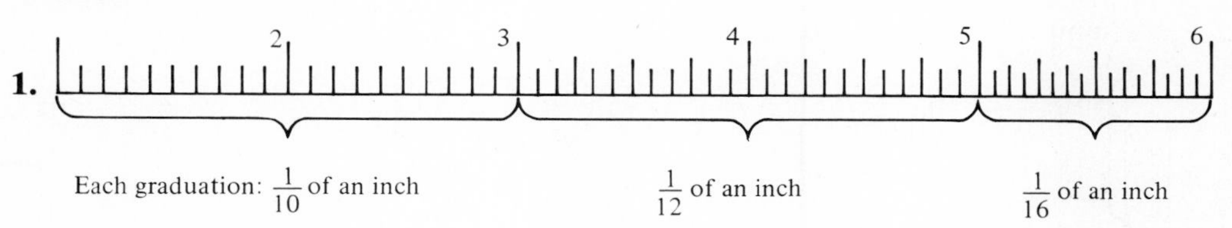

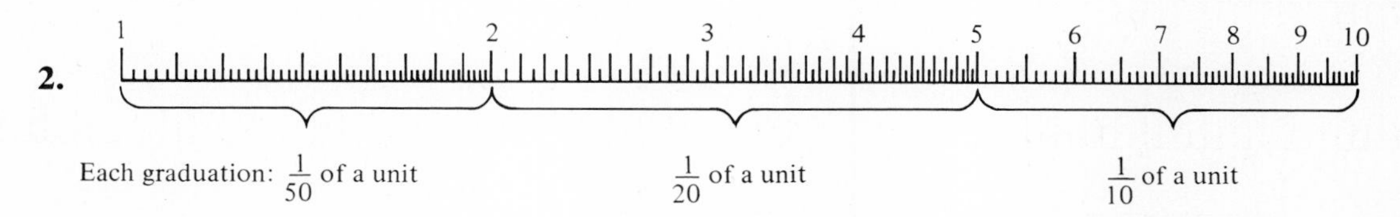

In addition to the difference in value of graduations, in this example the length of each graduation varies as well. Notice that the space between graduation marks gets smaller as you go from left to right.

Q1 Indicate whether each scale represents a uniform or nonuniform scale:

a.

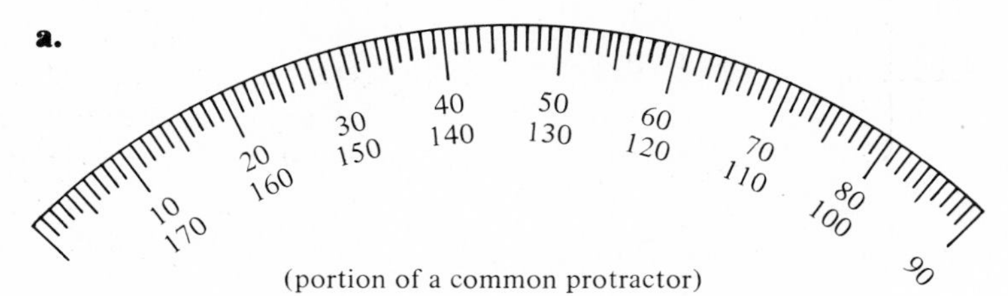

(portion of a common protractor) ____________

b. 2 3 4 5 ____________

\# # # • # # # • # # # • # # # • # # # • # # # • # # # • # # # • # # #

A1 **a.** uniform **b.** nonuniform

---

2 When using uniform or nonuniform scales to determine a measurement, it is necessary to determine the value of the graduations of the scale being used.

Example: What is the length of the nail below?

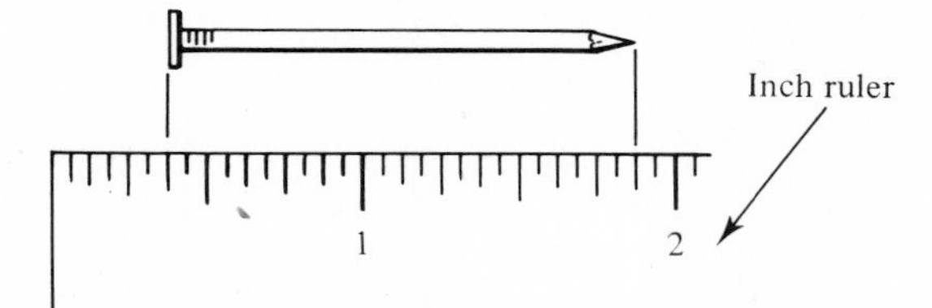

Solution:

Step 1: Count the number of graduations per inch. There are 16; hence, each graduation represents $\frac{1}{16}$ of an inch.

Step 2: Count the total number of graduations in the length of the nail. There are 24; hence, the length of the nail is $\frac{24}{16}$ or $1\frac{1}{2}$ inches long.

Q2 Determine the lengths of the items below:

a. ______ unit **b.** ______ unit **c.** ______ unit

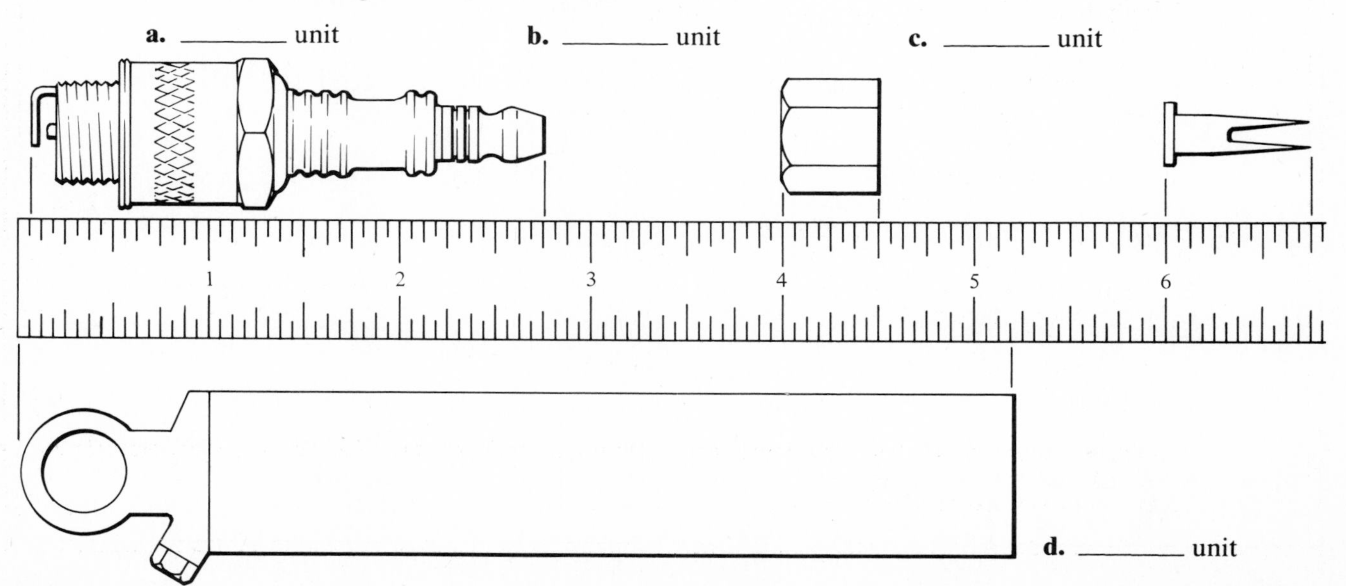

d. ______ unit

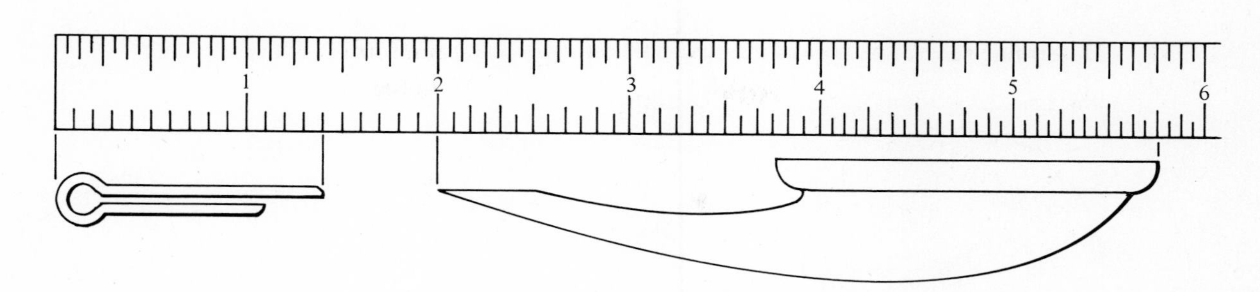

e. ________ inches **f.** ________ inches

\# \# \# • \# \# \# • \# \# \# • \# \# \# • \# \# \# • \# \# \# • \# \# \# • \# \# \# • \# \# \#

A2 **a.** $2\frac{11}{16}$ **b.** $\frac{1}{2}$ **c.** $\frac{3}{4}$ **d.** $5\frac{3}{16}$ **e.** 1.4 **f.** $3\frac{3}{4}$

3 A convenient method to determine the value of each graduation is to divide the value between two known graduation marks by the number of graduations (spaces).

Example 1: What is the value of each graduation?

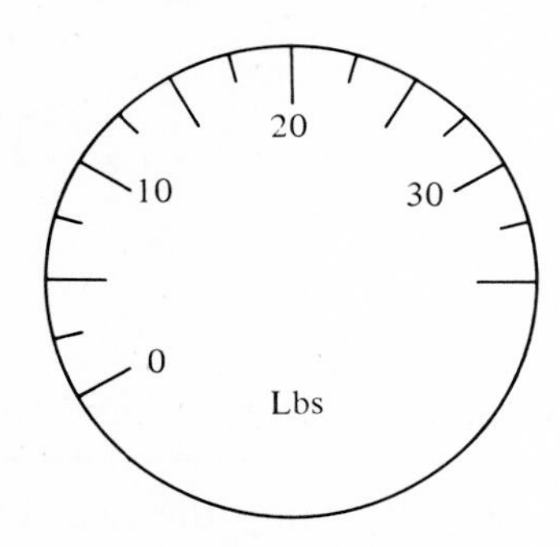

Solution: Between 0 and 10 there are 4 graduations. Count the spaces between 0 and 10, *not* the marks. Therefore, each graduation represents $\frac{10}{4}$ or 2.5 lbs.

Example 2: What is the value of each graduation?

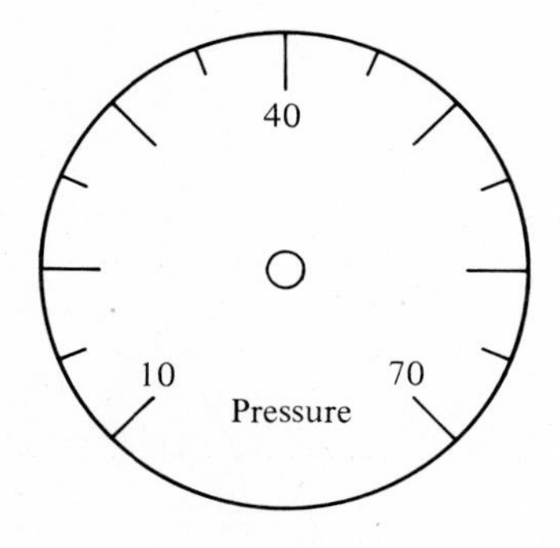

Solution: Between 10 and 40 there are 6 graduations (spaces). Therefore, each graduation represents a pressure of $\frac{40-10}{6}$ or 5 lbs.

Q3 Determine the value of each graduation:

a. ________ gallons

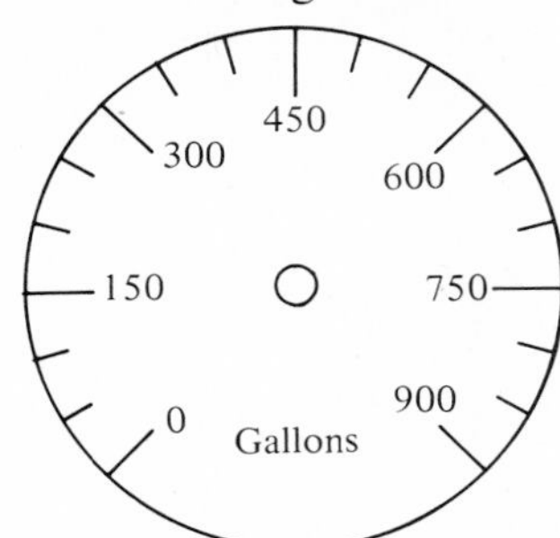

b. ________ unit

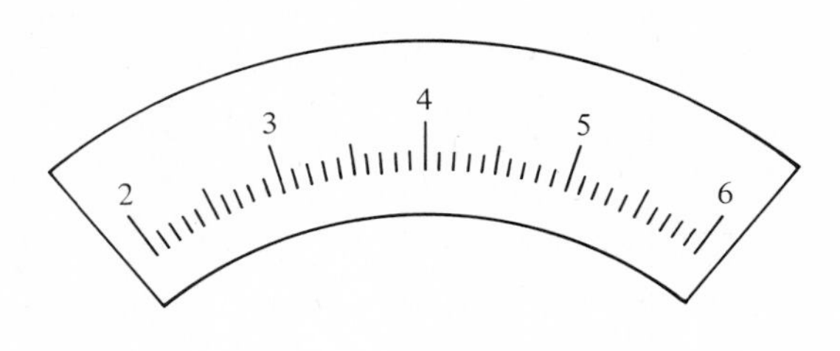

c. ________°

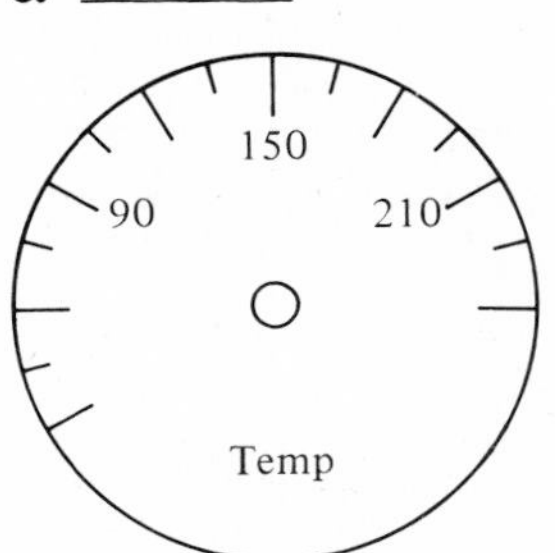

d. ________ unit

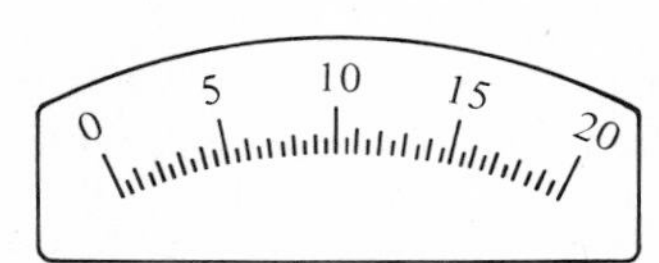

\# # # • # # # • # # # • # # # • # # # • # # # • # # # • # # # • # # #

A3 **a.** 50: $\frac{150}{3}$ **b.** 0.1: $\frac{1}{10}$

c. 15: $\frac{150-90}{4}$ **d.** 0.5: $\frac{5}{10}$ or $\frac{1}{2}$

4 When obtaining a reading from a dial, gauge, or scale, be sure to first determine the value of each graduation.

Example: What is the reading on the gauge below?

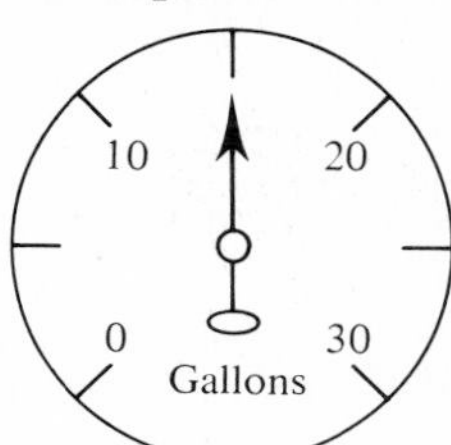

Solution: Each graduation represents 5 gallons. Hence, the reading is 15 gallons.

Q4 Determine the readings:

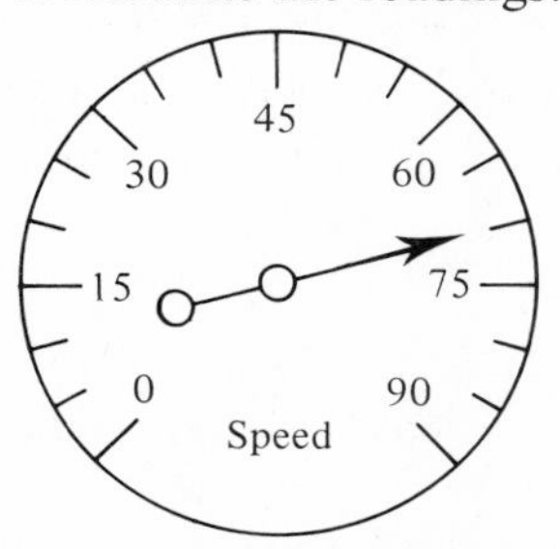

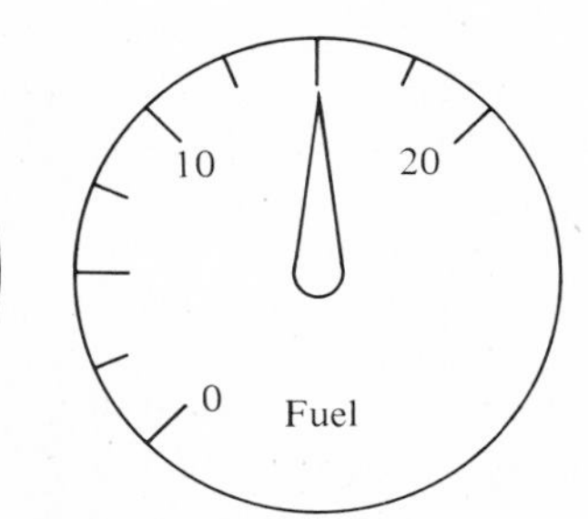

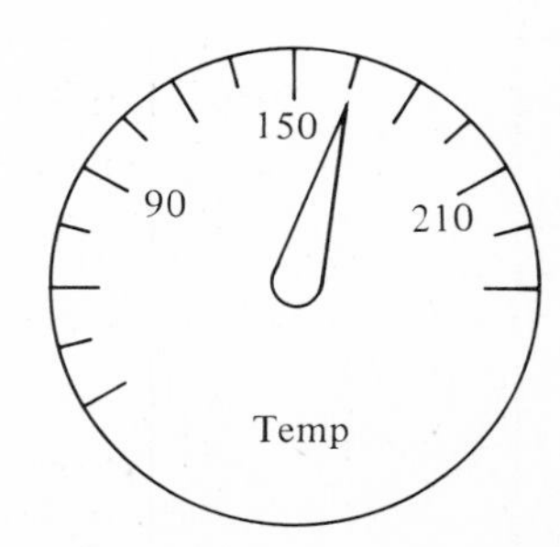

a. ________ rpm **b.** ________ lbs **c.** ________ gallons **d.** ________°

\# # # • # # # • # # # • # # # • # # # • # # # • # # # • # # # • # # #

A4 **a.** 70: Each graduation represents $\frac{15}{3} = 5$ rpm.

b. 25: Each graduation represents $\frac{40}{8} = 5$ lbs.

c. 15: Each graduation represents $\frac{10}{4} = 2.5$ gallons.

d. 165: Each graduation represents $\frac{60}{4} = 15°$.

Q5 Determine the readings:

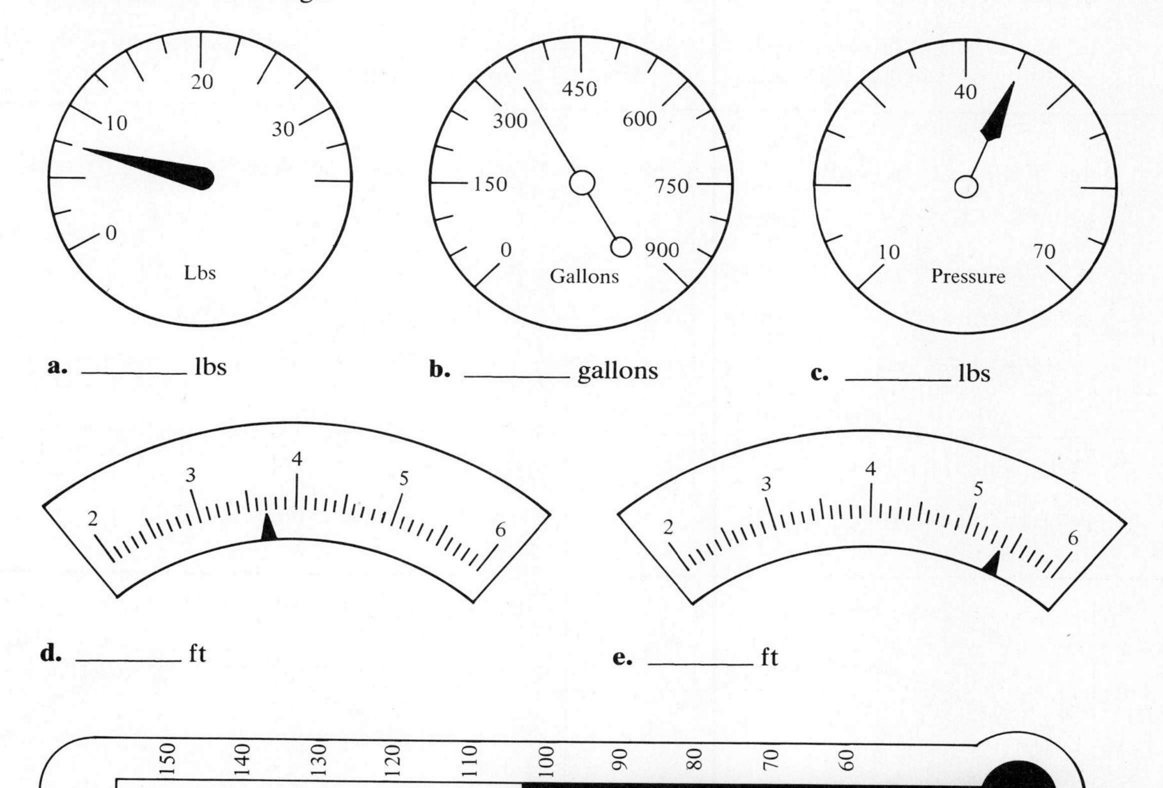

a. ________ lbs **b.** ________ gallons **c.** ________ lbs

d. ________ ft **e.** ________ ft

f. ________°

\# \# \# • \# \# \# • \# \# \# • \# \# \# • \# \# \# • \# \# \# • \# \# \# • \# \# \# • \# \# \#

A5 **a.** 7.5 **b.** 350 **c.** 45 **d.** 3.7 **e.** 5.4 **f.** 103

5 Reading nonuniform scales requires special care since graduations have different values depending on their location on the scale. On the ohmmeter (a device for measuring electrical resistance) observe the following:

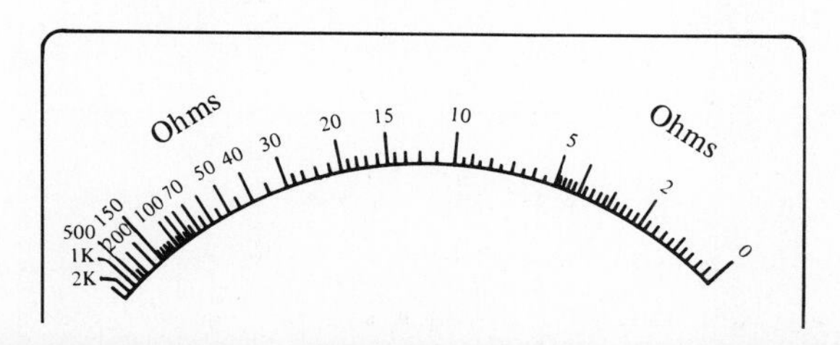

1. Between 0 and 5 each graduation represents $\frac{1}{5}$ or 0.2 ohms.
2. Between 5 and 10 each graduation represents $\frac{10-5}{10}=\frac{1}{2}$ or 0.5 ohms.
3. Between 10 and 20 each graduation represents $\frac{20-10}{10}=1$ ohm. The same result can be obtained by $\frac{15-10}{5}=1$ ohm.
4. Between 20 and 30 each graduation represents $\frac{30-20}{5}=2$ ohms.
5. The values of graduations between 30 and 100, 100 and 150, 150 and 200, 200 and 300, 300 and 500, 500 and 1 K (thousand), and 1 K and 2 K are all different.

Q6 Refer to the scale in Frame 5. What is the value of each graduation between:

a. 30 and 100? ________ ohms **b.** 100 and 150? ________ ohms

c. 150 and 200? ________ ohms **d.** 200 and 500? ________ ohms

e. 500 and 1 K? ________ ohms **f.** 1 K and 2 K ________ ohms

• # # # • # # # • # # # • # # # • # # # • # # # • # # # • # #

A6 **a.** 5 **b.** 10 **c.** 50 **d.** 100 **e.** 500 **f.** 1000

Q7 Determine the readings:

a. ________ ohms

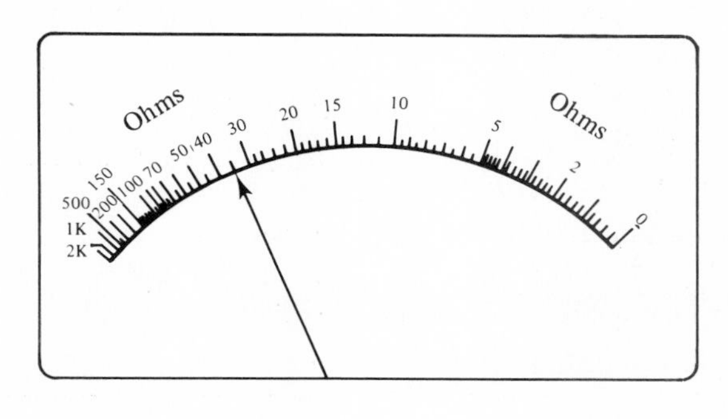

b. ________ ohms

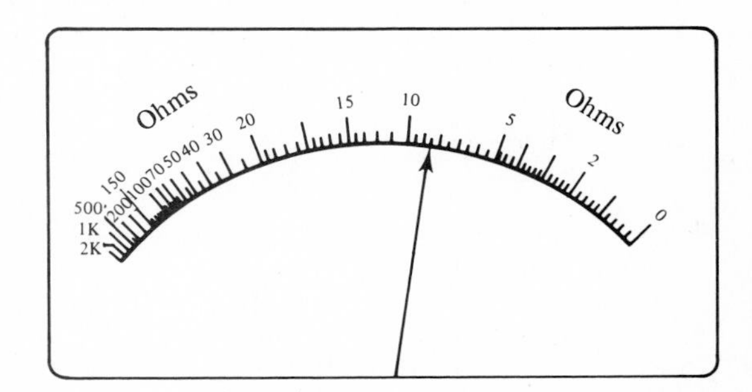

c. ________ ohms

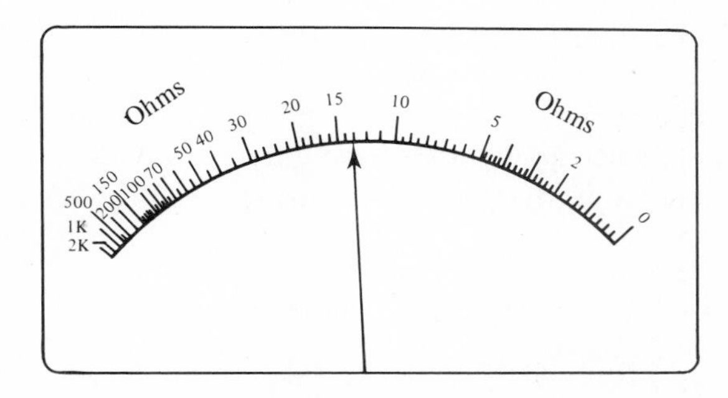

d. ________ ohms

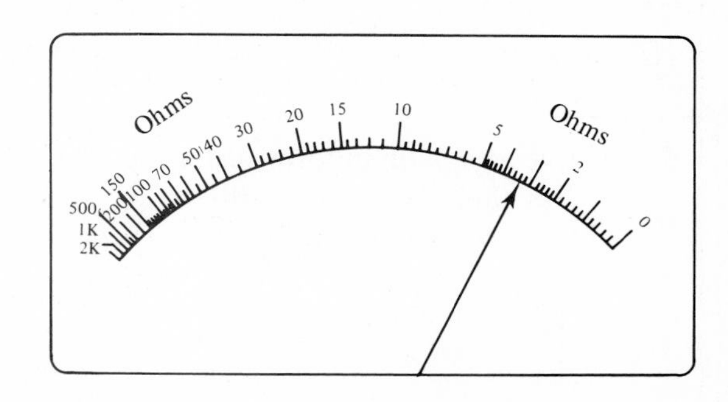

• # # # • # # # • # # # • # # # • # # # • # # # • # # # • # #

A7 **a.** 35 **b.** 8.5 **c.** 13 **d.** 3.4

6 The diagram below shows the AC (alternating current) scale which is typical of the one found on a volt-ohm milliameter (VOM). It is used to measure AC voltage in electrical circuits. Each graduation represents $\frac{1}{5}$ or 0.2 volts.

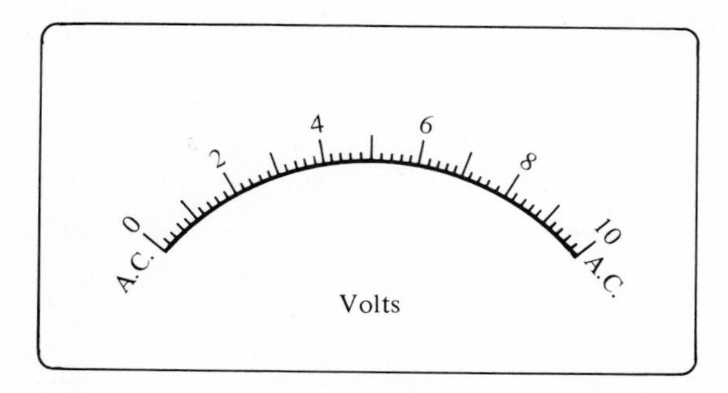

Q8 Indicate the reading of the scales:

a. ______ volts **b.** ______ volts

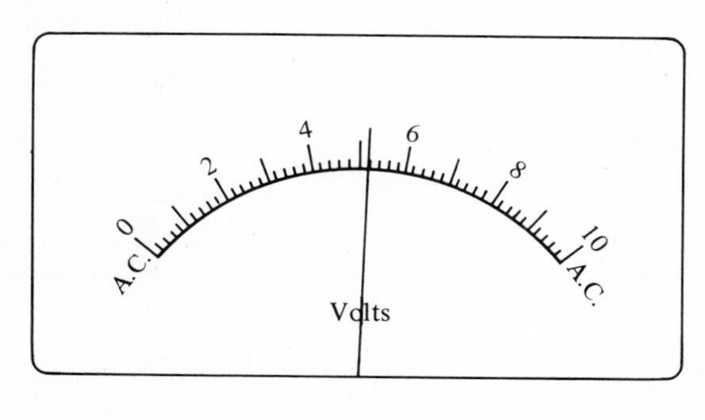

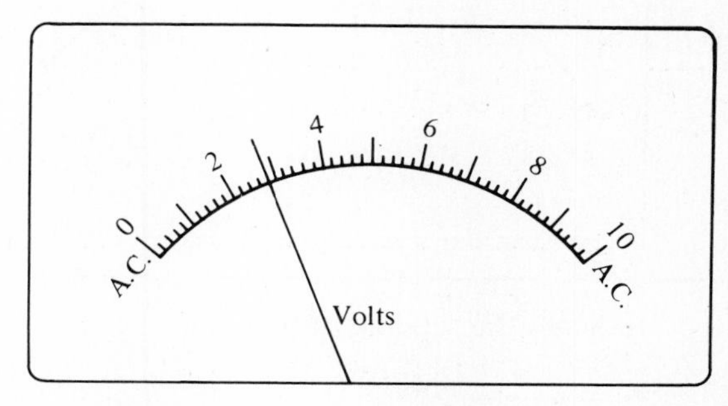

\# \# \# • \# \# \# • \# \# \# • \# \# \# • \# \# \# • \# \# \# • \# \# \# • \# \# \# • \# \# \#

A8 **a.** 5.2 **b.** 2.8

7 The diagram below shows an 2.5 AC volt scale (VAC) which is typical of the one found on a VOM. Each graduation represents 0.1 of a volt.

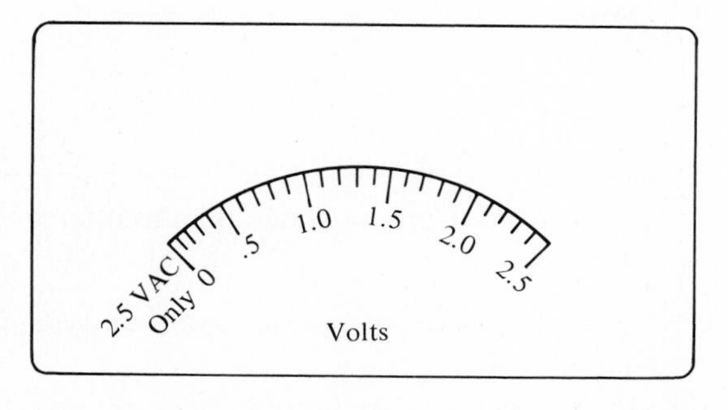

Q9 Indicate the reading of the scale:

a. ________ volts

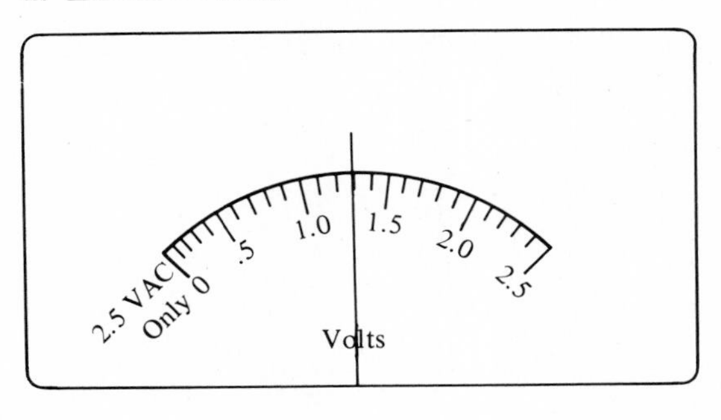

b. ________ volts

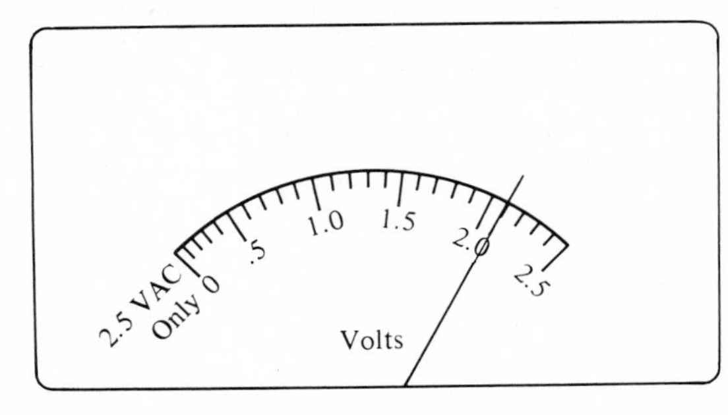

\# \# \# • \# \# \# • \# \# \# • \# \# \# • \# \# \# • \# \# \# • \# \# \# • \# \# \# • \# \# \#

A9 **a.** 1.3 **b.** 2.1

8 The diagram below shows a DC (direct current) scale which is typical of the one found on a VOM. Each graduation represents 5 volts.

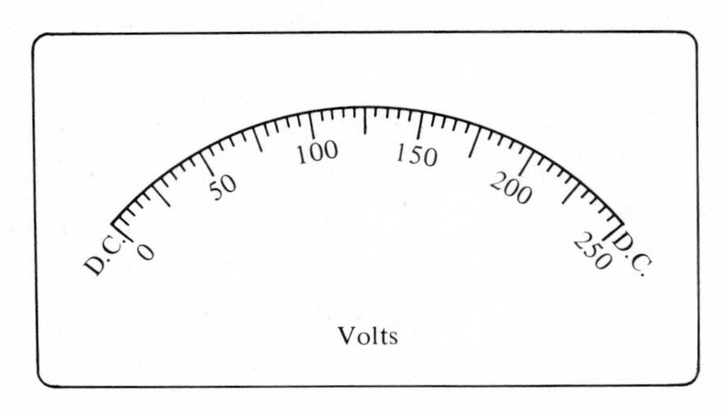

Q10 Indicate the reading of the scale:

a. ________ volts

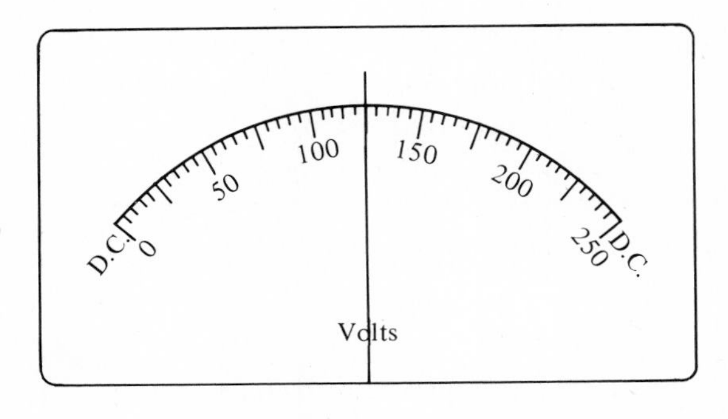

b. ________ volts

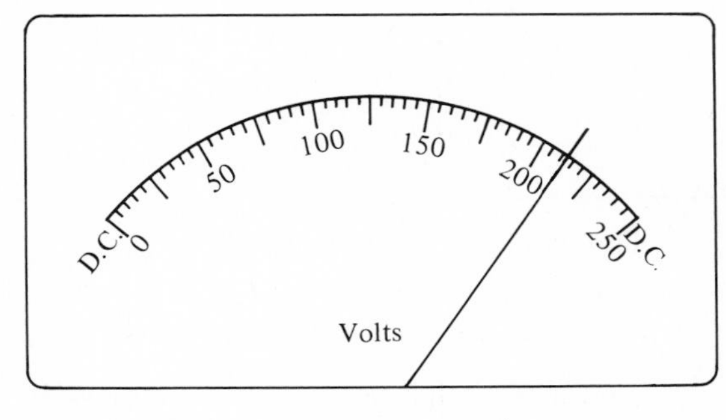

\# \# \# • \# \# \# • \# \# \# • \# \# \# • \# \# \# • \# \# \# • \# \# \# • \# \# \# • \# \# \#

A10 **a.** 125 **b.** 212

9 Usually more than one scale is present on the face of a VOM. Care must be taken to interpret each scale correctly.

1. Scale 1 is similar to the one in Frame 5. It is nonuniform; hence, the value of graduations vary.
2. Scales 2–5 are uniform, but the value of graduations on each scale is different from the others.
3. Scale 6 is nonuniform.

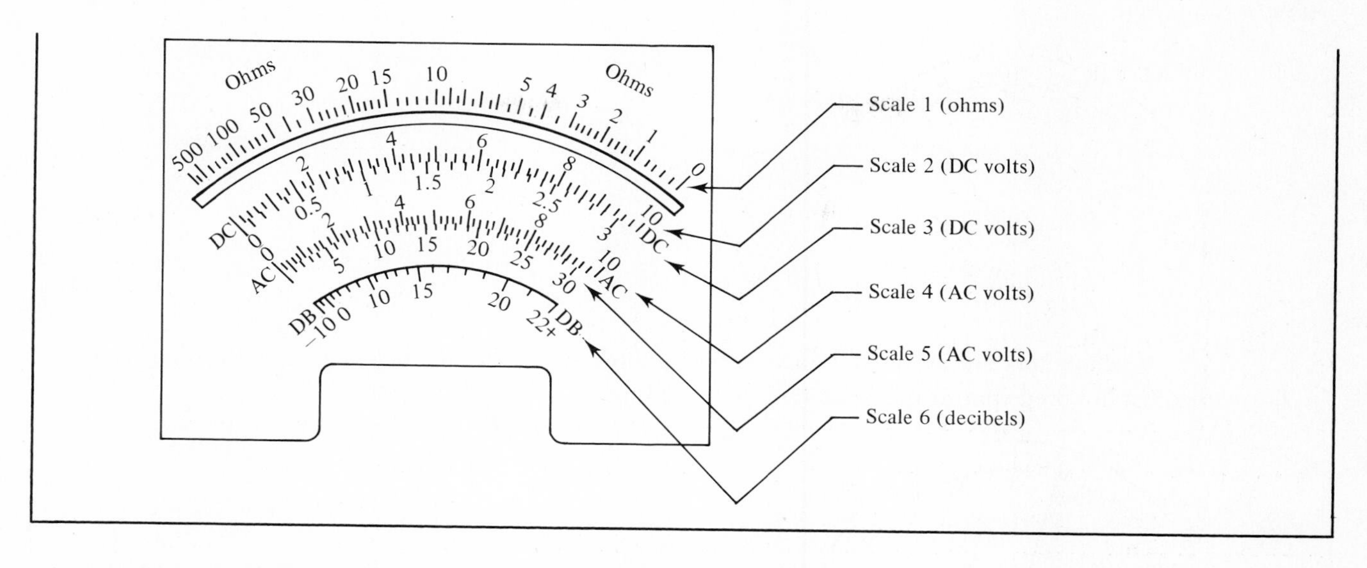

Q11 Determine the readings on the given scale (to the nearest whole unit or tenth of a unit):

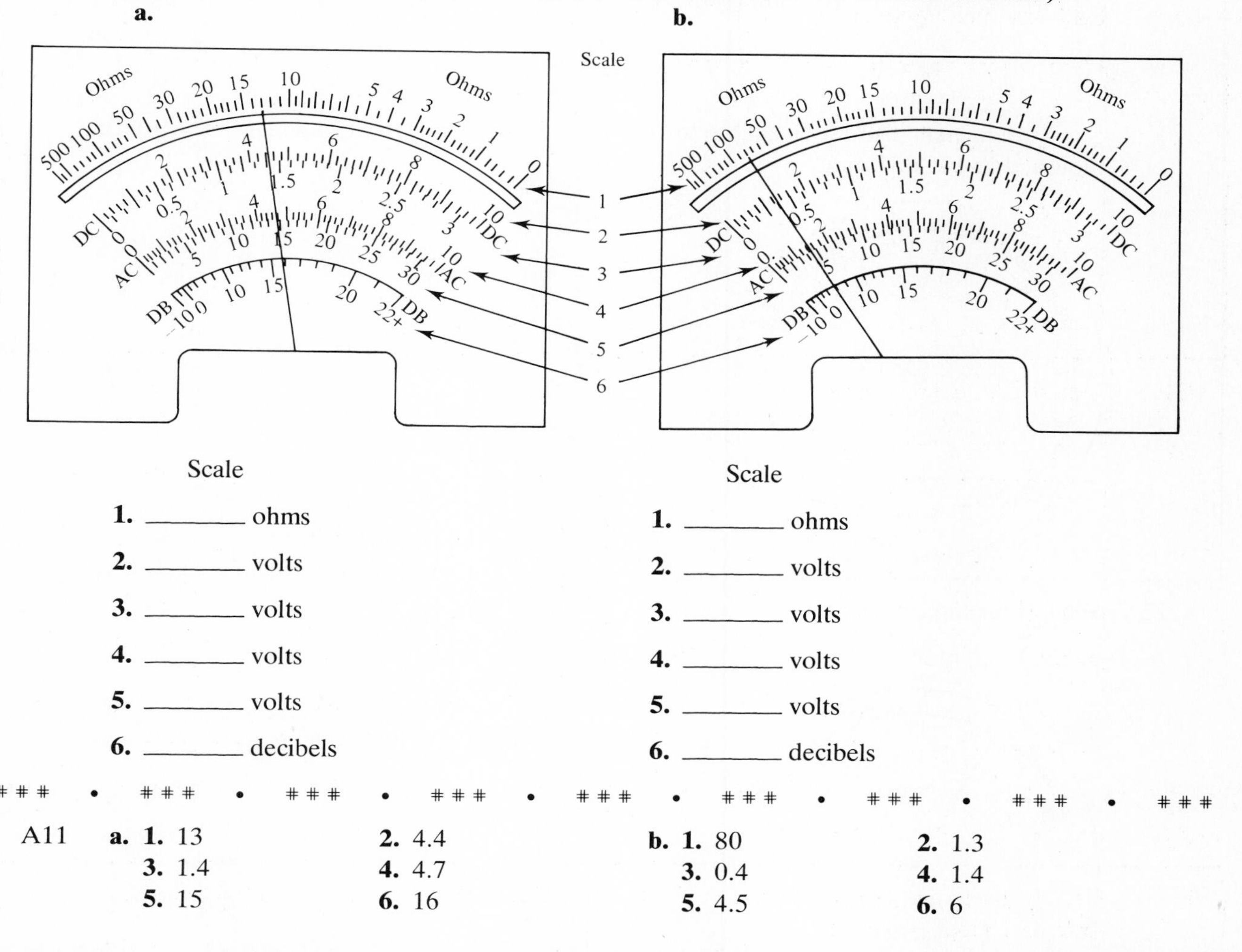

a.

Scale

1. ________ ohms
2. ________ volts
3. ________ volts
4. ________ volts
5. ________ volts
6. ________ decibels

b.

Scale

1. ________ ohms
2. ________ volts
3. ________ volts
4. ________ volts
5. ________ volts
6. ________ decibels

• ### • ### • ### • ### • ### • ### • ### •

A11 **a.** **1.** 13 **2.** 4.4 **3.** 1.4 **4.** 4.7 **5.** 15 **6.** 16

b. **1.** 80 **2.** 1.3 **3.** 0.4 **4.** 1.4 **5.** 4.5 **6.** 6

10 A *tachometer* is used to measure the number of revolutions the shaft of a motor makes with respect to a unit of time, usually minutes. Thus, a reading from a tachometer is generally in revolutions per minute (rpm). The reading from the tachometer shown would be 55 hundred or 5500 rpm.

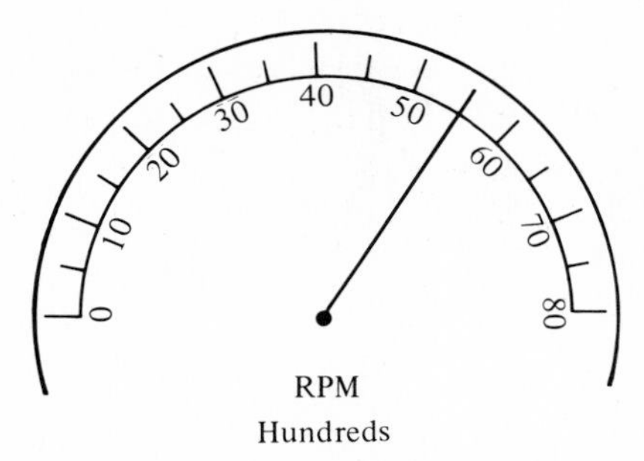

Tachometers are available in a variety of calibrations. The one below permits readings to the nearest hundred rpm. It reads 3300 (33 hundred) rpm.

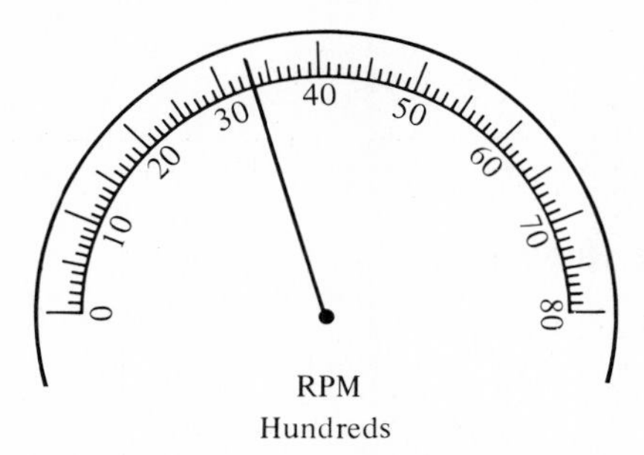

Tachometers also are available for higher speed motors. The tachometer is registering 105 thousand or 105,000 rpm.

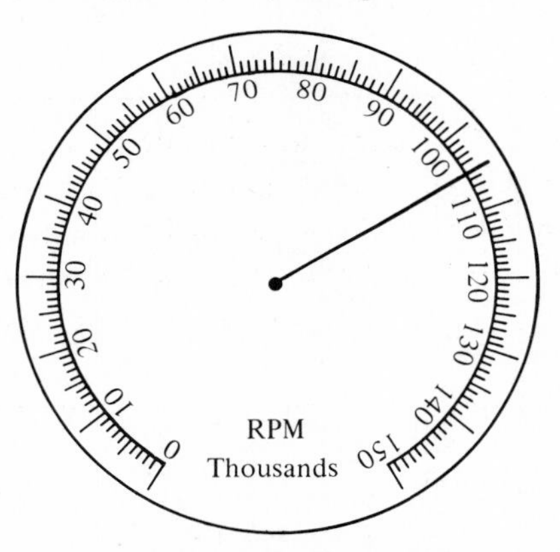

Q12 What is the indicated rpm?

a. _______ rpm

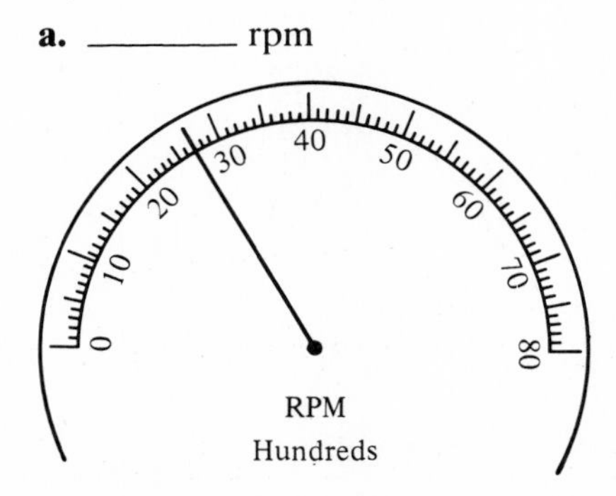

b. _______ rpm

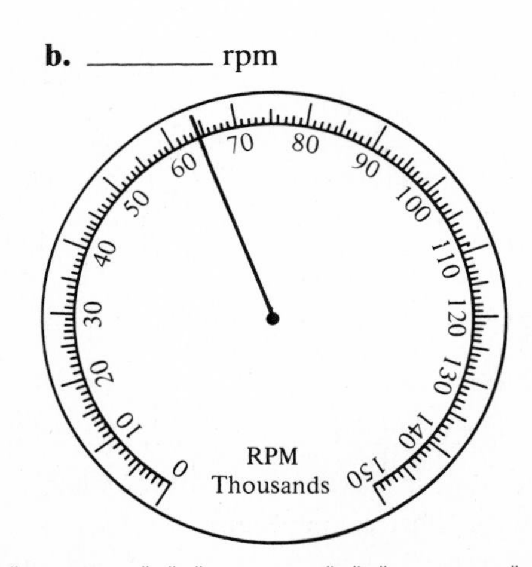

\# \# \# • \# \# \# • \# \# \# • \# \# \# • \# \# \# • \# \# \# • \# \# \# • \# \# \# • \# \# \#

A12 **a.** 2700 (27 hundred) **b.** 64,000 (64 thousand)

11 Below is the face of a Dwell-Tachometer Points Tester which may be used in timing an automobile engine. Again, care must be exercised in reading any given scale correctly.

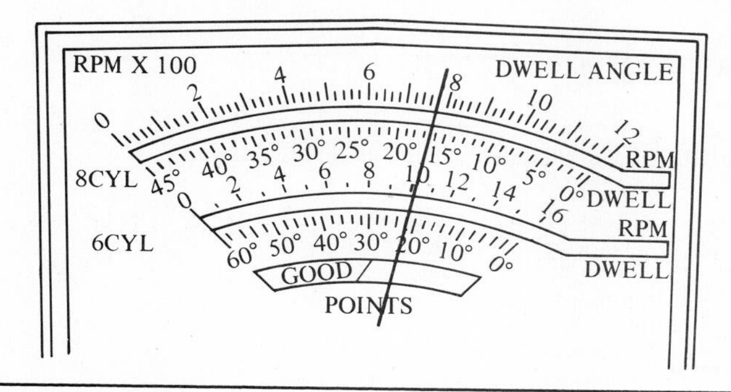

Q13 Refer to the diagram of Frame 11:

a. What is the rpm if an 8-cylinder engine is being tested? _______

b. What is the rpm if a 6-cylinder engine is being tested? _______

c. What is the dwell of the 8-cylinder engine? _______

d. What is the dwell of the 6-cylinder engine? _______

\# \# \# • \# \# \# • \# \# \# • \# \# \# • \# \# \# • \# \# \# • \# \# \# • \# \# \# • \# \# \#

A13 **a.** 780 rpm **b.** 1020 rpm **c.** 16.5° **d.** 22°

12 It would be impractical to attempt to illustrate all of the possible dials, gauges, meters, etc., which might be unique to a specific technical area. However, in each case the techniques discussed here should enable you to analyze the value of each graduation on the scale of the instrument to obtain an accurate reading. The exercise which accompanies this section and the sample test at the end of the chapter will contain scales which might not have been used here. Nevertheless, you should be able to read the scales correctly.

This completes the instruction for this section.

10.1 EXERCISE

1. Determine the following measurements with the inch ruler provided:

a. **b.**

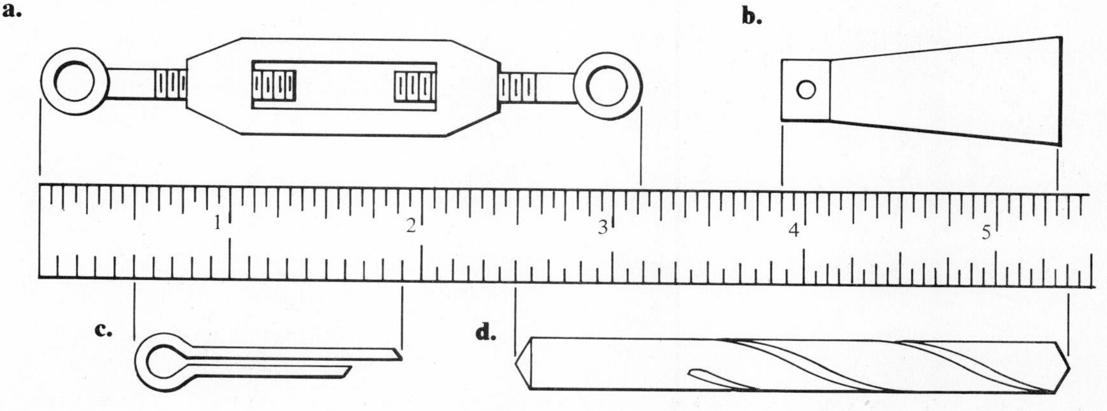

e.

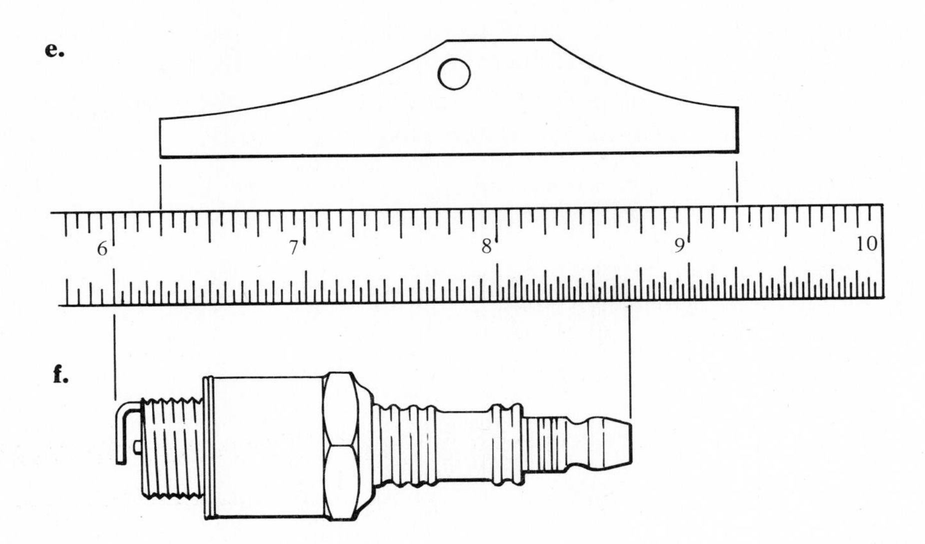

f.

2. Determine the following measurements with the metric ruler provided (the numbers indicate centimetres):

a.

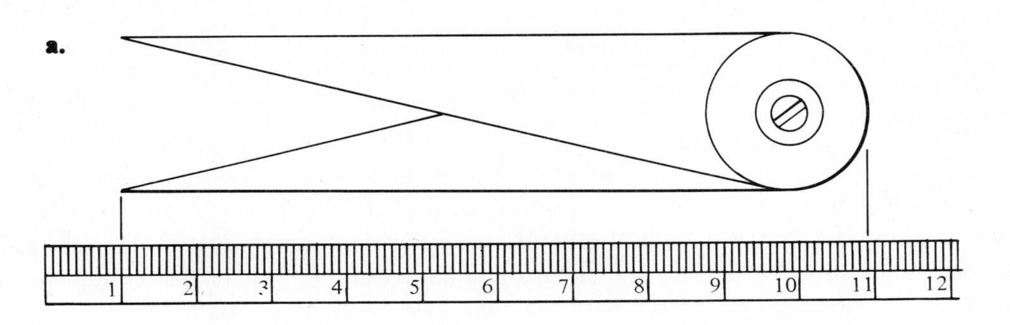

b.

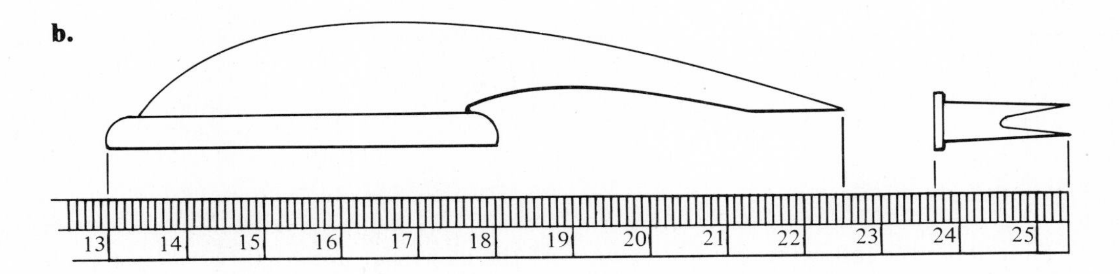

3. Determine the readings on the following scales (do not attach a unit of measurement):

a. **b.** **c.**

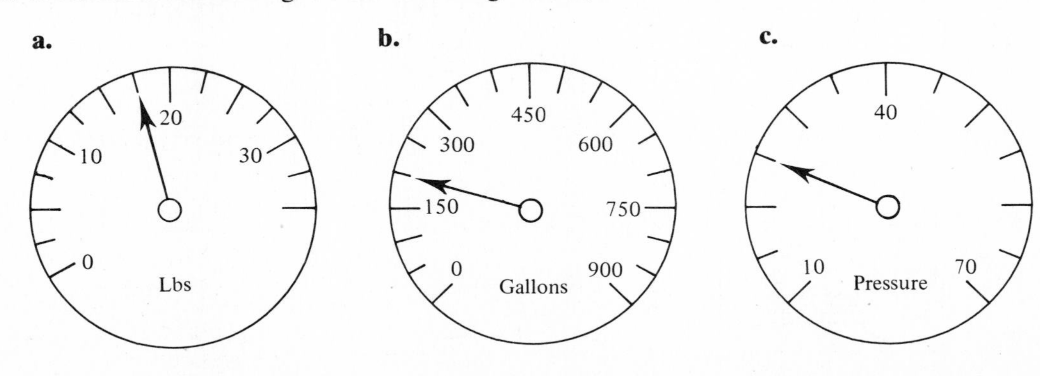

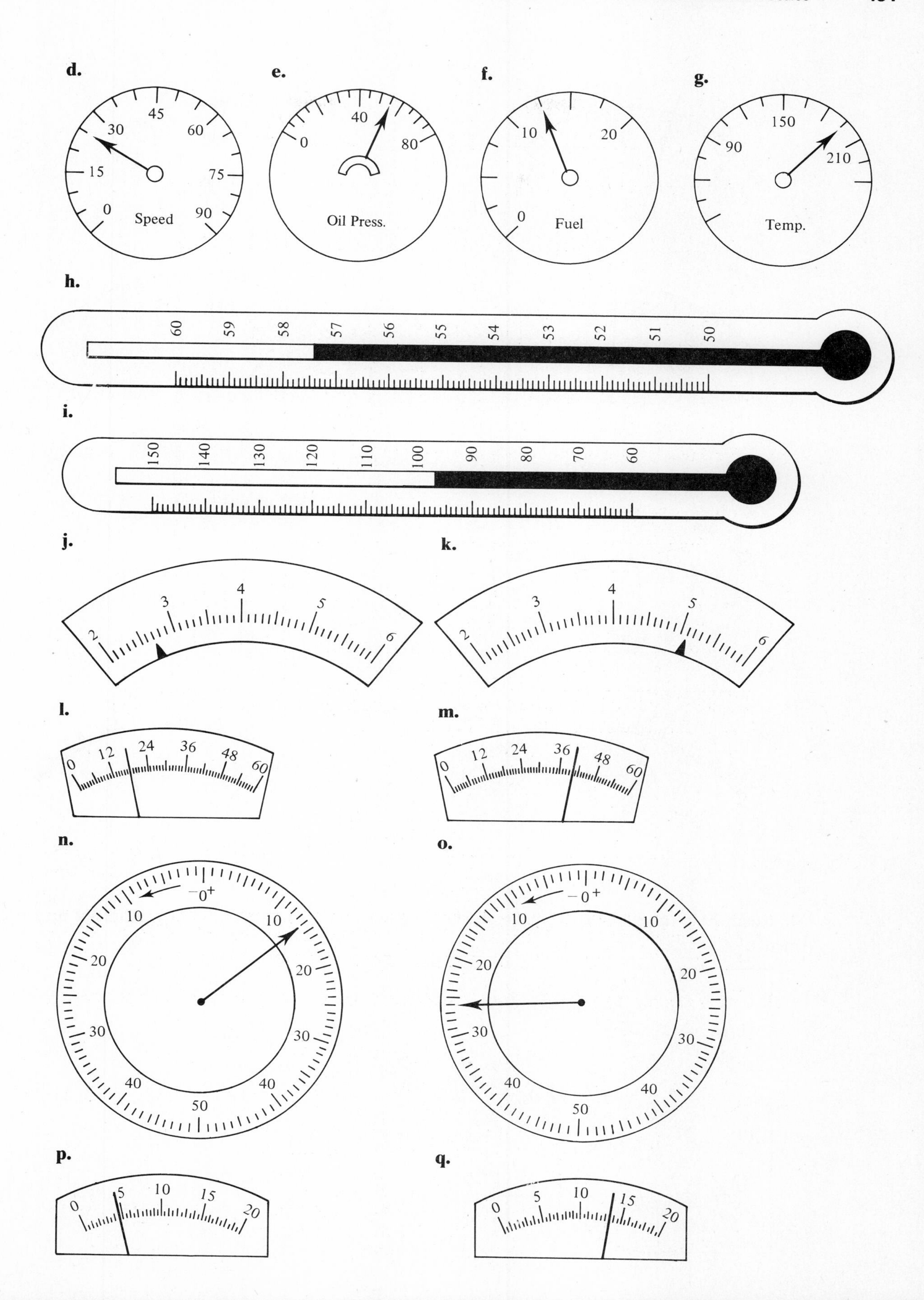
d.
e.
f.
g.
45
30
60
15
75
0
90
Speed
40
0
80
Oil Press.
10
20
0
Fuel
150
90
210
Temp.
h.
60
59
58
57
56
55
54
53
52
51
50
i.
150
140
130
120
110
100
90
80
70
60
j.
k.
2
3
4
5
6
l.
m.
0
12
24
36
48
60
n.
o.
−0+
10
20
30
40
50
p.
q.
0
5
10
15
20

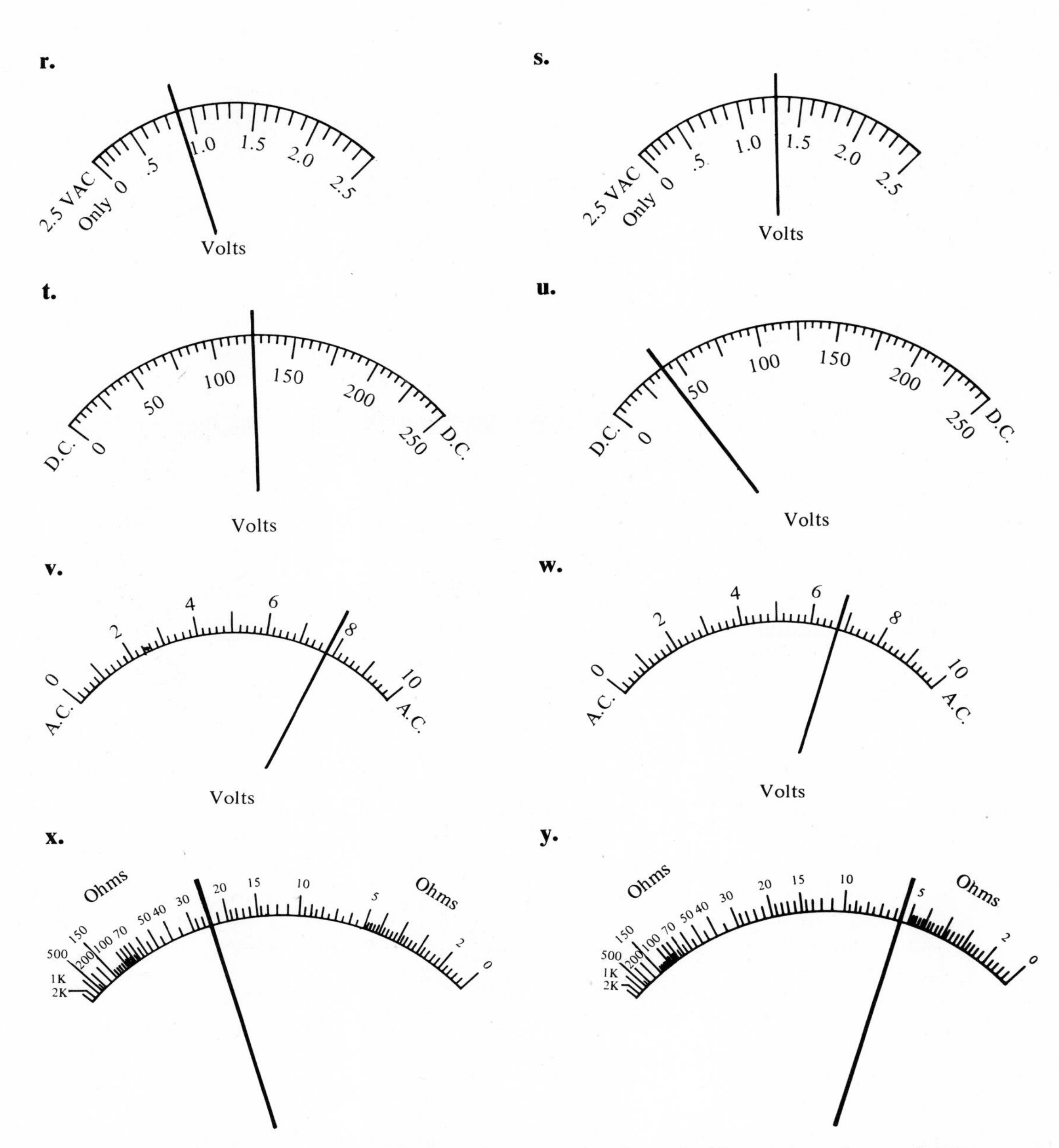

4. Read the given scale to the nearest whole or tenth of a unit (do not attach a unit of measurement):

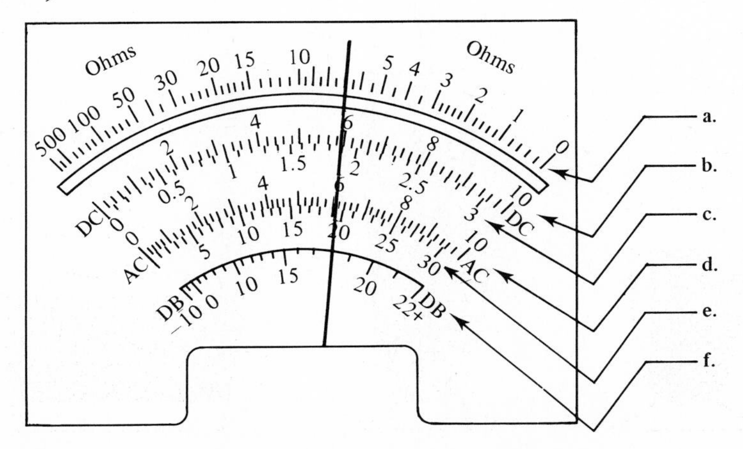

5. Determine the rpm:

a.

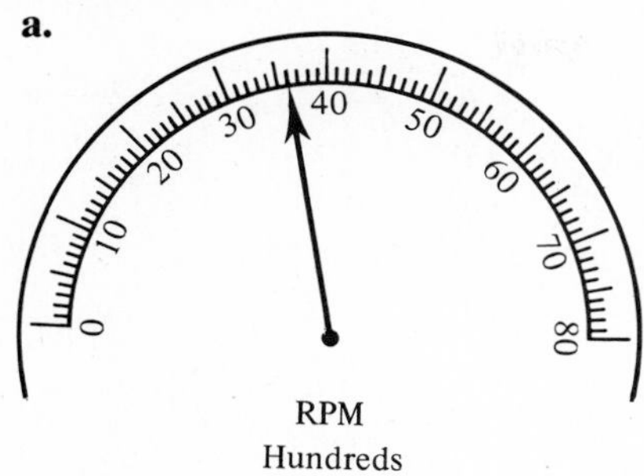

b.

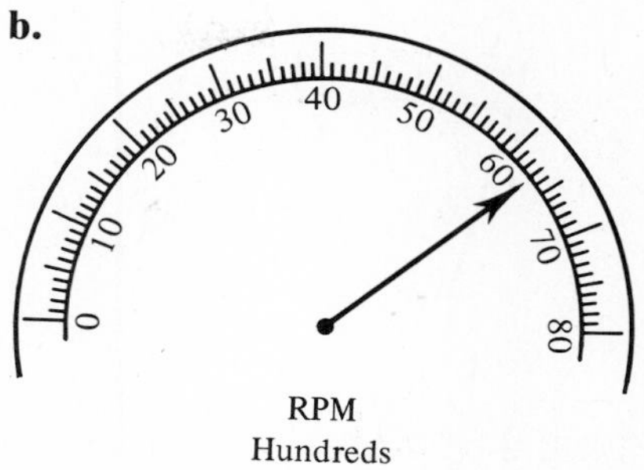

c.

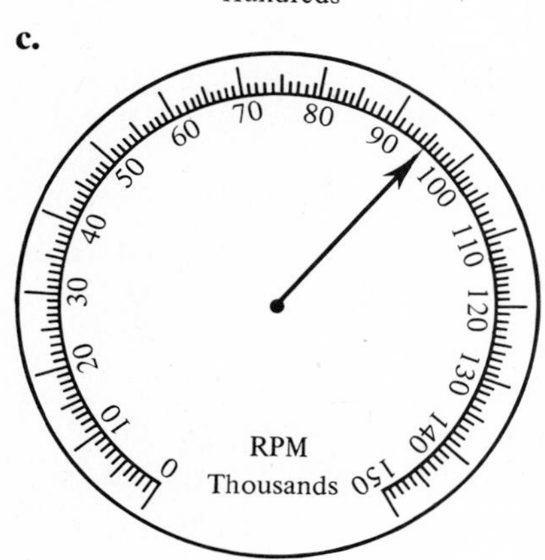

d.

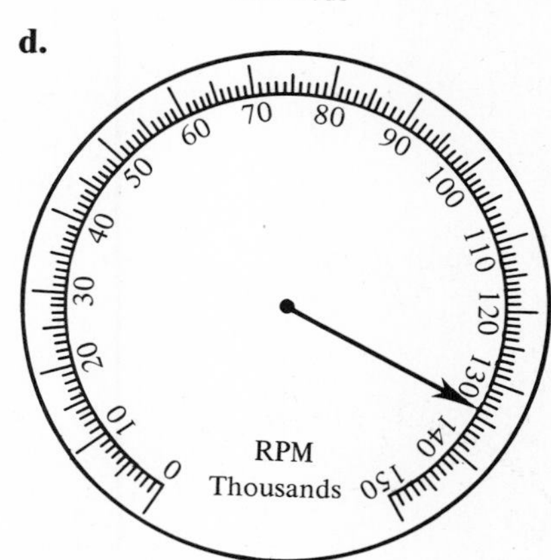

6. Read the given scale:

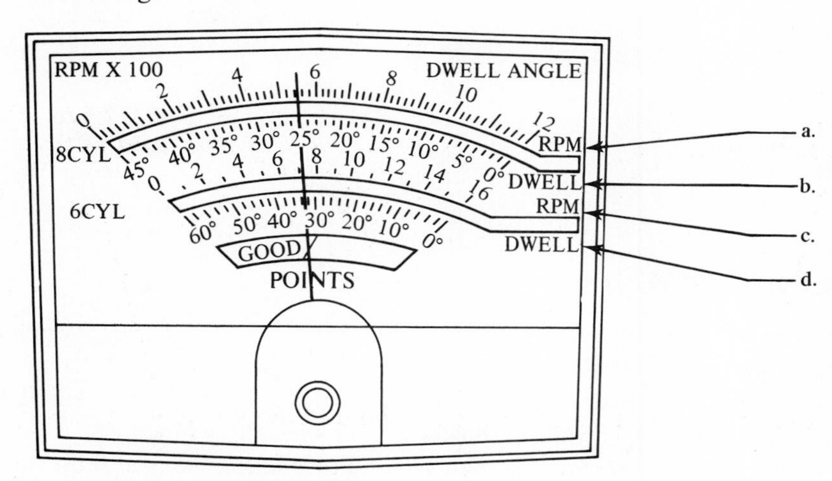

***7.** Determine the measurement:

a.

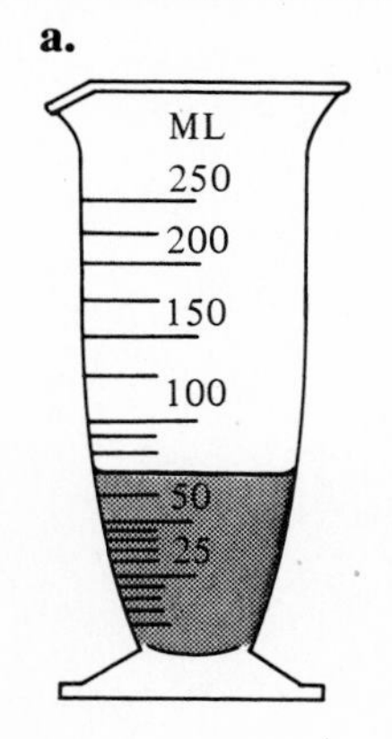

b.

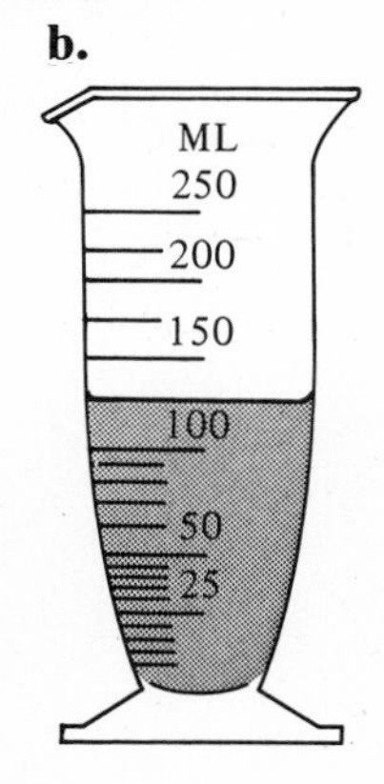

c.

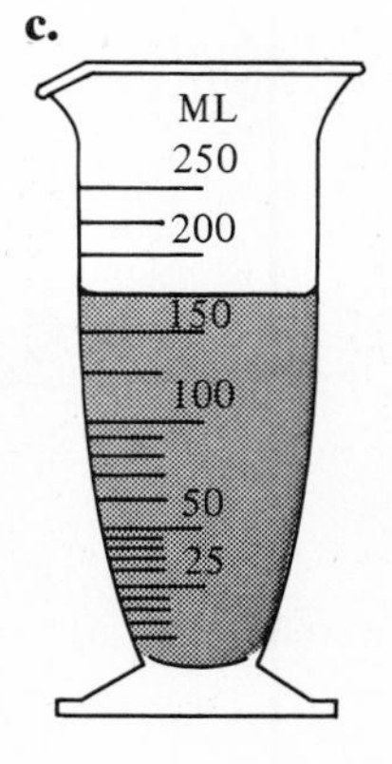

d.

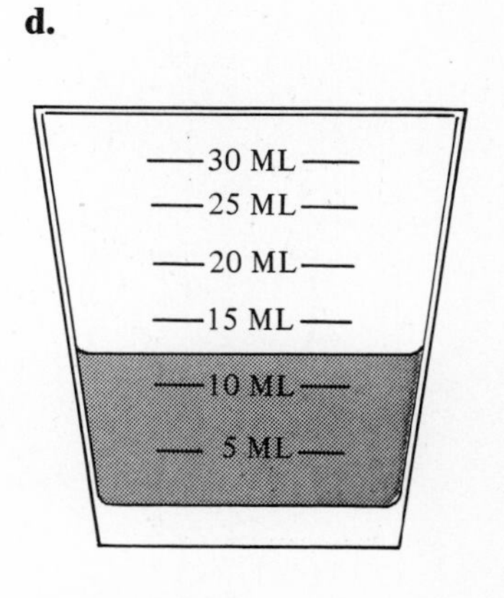

*8. Read the given scale:

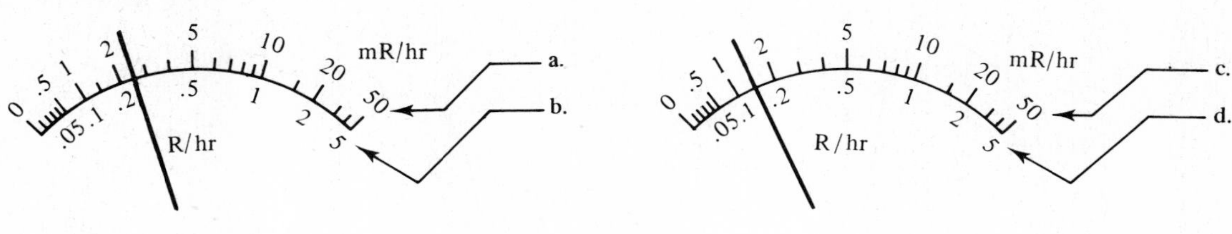

10.1 EXERCISE ANSWERS

1. a. $3\frac{1}{8}$ in. b. $1\frac{7}{16}$ in. c. 1.4 in.
   d. $2\frac{7}{8}$ in. e. 3 in. f. $2\frac{11}{16}$ in.
2. a. 9.9 cm or 99 mm b. 9.5 cm or 95 mm c. 1.7 cm or 17 mm
3. a. 17.5 b. 200 c. 25 d. 25
   e. 55 f. 12.5 g. 195 h. 57.4
   i. 97 j. 2.7 k. 5.1 l. 18
   m. 40 n. $^{+}14$ o. $^{-}26$ p. 4.6
   q. 13.5 r. 0.9 s. 1.3 t. 125
   u. 40 v. 7.8 w. 6.8 x. 24
   y. 5.5
4. a. 7 b. 6 c. 1.9
   d. 6.1 e. 19 f. 18
5. a. 3600 rpm b. 6300 rpm c. 95000 rpm d. 134000 rpm
6. a. 550 rpm b. 25° c. 700^{+} rpm d. 33°
*7. a. 70 ml b. 125 ml c. 175 ml d. 12.5 ml
*8. a. 2.5 mR/hr b. 0.25 R/hr c. 1.5 mR/hr d. 0.15 R/hr

10.2 MEASURING INSTRUMENTS

1 This section will discuss a variety of measuring instruments. These instruments contain uniform scales; hence, your experience in reading such scales in Section 1 of this chapter will be helpful.

The English *micrometer caliper* (mike) illustrated is especially valuable in measuring small distances where great accuracy is required. When the end of the thimble is at 0 on the sleeve, the end of the spindle should just touch the anvil. The object to be measured is placed between the anvil

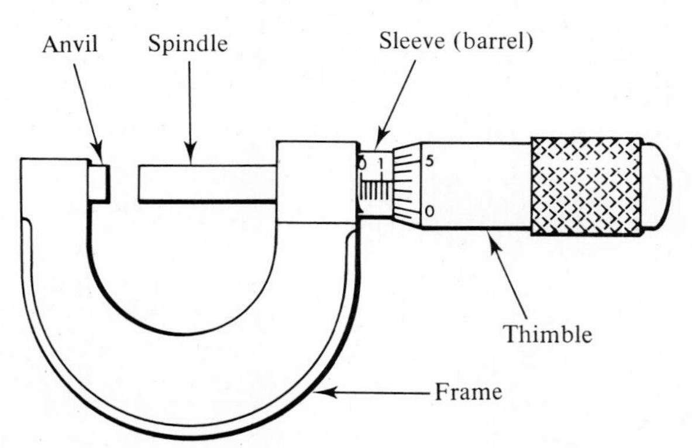

and the spindle. *One complete turn of the thimble changes the opening of the micrometer by* $\frac{1}{40}$ *of an inch.* (Note: $\frac{1}{40}$ in. $= 0.025$ in.). The sleeve is graduated on a line along the length of the spindle into divisions of $\frac{1}{40}$ inch each. Every fourth division (graduation mark) is labeled 1, 2, 3, etc. Hence, the numbered graduation marks represent tenths of an inch. (Note: $4 \times 0.025 = 0.1$)

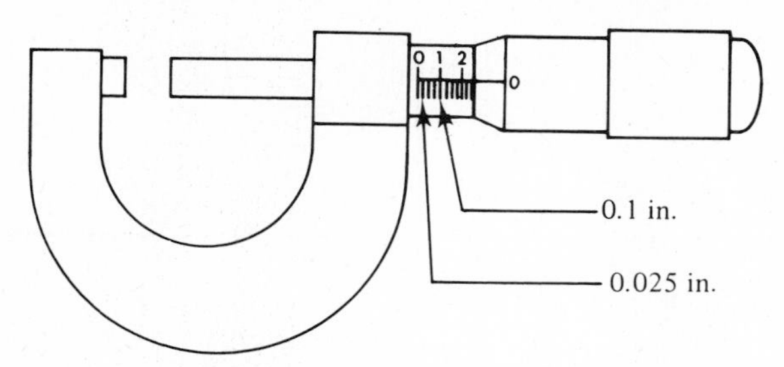

In the diagram, the thimble has been turned away from the 0 point. To determine the reading on the sleeve:

1. Note that 2 numbered graduation marks are exposed ($2 \times 0.1 = 0.2$), and
2. 2 small graduations are also exposed ($2 \times 0.025 = 0.050$). Therefore,

2 numbered graduation marks = 0.200 in.
2 small graduations (sleeve) = 0.050 in.
Reading = 0.250 in.

Q1 Determine the readings (only the exposed portion of the sleeve is shown):

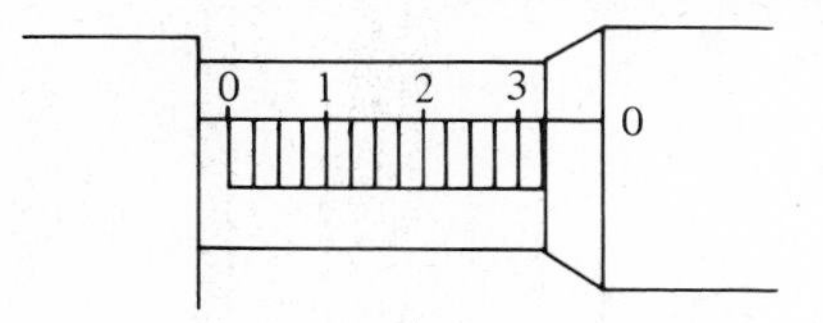

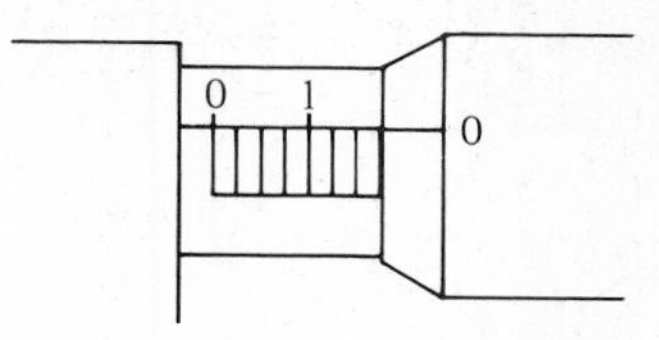

a. ______ **b.** ______

\# # # • # # # • # # # • # # # • # # # • # # # • # # # • # # # • # # #

A1 **a.** 0.325 in.:
0.3
+0.025
0.325

b. 0.175 in.:
0.1
+0.075
0.175

2 The thimble has a beveled end that is divided into 25 equal divisions. *A turn from one of these graduation marks to the next moves the spindle* $\frac{1}{25}$ *of 0.025 in.*, or *0.001 in.* To determine the reading indicated, examine the enlarged portion of the micrometer in the insert on the next page.

1. Notice that the sleeve shows 1 numbered graduation mark (1×0.1), and
2. 3 small graduations (sleeve) indicate 0.075 (3×0.025), and
3. the third division from 0 on the beveled edge of the thimble is on the centerline of the sleeve which indicates 0.003 (3×0.001). Thus,

1 numbered graduation marks = 0.100 in.
3 small graduation marks (sleeve) = 0.075 in.
3 graduation marks (thimble) = 0.003 in.
Complete reading = 0.178 in.

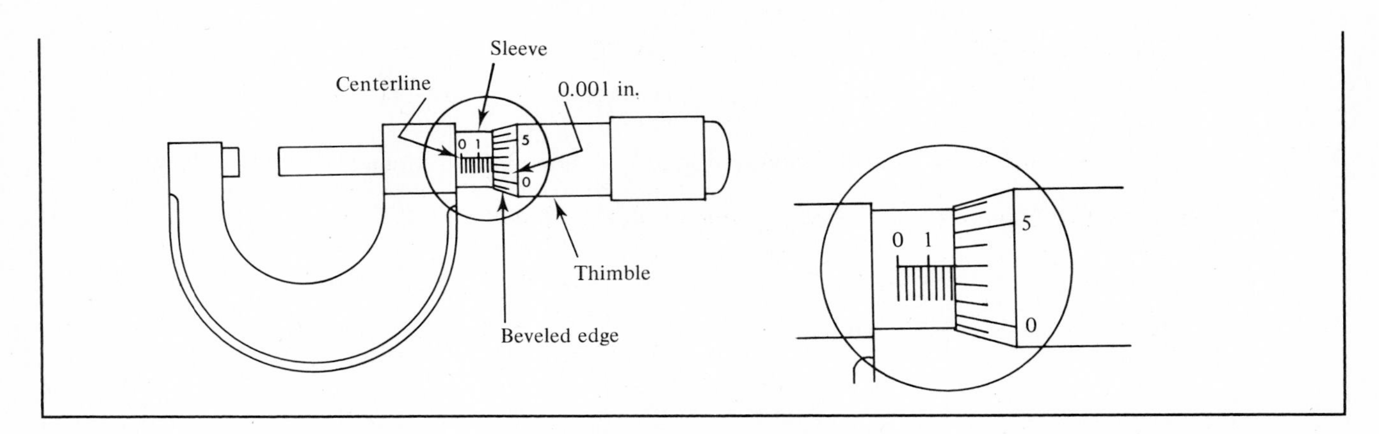

Q2 What is the reading on this English micrometer? ________

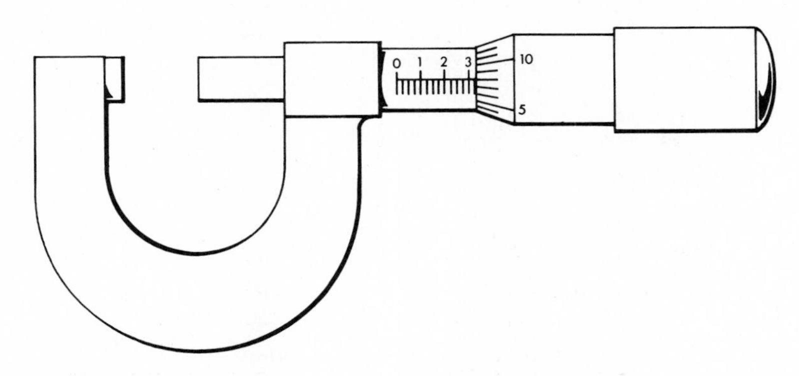

\# \# \# • \# \# \# • \# \# \# • \# \# \# • \# \# \# • \# \# \# • \# \# \# • \# \# \# • \# \# \#

A2 0.333 in.:

$$\begin{array}{r} 0.300 \\ 0.025 \\ +0.008 \\ \hline 0.333 \end{array}$$

Q3 Determine the readings on the following English micrometers:

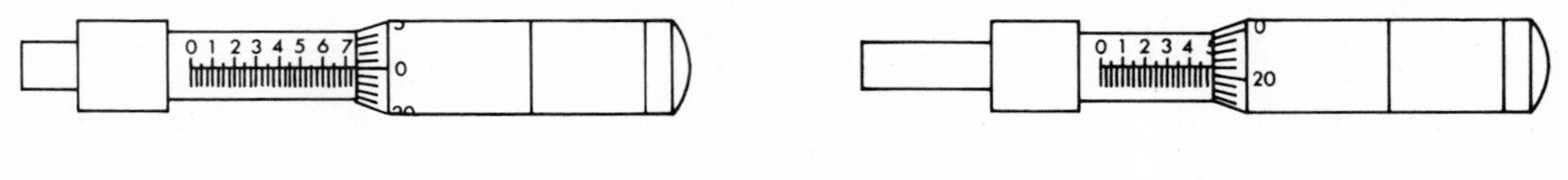

a. ________ **b.** ________

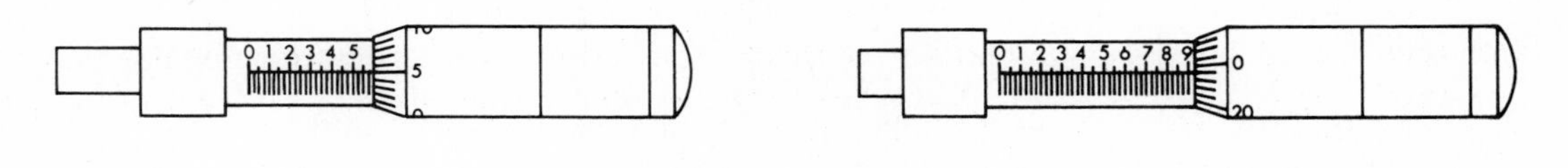

c. ________ **d.** ________

\# \# \# • \# \# \# • \# \# \# • \# \# \# • \# \# \# • \# \# \# • \# \# \# • \# \# \# • \# \# \#

A3 **a.** 0.750 in.:

$$\begin{array}{r} 0.700 \\ +0.050 \\ \hline 0.750 \end{array}$$

b. 0.496 in.:

$$\begin{array}{r} 0.400 \\ 0.075 \\ +0.021 \\ \hline 0.496 \end{array}$$

c. 0.580 in.:

$$\begin{array}{r} 0.500 \\ 0.075 \\ +0.005 \\ \hline 0.580 \end{array}$$

d. 0.924 in.:

$$\begin{array}{r} 0.900 \\ +0.024 \\ \hline 0.924 \end{array}$$

Q4 Determine the readings on the following English micrometers:

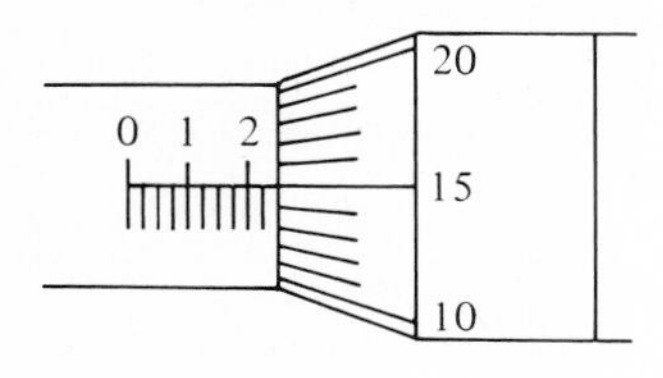

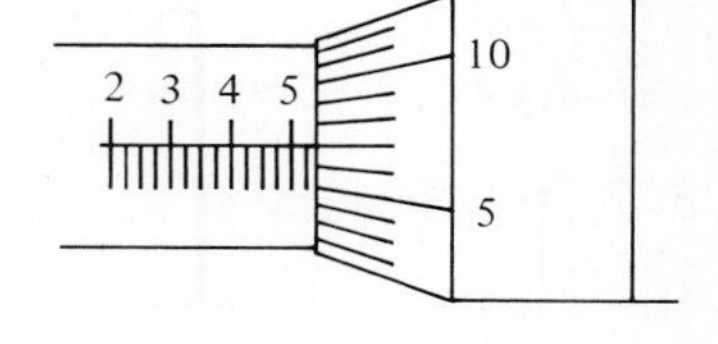

a. ______ **b.** ______

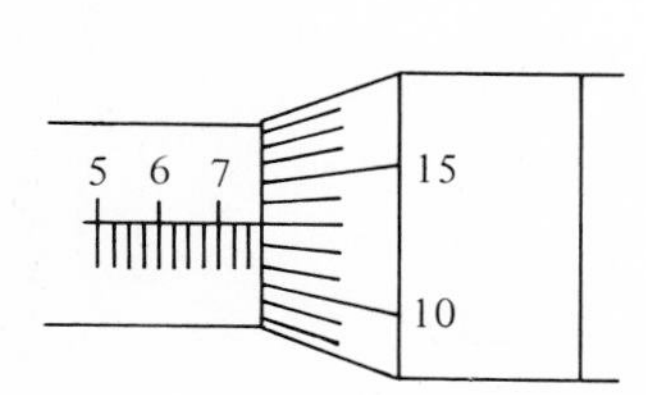

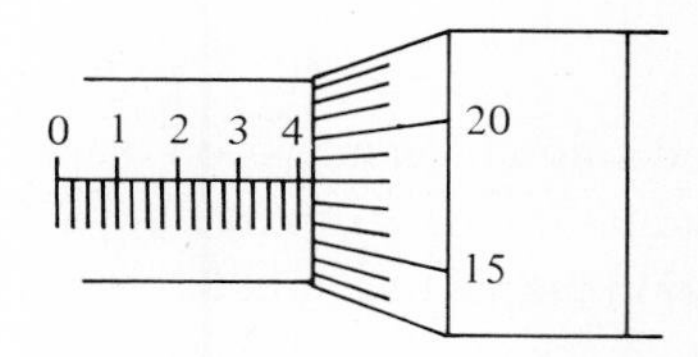

c. ______ **d.** ______

• ### • ### • ### • ### • ### • ### • ### •

A4 **a.** 0.240 in. **b.** 0.532 in. **c.** 0.763 in. **d.** 0.418 in.

---

3 A micrometer caliper that reads to thousandths of an inch (see previous examples) can be made to read to ten-thousandths of an inch by putting a *vernier* on the sleeve so that 10 divisions on the vernier correspond to 9 divisions on the thimble. For example,

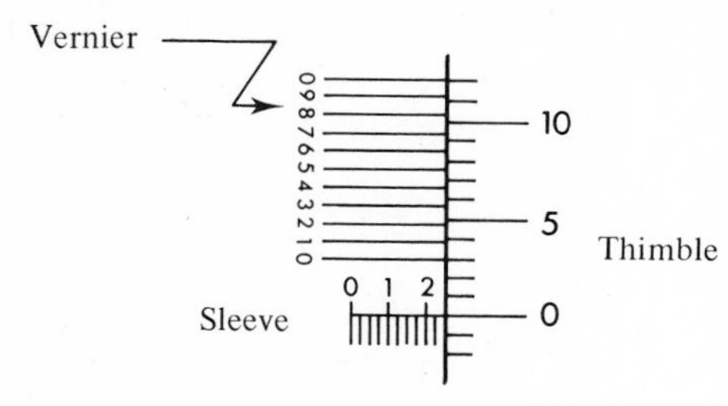

The 10 graduations on the vernier occupy the same space as 9 graduations on the thimble. The difference between one of the 10 spaces (graduations) on the sleeve and one of the 9 spaces on the thimble is $\frac{1}{10}$ of a space on the thimble or $\frac{1}{10{,}000}$ inch (0.0001 in.) in the micrometer reading. When the micrometer is opened, the thimble is turned to the left and each space on the thimble represents 0.001 in. Therefore, when the thimble is turned so that the 5 (on the thimble) and the 2 (on vernier) coincide, the micrometer has been opened $\frac{2}{10}$ of 0.001 or 0.0002 in.

If the thimble is turned further, so that 10 (on the thimble) coincides with 7 (on the vernier), the micrometer has been opened 0.0007 in. (see the vernier scale at the top of page 498). To read a micrometer containing a vernier (which permits a reading to ten-thousandths), read the "mike" in the usual way to thousandths, then observe the number of graduations on the vernier until a line (on the vernier) coincides with a line on the thimble. If it is the line marked 1, add 0.0001, if it is the line marked 2, add 0.0002, etc.

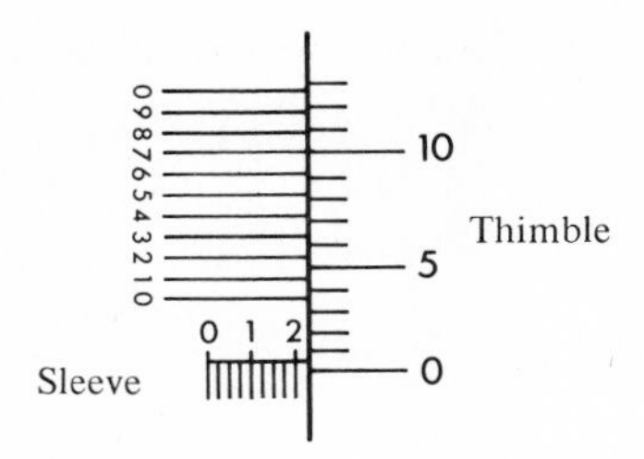

Example: Determine the reading on the English micrometer below which contains a vernier scale.

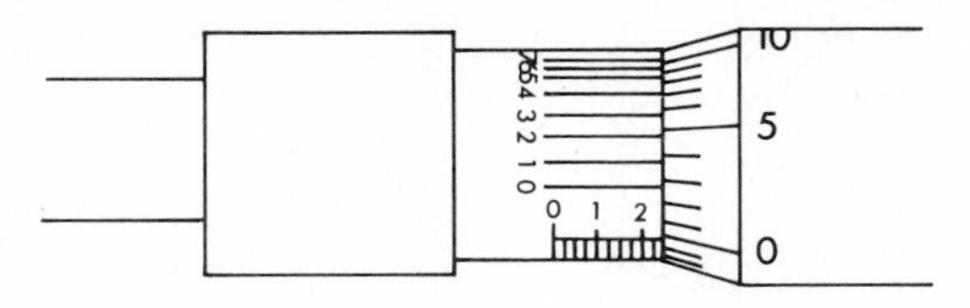

Solution:

1. 2 numbered graduation marks on the sleeve $= 0.2000$
2. 1 small graduation mark on the sleeve $= 0.0250$
3. 24 graduation marks on thimble opposite centerline on sleeve $= 0.0240$
4. Graduation mark numbered 4 is first line on vernier opposite a line on thimble $= 0.0004$

Reading $= 0.2494$ in.

Q5 Determine the reading to ten-thousandths on the English micrometer below. ________

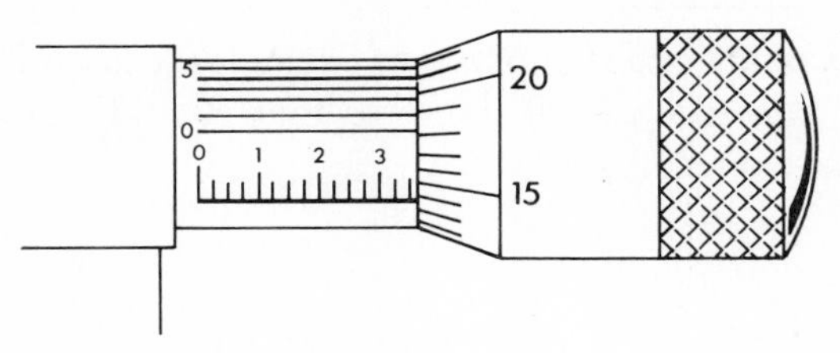

\# \# \# • \# \# \# • \# \# \# • \# \# \# • \# \# \# • \# \# \# • \# \# \# • \# \# \# • \# \# \#

A5 0.3634 in.:

0.3000	(3×0.1)
0.0500	(2×0.025)
0.0130	(13×0.001)
+0.0004	
0.3634	

4 When reading an actual micrometer, you would obtain the reading to thousandths from the sleeve and thimble, then you would more than likely need to rotate the micrometer slightly so that you could examine the vernier scale. The illustration shows the micrometer as it would appear to your eyes.

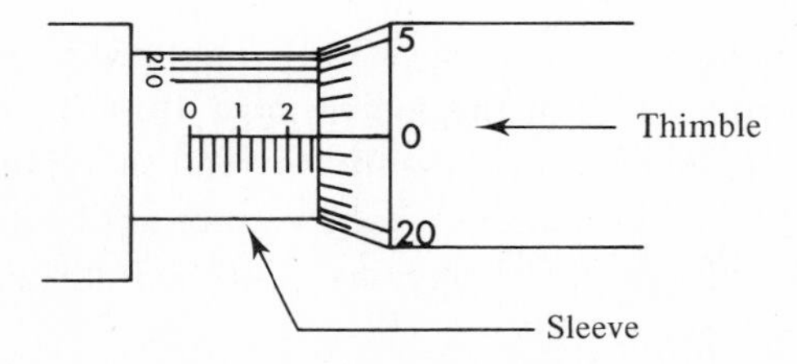

Then, rotating the micrometer slightly, you can imagine the top view looking something like:

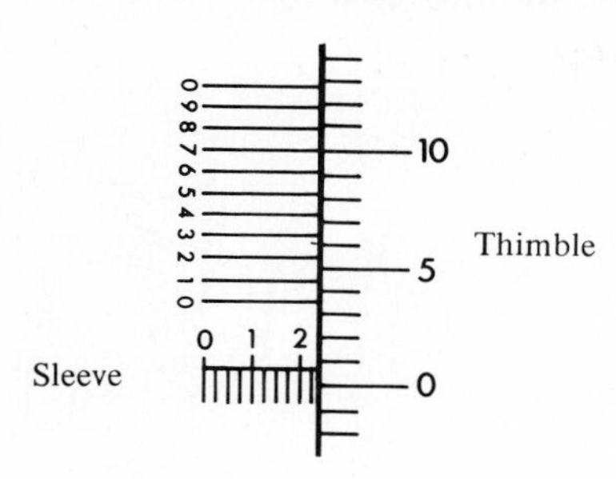

The reading on the micrometer mock-up would be

$$\begin{array}{r} 0.2000 \\ 0.0250 \\ 0.0000 \\ +0.0007 \\ \hline 0.2257 \text{ in.} \end{array}$$

Q6 Indicate the measurements obtained from the following mock-ups of an English micrometer graduated in ten-thousandths:

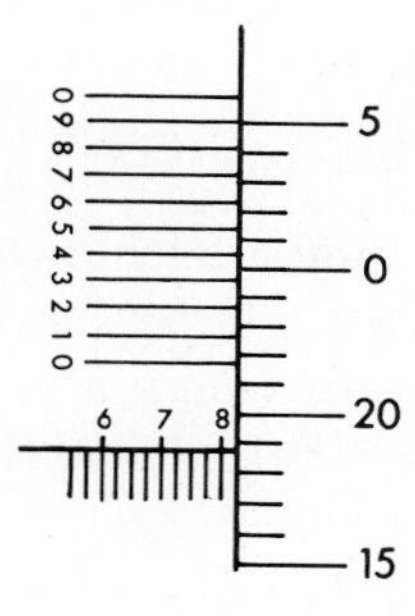

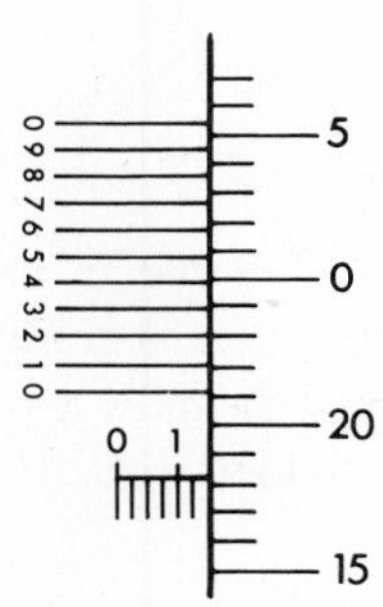

a. ________ **b.** ________

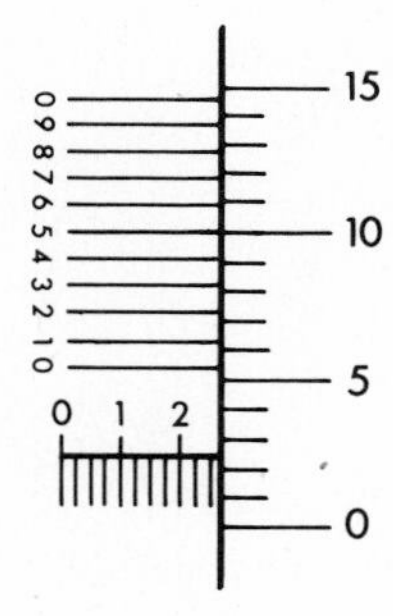

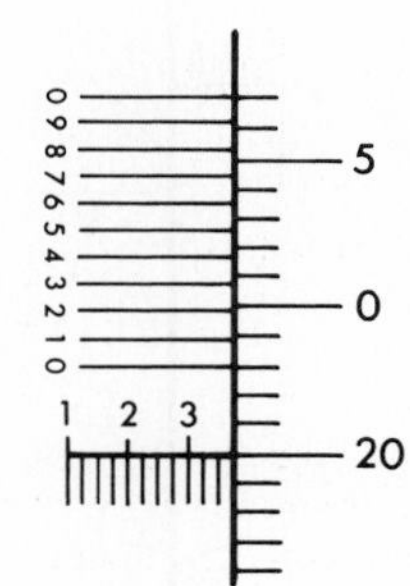

c. ________ **d.** ________

\# # # • # # # • # # # • # # # • # # # • # # # • # # # • # # # • # # #

A6 **a.** 0.8189 in.:

$$\begin{array}{r} 0.8000 \\ 0.0000 \\ 0.0180 \\ +0.0009 \\ \hline 0.8189 \end{array}$$

b. 0.1432 in.:

$$\begin{array}{r} 0.1 \\ 0.025 \\ 0.018 \\ +0.0002 \\ \hline 0.1432 \end{array}$$

c. 0.2525 in.

d. 0.3700 in.

5 Micrometers are available in many styles and forms, some designed for specific purposes. The illustrations indicate but a few.

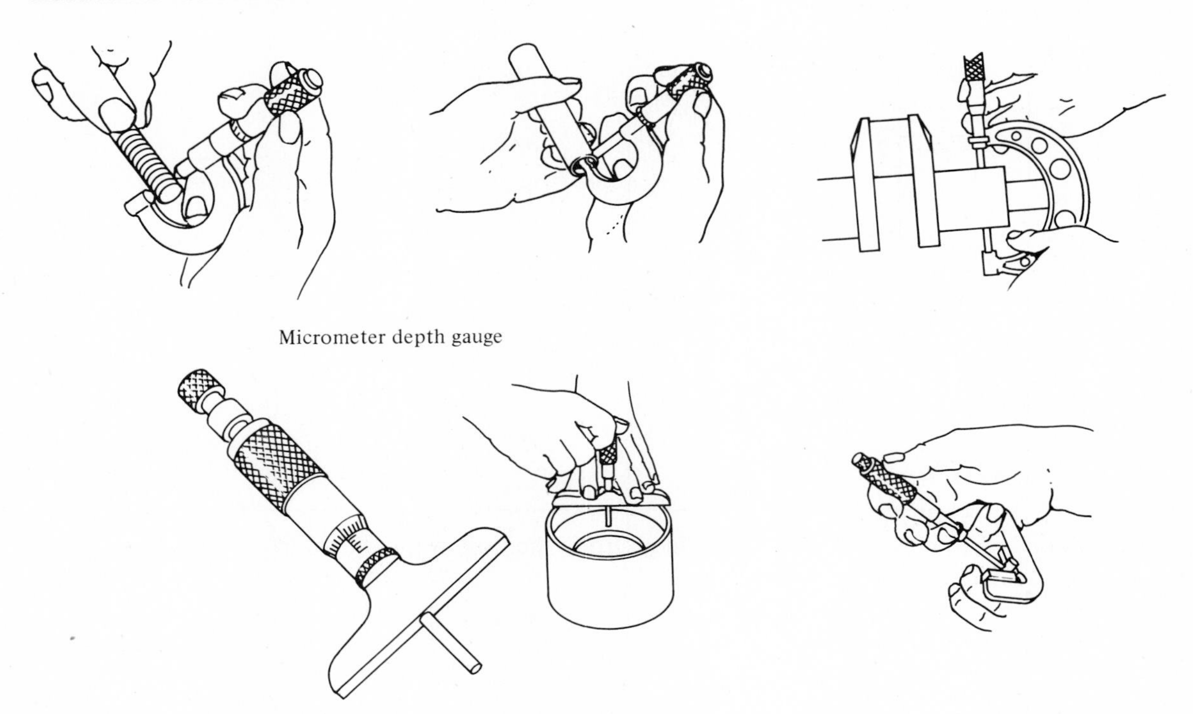

Micrometers are also available in metric units. The figure shows a metric micrometer. The sleeve of most metric micrometers is graduated in millimetres. On the micrometer, the sleeve is also graduated in 0.5 millimetres. *It takes two complete turns of the thimble for the spindle to move one millimetre*. The thimble is divided into 50 equal graduations, each indicating 0.01 millimetres.

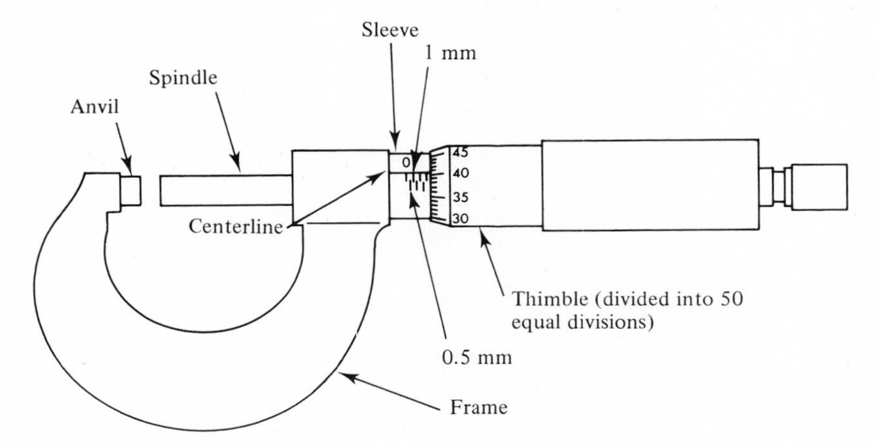

To read a metric micrometer:

1. Determine the number of exposed millimetre graduation marks (each represents 1 mm).
2. Add 0.50 mm if a half millimetre graduation mark is exposed.
3. Obtain the number of 0.01 mm from the thimble by reading the graduation mark (on the thimble) which is most nearly in line with the centerline (on the sleeve). Each graduation is multiplied by 0.01 mm.
4. Determine the sum of all individual readings.

Example: Determine the reading from the micrometer illustrated above in this frame.

Solution:

1. 3 exposed mm graduation marks (3 × 1 mm) = 3.00 mm
2. 0 exposed 0.5 mm = 0.00
3. The centerline appears to be $\frac{1}{2}$ way between 40 and 41 (on the thimble). Hence, record 0.40 mm* (40 × 0.01) = 0.40

Complete reading = 3.40 mm

*Round off to nearest even value.

Q7 Determine the readings on the metric micrometer below:

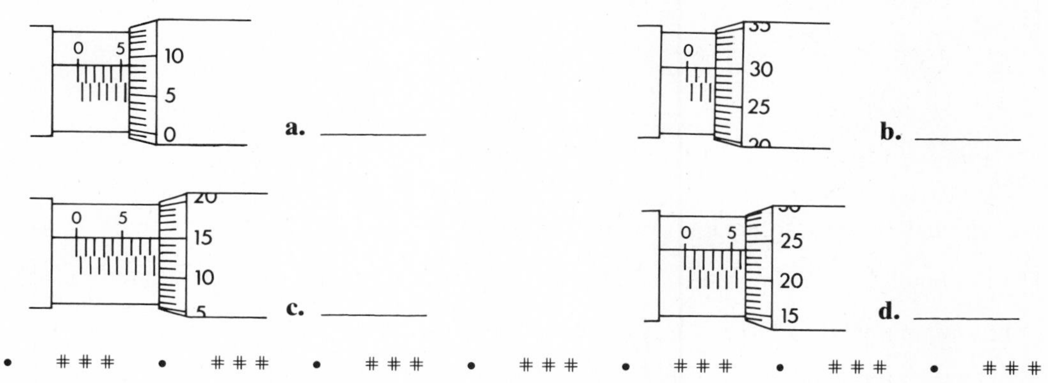

\# \# \# • \# \# \# • \# \# \# • \# \# \# • \# \# \# • \# \# \# • \# \# \# • \# \# \# • \# \# \#

A7 **a.** 5.59 mm:

5.00
0.50
+0.09
5.59

b. 2.80 mm:

2.00
0.50
+0.30
2.80

c. 8.65 mm

d. 6.24 mm

6 The scales on different models of English and metric micrometers will vary slightly. However, the principles discussed here will still apply. (Note: It is highly recommended that the student should have the opportunity to handle and manipulate both the English and metric micrometer.)

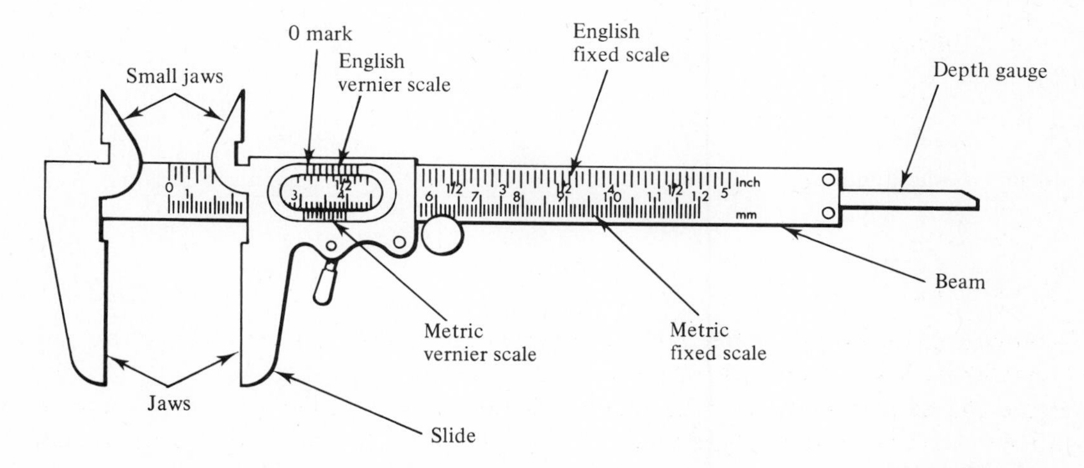

Numerous other measuring instruments exist which assist the technician. Many also involve reading scales. One such instrument is the *vernier caliper* illustrated here. This caliper can be used to make inside, outside, and depth measurements in both English and/or metric units.

1. The English fixed scale is graduated in $\frac{1}{16}$ in. The graduations on the English vernier scale represent $\frac{1}{128}$ in.
2. The metric part of the instrument will be discussed in Frame 7.

Example: Read the measurement on the vernier caliper below in English units.

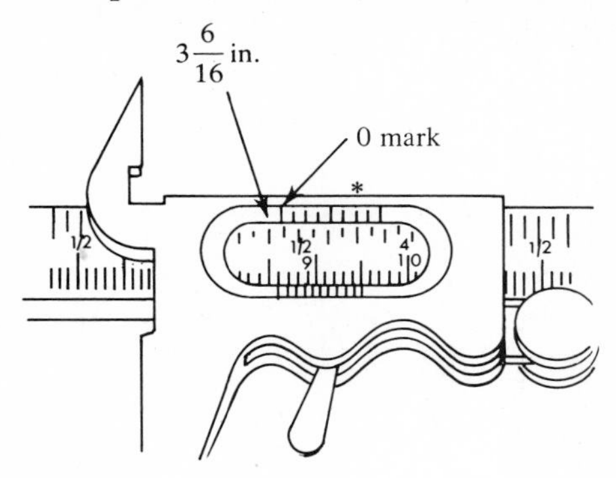

Solution:

1. The graduation mark to the left of the 0 mark represents $3\frac{6}{16}$ inches.
2. The graduation mark (*) on the vernier that most nearly lines up with a graduation mark on the fixed scale is 6 which represents $\frac{6}{128}$ inches. Therefore,

$$3\frac{6}{16} = 3\frac{3}{8} = 3\frac{24}{64}$$
$$+\frac{6}{128} = \frac{3}{64} = \frac{3}{64}$$
$$3\frac{27}{64}$$

Hence, the complete reading is $3\frac{27}{64}$ in.

Q8 Read the measurement from the vernier caliper in English units:

a.

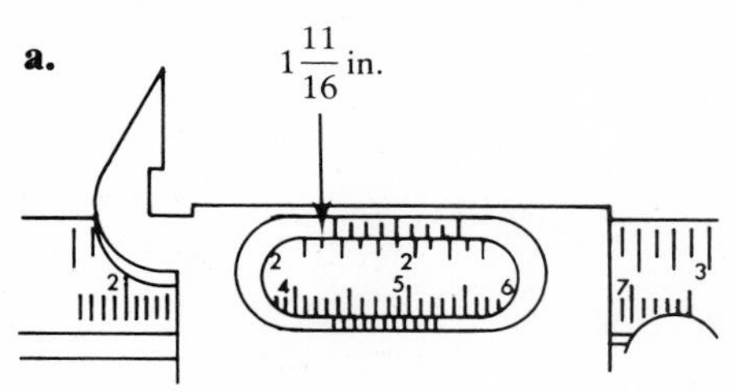

b.

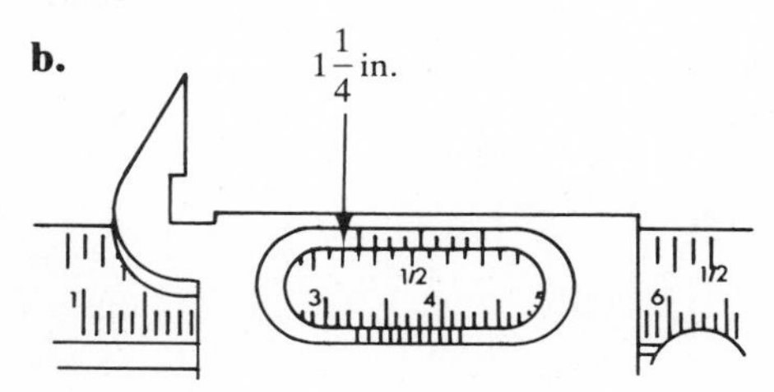

________ ________

• ### • ### • ### • ### • ### • ### • ### •

A8 **a.** $1\frac{23}{32}$ in.: $1\frac{11}{16} = 1\frac{88}{128}$

$$+\frac{4}{128} = \frac{4}{128}$$
$$1\frac{92}{128}$$

b. $1\frac{5}{16}$ in.: $1\frac{1}{4} + \frac{1}{16}$

7 When reading the vernier caliper in metric units the lower part of the beam is used.

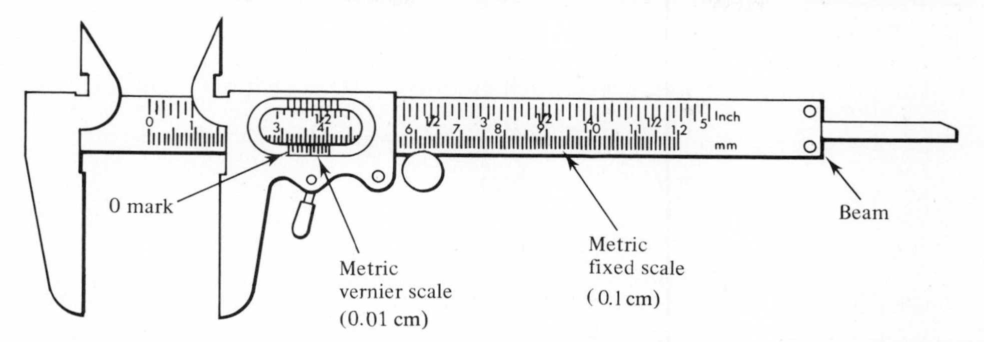

Examine the following illustration which shows the metric fixed scale and the metric vernier scale in greater detail.

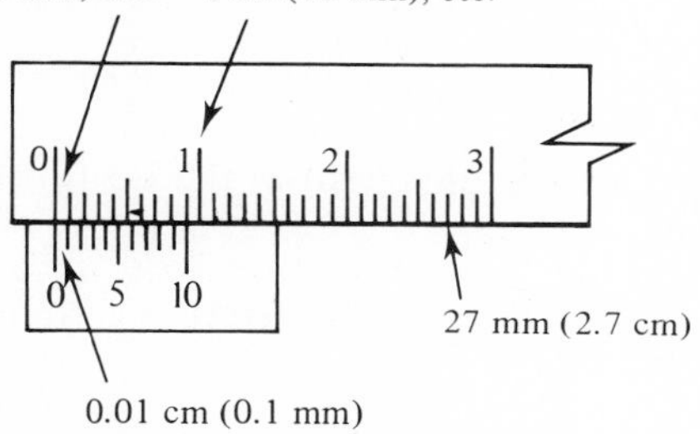

Example: Read the measurement on the vernier caliper below in metric units.

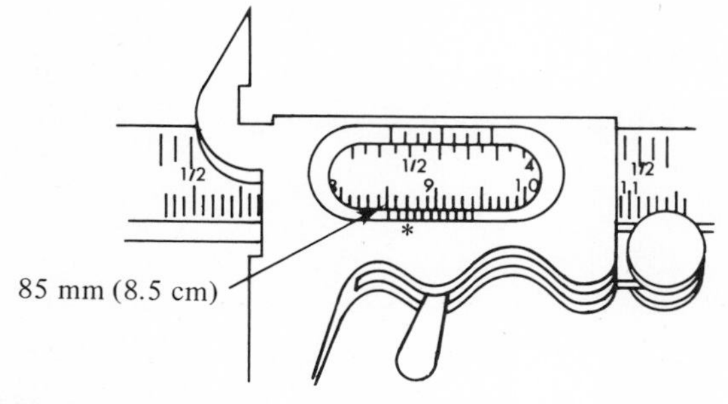

Solution:

1. From the fixed scale is read 85 mm.
2. The graduation mark (*) on the vernier which most nearly lines up with a graduation mark on the fixed scale represents 0.2 mm. Therefore, the complete reading is 85.2 mm (85 + 0.2). This reading could be converted to 8.52 cm.

Q9 Read the measurement on the vernier caliper below in metric units:

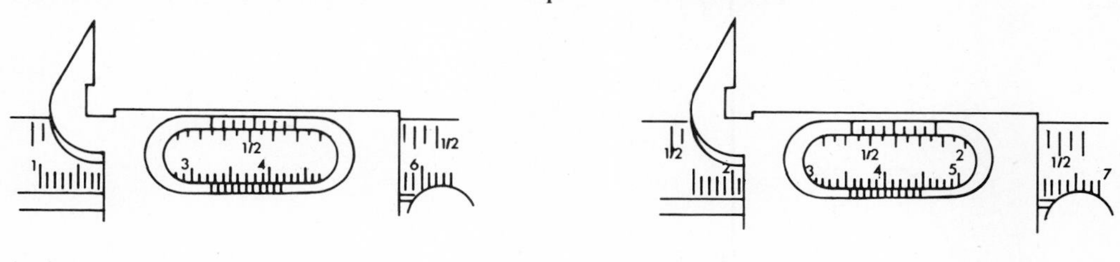

a. ________ **b.** ________

• # # # • # # # • # # # • # # # • # # # • # # # • # # # • # #

A9 **a.** 32.5 mm (3.25 cm) **b.** 35.5 mm

8 Other styles of vernier calipers exist. For example, in the type illustrated, each inch on the main scale is divided into 40 graduations; hence, each graduation is $\frac{1}{40}$ inch (0.025 in.). Every fourth is numbered 1, 2, 3, etc. Each numbered graduation mark represents 0.1 in., 0.2 in., etc.

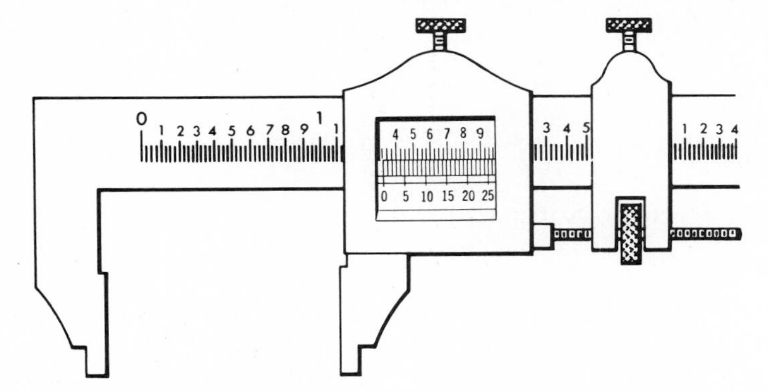

The first vernier number that coincides with a fixed scale number (from the beam) shows the number of 0.001 inch to be added to the reading on the main scale.

Example: What is the reading shown on the English vernier caliper illustrated in this frame?

Solution:

1. From the main scale, it can be determined that the measurement is at least 1.3 inches.
2. Also, one small graduation to the right of 1.3 inches is exposed; hence, the measurement is 0.025 inches more.
3. On the vernier, the fourth graduation mark coincides with a graduation mark on the main scale; hence 0.004 inches must also be added. Therefore, the complete reading is 1.329 in.

$$\begin{array}{r} 1.300 \\ 0.025 \\ +0.004 \\ \hline 1.329 \end{array}$$

(Note: Since the techniques involved in reading this English vernier caliper are identical to reading an English micrometer with a vernier, no additional practice will be included.)

9 You are familiar with the ordinary protractor. For more accurate work, a vernier scale can be added which will allow measurements to the nearest minute of a degree.

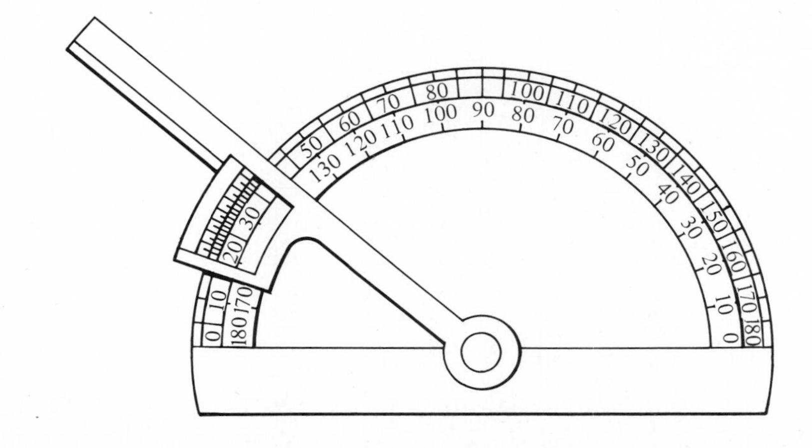

Recall that if a degree is divided into 60 equal parts, each part is called a minute. That is, 1° (1 degree) = 60′ (60 minutes). Since instruments of this kind vary a great deal depending on the manufacturer, the illustration was included only to indicate another adaptation of the use of scales to assist the technician when more precise measurements are needed. With the experiences provided in this chapter, you should be prepared to adapt your knowledge to the situation and the availability of a particular measuring instrument.

This completes the instruction for this section.

10.2 EXERCISE

1. Determine the readings on the following English micrometers:

a.

b.

c.

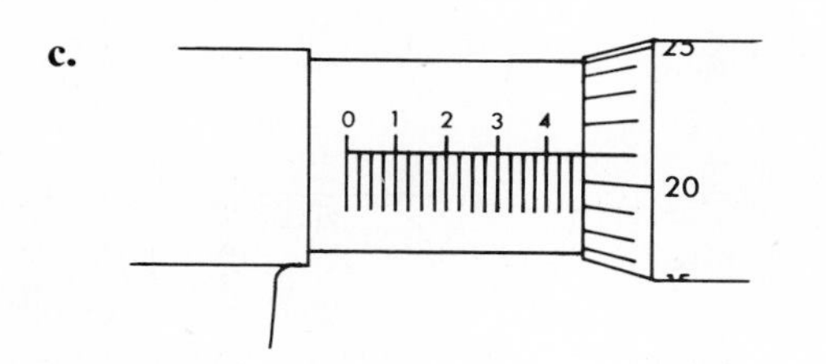

d.

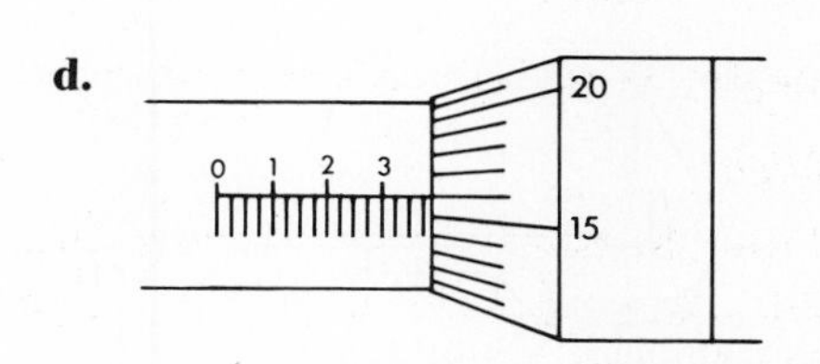

e.

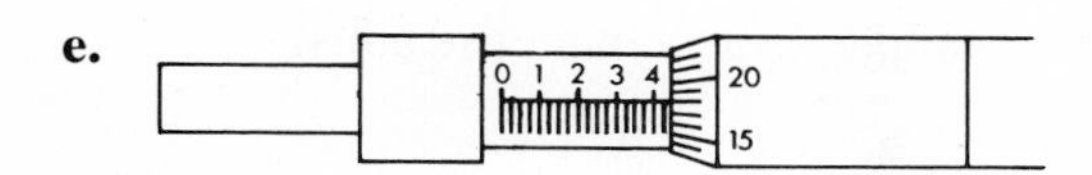

f.

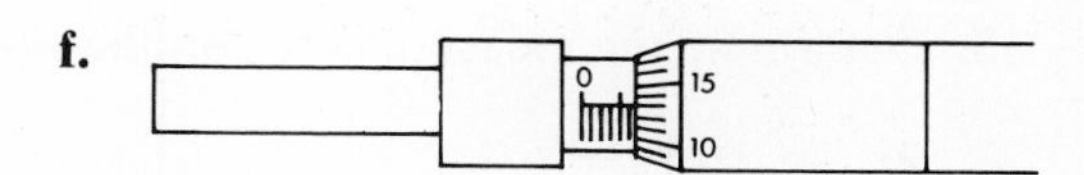

2. Determine the readings on the following English micrometers containing a vernier scale:

a.

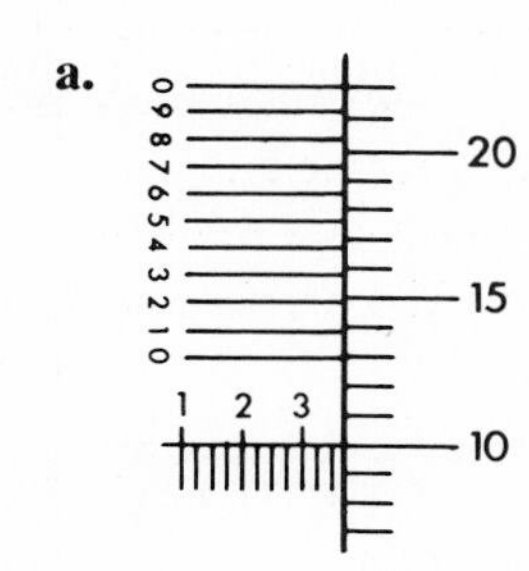

b.

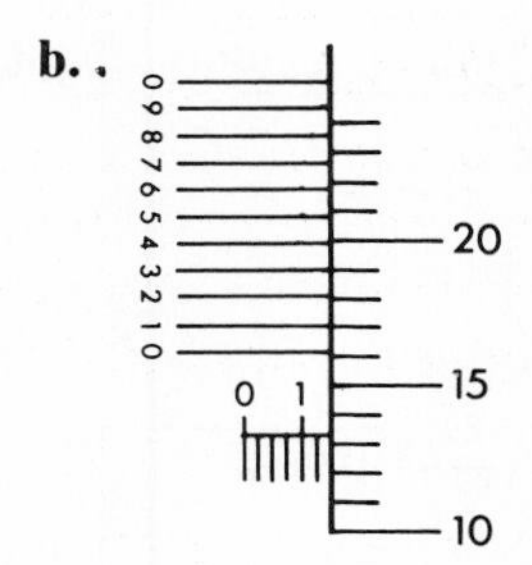

c.

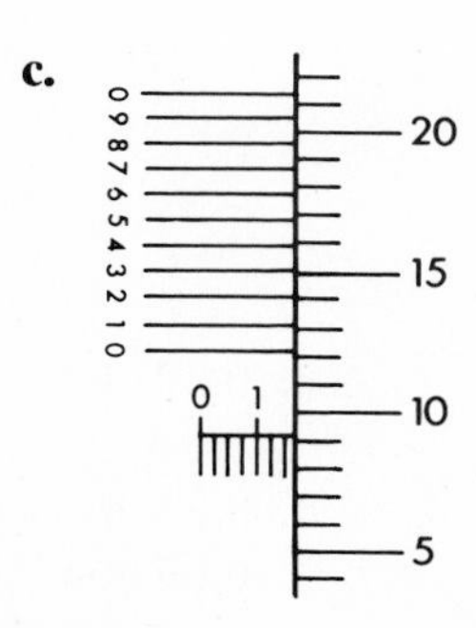

d.

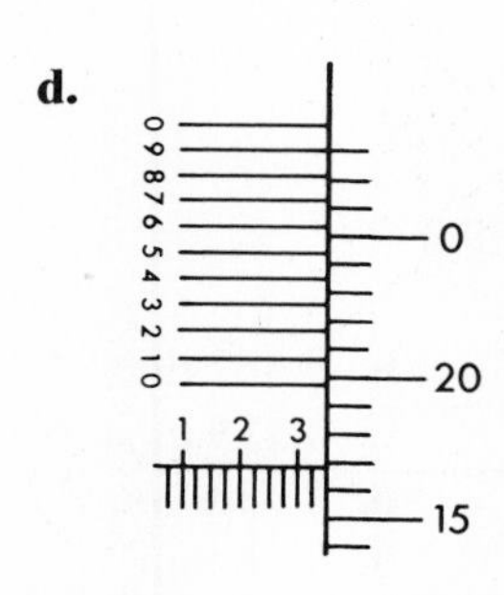

3. Determine the measurements shown on the following metric micrometers:

a.

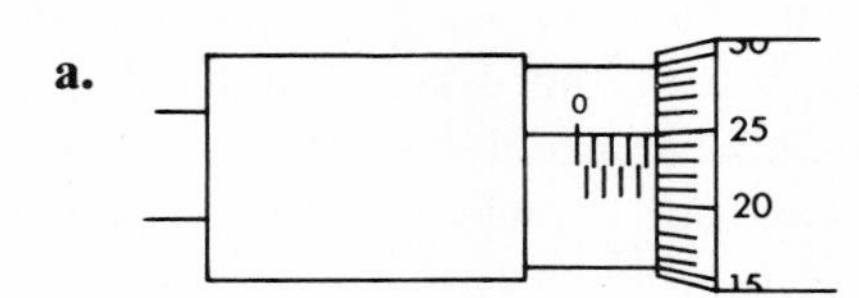

b.

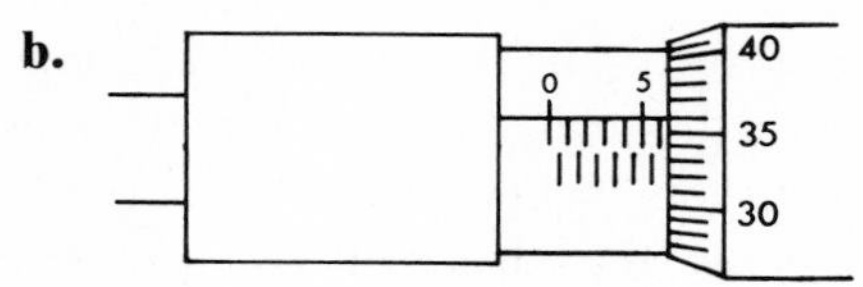

c.

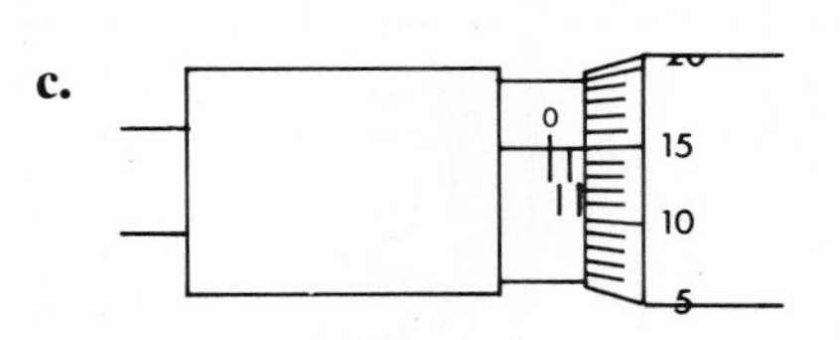

d.

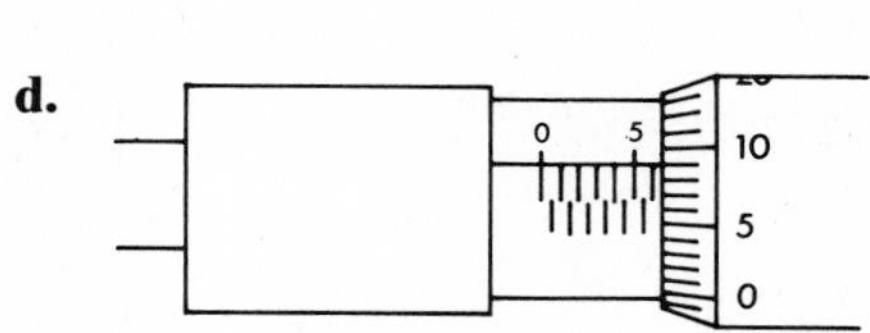

4. Determine the measurements on the vernier calipers in English units:

a.

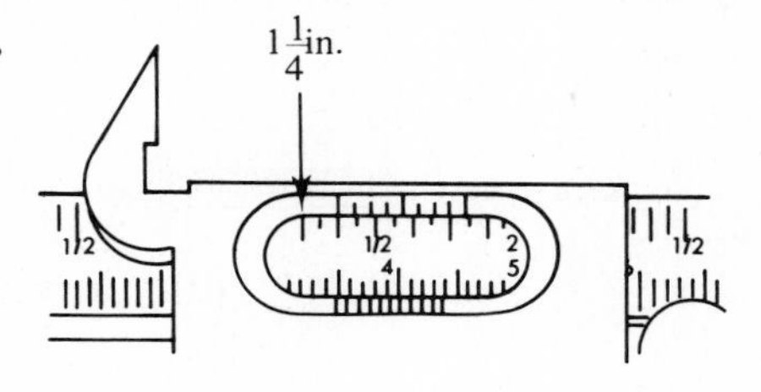

b.

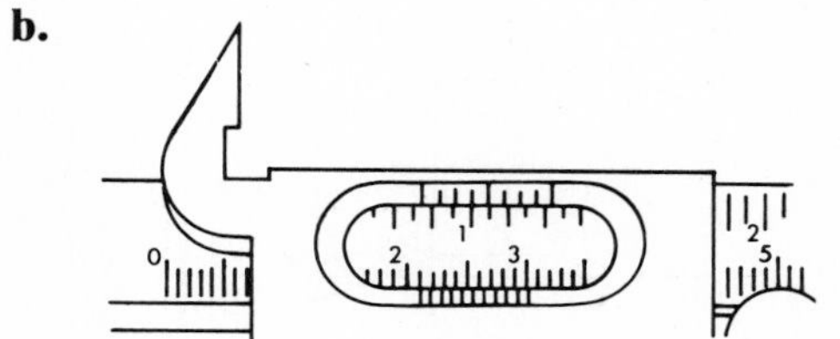

c.

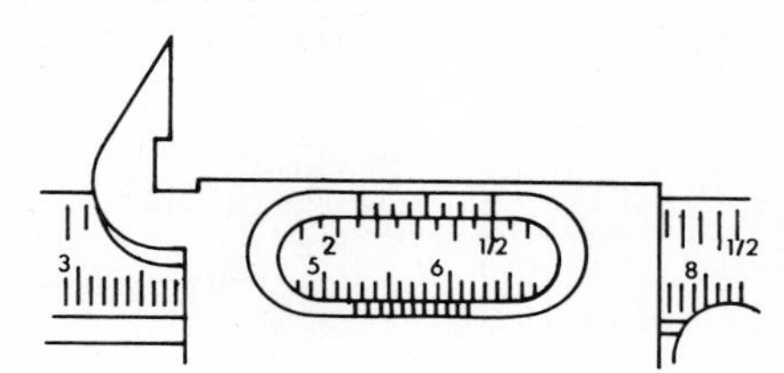

d.

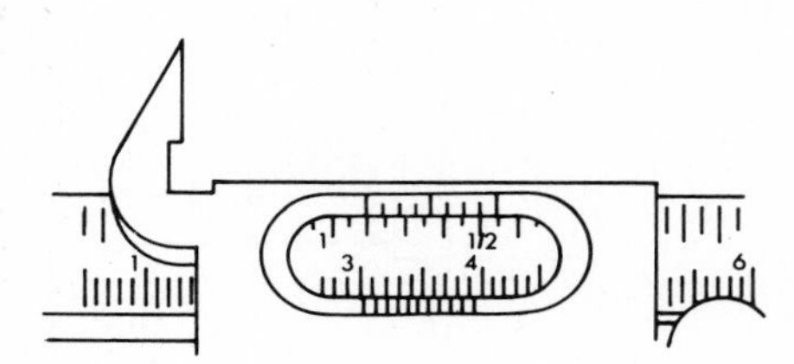

5. Determine the measurements on the vernier calipers below in metric units:

a.

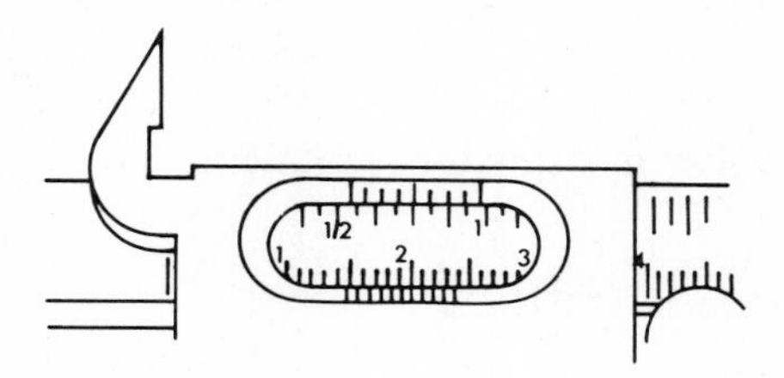

b.

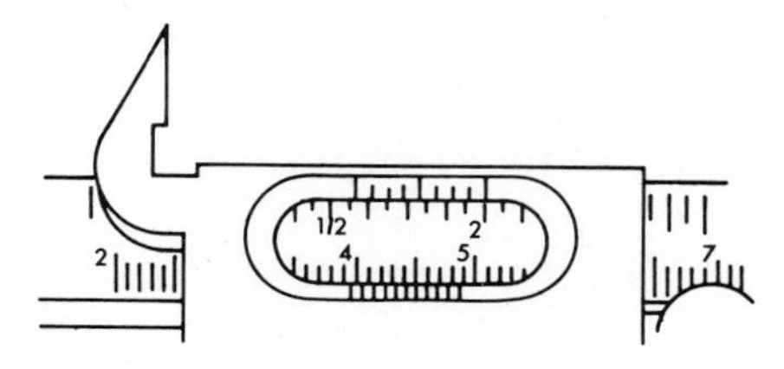

c.

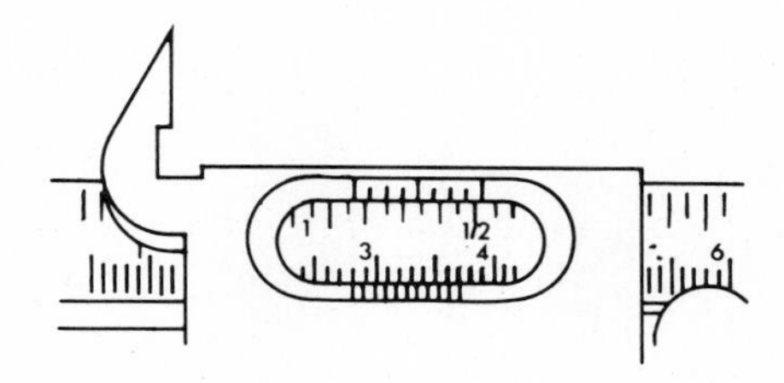

d.

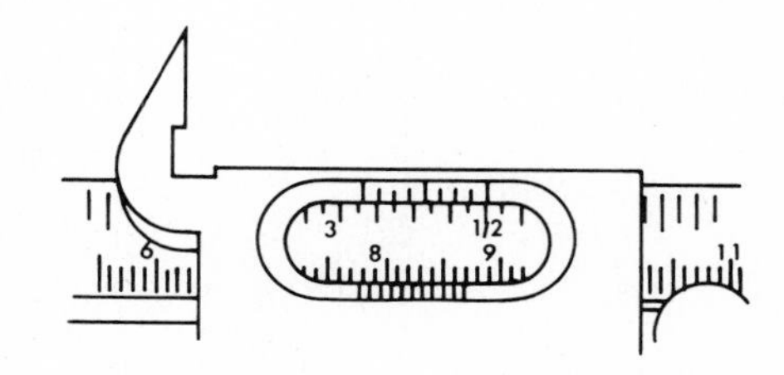

10.2 EXERCISE ANSWERS

1. a. 0.664 in. **b.** 0.562 in. **c.** 0.471 in. **d.** 0.391 in.
e. 0.443 in. **f.** 0.138 in.

2. a. 0.3600 in. **b.** 0.1383 in. **c.** 0.1592 in. **d.** 0.3419 in.

3. a. 4.25 mm **b.** 6.36 mm **c.** 1.65 mm **d.** 6.09 mm

4. a. $1\frac{3}{8}$ in. **b.** $\frac{107}{128}$ in. **c.** $2\frac{1}{16}$ in. **d.** $1\frac{1}{8}$ in.

5. a. 14.6 mm **b.** 39.7 mm **c.** 28.0 mm **d.** 77.8 mm

CHAPTER 10 SAMPLE TEST

At the completion of Chapter 10 you should be able to work the following problems.

10.1 UNIFORM AND NONUNIFORM PROBLEMS

1. Determine the following measurements with the inch ruler provided:

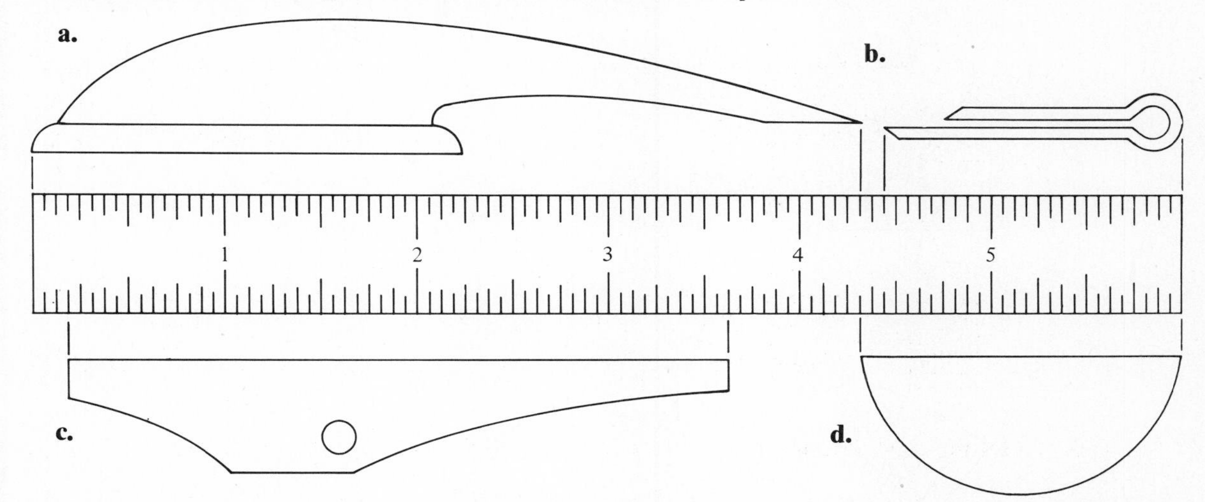

2. Determine the following measurements with the metric ruler provided (the numbers indicate centimetres):

a. b.

1 2 3 4 5 6 7 8 9 10 11 12 13 14 15 16

3. Determine the readings on the following scales (do not attach a unit of measurement):

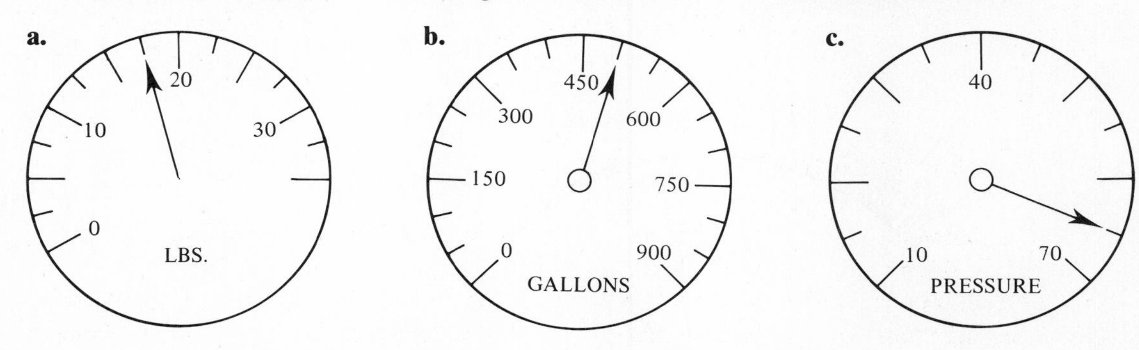

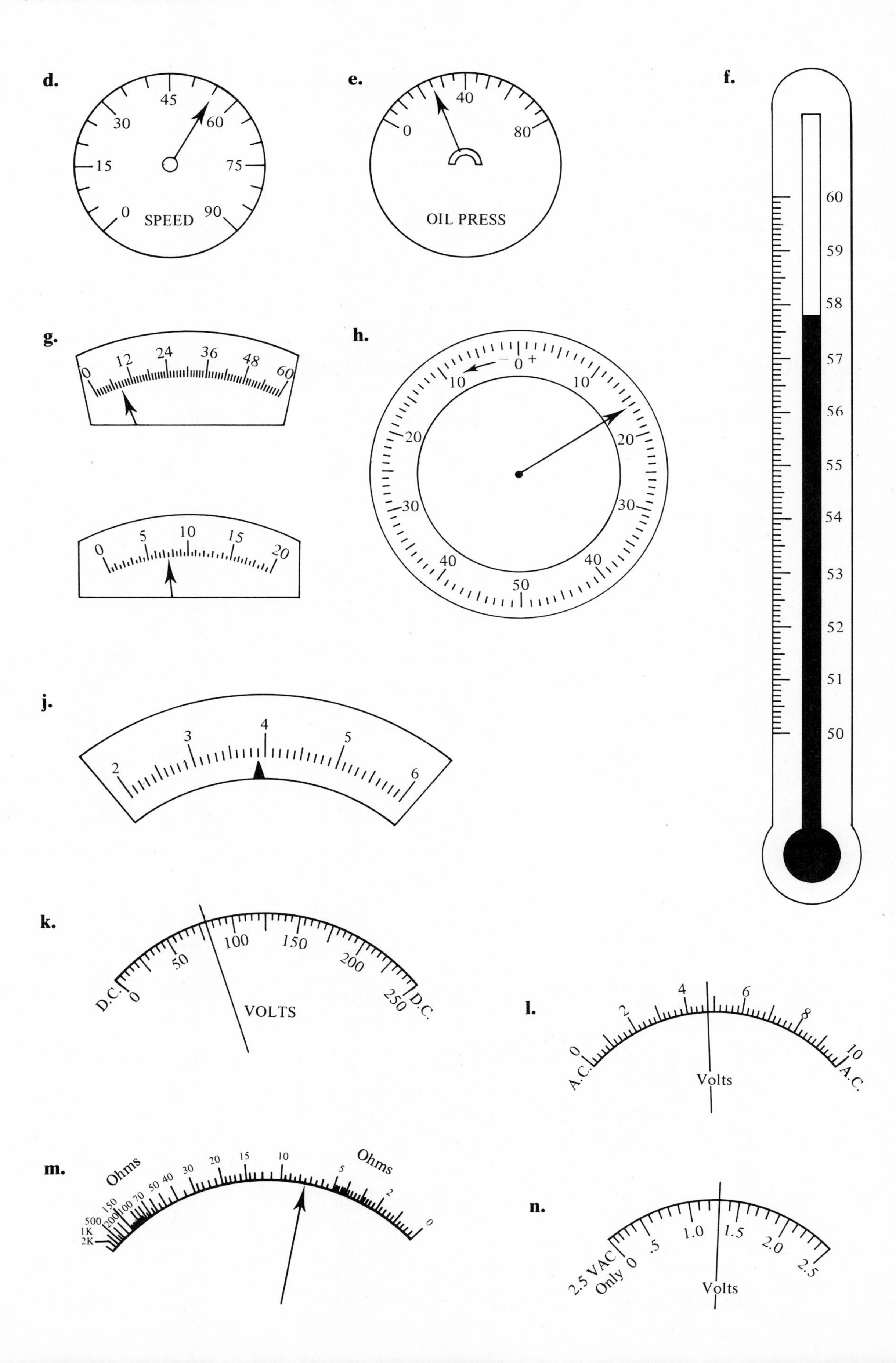
d.
0
15
30
45
60
75
90
SPEED
e.
0
40
80
OIL PRESS
f.
50
51
52
53
54
55
56
57
58
59
60
g.
0
12
24
36
48
60
0
5
10
15
20
h.
0
10
10
20
20
30
30
40
40
50
j.
2
3
4
5
6
k.
D.C. 0
50
100
150
200
250 D.C.
VOLTS
l.
A.C. 0
2
4
6
8
10 A.C.
Volts
m.
Ohms
2K
1K
500
200
150
100
70
50
40
30
20
15
10
5
2
0
Ohms
n.
2.5 VAC Only 0
.5
1.0
1.5
2.0
2.5
Volts

4. Read the given scale (do not attach a unit of measurement):

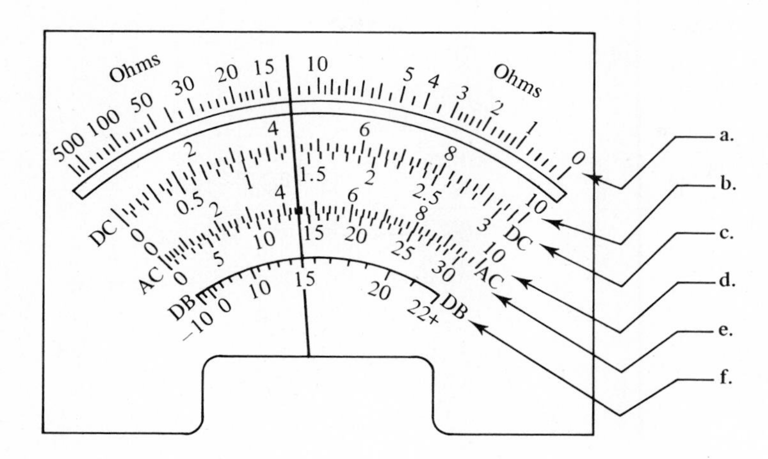

5. Determine the rpm:

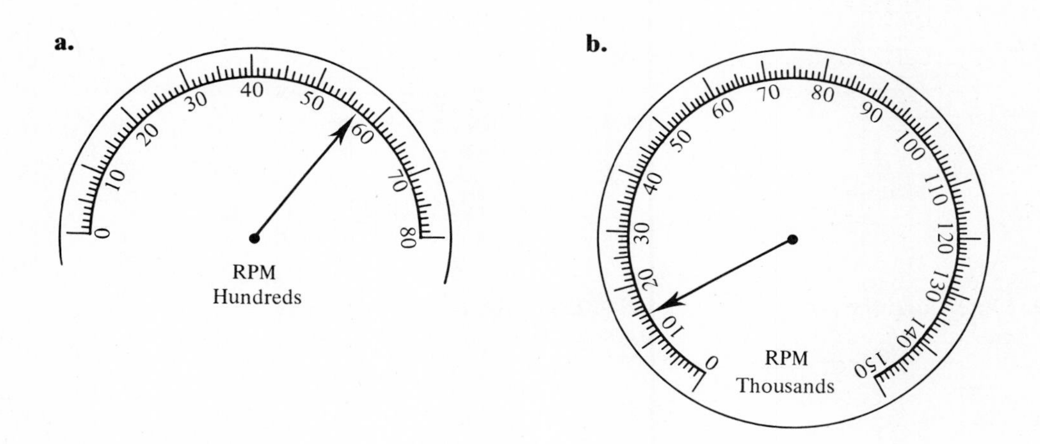

6. Read the given scale:

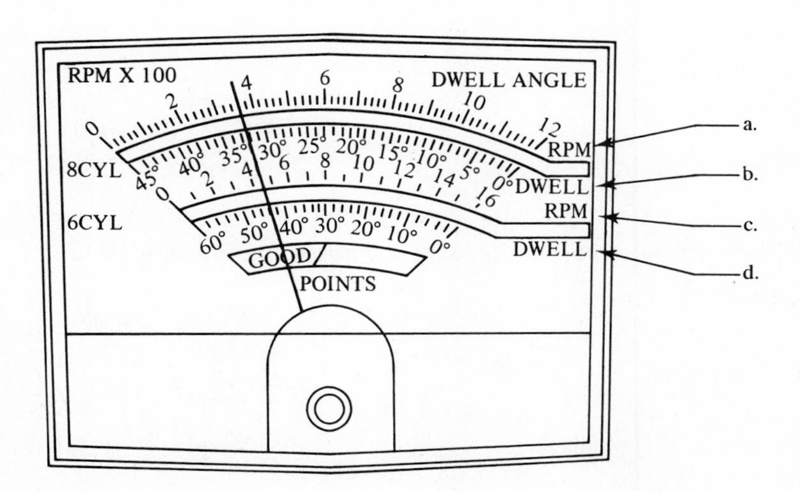

10.2 MEASURING INSTRUMENTS

7. Determine the readings on the following English micrometers:

a.

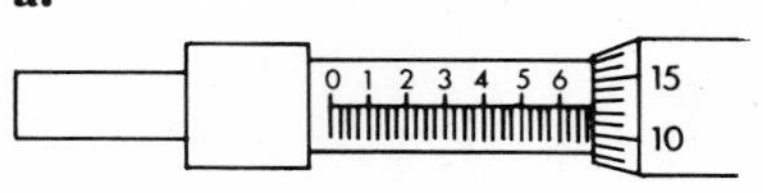

b.

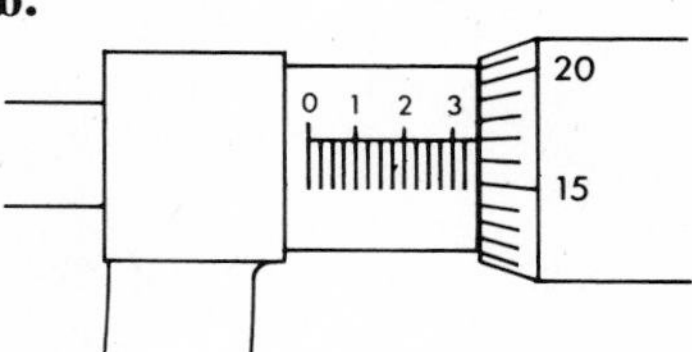

c.

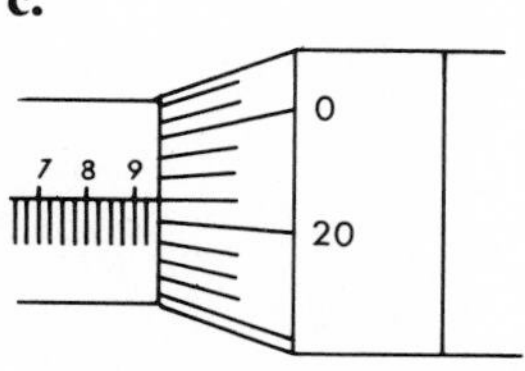

8. Determine the readings on the following English micrometers containing a vernier scale:

a.

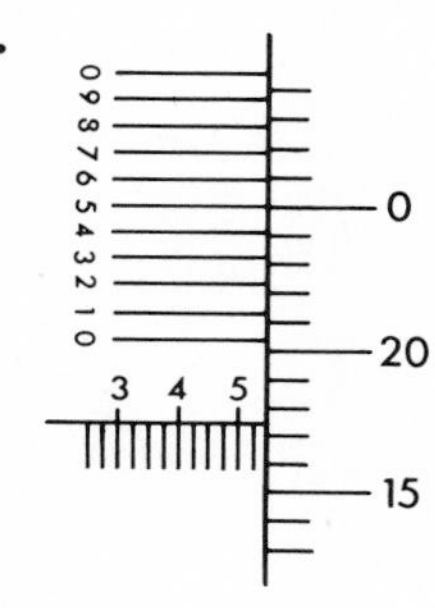

b.

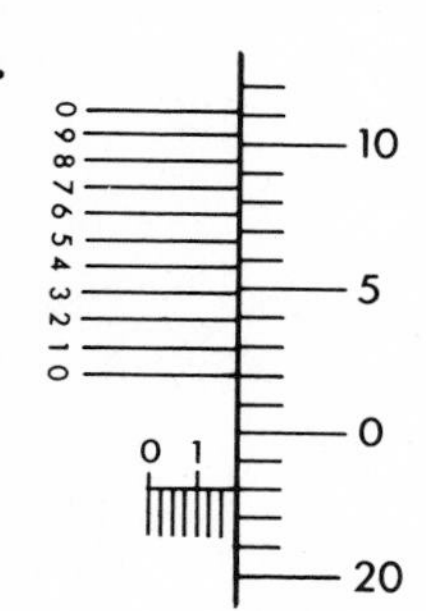

9. Determine the measurement on the following metric micrometer.

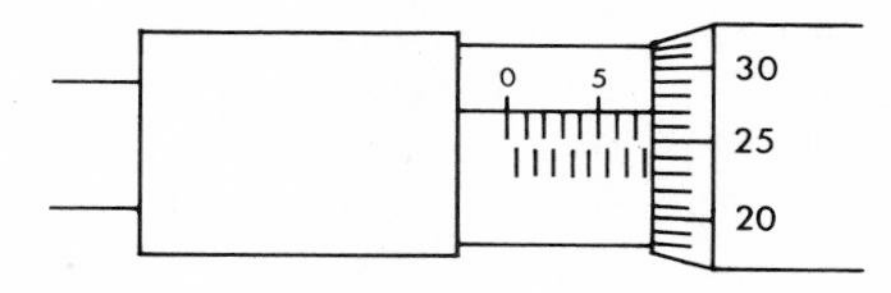

10. Determine the measurements on the vernier calipers below in English Units:

a.

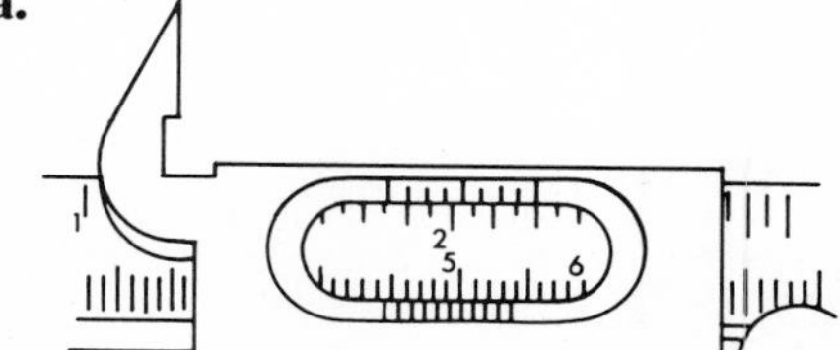

b.

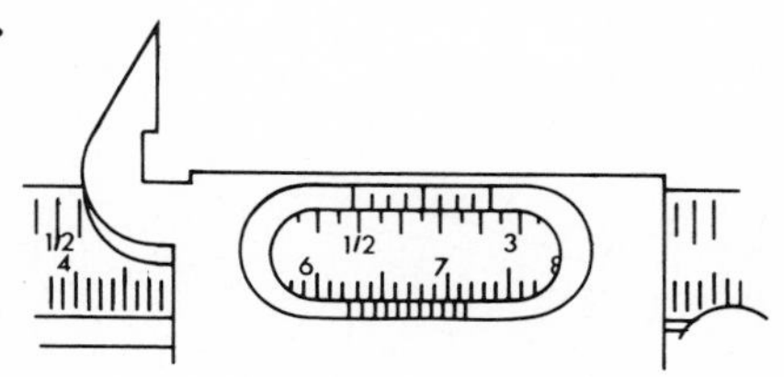

11. Determine the measurements on the vernier calipers below in metric units:

a.

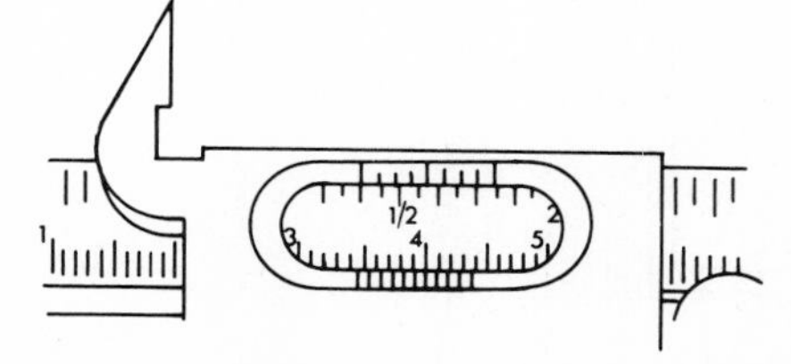

b.

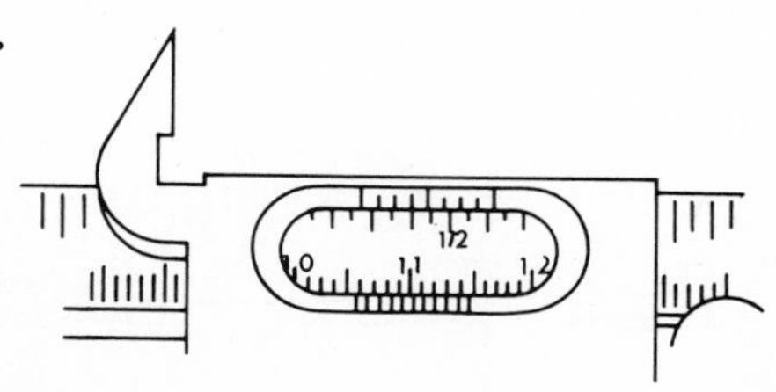

CHAPTER 10 SAMPLE TEST ANSWERS

1. **a.** $4\frac{5}{16}$ in. **b.** $1\frac{9}{16}$ in. **c.** $3\frac{7}{16}$ in. **d.** $1\frac{11}{16}$ in.
2. **a.** 7.6 cm (76 mm) **b.** 3.4 cm (34 mm)
3. **a.** 17.5 **b.** 500 **c.** 65 **d.** 55
   e. 25 **f.** 57.8 **g.** 8 **h.** 16
   i. 7.5 **j.** 3.9 **k.** 80 **l.** 4.8
   m. 7.5 **n.** 1.3
4. **a.** 13 **b.** 4.4 **c.** 1.4 **d.** 4.5
   e. 14 **f.** 15
5. **a.** 5,700 rpm **b.** 14,000 rpm
6. **a.** 360 rpm **b.** 32° **c.** 450 rpm **d.** 44°
7. **a.** 0.687 in. **b.** 0.342 in. **c.** 0.946 in.
8. **a.** 0.5425 in. **b.** 0.1730 in.
9. 7.77 mm (0.777 cm)
10. **a.** $1\frac{105}{128}$ in. **b.** $2\frac{61}{128}$ in.
11. **a.** 34.6 mm (3.46 cm) **b.** 105.8 mm (10.58 cm)

CHAPTER 11

INTERPRETATION AND CONSTRUCTION OF CIRCLE, BAR, AND LINE GRAPHS

11.1 CHOOSING THE TYPE OF GRAPH

1 Frequently people are asked to read and understand material that contains many numbers. The method of presentation of those numbers determines the ease with which the number relationships are understood. Written paragraphs containing numbers are difficult to understand unless there are very few numbers. Tables in which numbers are located in a uniform pattern are easily read, and graphs will often show relationships much more forcefully and possibly help the reader recognize and remember relationships.

A graph is used to present information artistically and accurately. A few mathematical principles are used which help to ensure that the graphs convey the truth. If these principles are ignored, the relationships between the numbers are distorted and the reader is left with an inaccurate impression of the facts. A person should be able to recognize distortions of graphs that he reads and avoid introducing distortions in graphs that he constructs.

Three types of graphs are used so frequently that everyone should be able to interpret and construct them. They are the *circle graph*, the *bar graph*, and the *line graph*. The name of the type of graph describes its character. Numbers are represented by sectors* of a circle on a circle graph, by bars in a bar graph, and by a continuous or broken line on a line graph.

*A sector is a section of a circle shaped like a piece of pie.

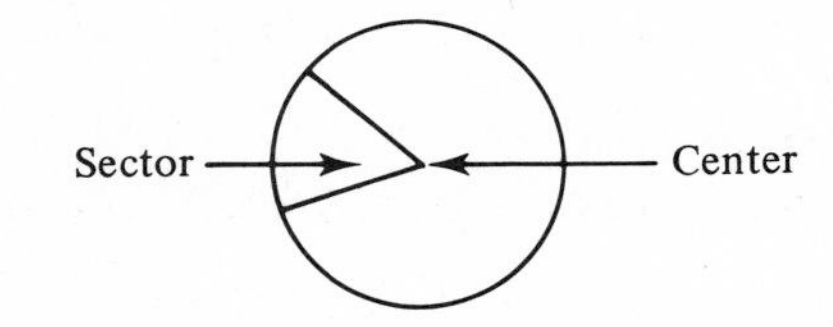

Q1 Examine the differences in the following graphs and identify each type of graph by writing its name on the line under the graph.

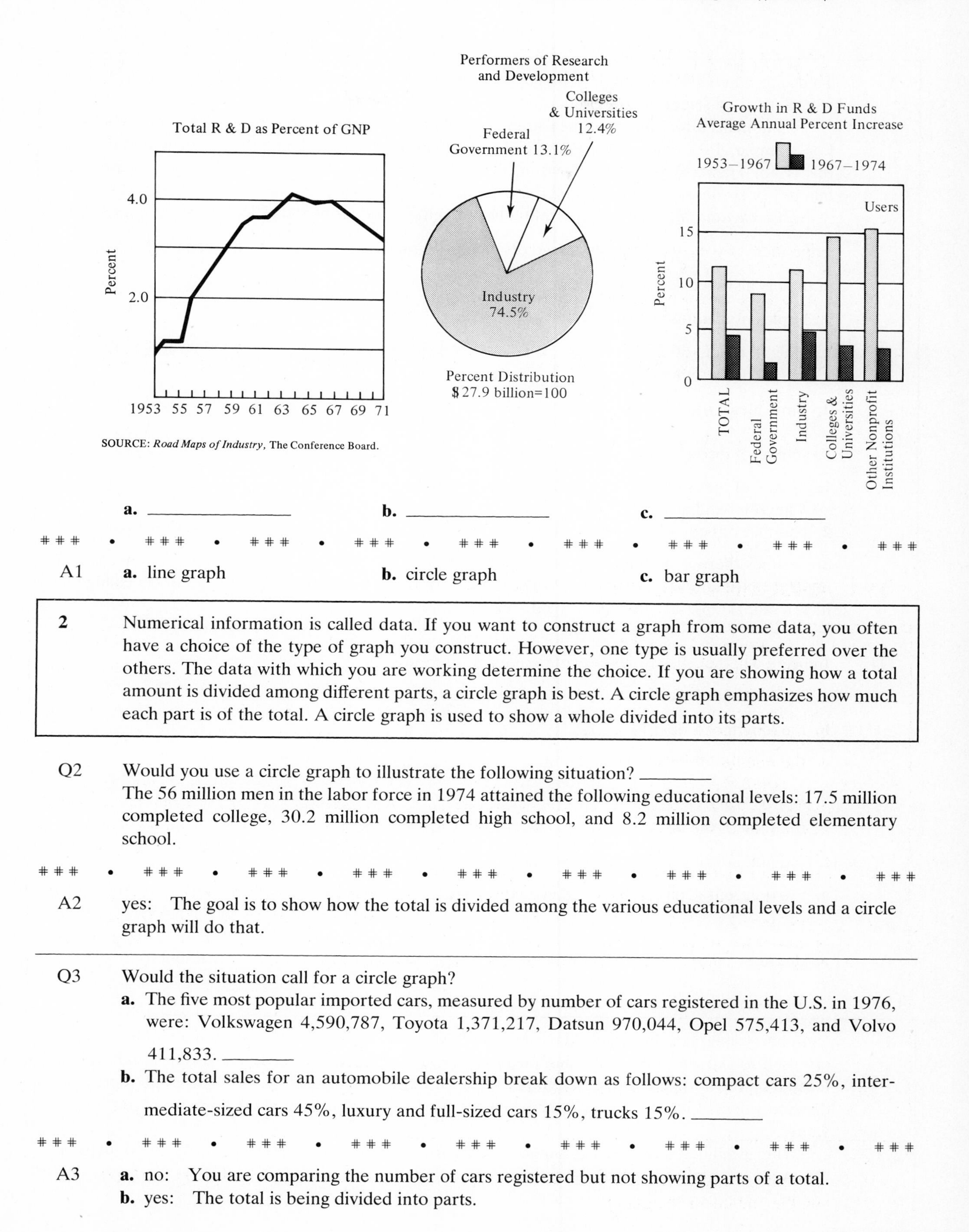

a. ______________ **b.** ______________ **c.** ______________

• ### • ### • ### • ### • ### • ### • ### •

A1 **a.** line graph **b.** circle graph **c.** bar graph

---

2 Numerical information is called data. If you want to construct a graph from some data, you often have a choice of the type of graph you construct. However, one type is usually preferred over the others. The data with which you are working determine the choice. If you are showing how a total amount is divided among different parts, a circle graph is best. A circle graph emphasizes how much each part is of the total. A circle graph is used to show a whole divided into its parts.

Q2 Would you use a circle graph to illustrate the following situation? ________
The 56 million men in the labor force in 1974 attained the following educational levels: 17.5 million completed college, 30.2 million completed high school, and 8.2 million completed elementary school.

• ### • ### • ### • ### • ### • ### • ### •

A2 yes: The goal is to show how the total is divided among the various educational levels and a circle graph will do that.

Q3 Would the situation call for a circle graph?

a. The five most popular imported cars, measured by number of cars registered in the U.S. in 1976, were: Volkswagen 4,590,787, Toyota 1,371,217, Datsun 970,044, Opel 575,413, and Volvo 411,833. ________

b. The total sales for an automobile dealership break down as follows: compact cars 25%, intermediate-sized cars 45%, luxury and full-sized cars 15%, trucks 15%. ________

• ### • ### • ### • ### • ### • ### • ### •

A3 **a.** no: You are comparing the number of cars registered but not showing parts of a total.
b. yes: The total is being divided into parts.

3 To decide whether a bar graph or a line graph should be constructed, it is necessary to determine if your data are discrete or continuous.

Continuous data consist of measurements that can be made more precise if a person chooses. For example, if the data are the amount of time spent on a project, it might be measured in days. If this is not satisfactory for some reason, time could be measured in hours or, more precisely, in minutes or seconds. Therefore, time is a continuous variable.

Discrete data allow no interpretation between the value of the data. For example, in a football game there may be 2 field goals scored or 3, but it is impossible to score $2\frac{1}{2}$ field goals. The number of field goals is a discrete variable.

Examples of continuous data:

1. Height: 60 in., 60.25 in., 6.2546 in. (the number of decimal places is dependent on the precision of the measurement)
2. Time measured in hours: 4 hours, $4\frac{1}{2}$ hours, 4.2576 hours

Examples of discrete data:

1. Makes of automobiles: Ford, Chevrolet, VW (there is no interpretation of halfway between a Chevrolet and a Ford)
2. Number of people in a classroom: 42, 16, 4 (you would not count people in fractional parts)

Sometimes there is no natural order for discrete data. For example, it makes no difference which make of automobile is listed first. When there is an order to discrete data, the numbers change in incremental steps, usually increasing by one each time.

Q4 Indicate whether the variable is continuous or discrete:

a. the number of men required to do a job __________

b. the length of a board __________

c. the number of inches of rainfall in a year __________

d. the inside diameter of a hole in a block __________

e. the time of the day __________

f. the number of cars produced per day in a manufacturing plant __________

g. the different department stores in a city __________

• ### • ### • ### • ### • ### • ### • ### •

A4 **a.** discrete **b.** continuous **c.** continuous **d.** continuous
e. continuous **f.** discrete **g.** discrete

4 Both bar graphs and line graphs have a vertical axis and a horizontal axis. Both bar and line graphs are used to show a numerical value measured on the scale on the vertical axis for various values located on the horizontal axis. Bar graphs are used when the horizontal axis of the graph represents discrete values. Line graphs are used when the horizontal axis of the graph represents continuous values.

The graph below shows the number of cars, trucks, and buses registered per 1,000 population. On the horizontal axis are the countries United States, Canada, France, and West Germany. The countries are discrete data, so a bar graph is used.

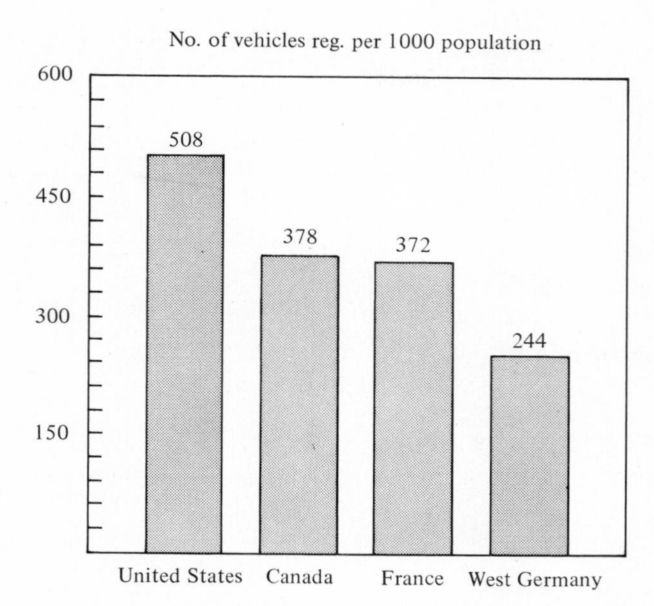

The horizontal axis on the following figure represents time measured in months. Since time is a continuous variable, the graph should be readable between numbers on the horizontal axis. Therefore, a line graph is constructed.

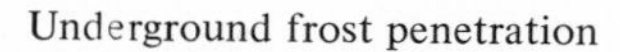

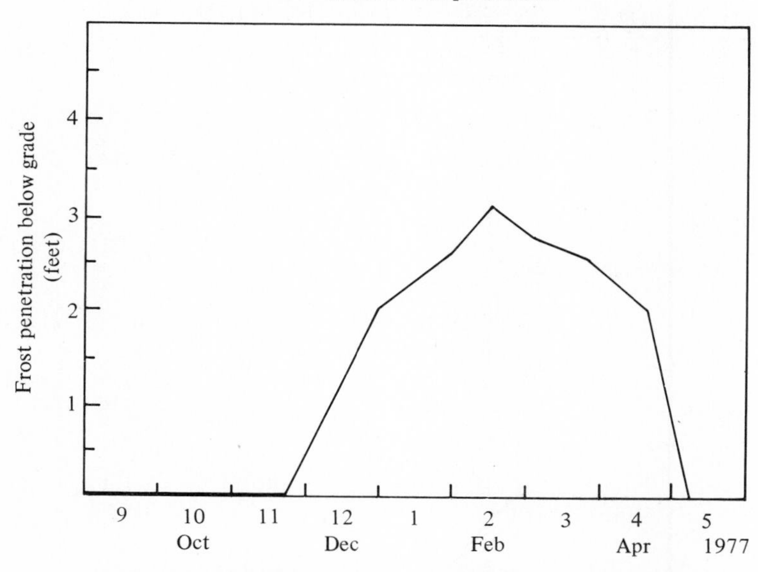

In order to tell what kind of graph to construct, first determine the type (continuous or discrete) of data on the horizontal axis.

Q5 Suppose the pressure in a boiler is to be compared throughout the day on a graph. Are the data on the horizontal axis continuous or discrete? ___________

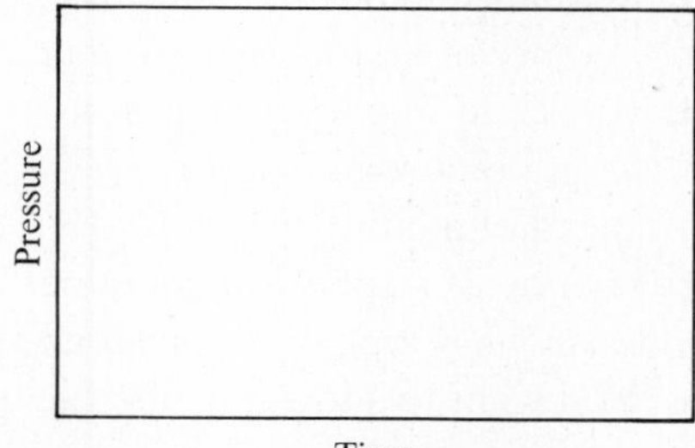

\# # # • # # # • # # # • # # # • # # # • # # # • # # # • # # # • # # #

A5 continuous: Time changes continuously throughout the day.

Q6 What kind of graph would you use in Q5? ____________

• # # # • # # # • # # # • # # # • # # # • # # # • # # # • # #

A6 line graph

Q7 Suppose that a graph is to be made that shows the maximum legal weight allowed in a truck with various numbers of axles.

a. Is the number of axles continuous or discrete? ____________

b. What type of graph would be drawn? ____________

Weight

Number of axles

• # # # • # # # • # # # • # # # • # # # • # # # • # # # • # #

A7 **a.** discrete: A truck would not have one-half axle.
b. bar graph

Q8 Assume that the variable listed is on the horizontal axis of a graph. Indicate whether the data are continuous or discrete and what type of graph would be drawn:

Data	*Type data*	*Type graph*
a. pressure measured in lb/in.^{2}	____________	____________
b. months	____________	____________
c. voltage measured in volts	____________	____________
d. car model: Ford, Chevrolet, etc.	____________	____________
e. number of customers in a garage	____________	____________

• # # # • # # # • # # # • # # # • # # # • # # # • # # # • # #

A8 **a.** continuous, line graph
b. continuous, line graph
c. continuous, line graph
d. discrete, bar graph
e. discrete, bar graph

This completes the instruction for this section.

11.1 EXERCISE

1. Indicate whether each of the following variables is continuous or discrete:
   a. number of tires on a truck
   b. resistance in a wire measured in ohms
   c. number of employees of a manufacturing plant
   d. votes cast in a union election
   e. length of a road in miles
   f. speed of a lathe in rpm
   g. volume of a spherical container in cm^3

2. Indicate what type of graph should be used to illustrate each of the following sets of data:
   a. You want to show how the fuel usage increases as you increase in speed from 0 to 90 mph.
   b. An oil is made up of pure oil plus certain percentages of 3 additives. You want to show the relationships of the four parts of the whole.

c. A service station manager wants to show how much of the income from his station comes from gasoline sales, automobile repair, car washes, and towing.
d. You want to show how the strength of a metal decreases as the temperature goes up.
e. You want to compare the accident frequency of manufacturing plants, construction sites, refuse collection, and service station garages. You will measure it in number of disabling injuries in 1,000,000 man-hours.

11.1 EXERCISE ANSWERS

1. a. discrete **b.** continuous **c.** discrete **d.** discrete
e. continuous **f.** continuous **g.** continuous
2. a. line graph **b.** circle graph **c.** circle graph **d.** line graph
e. bar graph

11.2 CIRCLE GRAPHS

1 A circle graph has the general appearance of a pie cut into uneven wedges. The geometric name of each wedge is a *sector*. The angle that is located at the center of the circle in each sector is called the *central angle* of the sector.

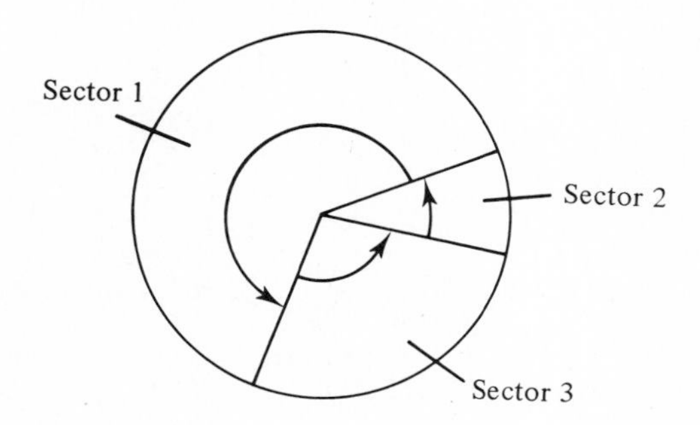

To interpret a circle graph you must know that the complete circle represents the total amount and that each sector represents a part of the total amount in the same ratio as the measure of its central angle is to 360°.

Q1 If a circle graph has a sector with a central angle of 90°, what percent of the total does that sector represent?

____

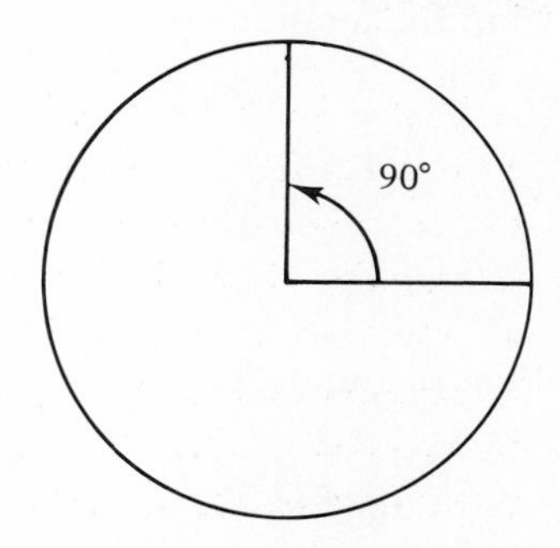

\# # # • # # # • # # # • # # # • # # # • # # # • # # # • # # # • # # #

A1 25%: $\frac{90}{360}=\frac{1}{4}=25\%$

2	Just as the sum of the parts must equal the total amount represented, the sum of the percent of all sectors must be 100% and the sum of the central angles must be 360°.

Q2 A circle graph showing the personal expenses of the family of an electrical technician is found on the right. Find the percent for the category "miscellaneous."

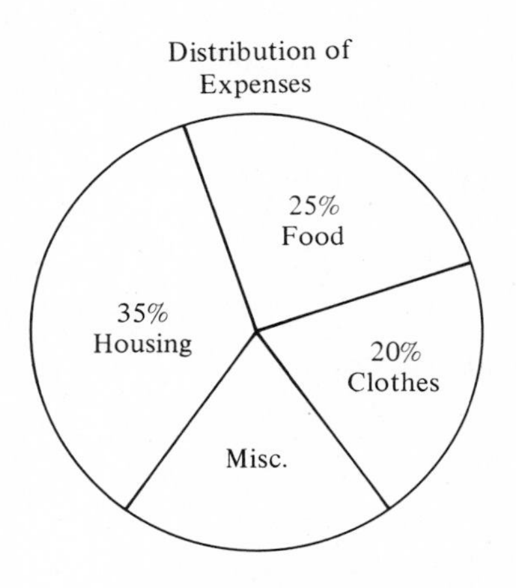

\# \# \# • \# \# \# • \# \# \# • \# \# \# • \# \# \# • \# \# \# • \# \# \# • \# \# \# • \# \# \#

A2 20%: $35\% + 25\% + 20\% = 80\%$
$100\% - 80\% = 20\%$ for miscellaneous

3 Usually the total amount that the whole circle represents is given in the title of the circle graph. To find the amount that each sector represents, multiply the total amount represented times the percent of the total included in the sector. For example, in the circle graph shown, to find the number of dollars used for housing in a family with income \$11,500, we would find $\$11{,}500 \times 35\% = \$11{,}500 \times 0.35 = \$4{,}025$.

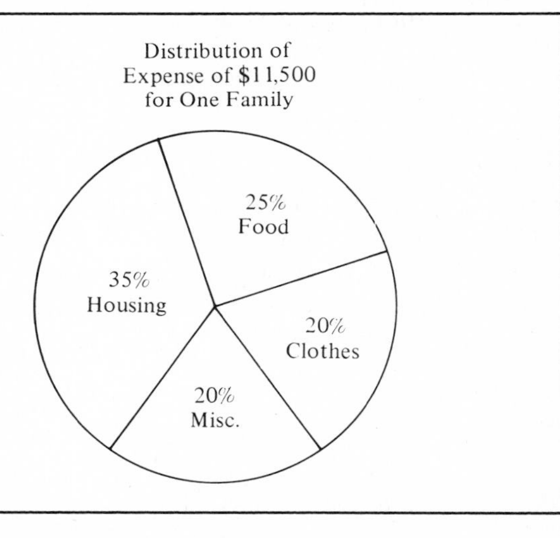

Q3 **a.** Find the amount spent on food for the family whose expenses are shown in the preceding circle graph.

b. Find the amount spent on clothes for the same family.

\# \# \# • \# \# \# • \# \# \# • \# \# \# • \# \# \# • \# \# \# • \# \# \# • \# \# \# • \# \# \#

A3 **a.** \$2,875: $\$11{,}500 \times 25\% = \$11{,}500 \times 0.25 = \$2{,}875$
b. \$2,300: $\$11{,}500 \times 20\% = \$11{,}500 \times 0.20 = \$2{,}300$

4 To construct a circle graph you must find the number of degrees to use for the central angle for each sector. This can be done with a proportion.

$$\frac{X}{360°} = \frac{\text{amount of part}}{\text{amount of total}} \quad \text{where } X = \text{number of degrees in the sector}$$

If you substitute in the proportion and solve for X, you will have the degrees in the central angle of the sector. To actually draw this, you will need a protractor.

For example, to find the number of degrees for the gas and oil sector of a circle graph of the information below you would proceed as shown.

Cost of Used-Car Ownership for Two-Year Period

Initial cost of car	\$1,500
Gas and oil	850
Upkeep and repairs	750
Insurance	500
Total	\$3,600

To find central angle of gas and oil sector:

$$\frac{X}{360°} = \frac{850}{3{,}600}$$

$$X = \frac{850}{\cancel{3{,}600}_{10}} \times \frac{\cancel{360°}^{1}}{1}$$

$$X = 85°$$

Q4 **a.** Determine the central angle for initial cost of the car, upkeep and repairs, and insurance for the information given in Frame 4.
b. Construct a circle graph by using the results of part **a** and the example in Frame 4. Be sure to write a title and label each sector.

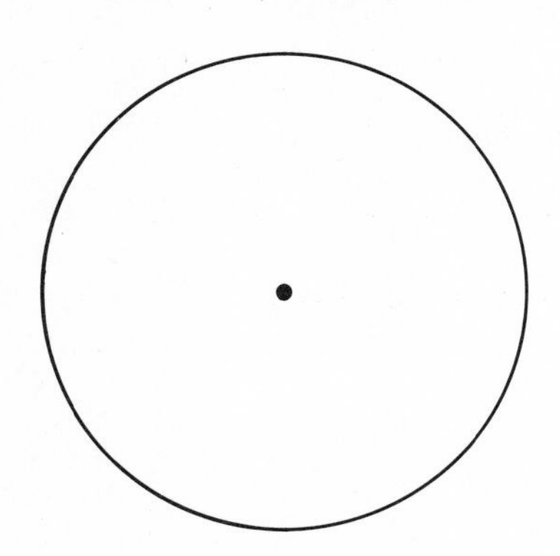

\# \# \# • \# \# \# • \# \# \# • \# \# \# • \# \# \# • \# \# \# • \# \# \# • \# \# \# • \# \# \#

A4 **a.**

$$\frac{X}{360°} = \frac{1{,}500}{3{,}600} \qquad \frac{X}{360°} = \frac{750}{3{,}600} \qquad \frac{X}{360°} = \frac{500}{3{,}600}$$

$$X = \frac{1{,}500}{3{,}600} \cdot 360° \qquad X = \frac{750}{3{,}600} \cdot 360° \qquad X = \frac{500}{3{,}600} \cdot 360°$$

$$X = 150° \qquad X = 75° \qquad X = 50°$$

b. Cost of Used-Car Ownership for Two-Year Period

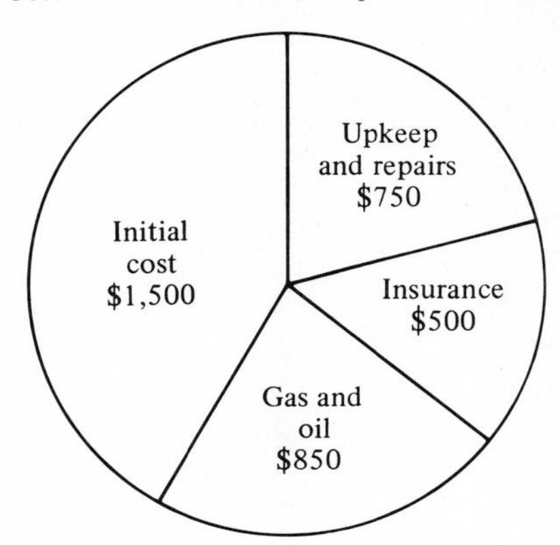

5 Suppose you know the percent of the total for each sector. To find the number of degrees in each sector you may use two methods. The first is based on proportions, and the second is done with percents.

Example: The highest level of education is given for men in 1974. Find the number of degrees in the sector of a circle graph representing elementary school.

Elementary school	15%
High School	55%
College	30%

Solution 1: $\dfrac{15\%}{100\%} = \dfrac{x}{360°}$

$$1.00x = (0.15)(360)$$
$$x = 54°$$

Solution 2: 15% of 360° must be in the sector representing elementary school, therefore $(0.15)\,(360°) = 54°$

Q5 **a.** Use the data of Frame 5 to find the number of degrees in the sector representing high school.

b. Find the number of degrees in the sector representing college.

• # # # • # # # • # # # • # # # • # # # • # # # • # # # • # #

A5 **a.** 198°: 55% of 360° = (0.55)(360°) = 198°
b. 108°: 30% of 360° = (0.30)(360°) = 108°

This completes the instruction for this section.

11.2 EXERCISE

Use the accompanying circle graph which shows the sales of trucks in 1974 in the U.S., broken down by gross vehicular weight, to answer problems 1–5.

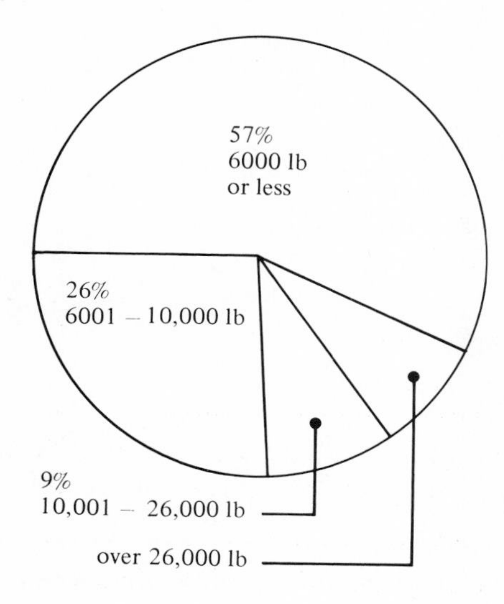

Source: Motor Vehicle Manufacturers Association of the U.S., Inc.

1. Determine the percent of trucks sold in the over 26,000 lb class.
2. Find the number of trucks sold in the 6,000 lb or less class. Round off to the 100,000's place.
3. Use your answer to problem 1 to determine the number of trucks sold in the over 26,000 lb class. Round off to the 100,000's place.
4. Use your answer to problem 2 to find the number of degrees in the central angle of the sector that represents sales of trucks with gross vehicular weight 6,000 lbs or less. Round off to the nearest degree. (Do not measure.)
5. Determine the number of degrees in the central angle of the sector that represents the sales of trucks in the 6,001–10,000 lb class. Round off to the nearest degree. (Do not measure.)

Use the following pie graph for problems 6–7.

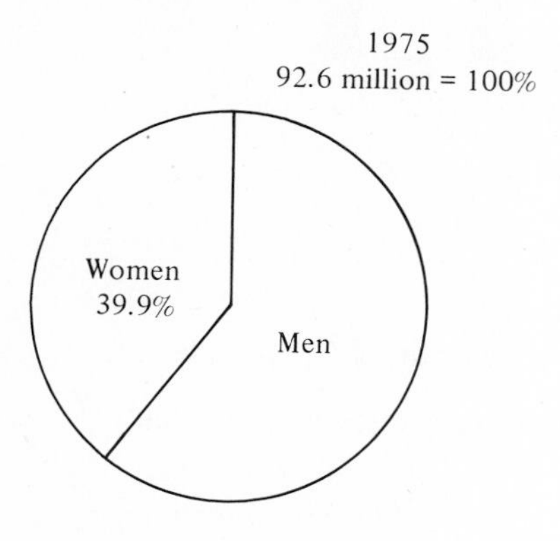

6. Use the graph to find the total number of women in the labor force in 1975. Round off to the nearest tenth of a million.

7. Find the number of degrees in the sector representing women in the labor force. Round off to the nearest degree. (Do not measure.)

11.2 EXERCISE ANSWERS

1. 8%

2. 1,500,000

3. 200,000

4. 200°: $\frac{x}{360°} = \frac{1,500,000}{2,700,000}$

$x = 200°$

5. 94°: 26% of 360°

6. 36.9 million (36,900,000)

7. 144°

11.3 BAR GRAPHS

1 A bar graph is useful when a quick visual impression of the relative sizes of various quantities is to be communicated. If the actual numbers are desired corresponding to the heights of the bars, they should be obtainable on a scale given as a part of the graph. The values will, by necessity, be approximate. The following bar graph is appropriate for the data being considered because the occupational groups are discrete data.

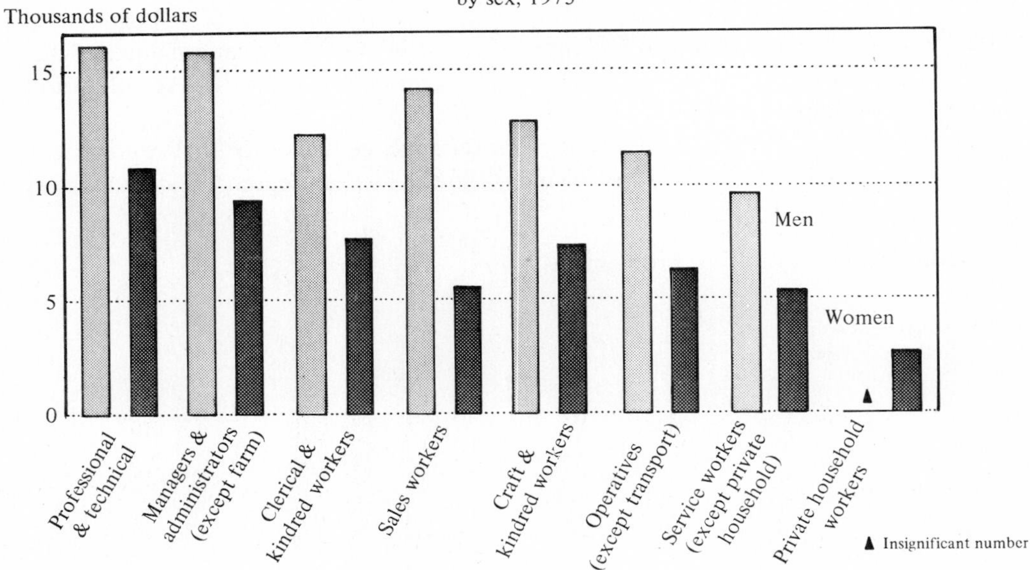

Example: Find the difference in earnings for a male sales worker and a female sales worker.

Solution: A male sales worker makes $14,000 and a female sales worker makes approximately $5,500. The difference is $14,000−$5,500=$8,500.

Q1 Use the graph in Frame 1 to answer the following questions:

a. What was the average earnings of a professional or technical male? ________

b. What was the average earnings of a professional or technical female? ________

c. What was the difference in average earnings between male and female professional or technical workers? __________

d. (1) Which male occupation group (other than private household workers) had the lowest average income? __________ (2) The highest average income? __________

e. (1) Which female occupational group had the lowest average income? __________ (2) The highest average income? __________

f. Within which occupational group is there the largest difference between male and female workers' average income? __________

\# # # • # # # • # # # • # # # • # # # • # # # • # # # • # # # • # # #

A1 **a.** $16,000 **b.** $10,500 **c.** $5,500
d. (1) service workers (2) professional and technical
e. (1) private household workers (2) professional and technical workers
f. sales workers

2 A bar consists of a series of vertical or horizontal bars drawn according to scale. When constructing a bar graph one must be careful not to introduce a distortion in the impression left with the reader. The visual impression should match the actual relationships between the numbers being represented. An example will show how a bar graph can leave the wrong impression when not constructed correctly. Study the data in the table and the bar graph and see if you can locate the distortion.

Average Weekly Earnings	
Ardis	$300
Baker	$400
Chambers	$600

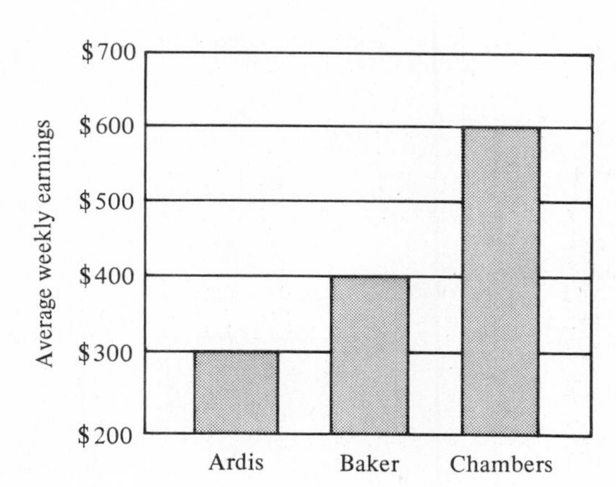

The relationships that exist in the chart (Chambers making twice as much as Ardis, Baker making $\frac{2}{3}$ of Chambers) are not reflected in the height of the bars. In the graph it appears that Chambers makes 4 times the earnings of Ardis and Baker makes $\frac{1}{2}$ the earnings of Chambers. Although the top of the bar corresponds to the correct dollar amounts, the visual impression (some say this is what people usually remember) is inaccurate. The graph shown is actually only the top portion of a graph as it should be drawn. Compare the two graphs on the next page.

Notice that starting the vertical scale at zero instead of a larger number causes the heights of the bars to be in the same proportion as the numbers they represent. Using any other number as the smallest value on the vertical scale will distort the impression. Therefore, remember—*always start the vertical scale on a bar graph at the number zero.* Removing a section of a bar or removing a

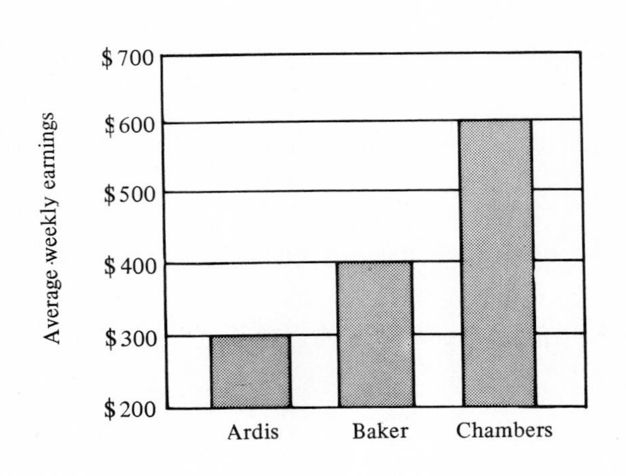

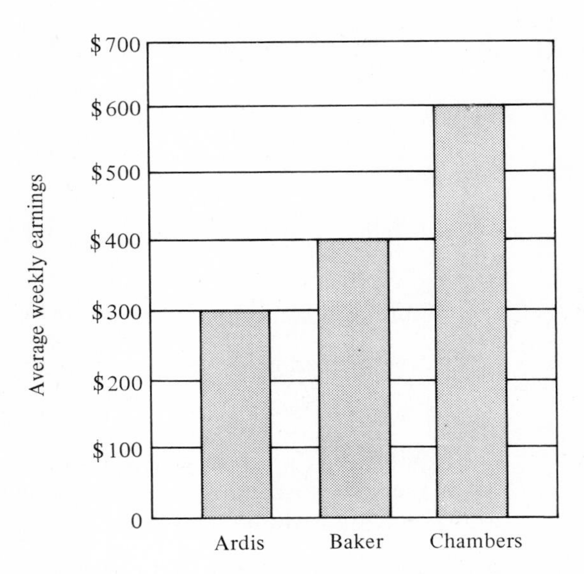

section of the vertical scale also distorts the ratios of the bars of a bar graph. The omissions are usually indicated with a jagged line. If you see such a jagged line, be on the lookout for a distorted graph.

Q2 For each of the graphs below (1) indicate the actual ratio of the heights of the bars and the ratio of the numbers represented by the bars, and (2) what in the construction of the graph has caused the distortion?

a.

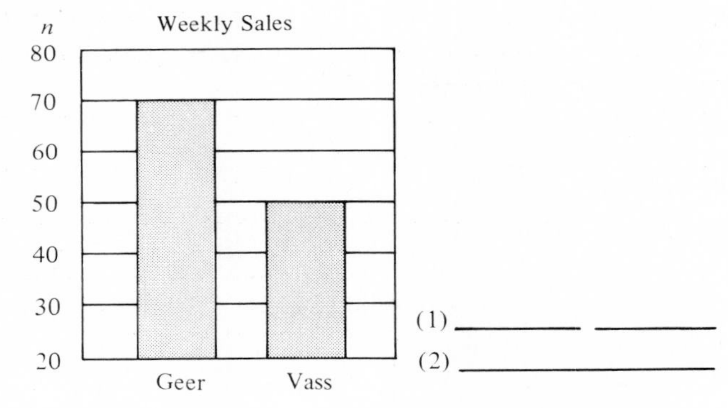

b.

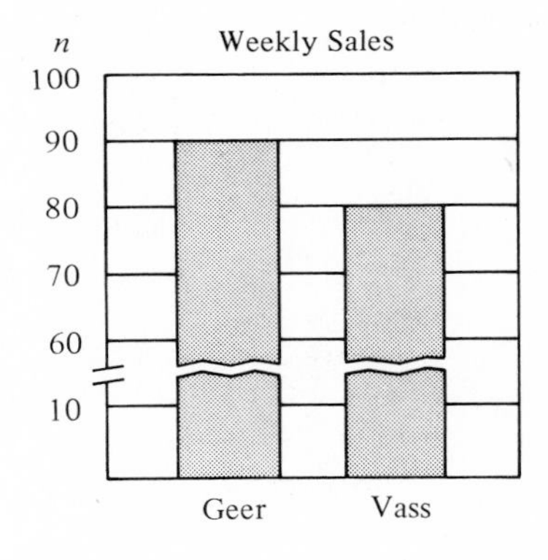

(1) ________ ________

(2) ________________

c.

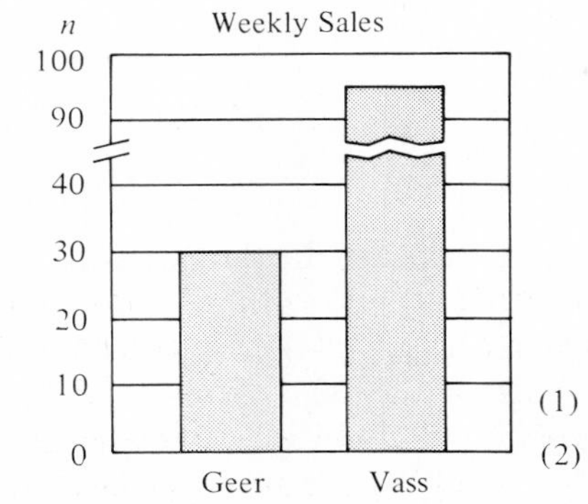

(1) ________ ________

(2) ________________

\# # # • # # # • # # # • # # # • # # # • # # # • # # # • # # # • # # #

A2 **a.** (1) $\frac{5}{3}, \frac{70}{50} = \frac{7}{5}$

(2) vertical scale starts at 20 instead of 0

b. (1) $\frac{5}{4}, \frac{90}{80} = \frac{9}{8}$

(2) section removed from bars

c. (1) $\frac{3}{5\frac{1}{2}} = \frac{6}{11}, \frac{30}{95} = \frac{6}{19}$

(2) section removed from bar

Q3 For each of the bar graphs in Q2, indicate whether the distortion favors Geer or Vass.

a. ________ **b.** ________ **c.** ________

• ### • ### • ### • ### • ### • ### • ### •

A3 **a.** Geer: The graph makes it look like he sold $\frac{5}{3}(\doteq 1.7)$ as much as Vass, but he only sold $\frac{7}{5}(= 1.4)$ as much.

b. Geer: The graph makes it look like he sold $\frac{5}{4}(\doteq 1.3)$ as much as Vass, but he only sold $\frac{9}{8}(\doteq 1.1)$ as much.

c. Geer: The graph makes it look like he sold $\frac{6}{11}(\doteq 0.55)$ as much as Vass, but he sold only $\frac{6}{19}(\doteq 0.32)$ as much.

---

3 The most important part of drawing a bar graph is determining the scale. You must be able to represent the largest number by a bar that fits into the available space. Suppose that you were to graph the data from the chart below on the space provided.

Imported Cars Registered in the U.S.

Volkswagen	4,590,000
Toyota	1,370,000
Datsun	970,000
Opel	570,000
Volvo	410,000

SOURCE: *Automotive News*, April 1975.

The number of Volkswagens would have to be represented by a bar no longer than 10 units because there are 10 spaces available. *To find the scale, divide the amount to be represented by the longest bar by the number of available spaces.* (If your paper is unlined, divide by the number of inches or centimetres available.) Round the resulting value up to a *convenient larger* value. Do not ever round down; for if you do, the longest bar will be longer than the height available. In the example given:

$4{,}590{,}000 \div 10 = 459{,}000$

Since the number would be inconvenient to use, it is rounded *up* to 500,000. This makes the vertical axis easy to read.

Before you are given an opportunity to complete the bar graph of this example, you will be given a chance to practice determining the scale for various graphs.

Q4 Suppose that the largest number to be represented was 426 and you wanted to draw your graph on a location that had 10 vertical spaces. What would you use for a scale?

• ### • ### • ### • ### • ### • ### • ### •

A4 50 units per space: $426 \div 10 = 42.6$ rounded upward would be 50. (45 would not be incorrect.)

Q5 If the longest bar on a bar graph was to represent 45,050 and you had 7 inches in the vertical direction available, what would you use for a scale?

• ### • ### • ### • ### • ### • ### • ### •

A5 7,000 per inch: $45{,}050 \div 7 = 6{,}436$ rounded up to 7,000. Therefore, use 7,000 per inch. In this case 8,000, or even 10,000, would not be incorrect. Use of 10,000 would cause your bars to be much shorter, and you would not be using as much of your available space.

Q6 The longest bar should represent $44, and there are 6 vertical spaces available. What should be used for the vertical scale?

• ### • ### • ### • ### • ### • ### • ### •

A6 $8 per space: $\$44 \div 6 = \$7.33$ rounded upward would be $8. It would also be correct to use $10 per space.

Q7 Use the scale of 500,000 per vertical space and complete the bar graph in Frame 3. Be sure that both scales are labeled and that a title is included.

• ### • ### • ### • ### • ### • ### • ### •

A7 (*Note*: The lengths of each bar must be approximated as nearly as possible.)

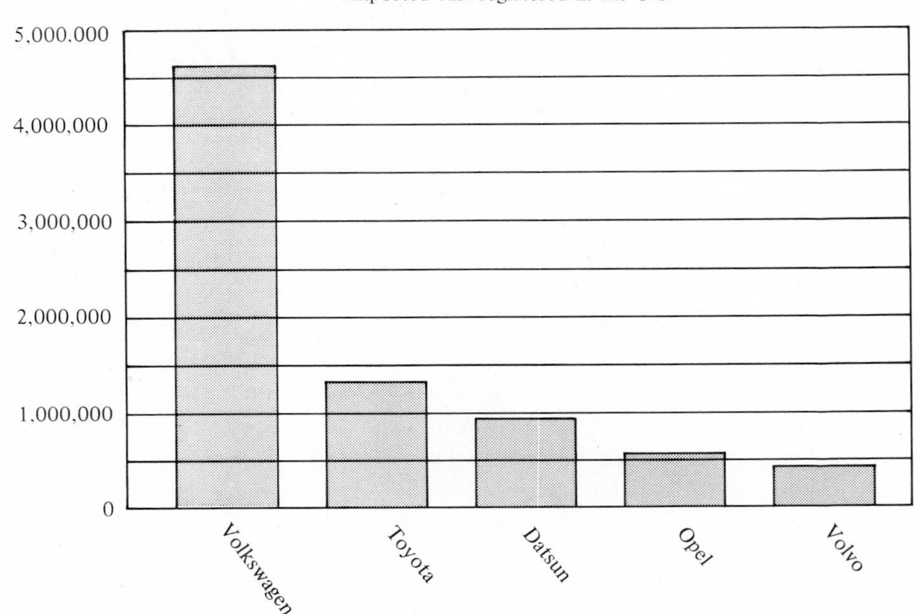

Q8 Use the space provided to draw a bar graph of the data given here.

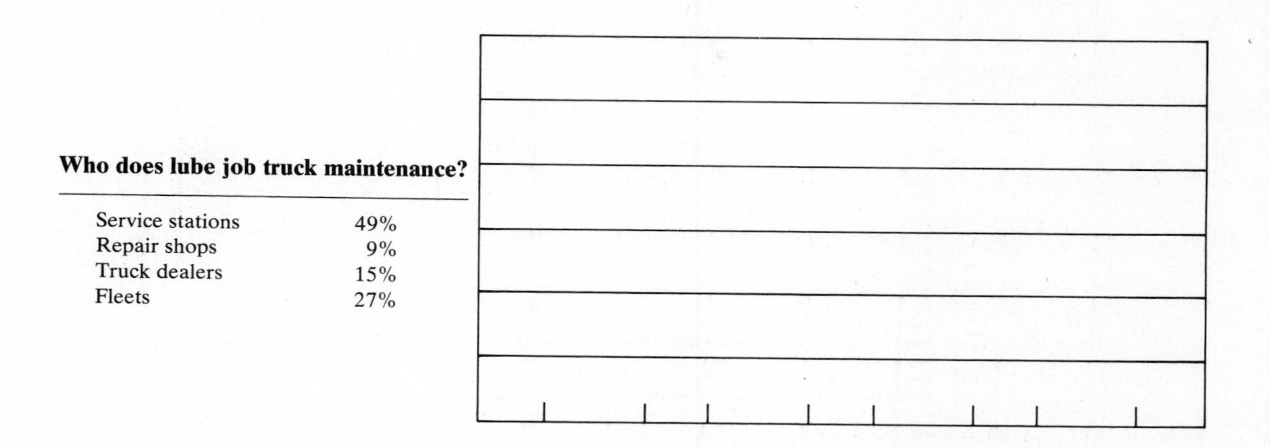

Who does lube job truck maintenance?

Service stations	49%
Repair shops	9%
Truck dealers	15%
Fleets	27%

\# # # • # # # • # # # • # # # • # # # • # # # • # # # • # # # • # # #

A8 The scale is determined by $49 \div 6 = 8.2$ which is rounded up to the convenient number 10.

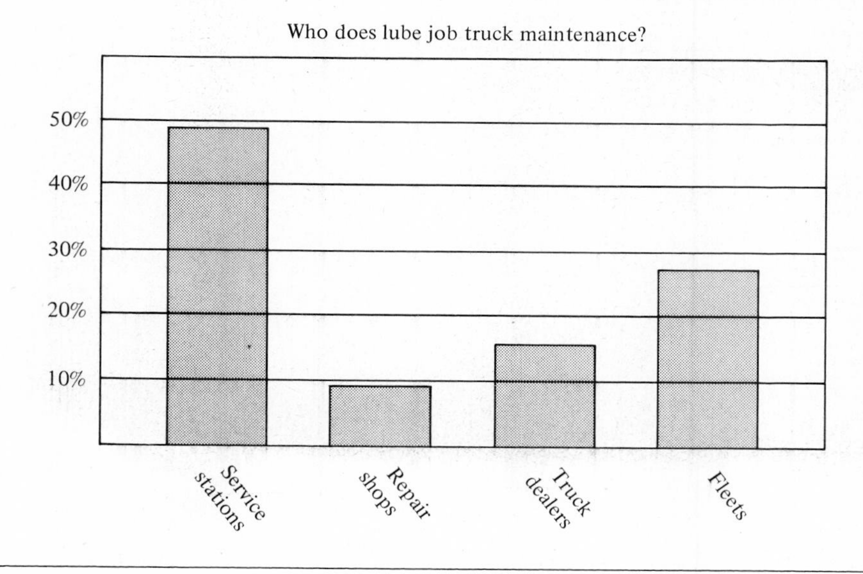

This completes the instruction for this section.

11.3 EXERCISE

A distortion is sometimes introduced into a bar graph by removing a section of a bar. Since this type of graph appears often, you should read the graph very critically. The questions following the graph will assist your thinking. The graph is more "sophisticated" than the ones that you have been reading, but the same principles apply.

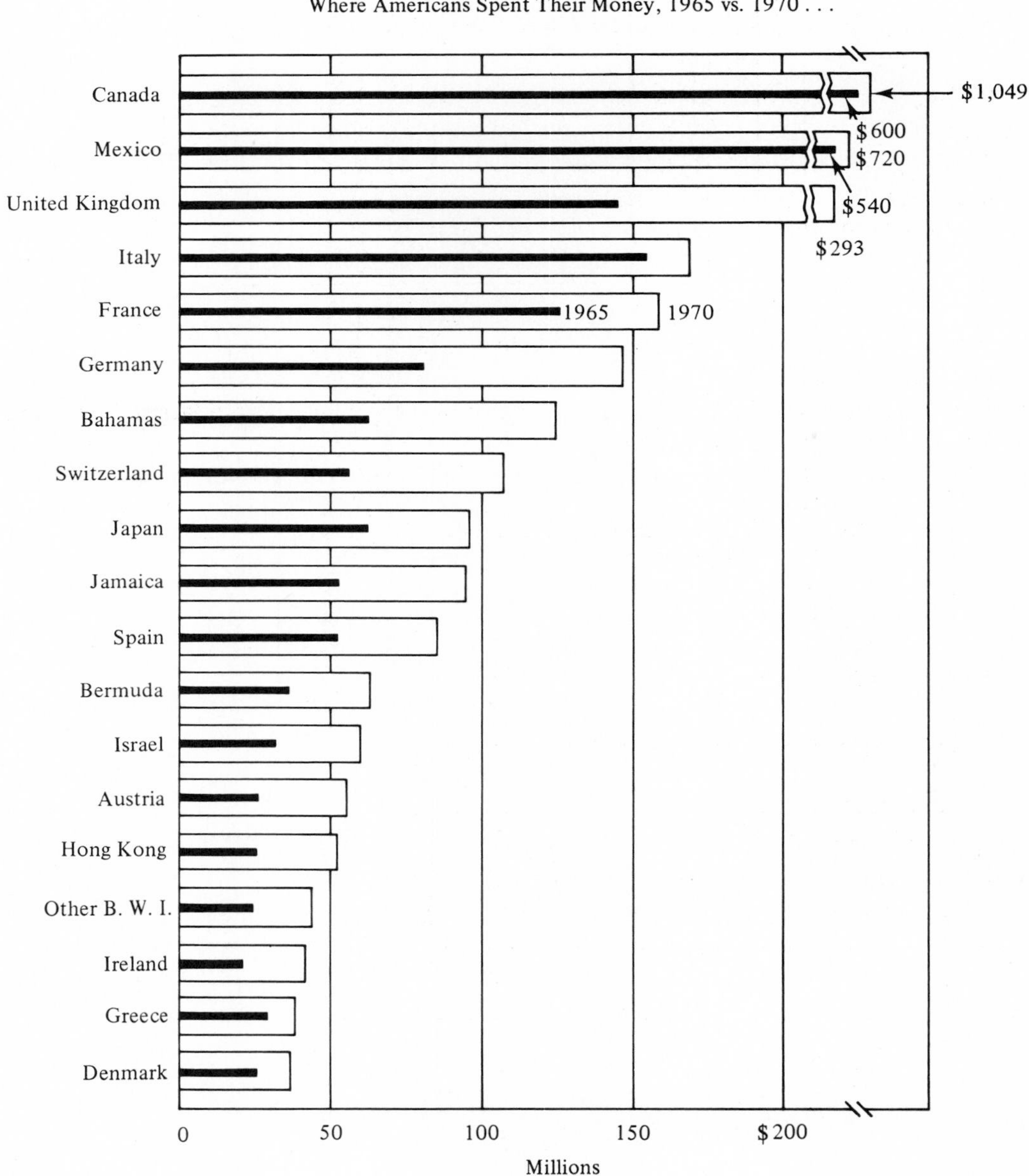

SOURCE: *Road Maps of Industry*, The Conference Board.

1. How much money was spent in France in 1970?
How much money was spent in France in 1965?
3. How much money was spent in Spain in 1965?
4. The amount spent in Italy in 1970 was approximately how many times the amount spent in Austria in 1970?

5. Does the ratio of the apparent length of the bar for 1970—Canada to the bar for 1970—France equal the ratio of the dollar amounts?
6. Was the same length of bar omitted from the first three bars?
7. The 1970 bar for Mexico should be approximately how many times as long as the 1970 bar for Switzerland?
8. Should the length of the 1965 bar for Canada actually reach as close to the end of the separated section of the bar as is shown on the graph?

Proceed as the number items indicate to construct a graph of the data in the chart.

Six Largest Employee organizations

Teamsters	1,800,000
Automobile workers	1,500,000
Steel workers	1,200,000
National Education Association	1,100,000
Electrical Workers (BEW)	900,000
Machinists	850,000

9. What kind of graph is appropriate, and why?
10. Determine a scale.
11. Label the horizontal and vertical scales. Arrange organizations from largest to smallest.
12. Draw the bars.
13. Write an appropriate title.

11.3 EXERCISE ANSWERS

1. 160,000,000
2. 125,000,000
3. 50,000,000
4. 3 times
5. no
6. no
7. almost 7 times
8. no
9. bar graph, Organizations are discrete data.
10. 300,000 per space: $1{,}800{,}000 \div 7 = 257{,}000$, so round to 300,000 per space

11, 12, and **13** are shown below.

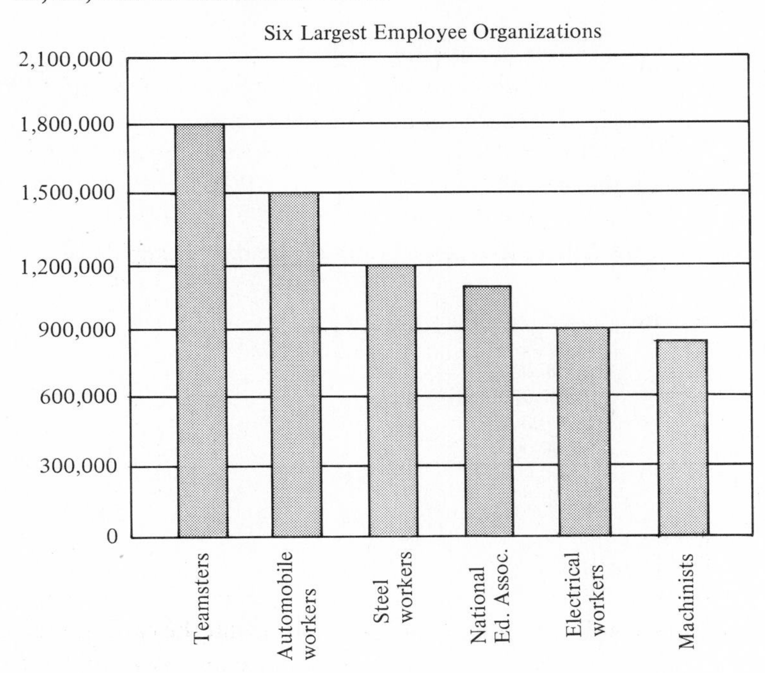

11.4 LINE GRAPHS

1 Line graphs are used when the horizontal scale indicates *continuous* data. The line graph is particularly useful for showing a continuous trend over a period of time. Time is then located on the horizontal axis. One can interpret a line graph for values in between the numbers shown on the horizontal scale.

The following line graph shows the distances that a car has traveled since the beginning of its trip at 12:00.

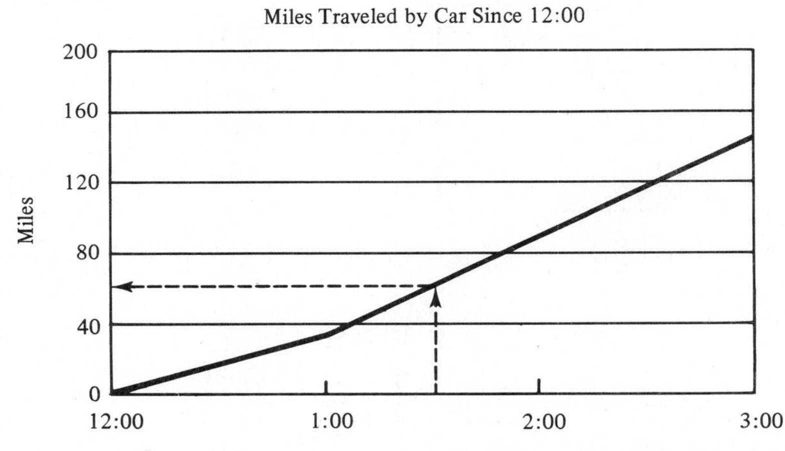

The distance traveled at a given time is found by locating the time on the horizontal scale, moving vertically to the line, then moving horizontally to read the height of the intersection point on the vertical scale. The dashed line on the graph shows how to locate the distance the car had traveled at 1:30. Reading the vertical scale corresponding to 1:30 on the horizontal scale, we see that the car had traveled 60 miles.

Q1 Use the line graph in the previous example to approximate the distance traveled at:

a. 12:30 ________ **b.** 1:15 ________ **c.** 2:45 ________

• ### • ### • ### • ### • ### • ### • ### •

A1 **a.** 15 miles
b. 45 miles
c. 135 miles

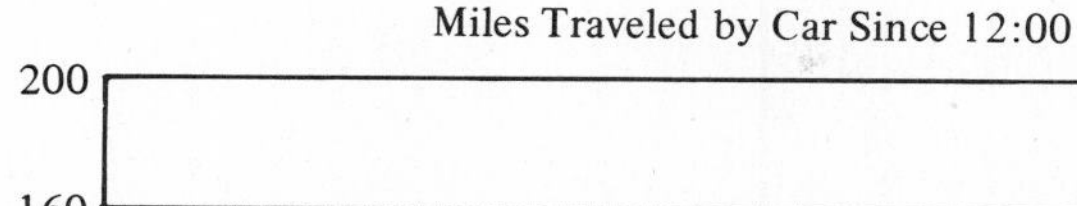

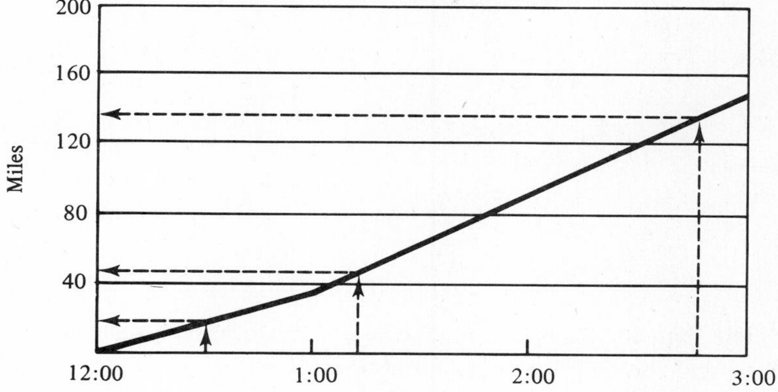

(answers may vary by 5 miles and still be correct)

---

2 Notice the change in the slope (slant) of the line at 1:00. This indicates that the car is traveling faster, such as would happen if the car left the city streets and entered an expressway at 1:00. The car traveled from 0 miles at 12:00 to 30 miles at 1:00 for an average of 30 miles per hour. From 1:00 to 2:00 the car traveled from 30 miles to 90 miles for an average of 60 miles per hour. From this example you can see that the steeper the slope of the line of the graph, the faster the variable on the vertical scale is changing.

A line graph is useful for reading particular values, showing trends that are increasing or decreasing, and for showing the rate of change of a quantity.

Q2 The line graph shows the total amount of money collected by the end of each day in a United Fund campaign. Find the following:

a. Money collected by the 4th day.

____

b. Money collected by the 10th day.

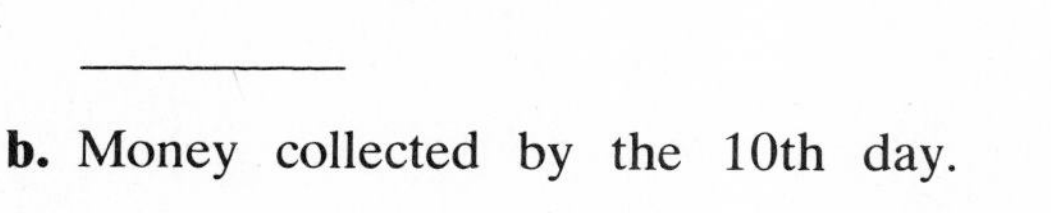

____

c. Is money coming in faster on the 3rd day or the 7th? ________

d. Is money coming in faster on the 7th day or the 13th? ________

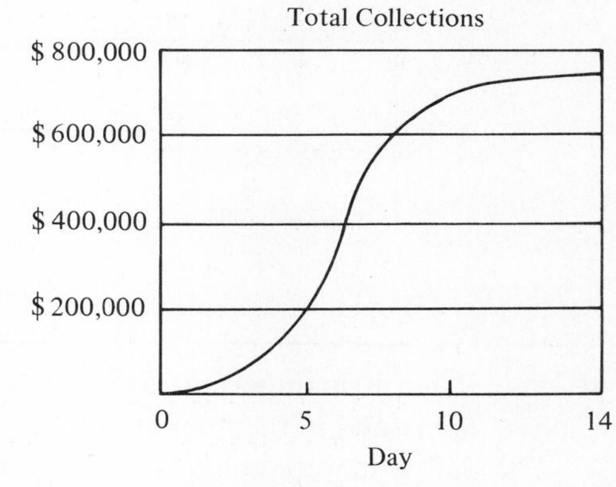

• ### • ### • ### • ### • ### • ### • ### •

A2 **a.** $100,000 **b.** $700,000 **c.** 7th **d.** 7th

---

3 When constructing a line graph, determine the vertical scale in the same manner as the bar graph. (Divide the largest number to be represented by the number of spaces available.) Lay off the horizontal scale corresponding to the appropriate variable. Time is located on the horizontal scale. Locate points corresponding to the given data and then connect the points with straight lines from left to right.

Q3 Construct a line graph from the given data.

Temperatures for a 24-hour period

12:00 midnight	50 °F
3:00	42 °F
8:00	52 °F
12:00 noon	68 °F
2:00	74 °F
5:00	70 °F
8:00	60 °F
12:00 midnight	45 °F

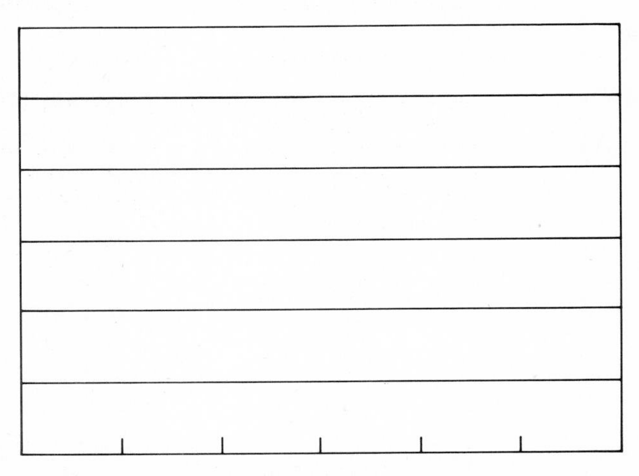

(*Note*: The data for the horizontal scale are not evenly spaced. Therefore, let each space on the horizontal scale be an increment of 4 hours.)

• # # # • # # # • # # # • # # # • # # # • # # # • # # # • # #

A3 The scale is determined by $74 \div 6 = 12\frac{1}{3}$, which is rounded to 15.

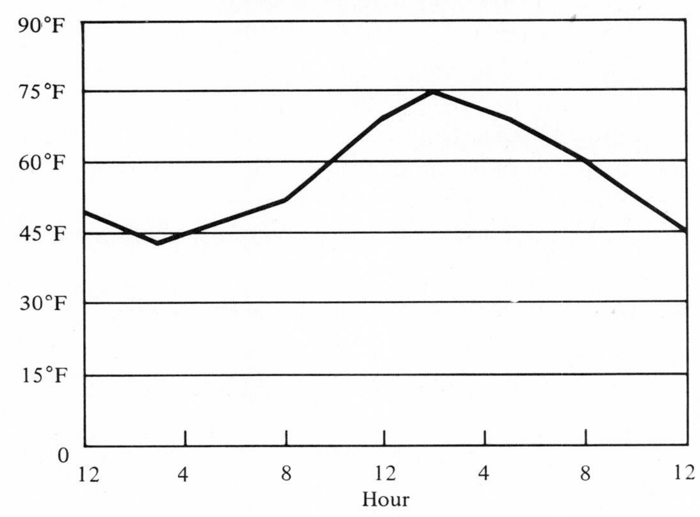

4 Often it is important to know a value in between values of a chart. For example, one might be interested in the temperature at 10:00 A.M. in the situation presented in Q3. The line graph allows you to approximate this temperature quite accurately. The graph of A3 indicates that the temperature at 10:00 A.M. was 60°; 60° is a much more reasonable answer than giving the temperature of either of the two closest times that are in the chart. Approximation between two known values is called *interpolation*. Interpolation is used a great deal in mathematics and everyday life. There are some hazards in interpolation which you should be aware of. The example at the top of page 533 will illustrate.

Notice that the first line graph is good for showing the general trend over the year. It shows the significant decrease in production in July and August when car companies change production to the new models. It misses the sudden changes in production that the more sensitive second graph illustrates. Answer the questions following this frame to see the accuracy of interpolation.

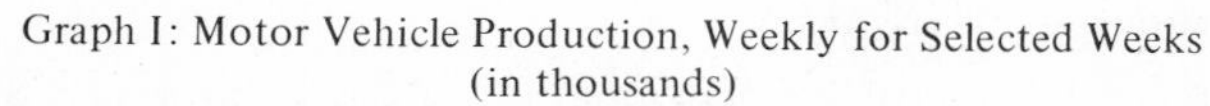

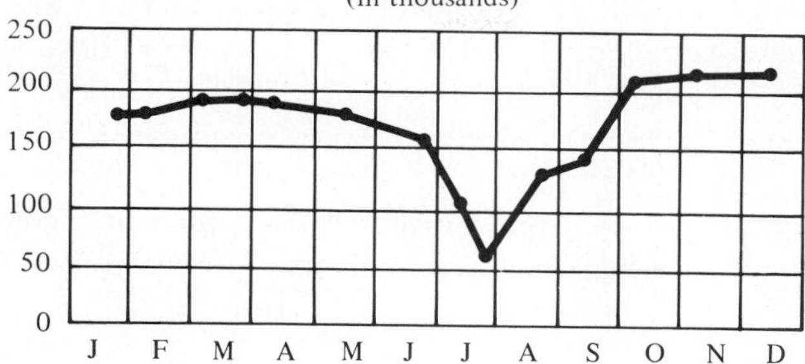

Graph II: Motor Vehicle Production, Weekly (in thousands)

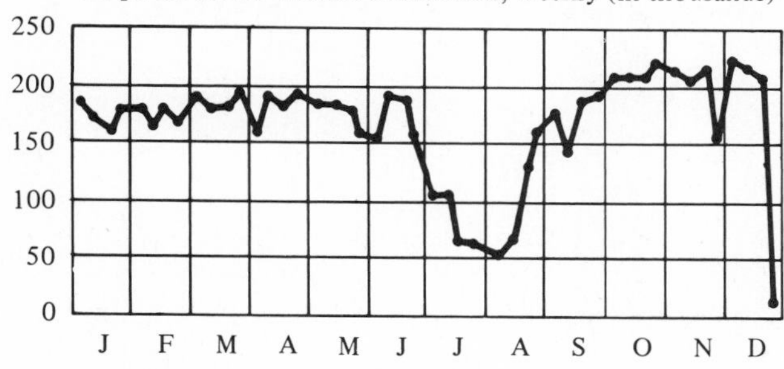

Q4 Use the two graphs in the previous frame to answer the questions below:

a. Find the production for the third week in March.

Graph I—interpolation __________ Graph II __________

b. Find the production for the fourth week in November.

Graph I—interpolation __________ Graph II __________

c. What was the amount of the lowest weekly production of the year?

Graph I __________ Graph II __________

d. Which graph would be most useful for predicting the week of the Easter holiday?

____

e. What week would you say included the Easter holiday? ____________________

• # # # • # # # • # # # • # # # • # # # • # # # • # # # • # #

A4

a. Graph I, 190,000; graph II, 180,000

b. Graph I, 220,000; graph II, 150,000
This fluctuation was probably due to Thanksgiving.

c. Graph I, 70,000; graph II, 10,000: in December

d. Graph II

e. first week in April

5 From the preceding questions you can tell that interpolation will be inaccurate when there are rapid changes that are not reflected in the graph. In situations where the data change smoothly, interpolation is more accurate.

Another type of approximation is much less likely to be accurate but is still useful. Approximating beyond the available data is called *extrapolation.* If a person had only graph I in Frame 4 available to him and wanted to know the most likely production for the fourth week of December, at the end of the year, he would probably have predicted about 220,000. (Check this yourself.) Such a prediction would then have been disastrous, however, because that was the week of the Christmas holidays and production went to 10,000 (check graph II). A prediction on the other end of the year for the first week in January would have been much more accurate. Extrapolating on the curve of graph I would have given an approximation of about 180,000, whereas from graph II we see that it was actually 190,000.

Q5 Over-advancing engine timing is frowned on by engine manufacturers since it increases power at the expense of greatly increasing heat. The chart below contains data recorded from tests on an engine.

Increase in heat and Power from Timing Over-advanced

Timing Over-advanced	*Increase in heat*	*Increase in Power*
2°	3.0%	1.0%
4°	4.8%	1.5%
6°	7.0%	2.5%
8°	8.8%	3.2%

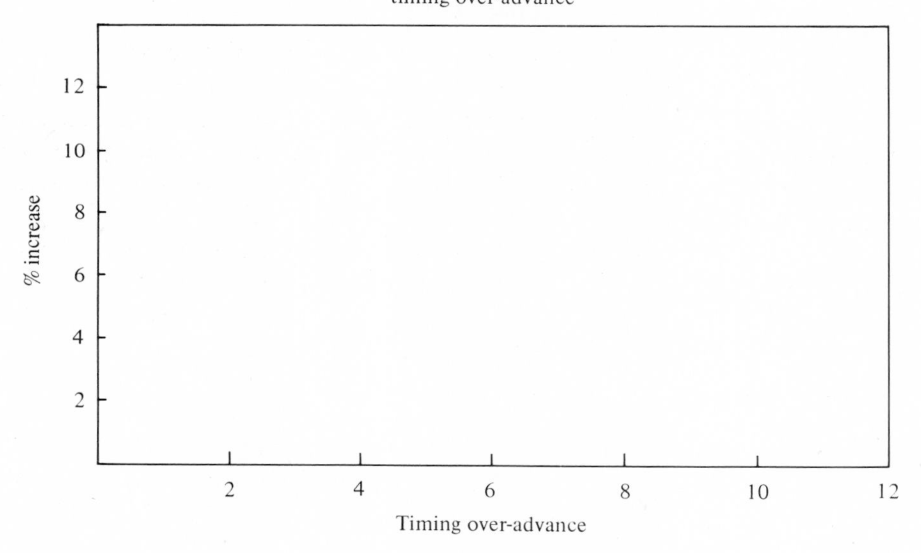

a. Use the data in the chart above to draw a line graph of the percent increase in heat. Use a solid line.

b. Use the data in the chart above to draw a line graph of the percent increase in power. Use a dashed line.

c. Extend the graph of the increase in heat (with a dotted line) to extrapolate the percent of heat increase when the timing is over-advanced 10°. ________

d. Extend the graph of the increase in power (with a dotted line) to extrapolate the percent of power increase when the timing is over-advanced 10°. ________

\# \# \# • \# \# \# • \# \# \# • \# \# \# • \# \# \# • \# \# \# • \# \# \# • \# \# \# • \# \# \#

A5 **a, b**

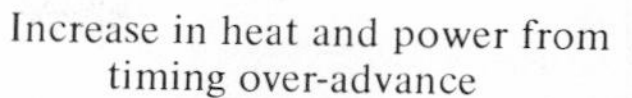

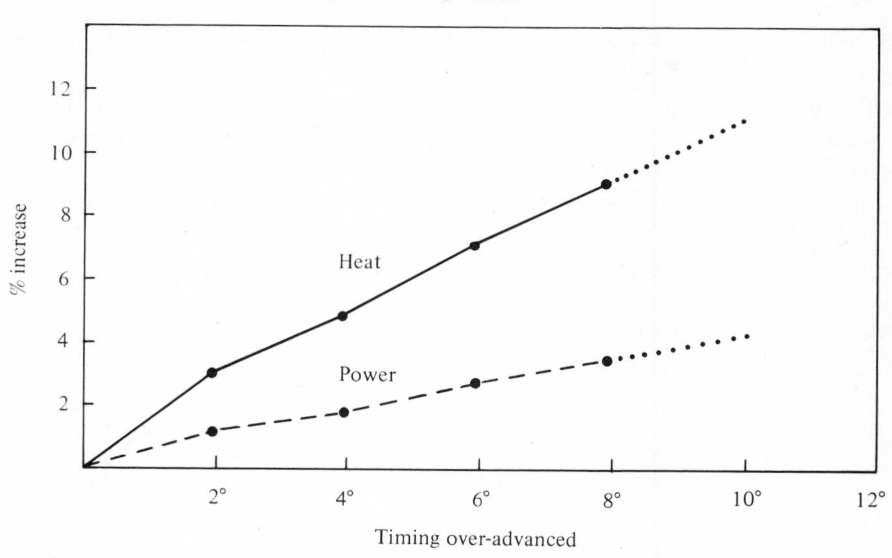

c. 10.8%: The extrapolated value is read from the dotted line in the graph above.
d. 4.0%: The extrapolated value is read from the dotted line in the graph above.

6 The danger of extrapolation is illustrated by a graph of the same information contained in Q5, augmented with additional data which were determined experimentally. The line graph indicating heat increase continued to climb, and the extrapolated value of 10.8% at 10° over-advanced timing is very close to the actual value of 11%.

However, the actual value of the line graph representing power dropped and the extrapolated value of 4.0% increase is much higher than the actual value of 1.8% increase.

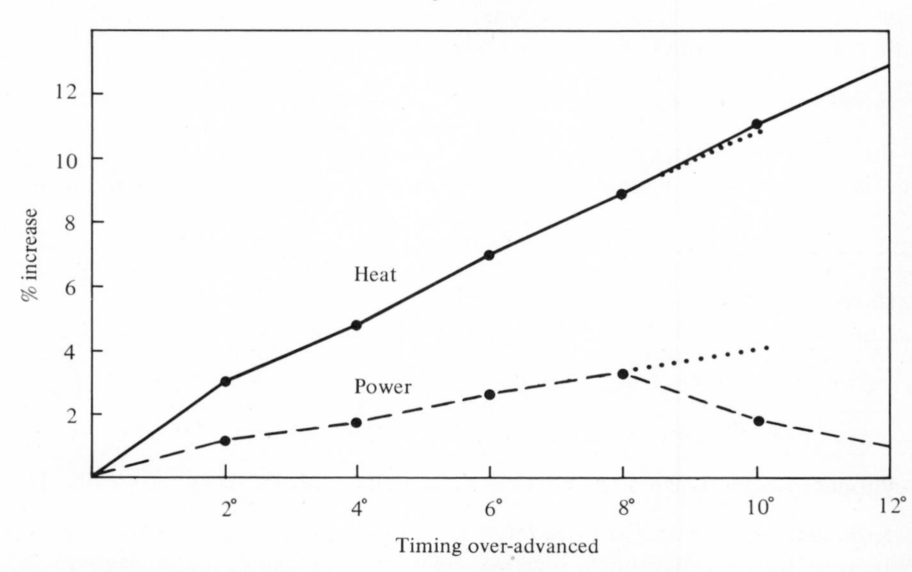

Q6 Which is most likely to be accurate, extropolation or interpolation? __________

• ### • ### • ### • ### • ### • ### • ### •

A6 interpolation: There is less likelihood of large fluctuations between data points than changes in direction of the curve beyond the data.

7 One of the tools available to a technician to analyze the operation of an engine is an *oscilloscope* (or scope). When connected to an operating engine it displays the pattern of the voltage fluctuations on a display screen similar to a television screen. These patterns are actually line graphs which are mechanically produced. Properly tuned engines have recognizably similar patterns, and variations from these patterns signal problems in such areas as spark duration, necessary voltage, coil and condenser action, secondary wire condition, etc. A typical pattern is shown. The vertical axis measures voltage in kilovolts (kV) and generally has a scale that extends above 20 kV, although the highest the graph goes under normal conditions is 5–10 kV.

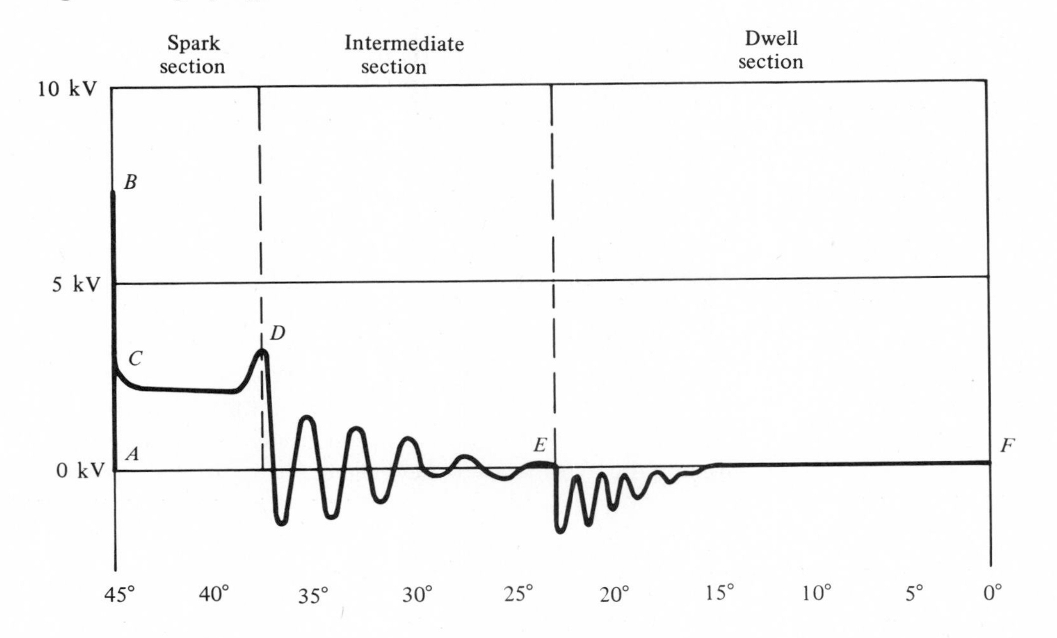

The horizontal scale is a measurement of time, even though the scale is not marked in units of time. As the distributor turns one revolution (360°) all of the cylinders of the car fire. In an eight-cylinder car the distributor turns $\frac{1}{8}$ (360°) or 45° between the time one cylinder fires and the next cylinder fires. Therefore, in order to display the voltage change from one cylinder, the horizontal scale is 45° long for an eight-cylinder car and is marked in degrees of turning of the distributor.

Q7 **a.** What is measured on the vertical scale of an oscilloscope? ________

b. What unit of measurement is used on the horizontal scale of an oscilloscope? ________

• ### • ### • ### • ### • ### • ### • ### • ### •

A7 **a.** voltage **b.** degrees

Q8 **a.** With a six-cylinder engine, as the distributor turns 360°, each of the six cylinders fire. How many degrees does the distributor turn per cylinder? ________

b. How many degrees does the distributor turn per cylinder for a four-cylinder engine? ________

• ### • ### • ### • ### • ### • ### • ### • ### •

A8 **a.** 60°: 360° ÷ 6 **b.** 90°: 360° ÷ 4

8 To become proficient at analyzing an ignition system with an oscilloscope, you would have to study material designed more specifically for that purpose, as well as experiment with a scope in a laboratory. However, we will present a few of the fundamental notions of scope patterns here in order to show the relationship they have to line graphs.

The pattern for one cylinder has three sections. In the illustration, the *firing section* starts with the firing of the plug at point A and extends until the spark extinguishes at point D. The *intermediate section* starts at D with gradually diminishing oscillations because of coil and condenser energy dissipation. The intermediate section ends at E when the points close. The *dwell section* represents the period of time during the ignition cycle in which the breaker points are closed. There is a short, downward line at E, followed by small oscillations, all below the zero line which indicates negative voltage.

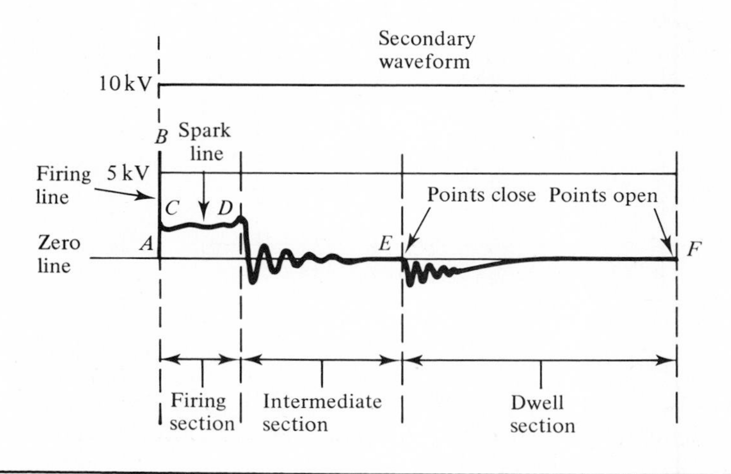

Q8 Use the oscilloscope pattern in Frame 8 to answer the following:

a. At what point on the pattern does the spark plug firing section begin? ______

b. At the beginning of the pattern in the firing section, the voltage spikes up to what amount? ______

c. In the firing section the pattern becomes fairly level at about what voltage? ______

d. What is the pattern in the intermediate section? ________________________

e. Is the voltage always positive in the intermediate section? ______

f. The distributor contacts are open for one-half of the pattern for each cylinder. How many degrees does the distributor turn while the points are open for a six-cylinder engine? ______

g. Is the voltage mostly positive or mostly negative in the dwell section? ____________

• ### • ### • ### • ### • ### • ### • ### • ### •

A8 **a.** A **b.** 7 kV **c.** 2 kV

d. initial oscillations gradually diminishing

e. no

f. 30°: A six-cylinder engine has a 60° cycle (see problem 8).

g. negative

This completes the instruction for this section.

11.4 EXERCISE

A line graph representing the miles per gallon which a particular automobile obtains at various values of air to fuel ratios is given below.

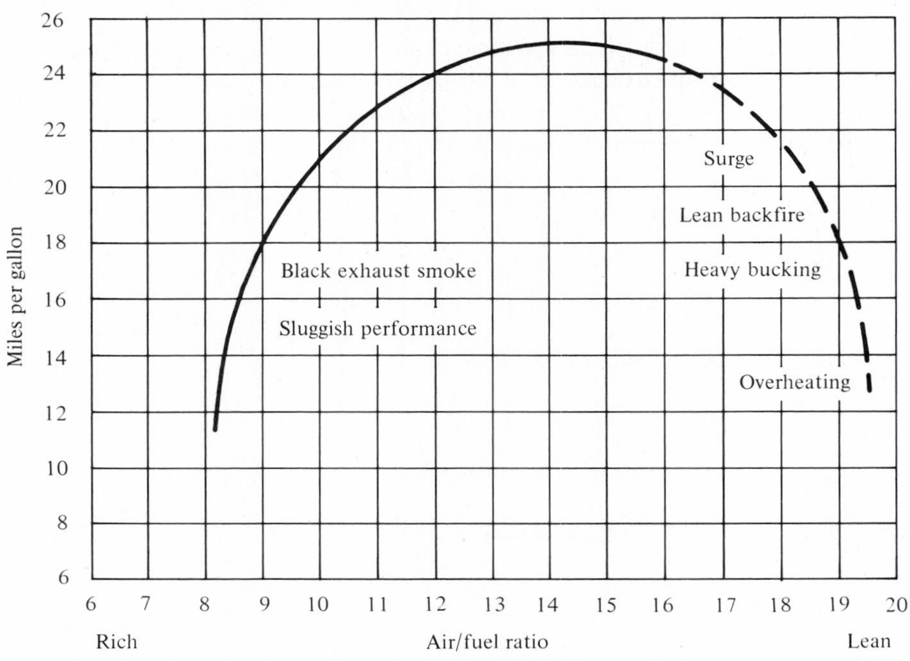

1. If the air to fuel ratio is above 15 to 1, excess air will be left after the burning process is completed, but all gasoline will be utilized. This results in maximum economy. What is the miles per gallon at a ratio of 15 to 1?
2. Between what two air to fuel ratios does the mpg stay above 24 mpg?
3. How rich can the air to fuel ratio become before the mpg falls below 18 mpg?
4. The dotted line is based on extrapolation. When the air/fuel ratio is as lean as 19 to 1, what is the predicted mpg?

As the altitude above sea level increases, an engine loses power. A few values are given in the chart below relating altitude to the percent of power loss.

Altitude	*Power Loss*
0	0
3,000 ft	9%
8,000 ft	24%

5. Draw a line graph of the data above.
6. Use your graph to find the power loss in:
   a. Ann Arbor Michigan 850 ft
   b. Denver, Colorado 5,000 ft
   c. Pittsburgh, Pa. 1,000 ft
   d. Milner Pass, Co. 10,000 ft

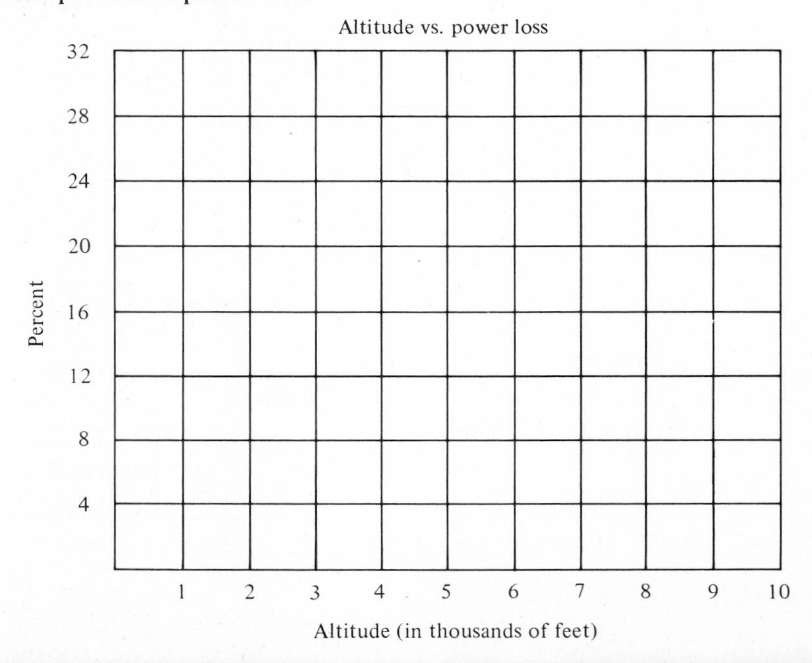

7. Suppose an engine was tested and found to have 340 horsepower at sea level in Miami, Florida. What would be its horsepower in Denver, Colorado?
8. What is the name of the instrument used to monitor and display a graph of the electrical voltage in an engine?
9. What units are used on the horizontal and vertical axis of an oscilloscope?
10. **a.** At what points on the following oscilloscope pattern do the points open?
    b. At what point do the points close?

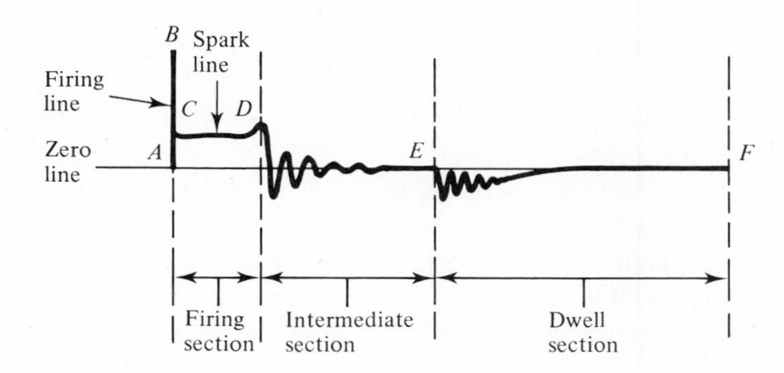

11.4 EXERCISE ANSWERS

1. 25 mpg
2. 12 to 1 and 16.5 to 1
3. 9 to 1
4. 18 mpg
5.

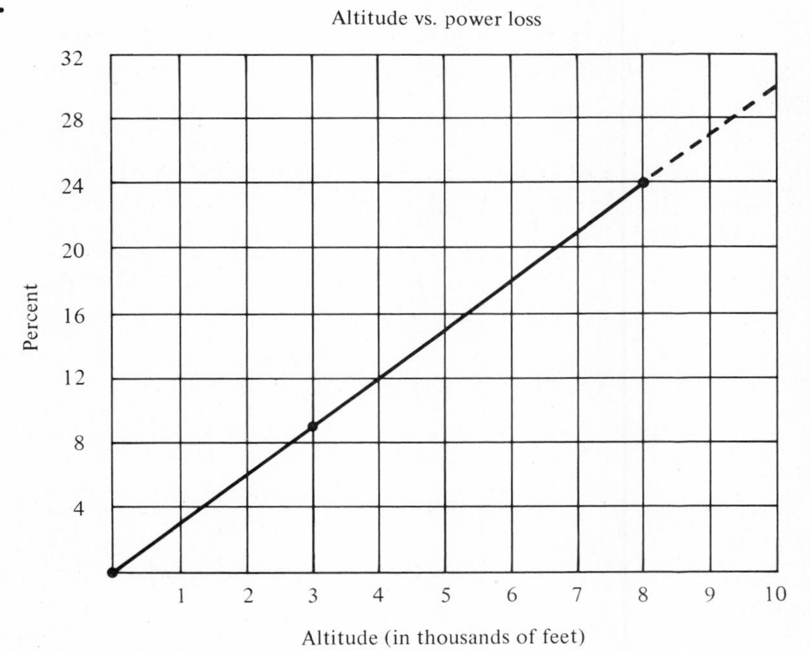

6. a. 2% **b.** 15% **c.** 3% **d.** 30%
7. 289 hp
8. oscilloscope
9. degrees, kilovolts
10. a. A and F **b.** E

CHAPTER 11 SAMPLE TEST

At the completion of Chapter 11 you should be able to work the following problems.

11.1 CHOOSING THE TYPE OF GRAPH

1. Identify each of the following variables as continuous or discrete:
   a. number of revolutions per minute that a gear is turning
   b. number of loss-of-time accidents in a manufacturing plant for a month
   c. air/fuel ratio from a carburetor
   d. percent of carbon monoxide in exhaust fumes
   e. number of telephone calls entering a telephone exchange in an hour.
2. Choose the type of graph (bar graph or circle graph) that would be used to illustrate the following data:
   a. You want to compare the average brake horsepower for the top of the line engine for three major automobile manufacturers.
   b. You want to show where trucks are serviced by showing the percent that are serviced in service stations, garages, truck dealers, and fleet service.
   c. You want to break down the total piecework production of a plant for an 8-hour shift among the 5 production workers on that shift.
3. Choose the type of graph (line graph or bar graph) that would be used to illustrate the following data:
   a. You want to show the total sales for Ford, Chrysler, and General Motors last year.
   b. You want to show the change in wind resistance of an automobile as the speed increases.
   c. You want to show how the level of carbon monoxide in the exhaust is related to the richness of leanness of the air to fuel ratio.

11.2 CIRCLE GRAPHS

Distribution of Foreign Direct Investments in the U.S. by Country of Ownership, 1971 ($15 billion)

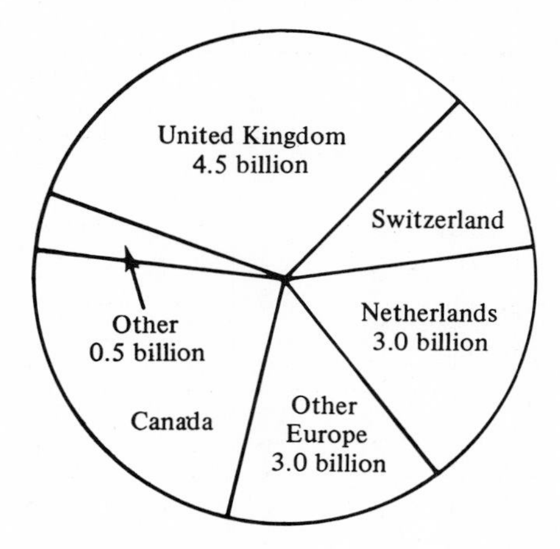

4. If Switzerland controls 10% of the foreign investments in the United States, what is its investment in the United States?
5. Find the dollars invested by Canada in the United States.
6. What is the sum of all the central angles of the sectors of a circle?
7. Find the number of degrees in the central angle of the sector representing Switzerland.
8. Find the number of degrees in the central angle of the sector representing Netherlands.

11.3 BAR GRAPHS

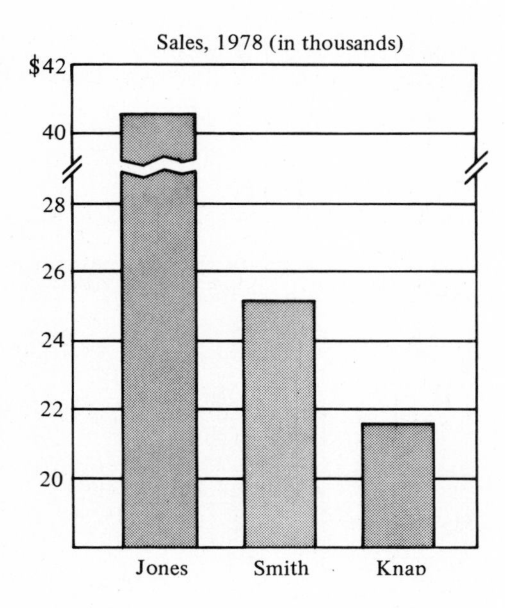

Use the graph to answer questions 9–11.

9. Give two reasons why the bar graph is a distortion of the data on which it is based.
10. What were the total sales of Smith in the graph?
11. What were the total sales of Knap in the graph?
12. If you were constructing a bar graph and had to draw the longest bar representing 108 million on a graph with six vertical spaces, what scale would you use?
13. Graph the following data.

Retirement Income/Month	
Atwood	$1,200
Baker	600
Calson	420
Dunlap	2,000
Ehen	800

11.4 LINE GRAPHS

A production manager recorded the total production at several times during an 8-hour shift. The shift started at 8:00 A.M. and ended at 5:00 P.M. with an hour off for lunch.

Total Production Until a Given Time in a Shift

Time	*Production Units*
8:00 A.M.	0
9:00 A.M.	220
10:00 A.M.	400
12:00 noon	630
1:00 P.M.	630
2:00 P.M.	730
4:00 P.M.	850

14. Use the data to construct a line graph of the total production at various times during the shift.

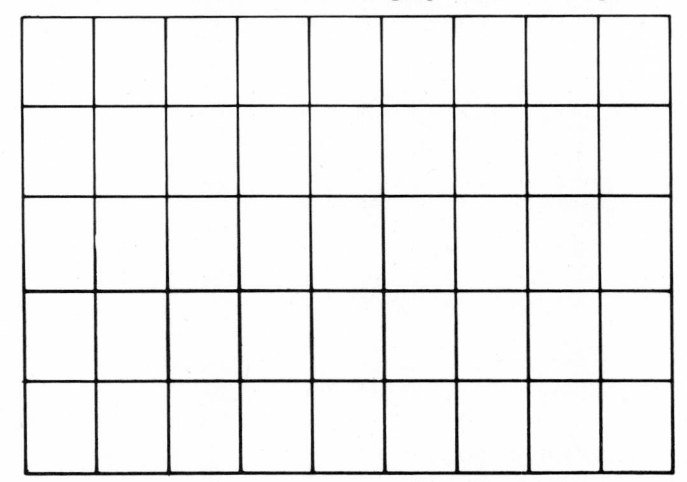

15. From the graph you constructed in problem 14, read the total production at 11:00 A.M.

16. The process of approximating the production for 11:00 A.M. between the two known values of the production at 10:00 A.M. and noon is called ______________ (interpolation or extrapolation).

17. Predict from your graph what the production will be at the end of the shift at 5:00 P.M.

18. The process of approximating the production at 5:00 P.M. beyond the known values is called ______________ (interpolation or extrapolation).

19. How could you explain the leveling off of the production from noon to 1:00 P.M.?

20. Is production increasing at a faster or slower rate at 9:00 A.M. than it is at 2:00 P,M.? Explain why this is reasonable.

CHAPTER 11 SAMPLE TEST ANSWERS

1. a. continuous **b.** discrete **c.** continuous **d.** continuous
e. discrete

2. a. bar graph **b.** circle graph **c.** circle graph

3. a. bar graph **b.** line graph **c.** line graph

4. 1.5 billion

5. 2.5 billion

6. 360°

7. 36°

8. 72°

9. **1.** The vertical scale has been broken between 28,000 and 40,000.
2. The vertical scale does not begin at zero.

10. $25,000 **11.** $21,500 **12.** 1 space = 20 million

13.

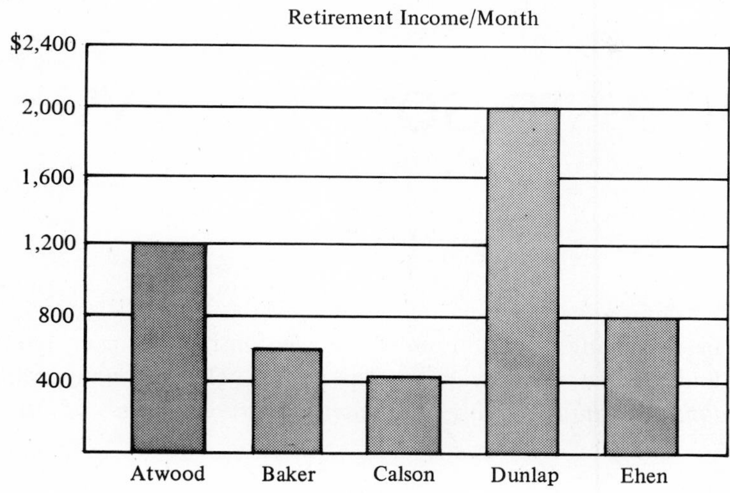

14.

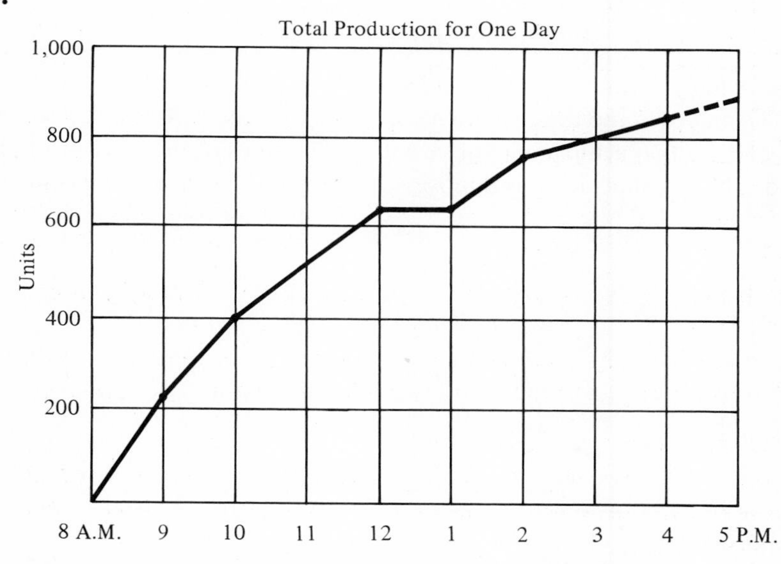

15. 520 **16.** interpolation **17.** 900
18. extrapolation **19.** lunch break
20. faster rate, People are fresh in the morning and tired later in the day.

CHAPTER 12

RIGHT-ANGLE TRIGONOMETRY

In this chapter, we will be restricting our attention to the relationship of angles and sides of right triangles. An understanding of these triangle relationships will be helpful in many technical occupations. *The use of a calculating machine is highly recommended during the study of this chapter.*

12.1 TRIGONOMETRIC RATIOS

1 A right triangle is one which contains one right angle (90°). The vertices of the triangle are labeled with capital letters. The angles of the triangle are named by listing the vertices. It is common to label the right angle as C, and the two acute angles as A and B. The sides of the triangle are labeled with small letters a, b, and c. Notice that side c is opposite $\measuredangle C$ (angle C), side b is opposite $\measuredangle B$, and side a is opposite $\measuredangle A$.

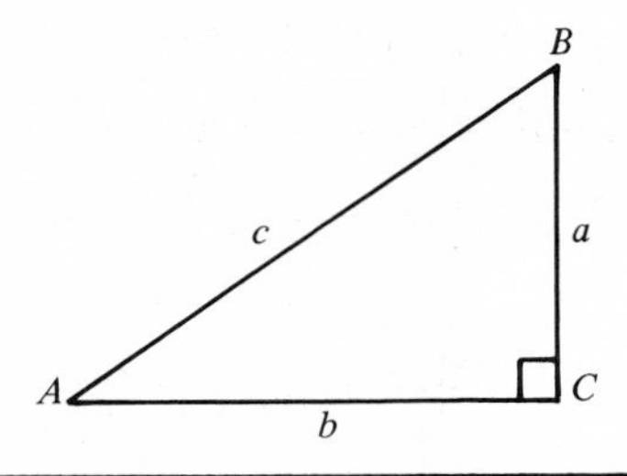

Q1 Label the right sides of the triangle below.

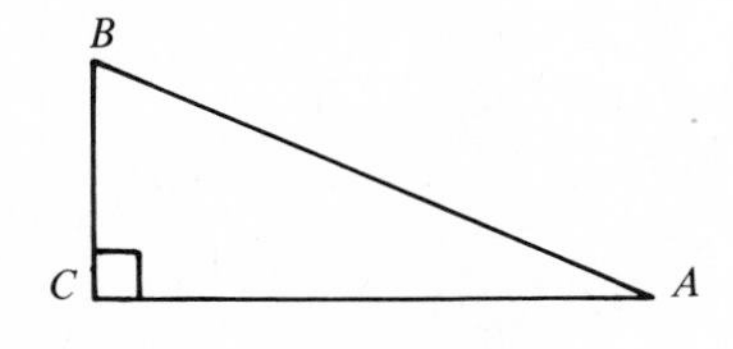

\# # # • # # # • # # # • # # # • # # # • # # # • # # # • # # # • # # #

A1 B

a c

C b A

2 The sides of a right triangle are often named with reference to a specific angle. For example, side a is the side *opposite* angle A, while side b is the side *adjacent* (next to) angle A. The side across from the right angle is always the *hypotenuse*.

With reference to $\measuredangle B$, the sides are labeled in a similar manner.

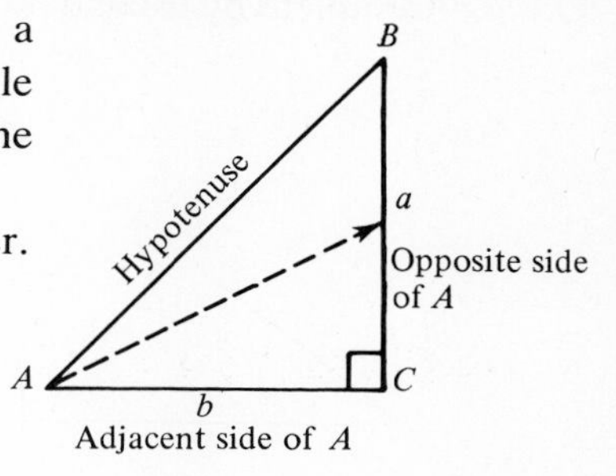

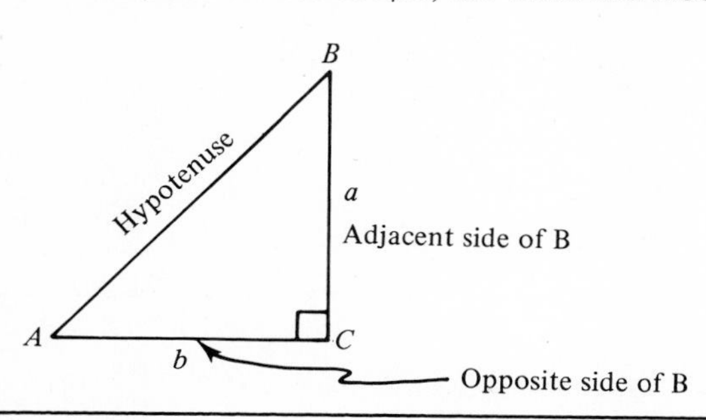

Q2 Refer to the figure.

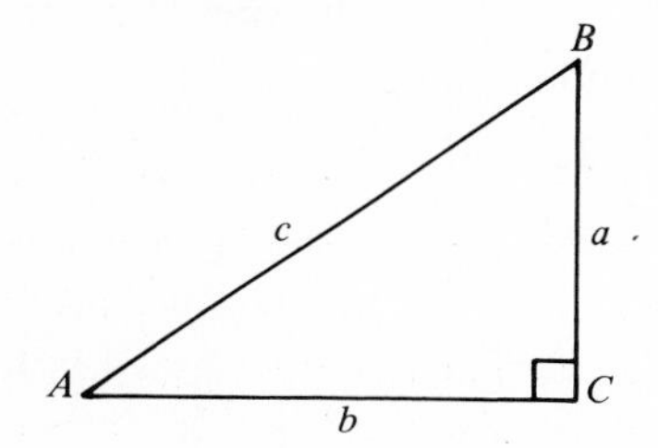

With reference to $\measuredangle A$:

a. Side a is the ______________ side.

b. Side b is the ______________ side.

c. Side c is the ______________ .

With reference to $\measuredangle B$:

d. Side ______ is the adjacent side.

e. Side ______ is the opposite side.

f. Side ______ is the hypotenuse.

• ### • ### • ### • ### • ### • ### • ### •

A2 **a.** opposite **b.** adjacent **c.** hypotenuse
d. a **e.** b **f.** c

3 In any right triangle, six trigonometric ratios are defined. Each will be discussed in subsequent frames. The *sine* of $\measuredangle A$ is defined to be the numerical value of the ratio $\frac{a \text{ (opposite side)}}{c \text{ (hypotenuse)}}$. The sine of $\measuredangle A$ may be shortened to sin A. The sine of $\measuredangle B$ (sin B) is defined to be the ratio $\frac{b \text{ (opposite side)}}{c \text{ (hypotenuse)}}$. That is, the sine of either acute angle in a right triangle is the ratio of the side opposite (opp) to the hypotenuse (hyp).

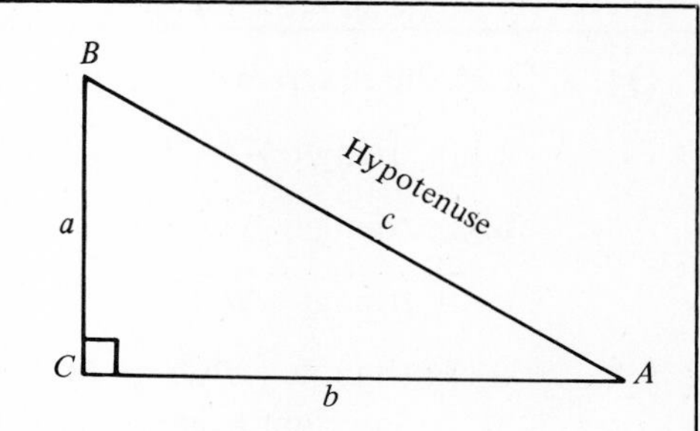

$$\sin A = \frac{\text{opp}}{\text{hyp}} = \frac{a}{c}$$ (Note: sin A means sin $\measuredangle A$)

$$\sin B = \frac{\text{opp}}{\text{hyp}} = \frac{b}{c}$$

Q3 Complete by using data from the accompanying triangle:

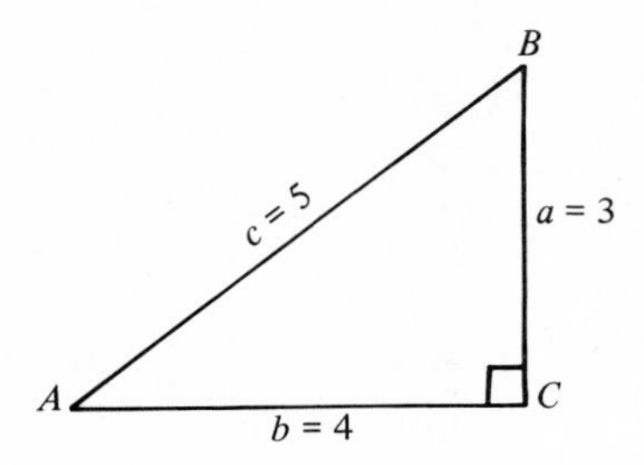

a. $\sin A =$ ________ **b.** $\sin B =$ ________

• ### • ### • ### • ### • ### • ### • ### •

A3 **a.** $\frac{3}{5}$ or 0.6 **b.** $\frac{4}{5}$ or 0.8

---

4 You should observe that *the sine of an angle depends on the size of the angle, not the size of the triangle*. In two similar triangles (same shape, not necessarily the same size), the sine of corresponding angles will be equal. Triangles ACB, AED, and AGF are similar triangles.

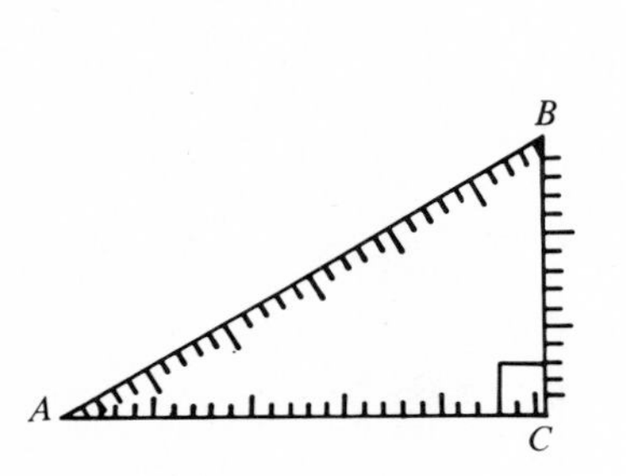

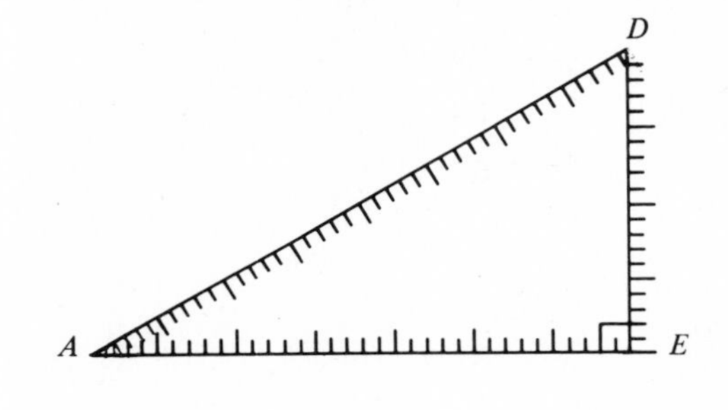

In triangle ABC,

$$\sin A = \frac{15}{30} = \frac{1}{2} = 0.5$$

In triangle ADE,

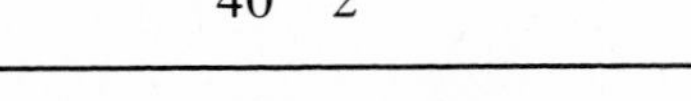

$$\sin A = \frac{20}{40} = \frac{1}{2} = 0.5$$

Q4 Use the diagrams in Frame 4:

a. In triangle AFG, $\sin A =$ ________.

b. In triangle ABC, $\sin B =$ ________.

c. In triangle ADE, $\sin D =$ ________.

d. In triangle AFG, $\sin F =$ ________.

e. In any right triangle, the sine of angle depends on the size of the ________, not the size of the ________.

• ### • ### • ### • ### • ### • ### • ### •

A4 **a.** $\frac{1}{2}$ or 0.5 **b.** $\frac{26}{30}$ or $\frac{13}{15}$ or $0.8\overline{6}$

c. same as b **d.** same as b **e.** angle, triangle

5 The second trigonometric ratio we will discuss is the *cosine* (cos) of an angle. In a right triangle, the cosine of either acute angle is defined to be the ratio of the adjacent side to the hypotenuse. Given triangle ABC

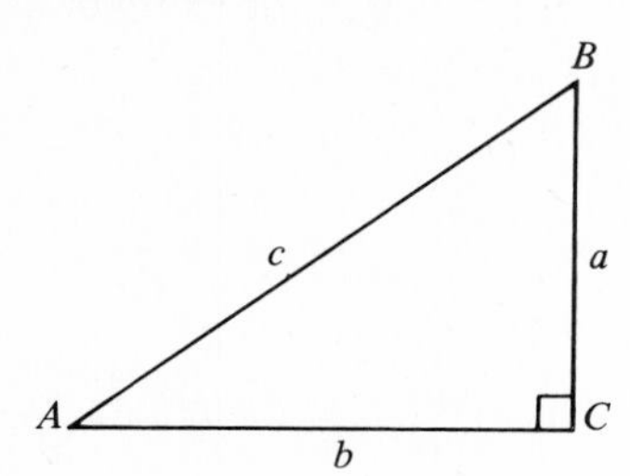

$$\cos A = \frac{\text{adjacent side (adj)}}{\text{hypotenuse (hyp)}} = \frac{b}{c}$$

$$\cos B = \frac{\text{adj}}{\text{hyp}} = \frac{a}{c}$$

Q5 Complete by using data from the accompanying triangle:

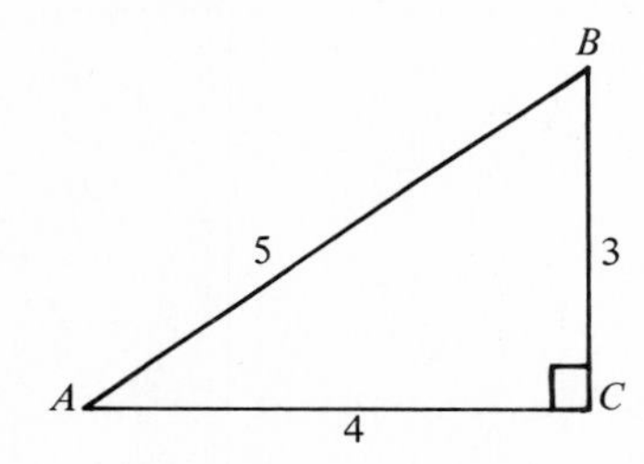

a. $\cos A =$ ________ **b.** $\cos B =$ ________

\# \# \# • \# \# \# • \# \# \# • \# \# \# • \# \# \# • \# \# \# • \# \# \# • \# \# \# • \# \# \#

A5 **a.** $\frac{4}{5}$ or 0.8 **b.** $\frac{3}{5}$ or 0.6

Q6 Complete by using data from the accompanying triangle (leave answer in fraction form):

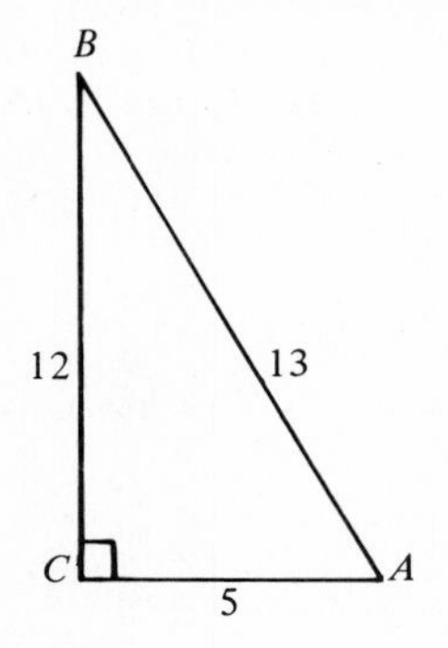

a. $\sin A =$ ________ **b.** $\sin B =$ ________

c. $\cos A =$ ________ **d.** $\cos B =$ ________

True or false:

e. $\sin A = \cos B$ **f.** $\cos A = \sin B$

\# \# \# • \# \# \# • \# \# \# • \# \# \# • \# \# \# • \# \# \# • \# \# \# • \# \# \# • \# \# \#

A6 **a.** $\frac{12}{13}$ **b.** $\frac{5}{13}$ **c.** $\frac{5}{13}$ **d.** $\frac{12}{13}$ **e.** true **f.** true

6 A third trigonometric ratio is the *tangent* (tan) of an angle. In a right triangle, the tangent of either acute angle is the ratio of the side opposite to the side adjacent. In triangle ABC

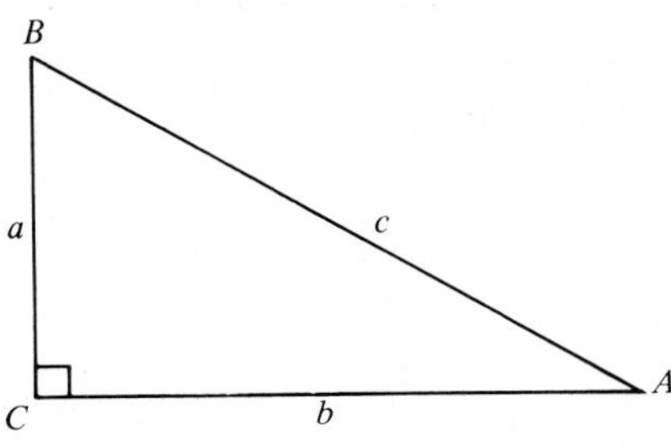

$\tan A = \frac{\text{opp}}{\text{adj}} = \frac{a}{b}$

$\tan B = \frac{\text{opp}}{\text{adj}} = \frac{b}{a}$

Q7 Complete by using data from the accompanying triangle (leave answers in fraction form):

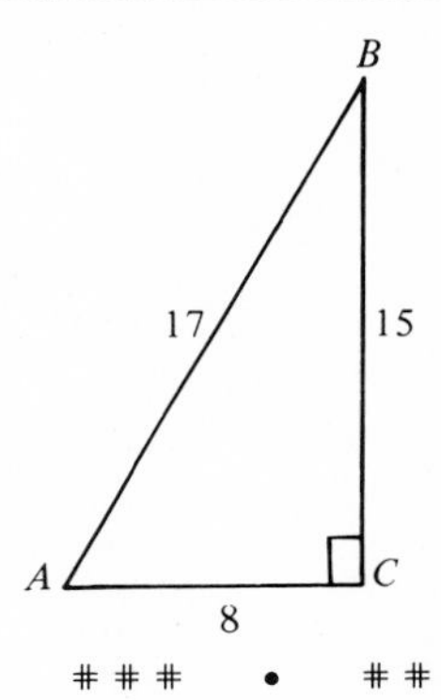

a. $\tan A =$ ________ **b.** $\tan B =$ ________

• # # # • # # # • # # # • # # # • # # # • # # # • # # # • # #

A7 **a.** $\frac{15}{8}$ **b.** $\frac{8}{15}$

Q8 Using opp, adj, and hyp, complete the following. In a right triangle,

a. sine of an acute angle = ________.

b. cosine of an acute angle = ________.

c. tangent of an acute angle = ________.

• # # # • # # # • # # # • # # # • # # # • # # # • # # # • # #

A8 **a.** $\frac{\text{opp}}{\text{hyp}}$ **b.** $\frac{\text{adj}}{\text{hyp}}$ **c.** $\frac{\text{opp}}{\text{adj}}$

Q9 Use the accompanying triangle to determine the following (give answers rounded off to hundredths):

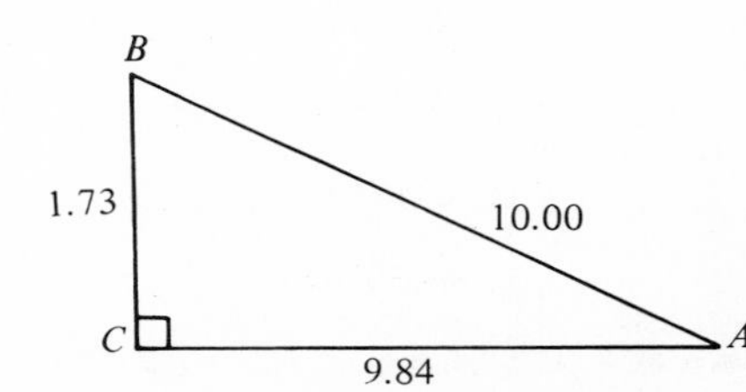

a. $\sin A =$ _______ **b.** $\sin B =$ _______ **c.** $\cos A =$ _______

d. $\cos B =$ _______ **e.** $\tan A =$ _______ **f.** $\tan B =$ _______

\# \# \# • \# \# \# • \# \# \# • \# \# \# • \# \# \# • \# \# \# • \# \# \# • \# \# \# • \# \# \#

Q9 **a.** 0.17 **b.** 0.98 **c.** 0.98 **d.** 0.17 **e.** 0.18 **f.** 5.69

7 The other three trigonometric ratios are reciprocals of either sin, cos, or tan. In any right triangle, the *cosecant* (csc) is the reciprocal of the sine. That is,

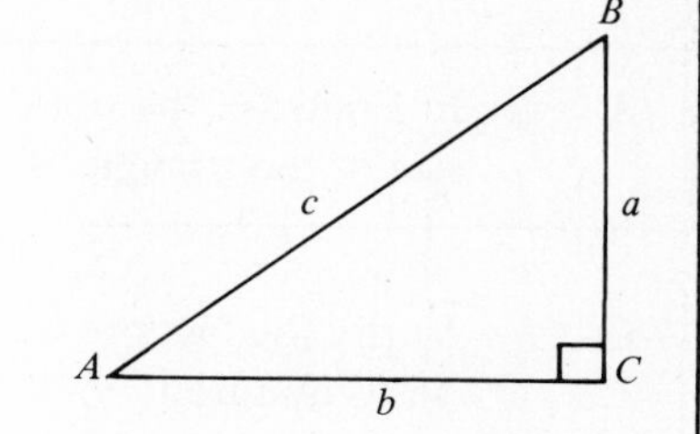

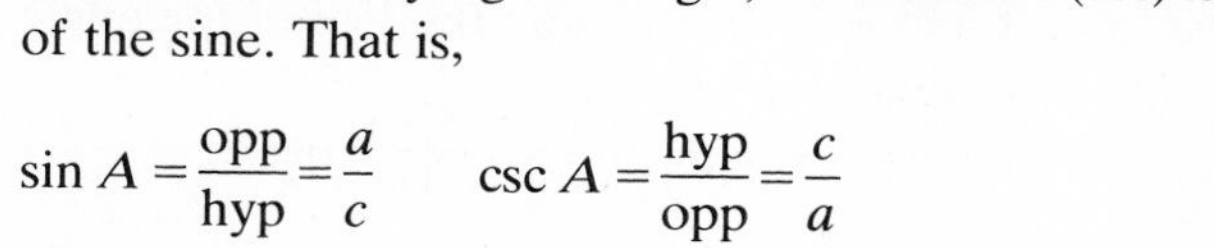

$$\sin A = \frac{\text{opp}}{\text{hyp}} = \frac{a}{c} \qquad \csc A = \frac{\text{hyp}}{\text{opp}} = \frac{c}{a}$$

$$\sin B = \frac{\text{opp}}{\text{hyp}} = \frac{b}{c} \qquad \csc B = \frac{\text{hyp}}{\text{opp}} = \frac{c}{b}$$

The *secant* (sec) is the reciprocal of the cosine. That is,

$$\cos A = \frac{\text{adj}}{\text{hyp}} = \frac{b}{c} \qquad \sec A = \frac{\text{hyp}}{\text{adj}} = \frac{c}{b}$$

$$\cos B = \frac{\text{adj}}{\text{hyp}} = \frac{a}{c} \qquad \sec B = \frac{\text{hyp}}{\text{adj}} = \frac{c}{a}$$

Finally, the *cotangent* (cot) is the reciprocal of the tangent. That is,

$$\tan A = \frac{\text{opp}}{\text{adj}} = \frac{a}{b} \qquad \cot A = \frac{\text{adj}}{\text{opp}} = \frac{b}{a}$$

$$\tan B = \frac{\text{opp}}{\text{adj}} = \frac{b}{a} \qquad \cot B = \frac{\text{adj}}{\text{opp}} = \frac{a}{b}$$

These relationships should be memorized since they form the basis for successful completion of this chapter. It will be helpful to memorize the ratios for sine, cosine, and tangent in terms of opp, adj, and hyp and then remember the reciprocal relationships:

cosecant is reciprocal of sine
secant is reciprocal of cosine
cotangent is reciprocal of tangent

The following exercise is intended to assist you in memorizing these ratios.

Q10 Use the triangle illustrated to complete the following:

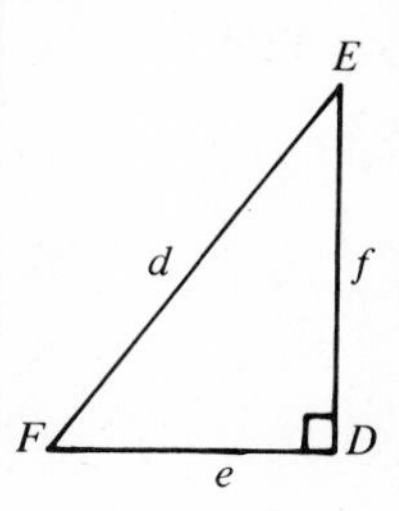

a. $\sin F =$ _______ **b.** $\csc F =$ _______ **c.** $\cos F =$ _______

d. $\sec F =$ _______ **e.** $\tan F =$ _______ **f.** $\cot F =$ _______

g. $\sin E =$ _______ **h.** $\csc E =$ _______ **i.** $\cos E =$ _______

j. $\sec E =$ _______ **k.** $\tan E =$ _______ **l.** $\cot E =$ _______

\# \# \# • \# \# \# • \# \# \# • \# \# \# • \# \# \# • \# \# \# • \# \# \# • \# \# \# • \# \# \#

A10 **a.** $\frac{f}{d}$ **b.** $\frac{d}{f}$ **c.** $\frac{e}{d}$ **d.** $\frac{d}{e}$

e. $\frac{f}{e}$ **f.** $\frac{e}{f}$ **g.** $\frac{e}{d}$ **h.** $\frac{d}{e}$

i. $\frac{f}{d}$ **j.** $\frac{d}{f}$ **k.** $\frac{e}{f}$ **l.** $\frac{f}{e}$

8	In Frame 4, the point was made that the sine of an angle depends on the size of the angle, not the size of the triangle. This fact is also true for all trigonometric ratios.

Q11 Verify the fact presented in Frame 8 by completing the following problems (express results rounded off to hundredths):

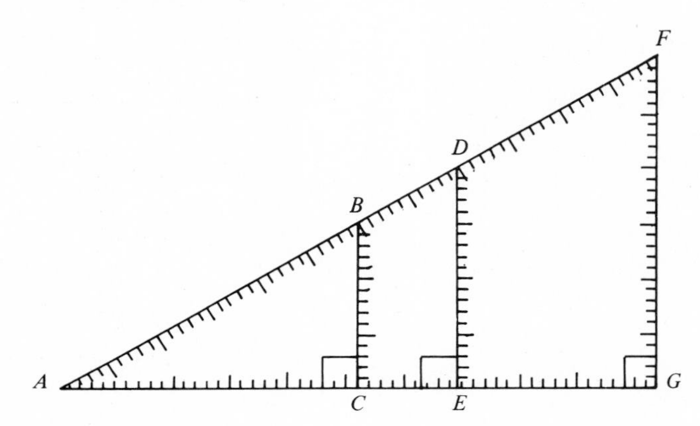

$AB = 30$, $AD = 40$, $AF = 60$, $BC = 15$, $DE = 20$, $FG = 30$, $AC = 26$, $AE = 34.6$, $AG = 52$
[*Note*: The notation $m(\overline{AB})$ is read "the measure of the line segment AB." In this chapter, the notation AB will be used in place of $m(\overline{AB})$; that is, $AB = m(\overline{AB})$.]

a. In triangle ABC, $\cos B =$ ______. **b.** In triangle ADE, $\cos D =$ ______.

c. In triangle ADE, $\tan D =$ ______. **d.** In triangle AFG, $\tan F =$ ______.

e. In triangle ABC, $\csc A =$ ______. **f.** In triangle AFG, $\csc A =$ ______.

g. In triangle ADE, $\sec A =$ ______. **h.** In triangle AFG, $\sec A =$ ______.

i. In triangle ABC, $\cot B =$ ______. **j.** In triangle ADE, $\cot D =$ ______.

• ### • ### • ### • ### • ### • ### • ### •

A11 **a.** $\frac{15}{30} = \frac{1}{2} = 0.5$ **b.** $\frac{20}{40} = \frac{1}{2} = 0.5$ **c.** $\frac{34.6}{20} = \frac{173}{100} = 1.73$

d. $\frac{52}{30} = \frac{26}{15} = 1.73$ **e.** $\frac{30}{15} = 2$ **f.** $\frac{40}{20} = 2$

g. $\frac{40}{34.6} = \frac{200}{173} = 1.16$ **h.** $\frac{60}{52} = \frac{15}{13} = 1.15$* **i.** $\frac{15}{26} = 0.58$

j. $\frac{20}{34.6} = \frac{100}{173} = 0.58$

*Slight difference is due to the fact that original measurement of AG is approximate.

Q12 Find the values of the following trigonometric ratios, using the accompanying triangle. Write all answers in fraction form:

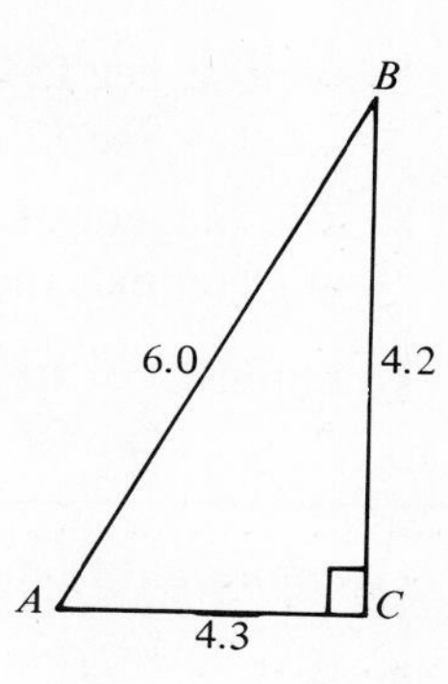

a. $\sin A =$ _______ **b.** $\cos B =$ _______ **c.** $\sec B =$ _______

d. $\csc A =$ _______ **e.** $\tan A =$ _______ **f.** $\tan B =$ _______

g. $\cos A =$ _______ **h.** $\sin B =$ _______ **i.** $\cot A =$ _______

j. $\cot B =$ _______ **k.** $\sec A =$ _______ **l.** $\csc B =$ _______

• # # # • # # # • # # # • # # # • # # # • # # # • # # # • # #

A12 **a.** $\frac{4.2}{6.0}$ or $\frac{7}{10}$ **b.** $\frac{4.2}{6.0}$ or $\frac{7}{10}$ **c.** $\frac{6.0}{4.2}$ or $\frac{10}{7}$ **d.** $\frac{6.0}{4.2}$ or $\frac{10}{7}$

e. $\frac{4.2}{4.3}=\frac{42}{43}$ **f.** $\frac{4.3}{4.2}=\frac{43}{42}$ **g.** $\frac{4.3}{6.0}=\frac{43}{60}$ **h.** $\frac{4.3}{6.0}=\frac{43}{60}$

i. $\frac{4.3}{4.2}=\frac{43}{42}$ **j.** $\frac{4.2}{4.3}=\frac{42}{43}$ **k.** $\frac{6.0}{4.3}=\frac{60}{43}$ **l.** $\frac{4.3}{6.0}=\frac{43}{60}$

Q13 Use the results of Q12 to complete the following:

a. $\sin A = \cos$ _______ **b.** sec _______ $= \csc B$

c. $\tan A = \cot$ _______ **d.** $\cos A = \sin$ _______

e. cot _______ $= \tan B$ **f.** $\csc A = \sec$ _______

• # # # • # # # • # # # • # # # • # # # • # # # • # # # • # #

A13 **a.** B **b.** A **c.** B **d.** B **e.** A **f.** B

---

9 The sum of the three angles of any triangle is always equal to 180°. That is,

$m\measuredangle A + m\measuredangle B + m\measuredangle C = 180°$

(*Note*: "$m\measuredangle A$" is read "the measure of angle A." In this chapter, $C = 90°$ will mean $m\measuredangle C = 90°$, $B = 45°$ will mean $m\measuredangle B = 45°$, etc.)

If the sum of two (or more) angles is 90°, the angles are called *complementary* angles. The two acute angles in a right triangle are complementary angles. One angle is said to be the complement of the other. In other words, if $C = 90°$, $A + B + 90° = 180°$. Therefore, $A + B = 90°$.

The results of Q13 can now be stated as: In a right triangle,

1. The sine of one of the acute angles is equal to the cosine of its complement (the other acute angle). Likewise, the cosine of one acute angle is equal to the sine of its complement.

Examples: complementary angles

$\sin 30° = \cos 60°$
$\cos 17° = \sin 73°$, etc.

(*Note*: Stated in another way, "sine and cosine of complementary angles are equal.")

2. The secant of an angle equals the cosecant of its complement. Likewise, the cosecant of an angle equals the secant of its complement.

Examples: sec 15° = csc 75°
csc 32° = sec 58°, etc.

3. The tangent of an angle equals the cotangent of its complement. Likewise, the cotangent of an angle equals the tangent of its complement.

Examples: cot 40° = tan 50°
tan 57° = cot 33°, etc.

Q14 Given a right triangle where A and B are the acute angles:

a. If $A = 35°$, $B =$ ________.
b. If $B = 11°$, $A =$ ________.
c. If $A = 28°$, $B =$ ________.
d. cot 11° = tan ________.
e. sin 28° = ________ 62°.
f. sec 35° = ________.

• ### • ### • ### • ### • ### • ### • ### •

A14 **a.** 55° **b.** 79° **c.** 62° **d.** 79° **e.** cos **f.** csc 55°

10 The relationships illustrated in Frame 9 are the basis for the organization and use of the tables which will be presented in the next section.

This completes the instruction for this section.

12.1 EXERCISE

1. Using opp, adj, and hyp, define all six trigonometric ratios of an acute angle A in a right triangle.
2. Label the sides of triangle GHI using small letters (g, h, i) and indicate:

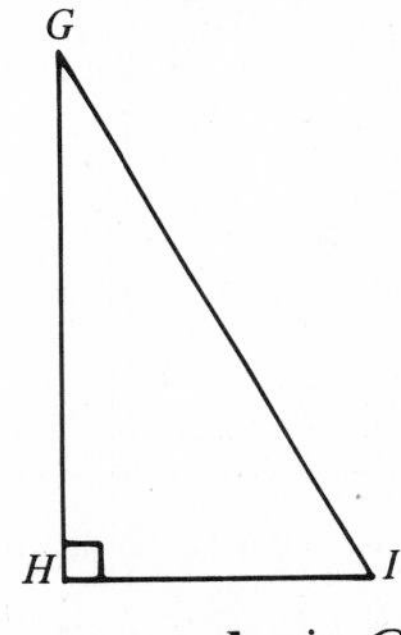

a. sin I **b.** cos I **c.** tan I **d.** sin G
e. cos G **f.** tan G **g.** csc I **h.** sec I
i. cot I **j.** csc G **k.** sec G **l.** cot G

3. Given triangle ABC, calculate the following ratios (round off to hundredths):

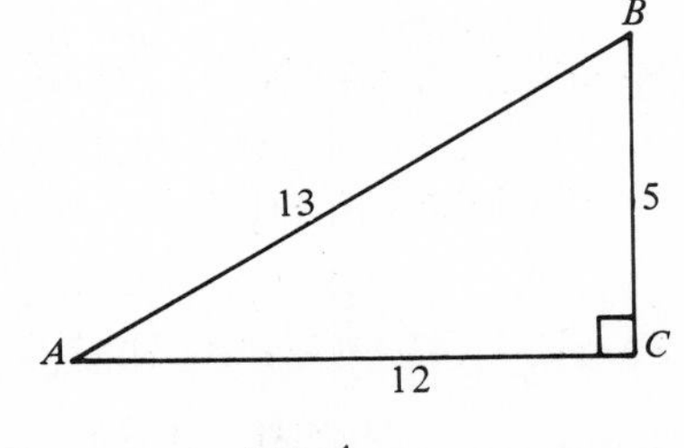

a. sin A **b.** cos B **c.** tan A
d. cot B **e.** sec A **f.** csc B

4. True or false:
   a. If ∡A and ∡B are complementary, the sine of one is equal to the cosine of the other.
   b. The csc A is equal to the reciprocal of the sin A.

c. The trigonometric ratios of a given angle depend on the size of the triangle.
d. The tangent and cotangent of complementary angles are equal.
e. The tangent and cotangent of equal angles are reciprocals of each other.

12.1 EXERCISE ANSWERS

1. $\sin A = \frac{\text{opp}}{\text{hyp}}$ $\csc A = \frac{\text{hyp}}{\text{opp}}$ $\cos A = \frac{\text{adj}}{\text{hyp}}$

$\sec A = \frac{\text{hyp}}{\text{adj}}$ $\tan A = \frac{\text{opp}}{\text{adj}}$ $\cot A = \frac{\text{adj}}{\text{opp}}$

2. a. $\frac{i}{h}$ **b.** $\frac{g}{h}$ **c.** $\frac{i}{g}$ **d.** $\frac{g}{h}$ **e.** $\frac{i}{h}$ **f.** $\frac{g}{i}$

g. $\frac{h}{i}$ **h.** $\frac{h}{g}$ **i.** $\frac{g}{i}$ **j.** $\frac{h}{g}$ **k.** $\frac{h}{i}$ **l.** $\frac{i}{g}$

3. a. 0.38 **b.** 0.38 **c.** 0.42 **d.** 0.42 **e.** 1.08 **f.** 1.08

4. a. true **b.** true **c.** false **d.** true **e.** true

12.2 TRIGONOMETRIC TABLES

1 Two triangles which appear frequently in technical work are the 30-60-90 triangle and the 45-45-90 triangle. Numerical values can be calculated for trigonometric ratios of the acute angles of these triangles.

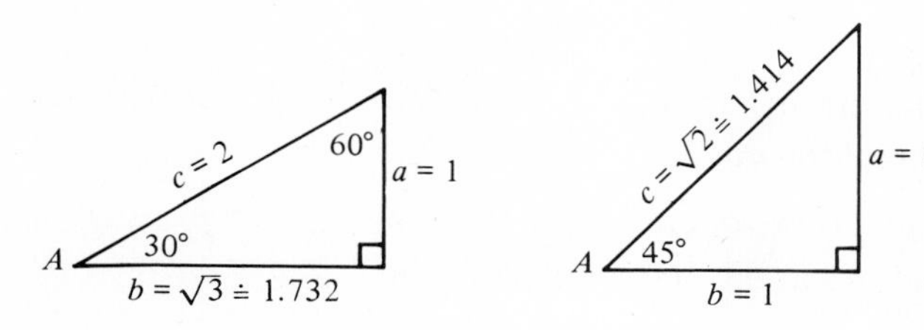

Example 1: Calculate the sin 30° correct to three decimal places.

Solution: $\sin 30° = \frac{\text{opp}}{\text{hyp}} = \frac{1}{2} = 0.500$

Example 2: Calculate the cos 30° correct to three decimal places.

Solution: $\cos 30° = \frac{\text{adj}}{\text{hyp}} = \frac{1.732}{2} = 0.866$

Example 3: Calculate the csc 45° correct to three decimal places.

Solution: $\csc 45° = \frac{\text{hyp}}{\text{opp}} = \frac{1.414}{1} = 1.414$

(*Note*: The results of Examples 2 and 3 are approximate since the values 1.732 and 1.414 are approximations.)

Q1 Use the triangles given in Frame 1 to complete the table below (give answers correct to three decimal places):

Angle	*sin*	*cos*	*tan*	*cot*	*sec*	*csc*
30°						
45°						
60°						

\# # # • # # # • # # # • # # # • # # # • # # # • # # # • # # # • # # # • # # #

A1

Angle	*sin*	*cos*	*tan*	*cot*	*sec*	*csc*
30°	0.500	0.866	0.577	1.732	1.155	2.000
45°	0.707	0.707	1.000	1.000	1.414	1.414
60°	0.866	0.500	1.732	0.577	2.000	1.555

2 A table of numerical values for every acute angle in a right triangle can be computed. Table X in the Appendix is such a table. It shows the sine, cosine, tangent, cotangent, secant, and cosecant of acute angles from 0–90°. The table's construction is based on the following fact developed in Section 12.1. If $\measuredangle A$ and $\measuredangle B$ are complementary:

1. $\sin A = \cos B$ and $\sin B = \cos A$
2. $\sec A = \csc B$ and $\sec B = \csc A$
3. $\tan A = \cot B$ and $\tan B = \cot A$

For the following, refer to the table of trigonometric ratios in the Appendix (Table X).

1. To find the trigonometric ratios of angles from 0–45°: Read the table from the *top down*, using the values of the angles at the *left* and the headings at the *top* of the table.

Example 1: Find the cos 17°.

Solution: Locate the column headed at the *top* by cos. Follow *down* until you reach the row opposite 17° on the *left*. Read .9563. Therefore, cos 17° = 0.9563.

2. To find the trigonometric ratios of angles from 45–90°: Read from the *bottom up*, using the values of the angles at the *right* and the headings at the *bottom* of the table.

Example 2: Find the csc 65°.

Solution: Locate the column headed at the *bottom* by csc. Follow *up* until you reach the row opposite 65° on the *right*. Read 1.1034. Therefore, csc 65° = 1.1034.

Q2 Use Table X to find:

a. tan 34° = ______ **b.** sin 80° = ______

\# # # • # # # • # # # • # # # • # # # • # # # • # # # • # # # • # # # • # # #

A2 **a.** 0.6745: Locate tan at *top*, follow *down* opposite 34° on *left*.
b. 0.9848: Locate sin at *bottom*, follow *up* opposite 80° on *right*.

Q3 Use Table X to find:

a. cos 49° = ______ **b.** sec 10° = ______ **c.** sin 45° = ______

d. tan 45° = ______ **e.** csc 4° = ______ **f.** cot 55° = ______

g. cos 80° = ________ **h.** cos 20° = ________ **i.** sec 77° = ________

j. csc 80° = ________

• ### • ### • ### • ### • ### • ### • ### •

A3 **a.** 0.6561 **b.** 1.0154 **c.** 0.7071 **d.** 1.0000
e. 14.3356 **f.** 0.7002 **g.** 0.1736 **h.** 0.9397
i. 4.4454 **j.** 1.0154

---

3 Two important observations should be made about the use of the table.

1. Trigonometric ratios of a 45° angle can be obtained in two ways: down and to the left or up and to the right. Since 45° appears at the bottom of the table, it is easier to go up to the right.
2. To find trigonometric ratios of angles greater than 45, you can use the fact that
   - **a.** the *sine* and *cosine* of complementary angles are equal
   - **b.** the *secant* and *cosecant* of complementary angles are equal
   - **c.** the *tangent* and *cotangent* of complementary angles are equal

Examples: **1.** sin 76° = cos 14° = 0.9703
2. cos 57° = sin 23° = 0.3907
3. sec 68° = csc 22° = 2.6695
4. csc 85° = sec 15° = 1.0353
5. tan 45° = cot 45° = 1.0000
6. cot 51° = tan 39° = 0.8098

Q4 Use Table X in the manner described in Frame 3 to find:

a. csc 50° = ________ = ________ **b.** cot 74° = ________ = ________

c. sin 89° = ________ = ________ **d.** cos 60° = ________ = ________

e. tan 65° = ________ = ________ **f.** sec 82° = ________ = ________

• ### • ### • ### • ### • ### • ### • ### •

A4 **a.** sec 40° = 1.3054 **b.** tan 16° = 0.2867 **c.** cos 1° = 0.9998
d. sin 30° = 0.5000 **e.** cot 25° = 2.1445 **f.** csc 8° = 7.1853

---

4 There is another use of the table which is very valuable. We know that in a right triangle, sin *A* and csc *A* are reciprocals. If *A* is 25°, sin 25° = 0.4226, and csc 25° = 2.3662. To show that 0.4226 and 2.3662 are reciprocals, observe that 0.4226 and $\frac{1}{0.4226}$ are reciprocals. By division, $\frac{1}{0.4226} =$ 2.3662. Therefore, $\frac{1}{\sin 25°} = \csc 25° = 2.3662$.

Other examples of this technique using the reciprocal relationships are:

1. $\frac{1}{\cos 75°} = \sec 75° = 3.8637$ (Note: sec 75° = csc 15°)

2. $\frac{1}{\tan 15°} = \cot 15° = 3.7321$

3. $\frac{1}{\csc 55°} = \sin 55° = 0.8192$ (Note: sin 55° = cos 35°)

Q5 Use the technique of Frame 4 to find:

a. $\frac{1}{\sin 15°}$ = ________ = ________

b. $\frac{1}{\cot 50°}$ = ________ = ________

c. $\frac{1}{\sec 20°}$ = ________ = ________

d. $\frac{1}{\tan 60°}$ = ________ = ________

e. $\frac{1}{\csc 10°}$ = ________ = ________

f. $\frac{1}{\cos 45°}$ = ________ = ________

• # # # • # # # • # # # • # # # • # # # • # # # • # # # • # #

A5 **a.** $\csc 15° = 3.8637$ **b.** $\tan 50° = 1.1918$ (Note: $\tan 50° = \cot 40°$) **c.** $\cos 20° = 0.9397$

d. $\cot 60° = 0.5774$ **e.** $\sin 10° = 0.1736$ **f.** $\sec 45° = 1.4142$

5 Table X can be used to find the angle if the trigonometric ratio is known.

Example 1: Find A if $\tan A = 1.4281$.

Solution: Look in the column headed at the top by tan. 1.4281 does not appear. Hence, look in the column headed by tan at the bottom. 1.4281 appears opposite 55° on the right. Therefore, $A = 55°$.

Example 2: If $\cos B = 0.9455$, find B.

Solution: Look in the column headed by cos at the top. 0.9445 appears opposite 19° on the left. Therefore, $B = 19°$.

Q6 Find the approximate angle:

a. $\sin A = 0.2250$

$A =$ ________

b. $\tan B = 1.9626$

$B =$ ________

c. $\cos B = 0.1908$

$B =$ ________

d. $\csc A = 11.4737$

$A =$ ________

• # # # • # # # • # # # • # # # • # # # • # # # • # # # • # #

A6 **a.** 13° **b.** 63° **c.** 79° **d.** 5°

THE REMAINING PORTION OF THIS SECTION IS OPTIONAL

6 If greater accuracy is required, tables are available which give the ratios to more than 4 decimal places. Tables are also available which will give the trigonometric ratios to subdivisions of a degree. A degree can be divided into 60ths where $\frac{1}{60}$th of a degree is 1 minute (1′). Hence, $1° = 60'$ and $1' = \frac{1}{60}^{\circ}$.

The accompanying table is a portion of a trigonometric table showing ratios of angles to the nearest 10′. Such a table is read in exactly the same manner as Table X.

ANGLE	SIN	COS	TAN	COT	SEC	CSC	
18° 00′	0.3090	0.9511	0.3249	3.078	1.051	3.236	**72° 00′**
10′	118	502	281	047	052	207	71° 50′
20′	145	492	314	3.018	053	179	40′
30′	0.3173	0.9483	0.3346	2.989	1.054	3.152	30′
40′	201	474	378	960	056	124	20′
18° 50′	228	465	411	932	057	098	10′
19′ 00°	0.3256	0.9455	0.3443	2.904	1.058	3.072	**71° 00′**
10′	283	446	476	877	059	046	70° 50′
20′	311	436	508	850	060	3.021	40′
30′	0.3338	0.9426	0.3541	2.824	1.061	2.996	30′
40′	365	417	574	798	062	971	20′
19° 50′	393	407	607	773	063	947	10′
20° 00′	0.3420	0.9397	0.3640	2.747	1.064	2.924	**70° 00′**
10′	448	387	673	723	065	901	69° 50′
20′	475	377	706	699	066	878	40′
30′	0.3502	0.9367	0.3739	2.675	1.068	2.855	30′
40′	529	356	772	651	069	833	20′
20° 50′	557	346	805	628	070	812	10′
21° 00′	0.3584	0.9336	0.3839	2.605	1.071	2.790	**69° 00′**
10′	611	325	872	583	072	769	68° 50′
20′	638	315	906	560	074	749	40′
30′	0.3665	0.9304	0.3939	2.539	1.075	2.729	30′
40′	692	293	0.3973	517	076	709	20′
21° 50′	719	283	0.4006	496	077	689	10′
22° 00′	0.3746	0.9272	0.4040	2.475	1.079	2.669	**68° 00′**
10′	773	261	074	455	080	650	67° 50′
20′	800	250	108	434	081	632	40′
30′	0.3827	0.9239	0.4142	2.414	1.082	2.613	30′
40′	854	228	176	394	084	595	20′
22° 50′	881	216	210	375	085	577	10′
23° 00′	0.3907	0.9205	0.4245	2.356	1.086	2.559	**67° 00′**
10′	934	194	279	337	088	542	66° 50′
20′	961	182	314	318	089	525	40′
30′	0.3987	0.9171	0.4348	2.300	1.090	2.508	30′
40′	0.4014	159	383	282	092	491	20′
23° 50′	041	147	417	264	093	475	10′
24° 00′	0.4067	0.9135	0.4452	2.246	1.095	2.459	**66° 00′**
	COS	SIN	COT	TAN	CSC	SEC	ANGLE

Q7 Use the table in Frame 6 to find:

a. tan 20° 30′ = ________ **b.** csc 66° 40′ = ________

\# \# \# • \# \# \# • \# \# \# • \# \# \# • \# \# \# • \# \# \# • \# \# \# • \# \# \# • \# \# \#

A7 **a.** 0.3739 **b.** 1.089

7 A procedure called *interpolation* can be used to obtain the ratio of values not immediately found in the table.

Example: Find sin 19° 44′.

Solution: The table in Frame 6 provides:

sin 19° 40′ = 0.3365 and
sin 19° 50′ = 0.3393

The solution can be completed as follows:

$$10' \left[4' \left[\begin{array}{l} \sin 19^\circ 40' = 0.3365 \\ \sin 19^\circ 44' = \underline{\quad ? \quad} \end{array} \right] x \right] 0.0028 \quad \sin 19^\circ 50' = 0.3393$$

The following proportion provides a method for finding x.

$$\frac{4}{10} = \frac{x}{0.0028}$$

Solving the proportion for x:

$$0.0028 \cdot 0.4 = x$$
$$0.00112 = x$$

This value can now be added to the value of sin 19° 40′.

$$\begin{array}{r} 0.3365 \\ +0.00112 \\ \hline 0.33762 \end{array}$$

Therefore sin 19° 44′ = 0.33762.

This procedure has not been presented for mastery but simply to indicate a manner in which the use of trigonometric tables may be extended. With the increasing popularity and accessibility of electronic minicomputers, the use of trigonometric tables and interpolation is declining. However, knowledge of these fundamental procedures will never be a hindrance.

Q8 Use the method of interpolation to find (use the table from Frame 6):

a. sin 18° 35′ **b.** csc 67° 48′

• # # # • # # # • # # # • # # # • # # # • # # # • # # # • # #

A8 **a.** 0.3187:

$$10' \left[5' \left[\begin{array}{l} \sin 18^\circ 30' = 0.3173 \\ \sin 18^\circ 35' = \underline{\quad ? \quad} \end{array} \right] x \right] 0.0028 \quad \sin 18^\circ 40' = 0.3201$$

$$\frac{x}{0.0028} = \frac{5}{10}$$

$$x = 0.0014 \quad (0.0028 \times 0.5)$$

$$\sin 18^\circ 35' = 0.3173 + 0.0014$$

b. 1.0802:

$$10'\left[2'\left[\begin{matrix}\csc 67^\circ 40' = 1.081 \\ \csc 67^\circ 48' = \underline{\quad ? \quad} \\ \csc 67^\circ 50' = 1.080\end{matrix}\right]x\right]0.001$$

$$\frac{x}{0.001} = \frac{2}{10}$$

$$x = 0.0002 \quad (0.001 \times 0.2)$$

$$\csc 67^\circ 48' = 1.080 + 0.0002$$

or

$$10'\left[8'\left[\begin{matrix}\csc 67^\circ 40' = 1.081 \\ \csc 67^\circ 48' = \underline{\quad ? \quad}\end{matrix}\right]x \atop \csc 67^\circ 50' = 1.080\right]0.001$$

$$\frac{x}{0.001} = \frac{8}{10}$$

$$x = 0.0008 \quad (0.001 \times 0.8)$$

$$\csc 67^\circ 48' = 1.081 - 0.0008$$

Q9 Use the method of interpolation to find (use the table from Frame 6):

a. cos 20° 23′

b. tan 23° 17′

c. cot 69° 32′

d. sec 71° 16′

\# \# \# • \# \# \# • \# \# \# • \# \# \# • \# \# \# • \# \# \# • \# \# \# • \# \# \# • \# \# \#

A9 **a.** 0.9374: $0.9367 + 0.0010 \times 0.7$ or $0.9377 - 0.0010 \times 0.3$

b. 0.43035: $0.4279 + 0.0035 \times 0.7$ or $0.4314 - 0.0010 \times 0.3$

c. 0.37126: $0.3706 + 0.0033 \times 0.2$ or $0.3739 - 0.0033 \times 0.8$

d. 3.1136: $3.098 + 0.026 \times 0.6$ or $3.124 - 0.026 \times 0.4$

This completes the instruction for this section.

12.2 EXERCISE

1. Determine the following complementary relationships:

a. sin 75° = cos ______ **b.** cot 69° = tan ______ **c.** tan 52° = cot ______

d. sec 86° = csc ______ **e.** csc 45° = sec ______ **f.** cos 71° = sin ______

2. Find the values of the following (use Table X):

a. cos 23° **b.** sin 47° **c.** tan 70°

d. csc 39° **e.** sec 1° **f.** cot 74°

3. Find the values of the following (use Table X):

a. $\frac{1}{\cos 12^\circ}$ **b.** $\frac{1}{\sin 45^\circ}$ **c.** $\frac{1}{\tan 33^\circ}$

d. $\dfrac{1}{\sec 68°}$ **e.** $\dfrac{1}{\csc 56°}$ **f.** $\dfrac{1}{\cot 17°}$

4. Find A (use Table X):

a. $\tan A = 1.2349$ **b.** $\sin A = 0.5299$ **c.** $\cot A = 6.3138$

d. $\cos A = 0.6947$ **e.** $\sec A = 1.4142$ **f.** $\csc A = 2.2812$

*__5.__ Find A (use Table X):

a. $\dfrac{1}{\sin A} = 1.3902$ **b.** $\dfrac{1}{\sec A} = 0.9613$

12.2 EXERCISE ANSWERS

1. a. 15° **b.** 21° **c.** 38° **d.** 4° **e.** 45° **f.** 19°

2. a. 0.9205 **b.** 0.7314 **c.** 2.7475 **d.** 1.5890 **e.** 1.0002 **f.** 0.2867

3. a. 1.0223: sec 12° **b.** 1.4142: csc 45°

c. 1.5399: cot 33° **d.** 0.3746: cos 68° or sin 22°

e. 0.8290: sin 56° or cos 34° **f.** 0.3057: tan 17°

4. a. 51° **b.** 32° **c.** 9° **d.** 46° **e.** 45° **f.** 26°

*__5. a.__ 46°: $\csc A = 1.3902$ **b.** 16°: $\cos A = 0.9613$

12.3 SOLVING RIGHT TRIANGLES

1 If two of the three sides of a right triangle are known, the third side can be obtained by use of the Pythagorean theorem. It states: In a right triangle ABC, the square of the hypotenuse is equal to the sum of the squares of the legs. In symbols, for triangle ABC:

$$a^2 + b^2 = c^2$$

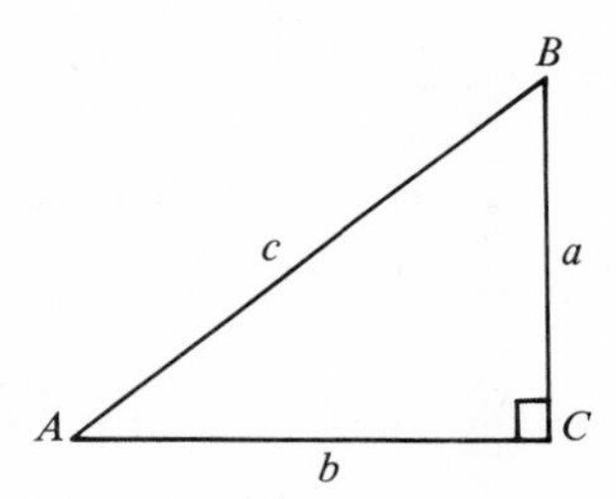

Example 1: In a right triangle ABC, find c if $a = 1$ and $b = 1$.

Solution:

$$\begin{aligned} a^2 + b^2 &= c^2 \\ 1^2 + 1^2 &= c^2 \\ 1 + 1 &= c^2 \\ 2 &= c^2 \\ \sqrt{2} &= c \end{aligned}$$

Using Table IX in the Appendix, $\sqrt{2}$ may be given as 1.414 (rounded off to thousandths). Therefore, $c = \sqrt{2} = 1.414$.*

* $\sqrt{2} \doteq 1.414$. In this chapter, we will use the = symbol even though the $\doteq$ symbol is more appropriate in many cases.

Example 2: In a right triangle ABC, find a if $b=1$ and $c=2$.

Solution:

$$a^2+b^2=c^2$$
$$a^2+1^2=2^2$$
$$a^2+1=4$$
$$a^2=3$$
$$a=\sqrt{3}$$

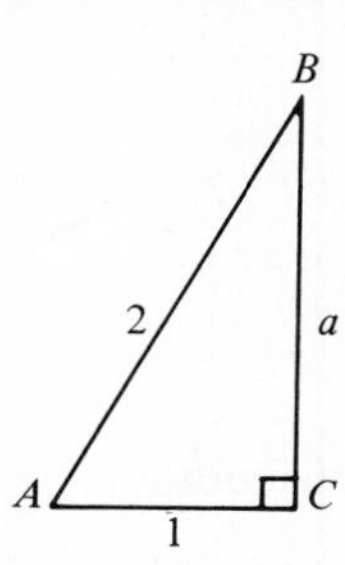

Using Table IX, $\sqrt{3}=1.732$. Therefore, $a=\sqrt{3}=1.732$.

Q1 Find the missing side:

a.

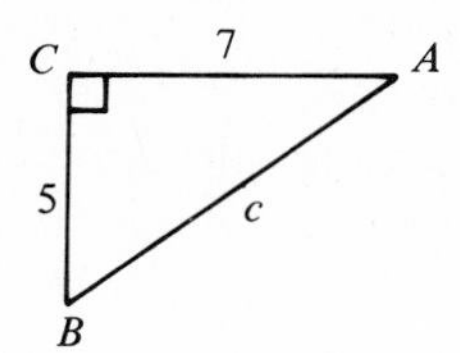

b.

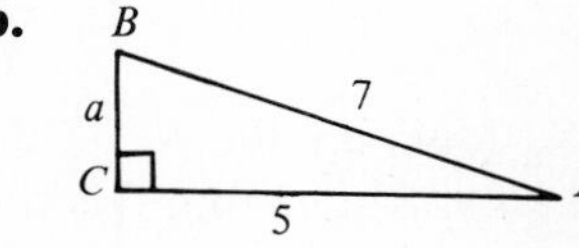

\# \# \# • \# \# \# • \# \# \# • \# \# \# • \# \# \# • \# \# \# • \# \# \# • \# \# \# • \# \# \#

A1 **a.** $\sqrt{74}=8.602$:

$$a^2+b^2=c^2$$
$$5^2+7^2=c^2$$
$$25+49=c^2$$
$$74=c^2$$
$$\sqrt{74}=c$$

b. $\sqrt{24}=4.899$:

$$a^2+b^2=c^2$$
$$a^2+5^2=7^2$$
$$a^2+25=49$$
$$a^2=24$$
$$a=\sqrt{24}$$

(*Note*: For a more complete review, see Section 5.5.)

2 In any right triangle with one side and one acute angle known, we can find the lengths of the other sides and the size of the other acute angle.

Example: Given triangle ABC, find b, c, and B.

Solution:

$$A+B=90°$$
$$60°+B=90°$$
$$B=30°$$

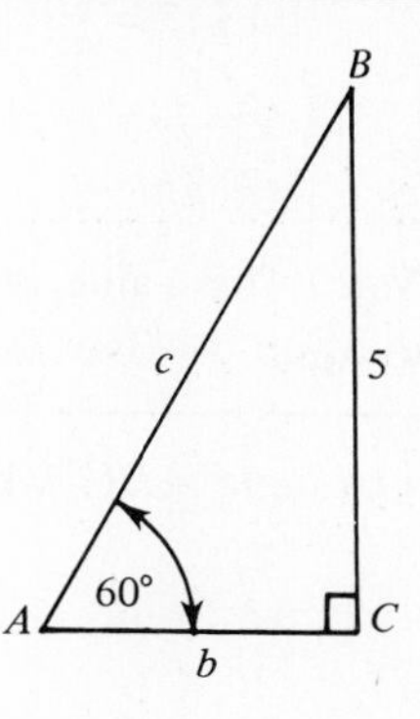

To find b, we can use one of the following trigonometric equations:

$$\tan 60° = \frac{5}{b}\ \frac{\text{(opp)}}{\text{(adj)}} \qquad \cot 60° = \frac{b}{5}\ \frac{\text{(adj)}}{\text{(opp)}}$$

$$\tan 30° = \frac{b}{5}\ \frac{\text{(opp)}}{\text{(adj)}} \qquad \cot 30° = \frac{5}{b}\ \frac{\text{(adj)}}{\text{(opp)}}$$

Either $\cot 60° = \frac{b}{5}$ or $\tan 30° = \frac{b}{5}$ will be easier to solve (since the unknown is in the numerator).

Using $\tan 30° = \frac{b}{5}$, look up the tan 30° in Table X; $\tan 30° = 0.5774$. Hence,

$$\tan 30° = \frac{b}{5}$$
$$0.5774 = \frac{b}{5}$$
$$5(0.5774) = b$$
$$2.887 = b \quad \text{(rounded off to thousandths)}$$

To find c, the Pythagorean theorem can be used:

$$a^2 + b^2 = c^2$$
$$5^2 + (2.887)^2 = c^2$$
$$25 + 8.335 = c^2$$
$$33.335 = c^2$$

Using Table IX, look up the $\sqrt{33}$. If greater accuracy is required, a calculator can be used. Therefore, $5.745 = c$. Notice that this result is slightly less than the true value of c, since $\sqrt{33}$ is less than $\sqrt{33.335}$.

The value of c could have been obtained using a trigonometric equation. For example:

$$\sin 60° = \frac{5\,(\text{opp})}{c\,(\text{hyp})}$$
$$\csc 60° = \frac{c\,(\text{hyp}}{5\,(\text{opp})}$$
$$\cos 30° = \frac{5\,(\text{adj})}{c\,(\text{hyp})}$$
$$\sec 30° = \frac{c\,(\text{hyp})}{5\,(\text{adj})}$$

Using $\sec 30° = \frac{c}{5}$

$$1.1547 = \frac{c}{5}$$
$$5(1.1547) = c$$
$$5.7735 = c$$

(*Note*: This value of c is more accurate than the value obtained earlier. Also, using a trigonometric equation involves less arithmetic than using the Pythagorean theorem.)

Q2 In triangle ABC, what is B?

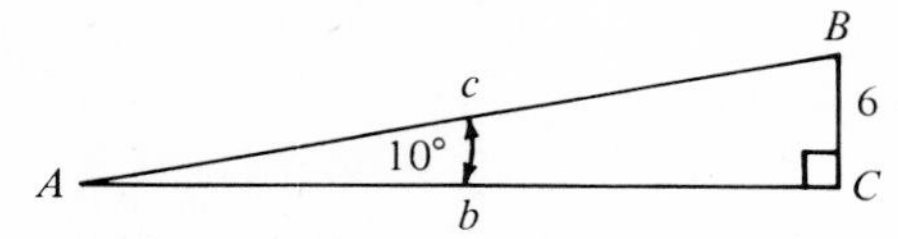

\# \# \# • \# \# \# • \# \# \# • \# \# \# • \# \# \# • \# \# \# • \# \# \# • \# \# \# • \# \# \#

A2 80°: $90° - 10°$

Q3 Using the triangle of Q2, complete the following:

a. $\tan 10° =$ ________ **b.** $\cot 10° =$ ________

c. $\tan 80° =$ ________ **d.** $\cot 80° =$ ________

\# \# \# • \# \# \# • \# \# \# • \# \# \# • \# \# \# • \# \# \# • \# \# \# • \# \# \# • \# \# \#

A3 **a.** $\frac{6}{b}$ **b.** $\frac{b}{6}$ **c.** $\frac{b}{6}$ **d.** $\frac{6}{b}$

Q4 Of the four trigonometric equations in Q3, which two would be easiest to solve? ________________ or ________________

\# # # • # # # • # # # • # # # • # # # • # # # • # # # • # # # • # # #

A4 $\cot 10° = \frac{b}{6}$ or $\tan 80° = \frac{b}{6}$ (either order): These two are chosen since the variable is in the numerator.

Q5 Use $\cot 10° = \frac{b}{6}$ and solve for b (round off to hundredths).

\# # # • # # # • # # # • # # # • # # # • # # # • # # # • # # # • # # #

A5 $b = 34.03$: $5.6713 = \frac{b}{6}$

Q6 Using the accompanying triangle, complete the following:

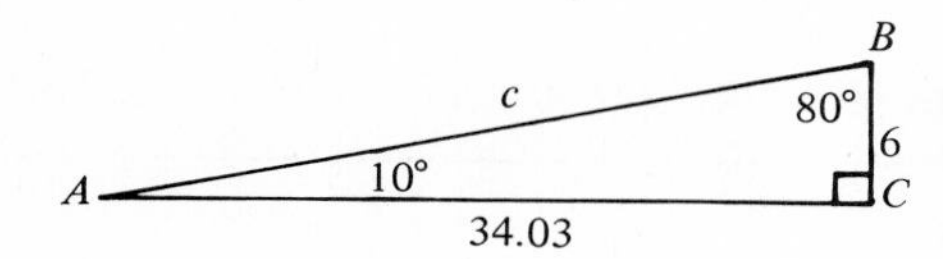

a. $\sin 10° =$ ________ **b.** $\csc 10° =$ ________

c. $\cos 80° =$ ________ **d.** $\sec 80° =$ ________

\# # # • # # # • # # # • # # # • # # # • # # # • # # # • # # # • # # #

A6 **a.** $\frac{6}{c}$ **b.** $\frac{c}{6}$ **c.** $\frac{6}{c}$ **d.** $\frac{c}{6}$

Q7 Use $\csc 10° = \frac{c}{6}$ or $\sec 80° = \frac{c}{6}$ to solve for c (round off to hundredths):

\# # # • # # # • # # # • # # # • # # # • # # # • # # # • # # # • # # #

A7 $c = 34.55$: $5.7588 = \frac{c}{6}$

---

3 Finding the missing sides and angles of a right triangle is called "solving the right triangle."

Example: Solve the right triangle.

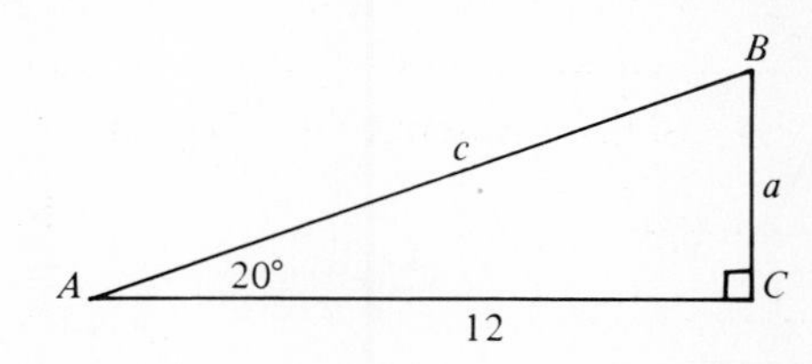

Solution: $B = 90° - 20° = 70°$

To solve for side a, an equation containing two other known values must be found. Such an equation is:

$$\tan 20° = \frac{a}{12}$$

$$0.3640 = \frac{a}{12}$$

$$12(0.3640) = a$$

$$4.368 = a$$

To solve for side c, an equation containing c and two other known values must be found. Such an equation is:

$$\csc 20° = \frac{c}{4.368}$$

$$2.9238 = \frac{c}{4.368}$$

$$4.368(2.9238) = c$$

$$12.771 = c \quad \text{(rounded off to thousandths)}$$

Therefore,

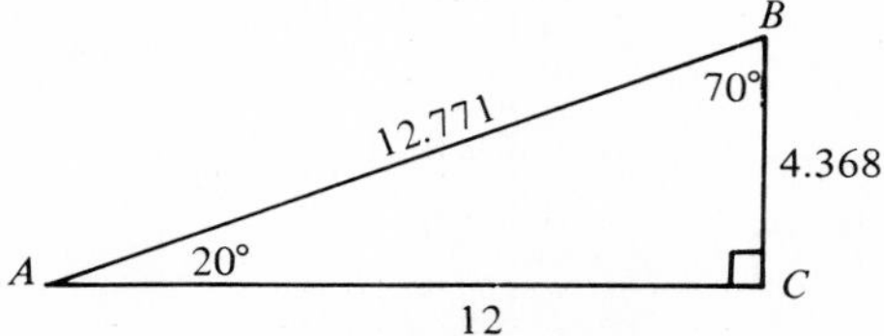

(*Note*: Answers may vary slightly depending on the number of digits used in the calculation.)

Q8 Solve the accompanying triangle (round off to tenths).

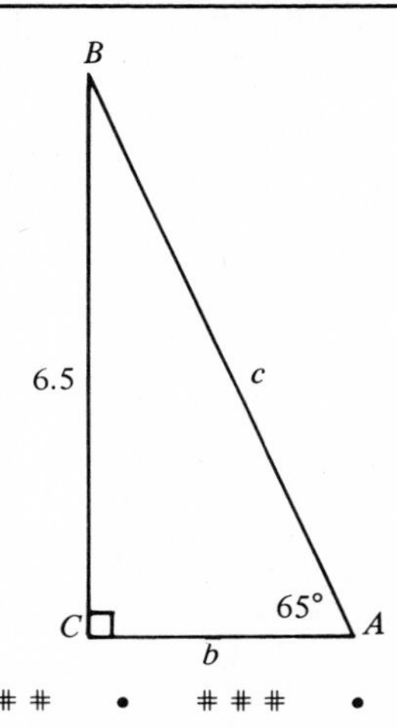

\# \# \# • \# \# \# • \# \# \# • \# \# \# • \# \# \# • \# \# \# • \# \# \# • \# \# \# • \# \# \#

A8

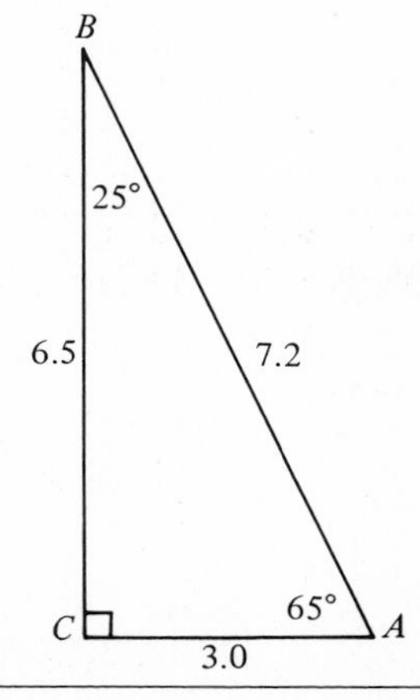

: $B = 90° - 65°$

(Note: the solution shown is not the only possibility.)

$$\cot 65° = \frac{b}{6.5} \qquad \csc 65° = \frac{c}{6.5}$$

$$0.4663 = \frac{b}{6.5} \qquad 1.1034 = \frac{c}{6.5}$$

Q9 Solve the accompanying triangle (round off to tenths).

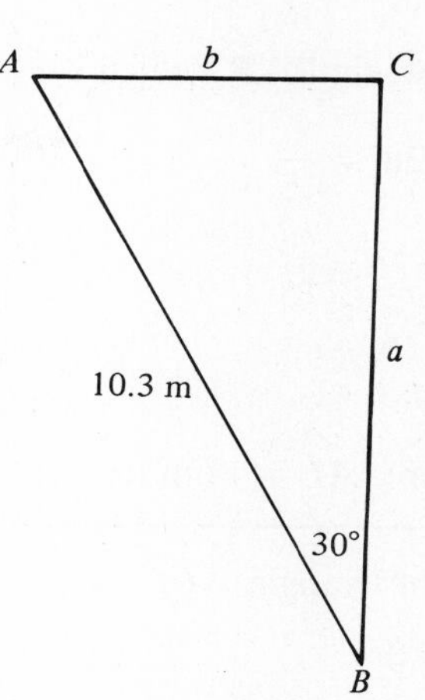

\# \# \# • \# \# \# • \# \# \# • \# \# \# • \# \# \# • \# \# \# • \# \# \# • \# \# \# • \# \# \#

A9

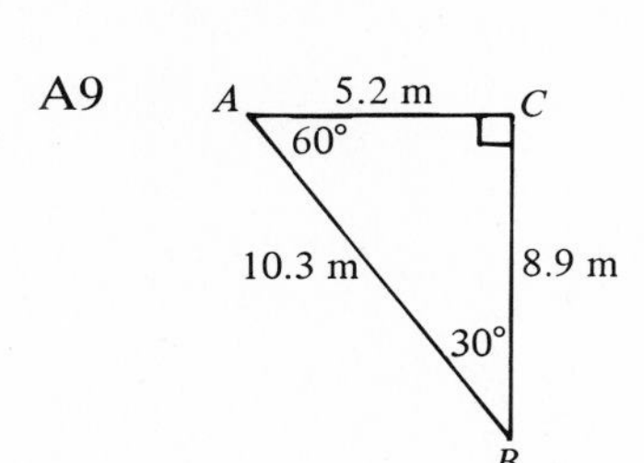

: $A = 90° - 30°$

$$\sin 30° = \frac{b}{10.3} \qquad \sin 60° = \frac{a}{10.3}$$

$$0.5 = \frac{b}{10.3} \qquad 0.866 = \frac{a}{10.3}$$

Q10 Solve the accompanying triangle (round off to tenths).

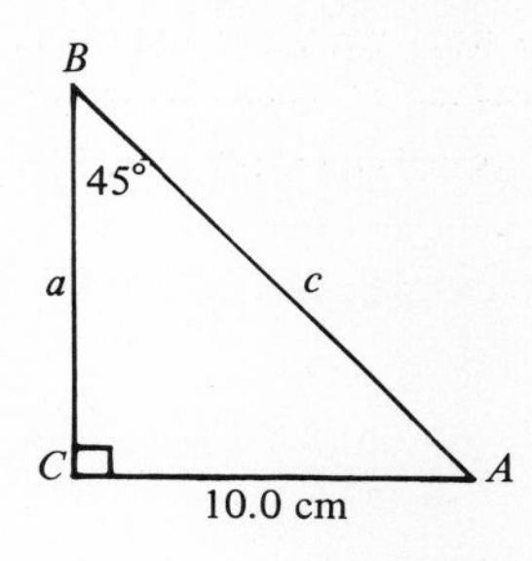

\# \# \# • \# \# \# • \# \# \# • \# \# \# • \# \# \# • \# \# \# • \# \# \# • \# \# \# • \# \# \#

A10

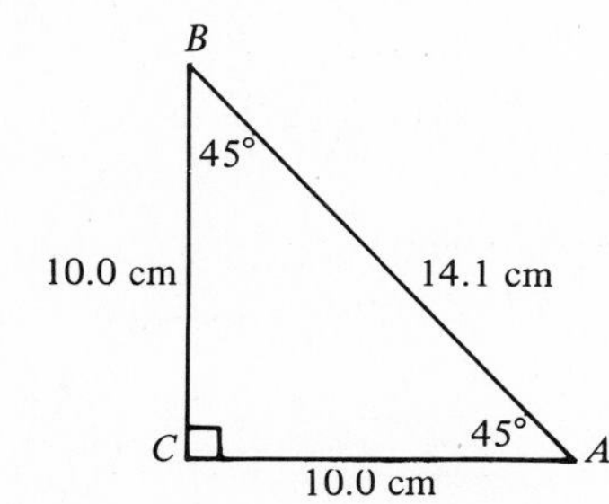

: In a 45-45-90 triangle, the sides of the 90° angles are equal.

$$\csc 45° = \frac{c}{10}$$

$$1.4142 = \frac{c}{10}$$

4 Frequently, the information about a right triangle is given in verbal or written form. It is then necessary to sketch a triangle indicating the given information in the appropriate places in order to solve for the desired part.

Example: In a right triangle ABC, $A = 28°$ and $AC = 15$ m. Find AB.

Solution: Sketch a right triangle and indicate the given information on it.

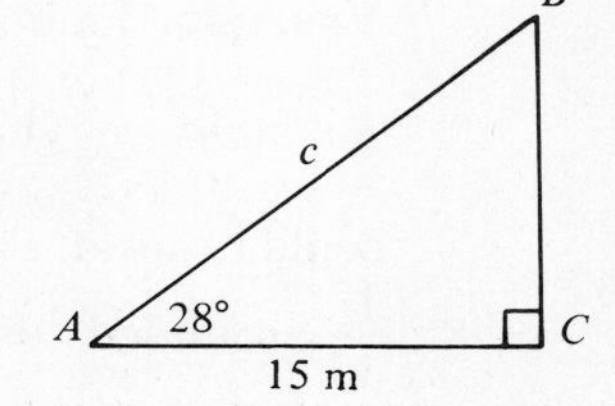

Now, find AB (which is c).

$$\sec 28° = \frac{c}{15}$$

$$1.1326 = \frac{c}{15}$$

$$15(1.1326) = c$$

$$16.989 = c$$

Therefore, $AB = 17$ m (rounded off to the nearest metre).

Q11 In a right triangle ABC, $B = 10°$ and $BC = 10.6$ mm. Find AC rounded off to the nearest tenth.

\# \# \# • \# \# \# • \# \# \# • \# \# \# • \# \# \# • \# \# \# • \# \# \# • \# \# \# • \# \# \#

A11 $AC = 1.9$ mm:

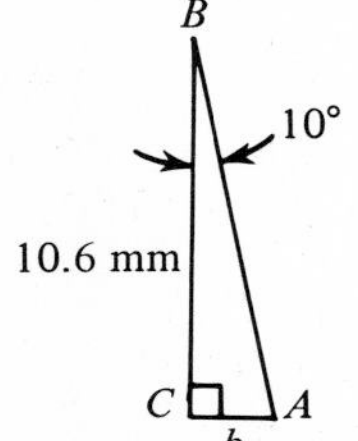

$$\tan 10° = \frac{b}{10.6}$$

$$0.1763 = \frac{b}{10.6}$$

Q12 In a right triangle ABC, $A = 45°$ and $c = 50$ km. Find a to the nearest kilometre.

\# \# \# • \# \# \# • \# \# \# • \# \# \# • \# \# \# • \# \# \# • \# \# \# • \# \# \# • \# \# \#

A12 $a = 35$ km:

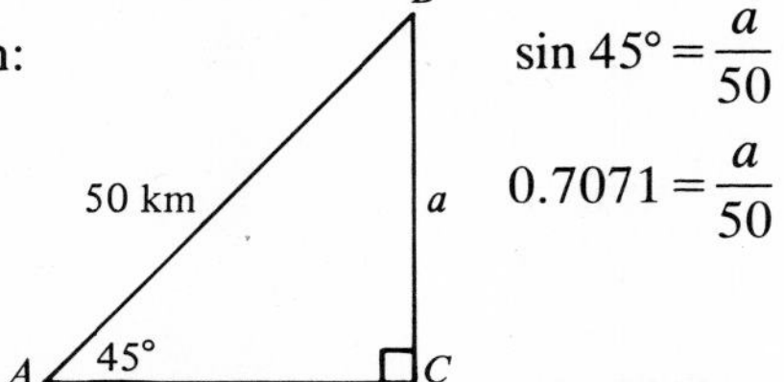

$$\sin 45° = \frac{a}{50}$$

$$0.7071 = \frac{a}{50}$$

5 If two sides of a right triangle are known, either acute angle can be found by using an appropriate trigonometric equation.

Example. Find A.

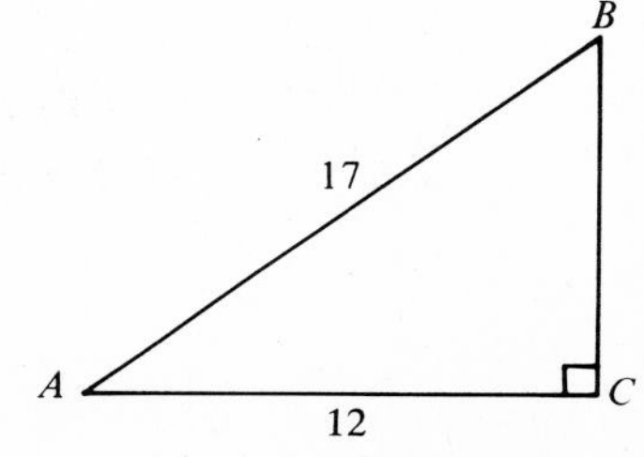

Solution: $\cos A = \frac{12}{17}$ or $\sec A = \frac{17}{12}$. Either trigonometric equation could be used. We will use $\cos A = \frac{12}{17}$. First, calculate the value of $\frac{12}{17}$ rounded off to ten-thousandths (4 decimal places). $\frac{12}{17} = 0.7059$. Hence, $\cos A = 0.7059$.

Next, attempt to locate 0.7059 in a cos column in Table X. You will find that

$$\left.\begin{array}{r} \cos 44° = 0.6947 \\ \cos A = 0.7059 \\ \cos 45° = 0.7071 \end{array}\right] \begin{array}{l} 12 \\ 12 \end{array}$$

Since 0.7059 is at least halfway between 0.6947 and 0.7071, we will estimate that $A = 45°$ to the nearest whole degree.* Greater accuracy could be obtained by using a trigonometric calculator, interpolating, or by using a table similar to the one illustrated in Section 12.2. For our purposes, estimating to the nearest whole degree will be satisfactory.

*Had 0.7059 been less than halfway between 0.6947 and 0.7071, then we would have estimated A to be 44°.

Q13 In a right triangle, $a = 6.0$ and $b = 4.5$. Find B to the nearest degree. Sketch a triangle if necessary.

\# \# \# • \# \# \# • \# \# \# • \# \# \# • \# \# \# • \# \# \# • \# \# \# • \# \# \# • \# \# \#

A13 $B = 37°$:

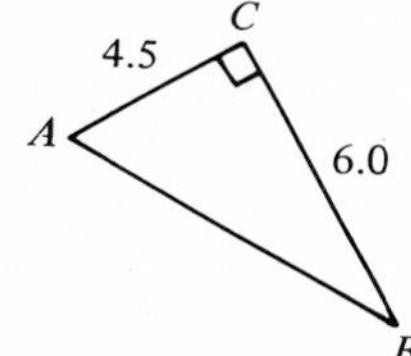

$$\tan B = \frac{4.5}{6.0} = 0.7500$$

$$\tan 36° = 0.7265$$
$$\tan B = 0.7500$$
$$\tan 37° = 0.7536$$

This completes the instruction for this section.

12.3 EXERCISE

1. Solve the triangle (give answers to nearest tenth):

a.

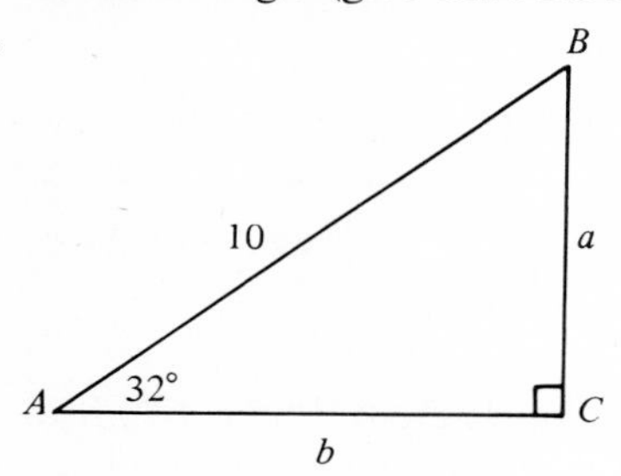

b.

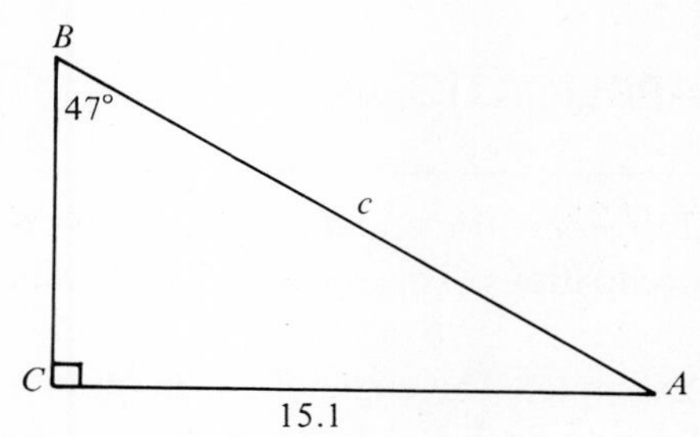

c.

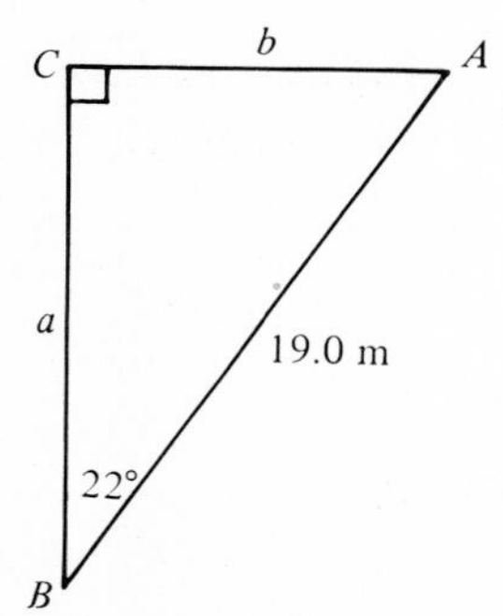

d.

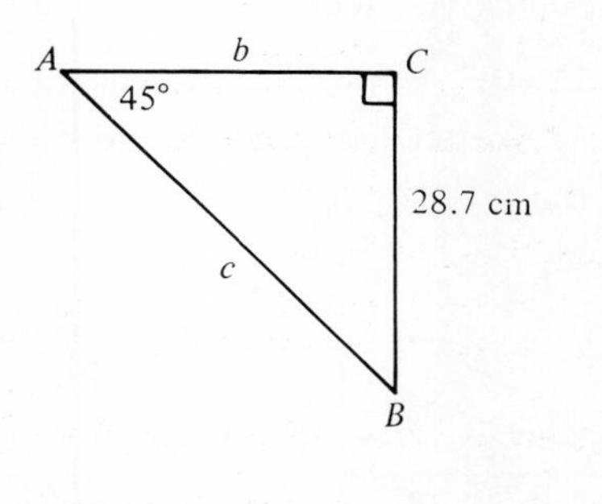

2. In a right triangle ABC:
   a. $A = 30°$ and b $(AC) = 24.3$ dm. Find c (AB) to the nearest tenth of a decimetre.
   b. $a = 15$ units and $b = 30$ units. Find A to the nearest degree.
   c. $B = 73°$ and a $(BC) = 45$ cm. Find b (AC) to the nearest centimetre.
   d. $A = 81°$ and c $(AB) = 56.0$ m. Find a (BC) to the nearest tenth of a metre.

12.3 EXERCISE ANSWERS

1. a.

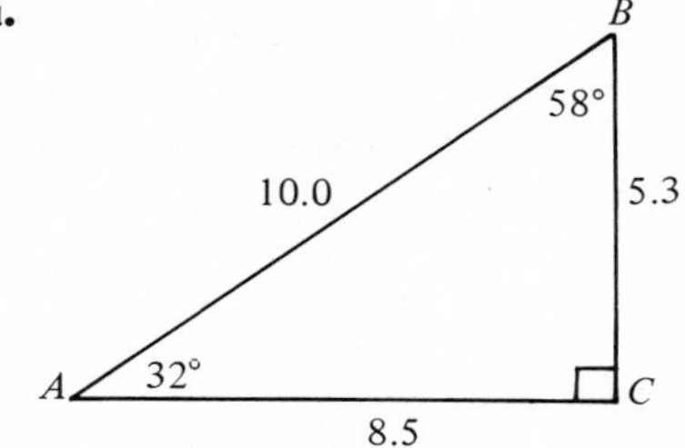

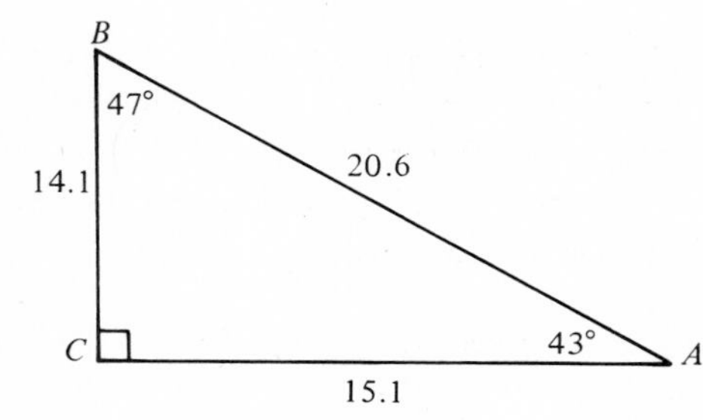

c.

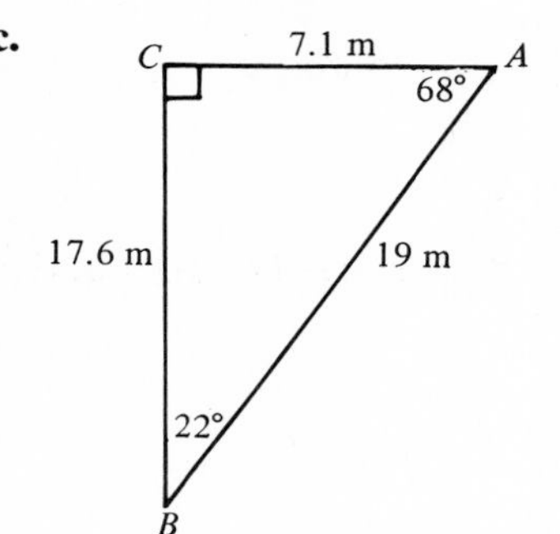

d.

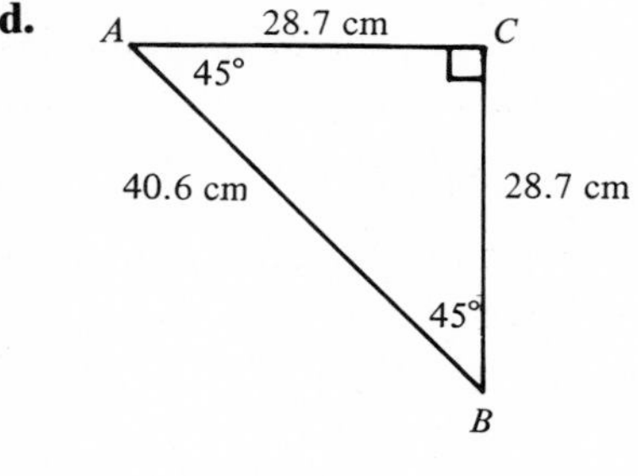

2. a. $c = 28.1$ dm
   b. $A = 27°$
   c. $b = 147$ cm
   d. $a = 55.3$ m

12.4 APPLICATIONS

1 Trigonometric equations can be used to solve a wide variety of problems which occur in technical occupations. This section will examine numerous such examples.

Example: The angle of elevation of a loading ramp is to be no greater than 18°. If the height of the loading dock is 4 feet above the road bed, what is the minimum length of the base of the loading ramp?

Solution: Sketch a diagram of the situation.

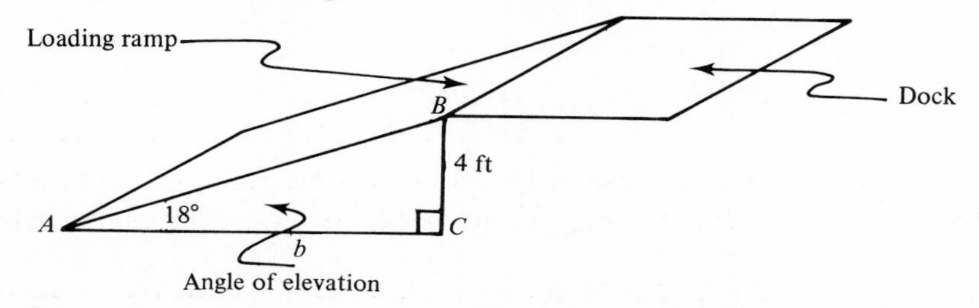

Write and solve an appropriate trigonometric equation containing side b and two known values.

$$\cot 18° = \frac{b}{4}$$

$$3.0777 = \frac{b}{4}$$

$$4(3.0777) = b$$

$$12.3108 = b$$

Therefore, the minimum length of the base of the loading ramp is 12.3 feet (rounded off to the nearest tenth of a foot).

Q1 The plans for a 24-foot wide garage call for the roof rafters to be at an 18° angle of elevation. There is also to be an 8-inch overhang.

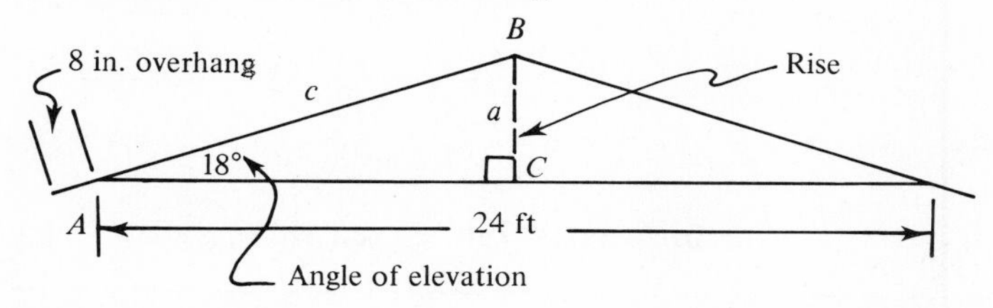

a. What is the minimum length of 2×6 stock from which each rafter can be cut (round off to nearest tenth of a foot)? (Hint: Determine the length of AC.)

b. What is the rise of the roof (round off to nearest tenth of a foot)?

• ### • ### • ### • ### • ### • ### • ### •

A1 **a.** 13.3 ft (≐ 13 ft 4 in.): $AC = 12$ ft $\quad 8 \text{ in.} = \frac{2}{3} \text{ ft} = 0.7 \text{ ft}$

$$\sec 18° = \frac{c}{12} \qquad \text{Rafter} = 12.6 + 0.7$$

$$1.0515 = \frac{c}{12}$$

$$12.6 = c$$

b. 3.7 ft (≐ 3 ft 8 in.): $\sin 18° = \frac{a}{12}$

$$0.3090 = \frac{a}{12}$$

$$3.7 = a$$

2 As you can see, a sketch of the situation is a great help in analyzing the problem. In the following problems, you will be asked to practice solving problems by first sketching and labeling an appropriate diagram.

Q2 A four-foot-square cupola is to be placed atop of the garage roof of Q1.

a. Label the sketch.

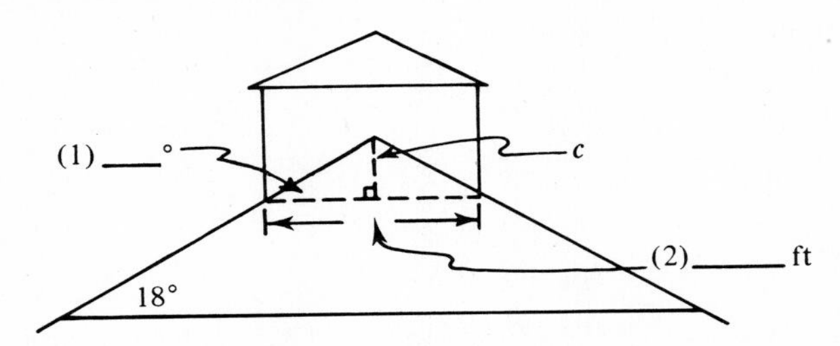

b. Determine the height of the triangle, c, (to the nearest tenth of a foot) which would have to be cut from the cupola in order for it to rest snugly on the roof.

\# \# \# • \# \# \# • \# \# \# • \# \# \# • \# \# \# • \# \# \# • \# \# \# • \# \# \# • \# \# \#

A2 **a.** (1) 18° (2) 4 ft **b.** 0.6 ft (≐7 in.): $\tan 18° = \frac{c}{2}$

Q3 A staircase rises vertically 5 feet for every 8 feet it progresses horizontally. Draw sketches to help in answering these questions.

a. What is the angle of elevation (nearest whole degree) of the staircase?

b. If the foot of the stairs is 12.0 feet from the vertical drop of the staircase, what is the vertical drop (nearest tenth of a foot)?

\# \# \# • \# \# \# • \# \# \# • \# \# \# • \# \# \# • \# \# \# • \# \# \# • \# \# \# • \# \# \#

A3 **a.** 32°:

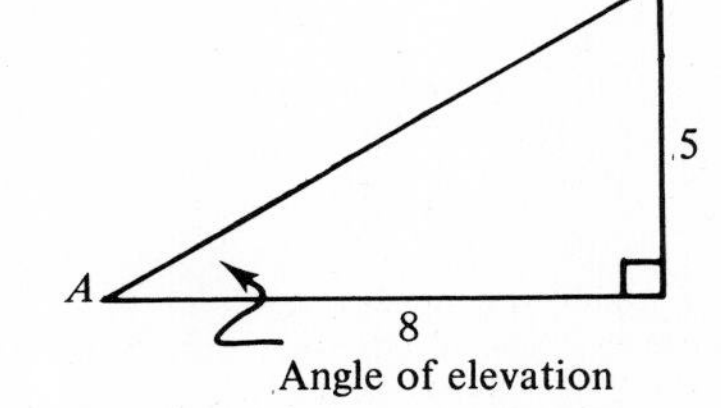

$\tan A = \frac{5}{8}$

$\tan A = 0.625$

b. 7.5 ft:

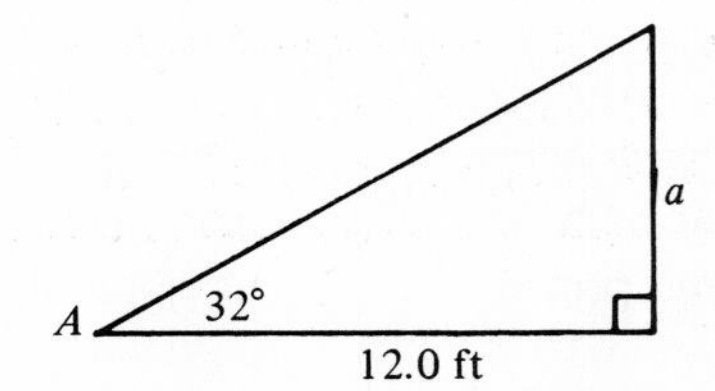

$\tan 32° = \frac{a}{12}$

$0.6249 = \frac{a}{12}$

Q4 The height of a radio station's antenna must be determined. At a point 100 feet from the base of the antenna the angle of elevation to the top of the antenna is 52°.

a. How tall (nearest foot) is the antenna?

b. How long (nearest foot) is a guy wire connected 5 feet from the top of the antenna to a point 100 feet from the base of the antenna?

\# # # • # # # • # # # • # # # • # # # • # # # • # # # • # # # • # # #

A4 **a.** 128 ft:

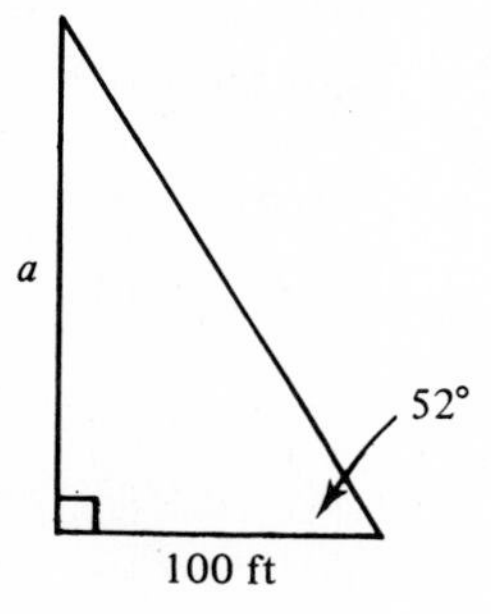

b. 159 ft:

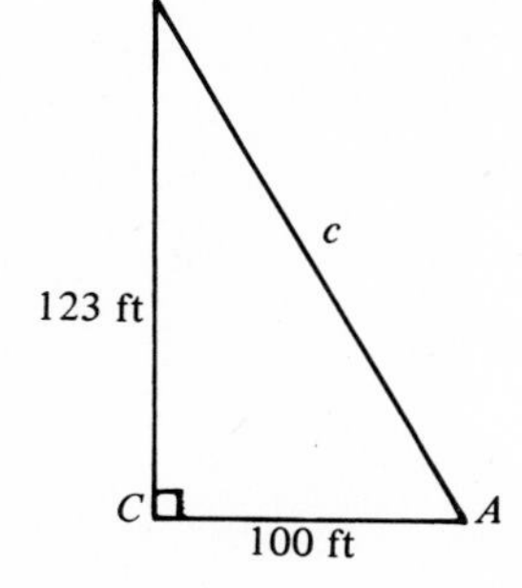

$$\tan A = 1.23$$
$$A = 51°$$
$$\sec 51° = \frac{c}{100}$$

3	It would not be possible to illustrate all possible applications of right-angle trigonometry to the wide variety of technical occupations. However, the procedures developed here will permit you to solve problems involving right triangles which occur in your specific area of interest. In the exercise that follows, problems have been selected to cover a wide range of interest.

This completes the instruction for this section.

12.4 EXERCISE

(Round off all answers to the nearest tenth unless otherwise indicated.)

1. a. What is the "angle of fitting" to the nearest whole degree?

b. What is the "run"?

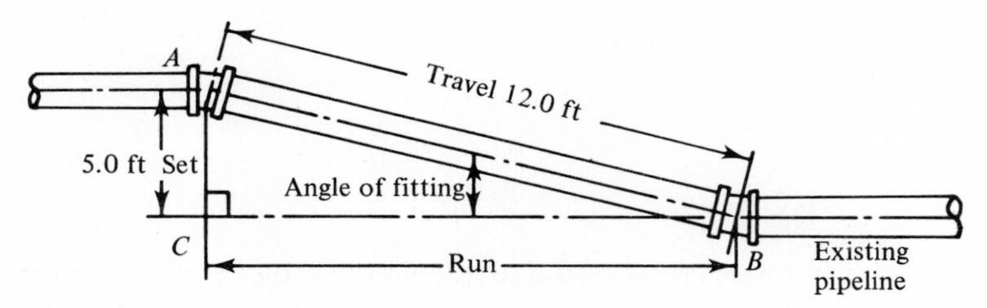

2. Find BC.

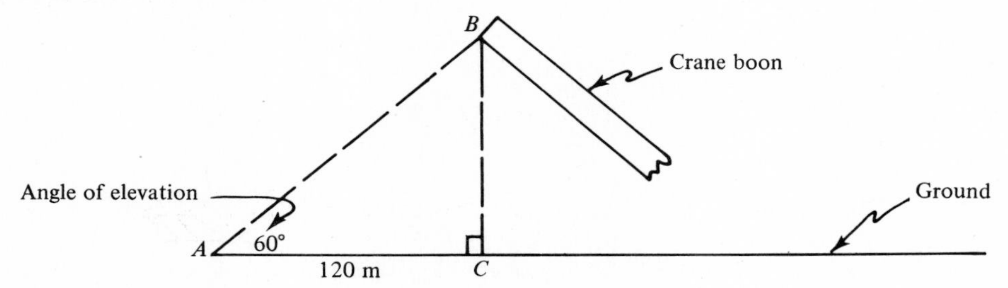

3. Find AB.

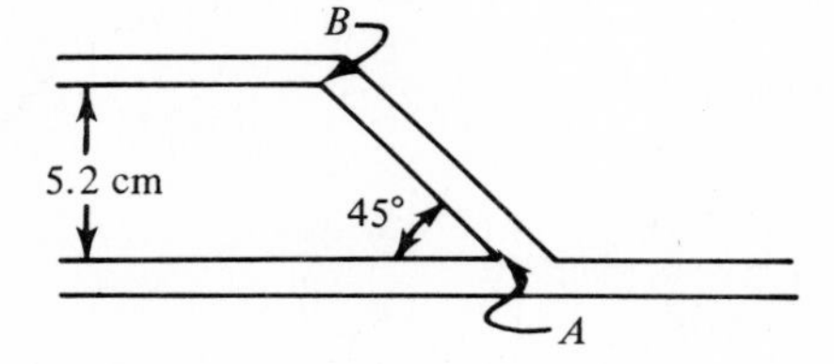

4. Find the height of the tree.

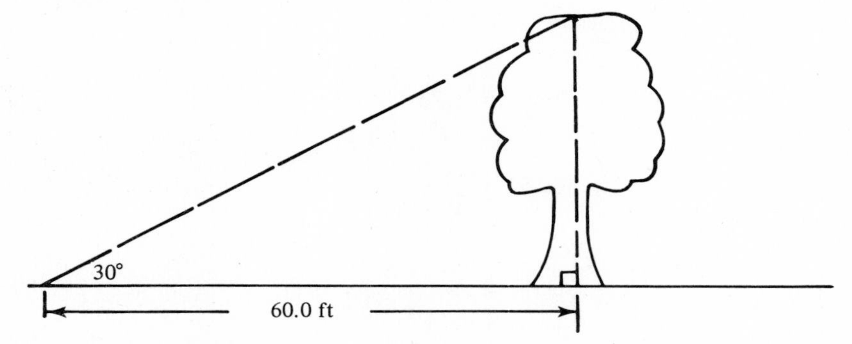

5. Find A (nearest degree).

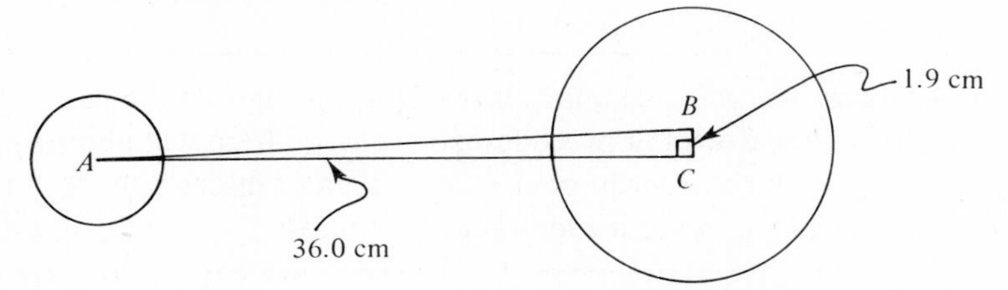

6. The most efficient operating angle of elevation for a conveyor used to raise feed to a storage bin is 31°. If the feed is to be elevated 60 feet, what length of conveyor is needed?

7. A millwright is to weld supports for a 30-metre conveyor so that it will operate at a 17° angle. What should be the length of the supports?

8. The template shown must be machined. What is the length of AC? (Hint: $AC = 2AB$.)

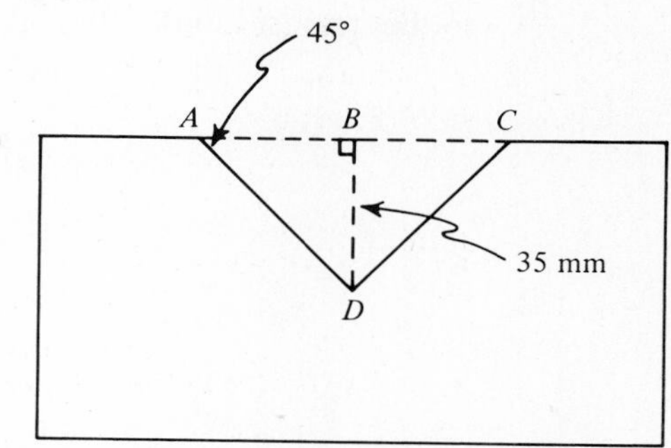

9. A pipe fitter is to connect the two pipes illustrated below. Determine (1) the "travel" and (2) the "run" of the pipe connection. Subtract 1 inch from the "travel" to allow for the pipe fittings.

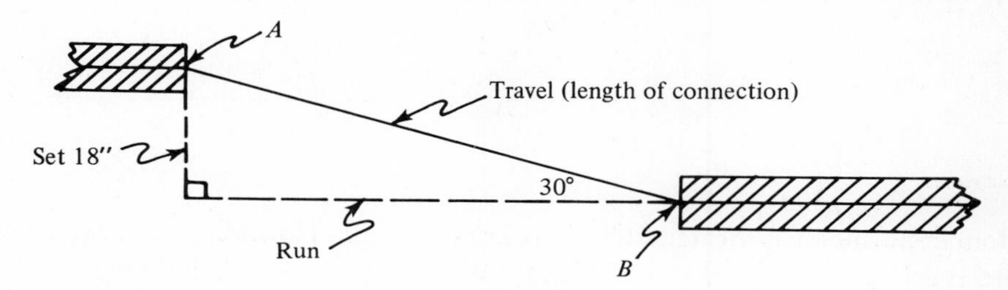

10. A gauge to check the diameter of a crankshaft journal is made according to the dimensions shown in the figure. Find BC.

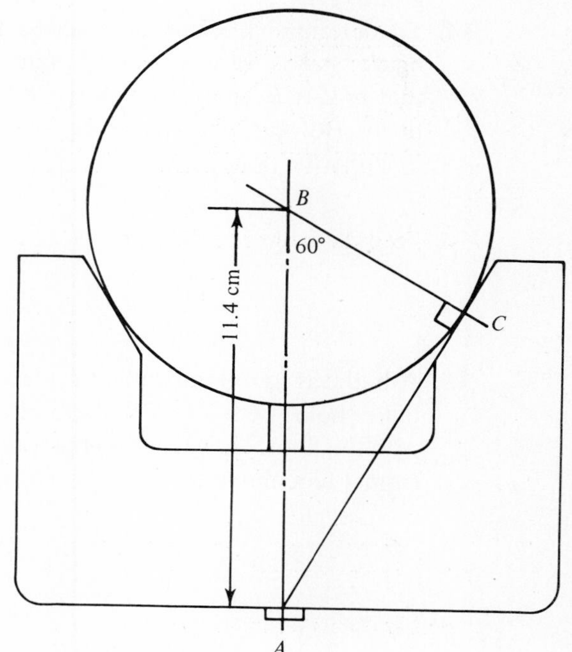

11. In order to machine the part shown, the machinist must determine $m\measuredangle A$ first. Find $m\measuredangle A$ to the nearest whole degree.

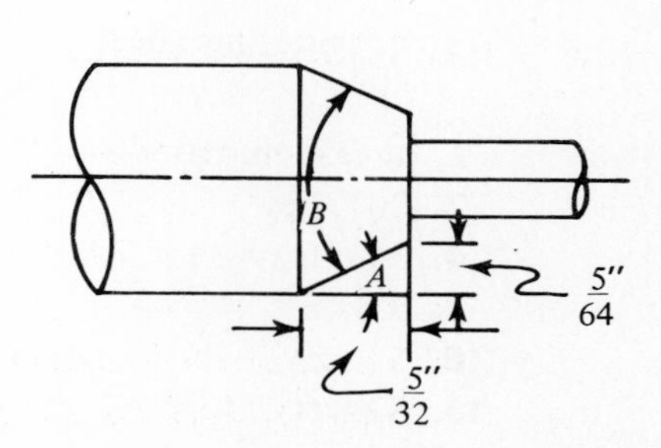

12. In order to locate the center of the drill hole, find (1) the value of x and (2) the value of y (round off to the nearest hundredth).

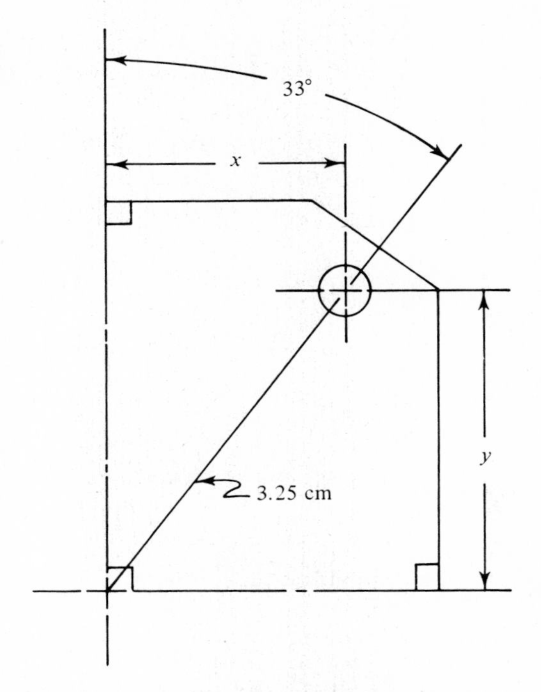

13. The bottom of a window is located 80 feet from the ground. How long a fire extension ladder is needed to reach the window when the ladder is set at a recommended angle of 75° with the ground?

14. To determine the distance across the pond in the figure, stakes were located at points A, B, and C so that $m\measuredangle ACB$ and $m\measuredangle CBA$ are 90° and 60°, respectively. BC was measured to be 40 metres. How far is it across the pond?

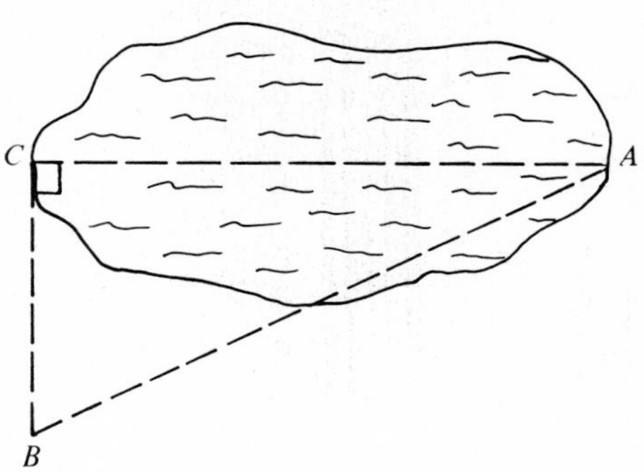

15. A bullet is found imbedded in the wall of a room. The vertical distance from the floor to the bullet hole is 6 ft 6 in. The path of the bullet in the wall is inclined upward at an angle of 18° with the floor. How far from the wall was the gun when fired if the gun was estimated to be fired from 4 feet above the floor?

12.4 EXERCISE ANSWERS

1. a. 25° **b.** 10.9 ft (≐ 10 ft 11 in.)
2. 207.9 m
3. 7.4 m
4. 34.6 ft (≐ 34 ft 7 in.)
5. 3°
6. 116.5 ft (116 ft 6 in.)
7. 8.8 m
8. 70 mm
9. (1) 35 in. (2) 31.2 in.
10. 5.7 cm
11. 27°
12. (1) 1.77 cm (2) 2.73 cm
13. 82.8 ft or longer
14. 69.3 m
15. 7.7 ft (≐ 7 ft 8 in.)

CHAPTER 12 SAMPLE TEST

At the completion of Chapter 12 you should be able to work the following problems. (Note: Use Tables IX and X whenever necessary. These tables will be provided when you are taking a post test on this chapter.)

12.1 TRIGONOMETRIC RATIOS

1. Using opp, adj, and hyp, define all six trigonometric ratios of an acute angle A in a right triangle.
2. Label the sides of triangle DEF appropriately and indicate:

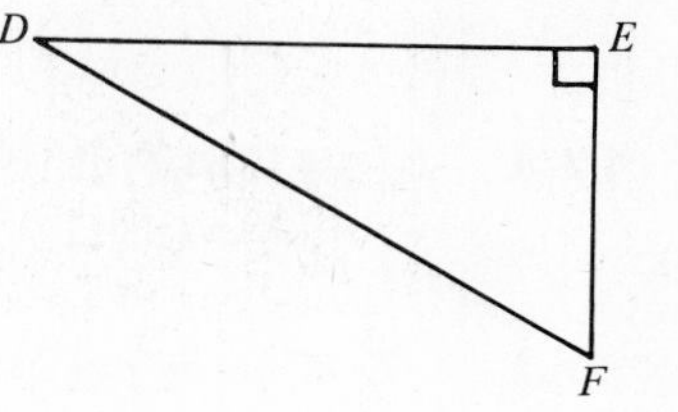

a. cos F **b.** sin D **c.** cot D **d.** csc F
e. sec F **f.** tan D **g.** cot F **h.** cos D
i. sin F **j.** csc D **k.** tan F **l.** sec D

12.2 TRIGONOMETRIC TABLES (Use Table X)

3. **a.** $\cos 60° = \sin$ ______ = ______ **b.** $\sec 75° = \csc$ ______ = ______
   c. $\cot 52° = \tan$ ______ = ______ **d.** $\sin 88° = \cos$ ______ = ______
   e. $\tan 47° = \cot$ ______ = ______ **f.** $\csc 59° = \sec$ ______ = ______
4. Find A:
   a. $\cos A = 0.9063$ **b.** $\csc A = 1.3054$ **c.** $\cot A = 28.6363$
   d. $\sin A = 0.8192$ **e.** $\sec A = 1.0014$ **f.** $\tan A = 0.2867$

12.3 SOLVING RIGHT TRIANGLES (Use Table IX and Table X)

5. Solve the triangle (give answers to nearest tenth):

a.

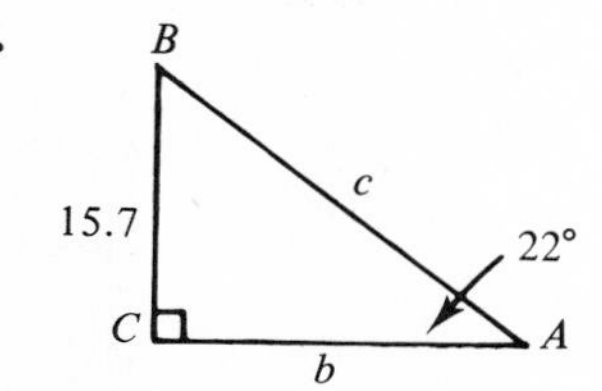

b.

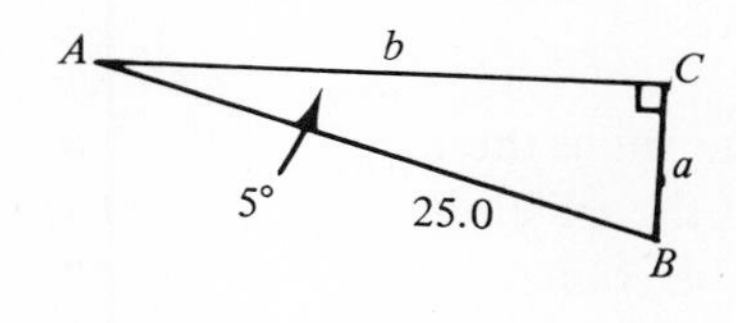

c.

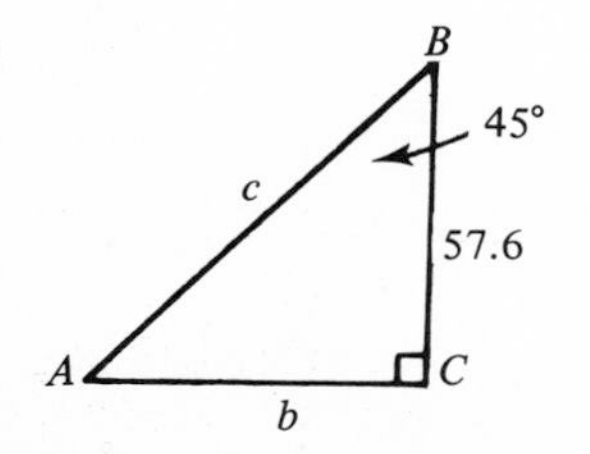

d.

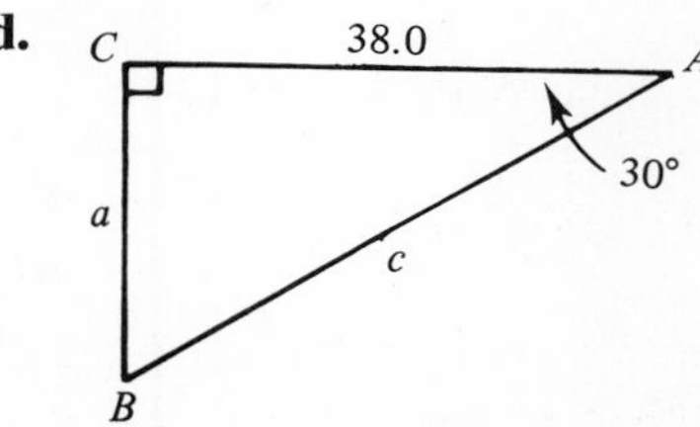

6. In triangle ABC:

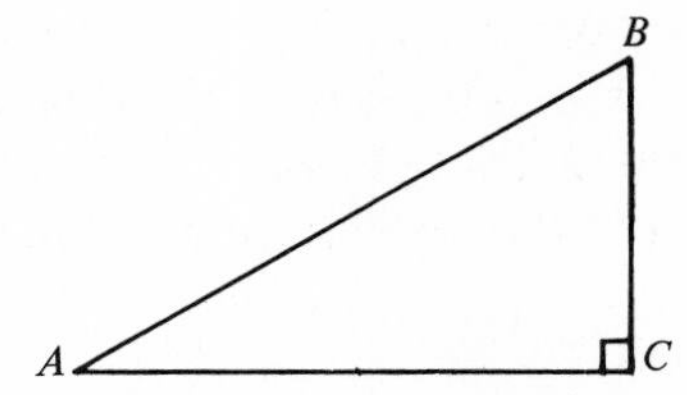

a. $a = 7$ m and $c = 12$ m. Find B to the nearest degree.
b. $A = 37°$ and a $(BC) = 50.0$ ft. Find b (AC) to the nearest tenth of a foot.
c. $B = 45°$ and a $(BC) = 17.5$ cm. Find b (AC) to the nearest tenth of a centimetre.
d. $a = 15.0$ in. and $b = 17.0$ in. Solve the triangle. Give sides to the nearest tenth of an inch and the angles to the nearest degree.

12.4 APPLICATIONS (Give all answers rounded off to the nearest tenth unless otherwise indicated)

7. (1) Find the perpendicular distance (OC) from the center to the side of a regular hexagon with 5″ sides and (2) determine the value of x. (Hint: $AC = 2.5″$ and $m\measuredangle AOB = 60°$; therefore, $m\measuredangle AOC = 30°$.)

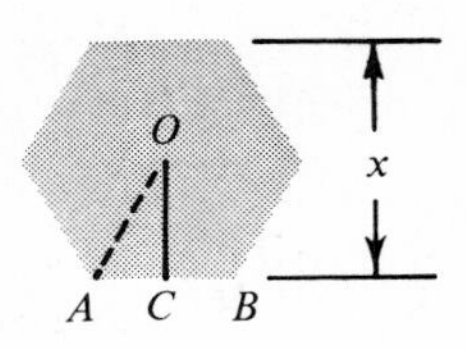

8. (1) Find the angle of elevation of the rafters to the nearest whole degree and (2) the length of the rafters of a garage 24 ft wide with a 10-inch overhang.

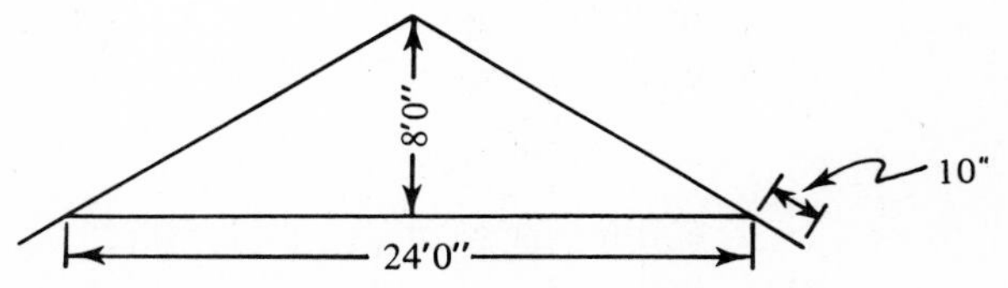

9. A concrete ramp is to be built with a 20° angle of elevation to a dock 4 feet 9 inches high. How far from the dock should the concrete forms be started?

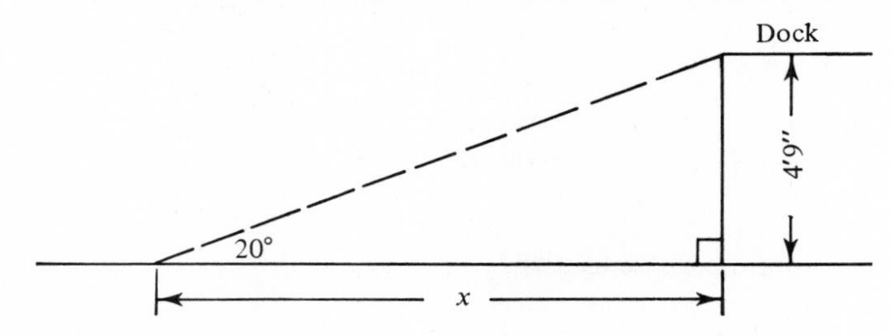

10. Determine the length of the connection between the new pipeline and the existing pipeline. Subtract 2.5 centimetres from the length of the connection to allow for the fittings on each end.

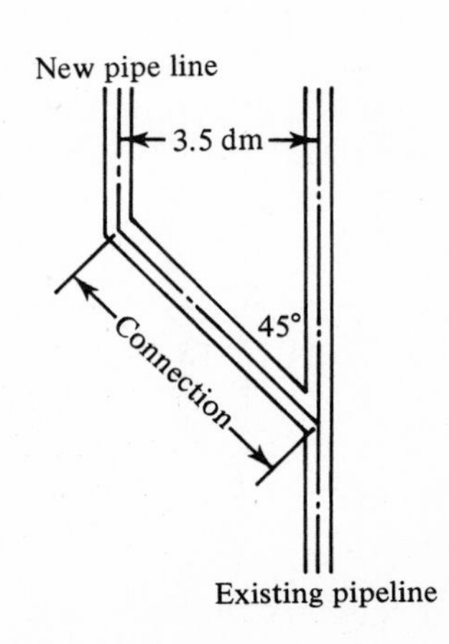

11. A guy wire attached to the top of a pole to a point 40.0 ft from the base of the pole makes an angle of elevation of 80°. (1) How long is the guy wire (disregard the extra length needed to secure the wire to the pole and to the stake in the ground) and (2) how high is the pole?

12. A road bed rises 3.5 metres in every 100 metres in length. What is the angle of elevation to the nearest degree?

CHAPTER 12 SAMPLE TEST ANSWERS

1. $\sin A = \dfrac{\text{opp}}{\text{hyp}}$ $\quad\cos A = \dfrac{\text{adj}}{\text{hyp}}$ $\quad\tan A = \dfrac{\text{opp}}{\text{adj}}$

$\csc A = \dfrac{\text{hyp}}{\text{opp}}$ $\quad\sec A = \dfrac{\text{hyp}}{\text{adj}}$ $\quad\cot A = \dfrac{\text{adj}}{\text{opp}}$

2. a. $\dfrac{d}{e}$ **b.** $\dfrac{d}{e}$ **c.** $\dfrac{f}{d}$ **d.** $\dfrac{e}{f}$

e. $\dfrac{e}{d}$ **f.** $\dfrac{d}{f}$ **g.** $\dfrac{d}{f}$ **h.** $\dfrac{f}{e}$

i. $\dfrac{f}{e}$ **j.** $\dfrac{e}{d}$ **k.** $\dfrac{f}{d}$ **l.** $\dfrac{e}{f}$

3. a. sin 30° = 0.5000 **b.** csc 15° = 3.8637 **c.** tan 38° = 0.7813
d. cos 2° = 0.9994 **e.** cot 43° = 1.0724 **f.** sec 31° = 1.1666

4. a. 25° **b.** 50° **c.** 2° **d.** 55° **e.** 3° **f.** 16°

5. a.

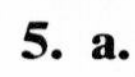

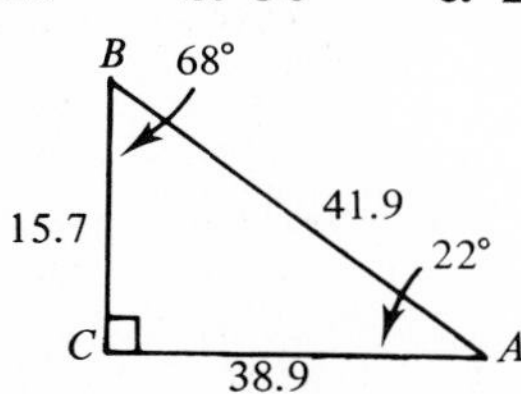

b.

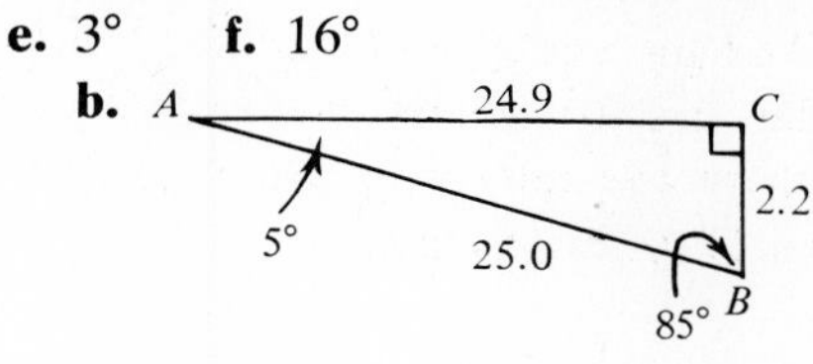

c.

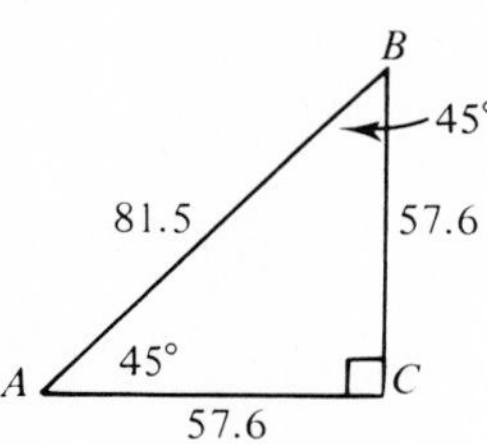

d.

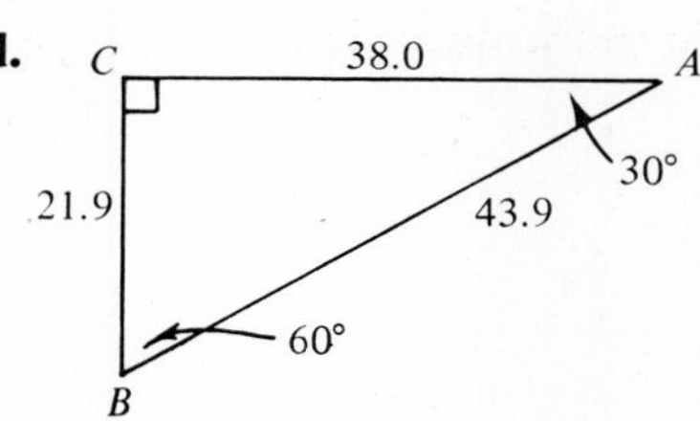

6. a. 54° **b.** 66.4 ft (≐ 66 ft 5 in.)
c. 17.5 cm **d.**

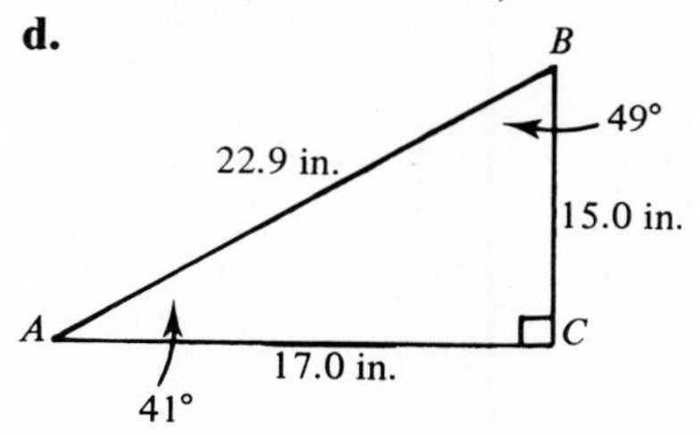

7. (1) 4.3 in. (2) 8.6 in. **8.** (1) 34° (2) 15.2 ft (≐ 15 ft 3 in.)
9. 13.1 ft (≐ 13 ft 1 in.)
10. 4.65 dm (4.7 to nearest decimetre): 2.5 cm = 0.25 dm
11. (1) 230.4 ft (2) 226.9 ft **12.** 2°

APPENDIX

Table I Multiplication Facts

1	**2**	**3**	**4**	**5**	**6**	**7**	**8**	**9**	**10**	**11**	**12**	**13**	**14**	**15**	**16**	**17**	**18**	**19**	**20**	**21**	**22**	**23**	**24**	**25**
2	**4**	6	8	10	12	14	16	18	20	22	24	26	28	30	32	34	36	38	40	42	44	46	48	50
3	6	**9**	12	15	18	21	24	27	30	33	36	39	42	45	48	51	54	57	60	63	66	69	72	75
4	8	12	**16**	20	24	28	32	36	40	44	48	52	56	60	64	68	72	76	80	84	88	92	96	100
5	10	15	20	**25**	30	35	40	45	50	55	60	65	70	75	80	85	90	95	100	105	110	115	120	125
6	12	18	24	30	**36**	42	48	54	60	66	72	78	84	90	96	102	108	114	120	126	132	138	144	150
7	14	21	28	35	42	**49**	56	63	70	77	84	91	98	105	112	119	126	133	140	147	154	161	168	175
8	16	24	32	40	48	56	**64**	72	80	88	96	104	112	120	128	136	144	152	160	168	176	184	192	200
9	18	27	36	45	54	63	72	**81**	90	99	108	117	126	135	144	153	162	171	180	189	198	207	216	225
10	20	30	40	50	60	70	80	90	**100**	110	120	130	140	150	160	170	180	190	200	210	220	230	240	250
11	22	33	44	55	66	77	88	99	110	**121**	132	143	154	165	176	187	198	209	220	231	242	253	264	275
12	24	36	48	60	72	84	96	108	120	132	**144**	156	168	180	192	204	216	228	240	252	264	276	288	300
13	26	39	52	65	78	91	104	117	130	143	156	**169**	182	195	208	221	234	247	260	273	286	299	312	325
14	28	42	56	70	84	98	112	126	140	154	168	182	**196**	210	224	238	252	266	280	294	308	322	336	350
15	30	45	60	75	90	105	120	135	150	165	180	195	210	**225**	240	255	270	285	300	315	330	345	360	375
16	32	48	64	80	96	112	128	144	160	176	192	208	224	240	**256**	272	288	304	320	336	352	368	384	400
17	34	51	68	85	102	119	136	153	170	187	204	221	238	255	272	**289**	306	323	340	357	374	391	408	425
18	36	54	72	90	108	126	144	162	180	198	216	234	252	270	288	306	**324**	342	360	378	396	414	432	450
19	38	57	76	95	114	133	152	171	190	209	228	247	266	285	304	323	342	**361**	380	399	418	437	456	475
20	40	60	80	100	120	140	160	180	200	220	240	260	280	300	320	340	360	380	**400**	420	440	460	480	500
21	42	63	84	105	126	147	168	189	210	231	252	273	294	315	336	357	378	399	420	**441**	462	483	504	525
22	44	66	88	110	132	154	176	198	220	242	264	286	308	330	352	374	396	418	440	462	**484**	506	528	550
23	46	69	92	115	138	161	184	207	230	253	276	299	322	345	368	391	414	437	460	483	506	**529**	552	575
24	48	72	96	120	144	168	192	216	240	264	288	312	336	360	384	408	432	456	480	504	528	552	**576**	600
25	50	75	100	125	150	175	200	225	250	275	300	325	350	375	400	425	450	475	500	525	550	575	600	**625**

Table II English Weights and Measures

Units of Length

1 foot (ft or ′) = 12 inches (in. or ″)
1 yard (yd) = 3 feet
1 rod (rd) = $5\frac{1}{2}$ yards = $16\frac{1}{2}$ feet
1 mile(mi) = 1,760 yards = 5,280 feet

Units of Area

1 square foot = 144 square inches
1 square yard = 9 square feet
1 acre = 43,560 square feet
1 square mile = 640 acres

Units of Weight

1 pound (lb) = 16 ounces (oz)
1 ton = 2,000 pounds

Units of Capacity (Volume) Liquid

1 pint (pt) = 16 fluid ounces (fl oz)
1 quart (qt) = 2 pints
1 gallon (gal) = 4 quarts = 8 pints

Dry

1 quart (qt) = 2 pints (pt)
1 peck (pk) = 8 quarts
1 bushel (bu) = 4 pecks = 32 quarts

Units of Capacity

1 cubic foot = 1,728 cubic inches
1 cubic yard = 27 cubic feet
1 gallon = 231 cubic inches
1 cubic foot = 7.48 gallons

Table III Metric Weights and Measures

Units of Length

1 millimetre (mm) = 1000 micrometres (μm)
1 centimetre (cm) = 10 millimetres
1 decimetre (dm) = 10 centimetres
1 metre (m) = 10 decimetres
1 dekametre(dam) = 10 metres
1 hectometre (hm) = 10 dekametres
1 kilometre (km) = 10 hectometres

Units of Weight

1 milligram (mg) = 1000 micrograms (μg)
1 centigram (cg) = 10 milligrams
1 decigram (dg) = 10 centigrams
1 gram (g) = 10 decigrams
1 dekagram (dag) = 10 grams
1 hectogram (hg) = 10 deckagrams
1 kilogram (kg) = 10 hectograms
1 megagram (Mg) = 1000 kilograms (metric ton)

Units of Area

1 square centimetre = 100 square millimetres
1 square decimetre = 100 square centimetres
1 square metre = 100 square decimetres
1 square kilometre = 1 000 000 square metres
1 hectare (ha) = 10 000 square metres
1 are (a) = 100 square metres

Units of Capacity (Volume)

1 millilitre (ml) = 1 cubic centimetre (cc or cm^3)
1 centilitre (cl) = 10 millilitres
1 decilitre (dl) = 10 centilitres
1 litre (l) = 10 decilitres
= 1000 cubic centimetres
1 deckalitre (dal) = 10 litres
1 hectolitre (hl) = 10 dekalitres
1 kilolitre (kl) = 10 hectolitres
1 megalitre (Ml) = 1000 kilolitres

Table IV Metric and English Conversion

Length

English	Metric
1 inch =	2.54 cm
1 foot =	30.5 cm
1 yard =	91.4 cm
1 mile =	1610 m
1 mile =	1.61 km
0.0394 in. =	1 mm
0.394 in. =	1 cm
39.4 in. =	1 m
3.28 ft =	1 m
1.09 yd =	1 m
0.621 mi =	1 km

Capacity

English	Metric
1 gallon =	3.79 l
1 quart =	0.946 l
0.264 gal =	1 l
1.05 qt =	1 l

Weight

English	Metric
1 ounce =	28.3 g
1 pound =	454 g
1 pound =	0.454 kg
0.0353 oz =	1 g
0.00220 lb =	1 g
2.20 lb =	1 kg

Area

English	Metric
1 square inch =	6.45 square centimetres
1 square foot =	929 square centimetres
1 square yard =	8361 square centimetres
1 square rod =	25.3 square metres
1 acre =	4047 square metres
1 square mile =	2.59 square kilometres
10.76 square feet =	1 square metre
1,550 square inches =	1 square metre
0.0395 square rod =	1 square metre
1.196 square yards =	1 square metre
0.155 square inch =	1 square centimetre
247.1 acres =	1 square kilometre
0.386 square mile =	1 square kilometre

Volume

English	Metric
1 cubic inch =	16.39 cubic centimetres
1 cubic foot =	28.317 cubic centimetres
1 cubic yard =	0.7646 cubic metres
0.06102 cubic inch =	1 cubic centimetre
35.3 cubic feet =	1 cubic metre

Table V Decimal Fraction and Common Fraction Equivalents

$\frac{1}{64} = 0.015625$	$\frac{9}{32} = 0.28125$	$\frac{17}{32} = 0.53125$	$\frac{25}{32} = 0.78125$
$\frac{1}{32} = 0.03125$	$\frac{19}{64} = 0.296875$	$\frac{35}{64} = 0.546875$	$\frac{51}{64} = 0.796875$
$\frac{3}{64} = 0.046875$	$\frac{5}{16} = 0.3125$	$\frac{9}{16} = 0.5625$	$\frac{4}{5} = 0.8$
$\frac{1}{16} = 0.0625$	$\frac{21}{64} = 0.328125$	$\frac{37}{64} = 0.578125$	$\frac{13}{16} = 0.8125$
$\frac{5}{64} = 0.078125$	$\frac{1}{3} = 0.33\bar{3}$	$\frac{19}{32} = 0.59375$	$\frac{53}{64} = 0.828125$
$\frac{3}{32} = 0.09375$	$\frac{11}{32} = 0.34375$	$\frac{3}{5} = 0.6$	$\frac{27}{32} = 0.84375$
$\frac{7}{64} = 0.109375$	$\frac{23}{64} = 0.359375$	$\frac{39}{64} = 0.609375$	$\frac{55}{64} = 0.859375$
$\frac{1}{8} = 0.125$	$\frac{3}{8} = 0.375$	$\frac{5}{8} = 0.625$	$\frac{7}{8} = 0.875$
$\frac{9}{64} = 0.140625$	$\frac{25}{64} = 0.390625$	$\frac{41}{64} = 0.640625$	$\frac{57}{64} = 0.890625$
$\frac{5}{32} = 0.15625$	$\frac{2}{5} = 0.4$	$\frac{21}{32} = 0.65625$	$\frac{29}{32} = 0.90625$
$\frac{11}{64} = 0.171875$	$\frac{13}{32} = 0.40625$	$\frac{2}{3} = 0.66\bar{6}$	$\frac{59}{64} = 0.921875$
$\frac{3}{16} = 0.1875$	$\frac{27}{64} = 0.421875$	$\frac{43}{64} = 0.671875$	$\frac{15}{16} = 0.9375$
$\frac{1}{5} = 0.2$	$\frac{7}{16} = 0.4375$	$\frac{11}{16} = 0.6875$	$\frac{61}{64} = 0.953125$
$\frac{13}{64} = 0.203125$	$\frac{29}{64} = 0.453125$	$\frac{45}{64} = 0.703125$	$\frac{31}{32} = 0.96875$
$\frac{7}{32} = 0.21875$	$\frac{15}{32} = 0.46875$	$\frac{23}{32} = 0.71875$	$\frac{63}{64} = 0.984375$
$\frac{15}{64} = 0.234375$	$\frac{31}{64} = 0.484375$	$\frac{47}{64} = 0.734375$	
$\frac{1}{4} = 0.25$	$\frac{1}{2} = 0.5$	$\frac{3}{4} = 0.75$	
$\frac{17}{64} = 0.265625$	$\frac{33}{64} = 0.515625$	$\frac{49}{64} = 0.765625$	

Table VI Formulas from Geometry

	PERIMETER	AREA
Rectangle	$p = 2l + 2w$	$A = lw$
Square	$p = 4s$	$A = s^2$
Parallelogram	$p = 2a + 2b$	$A = bh$
Trapezoid		$A = \frac{1}{2}h(b_1 + b_2)$
Triangle	$p = a + b + c$	$A = \frac{1}{2}bh$

The sum of the measures of the angles of a triangle = 180°. In a right triangle:

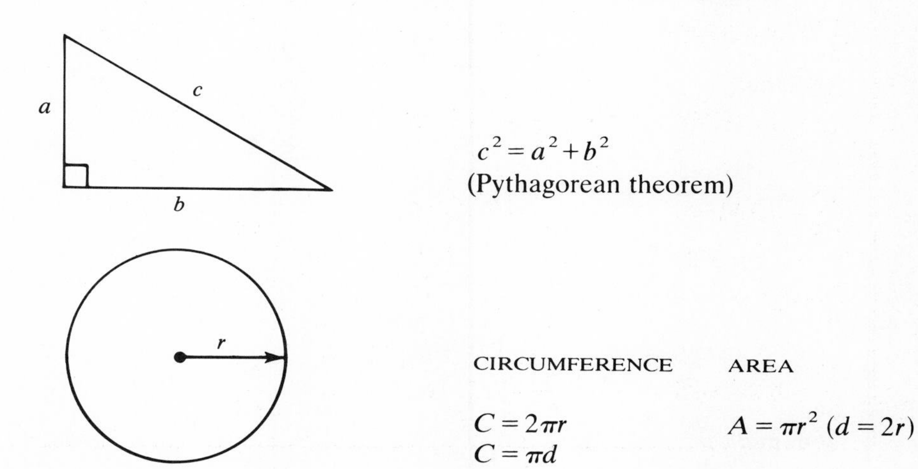

$c^2 = a^2 + b^2$
(Pythagorean theorem)

Circle

CIRCUMFERENCE	AREA
$C = 2\pi r$	$A = \pi r^2$ $(d = 2r)$
$C = \pi d$	

VOLUME

Triangular prism

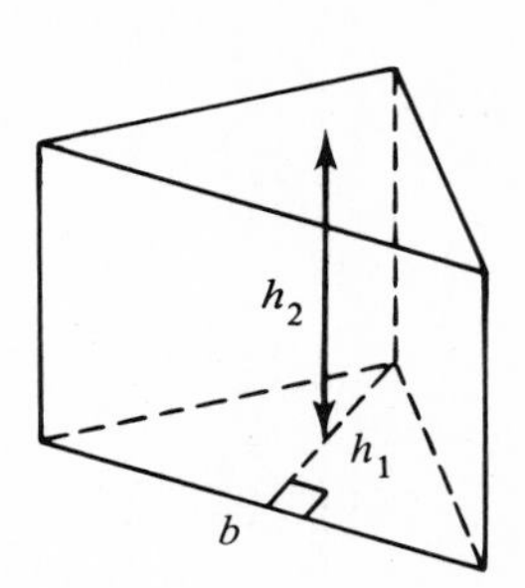

$V = \frac{1}{2}bh_1h_2$

or

$V = Bh$, where B is the area of the base

Cube

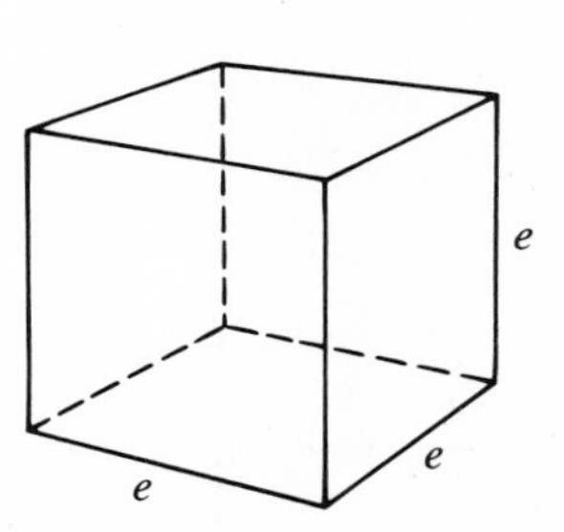

$V = e^3$

Rectangular prism

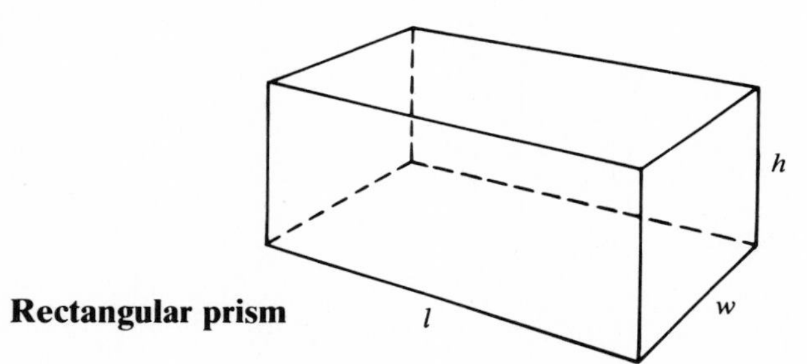

$V = lwh$

or

$V = Bh$, where B is the area of the base

Cylinder

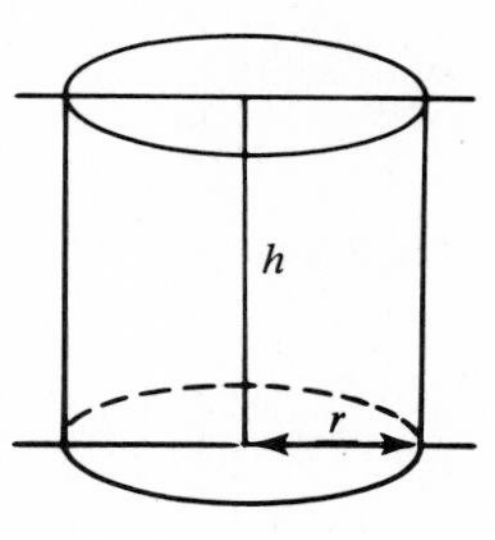

$V = \pi r^2 h$

Pyramid

h

B (area of the base)

$V = \frac{1}{3}Bh$

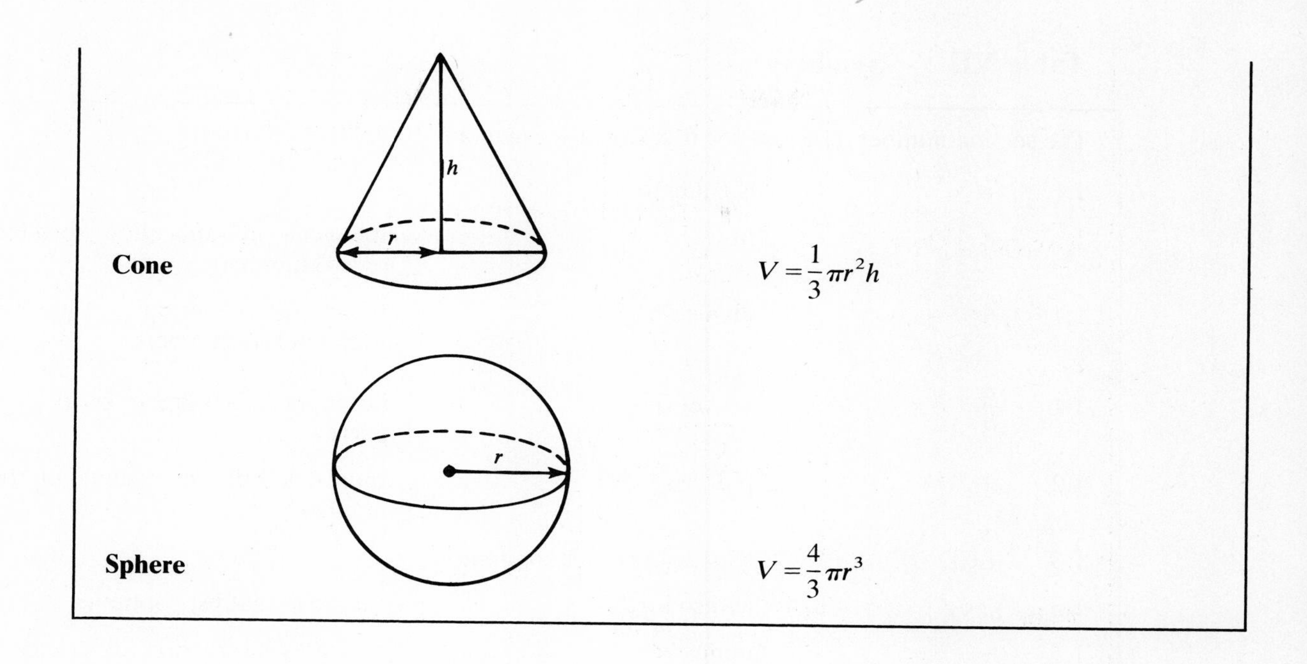
Cone
h
r
$V = \frac{1}{3}\pi r^2 h$
Sphere
r
$V = \frac{4}{3}\pi r^3$

Table VII Symbols

The section number indicates the first use of the symbol.

1.1	$=$	"is equal to"	
1.1	$\times$	"times"	Indicates multiplication (product) of two numbers
1.1	$\div$	"divided by"	Indicates division (quotient) between two numbers
1.1	$\frac{a}{b}$	"a over b"	Fraction indicating a of b equal parts
1.2	$+$	"plus"	Indicates addition (sum) of two numbers
1.3	LCD	"least common denominator"	
1.3	, . . .	"and so forth"	Indicates that list continues
1.5	$-$	"minus"	Indicates subtraction (difference) of two numbers
1.7	$''$	"inch or inches"	
1.7	$'$	"foot or feet"	
2.1	.	"point" or "and"	Decimal point used to separate the whole-number part of a number from the decimal (fraction) part of the number
2.5	$0.\overline{ab}$	"bar"	Indicates that the group of numbers under the bar repeat endlessly
2.6	@	"at"	Means for each item
3.1	%	"percent"	
4.1	l_1 and l_2	"l sub-one and l sub-two"	Subscripts used to distinguish between two lines
4.1	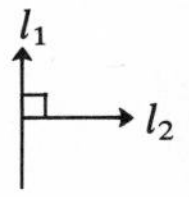		Indicates that l_1 and l_2 are perpendicular
4.1	$\parallel$	"is parallel to"	
4.1	$\perp$	"is perpendicular to"	
4.1	$\overleftrightarrow{AB}$	"line AB"	
4.1	$\overline{AB}$	"line segment AB"	
4.1	$\overrightarrow{AB}$	"ray AB"	
4.2	$\measuredangle$	"angle"	
4.2	$^\circ$	"degree"	Unit of measure of an angle
4.2	$m\measuredangle A$	"measure of angle A"	
4.3	R and R'	"R and R prime"	Notation to distinguish between two different radii
5.2	$\doteq$	"is approximately equal to"	

5.3	π	"pi"	Constant whose value is generally given as approximately equal to 3.14 or $\frac{22}{7}$
5.3	$\neq$	"is not equal"	
5.3	$a(b) = (a)(b) = (a)b = ab$		Product of a and b
5.4	$a \cdot b$	"times"	Raised dot to indicate the product of a and b
5.4	a^b	"a to the bth power"	Exponential notation
5.5	$\sqrt{\ }$	"radical" or "square root of"	
7.1	$^+$	"positive"	Raised plus sign to denote a positive number
7.1	$^-$	"negative"	Raised minus sign to denote a negative number
7.2	^-a	"opposite of a"	Raised minus sign to indicate the opposite of a number
7.2	( )	"quantity"	Parentheses used as a grouping symbol
9.1	$a:b$	"a to b" or "a is to b"	Colon used in expressing a ratio of two numbers or quantities
9.3	/	"per"	km/hr indicates km per hr

Table VIII Fraction, Decimal, and Percent Equivalents

$\frac{1}{16} = 0.06\frac{1}{4}$ or $0.0625 = 6\frac{1}{4}\%$ or 6.25%

$\frac{1}{12} = 0.08\frac{1}{3} = 8\frac{1}{3}\%$

$\frac{1}{10} = 0.1 = 10\%$

$\frac{1}{9} = 0.11\frac{1}{9} = 11\frac{1}{9}\%$

$\frac{1}{8} = 0.12\frac{1}{2}$ or $0.125 = 12\frac{1}{2}\%$ or 12.5%

$\frac{1}{7} = 0.14\frac{2}{7} = 14\frac{2}{7}\%$

$\frac{1}{6} = 0.16\frac{2}{3} = 16\frac{2}{3}\%$

$\frac{1}{5} = 0.2 = 20\%$

$\frac{1}{4} = 0.25 = 25\%$

$\frac{3}{10} = 0.3 = 30\%$

$\frac{1}{3} = 0.33\frac{1}{3} = 33\frac{1}{3}\%$

$\frac{3}{8} = 0.37\frac{1}{2}$ or $0.375 = 37\frac{1}{2}\%$ or 37.5%

$\frac{2}{5} = 0.4 = 40\%$

$\frac{1}{2} = 0.5 = 50\%$

$\frac{3}{5} = 0.6 = 60\%$

$\frac{5}{8} = 0.62\frac{1}{2}$ or $0.625 = 62\frac{1}{2}\%$ or 62.5%

$\frac{2}{3} = 0.66\frac{2}{3} = 66\frac{2}{3}\%$

$\frac{7}{10} = 0.7 = 70\%$

$\frac{3}{4} = 0.75 = 75\%$

$\frac{4}{5} = 0.8 = 80\%$

$\frac{5}{6} = 0.83\frac{1}{3} = 83\frac{1}{3}\%$

$\frac{7}{8} = 0.87\frac{1}{2}$ or $0.875 = 87\frac{1}{2}\%$ or 87.5%

Table IX Powers and Roots

No.	Square	Cube	Square Root	Cube Root	No.	Square	Cube	Square Root	Cube Root
1	1	1	1.000	1.000	51	2,601	132,651	7.141	3.708
2	4	8	1.414	1.260	52	2,704	140,608	7.211	3.732
3	9	27	1.732	1.442	53	2,809	148,877	7.280	3.756
4	16	64	2.000	1.587	54	2,916	157,464	7.348	3.780
5	25	125	2.236	1.710	55	3,025	166,375	7.416	3.803
6	36	216	2.449	1.817	56	3,136	175,616	7.483	3.826
7	49	343	2.646	1.913	57	3,249	185,193	7.550	3.848
8	64	512	2.828	2.000	58	3,364	195,112	7.616	3.871
9	81	729	3.000	2.080	59	3,481	205,379	7.681	3.893
10	100	1,000	3.162	2.154	60	3,600	216,000	7.746	3.915
11	121	1,331	3.317	2.224	61	3,721	226,981	7.810	3.936
12	144	1,728	3.464	2.289	62	3,844	238,328	7.874	3.958
13	169	2,197	3.606	2.351	63	3,969	250,047	7.937	3.979
14	196	2,744	3.742	2.410	64	4,096	262,144	8.000	4.000
15	225	3,375	3.873	2.466	65	4,225	274,625	8.062	4.021
16	256	4.096	4.000	2.520	66	4,356	287,496	8.124	4.041
17	289	4,913	4.123	2.571	67	4,489	300,763	8.185	4.062
18	324	5,832	4.243	2.621	68	4,624	314,432	8.246	4.082
19	361	6,859	4.359	2.668	69	4,761	328,509	8.307	4.102
20	400	8,000	4.472	2.714	70	4,900	343,000	8.367	4.121
21	441	9.261	4.583	2.759	71	5,041	357,911	8.426	4.141
22	484	10,648	4.690	2.802	72	5,184	373,248	8.485	4.160
23	529	12,167	4.796	2.844	73	5,329	389,017	8.544	4.179
24	576	13,824	4.899	2.884	74	5,476	405,224	8.602	4.198
25	625	15,625	5.000	2.924	75	5,625	421,875	8.660	4.217
26	676	17,576	5.099	2.962	76	5,776	438,976	8.718	4.236
27	729	19,683	5.196	3.000	77	5,929	456,533	8.775	4.254
28	784	21,952	5.292	3.037	78	6,084	474,552	8.832	4.273
29	841	24,389	5.385	3.072	79	6,241	493,039	8.888	4.291
30	900	27,000	5.477	3.107	80	6,400	512,000	8,944	4.309
31	961	29,791	5.568	3.141	81	6,561	531,441	9.000	4.327
32	1,024	32,768	5.657	3.175	82	6,724	551,368	9.055	4.344
33	1,089	35,937	5.745	3.208	83	6,889	571,787	9.110	4.362
34	1,156	39,304	5.831	3.240	84	7,056	592,704	9.165	4.380
35	1,225	42,875	5.916	3.271	85	7,225	614,125	9.220	4.397
36	1,296	46,656	6.000	3.302	86	7,396	636,056	9.274	4.414
37	1,369	50,653	6.083	3.332	87	7,569	658,503	9.327	4.431
38	1,444	54,872	6.164	3.362	88	7,744	681,472	9.381	4.448
39	1,521	59,319	6.245	3.391	89	7.921	704,969	9.434	4.465
40	1,600	64,000	6.325	3.420	90	8,100	729,000	9.487	4.481
41	1,681	68,921	6.403	3.448	91	8,281	753,571	9.539	4.498
42	1,764	74,088	6.481	3.476	92	8,464	778,688	9.592	4.514
43	1,849	79,507	6.557	3.503	93	8,649	804,357	9.644	4.531
44	1,936	85,184	6.633	3.530	94	8,836	830,584	9.695	4.547
45	2,025	91,125	6.708	3.577	95	9,025	857,375	9.747	4.563
46	2,116	97,336	6.782	3.583	96	9,216	884,736	9.798	4.579
47	2,209	103,823	6.856	3.609	97	9,409	912,673	9.849	4.595
48	2,304	110,592	6.928	3.634	98	9.604	941,192	9.899	4.610
49	2,401	117,649	7.000	3.659	99	9,801	970,299	9.950	4.626
50	2,500	125,000	7.071	3.684	100	10,000	1,000,000	10.000	4.642

Table X Trigonometric Ratios

ANGLE	SIN	COS	TAN	COT	SEC	CSC	
0°	0.0000	1.0000	0.0000	∞	1.0000	∞	**90°**
1	0.0175	0.9998	0.0175	57.2900	1.0002	57.2987	**89**
2	0.0349	0.9994	0.0349	28.6363	1.0006	28.6537	**88**
3	0.0523	0.9986	0.0524	19.0811	1.0014	19.1073	**87**
4	0.0698	0.9976	0.0699	14.3007	1.0024	14.3356	**86**
5°	0.0872	0.9962	0.0875	11.4301	1.0038	11.4737	**85°**
6	0.1045	0.9945	0.1051	9.5144	1.0055	9.5668	**84**
7	0.1219	0.9925	0.1228	8.1443	1.0075	8.2055	**83**
8	0.1392	0.9903	0.1405	7.1154	1.0098	7.1853	**82**
9	0.1564	0.9877	0.1584	6.3138	1.0125	6.3925	**81**
10°	0.1736	0.9848	0.1763	5.6713	1.0154	5.7588	**80°**
11	0.1908	0.9816	0.1944	5.1446	1.0187	5.2408	**79**
12	0.2079	0.9781	0.2126	4.7046	1.0223	4.8097	**78**
13	0.2250	0.9744	0.2309	4.3315	1.0263	4.4454	**77**
14	0.2419	0.9703	0.2493	4.0108	1.0306	4.1336	**76**
15°	0.2588	0.9659	0.2679	3.7321	1.0353	3.8637	**75°**
16	0.2756	0.9613	0.2867	3.4874	1.0403	3.6280	**74**
17	0.2924	0.9563	0.3057	3.2709	1.0457	3.4203	**73**
18	0.3090	0.9511	0.3249	3.0777	1.0515	3.2361	**72**
19	0.3256	0.9455	0.3443	2.9042	1.0576	3.0716	**71**
20°	0.3420	0.9397	0.3640	2.7475	1.0642	2.9238	**70°**
21	0.3584	0.9336	0.3839	2.6051	1.0711	2.7904	**69**
22	0.3746	0.9272	0.4040	2.4751	1.0785	2.6695	**68**
23	0.3907	0.9205	0.4245	2.3559	1.0864	2.5593	**67**
24	0.4067	0.9135	0.4452	2.2460	1.0946	2.4586	**66**
25°	0.4226	0.9063	0.4663	2.1445	1.1034	2.3662	**65°**
26	0.4384	0.8988	0.4877	2.0503	1.1126	2.2812	**64**
27	0.4540	0.8910	0.5095	1.9626	1.1223	2.2027	**63**
28	0.4695	0.8829	0.5317	1.8807	1.1326	2.1301	**62**
29	0.4848	0.8746	0.5543	1.8040	1.1434	2.0627	**61**
30°	0.5000	0.8660	0.5774	1.7321	1.1547	2.0000	**60°**
31	0.5150	0.8572	0.6009	1.6643	1.1666	1.9416	**59**
32	0.5299	0.8480	0.6249	1.6003	1.1792	1.8871	**58**
33	0.5446	0.8387	0.6494	1.5399	1.1924	1.8361	**57**
34	0.5592	0.8290	0.6745	1.4826	1.2062	1.7883	**56**
35°	0.5736	0.8192	0.7002	1.4281	1.2208	1.7434	**55°**
36	0.5878	0.8090	0.7265	1.3764	1.2361	1.7013	**54**
37	0.6018	0.7986	0.7536	1.3270	1.2521	1.6616	**53**
38	0.6157	0.7880	0.7813	1.2799	1.2690	1.6243	**52**
39	0.6293	0.7771	0.8098	1.2349	1.2868	1.5890	**51**
40°	0.6428	0.7660	0.8391	1.1918	1.3054	1.5557	**50°**
41	0.6561	0.7547	0.8693	1.1504	1.3250	1.5243	**49**
42	0.6691	0.7431	0.9004	1.1106	1.3456	1.4945	**48**
43	0.6820	0.7314	0.9325	1.0724	1.3673	1.4663	**47**
44	0.6947	0.7193	0.9657	1.0355	1.3902	1.4396	**46**
45°	0.7071	0.7071	1.0000	1.0000	1.4142	1.4142	**45°**
	COS	SIN	COT	TAN	CSC	SEC	ANGLE

GLOSSARY

1. When a bar is placed over a vowel, the vowel says its own name.
2. A curved mark over a vowel indicates the following sounds:
- **a.** ă as in at
- **b.** ĕ as in bed
- **c.** ĭ as in it
- **d.** ŏ as in ox
- **e.** ŭ as in rug

Acute (ŭ kūt′)
(angle) Angle whose measure is less than 90°.
(triangle) Triangle with three angles less than 90°.

Add (ăd) To combine into a sum.

Addend (ăd′ ĕnd) Value to be added to another.

Addition (ă dĭ′ shŭn) Act of adding.

Adjacent (ŭ jā′ sĕnt) **(sides of a quadrilateral)** Any two sides that intersect at a vertex.

Algebraic (ăl′ jŭ brā′ ĭk) **(open expression)** Expression in which the position of an unknown number is held by a letter (variable).

Altitude (ăl′ tĭ to͞od)
(of a parallelogram) Perpendicular distance between two parallel sides.
(of a pyramid) Line segment from the apex perpendicular to the base.
(of a right-circular cone) Length of the axis.
(of a right-circular cylinder) Length of the axis.
(of a trapezoid) Perpendicular distance between the two parallel sides.
(of a triangle) Line segment from the vertex perpendicular to the base.

Ampere (ăm′ pēr) Unit of intensity of electrical current produced by one volt acting through a resistance of one ohm.

Angle (ăng′ g'l) Two rays with a common endpoint.

Apex (ā′ pĕx) **(of a pyramid)** Common vertex of the faces which is not part of the base.

Arc (ărk) Part of a circle connecting two points on a circle.

Area (air′ ē ŭ) Measure (in square units) of the region within a closed curve (including polygons) in a plane.

Associative (ŭ sō′ shŭ tĭv) Indicates that the grouping of three numbers in an addition or multiplication can be changed without affecting the sum or product.

Axis (ăk′sĭs)
(of a cone) Line segment connecting the vertex to the center of the base.
(of a cylinder) Line segment connecting the centers of the bases.

Base (bās) Value that is being multiplied by a percent. That is, in $5\% \times 30$, 30 is the base.
(of a power) Number being raised to a power.
(of a triangle) Side opposite a vertex.

Bases (bās′ ĭz)
(of a cylinder) Two circular regions that are parallel.
(of a prism) Parallel faces.
(of a trapezoid) Parallel sides.

Bevel (bĕv′ ĕl) Angle which one surface or line makes with another when they are not at right angles.

Bisect (bī sĕkt′) **(a line segment)** To find a point which divides the segment into two equal segments.

Camber (kăm′ bur)
(negative) Front wheels inclined inward at the top.
(positive) Front wheels inclined outward at the top.

Celsius (sĕl′ sē ŭs) **(scale)** Temperature scale on which the interval between the freezing point and boiling point of water is divided into 100 parts or degrees, so that 0 °Celsius corresponds to 32 °Fahrenheit and 100 °Celsius to 212 °Fahrenheit.

Centi (sĕn′ tĭ) Latin prefix indicating one-hundredth.

Central (sĕn′ trăl) **(angle)** Angle with its vertex at the center of a circle.

Chord (kord) Line segment connecting two points on a circle.

Circle (sur′ kŭl) Set of all points in a plane whose distance from a given point (center) is equal to a positive number, r (radius).

Circular (sĭr′ kū lur) **(pitch)** Distance between centers of adjoining teeth on a gear.

Circumference (sur kŭm′ fur ĕns) **(of a circle)** Distance around.

Coefficient (kō ĕ fĭsh′ ŭnt) Any numeral or literal symbol placed before another symbol or combination of symbols as a multiplier.

Commutative (kŭ mū′ tŭ tĭv) Indicates that the order of two numbers in an addition or multiplication can be changed without affecting the sum or product.

Compass (kŭm′ pŭs) Instrument used to draw circles and compare distances.

Complementary (kŏm plŭ mĕn′ tŭ rē) **(angles)** Two angles whose measures added together result in a sum of 90°.

Composite (kŏm pŏz′ ĭt) **(number)** Whole number greater than O that has more than two whole number factors.

Cone (kōn) **(right-circular)** Circular region (base) and the surface made up of line segments connecting the circle with a point (vertex) located on a line through the center of the circle and perpendicular to the plane of the circle.

Constant (kŏn′ stănt) Number without a literal coefficient (numbers that do not vary in value).

Continuous (kŭn tĭn′ ū ŭs) **(data)** Measurements that can be made more precise if a person chooses, such as the length of a line.

Cosecant (kō sē′ kant) **(of an angle)** Reciprocal of the sine of the angle.

Cosine (kō′ sin) **(of an angle)** In a right triangle, the numerical value of the ratio of the length of the angle's adjacent side divided by the length of the hypotenuse of the right triangle.

Cotangent (kō tăn′ jĕnt) **(of an angle)** Reciprocal of the tangent of the angle.

Cube (kūb) Regular hexagon (polyhedron of six equal faces).

Cylinder (sĭl′ ĭn dur) **(right-circular)** Two circular regions with the same radius in parallel planes (bases) connected by line segments perpendicular to the planes of the two circles.

Data (dā′ tŭ) Something, actual or assumed, used as a basis of reckoning, such as sets of numbers.

Decagon (dĕk′ ŭ gŏn) Polygon with 10 sides.

Deci (dĕs′ ĭ) Latin prefix indicating one-tenth (the decimal number system is a system based on the use of 10 digits, where each position in a numeral has a value one-tenth the position to its left).

Decimal (dĕs′ ĭ măl) **(fraction)** Fraction whose denominator is a power (multiple) of 10; such as $0.5=\frac{5}{10}$, $2.67=\frac{267}{100}$, etc.

Degree (dŭ grē′) **(of angular measure)** $\frac{1}{360}$ of a complete revolution of a ray around its end-point.

Deka (dĕk′ ŭ) Greek prefix indicating a multiple of 10.

Denominate (dē nŏm′ ĭ năt) **(number)** Number expressed in terms of a standard unit of measure; such as 15 inches.

Denominator (dŭ nŏm′ ĭ nā tur) Bottom number in a fraction, such as 3 in $\frac{2}{3}$.

Dial (dī′ ăl) Graduated plate or face with a pointer for indicating something, as steam pressure.

Diameter (dī ăm′ ŭ tur) **(of a circle)** Twice the radius (of the circle).

Difference (dĭf′ ur ĕns) Result of a subtraction of two numbers, such as 5 in $7-2=5$.

Digit (dĭj′ ĭt) Any one of the symbols 0, 1, 2, 3, 4, 5, 6, 7, 8, or 9.

Discount (dĭs′ kount) Amount the price of an article is reduced. Frequently expressed as a percent of the list price.

Discrete (dĭs krēt′) **(data)** Allows no interpretation between the values of the data, such as the number of people in a room.

Divide (dĭ vīd′)

Dividend (dĭv′ ĭ dĕnd) Number being divided, such as 15 in $15 \div 3$.

Divisibility (dĭ vĭz′ ĭ bĭl′ ĭ tē)

Divisible (dĭ vĭz′ ĭ b'l)

Division (dĭ vĭ′ zhŭn) Act of dividing.

Divisor (dĭ vī′ zur) Number by which another number (dividend) is divided, such as 3 in $15 \div 3$.

Dodecagon(dō dĕk′ ŭ gŏn) Polygon with twelve sides.

Dodecahedron (dō′ dĕk ŭ hē′ drŭn) Polyhedron that has twelve faces.

Edges (ĕ′ jĕz) **(of a polyhedron)** Sides of the faces of the polyhedron.

Elements (ĕl′ ŭ mĕnts) Things that make up a set.

Equation (ē kwā′ zhŭn) Statement of equality between two values or quantities.

Equilateral (ē′ kwĭ lăt′ ur ŭl) **(triangle)** Triangle with three equal sides.

Equivalent (ŭ kwĭv′ ŭ lĕnt) **(expressions)** Two or more expressions that have the same evaluation for all replacements of the variable.

Exponent (ĕks′ pō nŭnt) Indicates the number of times the base is used as a factor. For example, in 2^3, 3 is the exponent and indicates that 2 (base) is being used three times as a factor. That is, $2^3 = 2 \cdot 2 \cdot 2$.

Extrapolation (ĕks′ tră pŭ lă shŭn) Approximating beyond the available data.

Extremes (ĕks trēmz′) **(of a proportion)** First and fourth terms of a proportion, such as 2 and 35 in $2:5 = 14:35$.

Faces (fās′ ĭz) **(of a polyhedron)** Polygonal regions that form the surface of the polyhedron.

Factor (făk′ tur) One of the values in a multiplication expression.

Fahrenheit (făr′ ĕn hīt) **(scale)** Temperature scale on which the interval between the freezing point and boiling point of water is divided into 180 parts or degrees, so that 32 °Fahrenheit corresponds to 0 °Celsius and 212 °Fahrenheit to 100 °Celsius.

Finite (fī′ nīt) **(set)** Set in which the elements can be counted and the count has a last number.

Fraction (frăk′ shŭn) Number which indicates that some whole has been divided into a number of equal parts and that a portion of the equal parts are represented.

Fulcrum (fŭl′ krŭm) Support about which a lever turns.

Gauge (gāj) Instrument for or means of measuring.

Gear (gēr) Toothed wheel.

Great circle (of a sphere) Intersection of the sphere and a plane containing the center of the sphere.

Hectare (hĕk′ tair) 10 000 square metres.

Hecto (hĕk′ tŭ) Greek prefix indicating a multiple of 100.

Hemisphere (hĕm′ ĭ sfēr) Half of a sphere.

Hexagon (hĕk′ sŭ gŏn) Polygon with six sides.

Hexahedron (hĕk′ sŭ hē′ drŭn) Polyhedron that has six faces.

Horizontal (hŏr ĭ zŏn′ tăl) **(line)** Line parallel to the horizon.

Hypotenuse (hī pŏt′ ŭ noos) Side opposite the right angle in a right triangle.

Icosahedron (ī′ kō sŭ hē′ drŭn) Polyhedron that has twenty faces.

Improper (ĭm prŏp′ ur) **(fraction)** Fraction that has a value greater than or equal to 1.

Infinite (ĭn′ fĭ nĭt) **(set)** Set whose count is unending.

Integer (ĭn′ tŭ jur) Any number that is in the following list: . . . , $^{-}2$, $^{-}1$, 0, 1, 2, ...

Interpolation (ĭn tur′ pŭ lā′ shŭn) Approximation between two known values.

Intersecting (ĭn′ tur sĕkt′ ĭng) **(lines)** Two (or more) lines that have one point in common.

Isosceles (ī sŏs′ ĕ lēz) **(triangle)** Triangle with at least two equal sides.

Joule (jool) Unit of work or energy expended in one second by an electric current of one ampere in a resistance of one ohm.

Kelvin (kĕl′ vĭn) **(scale)** Scale of absolute temperature, in which the zero is approximately $^{-}273.15$ °Celsius. The interval between the freezing point and boiling point of water is divided into 100 parts or degrees, so 273.15 Kelvin corresponds to 0 °Celsius and 373.15 Kelvin to 100 °Celsius.

Kilo (kĭl′ ŭ) Greek prefix indicating a multiple of 1000.

Kilogram (kĭl′ ŭ gràm) Base unit of weight in the metric system (1 kilogram = 1000 grams).

Lateral (lăt′ ur ŭl) **(surface of a cylinder)** Curved surface connecting the bases.

Least common denominator Smallest number that is exactly divisible by each of the original denominators of two or more fractions.

Lever (lē′ vur) Rigid piece capable of turning about one point (fulcrum).

Like terms Terms of an algebraic expression which have exactly the same literal coefficients (including exponents).

Line (lĭn) Set of points represented with a picture such as ⟷ (a line is always straight, has no thickness, and extends forever in both directions).

Linear (lĭn′ ē ur) **(measurement)** Measurement along a (straight) line.

(line) segment (sèg′ mènt) Portion of a line between two points (including its end points).

List (lĭst) **price** (prĭs) Original price.

Literal (lĭt′ ŭr ŭl) **coefficient** (kō′ ĕ fĭsh′ ŭnt) Letter factor of an indicated product of a number and one or more variables.

Literal (lĭt′ ŭr ŭl) **(equation)** Equation having more than one letter or variable.

Litre (lē′ tur) Base unit of volume or capacity in the metric system.

Mean (mēn) **(of a set of numbers)** Result obtained when all numbers of the set are added and the sum is divided by the number of numbers.

Means (mēnz) **(of a proportion)** Second and third terms of a proportion, such as 5 and 14 in $2:5 = 14:35$.

Median (mē′dē ăn) **(of a set of numbers)** Middle number of a set of numbers that have been arranged from smallest to largest; in a set with an even number of numbers, the median is one-half way between the two middle numbers.

Metre (mē′ tŭr) Base unit of length in the metric system.

Micrometre (mī′ krō mē′ tŭr) 0.000 001 metre or $\frac{1}{1\,000\,000}$ metre.

Micrometer (mī krŏm′ ŭ tur) **caliper** (kăl′ ĭ pur) Instrument for measuring minute distances.

Midpoint Point on a line segment that divides its length by 2.

Milli (mĭl′ ĭ) Latin prefix indicating one-thousandth.

Mixed number Understood sum of a whole number and a proper fraction.

Mode (mōd) **(of a set of numbers)** Number that occurs with the greatest frequency.

Multiple (mŭl′ tĭ p'l) Product of a quantity by a whole number.

Multiplication (mŭl′ tĭ plī kā′ shun) Act of multiplying.

Multiply (mŭl tĭ plī)

Natural (năt′ yū răl) **(number)** Any number that is in the following list: 1, 2, 3, ...

Negative (nĕg′ ŭ tĭv) **(number)** Any number that is located to the left of zero on a horizontal number line.

Net (nĕt) **(price)** List price minus discount.

Numeral (nōō′ mur ŭl) A symbol that represents a number, such as V, 𝍸, and 5, which are all numerals naming the number five.

Numerator (nōō′ mur ā′ tur) Top number in a fraction, such as 2 in $\frac{2}{3}$.

Numerical (nōō mĕr′ ĭ kŭl) **(coefficient)** Number factor of an indicated product of a number and a variable.

Obtuse (ŏb′ tōōs′)
(angle) Angle whose measure is between 90° and 180°.
(triangle) Triangle that has one angle greater than 90°.

Octagon (ŏk′ tŭ gŏn) Polygon with eight sides.

Octahedron (ŏk′ tŭ hē′ drŭn) Polyhedron that has eight faces.

Ohm (ōm) Unit of electrical resistance of a circuit in which a potential difference of one volt produces a current of one ampere.

Open sentence Equation that does not contain enough information to be judged as either true or false, such as $x + 2 = 7$.

Opposite (ŏp′ ŭ zĭt) **(sides of a quadrilateral)** Pairs of sides that do not intersect.

Opposites (ŏp′ ŭ zĭtz) Two numbers on a number line which are the same distance from zero, such as $^{-}5$ and 5, $^{-}6\frac{2}{3}$ and $6\frac{2}{3}$, etc.

Oscilloscope (ŏ sĭl′ ō skōp) Instrument for showing visually the changes in a varying current.

Parallel (păr′ ŭ lĕl) **(lines)** Two (or more) lines in the same plane that have no points in common.

Parallelogram (păr′ ŭ lĕl′ ŭ gram) Quadrilateral with opposite sides parallel.

Pentagon (pĕn′ tŭ gŏn) Polygon with five sides.

Pentahedron (pĕn′ tŭ hē′ drŭn) Polyhedron that has five faces.

Percent (pur sĕnt′) Fraction with a denominator of 100, such as $\frac{7}{100} = 0.07 = 7\%$.

Percentage (pur sĕn′ tĭj) Result of multiplying a percent times a number. That is, in $5\% \times 30 = 1.5$, the percentage is 1.5.

Perimeter (pĕ rĭm′ ŭ tur) **(of a polygon)** Total length of all the sides of the polygon.

Perpendicular (pur′ pĕn dĭk′ ū lur) **(lines)** Two intersecting lines that form a right angle.

Plane (plān) Flat surface (such as a table top) that extends infinitely in every direction.

Point Location in space that has no thickness.

Polygon (pŏl′ ĭ gŏn) Plane closed figure of three or more angles.

Polygonal (pŭ lĭg′ ŭ nŭl) **(region)** Polygon and its interior.

Polyhedra (pŏl′ ĭ hē′ dră) Plural of polyhedron.

Polyhedron (pŏl′ ĭ hē′ drŭn) Geometric solid where surfaces are polygonal regions.

Positive (pŏz′ ĭ tĭv) **(number)** Any number that is located to the right of zero on a horizontal number line.

Power (pow′ ur) Expression used when referring to numbers with exponents, such as 3^5, "the fifth power of three."

Prime (prīm) **(number)** Whole number greater than 1 divisible by exactly two whole-number factors, itself and 1.

Prism (prĭz'm) Polyhedron that has two faces which have the same size and shape and which lie in parallel planes, and the remaining faces are parallelograms.

Product (prŏd′ ŭkt) Result of a multiplication, such as 15 is the product of 3×5.

Proper (prŏp′ ur) **(fraction)** Fraction that has a value less than 1.

Proportion (prŭ por′ shun) Statement of equality between two ratios.

Protractor (prō trăk′ tur) Instrument used to measure angles.

Pulley (pŭll′ ē) Wheel used to transmit power by means of a band or belt.

Pyramid (pĭr′ ŭ mĭd) Polygonal region (base), the surface made up of line segments connecting the polygon with a point outside the plane of the base (apex), and all interior points.

Pythagoras (Pŭ thag′ or us) Greek philosopher and mathematician.

Pythogorean (pŭ thag′ ō rē′ ăn) **theorem** (thēr′ um) The sum of the squares of the legs of a right triangle is equal to the square of the hypotenuse.

Quadrilateral (kwŏd′ rĭ lăt′ ur ŭl) Polygon with four sides.

Quantity (kwŏn′ tĭ tē) Indicates an expression that has been placed within parentheses or that is being considered in its entirety.

Quotient (kwō′ shŭnt) Result of a division, such as 5 is the quotient of $15 \div 3$.

Radical (răd′ ĭ kŭl) **(sign)** The symbol $\sqrt{\ }$ (used to indicate square root).

Radii (rā′ dē ī) **(of a circle)** Plural of radius.

Radius (rā′ dē ŭs) **(of a circle)** Line segment connecting the center of a circle to any point on the circle.

Ratio (rā′ shō) Quotient of two quantities or numbers.

Rational (răsh′ ŭn ŭl) **(number)** Any number that can be written in the form $\frac{p}{q}$, where p and q are integers and $q \neq 0$.

Ray (rā) Part of the line on one side of a point which includes the endpoint.

Rectangle (rĕk′ tăng′ g'l) Parallelogram whose sides meet at right angles.

Reduce (rŭ dōōs′) Process of dividing both numerator and denominator of a fraction by some common factor.

Regular (rĕg′ ū lur)
(polygon) Polygon that has sides and angles of equal measure.
(polyhedron) Polyhedron that has faces which are regular polygons.

Remainder (rŭ mān′ dur) Amount left over when it is not possible to divide the dividend into an equal number of groups exactly.

Replacement set Set of permissible values of a variable in an algebraic expression.

Right
(angle) Any of the angles formed by two intersecting lines, where all four angles are equal (the measure of a right angle is 90°).
(triangle) Triangle with one angle of 90°.

Scalene (skā lēn′) **(triangle)** Triangle that has no pair of sides equal.

Secant (sē′ kănt) **(of an angle)** Reciprocal of the cosine of the angle.

Sector (sĕk′ tur) **(of a circle)** Figure enclosed by a central angle and its intercepted arc.

Semi-circle (sĕm′ ĭ sur kŭl) Arc of a circle in which the end points of the arc are on a diameter of the circle.

Set (sĕt) Well-defined (membership in the set is clear) collection of things.

Sine (sīn) **(of an angle)** In a right triangle, the numerical value of the ratio of the length of the angle's opposite side divided by the length of the hypotenuse of the right triangle.

(slant) height (hīt) **(of a right circular cone)** Length of a line segment from the vertex to a point on the circle of the base.

Sphere (sfēr) Set of points in space at a fixed distance (radius) from a given point (center).

Square (skwair) Rectangle whose sides are of equal length.

Subtract (sŭb trăkt′)

Subtraction (sŭb trăk′ shŭn) Act of subtracting.

Sum (sŭm) Result of an addition of two or more numbers.

Supplementary (sŭp′ lŭ mĕn′ tŭ rē) **(angles)** Two or more angles whose measures added together result in a sum of 180°.

Tachometer (tă kŏm′ ĭ tur) Instrument used to measure the number of revolutions the shaft of a motor makes with respect to a unit of time.

Tangent (tăn′ jĕnt)
(line) Line that intersects only one point on a circle.
(of an angle) In a right triangle, the numerical value of the ratio of the length of the angle's opposite side divided by the length of the angle's adjacent side.

Taper (tā′ pur) Gradual diminishing of thickness or width in an elongated object.

Terms (tŭrmz) Parts of an expression separated by an addition symbol.

Tetrahedron (tĕt′ rŭ hē′ drŭn) Polyhedron that has four faces.

Torque (tork) That which produces or tends to produce rotation.

Trapezoid (trăp′ ŭ zoid) Quadrilateral in which only one pair of opposite sides are parallel.

Triangle (trī′ ăng′ g'l) Polygon with three sides.

Truss (trŭs) Assemblage of members, such as beams, bars, rods, etc., forming a rigid framework.

Variable (vair′ ĭ ŭ b'l) Letter in an algebraic expression that can be replaced by any one of a set of many numbers.

Vernier (vur′ nĭ ur) **(scale)** Short scale made to slide along the divisions of a graduated instrument to indicate parts of divisions.

Vertex (vur′ tĕks) **(of an angle)** Common endpoint of two rays that form an angle.

Vertical (vur′ tĭ kăl) **(line)** Line perpendicular to the horizon.

Vertices (vur′tĭ sēz) Plural of vertex.
(of a polygon) Vertices of the sides of the polygon.
(of a polyhedron) Vertices of the faces of the polyhedron.

Volt (vōlt) Electromotive force which when steadily applied to a conductor whose resistance is one ohm will produce a current of one ampere.

Volume (vŏl′ yŭm) Measure (in cubic units) of the space within a closed solid figure.

Watt (wŏt) Unit of power equal to work done at a rate of one joule a second or to the rate of work represented by a current of one ampere under a pressure of one volt.

Whole (number) Any number that is in the following list: 0, 1, 2, 3, . . . (the next whole number is formed by adding 1 to the previous whole number).

INDEX